APPLIED THERMODYNAMICS

(SI UNITS)

Prof. D.K. Chavan

Assistant Professor,
Mechanical Engineering Department,
Marathwada Mitra Mandal's
College of Engineering
(M.M.C.O.E) Pune – 52

Ex. Assistant Professor
Mechanical Engineering Department,
M.I.T. Pune – 38

Prof. G.K. Pathak

Sr. Faculty Member,
Mechanical Engineering Department,
Maharashtra Institute of Technology
(M.I.T.) Pune

STANDARD BOOK HOUSE

Unit of: **RAJSONS PUBLICATIONS PVT. LTD.**

1705-A, Nai Sarak, P.B. No. 1074, Delhi - 110006
Ph.: +91-(011)-23265506 • Fax: +91-(011)-25555295
Show Room: 4262/3, First Lane, G-Floor, Gali Punjbian, Ansari Road, Darya Ganj,
New Delhi - 110002 Ph.: +91-(011)-43551085, +91-(011)-43551185
E-mail: sbh10@hotmail.com • www.standardbookhouse.com

Published by:

RAJINDER KUMAR JAIN

Standard Book House
Unit of: Rajsons Publications Pvt. Ltd.
1705-A, Nai Sarak, Delhi - 110006
Post Box: 1074,
Ph.: +91-(011)-23265506 Fax: +91-(011)-25555295

Showroom:

4262/3, First Lane, G-Floor, Gali Punjbian,
Ansari Road, Darya Ganj,
New Delhi - 110002
Ph.: +91-(011)-43551085, +91-(011)-43551185
E-mail: sbh10@hotmail.com
www.standardbookhouse.com

First Published: 2010

Price: **Rs. 220.00**

ISBN: 978-81-89401-37-5

Typeset by:

Amandeep Singh Bansal
Graphic Era (*An Orgnisation of Book Designing*)

Printed by:

Radha Press, Delhi 110031

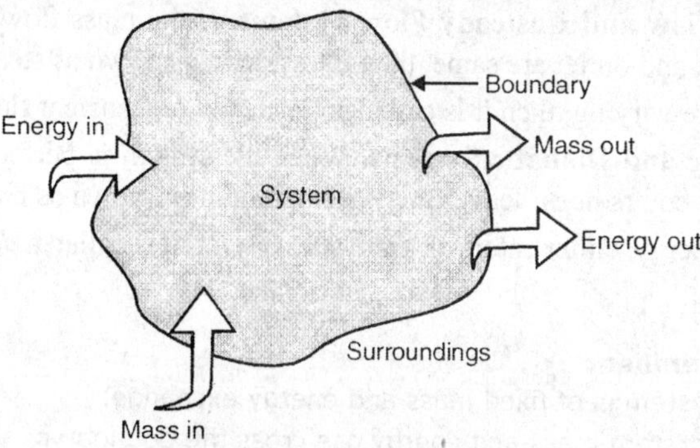

Fig. 1.5. *An open system.*

For example, consider a portion of steam plant as shown in Fig. 1.6. Mass of steam enters at section (1) and after expanding into the turbine, leaves at section (2). In this case work is produced when the turbine is driven. So, as shown, the mass, work and heat are crossing the system boundary.

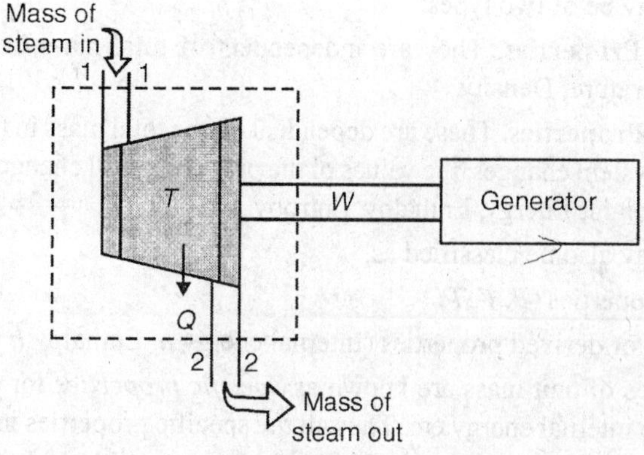

Fig. 1.6. *Example of an open system.*

(c) Isolated System. It is of fixed mass and energy and there is no mass or energy transfer across the system boundary. An isolated system is independent of all the changes that are taking place in the surroundings. (Fig. 1.7).

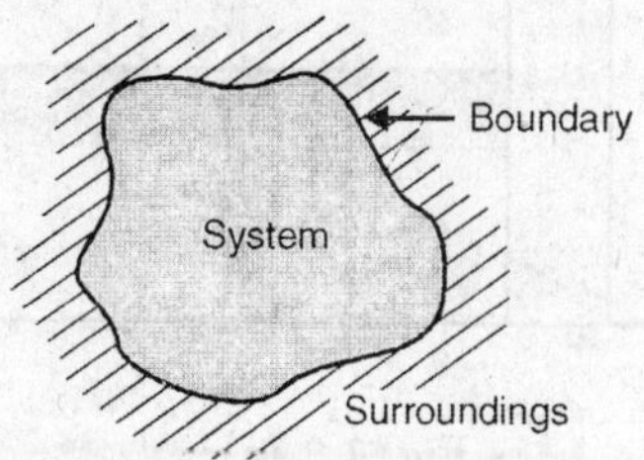

Fig. 1.7. *Isolated system (No mass or energy transfer).*

(d) Steady Flow and Unsteady Flow Systems. If the mass flow rate and energy flow rate at inlet and outlet are same, then the system is known as steady flow system.

If the rates are varying, then it is called an unsteady or transient flow system.

(e) Adiabatic and Diabatic Systems. When the system is perfectly insulated, so that neither heat enters nor it leaves the system, then it is known as *adiabatic system*.

When heat energy, either enters or leaves the system, it is called a *diabatic system*. It is not insulated.

Points to Remember :

- Closed system is of fixed mass and energy exchange.
- In open system mass and energy can cross the boundary.
- Isolated system is of fixed mass and energy. No energy or mass transfer occurs.

1.3 THERMODYNAMIC PROPERTIES, PROCESSES AND CYCLES

Every system has certain characteristics such as Pressure, Volume. Temperature, Density, Internal energy, Enthalpy etc. by which its physical condition may be described. Such characteristics are called as *Properties* of the system.

Properties may be of two types:

1. Intensive Properties. They are independent of total mass in the system e.g., Pressure, Temperature, Density.

2. Extensive Properties. These are dependent on the total mass in the system. When the mass of the system changes, the values of the properties will change proportionately. e.g., Volume, Weight, Energy, Enthalpy, Entropy.

Properties may also be classified as,

1. Primary properties (P, V, T)

2. Secondary or derived properties (Internal energy u, Enthalpy h and Entropy s)

The properties of unit mass are known as *Specific properties* for example specific volume, specific internal energy etc. Thus all the specific properties indicate the values for unit mass and they are independent of total mass, so *all the specific properties are intensive properties*.

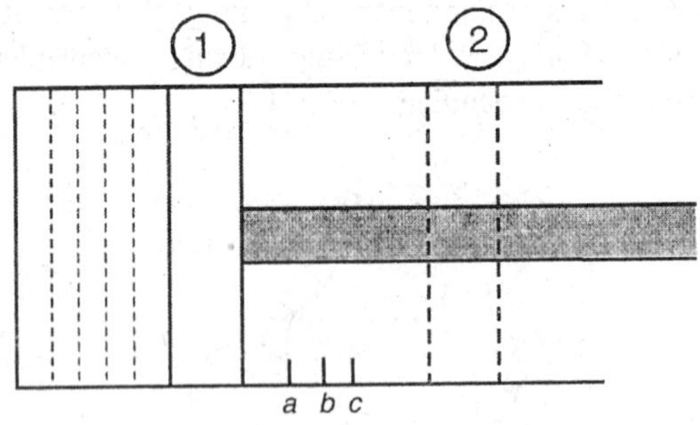

Fig. 1.8 (a)

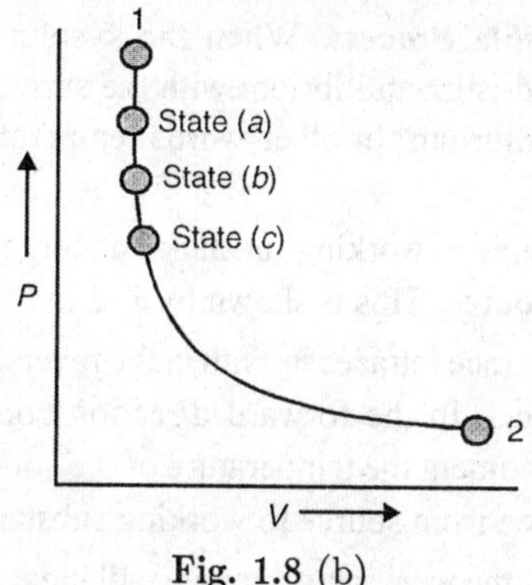

Fig. 1.8 (b)

When all the properties of a system have definite values (such as P, V, T) then the system is said to exist at a *Definite state* and the properties of the state are said to be *State functions* or *point functions.* Any operation in which one or more properties of the system changes, is called *Change of state,* and the series of states (and substates like *a*, *b*, *c*, etc. as shown in Fig. 1.9) passed during a change of state from state (1) to state (2) is called the *path* of change of state. The properties of the states and substances are called as *Path functions.*

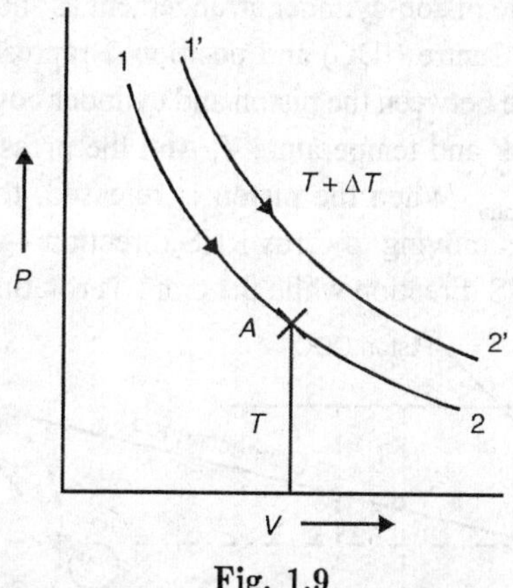

Fig. 1.9

When the path is completely specified, then the change of state is called a *Process.*

A *Process* is defined as the transformation of the system from one fixed state to another fixed state.

When any one of the properties changes, the working substance or system is said to have undergone a process.

One way of classifying a thermodynamic process is—Reversible process and Irreversible process.

(A) Reversible Process. The system will undergo a reversible process in three different conditions:

(i) Thermally Reversible Process. When the system has uniform temperature throughout the process and is in equilibrium with the surroundings, then the system is said to be in thermal equilibrium. In other words temperature gradient in the system does not exist.

Let us consider an example of working substance undergoing a process, during which it receives heat from the source. This is shown by 1–2 in Fig. 1.9.

When the working substance retraces its path in the reverse direction, then the process is called *Reversible process.* In the forward direction consider a point A where the temperature is T. At that moment the temperature of the source will be $(T + \Delta T)$, so that the heat transfer takes place from source to working substance.

In the reverse direction, the working substance will have to reject heat. At point A the working substance at a temperature of T has to transfer heat to source at $(T + \Delta T)$ that is not possible and the process cannot be retraced back and the process will not be reversible. To have the process reversible, the temperature gradient must be zero i.e. $\Delta T = 0$. In a thermally reversible process, therefore, temperature gradient does not exists between the surroundings (source) and the system.

(ii) Mechanically Reversible Process. A system is said to be mechanically reversible, when there are no mechanical losses either internally or externally.

Let us consider a simple piston-cylinder arrangement as shown in Fig 1.10. Position 1 represents Inner Dead Centre (IDC) and position 2 represents Outer Dead Centre position (ODC). Let space between the piston and cylinder cover be filled with a gas at a pressure of P, volume V and temperature T. And the pressure on the other side of piston be atmospheric P_{atm}. When the piston is released, then because of pressure potential, the piston starts moving towards RHS direction — till ODC and after that piston moves towards LHS direction while the crank is rotating.

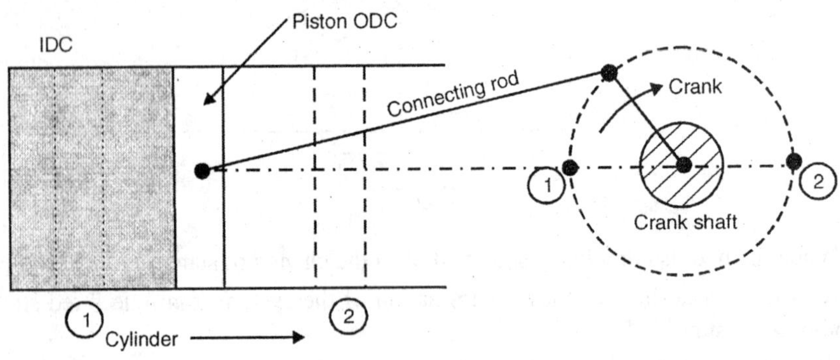

Fig. 1.10

If there are no mechanical losses the piston will reach to its original position. This type of process is called as *Mechanically Reversible Process.*

In actual practice this does not exist. Mechanical losses are mainly *frictional losses.* If the process is mechanically reversible, then the friction between the moving parts must be zero.

We can have other example as a tennis ball released from a certain height, then it will start bouncing ultimately it comes to rest. If there would not have been any mechanical losses, at the point of contact, then it would have been bouncing continuously. In practice this does not happen.

(iii) Chemically Reversible Process. The process is called chemically reversible process, when there are no chemical reactions during the process.

A system will be in thermodynamic equilibrium, when it is in thermal equilibrium, mechanical equilibrium, and chemical equilibrium simultaneously.

A system in thermodynamic equilibrium is incapable of any spontaneous change and it is in complete balance with its surroundings.

(B) Irreversible Process. When the process is not thermally reversible, mechanically reversible and chemically reversible, then it is called an *Irreversible Process.*

A process will be thermally irreversible, when the heat transfer with finite temperature difference takes place.

A process will be mechanically irreversible, when the friction exists during the process.

A process will chemically irreversible, when chemical reactions take place during the process.

The various thermodynamic processes (See Fig 1.10 a–c) which are used in engineering practice are discussed below.

1. Constant Pressure Process or Isobaric Process. A process during which pressure remains constant is called a constant pressure process. The law for the process is ($P = C$) and is represented on the PV-diagram by means of a horizontal line 1–2.

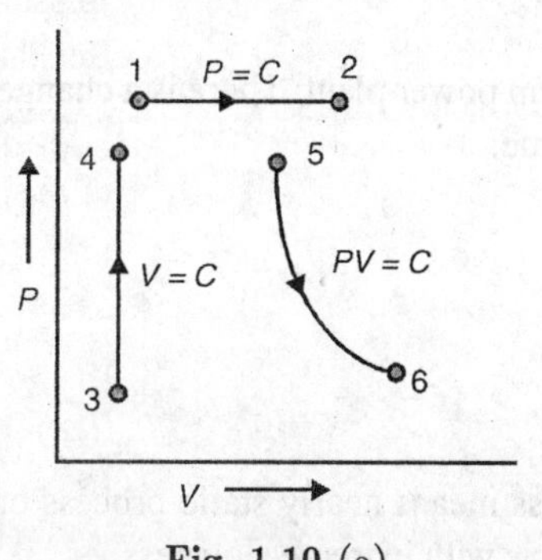

Fig. 1.10 (a)

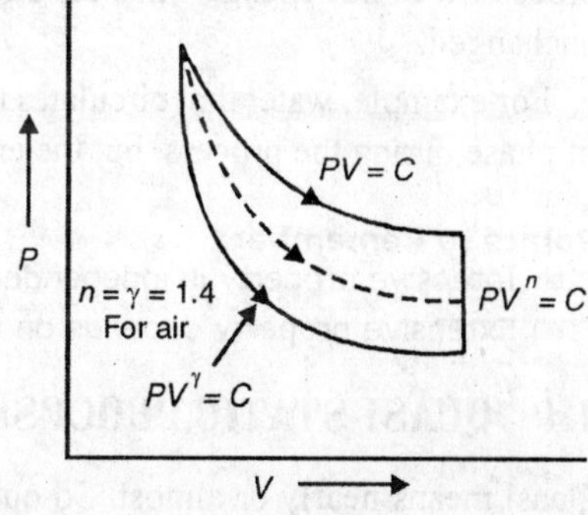

Fig. 1.10 (b)

2. Constant Volume Process or Isochoric Process. A process during which volume remains constant is called a constant volume process. The law for the process is $V = C$ and is represented by means of a vertical line 3–4 on the PV-diagram.

3. Constant Temperature or Isothermal Process. A process during which the temperature remains constant is called a constant temperature on isothermal process. The law for the process is $PV = C$ and is represented by the curve 5–6 on the PV-diagram. Also the law of the process may be given as $T = C$.

4. Adiabatic or Isentropic Process. A process during which heat exchange with the surroundings is zero, then it is called an *Adiabatic process.* A reversible adiabatic process is called as an Isentropic process, during which entropy remains constant ($S = C$). The law for the adiabatic process is $PV^{\gamma} = C$. Irreversible adiabatic process is called a *Throttling process*, during which enthalpy remains constant. It is given by law h = constant.

5. General Law Process or Polytropic Process. The law for the process is $PV^n = C$. In actual practice neither we get isothermal process, nor we get adiabatic process, but what we get is the process in between these two and in known as polytropic process [See Fig 1.10(b)].

Thermodynamic Cycle. It is defined as the series of state changes such that the final identical with the initial state.

or

When the system undergoes the series of processes such that the end states are same, the system is said to have undergone a thermodynamic cycle.

or

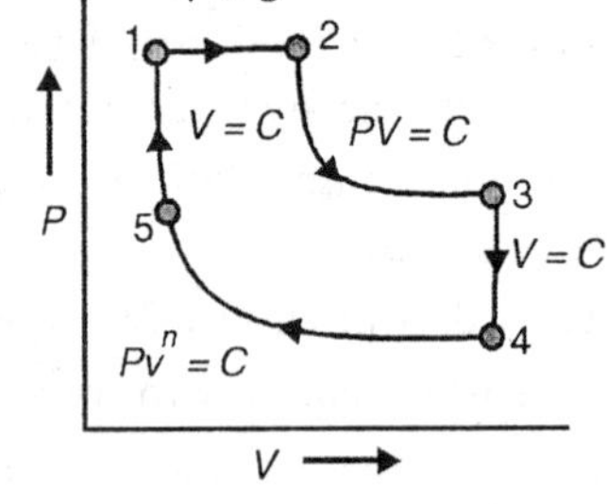

Fig. 1.10 (c)

Series of thermodynamic processes, the end states of which are identical, is called a *thermodynamic cycle.*

In a thermodynamic cycle, chemical composition of the working fluid during the process does not change Thus all the properties in the initial and final states remain unchanged.

For example, water that circulates through the steam power plant. There is a change of phase during the process, but the end states are same.

Points to Remember :

- Intensive property is independent of mass
- Extensive property depends on mass.

1.4 QUASI-STATIC PROCESS

Quasi means nearly or almost. So quasi-static process means nearly static process or nearly stationary process, or a process which proceeds with extreme slowness.

Quasi-static process proceeds from one equilibrium state to another equilibrium state till the end of the process i.e. all the states passed during the process are all equilibrium states.

In each state the process deviates from the equilibrium state by a very small amount and immediately attains its equilibrium.

Let us consider a system consisting of a gas in a cylinder fitted with a piston as shown in Fig. 1.11 (a). Let W be the weight kept on the piston which just balances the upward force exerted by the gas. The system is initially in the equilibrium state. During initial condition, let the properties be P_1 V_1 and t_1. If the weight is removed then there will be an unbalanced force between the system and the surroundings and the piston will move up till it hits the stops. Then again the system comes to another equilibrium state described by the properties P_2, V_2, t_2 but the intermediate states passed through by the system are non-equilibrium states. In Fig. 1.11 (b) points (1) and (2) are initial and final equilibrium states and since the system is passing through the non-equilibrium states, it is joined by dotted line. Such a process is called as *Irreversible Process.*

Now if the single weight on the piston is made up of many very small pieces of weights as shown in Fig. 1.11 (c) and these weights are removed one by one, then the departure of the state of the system from the thermodynamic equilibrium state will be very small. So, every state passed through by the system will be an equilibrium state.

A Quasi-static process is also called as *Reversible process* and is represented on the *PV*-diagram by a continuous line as shown in Fig. 1.11 (d).

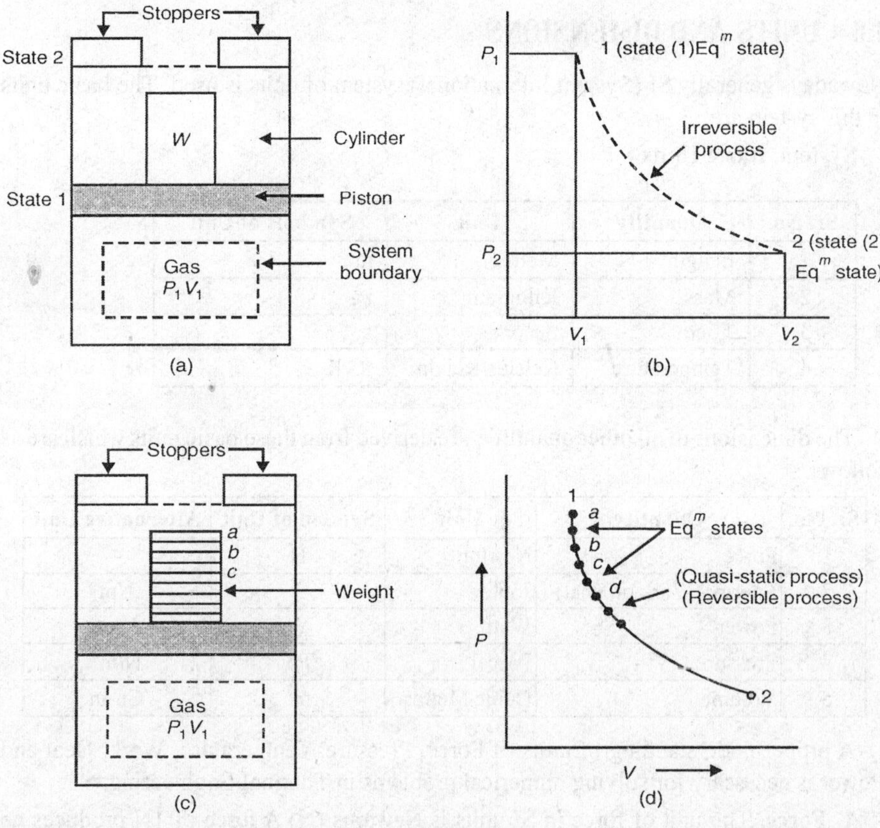

Fig. 1.11

In other words suppose when one weight say *a* is removed then the system will deviate from its equilibrium by a very small amount and immediately comes to an equilibrium state *a* (as shown in *PV*-diagram Fig. 1.12 (d) and when weight *b* is removed, the system again deviates by a small amount and immediately comes to another equilibrium state *b*. Thus when these small weights are removed one by one the system proceeds from one equilibrium state to another equilibrium state till end, such a process which passes through all the equilibrium states is known as quasistatic process.

1.5 WORKING SUBSTANCE

"*It is a fluid, in which energy can be stored and from which energy can be removed*".

or

"*It is a fluid which is used in thermodynamic systems, for converting one form of energy into the other*".

The working substance may be a liquid, vapour or a gas. Generally air and water are used as working substances, as they are freely available. However NH_3 and Freon-12 are also used depending upon specific applications.

1.6 UNITS AND DIMENSIONS

Nowadays generally SI (System International) system of units is used. The basic units in this system are:

System. Basic Units

Sr. No.	Quantity	Unit	Symbols of Unit
1.	Length	Metre	m
2.	Mass	Kilogram	kg
3.	Time	Seconds	s
4.	Temperature	Celsius/Kelvin	°C/K

The dimensions of all other quantities are derived from these basic units which are as follows.

Sr. No.	Quantity	Unit	Symbol of Unit	Alternative Unit
1.	Force	Newton	N	—
2.	Energy (Work or Heat)	Joule	J	Nm
3.	Power	Watt	W	J/sec
4.	Pressure	Pascal	Pa	N/m^2
5.	Volume	Cubic Metres	m^3	Cu-m

A proper understanding of units of Force, Pressure, Temperature, Work, Heat and Power is necessary for solving numerical problems in Thermal Engineering.

1. Force. The unit of force in SI units is Newtons (N) A force of 1N produces an acceleration of 1 m/sec^2 when applied to a mass of 1 kg.

We know that from Newton's Second law of motion $F \propto ma$

or $$F = \frac{ma}{g_c},$$ where g_c = constant of proportionality.

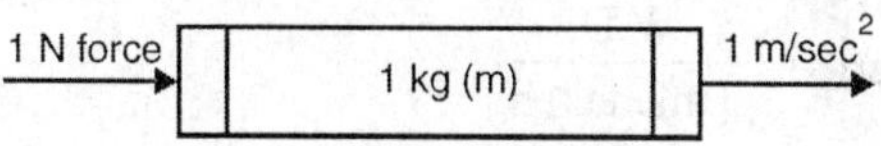

Fig. 1.12

As 1 N force produces an acceleration of 1 m/sec² in 1 kg mass, we have,

$$1 \text{ Newton} = \frac{1}{g_c} \times 1 \text{ kg} \times 1 \text{ m/sec}^2$$

$$\therefore \quad g_c = 1 \text{ in SI units and units will be}$$

$$g_c = \frac{1 \text{ kg-m}}{\text{N} - \text{sec}^2}$$

When $g_c = 1$, the product of mass and acceleration becomes the magnitude of force.

If the mass of 1 kg is allowed to fall freely under the action of standard gravitational force, it is accelerated at the rate of 9.806 m/sec² (9.81 m/sec²) and we have,

$$\text{Force} = 1 \text{ kg} \times 9.81 \text{ m/sec}^2 = 9.81 \text{ N}$$

∴ It follows that the weight of 1 kg mass equals 9.81 N and since the weight is the force,

$$\textbf{1 kgf} = \textbf{9.81 N}$$

2. Weight. Weight of a body is the force exerted on its mass due to gravity.

3. Density. Density is defined as mass per unit volume of the substance. It is denoted by ρ

$$\therefore \quad \rho = \frac{\text{mass}}{\text{volume}} = \frac{\text{kg}}{\text{m}^3}$$

Since unit of volume is m³ and mass is kg.

4. Work (mechanical). When a force F acts on a body and causes displacement through a distance dl in the direction of force. then the work is said to be done. This work is equal to the product of force and distance moved.

$$W = F \times dl$$

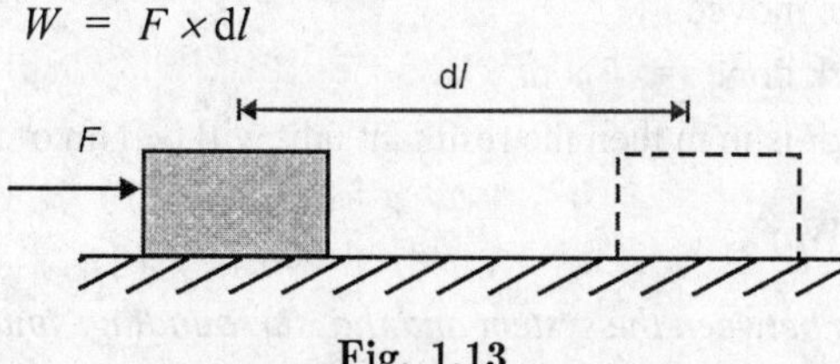

Fig. 1.13

If force F is in Newton and distance dl is in metre, then the resultant unit of W.D. will be Nm and 1 Nm = 1 joule.

5. Power. It is defined as the rate of doing work.

or $$\text{Power} = \frac{\text{W.D.}}{\text{Time taken}}$$

If the unit of W.D. is in joules and the time taken is in seconds, then the unit of Power is J/sec.

And the rate of doing work of 1 J/sec is called watt (W)

$$\therefore \quad \text{Power} = \frac{\text{J}}{\text{sec}} \text{ or W}$$

Earlier there used to be another unit of Power known as Horse Power (H.P.)

$$1 \text{ HP} = 746 \text{ watts} = 0.746 \text{ kW}$$

6. Pressure (P). Pressure is defined as the force per unit area. Thus if F is the force applied to an area A then,

$P = \frac{F}{A}$ and if F is in newton and A is in m^2 then the unit will be $\frac{\text{N}}{\text{m}^2}$.

This unit of pressure is called as pascal (Pa).

1.7 MECHANICAL AND THERMODYNAMIC WORK

Mechanical Work

$$\text{W.D.} = F \times dl$$

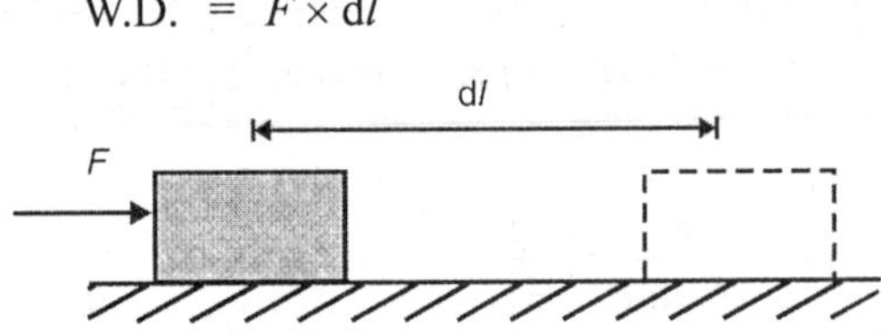

Fig. 1.14. *Mechanical work.*

When a force F acts on a body and causes a displacement through a distance in the direction of force, then the work is said to be done and this work is equal to the product of force and distance moved.

i.e. $$\text{Work done} = F \times dl$$

If F is in N, and dl is in m then the resultant unit will be Nm or Joule.

Thermodynamic Work

"It is an interaction between the system and the .surroundings and the work is said to be done by the system on surroundings, if the complete external effect is lifting up of the body".

Let us consider a battery and motor system as shown in Fig. 1.15. The motor is driving a fan. The system is doing work on the surroundings.

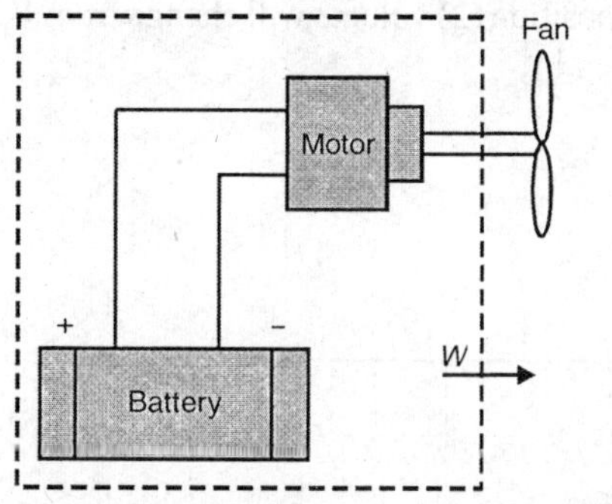

Fig. 1.15

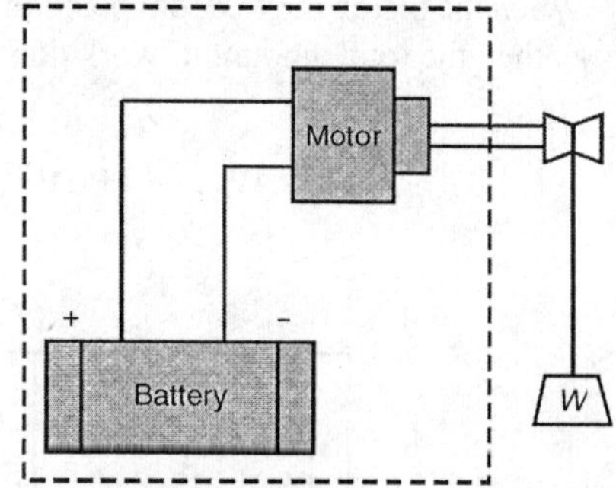

Fig. 1.16

When the fan is replaced by a pulley and weight as shown in Fig. 1.16, then the weight may be raised as the pulley is driven by the motor. Then the complete external effect is raising up the body, so the system is doing thermodynamic work.

When the work is done by the system, then it is taken as + ve, and when the work is done on the system, it is taken as –ve. (See Fig. 1. 17)

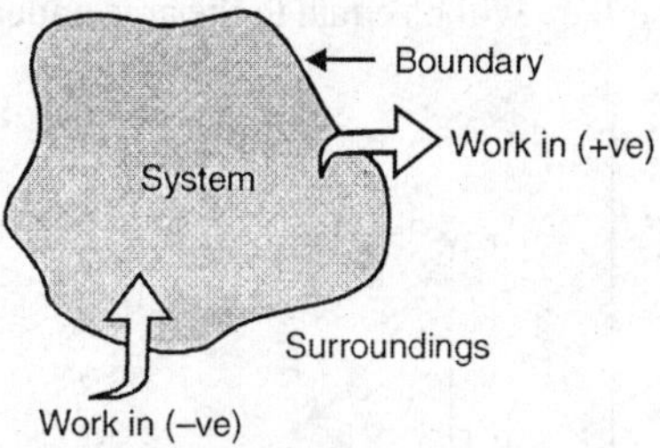

Fig. 1.17

1.8 *PdV*–WORK

Consider a system consisting of a gas in a cylinder fitted with a piston as shown in Fig. 1.19. During the initial condition of the piston i.e. when the piston is at position (1), the pressure inside the cylinder is P_1 and volume is V_1. Let the gas expands as the piston moves to position (2). Then the pressure falls to P_2 and volume will increase to V_2.

Now, consider a small movement of the piston dl during which pressure P is assumed to be constant. If a is the cross sectional area in m^2, then the force acting on the piston is given by

$$F = P \times a$$

And the small amount of work done by the gas on the piston in causing the displacement

dl = force × displacement.

$$\delta W = F \times \mathrm{d}l = P \times a \times \mathrm{d}l \qquad (\because F = P \times a)$$

$$\delta W - PdV'$$

where $dV = a \times \mathrm{d}l$ = small amount of displacement volume.

When the piston moves from position (1) to position (2) volume will change from V_1 to V_2 then the total amount of work done is given by,

$$W_{1-2} = \int_{V_1}^{V_2} PdV$$

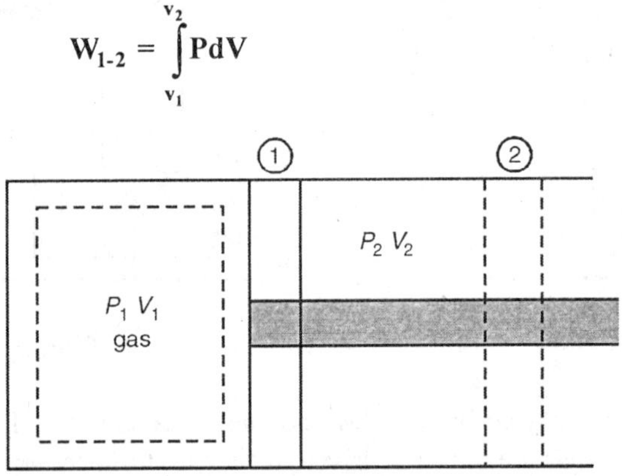

Fig. 1.18

The initial and final states of the system can be represented on PV-diagram as shown in Fig 1.19. The work done W_{1-2} will be equal to the area under the curve 1–2 on the PV-diagram.

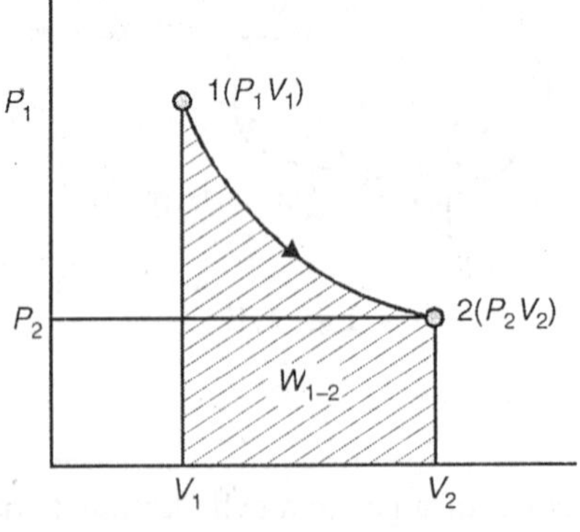

Fig. 1.19

Note. Here the piston is assumed to move very very slowly so that path 1–2 is Quasi-static (i.e. nearly static) and every state passed through is an equilibrium state.

1.9 WORK IS A PATH FUNCTION AND PROPERTIES ARE POINT FUNCTIONS

With reference to Fig. 1.20 it is possible to take a system from state (1) to state (2) by following many quasi-static paths such as *A*, *B*, *C* etc.

Now if we consider the path 1–*A*–2, then the work done during the process 1–*A*–2 will be equal to the area under the curve 1–*A*–2 and if we consider path 1–*B*–2, then the work done during the path 1–*B*–2 will be equal to the area under the curve 1–*B*–2.

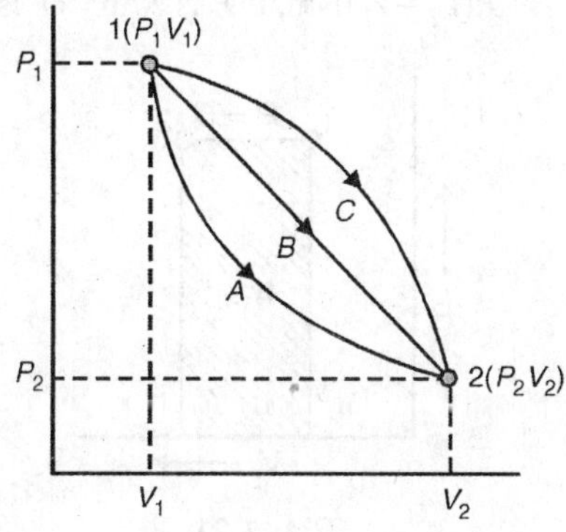

Fig. 1.20

So, the work done during the process directly depends upon the path and not on the end states. So, work is called Path function and δW is an inexact or imperfect differential.

i.e.
$$\int_1^2 \delta W \neq W_2 - W_1$$

But
$$\int_1^2 \delta W = W_{1-2} \text{ or } {}_1W_2$$

Thermodynamic properties are point functions, since for a given state, there is a definite value for each property. During change of states, the values of the properties will change and this change in values of the properties does not depend on path, but it depends upon end states and hence all the properties are Point functions.

The differentials of point functions are exact or perfect differentials and the integration is simply, e.g.

$$\int_1^2 dV = V_2 - V_1$$

Operator δ is used to denote inexact differentials, operator d is used to denote exact differentials.

1.10 EQUATIONS FOR WORK DONE IN VARIOUS PROCESSES

***PdV* work in various quasi-static processes.**

1. Constant pressure process (Isobaric process). Here pressure is constant.

We know that Work Done is given by,

$$W_{1-2} = \int_{V_1}^{V_2} PdV = P\int_{V_1}^{V_2} dV$$

$$= P(V_2 - V_1)\text{Nm, if } P \text{ is N/m}^2 \times V \text{ is m}^3$$

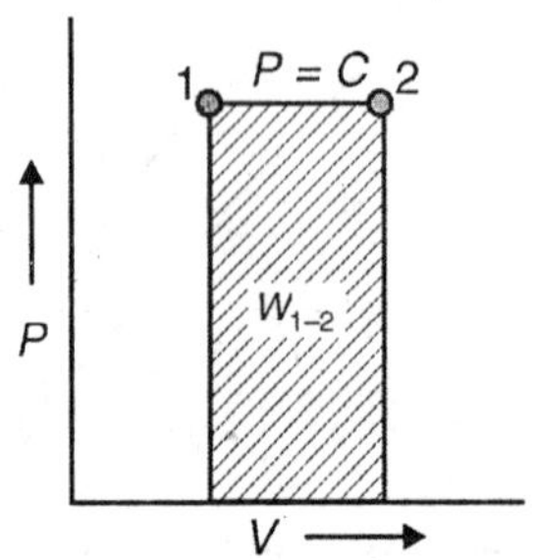

Fig. 1.21

2. Constant volume process. (Isochoric process)

$$W_{1-2} = \int PdV$$

$$= 0$$

During this process $V = C$ hence W.D. $= 0$

Since derivative of a constant is zero.

From PV-diagram area under the line 1–2 is zero.

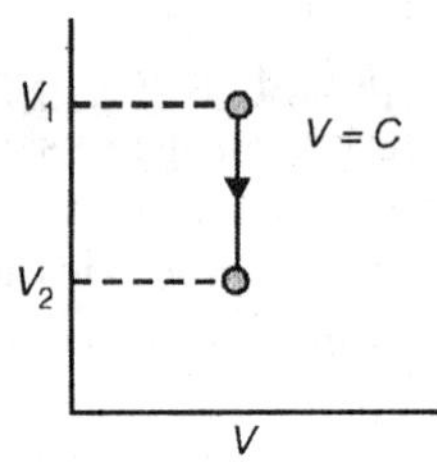

Fig. 1.22

3. Isothermal process. We know that work done in a non-flow system

$$W_{1-2} = \int_1^2 P.dV \qquad (1)$$

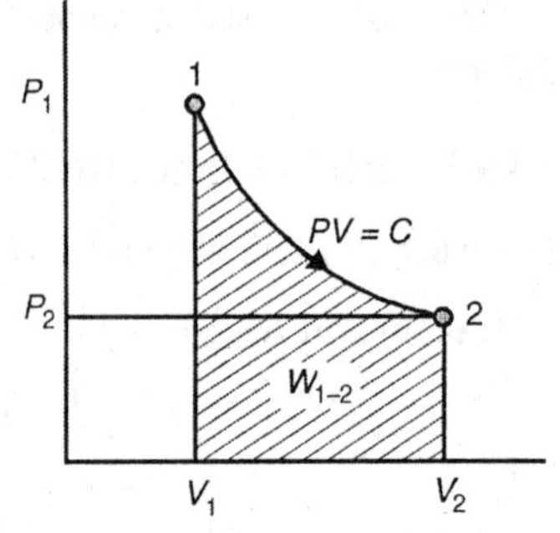

Fig. 1.23

The law for the process is,

$$PV = C$$

i.e. $PV = P_1V_1 = C$

$\therefore$ $P = \dfrac{P_1 V_1}{V}$

Substituting this in Eq. (1) we get,

$$\therefore \quad W_{1-2} = \int_1^2 \frac{P_1 V_1}{V} \cdot dV = P_1 V_1 \int_1^2 \frac{dV}{V}$$

$$\mathbf{W_{1-2} = P_1 V_1 \ln \frac{V_2}{V_1}}$$

4. Adiabatic process. The law for the process is $PV^\gamma = C$ where γ is an adiabatic index. We know that, the work down in non-flow system,

$$W_{1-2} = \int_1^2 PdV \qquad (2)$$

But for a polytropic process,

$$PV^\gamma = P_1 V_2^\gamma = P_2 V_2^\gamma = C$$

$$\therefore \quad P = \frac{C}{V^\gamma}$$

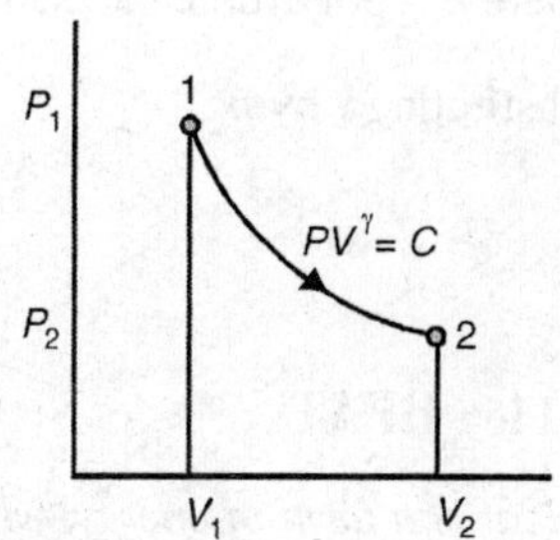

Fig. 1.24

Substituting in Eq. (2), we get,

$$W_{1-2} = \int_1^2 \frac{C}{V^\gamma} dV$$

$$W_{1-2} = \int_1^2 C V^{-\gamma} dV$$

$$= C\left[\frac{V^{-\gamma+1}}{-\gamma+1}\right]_1^2$$

$$= C\left[\frac{V_2^{-\gamma+1} - V_1^{-\gamma+1}}{-\gamma+1}\right] = \frac{CV_2^{-\gamma+1} - CV_1^{-\gamma+1}}{-\gamma+1}$$

$$= \frac{CV_2^{-\gamma+1}}{-\gamma+1} - \frac{CV_1^{-\gamma+1}}{-\gamma+1}$$

Substituting for C

$$= \frac{P_2V_2^{\gamma}.V_2^{-\gamma+1}}{-\gamma+1} - \frac{P_1V_1^{\gamma}.V_1^{-\gamma+1}}{-\gamma+1}$$

$$= \frac{P_2V_2^{\gamma-\gamma+1} - P_1V_1^{\gamma-\gamma+1}}{1-\gamma}$$

$$W_{1-2} = \frac{P_2V_2 - P_1V_1}{1-\gamma} \text{ or } \frac{P_1V_1 - P_2V_2}{\gamma-1} \quad (3)$$

Since $P_1V_1 = mRT_1$ and $P_2V_2 = mRT_2$

$$\mathbf{W_{1-2} = \frac{mR(T_2 - T_1)}{1 - r}}$$

and for unit mass
$$\mathbf{W_{1-2} = \frac{R(T_2 - T_1)}{1 - r}} \quad (4)$$

5. General law process or polytropic process. The law of the process is $PV^n = C$ where n is polytropic index. The work done for this process, $\int P.dV$ can be obtained by substituting γ by n.

$$\therefore \quad \mathbf{W_{1-2} = \frac{P_1 V_1 - P_2 V_2}{n-1}}$$

1.11 HEAT

"*Heat is a form of energy, which crosses the system boundary due to the temperature difference between the system and the surroundings.*"

When the two bodies, one hot and the other cold are kept in contact with each other, then the hot body loses heat and becomes colder and the cold body gains heat and becomes hotter and this process continues till the thermal equilibrium is reached. At this stage the two bodies will be at the same temperature.

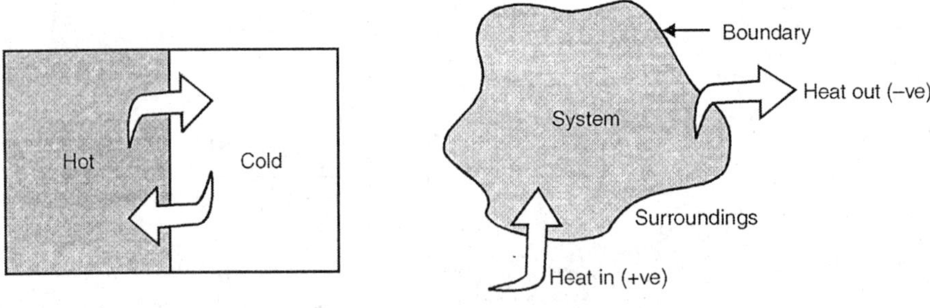

Fig. 1.25 **Fig. 1.26**

Like work, heat is a path function and we know that the differentials of path functions are imperfect differentials. If Q is the heat transfer, then the magnitude of heat transfer during the process 1–2 is given by,

$$\int_1^2 \delta Q = Q_{1-2} \text{ or } Q_2$$

$$\int_1^2 \delta Q \neq Q_2 - Q_1$$

Note. When heat flows into the system then it is taken as +ve and when heat flows out of the system then it is taken as –ve.

1.12 PRESSURE

It is defined as force per unit area i.e.

$$P = \frac{F}{A}$$

F = Force in newton (N)

A = Area is a m^2

Then the unit of pressure will be N/m^2, which is the basic unit of pressure, in SI units. This is also sometimes called as Pascal (Pa). Since this unit is very small, when compared to many engineering values, the units like, kPa, mPa, bar are used.

$$1 \text{ bar} = 10^5 \text{ N/m}^2 = 100 \text{ kN/m}^2 = 100 \text{ kPa}$$

Pressures are also measured in mm, or cm, of Hg or H_2O column. The pressure exerted by the atmosphere is known as atmospheric pressure and is denoted by 1 atm.

In various units, the value of 1 atm pressure are given by,

$$1 \text{ atm} = 760 \text{ mm of Hg} = 101325 \text{ N/m}^2 = 1.01325 \text{ bar.}$$
$$= 10.33 \text{ m of } H_2O = 1.033 \text{ kg/cm}^2$$

Another useful unit of pressure is 1 Torr and this is equal to Pressure exerted due to 1 mm of Hg = 133.34 N/m^2.

1.13 PRESSURE MEASUREMENT

Pressure is measured by using,

1. **Barometer.** This measures the atmospheric pressure.

2. **Bourdon pressure gauge.** This is used to measure pressure inside the pipes/containers etc.

3. **Manometers.** These are used to measure pressure, inside the pipes/vessels etc.

The atmospheric pressure acts both on the container and the gauge which measures the pressure inside the container.

When the pressure of the system is more (i.e. +ve pressure) than the atmosphere, then guage pressure is +ve and P_{abs} or P_{real} is given by,

$$P_{abs} = P_{aim} + P_{gauge}$$

If the pressure of the system is less (i.e. *s* –ve pressure) than the atmospheric pressure, its gauge pressure is –ve and it is known as vacuum and,

$$P_{abs} = P_{atm} - P_{vac}$$

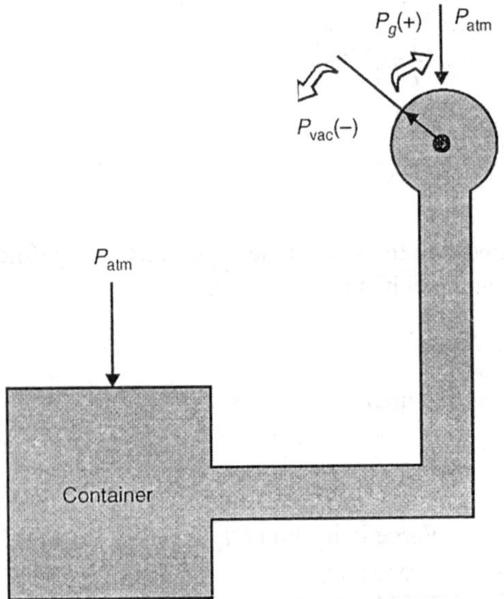

Fig. 1.27

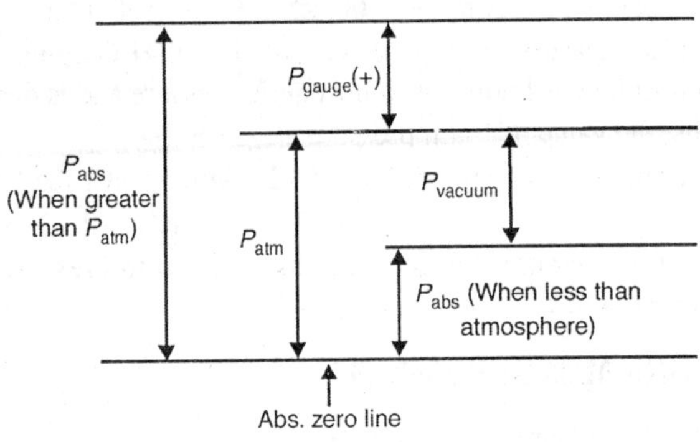

Fig. 1.28

Figure 1.28 shows the relationship between absolute and gauge pressure.

1.14 PRESSURE EXERTED DUE TO A COLUMN OF FLUID

Figure 1.29 shows a fluid of density ρ contained in container to a depth z or h. Now consider an elemental vertical cylinder of cross sectional area a.

Then we know that,

$$\text{Density} = \frac{m}{v}$$

$$\therefore \quad \text{Mass} = \text{density} \times \text{volume}$$

$$= \rho \times a \times Z$$

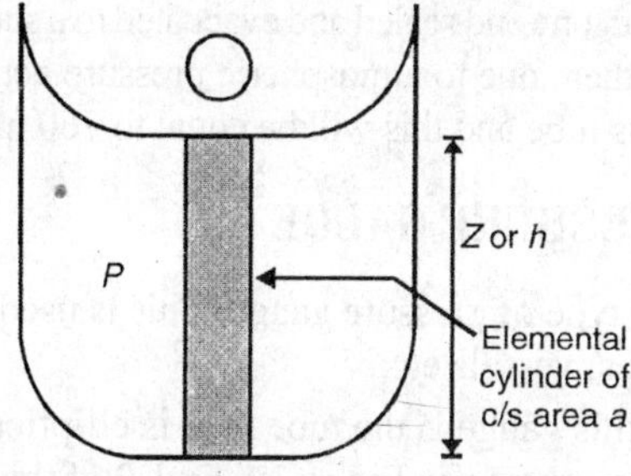

Fig. 1.29

But Force = mass × acceleration

$$= \rho . a . Z \times g$$

But pressure exerted due to a column of fluid at a depth Z

$$= \frac{\text{Force}}{\text{Area}} = \frac{\rho \cdot a \cdot Z \cdot g}{a}$$

$$\mathbf{P = \rho \; g \; Z}$$

Note.

Units
$$P = \frac{\text{kg}}{\text{m}^3} \times \frac{\text{m}}{\text{sec}^2} \times \text{m}$$

This we can write as, $= \text{kg} \times \frac{\text{m}}{\text{sec}^2} \times \frac{1}{\text{m}^2} = [\text{mass} \times \text{acceleration}] \times \frac{1}{\text{m}^2}$

$$P = \frac{\text{Force}}{\text{m}^2} = \frac{\text{N}}{\text{m}^2} \text{ resultant unit.}$$

1.15 BAROMETER

It is used to measure atmospheric pressure. Consider Hg in the container as shown in Fig. 1.31 (a) and (b). First of all, let a glass tube open at both the ends is immersed in the Hg as shown in Fig. 1.31 (a). Then the atmospheric pressure acting on the surface of Hg level is same as shown.

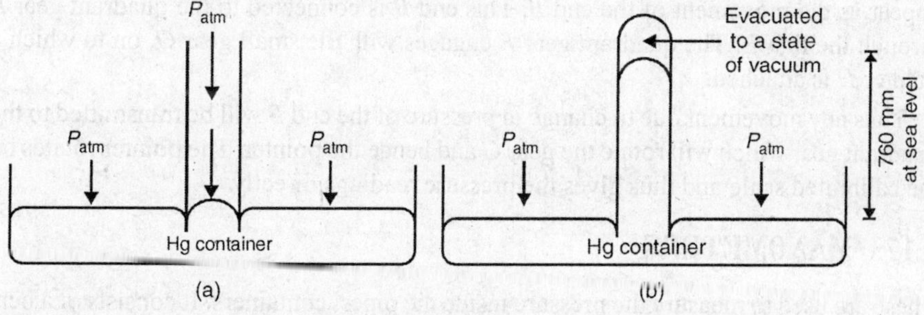

Fig. 1.30. *Barometer.*

Now if the glass tube with one end sealed and evacuated to a state of vacuum is immersed as shown in Fig. 1.31 (b) then, due to atmospheric pressure acting on the surface of Hg, the Hg will rise in the glass tube and this will be equal to 760 mm at sea level.

1.16 BOURDON PRESSURE GAUGE

This is the most common type of pressure gauge. This is used to measure the pressure inside the pipes, containers, vessels etc.

The basic element of this gauge is the tube *A*. It is elliptical in cross section and is bent into an arc of circle as shown in Fig. 1.32. End *B* of this tube is sealed, whereas open end *C* is connected to the connecting union *D* through which it can be mounted over the pipes or containers of whose pressures are to be measured.

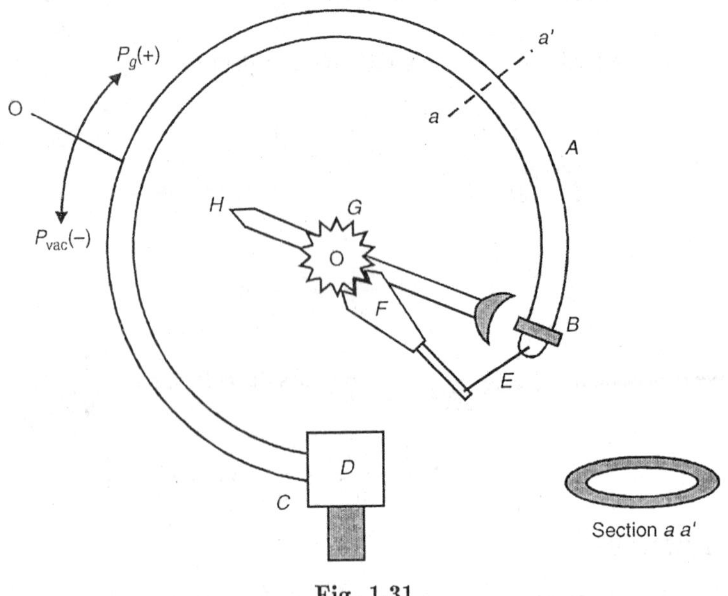

Fig. 1.31

If the pressure of the system is more than the atmosphere (i.e. +ve pressure) then the tube will tend to curl out. Conversely if the pressure of the system is less than atmosphere (i.e. –ve pressure) then the tube will tend to curl in. The change in pressure will therefore appear as the movement of the end *B*. This end *B* is connected to the quadrant gear *F* through the link *E*. The quadrant gear *F* engages with die small gear *G*, on to which a pointer *H* is attached.

Thus any movement due to change in pressure of the end *B* will be transmitted to the quadrant gear which will rotate the gear *G* and hence the pointer. The pointer rotates on the calibrated scale and thus gives the pressure readings directly.

1.17 MANOMETERS

These are used to measure the pressure inside the pipes, containers. It consists of a bent glass tube or a U-type glass tube which contains a liquid generally H_2O or Hg. One end of the manometer is connected to the container or pipe of whose pressure measurement is to be done and the other end is generally open to the atmosphere. The U-tube with one end closed and evacuated to state of vacuum indicates the absolute pressure directly as shown in Fig. 1.32(c).

Balancing the pressure values about the line 1–1′, we get for Fig. 1.32 (a).

$$P_{abs} = P_{abs} + \rho gZ = P_{abs} + P_{gauge}$$

For Fig. 1.32 (b) $P_{abs} + \rho gZ = P_{atm}$

i.e. $P_{abs} = P_{atm} - \rho gZ = P_{atm} - P_{gauge} = P_{atm} - P_{vac}$

For Fig. 1.32 (c) $P_{abs} = \rho gZ$

1.18 ZEROTH LAW OF THERMODYNAMICS (THERMAL EQUILIBRIUM)

When the two bodies one hot and the other cold, are placed in contact with each other, then the hot body loses heat and becomes colder and the cold body gains heat and becomes hotter, and this process continues till the thermal equilibrium is reached. In this state the two bodies will be at the same temperature.

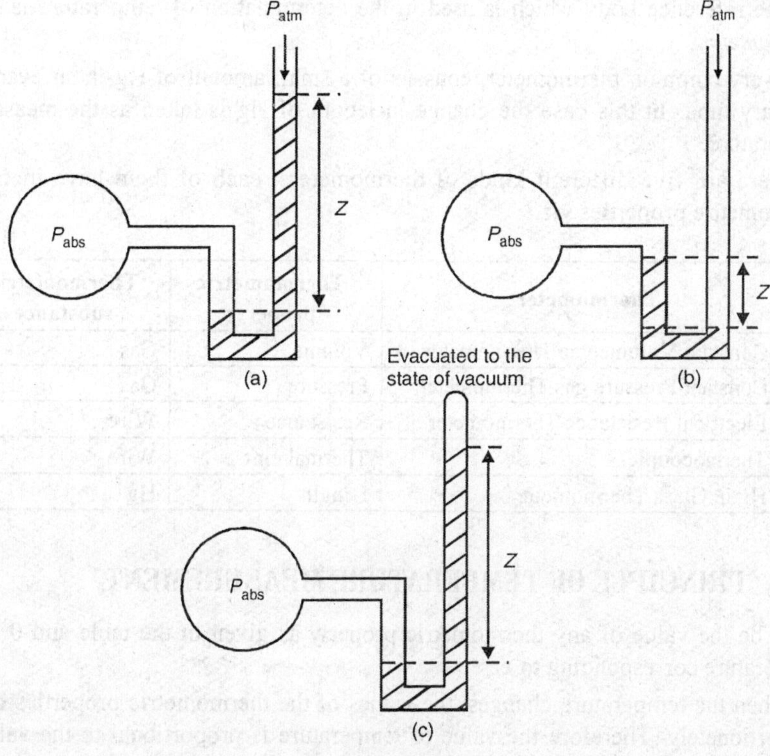

Fig. 1.32

From this a law is deduced known as *Zeroth law of thermodynamics.* It states that, "When a body *A* is in thermal equilibrium with the body *B* and also separately with the body C, then the bodies B and C will be in thermal equilibrium with each other. On the basis of this law, it is possible to compare the temperatures of B and C and it can be said that the temperatures of B and C are equal without actually making any contact between them. This law thus forms the basis of temperature measurement: by comparison method.

Thermal Equilibrium	**Thermal Equilibrium**	**Thermal Equilibrium**
A | *B*	*A* | *C*	*B* | *C*

If A is thermal equilibrium separately with B and C then B and C will also be in thermal equilibrium.

1.19 THERMOMETRIC PROPERTY AND THERMOMETERS

Temperature. It is the property of the system or of a body which distinguishes a hot body from cold one.

Thermometers. In order to obtain a quantitative measure of temperature, a reference body is used, and a certain physical characteristic of this body which changes with the temperature is selected. The selected characteristic is called the *thermometric property* and the reference body which is used in the determination of temperature is called *thermometer.*

A very common thermometer, consists of a small amount of Hg in an evacuated capillary tube. In this case the change in length of Hg is taken as the measure of temperature.

There are five different kinds of thermometers, each of them have their own thermometric properties viz.

Thermometer	Thermometric property	Thermometric substance
1. Constant Volume gas Thermometer	Volume	Gas
2. Constant Pressure gas Thermometer	Pressure	Gas
3. Electrical Resistance Thermometer	Resistance	Wire
4. Thermocouple	Thermal emf	Wire
5. Hg in Glass Thermometer	Length	Hg

1.20 PRINCIPLE OF TEMPERATURE MEASUREMENT

Let x be the value of any thermometric property as given in the table and θ be the temperature corresponding to x.

When the temperature changes, the values of the thermometric properties change proportionately. Therefore the value of temperature is proportional to the values of thermometric properties.

So we have, $\theta \propto x$

i.e. $$\theta = ax \qquad (1)$$

For Eqs (1) and (2) $\theta_1 = ax_1$

$$\theta_1 = ax_2 \quad (2)$$

where a is the constant of proportionality.

Dividing we get, $$\frac{\theta_1}{\theta_1} = \frac{x_1}{x_2} \quad (3)$$

Thus the ratio of temperature is equal to the ratio of values of properties, such as Length of Hg, volume, pressure etc.

Now, to construct a scale we need a reference point. The internationally accepted reference point is the *Triple point of water,* which is a temperature, at which all the 3-phases of water. i.e. ice, water and vapour, exists in equilibrium. And the value of triple point temperature is 273.16 K (or 0.01°C) at a pressure of 0.006 bar, where *K* stands for degree Kelvin.

∴ Equation (1) becomes,

$$273.16 = ax_t \Rightarrow \frac{273.16}{x_t} = a$$

where x_t is the value of the property at triple point.

Now susbtitute this value of a in Eq. (1)

i.e. $$\theta = ax \text{ we get, } \theta = \frac{273.16}{x_t} \cdot x$$

$$\theta = 273.16 \times \frac{x}{x_t}$$

Thus for any change in value of x (i.e. Thermometric property) the temperature θ can be found out. Thus from this equation. we can construct a scale and calibrate it to give us direct temperature readings. Here it is to be noted that x_t is the value of the property at triple point and is constant.

1.21 SCALE OF TEMPERATURE

A scale of temperature is based on some reference or fixed point to which an arbitrary value of temperature is assigned. Thus on a thermometer with some fixed or reference point, any temperature has a unique numerical value w. r. t. the fixed reference point.

Scales of Temperature

(1)(a) Celsius scale, (b) Fahrenheit scale.

(2) Absolute Temperature scale.

(3) Ideal Gas temperature scale.

1. (a) Celsius Scale. This scale is named after Anders the Celsius, who devised this scale. There are two fixed points on this scale. One is ice point and the other is the steam point. These two points are numbered as 0 and 100 on Celsius scale. The interval between them is divided into 100 equal parts and each division, is known as one degree Celsius and degree Celsius denoted as °C.

(b) Fahrenheit Scale. On this scale the ice point is numbered as 32 and steam point is numbered as 212. The interval between them is divided into 180 equal parts and each division, is known as 1 °F.

The temperature readings on one scale can be converted into a reading on the other scale by the following formula.

$$\frac{C}{100} = \frac{F-32}{180}$$

$$\therefore \quad C = \frac{100}{180}(F-32) = \frac{5}{9}(F-32)$$

and

$$F = \frac{9}{5}(C)+32$$

2. Absolute Temperature Scale or Kelvin Temperature Scale or Thermodynamic Temperature Scale. This is an internationally accepted scale. In this scale the temperature is measured from absolute zero temperature.

Absolute zero is the temperature at which all the vibratory, translatory and rotational motion of the molecules of a gas, is supposed to cease. A gas on cooling contracts in volume. In case of perfect gases, 'Charle's found that decreases in volume per degree centigrade decrease in temperature is 1/273. 16th of its volume at 0°C and pressure remaining constant. Thus the volume of gas will be zero at temperature –273.16°C. This temperature 273.16°C below or behind 0°C (or –273.16°C) is called the *absolute zero temperature.*

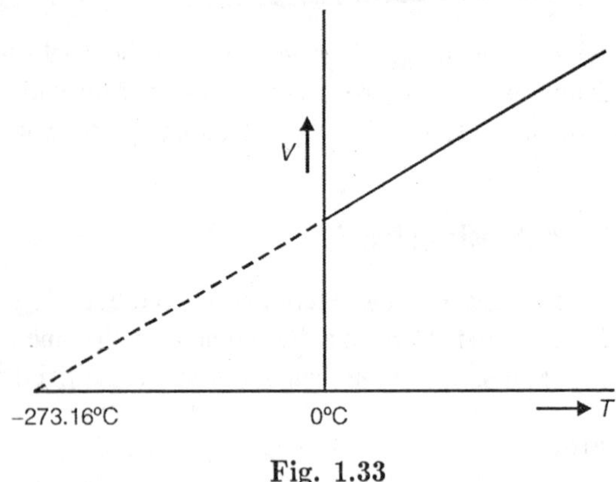

Fig. 1.33

In this scale there is one more reference point taken, the Triple point of water. This is a temperature at which all the three phases of water exists in equilibrium, and the value of triple point temperature is 0.01 °C or 273.16 K at a pressure of 0.006 bar. Here it is to be noted that, the ice point temperature on this scale is 273.15 K, and steam point temperature is 373.15 K and therefore to convert the temperature given in °C to K, just add 273.15.

$K = °C + 273.15$ or in general, $K = °C + 273$. Here it is to be noted that the value of Absolute zero on this scale is –273.16°C.

3. Ideal Gas-temperature Scale. This is similar to absolute temperature scale. This scale is constructed by using constant volume gas thermometer.

In a constant volume gas thermometer, gas will be used as the working substance and as all the gases behave as ideal gases at low pressure, thus scale is known as Ideal gas temperature scale. The value of absolute zero on this scale is same as that given by Absolute temperature scale i.e. – 273.16°C.

1.22 MICROSCOPIC AND MACROSCOPIC POINT OF VIEW

There are two points of views.

1. Microscopic or Statistical Thermodynamics.
2. Macroscopic or Classical Thermodynamics. From which the behaviour of matter or working of a system can be studied.

From microscopic point of view, it is considered that, the matter is not continuous, but it is made up of a large number of identical particles called molecules. For example, consider a gas in a cylinder as a system. Then this gas is made up of number of molecules. Each molecule of a gas, at a given instant, has certain position, energy, etc., and for each molecule these change very rapidly because of the collisions (strikings). The behaviour of the gas is described by summing up the behaviours of each molecule. And for summing up the behaviour of molecules statistical methods are employed and hence it is also called as *statistical thermodynamics.*

In macroscopic point of view a certain quantity of matter is considered, and the events that are taking place at the molecular level are not taken into account. For example, let us consider a system of an IC Engine consisting of a charge in the engine cylinder. At any instant the system has certain volume depending upon the position of the piston, this volume is easily measurable.

Another quantity to describe the system is the pressure of the gas inside the cylinder. A pressure gauge can be used to measure the same. Similarly temperature, chemical composition etc. may be described. Thus, in macroscopic point of view the system will be described by large scale properties.

Though the approach based on microscopic point of view seems to be different, from that based on macroscopic point of view, but there exists a relationship between them. The relationship between macroscopic and microscopic point of view lies in the fact that the macroscopic properties are in fact the average properties of a large number of microscopic characteristics.

Hence, when both the methods are applied to a particular system, they give the same result.

Points to Remember:

- The microscopic point of view postulates the existence of molecules, their movement and collisions and hence it is constantly being changed. Whereas the macroscopic properties can be measured and their existence can be felt by our senses.

1.23 ENERGY IN TRANSIT

When energy flows from one system (or one body) to another system (other body), then it is referred as energy in transit.

In general heat and work are called energies in transit. The energy transfer is due to some energy potential, e.g.

(i) Mechanical energy or work energy are transferred due to difference in pressure potential.
(ii) Electrical energy is transferred due to different in voltage potential.
(iii) Heat energy is transferred due to difference in temperature potential.

LIST OF FORMULAE

Note. (1) Unit of work in SI units is Nm or joules

(2) Unit of pressure in SI units is N/m² or Pa

(3) 1 bar = 10^5 N/m^2= 100 kN/m^2 = 100 kPa

(4) 1 atm = 760 mm of Hg

= 10.33 m of water

= 101325 N/m^2= 1.01325 bar

= 1.033 kg/cm^2

SOLVED EXAMPLES

Example 1.1 A fluid in a cylinder is at a pressure of 700 kN/m^2. It is expanded at constant pressure from a volume of 0.28 m^3 to a volume of 1.68 m^3. Determine the work done.

Data: $P = 700\text{ kN/m}^2 = 700 \times 10^3\text{ N/m}^2$

Solution

Expanded at constant pressure,

$$V_1 = 0.28\text{ m}^3 \qquad V_2 = 1.68\text{m}^3$$

We know that, work done $= \int_{V_1}^{V_2} PdV$

i.e. Work done $= W_{1-2} = P\ (V_2 - V_1)$ since $(P = C)$

Note : Units $\frac{N}{m^2} \times m^3 = Nm$

$$= 700 \times 10^3 (1.68 - 0.28) \text{ Nms}$$

$$= 700 \times 10^3 \times 1.4 \text{ Nm or Joules}$$

$$= 980000 \text{ J} = 9.8 \times 10^5 \text{ J}$$

$$\mathbf{W_{1\text{-}2} = 980\ kJ = 0.98\ MJ}$$

Fig. Ex. 1.1

Example 1.2 0.112 m³ of gas has a pressure of 138 kN/m². It is compressed to 690 kN/m², according to the law $PV^{1.4} = C$. Determine the new volume of the gas.

Data: The gas is compressed according to the $PV^{1.4} = C$

$P_1 = 138$ kN/m² $\quad P_2 = 690$ kN/m²

$V_1 = 0.112$ m³ $\quad V_2 = ?$

Solution

We know that polytropic process is given by $PV^n = C$

i.e. $$PV^n = P_1V_1^n = P_2V_2^n = C$$

$\therefore$ $$P_1V_1^{1.4} = P_2V_2^{1.4}$$

$$\frac{P_1}{P_2} = \left[\frac{V_2}{V_1}\right]^{1.4}$$

or $$\frac{V_2}{V_1} = \left[\frac{P_1}{P_2}\right]^{\frac{1}{1.4}}$$

or $$V_2 = V_1 \times \left[\frac{P_1}{P_2}\right]^{\frac{1}{1.4}}$$

$$= 0.112 \text{ m}^3 \times \left[\frac{138 \text{ kN/m}^2}{690 \text{ kN/m}^2}\right]^{1/1.4}$$

$$= 0.112 \text{ m}^3 \times (0.2)^{1/1.4} = 0.112 \times 0.3168$$

$$\mathbf{V_2 = 0.03555\ m^3}$$

Example 1.3 0.014 m³ gas at a pressure of 2070 kN/m² expands to a pressure of 207 kN/m², according to the law $PV^{1.35} = C$. Determine the work done by the gas during the expansion.

Data: Gas expands according to the law $PV^{1.35} = C$

$V_1 = 0.14\ m^3$ $\quad P_1 = 2070\ kN/m^2$

$P_2 = 207\ kN/m^2$ $\quad$ Work done $= ?$

Solution

The work done during a polytropic process is given by,

$$W_{1-2} = \frac{P_1V_1 - P_2V_2}{n-1} = \frac{P_2V_2 - P_1V_1}{1-n}$$

Since V_2 is not given, to determine V_2,
We know that since it is a polytropic process,

$$P_1V_1^n = P_2V_2^n$$

$$\frac{P_1}{P_2} = \left[\frac{V_2}{V_1}\right]^n$$

$$V_2 = V_1 \times \left[\frac{P_1}{P_2}\right]^{\frac{1}{n}} = 0.014\ m^3 \left[\frac{2070\ kN/m^2}{207\ kN/m^2}\right]^{\frac{1}{1.35}}$$

$$= 0.014 \times 5.5048\ m^3$$

$$\mathbf{V_2 = 0.0771\ m^3}$$

$$\text{Work done} = W_{1-2} = \frac{P_1V_1 - P_2V_2}{n-1}$$

$$= \frac{2070 \times 10^3\,N/m^2 \times 0.014\,m^3 - 207 \times 10^3\,N/m^2 \times 0.077\,m^3}{1.35-1}$$

$$= \frac{28980 - 15959.7}{0.35} = \frac{13020.3}{0.35} = 37200.857\ Nm$$

$$= 37.2 \times 10^3\ \text{Nm or Joules}$$

$$\mathbf{W_{1-2} = 37.2\ kJ}$$

Example 1.4 A gas is compressed hyperbolically (Isothermally) from a pressure and volume of 100 kN/m² and 0.056 m³ respectively to a volume of 0.007 m³. Determine the final pressure and the work done on the gas.

Data: Since the gas is compressed Isothermally, $PV = C$ i.e. $P_1V_1 = P_2V_2$

$$P_1 = 100\ kN/m^2 = 100 \times 10^3\ N/m^2$$

$P_2 = ?$ $\quad$ Work done $= ?$

$V_1 = 0.056\ m^3$ $\quad V_2 = 0.007\ m^3$

Solution

Since it is an Isothermal process $P_1V_1 = P_2V_2$

$$P_2 = \frac{P_1V_1}{V_2}$$

$$P_2 = \frac{100\times1000\ \text{N/m}^2\times0.056\ \text{m}^3}{0.007\ \text{m}^3}$$

$$P_2 = \frac{5600}{0.007}$$

$$P_2 = 8\times10^5\ \text{N/m}^2$$

$$\mathbf{P_2 = 8\times10^2\,kN/m^2 = 800\ kN/m^2}$$

We know that, Work done $= P_1V_1 \ln\dfrac{V_2}{V_1}$

$$= 100\times1000\times0.056\times\ln\left(\frac{0.007}{0.056}\right)$$

$$= 100\times1000\times0.056\times-2.079$$

$$W_{1-2} = -11644.973\ \text{Nm} = -11.65\ \text{kNm or kJ}$$

∴ The work done on the gas = 11.65 kJ

Example 1.5 A non-flow reversible, quasi-static process occurs for which pressure $P = [V^2 + 8/V]$ bar. Determine the work done if volume changes from 1 m³ to 3 m³.

Data: $P = [V_2 + 8/V]\ \text{bar} = [V^2 + 8/V]\times10^5\ \text{N/m}^2$

$V_1 = 1\text{m}^3,\quad V_2 = 3\ \text{m}^2,\quad$ work done = ?

Solution

We know that, work done in a closed system or displacement work done

$$W_{1-2} = \int PdV$$

$$W_{1-2} = \int_{V_1=1}^{V_2=3} PdV, \text{ subtitute for } P,$$

$$= \int_1^3\left[V^2+\frac{8}{V}\right]\times10^5\,dV = 10^5\int_1^3\left[V^2+\frac{8}{V}\right]dV$$

$$= 10^5\left[\frac{V^3}{3}+8\ln V\right]_1^3 = 10^5\left[\frac{3^3-1^3}{3}+8\ln\frac{3}{1}\right]$$

$$= 10^5\left[\frac{27-1}{3}+8\times 1.0986\right]=10^5\left[\frac{26}{3}+8.789\right]$$

$$\mathbf{W_{1\text{-}2} = 1745556.5\ Nm = 17.455\times 10^5\ Nm\ or\ Joules}$$

Example 1.6 In a certain thermodynamic process of an ideal gas, the volume changes from 0.2 m³ to 0.5 m³ while pressure change occurs according to the law $P = 1500 \left[\frac{V}{100}+1\right]$. Where P is in N/m² and V is in m³, then find out the work done by the gas in kJ.

Data: $V_1 = 0.2\,m^3$; $V_2 = 0.5\,m^3$ and $P=1500\left[\frac{V}{100}+1\right]$ N/m²

Solution

We know that, work done

$$= W_{1-2} = \int PdV;\ \text{substitute for } P$$

$$= \int_{0.2}^{0.5} 1500\left[\frac{V}{100}+1\right]dV = 1500\times\int_{0.2}^{0.5}\left[\frac{V}{100}+1\right]dV$$

$$1500\times\left[\frac{V^2}{2\times 100}+V\right]_{0.2}^{0.5} = 1500\times\left[\frac{0.5^2-0.2^2}{200}+(0.5-0.2)\right]$$

$$= 1500\times(0.00105+0.3) = 451.575\ \text{Nm}$$

$$\mathbf{W_{1\text{-}2} = 0.4515\ kJ}$$

Example 1.7 Show that for Van-der-Waals equation,

$$\left[P+\frac{a}{V^2}\right](V-b) = mRT$$

The work done is given by, $mRT\log_e\frac{V_2-b}{V_1-b} - a\left[\frac{1}{V_1}-\frac{1}{V_2}\right]$, where V_1 and V_2 are initial and final volumes.

Solution

From the given equation i.e.

$$\left[P + \frac{a}{V^2}\right](V - b) = mRT$$

$$P = \frac{mRT}{V-b} - \frac{a}{V^2}$$

$$\therefore \text{ Work done} = W_{1-2} = \int PdV = \int_1^2 \left[\frac{mRT}{V-b} - \frac{a}{V^2}\right] dV$$

$$= mRT \int_1^2 \frac{1}{V-b} dV - \int_1^2 \frac{a}{V^2} dV$$

$$\mathbf{W_{1-2} = mRT \log_e \frac{V_2 - b}{V_1 - b} - a\left[\frac{1}{V_1} - \frac{1}{V_2}\right]}$$

Example 1.8 A gas in a cylinder and piston arrangement comprise the system. Gas expands frictionlessly from 1.5 m^3 to 2 m^3 while receiving 2 Nm of work from a Paddle wheel. Pressure of gas remains constant at 6 N/m^2. Determine the net work done by the system.

Solution

Work done by the gas on the piston,

W_{gas} = Force × Displacement = $F \times dl = P \times A \times dl = P \times dV$

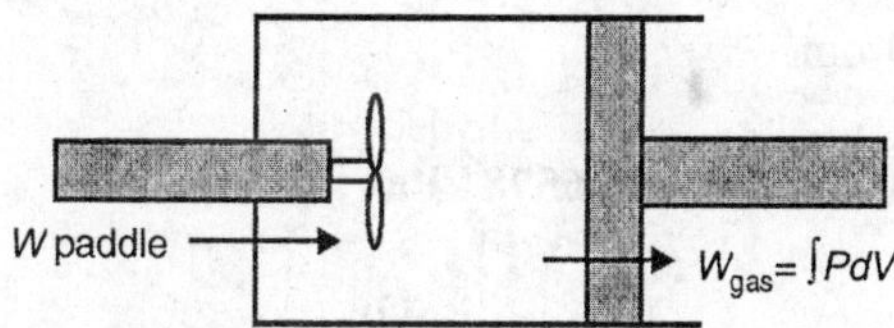

Fig. Ex. 1.8

The total work done by the gas, when the volume changes from V_1 to V_2 and the pressure remaining constant = $\int PdV$

i.e. $$W_{gas} = \int_1^2 PdV$$

$$= P \int_{1.5}^{2} dV \text{ since } P \text{ is constant}$$

$$= 6[V]_{1.5}^{2}$$

$$= 6[2-1.5] = 3 \text{ Nm}$$

But gas receives 2 Nm of work from the paddle wheel,

∴ **Net work done by the system = 3 – 2 = 1 Nm**

Example 1.9 Convert the following readings of pressure to kPa assuming that the barometer reads 760 mm of Hg

(a) 80 cm of Hg
(b) 40 cm of Hg vacuum
(c) 1.5 mt of water
(d) 5.2 bar

Solution

Let us assume ρ of Hg = 13596 kg/m³

and g = 9.806 m/sec², if the values are not given

Let Z_o = Barometric pressure = 760 mm of Hg

∴ Pressure exerted due to 760 mm of Hg column in the barometer,

$$P_{atm} = \rho gh \text{ or } \rho g Z_o = 13596 \times 9.806 \times \frac{760}{1000}$$

$$= 101325.01 \text{ N/m}^2 \text{ or Pa} = 101.325 \text{ kN/m}^2 \text{ or kPa}$$

(a) Pressure of 80 cm of Hg

For 76 cm = 101.325 kPa

∴ 80 cm = ?

$$\mathbf{\frac{80}{76} \times 101.325 = 106.6578 \text{ kPa}}$$

Or

We know that, pressure exerted due to a column of fluid = $\rho.g.Z$

$$= 13596 \times 9.806 \times 0.8 = 106657.8 \text{ N/m}^2$$

$$= 106.657 \text{ kPa}$$

(b) 40 cm of Hg vacuum

In this case, $P_{abs} = P_{atm} - P_{vac} = 76 \text{ cm} - 40 \text{ cm} = 36 \text{ cm}$

Since for 76 cm – 101.325 kPa

∴ 36 cms – ?

$$\frac{36}{76} \times 101.325 = 47.996 \text{ kPa}$$

(c) Pressure due to 1.5 mt of H_2O column on kPa

$$= \rho_w . g . Z_w = 1000 \times 9.806 \times 1.5$$

$$= \mathbf{14709\ N/m^2\ or\ Pa = 14709\ kPa}$$

(d) 5.2 bar $= 5.2 \times 10^5 \text{N/m}^2 \text{or Pa}$

$$= \mathbf{5.2 \times 10^2\ kPa = 520\ kPa}$$

Note. Pressure of 1 atm. equals 760 mm of Hg or 101325 N/m^2. These are standard values. To get this valve of 101325 N/m^2, we have to use ρ of Hg = 13596 kg/m^3 and g = 9.806 m/sec^2.

If we use ρ of Hg = 13.6 × 10^3 kg/m^3 and g = 9.81 m/sec^2 then we get P_{atm} = 101396 N/m^2.

Example 1.10 Convert the following readings of pressure into kPa (Abs) assuming barometric pressure of 750 mm of Hg.

(i) 55 cm of Hg (Abs)
(ii) 3.3 Atm (Gauge)
(iii) 3.2 m of H_2O (Gauge)
(iv) 4.6 bar (Abs)

Solution

For $$P_{ptm} = \rho_{Hg} \cdot g \cdot Z_o = \frac{13596 \times 9.81 \times 0.75}{1000}$$

$$= 100.32 \text{ kPa}$$

(i) 55 cm of Hg (Abs)

$$P_{abs} = 13596 \times 9.81 \times 0.55 = 73357 \text{ N/m}^2$$

$$= 73.35 \text{ kPa}$$

(ii) 3.3 atm (Gauge)

Since 1 atm = 101325 N/m^2

$\therefore$ 3.3 atm = $3.3 \times 101325 \text{ N/m}^2$

$$\therefore \quad P_{abs} = P_{atm} + P_{gauge} = 100.032 + \frac{3.3 \times 101325}{1000}$$

$$= 434.405 \text{ kPa}$$

(iii) 3.2 m of H_2O (Gauge)

$$P_{abs} = P_{atm} + P_{gauge} = 100.032 + \frac{100 \times 9.81 \times 3.2}{1000}$$

$$= 131.42 \text{ kPa}$$

(iv) 4.6 bar (Abs)

Since 1 bar $= 10^5 \text{N/m}^2 = 10^2 \text{ kN/m}^2$

$$\therefore \quad 4.6 \text{ bar} = 4.6\times100 \text{ kPa}$$
$$= \mathbf{460\ kPa}$$

Example 1.11 A manometer shows a reading of 50 cm of water. If atmospheric pressure is 763 mm of Hg. Find absolute pressure in kPa. Take ρ of Hg = 13.6 × 10³ kg/m³, g = 9.81 m/sec².

Solution

$$P_{abs} = P_{atm} + P_{gauge}$$

$$P_{atm} = \rho.g.Z_o = 13.6\times10^3 \frac{\text{kg}}{\text{m}^3}\times9.81\frac{\text{m}}{\text{sec}^2}\times0.765 \text{ m}$$

$$= 102063.24 \text{ N/m}^2 \text{ or Pa}$$

$$= 102.063 \text{ kPa}$$

and $P_{manometer}$ or $P_{gauge} = \rho_w.g.Z_w = 1000\times9.81\times0.50 = 4905 \text{ kPa} = 4.905 \text{ kPa}$

$$P_{abs} = P_{atm} + P_{gauge}$$

$$P_{abs} = 102.063 + 4.905 = 106.968 \text{ kPa}$$

Example 1.12 A vacuum gauge mounted on the condenser reads 70 cm of Hg. If atmospheric pressure is 101.325 kPa. Find absolute pressure in kPa.

Solution

We know that $P_{abs} = P_{atm} - P_{vac}$

To find $P_{vac} = \rho\cdot g\cdot Z = 13.6\times10^3\times9.81\times0.70$

$$= 93391.2 \text{ Pa} = 93.3912 \text{ kPa}$$

$$\mathbf{P_{abs} = 101.325 - 93.3912 = 7.9338\ kPa}$$

Example 1.13 A vacuum gauge connected to the tank reads 30 kPa at a location where a barometric reading is 755 mm of Hg. Determine the absolute pressure in the tank. Take density of Hg = 13590 kg/m³.

Solution

$$P_{vac} = 30 \text{ kPa}$$

$$Z_o = 755 \text{ mm of Hg}$$

$$P_{abs} = ?$$

$$P_{abs} = P_{atm} - P_{vac} \qquad (1)$$

To find P_{atm} $\quad P_{atm} = \rho_{Hg}\cdot g\cdot Z_o = 13590\times9.81\times0.755$

$$P_{atm} = 100655.01 \text{ N/m}^2$$

$$P_{atm} = 100.655 \text{ kPa} \quad (2)$$

$\therefore$ From (1) $\quad P_{abs} = P_{atm} - P_{vac} = 100.655 - 30$

$$\mathbf{P_{abs} = 70.655 \text{ kPa}}$$

Example 1.14 The pressure of a air in a pipe line measured with Hg–manometer having one arm open to the atmosphere. The level in the open arm is 562 mm higher than the arm connected to the air pipe. Calculate the air pressure. The barometer reads 760 mm of Hg, the acceleration due to gravity is 9.81 m/sec² and the density of Hg is 13596 kg /m³.

Data: $\quad z = 0.560$ m $\quad Z_o = 0.760$ m

$\quad g = 9.81$ m/sec² $\quad \rho_{Hg} = 13596$ kg/m²

Solution

Note: Since the plane 1–1′ is horizontal plane, and is in the same continuous static mass of liquid, the pressure heads at the marked (*) points in figure will be equal.

Therefore the pressure along a horizontal line 1–1′ is ,same. Balancing the pressure values about the line 1–1′ we get,

$$P_{abs} = P_{atm} + \rho \cdot g \cdot Z \quad (1)$$

But $\quad P_{atm} = \rho \times g \times Z_o$

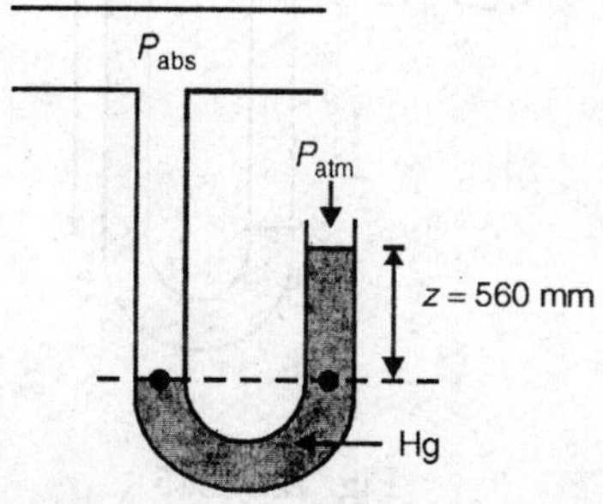

Fig. Ex. 1.14

$$= 13596 \times 9.81 \times 0.760$$

$$= 101366.3376 \text{ N/m}^2$$

$$= 101.37 \text{ kN/m}^2 = \text{kPa}$$

The pressure exerted due to the column of Hg of height $Z = \rho \times g \times Z$

$$= 13596 \times 9.81 \times 0.56 = 74690.98 \text{ N/m}^2 = \text{Pa}$$

$$= 74.69 \text{ kN/m}^2 = \text{kPa}$$

∴ From (1), (2) and (3)

$$\mathbf{P_{abs} = 101.37 + 74.69}$$
$$\mathbf{= 176.06\ kPa}$$

Example 1.15 A U-tube manometer contains a liquid having a density of 800 kg/m^3. When the manometer is connected to a gas pipe, the level in the open arm is 30 cm higher than the level in the column connected to the gas pipe. Find the absolute pressure of the gas in bar, if the barometric pressure is 75 cm of Hg.

Solution

We have, Density of liquid (ρ_{liq}) =800 kg/m^3

$Z = 30$ cm

∴ We have, gauge pressure

$$= \rho g Z$$
$$= 800 \times 9.81 \times 0.3$$
$$= 2354.4\ \text{N/m}^2 = 0.023544\ \text{bar}$$

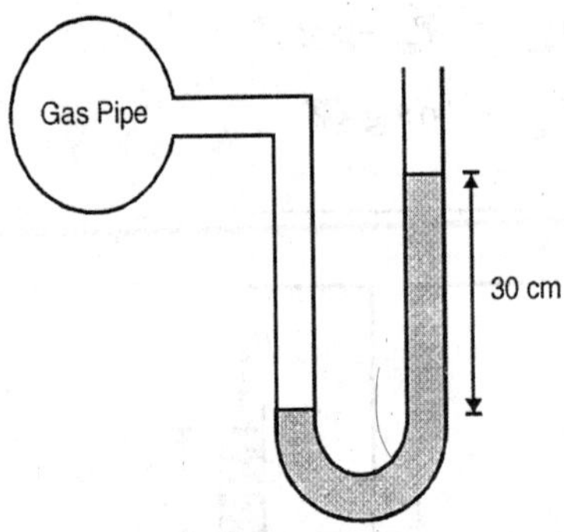

Fig. Ex. 1.15

Now $\quad P_{atm} = \rho.g.Z = 75$ cm of Hg

$$= 13.596 \times 10^3 \times 9.81 \times 0.75$$
$$= 99991.78\ \text{N/m}^2$$
$$= 0.9999\ \text{bar}$$

Also, $\quad P_{abs} = P_{gauge} + P_{atm} = 0.023544 + 0.9999$

$$\mathbf{P_{abs} = 1.234618\ bar}$$

Example 1.16 A U-tube manometer is connected to a gas pipe. The level of the liquid in the manometer arm open to the atm. is 17 cm lower than the level of liquid in the arm connected to the gas pipe. The liquid in the manometer has specific gravity of 0.8. Find the abs pressure of the gas if the barometer reads760 mm of Hg. Take $\rho_{Hg} = 13596$ and $g = 9.806$.

Data: $Z = 0.17$ m $P_{abs} = ?$

$Z_o = 0.76$ Sp. gr. $= 0.8$

Solution

Note: Specific gravity $= \dfrac{\rho \text{ of given liquid}}{\rho \text{ of water}}$

$\therefore$ ρ of given liquid $= 0.8 \times 10^3$ kg/m^3

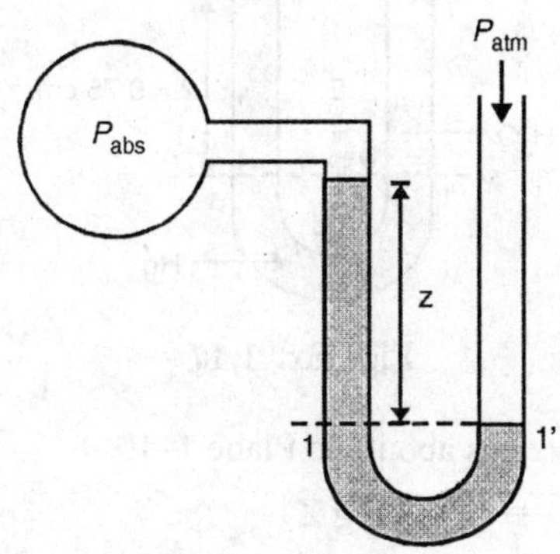

Fig. Ex. 1.16

Balancing the pressure values about the plane 1–1′

$$P_{abs} + \rho g Z = P_{atm}$$

$$P_{abs} = P_{atm} - \rho \cdot g \cdot Z$$

(a) $$P_{atm} = \rho \cdot g \cdot Z_o \qquad (1)$$

$$= 13596 \text{ kg/m}^3 \times 9.806 \text{ m/sec}^2 \times 0.7 \text{ m}$$

$$= 101325 \text{ N/m}^2$$

and $$\rho g Z = 0.8 \times 10^3 \times 9.806 \times 0.17 = 1333.616 \text{ N/m}^2$$

$\therefore$ $$P_{abs} = 101325 - 1333.616 = 99991.384 \text{ N/m}^2$$

$$\mathbf{P_{abs} = 0.999913 \text{ bar}}$$

Example 1.17 A U-tube manometer with one arm open to atmosphere is used to measure the pressure of steam flowing through a pipe. Hg is used in the manometer. The height of the column open to the atm is 9.75 cm greater than the Hg level in the arm connected to the steam pipe. Steam condenses in the manometer on the steam side.

The column of condensate is 3.4 cm. The atmospheric pressure is 76 cm of Hg. Find the absolute pressure of steam. Take,

$$g = 9.806 \text{ m/sec}^2 \quad \text{and} \quad \rho_{Hg} = 13596 \text{ kg/m}^3$$

$$P_{abs} = P \text{ of steam} = ?$$

Data: $P_{abs} = ?$ $\quad Z = 9.75$ cm

Solution

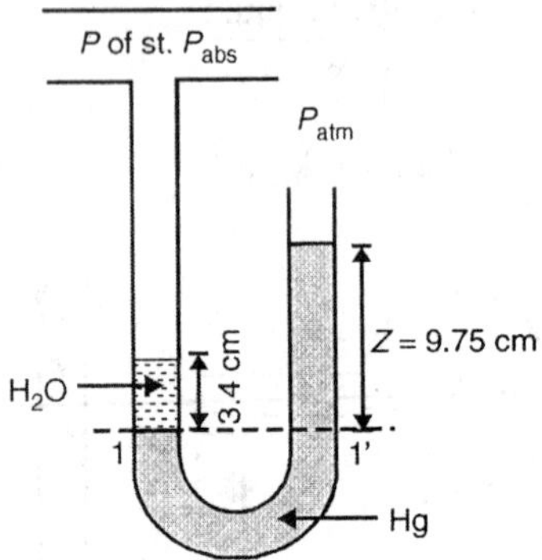

Fig. Ex. 1.17

Balancing the pressure values about the Plane 1–1′

$$P_{abs} + \rho_w \cdot g \cdot Z_w = P_{atm} + \rho\, g\, Z$$

Now to find,

(a) $$\rho_w \cdot g \cdot Z_w = 1000 \times 9.806 \times 0.034$$

$$= 333.404 \text{ N/m}^2 \qquad (2)$$

(b) $$P_{atm} = \rho \cdot g \cdot Z_o = 13596 \times 9.806 \times 0.76$$

$$= 101325 \text{ N/m}^2 \qquad (3)$$

(c) $$\rho \cdot g \cdot Z = 13596 \times 9.806 \times 0.0975$$

$$= 12998.932 \text{ N/m}^2 \qquad (4)$$

∴ Now substituting the values of Eqs (2), (3) and (4) in Eq. (1) we get,

$$P_{abs} + 333.404 = 101325 + 12998.932$$

$$P_{abs} = 114323.93 - 333.404 = 113990.53 \text{ N/m}^2$$

$$\mathbf{P_{abs} = 1.139905 \text{ bar}}$$

Example 1.18 A Hg manometer is used to measure the pressure of steam in the mains. (i.e. main steam pipe). Its one arm having higher level of Hg is open to atmosphere at

98 kPa. The other arm is connected to steam pipe. The manometer records a level difference of 60 cm of Hg. On the lower level of Hg, 4 cm of water column accumulates due to condensation of steam. Find the absolute pressure of steam in bar.

Data:

$$g = 9.7 \text{ m/sec}^2$$

$$\rho_{Hg} = 13.69 \text{ gm/cc} = 13.69 \times 10^3 \text{ kg/m}^3.$$

$$\rho_{water} = 1000 \text{ kg/m}^3$$

Solution

Take $g = 9.7 \text{ m/sec}^2$ and $\rho_{Hg} = 13.69$ gm/cc.

$$\rho_w = 1000 \text{ kg/m}^3$$

Now balancing pressure values the plane 1–1′

$$P_{abs} + \rho_w g Z_w = P_{atm} + \rho \cdot g \cdot Z \quad (1)$$

Now to find,

(a)

$$\rho_w g Z_w = 1000 \times 9.7 \times 0.04$$

$$= 388 \text{ N/m}^2$$

$$= 0.338 \text{ kN/m}^2 \text{ or kPa} \quad (2)$$

(b)

$$\rho \cdot g \cdot Z = 13.69 \times 10^3 \times 9.7 \times 0.6$$

$$= 79675.8 \text{ N/m}^2 = 79.6758 \text{ kPa}$$

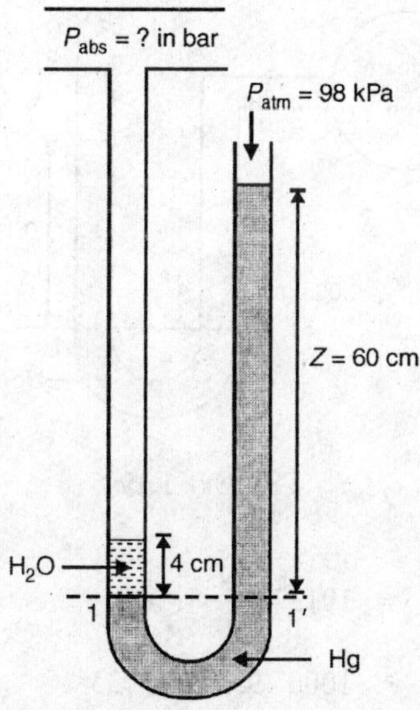

Fig. Ex. 1.18

∴ From (1)

$$P_{abs} = P_{atm} + \rho g Z - \rho_w g Z_w$$

$$= 98 + 79.68 - 0.388$$

$$= 177.68 - 0.388$$

$$= 177.292 \text{ kPa}$$

Since $\quad 1 \text{ bar} = 10^5 \text{ N/m}^2 = 100 \text{ kPa}$

$$\mathbf{P_{abs} = 1.77292 \text{ bar}}$$

Example 1.19 A U-tube manometer with one arm open to atmosphere is used to measure the pressure of steam flowing through a pipe. Mercury is used as a manomeric fluid. The height of the column of mercury in the open end is 10 cm greater than the mercury level in the arm connected to the steam pipe. Steam condenses in the manometer on the steam side. The column of condensate is 3.5 cm in height. The atmospheric pressure is 760 mm mercury. What is the absolute pressure of steam in the pipe in bar ?

Solution

Data: $\quad Z_o = 0.76 \text{ m} \qquad P_{abs} \text{ or } P_{steam} = ?$

$g = 9.806 \qquad \rho_{Hg} = 13596$

$$P_{abs} + \rho_w g Z_w = P_{atm} + \rho_{Hg}\, g\, Z_{Hg}$$

$$P_{atm} = \rho_{Hg}\, g\, Z_o$$

$$= 13596 \times 9.806 \times 0.76$$

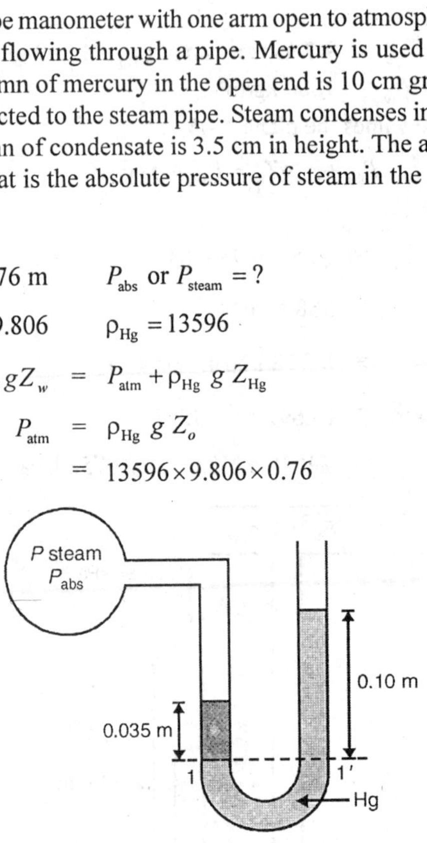

Fig. Ex. 1.19

$$= 101325 \frac{\text{N}}{\text{m}^2}$$

$$\rho_w g Z_w = 1000 \times 9.806 \times 0.035$$

$$= 333.404 \text{ Pa}$$

$$\rho_{Hg}\, g\, Z_{Hg} = 13596 \times 9.806 \times 0.10$$

$$= 12998.93 \text{ Pa}$$

$\therefore$ From (1) $\quad P_{abs} = P_{atm} + \rho_{Hg} g Z_{Hg} - \rho_w g Z_w$

$$= 101325 + 12998.9 - 333.404$$

[From (1), (2), (3) and (4)]

$$= 113990.52 \text{ Pa}$$

$$= 1.139 \text{ bar}$$

$$\mathbf{P_{abs} = 1.14 \text{ bar}}$$

Example 1.20 A U-tube manometer with one arm open to atmosphere is used to measure the pressure of kerosene vapour passing in a pipe. Hg level in the open arm is 10 cm greater than that in the arm connected to the kerosene pipe. There is a condensation of kerosene vapour on the Hg column of 5.1 cm height. If the atmospheric pressure is 75.5 cm of Hg. Find the absolute pressure of vapour in kPa. Assume specific gravity of kerosene as 0.8 and that of Hg as 13.6.

Solution

Balancing the pressure values @ the plane 1–1′

$$P_{\text{abs of Kero. vap}} + \rho_{kero} g Z_{kero} = P_{atm} + \rho_{Hg} g Z_{Hg} \quad (1)$$

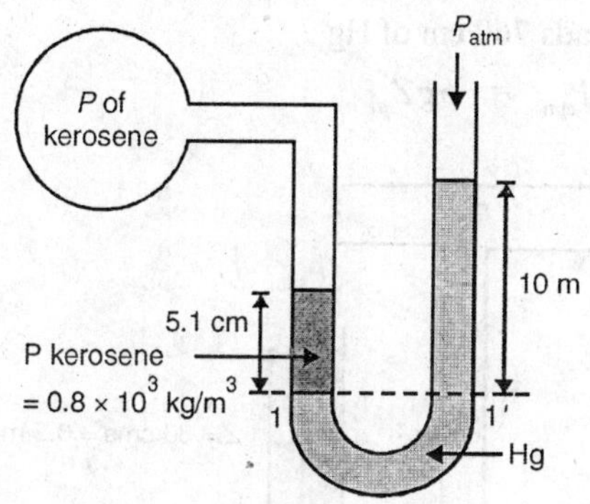

Fig. Ex. 1.20

Now to find

(a) $$\rho_{kero} \cdot g \cdot Z_{kero.} = 800 \times 9.81 \times 0.051 \text{ m}$$

$$= 400.02 \text{ N/m}^2 \quad (2)$$

(b) $$P_{atm} = \rho_{Hg} g Z_o$$

$$= 13600 \times 9.81 \times 0.755 \text{ m}$$

$$= 100726.08 \text{ N/m}^2 \quad (3)$$

(c) $$\rho_{Hg} g Z_{Hg} = 13600 \times 9.81 \times 0.1 \text{ m}$$

$$= 13341.6 \text{ N/m}^2 \quad (4)$$

Sustituting Eqs (2), (3) and (4) in Eq. (1) we get,

$$P_{abs} + 400.02 = 100729.08 + 13341.6$$

$$P_{abs} = 113670.68 \text{ N/m}^2$$

$$\mathbf{P_{abs} = 113.67068 \text{ kP}}$$

Example 1.21 A U-tube manometer having equal arms with one arm open to atmosphere is used to measure the pressure of a gas passing through a pipe line. The fluid used in the manometer has a specific gravity of 0.8, while the height of the fluid column in the arm open to the atmosphere is 30 cm greater than the height of fluid column in the arm connected to the gas pipe line. Determine, the absolute pressure of the gas in pipe line.

If the fluid used were Hg, what would be the difference of height of Hg-column in the 2-arms.

Take $\rho_{Hg} = 13596$ and $g = 9.806 \text{ m/sec}^2$

Solution

ρ of fluid used in the manometer = $0.8 \times 1000 \text{ kg/m}^3$

$$P_{abs} = P_{atm} + \rho g Z$$

Let the barometer reads 760 cm of Hg

$$P_{atm} = \rho g Z_o$$

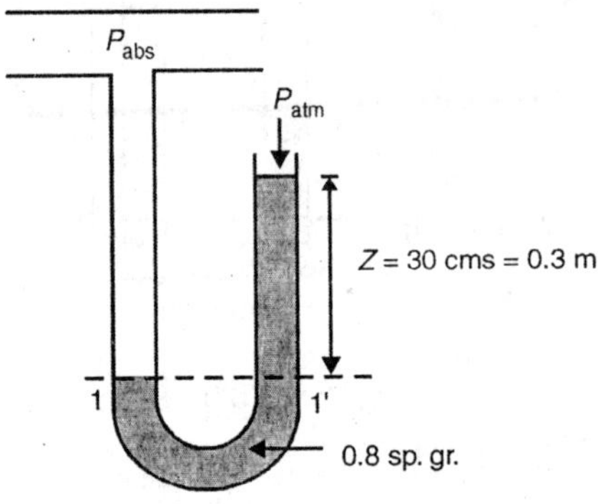

Fig. Ex. 1.21

$$= 13596 \times 9.806 \times 0.76$$

$$= 101325 \text{ N/m}^2$$

and $$\rho g Z = 0.8 \times 10^3 \times 9.806 \times 0.3$$

$$= 2353.44 \text{ N/m}^2$$

$\therefore$ $$P_{abs} = 101325 + 2353.44$$

$$= 1013678.44 \text{ N/m}^2$$

$$P_{abs} = \mathbf{1.0367844\ bar}$$

If Hg is used, then to find Z,

(i.e. to produce the same absolute pressure, we have to find the equivalent columnof Hg)

$$P_{abs} = P_{atm} + \rho g Z$$

$$103678.44 = 101325 + 13596 \times 9.806 \times Z$$

$$Z = \frac{103678.44 - 101325}{13596 \times 9.806}$$

$$= 0.01765 \text{ mt of Hg}$$

$$\mathbf{Z = 1.765.\ cm\ of\ Hg}$$

Example 1.22 A vertical composite liquid column with its upper end exposed to atmospheric pressure, comprise of 45 cms of Hg (sp. gr. 13.6); 65 cm of water and 80 cm of oil (sp. gr. 0.8). Calculate the pressure in bar,

(i) At the bottom of the column.

(ii) At the inter surface of oil and water.

(iii) At the inter surface of water and Hg.

Assume atmospheric pressure as 100 kPa and $g = 9.81$ m/sec^2

Solution

Since it is given $P_{atm} = 100$ kPa i.e. equal to 1 bar

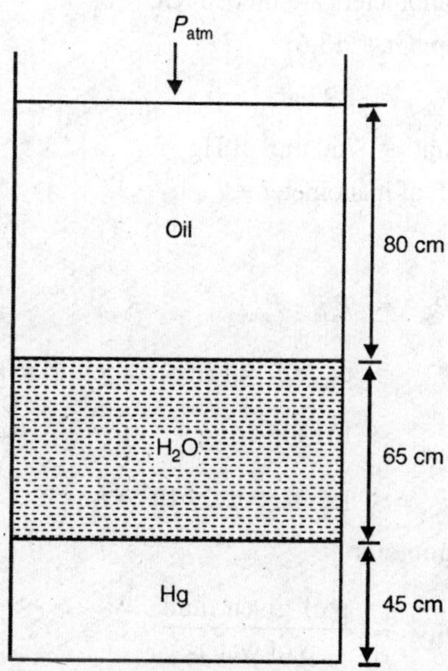

Fig. Ex. 1.22

(Since 10^5 N/m^2 = 1 bar = 100 kPa)

(i) $$P_{\text{at bottom}} = P_{\text{atm}} + P_{\text{oil}} + P_{\text{water}} + P_{\text{Hg}}$$

$$= 1 + \rho_{\text{oil}} g Z_{\text{oil}} + \rho_w g Z_w + \rho g Z$$

$$= 1 + \frac{0.8\times1000\times9.81\times0.8}{10^5} + \frac{1000\times9.81\times0.65}{10^5} + \frac{13.6\times10^3\times9.81\times0.45}{10^5}$$

Note: If ρ is in kg/m³, g is in m/sec², Z is in m, then the resultant unit will be N/m².

∴ To get in bar divide by 10^5

(i) $$P_{\text{at bottom}} = 1 + 0.062784 + 0.063765 + 0.6$$

$$\mathbf{P_{at\ bottom} = 1.72692\ bar}$$

(ii) Pressure at the interface of oil and water

$$= 1 + 0.062784 = 1.062784 \text{ bar}$$

(iii) Pressure at the inter surface of waterand Hg = 1 + 0.062784 + 0.063765

$$= 1.1265 \text{ bar}$$

Example 1.23 An air tank has three manometers connected to it. The fluids in them are oil (Specific Gravity = 0.8) water and mercury (Specific Gravity = 13.6). If the absolute pressure in the tank is 1.2 bar and the barometer reads 760 mm of Hg. Estimate the height of fluid in each of manometer.

Data: Fluids in the manometers are oil (Sp. Gr. = 0.8)

water and mercury (Sp. Gr. = 13.6)

$$P_{\text{abs}} = 1.2 \text{ bar}$$

Barometric reading = 760 mm of Hg

Height of fluid in each of manometer = ?

Solution

We know that, $$P_{\text{abs}} = P_{\text{atm}} + P_{\text{gauge}}$$

$$P_{\text{atm}} = \rho_{\text{Hg}} g Z_o = 13.6\times10^3\times9.81\times0.76$$

$$= 101396.16 \text{ N}$$

$$P_{\text{abs}} = 1.2 \text{ bar} = 1.2\times10^5 \text{ N/m}^2$$

Case I: For Oil Manometer

(a) $$\text{Sp. Gravity} = \frac{\rho \text{ of given fluid}}{\rho \text{ of water}}$$

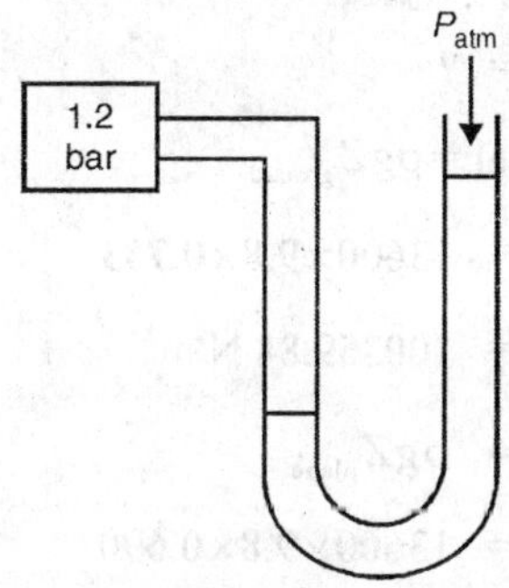

Fig. Ex. 1.23

$$0.8 = \frac{\rho_{oil}}{1000} \qquad \therefore \rho_{oil} = 800 \text{ kg/m}^3$$

$$\therefore P_{gauge} = \rho_{oil} g Z_{oil} = 800 \times 9.81 \times Z_{oil}$$

Subtituting these values in Eq. (1) we get,

$$1.2 \times 10^5 = 101396.6 + 800 \times 9.81 \times Z_{oil}$$

$$\mathbf{Z_{oil} = 2.37 \text{ m}}$$

Case II: For Water Manometer

$$P_{abs} = P_{atm} + \rho_w g Z_w$$

$$1.2 \times 10^5 = 101396.6 + 1000 \times 9.81 \times Z_w$$

$$\mathbf{Z_w = 1.89 \text{ m}}$$

Case III: For Hg Manometer

$$P_{abs} = P_{atm} + \rho_{Hg} g Z_{Hg}$$

$$1.2 \times 10^5 = 101396.6 + 13.6 \times 10^3 \times 9.81 \times Z_{Hg}$$

$$\mathbf{Z_{Hg} = 0.139 \text{ Hg}}$$

Example 1.24 A basic barometer can be used as an altitude measuring device in aeroplanes. The ground control reports a barometric reading of 753 mm of Hg, while the pilot's reading is 690 mm of Hg.

Estimate the altitude of the plane from the ground level if the average air density is 1.25 kg/m^3 and $g = 9.8$ m/sec^2.

Solution

Barometer reading at ground level $Z_{ground} = 0.753$ m

Pilot's Barometric reading $Z_{plane} = 0.690$ m

Altitude = ?

Pressure at the ground level $= \rho g Z_{ground}$

$$= 13600 \times 9.8 \times 0.753$$

$$= 100359.84 \text{ N/m}^2 \qquad (1)$$

Pressure at plane level $= \rho g Z_{plane}$

$$= 13600 \times 9.8 \times 0.690$$

$$= 91963.2 \text{ N/m}^2 \qquad (2)$$

∴ Change of pressure at the ground level and that of plane level,

$$\Delta P = 100359.84 - 91963.2$$

$$= 8396.64 \text{ N/m}^2$$

∴ This is the difference of pressure, now to find altitude for this much difference of pressure,

$$\Delta P = \rho_{air} g Z_{altitude}$$

$$8396.64 = 1.25 \times 9.8 \times Z_{altitude}$$

$$\mathbf{Z_{altitude} = 684.44 \text{ m}}$$

Example 1.25 A manometer connected between 2 pipes as shown in Fig. 1.26. Take,

$\rho_{water} = 1000 \text{ kg/m}^3$ $\rho_{Hg} = 13590 \text{ kg/m}^3$

Pressure at $A = 400$ kPa $g = 9.8 \text{ m/sec}^2$.

Find Pressure at B = ?

Solution

Note. If there are more than one U-turns, then start from point and as we go downwards take it as positive, and as we go upwards take it as negative. Then equate this pressure equation to pressure at the second point.

So, writing the pressure equation,

$$P_A + P_w \, g \, Z_w - \rho_{Hg} \, g \, Z_{Hg} + \rho_w \, g \, Z_w = P_B$$

$$400 \times 1000 + 1000 \times 9.8 \times 2.5 - 13590 \times 9.8 \times 0.6 + 1000 \times 9.8 \times 0.4 = P_B$$

$$400000 + 24500 - 79409.2 + 3920 = P_B$$

$$\mathbf{P_B = 348510.8 \text{ Pascals}}$$

$$\mathbf{P_B = 348.5108 \text{ kPa}}$$

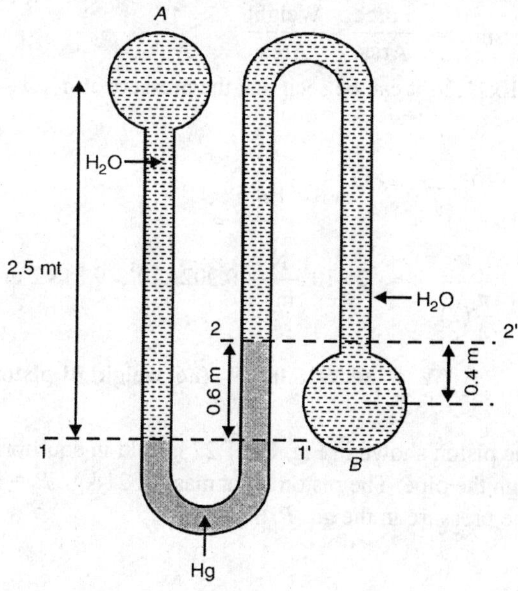

Fig. Ex. 1.25

Example 1.26 The cylinder and tubing shown in Fig. 1.26 contain oil of density 0.902 $\times$ 10^3 kg/m^3 for gauge reading of 2 bar. What is the total weight of piston and slab placed on it?

Solution

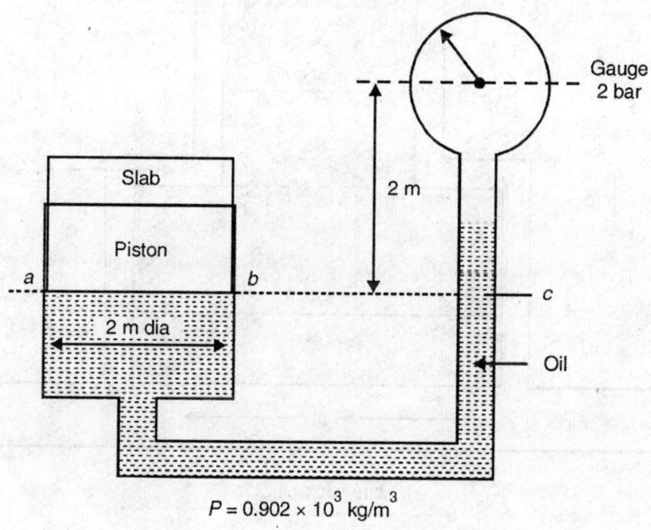

Fig. Ex. 1.26

We know that, Pressure $= \dfrac{\text{Force}}{\text{Area}} = \dfrac{\text{Weight}}{\text{Area}}$

And from Fig. Ex. 1.26 it can be seen that the pressure at $a - b - c$ level will be same.

$$\therefore \quad \frac{W}{A} = P_{gauge} + \rho_{oil} g h_{oil}$$

$$\therefore \quad \frac{W}{\frac{\pi}{4}(2)^2} = 2\times10^5 \frac{\text{N}}{\text{m}^2} + 0.902\times10^3 \times 9.81\times 2 \text{ m}$$

$$\mathbf{W = 683916.05 \text{ N} \quad (\text{i.e. weight of piston and slab})}$$

Example 1.27 The piston shown in Fig. Ex. 1.27 is held in equilibrium by pressure of gas flowing through the pipe. The piston has a mass of 21 kg. $P_I = 600$ kPa; $P_{II} = 170$ kPa. Determine the pressure in the gas P_{III}.

Solution

Total down ward force acting will be:

(i) Due to self weight.

(ii) Pressure P_I acting on 10 cm dia. piston.

(iii) Pressure P_{II} acting on 20 cm dia piston.

And this is balanced by pressure P_{III}.

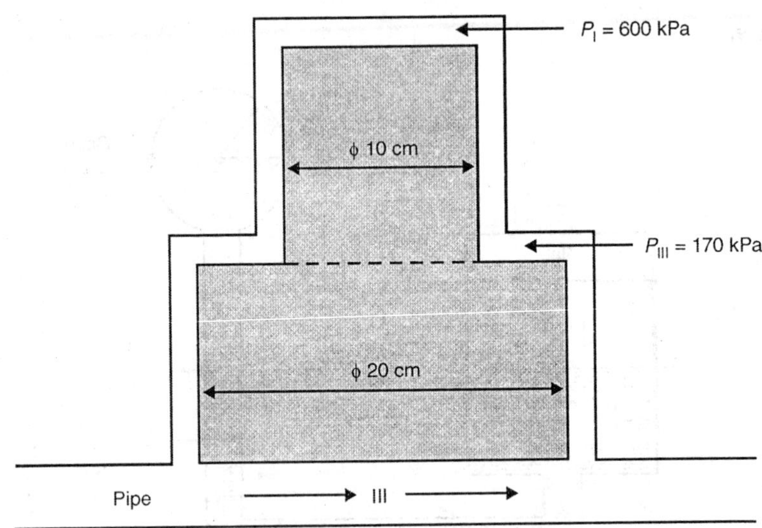

Fig. Ex. 1.27

$\therefore$ Force due to mass $= 21 \times 9.81 = 206.01 \text{ N} = 0.20601 \text{ kN}$

Then force I acting on 10 cm dia piston $= \text{Pr.} \times \text{Area} = 600 \times \frac{\pi}{4}(0.1)^2$

$= 4.712389$ kN

And force II acting on 20 cm dia piston =

$$170 \times \left(\frac{\pi}{4}\left(0.2^2 - 0.1^2\right)\right) = 4.0055306 \text{ kN}$$

$\therefore$ Total downward load/force $= 0.20601 + 4.712389 + 4.005306$
$= 8.9239296$ kN

This is balanced by P_{III}

$$\therefore \quad 8.9239296 = P_{III} \times \text{Area} = P_{III} \times \frac{\pi}{4}(0.2)^2$$

$$= P_{III} \times 0.0314$$

$$\mathbf{P_{III} = 284.0575 \text{ kPa}}$$

Example 1.28 The basic barometer can be used to measure the height of a building. If the barometric readings at top and bottom of a building are 730 and 755 mm of Hg respectively. Determine the height of the building. Assume an average air density of 1.18 kg/m^3.

Solution

The difference between pressure at top and bottom is equal to the air head of height equal to that of the building.

$$\therefore \quad P_1 - P_2 = \rho_{air} gh$$

$$(0.755 - 0.73) \times 13600 \times 9.81 = 1.18 \times 9.81 \times h$$

$$h = \frac{13600 \times 0.025}{1.18}$$

$$\mathbf{h = 288.14 \text{ m}}$$

Example 1.29 A large container *A* contains another small container *B*, both fitted with gauges. The gauge fitted to *A* reads 200 kPa and gauge fitted to *B* reads 120 kPa.

If barometer reads 750 mm of Hg, calculate the absolute pressure in the containers *A* and *B*.

Solution

Barometer reading h_{baro} = 750 mm of Hg=Z_o

$$P_{atm} = \rho \cdot g \cdot h_{baro}$$

$$= 13596 \times 9.806 \times 0.750$$

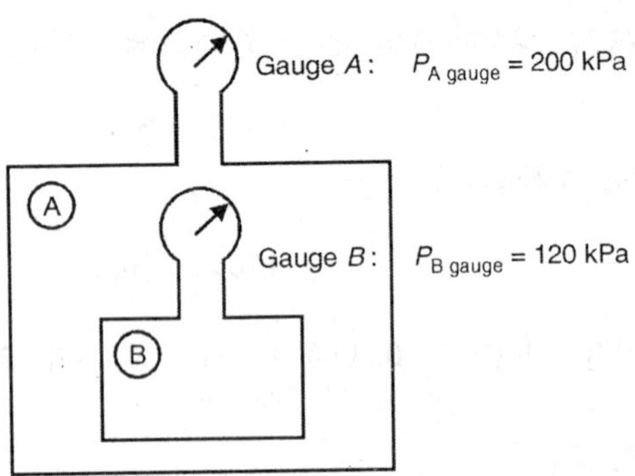

Fig. Ex. 1.29

$$= 100.03 \text{ kPa.}$$

$$P_{A\ abs} = P_{A\ gauge} + P_{atm}$$

$$= 200 + 100.03$$

$$= 300.03 \text{ kPa}$$

Note thet gauge *B* reads pressure relative to its surrounding presure i.e. gauge *B* is surrounded by pressure of 300.03 kPa.

$$P_{B\ abs} = P_{B\ gauge} + P_{atms} \text{ of } B$$

$$= 120 + 300.03$$

$$\mathbf{P_{B\ abs} = 420.03\ kPa}$$

Example 1.30 Convert the following temperatures into degree Fahrenheit.

(i) 40 °C

(ii) – 20 °C

Solution

(i) Using the relation $\frac{C}{100} = \frac{F-32}{180}$

$$\frac{40}{100} = \frac{F-32}{180}$$

$$F-32 = \frac{40 \times 180}{100} = 72$$

$$\mathbf{F = 72 + 32 = 104\ °F}$$

(ii) $$\frac{-20}{100} = \frac{F-32}{180}$$

$$F - 32 = \frac{-20 \times 180}{100} = -36$$

$$\mathbf{F = 32 - 36 = -4\ °F}$$

Example 1.31 Find the temperature which has the same value on both the centigrade and fahrenheit scales.

Solution

Using the relation $\dfrac{C}{100} = \dfrac{F-32}{180}$

Putting C = F, $\dfrac{C}{100} = \dfrac{C-32}{180}$

$\therefore$ $180\,C = 100\,C - 3200$

$\therefore$ $C = -40$

$\therefore$ $\mathbf{-40\ °C = -40\ °F}$

Example 1.32 A temperature scale of a certain thermometer is given by the relation $t = a \log_e P + b$, where a and b constants and P is the thermometric property of the fluid in the thermometer. If at the ice point and steam point, thermometric properties are found to be 1.5 and 7.5 respectively. What would be the temperature corresponding to the thermometric property 3.5 on Celsius Scale.

Solution

$$t = a \log_e P + b$$

$$100 = a \log_e 7.5 + b$$

$$0 = a \log_e 1.5 + b$$

Solving theses two equations we get,

$\therefore$ $a = 62.1336$ and $b = -25.168$

$\therefore$ For $P - 3.5$, $t = 62.1336 \log_e 3.5 - 25.168$

$$\mathbf{t = 52.670676\ °C}$$

Example 1.33 The temperature scale of a certain thermometer is given by $t = a \ln P + b$, when a and b are constants and P is the thermometric property. The temperature of ice point and steam point are 100 and 300 respectively on this scale. Experiment gives values of thermometric properties of $P_1 = 1.86$ and $P_2 = 6.81$ at ice point and steam point respectively. Evaluate the temperature corresponding to a reading of thermometric property at $P - 2.5$.

Solution

$$t = a \ln P + b \quad (1)$$

$$300 = a \ln 6.8 + b$$

$$100 = a \ln 1.86 + b \quad (2)$$

On subatracting we get, $200 = a \ln\left(\frac{6.81}{1.86}\right)$

$$a = \frac{200}{1.3} = 154 \quad (3)$$

Substituting this value of a in (2) we get,

$$100 = 154 \ln 1.86 + b$$

$\therefore$ $$b = 100 - 154 \ln 1.86 = 4.43$$

$\therefore$ For $$P = 2.5,\ t = 154 \ \ln 2.5 + 4.43$$

$$t = \mathbf{145.5^{o}}$$

Example 1.34 The pressure in a constant volume gas thermometer is measured as 32 mm of Hg above the atmospheric pressure at triple point of water, determine the temperature in °C, when pressure is 76 mm of Hg above atmospheric pressure. Barometric pressure is 752 mm of Hg.

Solution

Note: In a constant volume gas thermometer, volume remains constant but as the temperature changes, pressure change occurs.

$$\frac{\theta_1}{\theta_2} = \left(\frac{P_2}{P_1}\right) \text{ constant volume} \quad (1)$$

Given that $$\theta_1 = 273.16 \text{ K (triple point of water)}$$

$$P_1 = P_{abs} + P_{gauge} = 752 + 32$$

$$P_1 = 784 \text{ mm of Hg}$$

and $$P_2 = P_{atm} + P_{gauge} = 752 + 76 = 828 \text{ mm of Hg}$$

Substituting these in Eq. (1) we get

$$\frac{273.16}{\theta_2} = \frac{784}{828}$$

$$\theta_2 = 288.49 \text{ K}$$

$$\theta_2 = \mathbf{15.49°C}$$

Example 1.35 The temperature T on a thermometric scale is defined in terms of a property P by the relation,

$$T = a \ln P + b \text{ where } a \text{ and } b \text{ are constants.}$$

The temperatures of the ice point and steam points are assigned the numbers 32 and 212 respectively. Experiment gives values of P of 1.86 and 6.81 at the ice point and steam point respectively. Evaluate the temperature corresponding to a reading of P = 2.5 on the thermometer.

Solution

We have $T = a \ln P + b$

Substituting the values for ice point and steam point,

$$32 = a \ln 1.86 + b$$

and $$212 = a \ln 6.81 + b$$

Solving we get $a = 138.7,\ b = -54.07$

$\therefore$ Equation becomes $T = 138.7 \ \ln P - 54.07$

Now for $P = 2.5$, $T = 138.7 \ \ln 2.5 - 54.07$

$$= 73$$

THEORY QUESTIONS

1. What is Thermodynamics?
2. Define Thermodynamic system? Explain its types with suitable examples.
3. Discuss : (1) System, (2) Surrounding, (3) Boundary, (4) Universe.
4. What do you understand by thermodynamic system ? How these are classified ? Explain each with suitable sketch, or
   Explain
   1. Open system
   2. Closed system
   3. Isolated system

   and state the examples of each.
5. Distinguish between: Open system and closed system.
6. State with reasoning whether the following systems are closed, open or isolated.
   1. IC engine
   2. Refrigerator
   3. Pressure cooker.
7. Explain thermodynamic properties, processes and cycles.
8. Discuss :
   1. State
   2. Property

9. Explain thermodynamic properties.
10. Compare Intensive and Extensive properties.
11. What do you understand by state function and path function ?
12. Compare state function and path function.
13. Unit extensive property is an intensive property. Discuss.
14. Define the terms Process.
15. Explain 1. Cycle; 2. Process.
16. Define property and classify it with proper examples.
17. What is meant by Working Substance? Explain with an example.
18. Write a short note on Units and Dimensions.
19. What is Mechanical and Thermodynamic Work ? Explain.
20. What do you mean by thermodynamic work and mechanical work ?
21. What is PdV-work ? or Work done at the moving boundary ? or Displacement work.
22. Define heat and work and explain why neither is a property.
23. Prove that Work is a path function and Properties are point functions.
24. Define heat and work and explain why neither is a property.
25. 'Property is an exact differential'. Explain.
26. Derive the equations for work done in various quasi-static processes.
27. Write a note on Heat Energy.
28. Define the terms Work and heat.
29. Explain the similarities and differences between heat and work.
30. Define the property – Pressure. State the various units used in practice.
31. How pressure measurement is done ?
32. Write a note on pressure exerted due to a column of fluid.
33. Write a short note on Barometer.
34. Briefly explain why you cannot use a mercury barometer to measure pressure in a gravity free environment.
35. Write a note on Bourdon Pressure gauge.
36. Explain Bourdon's Pressure gauge with sketch.
37. Write a note on Manometers.
38. Define Thermal Equilibrium. State and explain the Zeroth law of Thermodynamics.
39. State and explain zeroth law of thermodynamics and explain how it leads to the concept of temperature.
40. What is meant by thermometric property. Enlist the temperature measuring devices.
41. Discuss various thermodynamic properties used for temperature measurement.
42. What is the principle of Temperature Measurement ?
43. What are the different Scales of Temperature ?
44. Write a note on Microscopic and Macroscopic point of view.
45. What do you understand by microscopic and macroscopic approach of thermodynamic study ? Explain.

46. What is the difference between the classical and statistical approaches to Thermodynamics.
47. Write a note on Energy in transit.
48. Explain the term Energy in transit.
49. What is a Quasi-static Process? Explain.
50. What do you mean by Quasi-static process.
51. Explain why a non-quasi equilibrium compression process requires a larger work input than the corresponding quasi equilibrium one, and, why a non-quasi equilibrium expansion process delivers less work than the corresponding quasi-equilibrium one.

PROBLEMS FOR PRACTICE

1. The pressure of a fluid of specific gravity 0.8 flowing in a horizontal pipe line in determined with a simple mercury U-tube manometer. The level of Hg surface in the right limb which is open to the atmosphere is 90 mm above the centre of the pipes. The level of Hg in the left limb which is connected to the pipe is 60 mm below the centre of pipe. Determine absolute pressure of liquid in the pipe is N/m². Assume standard P_{atm} to be equal to 10 m of water.

   (Ans. 117641.52 N/m²)
2. An inverted U-tube is connected across two pipes *A* and *B* carrying fluid as shown in Fig. P1.2. Calculate the pressure difference between the centre lines of the pipes.

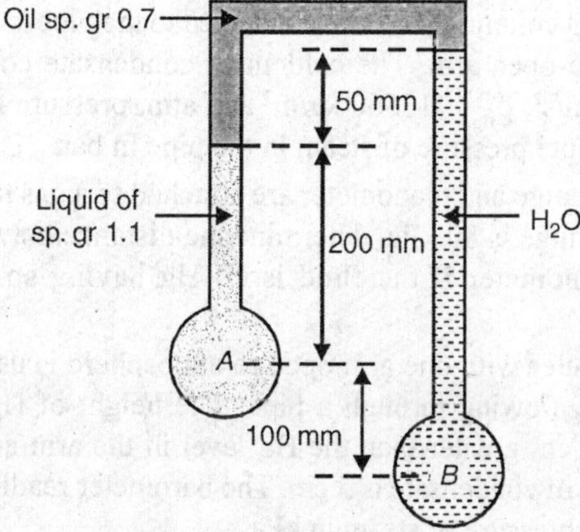

Fig. P1.2

3. In the arrangement shown in Fig. P1.3 P_3. Find the difference of pressure in the pipe *A* and *B*.

   (Ans. $P_A - P_B$ = 29135.7 N/m²)

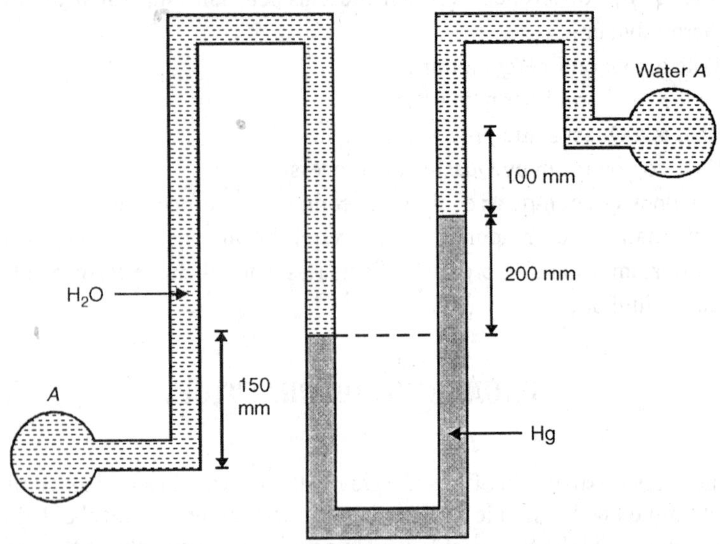

Fig. P1.3

4. A pressure gauge is attached to closed tank which contains some gas. The gauge reads 1.3 bar when barometer reads 75 cm of Hg. If the barometer changes to 72 cm of Hg, what will be the reading on the gauge for the same absolute pressure.
5. A U-tube manometer with one arm open to atmosphere is used to measure the pressure of the steam flowing through a pipe using mercury as manometric fluid. The level of the Hg column in the arm connected to the pipe is 125 mm below the level of Hg in the open arm. The column of condensate collected is 35 mm. Taking $g = 9.81$ m/s², $\rho_{Hg} = 13560$ kg/m³ and atm. pressure as 760 mm of Hg, estimate the absolute pressure of steam in the pipe in bar.
6. Both a Bourdon gauge and manometer are attached to a gas tank. If the reading on the pressure gauge is 80 kPa, determine the distance between the two fluid levels of the manometer if the fluid is (i) Hg having sp. gravity of 13.6, (ii) water.
7. A U-tube manometer with one arm open to atmosphere is used to measure the pressure of steam flowing through a pipe. The height of Hg column open to atmosphere is 15 cm greater than the Hg level in the arm connected to steam pipe. The column of condensate is 5 cm. The barometer reading is 70 cm of Hg. Find the absolute pressure of steam in kPa.
8. A water pump is required to supply water at 900 lit./min into a tank at a height of 20 m. Find the power required to drive the pump in kW.
9. Estimate the height of the atmospheric column to produce a pressure of 1 bar at the earth surface. Assume that the pressure P in Pascal and sp. volume V in m³/kg are related according to the law $PV^{1.39} = 2.25 \times 10^5$ and $g = 9.8$ m/s².

10. An air tank has three manometers connected to it. The fluids in them are oil (sp. gr. = 0.85), water and Hg. If the absolute pressure in the tank is 1.24 bar and the barometric height is 75 cm of Hg, estimate the height of fluid in each manometer.
11. A gas expands in a piston–cylinder device from 0.01 to 0.03 m³, the process being described by $P = aV^{-1} + b$ where P is in bar and V in m³. If $a = 0.06$ bar/m³ and work done by the gas on 'the piston is 10.6 kg, evaluate b.
12. In a piston cylinder arrangement the pressure is inversely proportional to the square of the volume. The initial pressure is 10 bar in the cylinder and initial volume is 0.1 m³. The volume is now changed so that the final pressure is 2 bar. Find the work done in kJ.

ꟸ ꟹ

2

Ideal Gases and Ideal Gas Processes

CHAPTER OBJECTIVES

After reading this chapter you will be able to learn the following

- Ideal Gas, Ideal Gas Laws, Equation of State, Specific Gas Constant and Universal Gas Constants, Specific Heat and Thier Relation
- Joules Experiment
- Work Done, Heat Transferred, Change of Internal Energy, Change of Enthalpy during various processes.

2.1 INTRODUCTION

An ideal gas is one, which obeys all the gas laws and the characteristic gas equation $PV = mRT$, at all temperatures and pressures.

But actually there is no perfect or Ideal gas in reality. But all the real gases like Air, O_2, H_2, N_2, He, behave as ideal gases, at very low pressures and high temperatures.

2.2 IDEAL GAS LAWS

The behaviour of ideal gas is governed by following gas laws:

1. Boyle's law.
2. Charle's law.
3. Avagadro's law.

1. Boyle's Law. Boyle experimentally established that, the volume of a given mass of a gas is inversely proportional to the absolute pressure, when the temperature is constant.

i.e. $$V \propto \frac{1}{P}, \text{ for particular temperature } T$$

$$\therefore \quad V = \frac{\text{Constant of proportionality}}{P} = \frac{C}{P}$$

or $\quad PV = C \quad$ where C is constant.

Also $\quad P_1V_1 = P_2V_2 = C.$

Consider a gas in a cylinder fitted with a frictionless piston. Let the initial properties be P_1 and V_1. Now when the gas is heated at constant temperature, then the gas expands and the pressure falls to P_2 and volume increases to V_2. This is represented on PV-diagram as shown in Fig. 2.1.

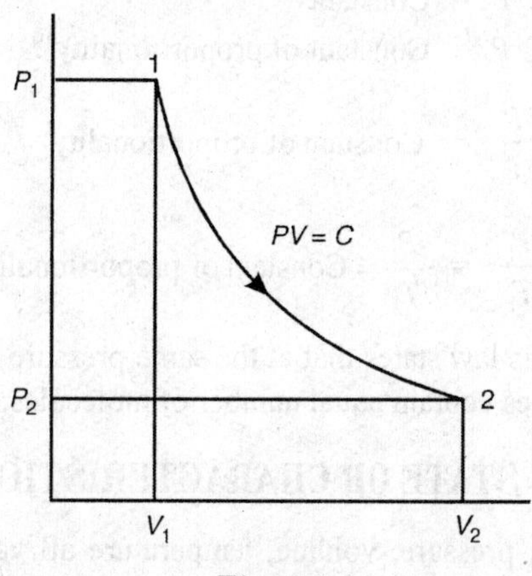

Fig. 2.1

2. Charle's Laws. (i) It states that, "when the pressure of a given mass of a gas is kept constant, then volume is directly to the absolute temperature". (Fig. 2.2)

i.e. when $\quad P = \text{Constant}$

Then $\quad V \propto T$

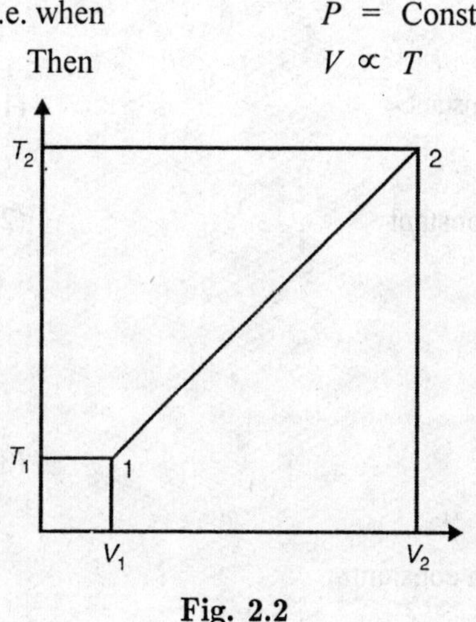

Fig. 2.2

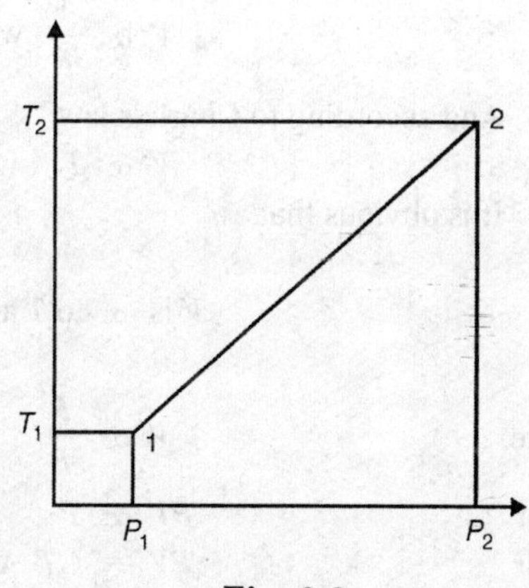

Fig. 2.3

or $\quad V = \text{Constant of proportionality}\, T$

or $$\frac{V}{T} = \text{Constant of proportionality}$$

or $$\frac{V_1}{T_1} = \frac{V_2}{T_2} = \text{Constant of proportionality}$$

(ii) It also states that, "when the volume of a given mass of a gas is kept constant, the pressure is directly proportional to the absolute temperature". (Fig. 2.3)

i.e. $$P \propto T$$

when $$V = \text{Constant}$$

or $$P = \text{Constant of proportionality } T$$

$$\frac{P}{T} = \text{Constant of proportionality}$$

or $$\frac{P_1}{T_1} = \frac{P_2}{T_2} = \text{Constant of proportionality}$$

3. Avagadro's Law. This law states that at the same pressure and temperature, equal volumes of different gases contain equal number of molecules.

2.3 EQUATION OF STATE OR CHARACTERISTIC GAS EQUATION

In engineering practice, pressure, volume, temperature all vary simultaneously. So, Boyle's law or Charle's law alone is not applicable, since one of three properties is kept constant. In order to establish a relationship between these three properties, Boyle's law and Charle's laws are combined together, which gives a general equation called characteristic gas equation.

According to Boyle's law,

$$V \propto \frac{1}{P} \text{ when } T = \text{Constant} \quad (1)$$

And according to Charle's law,

$$V \propto T \quad \text{when } P = \text{Constant} \quad (2)$$

It is obvious that,

$$V \text{ is } \propto \text{ to } T \text{ and } \frac{1}{P}$$

i.e. $$V \propto \frac{T}{P} \quad (3)$$

or $$PV \propto T$$

or $$PV = CT \quad \text{where } C \text{ is a constant}$$

or $$\frac{PV}{T} = C \quad (4)$$

or in general $$\frac{P_1V_1}{T_1} = \frac{P_2V_2}{T_2} = \frac{P_3V_3}{T_3} = ... = C$$

or since v = specific volume in $\frac{m^3}{kg}$

Then equation will be, $\frac{Pv}{T} = C$ (5)

When 1 kg of gas is considered, then the constant for the Eq. (4) is written as R and is called *Characteristic gas constant.*

$$\therefore \quad \frac{Pv}{T} = R \tag{6}$$

Now consider m kg of gas. Multiply both sides of Eq. (6) by m

$$\frac{P\cdot(m\cdot v)}{T} = m\,.\,R$$

$$m\,.\,v = V = \text{Total volume.}$$

$$\therefore \quad \mathbf{PV = mRT}$$

This is known as characteristic equation of an ideal gas.

Note that the value of R for air = $0.287\ \frac{kJ}{kg\text{-}K}$

2.4 UNIVERSAL GAS CONSTANT

Frequently it is required to express the equation of state, $PV = mRT$. (1) on mole basis. The mass of a substance m is equal to the product of number of moles n and the molecular weight M.

i.e. $$m = n\,.\,M \tag{1}$$

Hence susbtituting the value of m in Eq. (1) we get,

$$PV = n\,.\,M\,.\,RT \tag{2}$$

Now the molar specific volume is defined as the volume/unit mole and is denoted by $\bar{v}$,

$$\therefore \quad \bar{v} = \frac{V}{n}$$

Hence Eq. (2) becomes,

$$P\bar{v} = MRT \tag{3}$$

Now if we consider two gases, a and b occupying equal molar volumes at the same temperature and pressure, Eq. (3) becomes,

$$P_a.\bar{v}_a = M_aR_aT_a$$

and $$P_b.\bar{v}_b = M_b.R_b.T_b$$

According to Avagardo's law, perfect gases at the same temperature and pressure, occupy the same molar volumes.

i.e. $\bar{v}_a = \bar{v}_b$ when P_a and $T_a = T_b$

We have, $$\frac{P_a.\bar{v}_a}{T_a} = \frac{P_b.\bar{v}_b}{T_b}$$

$$\therefore \quad M_a.R_a = M_b.R_b = M.R$$

The product $M.R$ is called the *Universal Gas Constant* and it is denoted by $\bar{R}$

$\therefore$ Eq. (3) can be written as,

$$P\bar{v} = MRT$$

$$P\bar{V} = \bar{R}T$$

and Eq. (2) can be written as,

$$P\bar{V} = nMRT$$

$$P\bar{V} = n\bar{R}T$$

and the value of $\bar{R}$ for air $= 8.3143 \frac{\text{kJ}}{\text{kmol K}}$

2.5 JOULE'S EXPERIMENT TO PROVE $U = f(T)$

Internal Energy—Joule's Law

Joule's Law states that the internal energy of an ideal gas depends only on the temperature of gas and is independent of changes in pressure and volume i.e. $U = f(T)$.

Joule's Experiment

The apparatus for Joule's experiment is shown in Fig. 2.4. It consists of two copper vessels A and B. Connected by a valve V. Initially A contained high pressure air and B

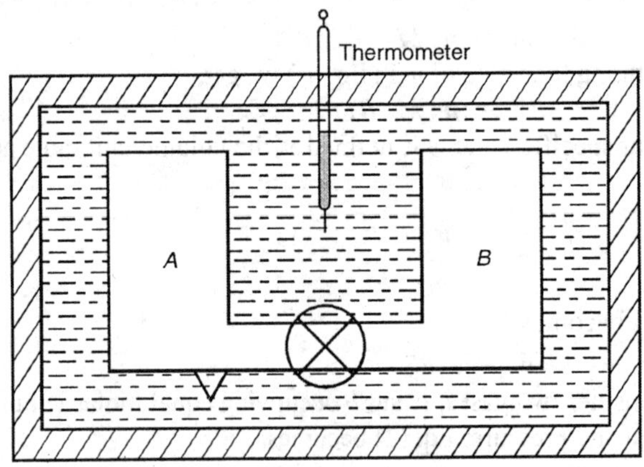

Fig. 2.2. *Joule's experiment.*

was evacuated. Both were placed in an insulated water bath with a thermometer to measure the temperature.

The initial temperature of water was recorded. The valve V was then opened and air is allowed to flow from vessel A to vessel B.

The operation is slow enough for air to attain equilibrium states. Eventually pressure in both the vessels will become equal and flow will be stopped. Joule observed that there was no change in the temperature of water during or after the process.

Since there was no change in temperature, there was no heat transfer between air and water. Therefore $\delta Q = 0$. Also no work is done during the process. Therefore $\delta W = 0$. By first law of thermodynamics for closed system we know that $\delta Q = dU + \delta W$.

Therefore $dU = 0$ or change in internal energy is zero. The internal energy therefore remains constant.

Again we observe that pressure and volume both changed during the process and only temperature remained constant. Joule therefore concluded that internal energy is the function of temperature only and independent of pressure and volume changes.

Thus, $U = f(T)$.

2.6 RELATIONS BETWEEN C_P AND C_V

We have studied specific heat at constant pressure and specific heat at constant volume in the First Law of Thermodynamics.

Since C_P and C_V are the properties of a system, there is a relationship between these.

We know that, specific enthalpy h is given by,

$$h = u + PV$$

$$dh = du + d(P \cdot v) \quad (1)$$

from the definition of,

$$C_p = \left(\frac{\delta q}{dT}\right)_p = \left(\frac{dh}{dT}\right)_p$$

$$\therefore \quad dh = C_p \cdot dT \quad (2)$$

and

$$C_v = \left(\frac{\delta q}{dT}\right)_v = \left(\frac{du}{dT}\right)_v$$

$$\therefore \quad du = C_v \cdot dT \quad (3)$$

Since,

$$PV = RT$$

$$d(PV) = d(RT)$$

$$d(PV) = R \cdot dT, \text{ since } R \text{ is constant} \quad (4)$$

Substituting Eqs (2), (3) and (4) in (1) we get,

$$C_p.dT = C_v dT + R.dT$$

$$\therefore \quad C_p = C_v + R \quad \text{(Since } dT \text{ is common in both sides)}$$

or $$\mathbf{C_p - C_v = R} \tag{5}$$

We also know that the ratio $\dfrac{C_p}{C_v} = \gamma$

$$\therefore \quad C_p = \gamma . C_v$$

Substituting in, $C_p - C_v = R$

We get,

$$\gamma . C_v - C_v = R$$

$$\mathbf{C_v = \frac{R}{\gamma - 1}} \tag{6}$$

and since $$C_p = \gamma . C_v ,$$

$$\mathbf{C_p = \frac{\gamma . R}{\gamma - 1}} \tag{7}$$

2.7 IDEAL GAS PROCESSES

2.7.1 Constant Volume Process

During this process, volume remains constant ($V = C$) and is represented on a PV-diagram by means of vertical line as shown.

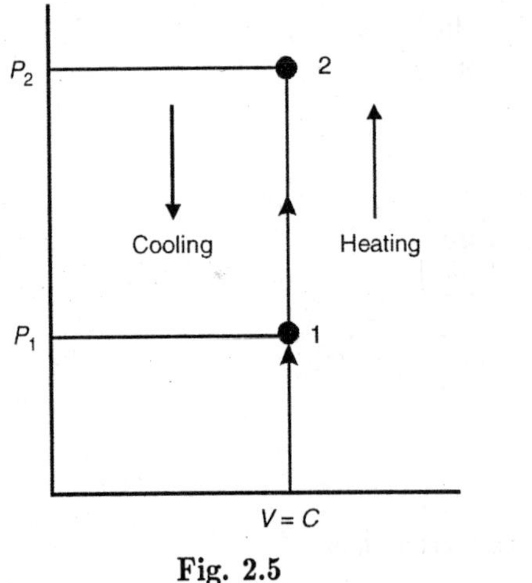

Fig. 2.5

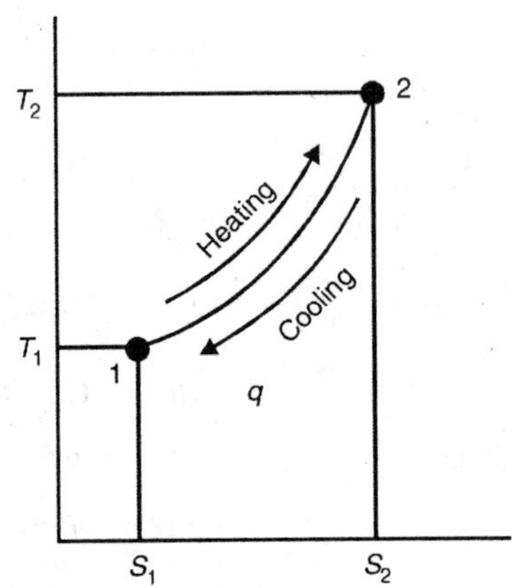

Fig. 2.6

When unit mass of a gas is heated in a closed vessel (i.e. $V = C$), then, since volume remains constant, no external work is done. But as temperature of the gas increases I.E. increases.

(a) Work done. Work done in a non-flow system = $\int PdV$.

Here, $V = C \quad \therefore \; dV = 0$

$$\therefore \; W_{1-2} = 0$$

(b) Heat supplied. From the I–Law for a closed system undergoing a process,

$$Q = \Delta U + W. \text{ In this case}$$

Since $W = 0$

$\therefore \quad Q = \Delta U$

or $\quad q = \Delta u$ (where q is kg)

$\therefore \quad \delta q = du = C_v . dT$

from the definition of C_v. For total mass,

$$\delta Q = dU = m \times C_v \times dT$$

2.7.2 Constant Pressure Process

During this process pressure remains constant ($P = C$) and this process is represented by means of a horizontal line on the PV-diagram as shown.

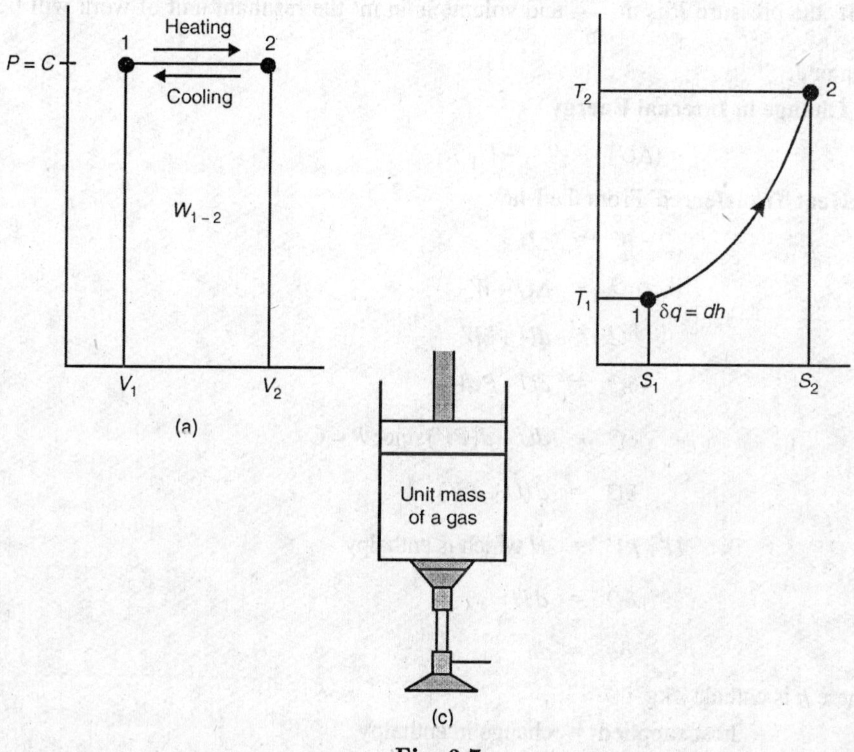

Fig. 2.7

When a unit mass of a gas is taken in a cylinder fitted with a frictionless piston and is heated. Then the piston moves up maintaining same pressure (*i.e.* $P = C$). As the volume of the gas increases, the work is done by the gas on the piston. As the temperature of the gas increases, Internal Energy increases. So, heat supplied in a constant pressure heating process is utilised for two purposes:

(i) For doing some external work.

(ii) For increasing the I.E. of the gas.

(a) Work Done

Work done in non-flow system.

$$W_{1-2} = \int_1^2 P dV$$

Since P is constant.

$$\therefore \quad W_{1-2} = P\int_1^2 dV$$

$$W_{1-2} = P(V_2 - V_1)$$

If the pressure P is in $\frac{\text{N}}{\text{m}^2}$ and volume is in m^3 the resultant unit of work will be N-m or J.

(b) Change in Internal Energy

$$(\Delta U) = U_2 - U_1$$

(c) Heat Transferred. From the I-law,

$$Q - W = \Delta U$$

$$Q = \Delta U + W$$

or

$$\delta Q = dU + \delta W$$

$$\delta Q = dU + PdV$$

$$\delta Q = dU + d(PV) \text{ since } P = C$$

$$\delta Q = d(U + PV)$$

$$U + PV = H \text{ which is enthalpy}$$

$$\delta Q = dH$$

or

$$\delta q = dh$$

where h is enthalpy/kg

i.e. heat supplied = change in enthalpy

2.7.3 Isothermal Process

During this process, temperature remains constant ($T = C$). The law for the process is $PV = C$ and is reprseented by means of a curve as shown on the PV-diagram. It is represented by means of a horizontal line on T-S diagram.

(a) Work Done. We know that work done in a non-flow system.

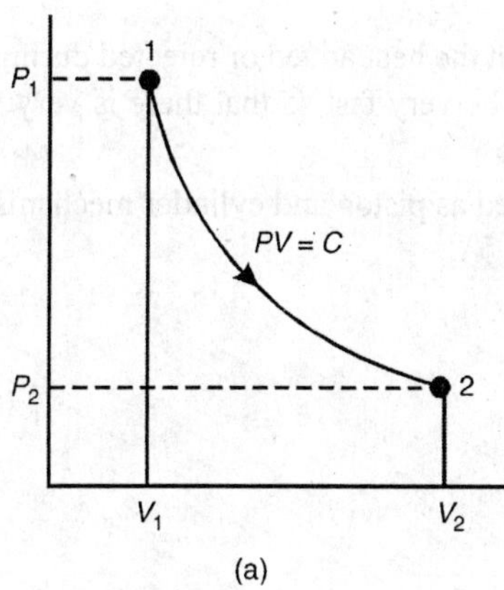

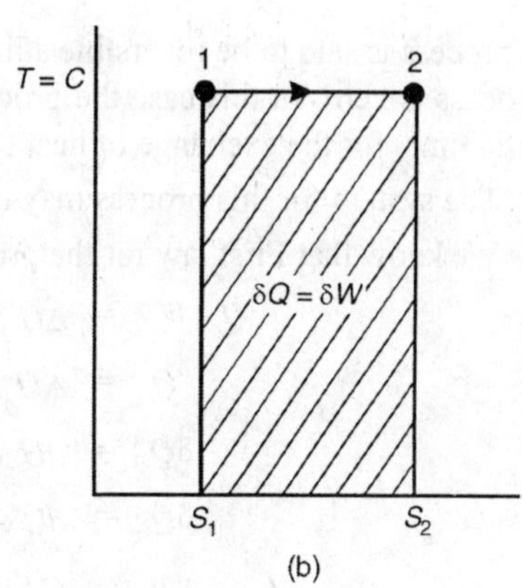

Fig. 2.8

$$W_{1-2} = \int_1^2 P.dV \tag{1}$$

The law for the process is,

$$PV = C$$

i.e. $$PV = P_1V_1 = P_2V_2 = C$$

$\therefore$ $$P = \frac{P_1V_1}{V}$$ Substitute this in (1) we get,

$$\mathbf{W_{1-2} = P_1V_1 \ln\frac{V_2}{V_1}}$$

$$\mathbf{W_{1-2} = P_1V_1 \ln\frac{P_1}{P_2}} \quad \text{since } \frac{V_1}{V_2} = \frac{P_1}{P_2}$$

(b) Change in IE. $dU = mC_v dT = 0 \; as \; T = C, \; dT = 0$

(c) Heat Transferred. From the First law for a closed system undergoing a process,

$$Q = \Delta U + W$$

$$\delta Q = dU + \delta W$$

$$\delta Q = 0 + \delta W$$

$$\delta Q = \delta W = P_1 V_1 \ln \frac{V_2}{V_1} = P_1 V_1 \ln \frac{P_1}{P_2}$$

2.7.4 Reversible Adiabatic Process

A process is said to be reversible adiabatic, when the heat added or rejected during the process is zero. In this case the process should be very fast so that there is very very little time for the exchange of heat to take place.

The system for this process may be considered as piston and cylinder mechanism.

We know that First law for the process is,

$$Q - W = \Delta U$$

$$Q = \Delta U + W$$

$$\delta Q = dU + \delta W$$

$$\delta Q = dU + PdV$$

$$0 = C_v . dT + PdV \qquad (1)$$

Since heat transfer is zero.

We also know that, for an ideal gas,

$$PV = RT$$

Diffrentiating this we get,

$$P.dV + VdP = R.dT$$

$$\frac{P.dV + VdP}{R} = dT \qquad (2)$$

Substituting the value of dT in Eq. (1)

We get,

$$O = C_v \frac{(P.dV + V.dP)}{R} + P.dV$$

$$= P \cdot dV \left(1 + \frac{C_v}{R}\right) + \frac{C_v}{R} \cdot VdP$$

$$= C_p \cdot \frac{P}{R} \cdot dV + \frac{C_v}{R} . V \cdot dP$$

Dividing this equation by PV we get,

$$O = \frac{C_p}{R} \cdot \frac{dV}{V} + \frac{C_v}{R} \cdot \frac{dP}{P}$$

Now dividing by C_v throughout, we get,

$$O = \frac{C_p}{C_v R} \cdot \frac{dV}{V} + \frac{1}{R} \cdot \frac{dP}{P}$$

$$= \frac{C_p}{C_V} \cdot \frac{dV}{V} + \frac{dP}{P}$$

$$= \gamma \cdot \frac{dV}{V} + \frac{dP}{P}$$

$$\therefore \quad \frac{dP}{P} = -\gamma \cdot \frac{dV}{V} \qquad (3)$$

Integration of this equation gives

$$\int_1^2 \frac{dP}{P} = -\gamma \int_1^2 \frac{dV}{V}$$

$$\mathbf{\log_e P = -\gamma . \log_e V}$$

$$\therefore \quad [\log_e P]_1^2 = -\gamma [\log_e V]_1^2$$

$$\therefore \quad \log_e P_2 - \log_e P_1 = -\gamma [\log_e V_2 - \log_e V_1]$$

$$\therefore \quad \log_e \frac{P_2}{P_1} = -\gamma \log_e \frac{V_2}{V_1}$$

$$\therefore \quad \log_e \frac{P_2}{P_1} = \log_e \left(\frac{V_2}{V_1}\right)^{-\gamma}$$

$$\therefore \quad \left(\frac{V_2}{V_1}\right)^{-\gamma} = \frac{P_2}{P_1}$$

$$\therefore \quad \left(\frac{V_1}{V_2}\right)^{\gamma} = \frac{P_2}{P_1}$$

$$\therefore \quad P_1 V_1^{\gamma} = P_2 V_2^{\gamma}$$

$$\text{or} \quad \mathbf{PV^{\gamma} = Constant} \qquad (4)$$

Combining this equation with ideal gas equation, we get,

$$\frac{P_1 V_1}{T_1} = \frac{P_2 V_2}{T_2}$$

$$\therefore \quad \frac{T_1}{T_2} = \frac{P_2 V_2}{P_1 V_1}$$

But $$\frac{P_2}{P_1} = \left(\frac{V_1}{V_2}\right)^{\gamma}$$

$\therefore$ $$\frac{T_2}{T_1} = \left(\frac{V_1}{V_2}\right)^{\gamma} \cdot \frac{V_2}{V_1} = \left(\frac{V_1}{V_2}\right)^{\gamma-1}$$

$$\mathbf{\frac{T_2}{T_1} = \left(\frac{V_1}{V_2}\right)^{\gamma-1}} \quad (5)$$

Similarly, $$\frac{V_2}{V_1} = \left(\frac{P_1}{P_2}\right)^{\frac{1}{\gamma}}$$

$\therefore$ $$\frac{T_2}{T_1} = \frac{P_2}{P_1} \times \left(\frac{P_1}{P_2}\right)^{\frac{1}{\gamma}}$$

$$= \left(\frac{P_2}{P_1}\right)^{\frac{\gamma-1}{\gamma}}$$

$\therefore$ $$\mathbf{\frac{T_2}{T_1} = \left(\frac{P_2}{P_1}\right)^{\frac{\gamma-1}{\gamma}}} \quad (6)$$

$\therefore$ $$\mathbf{\frac{T_2}{T_1} = \left(\frac{V_1}{V_2}\right)^{\gamma-1} = \left(\frac{P_2}{P_1}\right)^{\frac{\gamma-1}{\gamma}}} \quad (7)$$

Now for Adiabatic Process ($PV^{\gamma} = C$) and during this process neither heat enters the system nor heat leaves the system.

(a) Work done

Note: Derivation is just similar to work done in polytropic process, put γ in place of n.

$\therefore$ Work done, $$W_{1-2} = \int_1^2 PdV$$

$$W_{1-2} = \frac{P_2V_2 - P_1V_1}{1-\gamma}$$

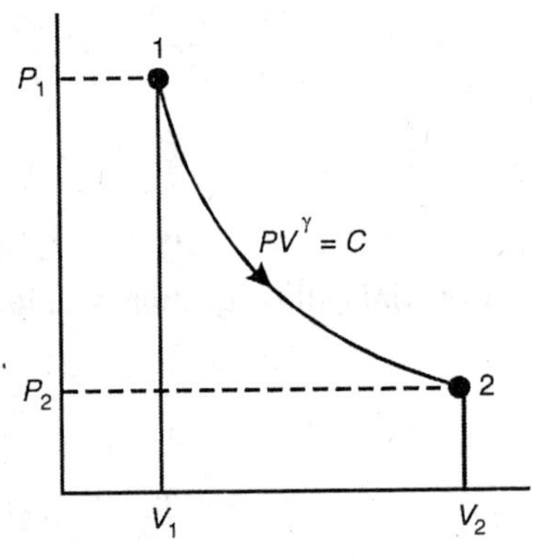

Fig. 2.9

$$W_{1-2} = \frac{P_1V_1 - P_2V_2}{\gamma - 1}$$

(b) Heat supplied. Here Heat supplied = 0 (Since neither heat enters nor leaves the system during adiabatic the process).

2.7.5 Polytropic Process

Polytropic Process is given by the law ($PV^n = C$) and is represented on the PV diagrams as shown.

(a) Work done. Work done in a non-flow process

$$W_{1-2} = \int PdV \quad (1)$$

For a polytropic process,

$$PV^n = P_1V_1^n = P_2V^n = C$$

$$\therefore \quad P = \frac{C}{V^n}$$

Substituting in Eq. (1), we get,

$$\int PdV = \int \frac{CdV}{V^n} = C\int V^{-n}dv$$

$$= \left[V^{1-n}\right]_1^2 = \left(V_2^{1-n} - V_1^{1-n}\right)$$

$$W_{1-2} = \frac{P_2V_2 - P_1V_1}{1-n} \text{ or } \frac{P_1V_1 - P_2V_2}{n-1} \quad (2)$$

Since, $P_1V_1 = mRT_1$

and $P_2V_2 = mRT_2$

$$W_{1-2} = \frac{mR(T_2 - T_1)}{1-n}$$

And for unit mass,

$$W_{1-2} = \frac{R(T_2 - T_1)}{1-n}$$

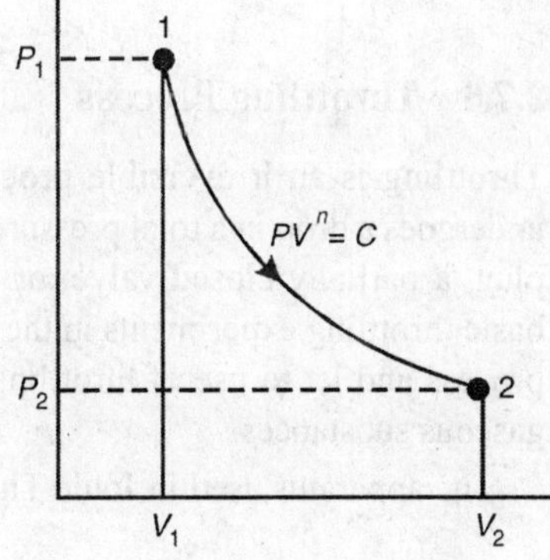

Fig. 2.10

(b) Heat supplied. From the I-law for closed system undergoing a process,

$$Q = \Delta U + W$$

For unit mass, $q = \Delta u + w$

In the differential form,

$$\delta q = du + \delta w$$

$$\delta q = du + Pdv$$

$$q_{1-2} = C_v(T_2 - T_1) + \frac{R(T_2 - T_1)}{1-n} \qquad \text{[From Eq. (3)]}$$

$$= \left[C_v + \frac{R}{1-n}\right](T_2 - T_1)$$

$$= \left[\frac{C_v - nC_v + C_p - C_v}{1-n}\right](T_2 - T_1) \text{ since } C_p - C_v = R$$

$$= \left[\frac{C_p - nC_v}{1-n}\right](T_2 - T_1)$$

Taking C_v common, we can write,

$$q_{1-2} = C_v\left[\frac{\gamma - n}{1-n}\right](T_2 - T_1), \text{ as } \frac{C_p}{C_v} = \gamma$$

$$\mathbf{q_{1-2} = C_n . (T_2 - T_1)}$$

where, $$C_n = C_v . \left[\frac{\gamma - n}{1-n}\right] \text{ or we can write as, } C_n = -C_v\left[\frac{\gamma - n}{n-1}\right]$$

i.e. Magnitude of C_n $$= C_v\left[\frac{\gamma - n}{n-1}\right] \text{ is known as Polytropic specific heat.} \qquad (5)$$

2.7.6 Throttling Process

Throttling is an irreversible process in which a fluid, flowing across a restriction, undergoes a drop in a total pressure, such a process occurs in the flow through a porous plug, a partially closed valve, or a small orifice. Joule and Thomson performed the basic throttling experiments in the period 1852–62, and their experiments clarified the process and let to use of throttling as a method for determining certain properties of gaseous substances.

The apparatus used in Joule Thomson experiment is showing in Fig. 2.11.

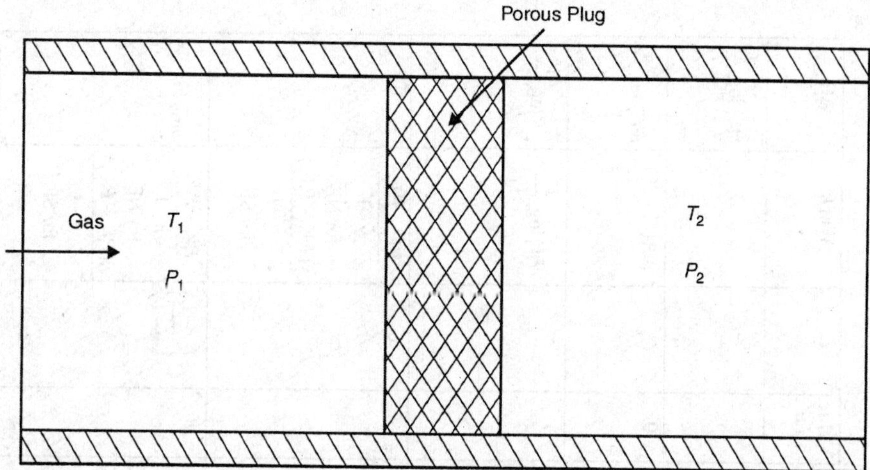

Fig. 2.11. *Throttling process, the Joule–Thomson porous plug experiment.*

A steady stream of gas flows through a porous plug contained in a horizontal tube. This system is open, is thermally insulated ($Q = 0$) and does not exchange work with its environment ($W = 0$). At sections 1 and 2, both the temperature and the pressure are measured. If the kinetic energy does not change significantly as the fluid passes through the porous plug, the steady flow energy equation reduces to

$$Q = \Delta P + \Delta K + \Delta H + W_{sf}$$

$$\therefore \quad O = 0 + 0 + \Delta H + 0$$

or $$\Delta H = 0$$

or $$H_2 - H_1 = 0$$

or $$H_2 = H_1 \text{ or } h_1 = h_2 \text{ or } t_1 = t_2$$

Hence, in an adiabatic throttling process the enthalpy remains constant. In actual experiments with different p_2, t_2 does not remain constant but is different than t_1.

2.8 EFFECT OF VARYING *N*

Since it is helpful to be able to sketch curves of various thermodynamic process on *PV* and *TS* planes about as they should appear. Figure 2.12 will be of assistance in learning to make these sketches. It shows also the effect of varying *n*. Note that the isentropic curve on the *PV* plane is steeper than the isothermal curve and that, on *TS*-plane, the constant volume curve is steeper than the constant pressure curve when both the curves are drawn for the same temperature limits.

Expansions or compressions are imagined to take place from some common point '1'.

Table 2.1 shows various quantities involved in various processes.

Table 2.1

Process	Relation between **P, V, T**	$\int_1^2 P.dv$	$-\int_1^2 V.dP$	ΔU	ΔH	Q	Value of n	C	Work	ΔH
Constant volume	$V_1 = V_2$ $\frac{P_1}{P_2} = \frac{T_1}{T_2}$	0	$V_1(P_2 - P_1)$	$mC_v(T_2 - T_1)$	$mC_p(T_2 - T_1)$	$mC_v(T_2 - T_1)$	∞	C_v	0	$mC_v \log_e \frac{T_2}{T_1}$
Constant pressure	$P_1 = P_2$ $\frac{T_2}{T_1} = \frac{V_2}{V_1}$	$P_1(V_2 - V_1)$	0	$mC_v(T_2 - T_1)$	$mC_p(T_2 - T_1)$	$mC_p(T_2 - T_1)$	0	C_p	$mP_1(V_2 - V_1)$	$mC_p \log_e \frac{T_2}{T_1}$
Isothermal	$T_1 = T_2$ $P_1V_1 = P_2V_2$	$P_1V_1 \log_e \frac{V_2}{V_1}$ or $P_1V_1 \log_e \frac{P_1}{P_2}$	$P_1V_1 \log_e \frac{V_2}{V_1}$ or $P_1V_1 \log_e \frac{P_1}{P_2}$	0	0	$P_1V_1 \log_e \frac{V_2}{V_1}$ or $P_1V_1 \log_e \frac{P_1}{P_2}$	1	∞	$P_1V_1 \log_e \frac{V_2}{V_1}$	$R \log_e \frac{V_2}{V_1}$
Reversible adiabatic	$P_1V_1^r = P_2V_2^r$ $\frac{T_2}{T_1} = \left(\frac{V_1}{V_2}\right)^{r-1}$ $= \left(\frac{P_2}{P_1}\right)^{\frac{r-1}{r}}$	$\frac{P_1V_1 - P_2V_2}{r - 1}$	$\frac{\gamma(P_1V_1 - P_2V_2)}{\gamma - 1}$	$mC_v(T_2 - T_1)$	$mC_p(T_2 - T_1)$	0	γ	0	$\frac{P_1V_1 - P_2V_2}{\gamma - 1}$ for NF $\frac{\gamma(P_1V_1 - P_2V_2)}{\gamma - 1}$ for SF	0
Polytropic	$P_1V_1^n = P_2V_2^n$ $\frac{T_2}{T_1} = \left(\frac{V_1}{V_2}\right)^{\frac{n-1}{1}}$ $= \left(\frac{P_2}{P_1}\right)^{\frac{n-1}{n}}$	$\frac{P_1V_1 - P_2V_2}{n - 1}$	$\frac{n}{n-1}(P_1V_1 - P_2V_2)$	$mC_v(T_2 - T_1)$	$mC_p(T_2 - T_1)$	$mC_v \cdot \frac{\gamma - 1}{n - 1}(T_2 - T_1)$ $= mC_n(T_2 - T_1)$	n	$C_v\left(\frac{\gamma - 1}{n - 1}\right)$ $= C_n$	$\frac{P_1V_1 - P_2V_2}{n - 1}$ for NF $\frac{n(P_1V_1 - P_2V_2)}{n - 1}$ for SF	$mC_n \log_e \frac{T_2}{T_1}$

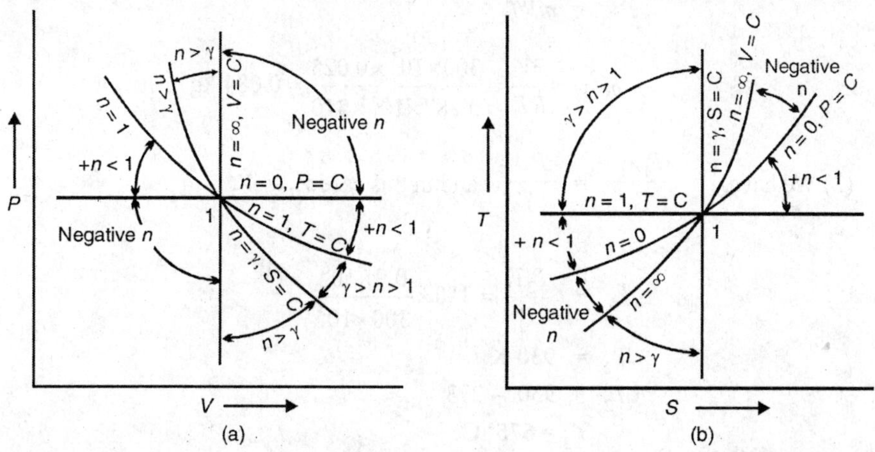

Fig. 2.12. *Effect of Varying n.*

SOLVED EXAMPLES

Example 2.1 A gas whose original pressure, volume and temperature were 150 kN/m²; 0.1 m³; and 30°C is compressed such that its new pressure is 750 kN/m² and its new temperature is 70°C. Determine the new volume of the gas.

Solution

By the characteristic equation i.e. $PV = mRT$

or $$\frac{P_1V_1}{T_1} = \frac{P_2V_2}{T_2}$$

$$\therefore \quad V_2 = \frac{P_1}{P_2}\cdot\frac{T_1}{T_2}V_1 = \frac{150}{750}\times\frac{343}{303}\times 0.1$$

$$V_2 = \mathbf{0.0226\ m^3}$$

Note. The temperatures must always be in degrees Kelvin.

Example 2.2 A quantity of gas has a pressure of 300 kN/m² when its volume is 0.025 m² and its temperature is 37°C. If the value of R is 0.287 kJ/kg K, determine the mass if the gas present. If the pressure of gas is now increased to 0.9 MN/m² while the volume remains constant what will be the new temperature of the gas.

Solution

(1) By characteristic gas equation,

$$PV = mRT$$

$$\therefore \quad m = \frac{PV}{RT} = \frac{300\times10^3\times0.025}{0.287\times10^3\times310} = 0.084 \text{ kg}$$

(2) We have, $\dfrac{P_1V_1}{T_1} = \dfrac{P_2V_2}{T_2}$ and in this case $V_1 = V_2$

$$\therefore \quad T_2 = \frac{P_2T_1}{P_1} = 310\times\frac{0.9\times10^6}{300\times10^3}$$

$$\mathbf{T_2 = 930\ K}$$

$$\therefore \quad T_2 = 930 - 273$$

$$\mathbf{T_2 = 675^\circ\ C}$$

Example 2.3 10 kg of air is heated at constant pressure, from a temperature of 100°C to 200°C. Calculate the heat added during the process and also the change in I.E. Take gas constant $R = 0.287\dfrac{\text{kJ}}{\text{kg-K}}$ and $g = 1.4$ for air.

Data: $m = 10 \text{ kg} \quad T_1 = 100°\text{C}, \quad T_2 = 200°\text{C}$

$$R = 0.287\frac{\text{kJ}}{\text{kg K}}; \ \gamma = 1.4$$

Solution

We know that, heat added during constant pressure process $= m\, C_P\left(T_2 - T_1\right)$ (1)

Also, $$C_V = \frac{R}{\gamma-1} = \frac{0.287}{1.4-1}$$

$$C_V = 0.7175$$

and $$C_P - C_V = R \quad (2)$$

$$\Rightarrow \quad C_P = R + C_V$$

$$C_P = 0.287 + 0.7175$$

$$C_P = 1.0045\frac{\text{kJ}}{\text{kg K}}$$

$\therefore$ Substituting (2) and (3) in (1) we get,

$$\text{Heat added} = m\, C_P\left(T_2 - T_1\right)$$

$$= 10 \times 1.0045(200-100)$$

Heat added = 1004.5 kJ

Charge in I.E. $= \Delta U = m\,C_V\,(T_2 - T_1) = 10 \times 0.7175(200-100)$

$$\mathbf{\Delta U = 717.5\ kJ}$$

Example 2.4 A mass of 0.8 kg of air at 1 bar and 25°C is contained in a gas tight frictionless piston cylinder device. The air is now compressed to a final pressure of 5 bar. During the process heat is transferred from the air such that the temperature inside the cylinder remains constant. Calculate the heat transfer and work done during the process and direction of each in the process.

Data :

$$m = 0.8\text{ kg} \qquad P_1 = 1\text{ bar} = 10^5\ \text{N/m}^2$$

$$T_1 = 298\text{ K} \qquad P_2 = 5\text{ bar} = 5\times 10^5\ \text{N/m}^2$$

$$T = C,\ \text{Isothermal process.}$$

Solution

∴ We know that, for an isothermal process,

$$Q = W = P_1 V_1 \ln\frac{P_1}{P_2} = mRT_1 \ln\frac{P_1}{P_2}$$

$$\mathbf{Q = W = -110.11\ kJ}$$

as Q is –ve ∴ W is also –ve

Example 2.5 1 kg of air at an initial condition of 3 bar and 175°C expands adiabatically withou friction to a final condition of 1 bar. Calculate,

(a) External work done during expansion.

(b) Change in I.E.

Take $\gamma = 1.4$ and $R = 0.287\dfrac{\text{kJ}}{\text{kg-K}}$

Data :

$$m = 1\text{ kg} \qquad P_1 = 3\times 10^5\frac{\text{N}}{\text{m}^2} = 3\times 10^2\ \text{kN/m}^2$$

$$T_1 = 175 + 273 = 448\text{ K} \qquad P_2 = 1\times 10^2\ \text{kN/m}^2$$

$$\gamma = 1.4,\ R = 0.287\frac{\text{kJ}}{\text{kg-K}}$$

Solution

(a) We know that the work done in a adiabatic expansion process is given by,

$$W_{1-2} = \frac{P_1V_1 - P_2V_2}{\gamma - 1} \qquad (1)$$

Here V_1 and V_2 are not given.

Since it is an adiabatic process,

$$P_1 V_1^{\gamma} = P_2 V_2^{\gamma}$$

$$\frac{P_1}{P_2} = \left(\frac{V_2}{V_1}\right)^{\gamma}$$

$$\left(\frac{P_1}{P_2}\right)^{\frac{1}{\gamma}} = \frac{V_2}{V_1}$$

$$\therefore \qquad V_2 = V_1 \times \left(\frac{P_1}{P_2}\right)^{\frac{1}{\gamma}} \qquad (2)$$

To find V_1, $\quad P_1 V_1 = mRT_1$

$$V_1 = \frac{mRT_1}{P_1} = \frac{1 \times 0.287 \times 448}{3 \times 10^2}$$

Note units : $\left[\dfrac{\text{kg}\dfrac{\text{kJ}}{\text{kg-K}} \times \text{K}}{\text{kN/m}^2} = \dfrac{\text{kJ}}{\text{kN/m}^2} = \dfrac{\text{kNm}}{\text{kN}} \times \text{m}^2 = \text{m}^3\right]$

$$V_1 = 0.429 \text{ m}^3$$

$\therefore$ From Eq. (2),

$$V_2 = 0.429 \times \left(\frac{300}{100}\right)^{\frac{1}{1.4}} = 0.429 \times (3)^{\frac{1}{1.4}}$$

$$\mathbf{V_2 = 0.940 \text{ m}^3}$$

$\therefore$ Now from Eq. (1), work done,

$$W_{1-2} = \frac{P_1 V_1 - P_2 V_2}{\gamma - 1}$$

$$= \frac{300 \times 0.429 - 100 \times 0.940}{1.4 - 1} = \frac{34.7}{0.4}$$

$$\mathbf{W_{1-2} = 86.75\ kJ}$$

(b) We also know from the I–law, that for a process, $Q = \Delta U + W$. Since it is adiabatic process $Q = 0$

$$\therefore \quad \mathbf{W = -\Delta U = 86.75\ kJ \Rightarrow \Delta U = -86.75\ kJ}$$

∴ I.E. decreases by 86.75 kJ

Example 2.6 Find the change of Internal energy and enthalpy of 3 kg of air, when the temperature changes from 25°C to 100°C.

Take $C_p = 1 \frac{\text{kJ}}{\text{kg-K}}$ and $C_v = 0.71 \frac{\text{kJ}}{\text{kg-K}}$

Solution

In the problem, the type of process is not given, but since temperature changes, it is not an isothermal process.

$$\therefore \quad \Delta U = m \cdot C_V \cdot (T_2 - T_1)$$

$$= 3\ \text{kg} \times 0.71 \frac{\text{kJ}}{\text{kg-K}} \times (373 - 298)\ \text{K}$$

$$= 3 \times 0.71 \times 75$$

$$\mathbf{\Delta U = 159.75\ kJ}$$

and change of enthalpy $\Delta H = m \cdot C_P \cdot (T_2 - T_1) = 3 \times 1 \times (75)$

$$\mathbf{\Delta H = 225\ kJ}$$

Example 2.7 0.44 kg of a gas having volume 0.28 m^3 and a pressure of 1.4 bar is compressed to a pressure of 14 bar according to $PV^{1.25} = C$. Find the amount of heat transferred during the process by calculating the work done and the change of internal energy.

Take $C_p = 1.041 \frac{\text{kJ}}{\text{kg-K}}$ and $C_v = 0.743 \frac{\text{kJ}}{\text{kg-K}}$

Data: $m = 0.44$ kg $\quad V_1 = 0.28\ \text{m}^3$

$P_1 = 1.4$ bar $= 140$ kPA $\quad P_2 = 14$ bar $= 1400$ kPa

Solution

$$PV^{1.25} = C \qquad Q = ?$$

$$W = ? \qquad \Delta U = ?$$

$$C_P = 1.041 \text{ and } C_V = 0.743 \frac{\text{kJ}}{\text{kg-K}}$$

$$\text{W. D.} = \frac{P_2 V_2 - P_1 V_1}{1-n} \qquad (1)$$

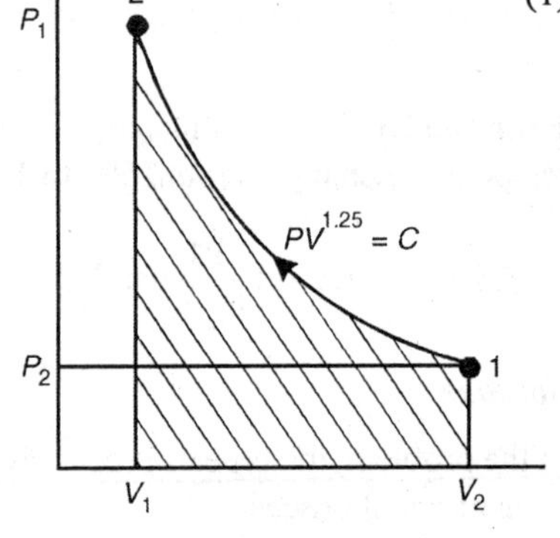

Fig. Ex. 2.7

Here only V_2 is not given and since it is a polytropic process,

$$P_1 V_1^n = P_2 V_2^n$$

$$\frac{P_1}{P_2} = \left(\frac{V_2}{V_1}\right)^n$$

i.e.

$$\frac{V_2}{V_1} = \left(\frac{P_1}{P_2}\right)^{\frac{1}{n}}$$

$$V_2 = V_1 \times \left(\frac{P_1}{P_2}\right)^{\frac{1}{n}} = 0.28 \times \left(\frac{140}{1400}\right)^{\frac{1}{1.25}}$$

$$= 0.28 \times (0.1)^{0.8}$$

$$V_2 = 0.044377 \text{ m}^3 \approx 0.0444 \text{ m}^3$$

Now work done from Eq. (1),

$$W_{1-2} = \frac{1400 \times 0.0444 - 140 \times 0.28}{1-1.25}$$

$$= \frac{62.16 - 39.2}{-0.25}$$

$$\mathbf{W_{1-2} = -91.84 \text{ kJ}}$$

W_{1-2} is – ve, so, work is done on the gas.

We know that heat transferred in a polytropic process is given by,

We know that, $$q = C_n \cdot (T_2 - T_1)$$

$$= C_V \cdot \left(\frac{\gamma - n}{1-n}\right)(T_2 - T_1)$$

$$= \frac{R}{\gamma-1}\left(\frac{\gamma-n}{1-n}\right)(T_2-T_1)$$

$$q = \frac{R}{(1-n)}\frac{(\gamma-n)}{(\gamma-1)}\times(T_2-T_1)$$

$$q = \frac{(\gamma-n)}{(\gamma-1)}\times\frac{R}{(1-n)}\times(T_2-T_1)$$

And total heat transfer,

$$Q = \frac{\gamma-n}{\gamma-1}\times\frac{mR\,(T_2-T_1)}{1-n}$$

i.e. $$Q = \frac{\gamma-n}{\gamma-1}\times \text{W.D. in a Polytropic process.}$$

Note. Remember this equation and derivation.

To find Q,

Let us find, $$\gamma = \frac{C_P}{C_V} = \frac{1.041}{0.743} = 1.401$$

$$Q = \frac{1.401-1.25}{1.104-1}\times -91.84 = \frac{0.15}{0.4}\times -91.84$$

$$\mathbf{Q = -34.583\ kJ}$$

$\therefore$ From the first law for a process,

$$Q-W = \Delta U$$

$$-34.583+91.84 = \Delta U$$

$\therefore$ $$\mathbf{\Delta U = 57.257\ kJ\ (Increases)}$$

Example 2.8 A closed system containing 2 kg of working substance undergoes a reversible constant pressure process, during which the I.E. of the system decreases by 80 kJ. If

$C_p = 1120\dfrac{\text{kJ}}{\text{kg-K}}$ and $C_v = 800\dfrac{\text{kJ}}{\text{kg-K}}$. Find the work done during the process.

Data : $m = 2$ kg $\quad P = C$

$\Delta U = -80$ kJ $\quad C_P = 1.120\dfrac{\text{kJ}}{\text{kg-K}}$

$$C_V = 0.8 \frac{\text{kJ}}{\text{kg-K}} \qquad W_{1-2} = ?$$

Solution

We know that work done in a constant pressure process,

$$W_{1-2} = P(V_2 - V_1) \tag{1}$$

Also from the I-law, $Q - W = \Delta U$ (2)

We know that heat supplied is utilised for,

(i) increasing I.E.,

(ii) doing external work and heat transfer,

$$\delta q = dh$$

or

$$\delta Q = dH = m\,C_P\,\Delta T \tag{3}$$

Also,

$$\Delta U = m \cdot C_V \cdot \Delta T \tag{4}$$

$$\frac{\Delta U}{m \cdot C_V} = \Delta T$$

$$\frac{-80 \text{ kJ}}{2 \text{ kg} \times 0.8 \frac{\text{kJ}}{\text{kg-K}}} = \Delta T$$

$$= \frac{-0.8}{1.6} = \Delta T$$

$$-50 \text{ K} = \Delta T \tag{5}$$

Fig. Ex. 2.8

$\therefore$ From (3),

$$\delta Q = m \cdot C_P \cdot \Delta T = 2 \times 1.12 \times (-50)$$

$$\delta Q \text{ or } Q = -112 \text{ kJ}$$

$\therefore$ From Eq. (2),

$$Q - W = \Delta U$$

$$Q - \Delta U = W$$

$$-112 + 80 = W$$

$\therefore$ **W = –32 kJ**

Example 2.9 A cylinder contains 0.12 m^3 of air at 1 bar and 100°C. The air is compressed to 0.03 m^3. The final pressure being 6 bar.

Determine:

(i) The value of index n.

(ii) Mass of air in the cylinder.

(iii) Increase in I.E.

Take $\gamma = 1.4$, $R = 0.287$ kJ/kg-K and $C_v = 0.72 \frac{\text{kJ}}{\text{kg-K}}$

Data:

$$V_1 = 0.12 \text{ m}^3 \qquad P_1 = 100 \text{ kPa}$$

$$T_1 = 100 + 273 = 373 \text{ K} \qquad V_2 = 0.03 \text{ m}^3$$

$$P_2 = 600 \text{ kPa}$$

Solution

(i) To find n.

So,

$$P_1 V_1^n = P_2 V_2^n$$

$$\frac{P_2}{P_1} = \left(\frac{V_1}{V_2}\right)^n$$

$$\left(\frac{600}{100}\right) = \left(\frac{0.12}{0.03}\right)^n$$

$$6 = (4)^n$$

$$\log_e 6 = n \log_e (4)$$

$$\frac{\log_e 6}{\log_e 4} = n$$

$$\frac{1.79175}{1.38629} = n$$

$$\mathbf{n = 1.29248 \approx 1.3}$$

(ii) To find mass of air in the cylinder.

We know that,

$$P_1 V_1 = mRT_1$$

$$100 \times 0.12 = m \times 0.287 \times 373$$

$$\therefore \quad \mathbf{m = 0.112096 \text{ kg}}$$

(iii) To find the increase in I.E.,

We have,

$$\Delta U = m \cdot C_V (T_2 - T_1)$$

Here T_2 is not known, so to find T_2, since it is a polytropic process,

$$\therefore \quad \frac{T_2}{T_1} = \left(\frac{V_1}{V_2}\right)^{n-1} = \left(\frac{P_2}{P_1}\right)^{n-1/n}$$

$$T_2 = T_1 \times \left(\frac{V_1}{V_2}\right)^{n-1}$$

$$= 373\times\left(\frac{0.12}{0.03}\right)^{1.3-1}$$

$$= 373\times(4)^{0.3} = 373\times1.516$$

$$T_2 = 565.36\text{ K}$$

$$\therefore \quad \Delta U = m\cdot C_V\cdot(T_2 - T_1)$$

$$= 0.112\text{ kg}\times0.72\,(565.36-373)$$

$$\mathbf{\Delta U = 15.51\ kJ\ (Increases)}$$

Example 2.10 Calculate the change of,

(i) IE and

(ii) enthalpy of 2 kg of air when the temperature changes from 20°C to 90°C. Take

$$C_p = 1\text{kJ/kg-K and } C_v = 0.71\frac{\text{kJ}}{\text{kg-K}}$$

Solution

(i) $$\Delta U = m\,C_V\,(T_2 - T_1) = 2\times0.71(90-20)$$

$$\mathbf{\Delta U = 99.4\ kJ}$$

(ii) $$\Delta H = m\cdot C_V\cdot(T_2 - T_1) = 2\times1\times(90-20)$$

$$\mathbf{\Delta H = 140\ kJ}$$

Example 2.11. 10 kg of air are heated in a rigid vessel from 20°C to 100°C. If the ratio of specific heats is 1.4, estimate the values of C_p and C_v, the change in internal energy and enthalpy.

Data: $T_1 = 20°\text{C}$, $T_2 = 100°\text{C}$

Solution

$$C_V = \frac{R}{\gamma-1}$$

and $$\gamma = 1.4 \text{ and } R = 0.287$$

$$\therefore \quad \mathbf{C_V = \frac{0.287}{0.4} = 0.7175\frac{kJ}{kg\text{-}K}}$$

$$\mathbf{C_P = R + C_V = 0.287 + 0.7175 = 1.0045\ kJ/kg\text{-}K}$$

We know that,

Change in internal energy,

$$dU = mC_V(T_2 - T_1)$$

$$= (10)\,0.7175\,(100-20)$$

dU = 574 kJ = 574 kJ

and change in enthalpy,

$$dH = mC_P(T_2 - T_1)$$

$$= (10)(1.0045)(100-80)$$

dH = 803.6 kJ

Example 2.12 10 kg of carbon monoxide at 40°C occupies 3 m^3. Determine the gas pressure in bar. An additional mass of CO is then very slowly added to raise the tank pressure to 10 bar. Assuming that the gas temperature remains constant, how much extra mass must have been added. Assume $R = 0.297$ KJ/kg-K.

Solution

We know that, $PV = mRT$

$$P = \frac{mRT}{V} = \frac{10 \times 0.297 \times 313}{3}$$

$$P = 309.87 \text{ kPa} = 3.0987 \text{ bar}$$

Since additional CO is added to the tank, $V = C$ and as given $T = C$.

$\therefore$ $m \propto$ Pressure.

$$\therefore \quad \frac{m_2}{m_1} = \frac{P_2}{P_1} = \frac{10}{3.0987}$$

$$\therefore \quad m_2 = m_1 \times \frac{10}{3.0987} = 10 \times \frac{10}{3.0987}$$

$$\mathbf{m_2 = 32.2768\ kg}$$

$\therefore$ Additional mass,

$$\mathbf{\Delta m = 32.2768 - 10 = 22.2768\ kg}$$

Example 2.13 A piston and cylinder machine contains 1 kg of air, initially $v = 0.8$ m^3/kg and $T = 290$ K. The air is then compressed in a slow frictionless process to a specific volume 0.2 m^3/kg and a temperature 580 K. The law for compression is $PV^{1.5} = 0.75$ with P in bar and V in m^3/kg. Determine work done and heat transferred during the process. Assume for air

$$C_p = 1.0\frac{\text{kJ}}{\text{kg-K}'} \text{ and } C_v = 0.743\frac{\text{kJ}}{\text{kg-K}} \text{ and } R = 0.287\frac{\text{kJ}}{\text{kg-K}}.$$

Data: $m = 1$ kg, $v_1 = 0.8$ m^3/kg, $T_1 = 290$ K, $v_2 = 0.2$ m^3/kg $T_2 = 580$ K

$W = ?$ $Q = ?$, law for the Process $= PV^{1.5} = C = 0.75$

Solution

We know that, for a Polytropic process,

$$W = \frac{P_2 V_2 - P_1 V_1}{1-n} = \frac{m\,R\,(T_2 - T_1)}{1-n}$$

$$= \frac{1 \times 0.287\,(580-290)}{1-1.5}$$

$$= \frac{83.23}{-0.5}\text{ kJ}$$

$$\mathbf{W = -166.46\ kJ}$$

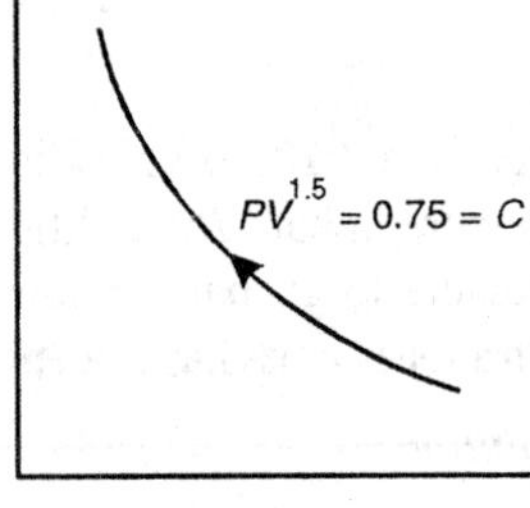

Fig. Ex. 2.13

The sign implies work is done on the air.

Also, $\Delta U = m \cdot C_V \cdot dT$

$$= 1 \times 0.743 \times (580 - 290)$$

$$\mathbf{\Delta U = 206.77\ kJ}$$

From I-law, for a process,

$$Q - W = \Delta U$$

$$Q = \Delta U + W$$

$$= 206.77 + (-166.46)$$

$$\mathbf{Q = 40.31\ kJ}$$

Example 2.14 2 kg of air at 150°C and 3 bar expands according to law $PV^{1.2} = C$ to final pressure of 1 bar. Calculate: heat supplied, work done and change in internal energy.

Take $R = 0.287$ kJ/kg – K and $= 1.41$.

Data: $P_1 = 3$ bar, $P_2 = 1$ bar, $T_1 = 150°\text{C} = 423$ K

Solution

(1) Non flow work done in polytropic process

$$= \frac{mR(T_2 - T_1)}{n-1}$$

We have, $$\frac{T_2}{T_1} = \left(\frac{P_2}{P_1}\right)^{\frac{n-1}{n}} = \left(\frac{1}{3}\right)^{\frac{0.2}{1.2}} = 0.832$$

$$T_2 = 423 \times 0.832 = 351.936 \text{ K}$$

$$\therefore \quad \text{W.D.} = \frac{0.2 \times 0.287 \times (423 - 351.936)}{1.2 - 1}$$

W.D. = 203.954 kJ

(2) Heat supplied in polytropic process,

$$Q = \left(\frac{\gamma - n}{\gamma - 1}\right) \times \text{W.D. in non flow process.}$$

$$= \frac{1.41 - 1.2}{1.41 - 1} \times 203.954$$

$\therefore$ **Q = 104.464 kJ**

(3) Change in internal energy.

$$\Delta U = Q - W$$

$$= 104.464 - 203.954$$

ΔU = – 99.49 kJ decrease

Alternatively :

(2) Change in internal energy,

$$\Delta U = m\, C_V \,(T_2 - T_1)$$

$$= 2 \times \frac{0.287}{1.41 - 1} \times (351.936 - 423)$$

$\therefore$ **ΔU = –99.49 kJ**

(3) Heat supplied

$$Q = W + \Delta U = 203.954 - 99.49$$

Q = 104.464 kJ

Example 2.15 One kg of certain gas undergoes a thermodynamic constant pressure process whereby the volume changes from 1 m^3 to 1.8 m^3 while the temperature changes from 50°C to 450°C. The specific heat at constant pressure is given by,

$$C_p = \left(2.5 + \frac{40}{T + 20}\right) \text{kJ/kg °C where } T \text{ is in °C.}$$

Find out (1) heat supplied, (2) work done, (3) change in internal energy and (4) change in enthalpy: Take $\gamma = 1.4$.

Solution

(1) Heat supplied in constant pressure process,

$$= \int m \,.\, C_P \, dT, \quad \text{as given } m = 1$$

$$= \int_{50}^{450} \left(2.5 + \frac{40}{T + 20}\right) dT$$

$$= \int_{50}^{450} 2.5 \, dT + \int_{50}^{450} \frac{0}{T + 20} dT$$

$$= 2.5[T]_{50}^{450} + 40[\ln(T + 20)]_{50}^{450}$$

$$= 2.5(450 - 50) + 40 \ln\left(\frac{470}{70}\right)$$

$$= 1000 + 76.169$$

Heat supplied in constant pressure process = 1076.169 kJ

(2) Change in internal energy is given by,

$$U_2 - U_1 = \int m \, C_V \, dT \text{ where } m = 1 \; C_P = \gamma \, C_V$$

$$= \frac{1}{\gamma} \int C_P \, dT = \frac{1076.169}{1.4}$$

$$\mathbf{U_2 - U_1 = 768.692 \text{ kJ}}$$

(3) Work done = Heat supplied – Change in internal energy

$$= 1076.169 - 768.692$$

Work done = 307.477 kJ

(4) Change in enthalpy,

= Heat supplied in constant pressure process

Change in enthalpy = 1076.169 kJ

Example 2.16 One kg of certain working substance undergoes a reversible constant pressure process at 1.2 bar during which its volume changes from 1 m^3 to 2.8 m^3 and temperature changes from 50°C to 370°C. The specific heat for the substance at constant pressure is given by,

$$C_p = \left(1.1 + \frac{40}{t + 30}\right) \text{kJ/kg °C where, '}t\text{' is in °C. Find out:}$$

(i) Heat supplied
(ii) Work done
(iii) Change in I.E.
(iv) Change in enthalpy.

Data: $P = C$ mass $= 1$ kg

$$P = 1.2 \text{ bar} = 1.2\times10^5 \frac{\text{N}}{\text{m}^2}$$

$$V_1 = 1 \text{ m}^3 \qquad V_2 = 1.8 \text{ m}^3$$

$$T_1 = 50°\text{C} \qquad T_2 = 370°\text{C}$$

Solution

Given, $$C_P = \left[1.1+\frac{40}{t+30}\right]\frac{\text{kJ}}{\text{kg °C}}$$

For $P = C$ process $\delta Q = dH$

$$\therefore \quad \delta Q = \text{Change in enthalpy} = \int_{T_1}^{T_2} mC_P\, dT$$

$$= \int_{T_1}^{T_2} 1\times\left(1.1+\frac{40}{t+30}\right)dT$$

$$= \int_{50}^{370} 1.1\, dT + \int_{50}^{370} \frac{40}{t+30}\, dT$$

$$= [1.1\,t]_{50}^{370} + 40\{\ln(t+30)\}_{50}^{370}$$

Change in enthalpy = 416.38 kJ

Also, for a constant pressure process,

Heat supplied = Change in enthalpy

$$\therefore \quad Q = 416.38 \text{ kJ}$$

Also, $$\text{W. D.} = P(V_2 - V_1) = 120(1.8-1)$$

$$\mathbf{W_{1-2} = 96\ kJ}$$

According to I-law for a process,

$$Q - W = \Delta U$$

$$\Rightarrow \quad \Delta U = 416.38 - 96$$

$$\mathbf{\Delta U = 320.38\ kJ}$$

Example 2.17 60 litres of hydrogen at 20°C and 1 bar is compressed adiabatically to 9.8 bar. It is then cooled at constant volume to a pressure P_3 such as to have temperature of 20°C and further expanded isothermally so as to reach initial state. Find:

(i) Work done during each process.

(ii) Change in internal energy during each process. Take

$$C_p = 14.4 \text{ kJ/kg-K}$$
$$C_v = 10.28 \text{ kJ/kg-K}$$

for hydrogen and show the process on PV and T-S plane.

Data:

$$V_1 = 0.06 \text{ m}^3 \left(H_2\right), T_1 = 20 + 273 = 293 \text{ K}, P_1 = 100 \text{ kPa}, PV^\gamma = C, P_2 = 980 \text{ kPa},$$

$$\left.\begin{aligned} C_P &= 14.4 \text{ kJ/kg-K} \\ P_3 = T_3 = 20 + 273 = 293 \text{ K}, Pv = C, \ C_V &= 10.28 \text{ kJ/kg-K} \end{aligned}\right\} \frac{C_P}{C_V} = \gamma = 1$$

Find : (i) W. D. during processes, (ii) during processes.

Solution

(i) Work done :

For process 1 – 2, $$W_{1-2} = \frac{P_2V_2 - P_1V_1}{1-\gamma} \quad (1)$$

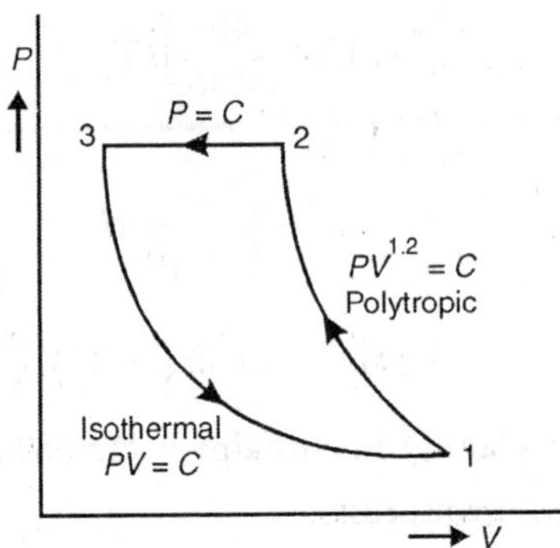

Fig. Ex. 2.17

To find V_2 and since process 1 – 2 is an adiabatic process,

$$P_1V_1^\gamma = P_2V_2^\gamma$$

$$\left(\frac{P_1}{P_2}\right)^{\frac{1}{\gamma}} \times V_1 = V_2$$

$$\therefore \quad \mathbf{V_2 = 0.01175 \ m^3}$$

$$\therefore \quad W_{1-2} = \frac{980 \times 0.01175 - 100 \times 0.06}{1 - 1.4}$$

$$\mathbf{W_{1-2} = -13.788\ kJ}$$

Process 2 – 3 : Constant Volume process $W_{2-3} = 0$

Process 3 – 1 : (Isothermal process) :

$$W_{3\ 1} = P_3 V_3 \ln \frac{V_1}{V_3} \qquad (2)$$

Note $\quad V_2 = V_3$

and for process 2 – 3,

$$\frac{P_2 V_2}{T_2} = \frac{P_3 V_3}{T_3}$$

$$\frac{P_2}{T_2} = \frac{P_3}{T_3}$$

$$\frac{980}{T_2} = \frac{P_3}{298\ (\text{i.e. } T_1)} \qquad (3)$$

For isothermal, $\quad P_1 V_1 = P_3 V_3$

$$100 \times 0.06 = P_3 \times 0.01175 \ (\text{as } V_2 = V_3)$$

$$\therefore \quad \mathbf{P_3 = 510.64\ kPa} \qquad (4)$$

From (3) and (4),

$$\mathbf{T_2 = 562.31\ K}$$

$$\therefore \quad W_{3-1} = P_3 V_3 \ln \frac{V_1}{V_3}$$

$$= 510.64 \times 0.01175 \times \ln \frac{0.06}{0.01175}$$

$$\mathbf{W_{3-1} = 9.7829\ kJ}$$

(ii) Change in I.E.:

For process 1 – 2, $\quad Q = 0$

$$\therefore \quad Q - W = \Delta U$$

$$0 - W = \Delta U$$

$$\therefore \quad \mathbf{\Delta U = 13.788\ kJ\ (i.e.\ U_2 - U_1)}$$

For process 2 – 3, $W = 0$

$$Q - W = \Delta U$$

$$Q = \Delta U = m \cdot C_V (T_3 - T_2)$$

$$P_1 V_1 = mRT_1$$

$$100 \times 0.06 = m (C_P = C_V) T_1$$

$$= m(14.4 - 10.28) \times 293$$

m = 0.00488 kg

$$\therefore \quad Q_{2-3} = \Delta U = 0.0048 \times 10.28(298 - 562.3)$$

$$\mathbf{\Delta U = Q_{2-3} = -13.27\ kJ}$$

for 3 – 1 $T = C$

$$\therefore \quad \Delta U = m \cdot C_V \cdot dT = 0 \text{ as } T = C.$$

Example 2.18 3 kg of air at a pressure of 150 kPa and temperature 360 K is compressed polytropically to 750 kPa according to law $PV^{1.2}$ = constant. The air is then cooled to initial temperature at constant pressure. The air is then brought to state (1) by following $PV = C$. Draw the cycle on PV-diagram and determine net work and heat.

Data: $P_1 = 150$ kPa, $P_2 = 750$ kPa, $P_2 = P_3$, $T_1 = 360$ K, $V_2 = ?$, $T_3 = T_1 = 360$ K

Solution

$V_1 = ?\ T_2 = ?\ V_3 = ?$

1 – 2 is a Polytropic process : $PV^{1.2} = C$

2 – 3 is a Constant pressure process

3 – 1 is a Isothermal process

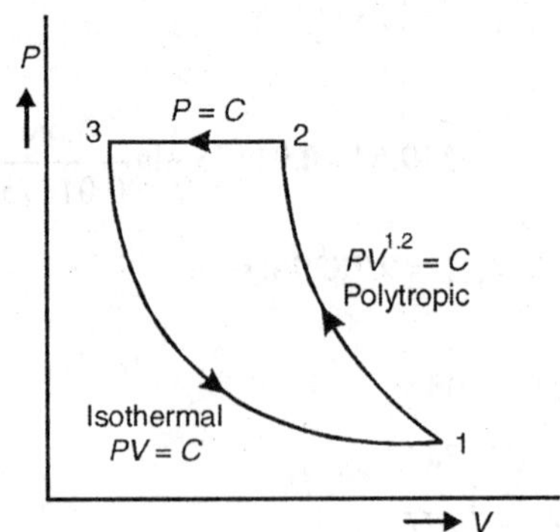

Fig. Ex. 2.18

We know that $PV = mRT$

$$P_1V_1 = mRT_1$$

$$\Rightarrow \quad V_1 = \frac{mRT_1}{P_1} = \frac{3\times0.287\times360}{150}$$

$$\mathbf{V_1 = 2.0664\ m^3}$$

1 – 2 adiabatic process

$$\frac{T_1}{T_2} = \left(\frac{P_1}{P_2}\right)^{\frac{n-1}{n}} \Rightarrow \frac{360}{T_2}$$

$$= \left(\frac{150}{750}\right)^{\frac{0.2}{1.2}} \Rightarrow T_2 = 470.76\ \text{K}$$

$$V_2 = \frac{mRT_2}{P_2} = \frac{3\times0.287\times470.76}{750}$$

$$= 0.540\ \text{m}^3$$

3 – 1 Isothermal process $PV = C$

$$P_3V_3 = P_1V_1 \Rightarrow V_3 = \frac{P_1V_1}{P_3}$$

$$= \frac{150\times2.0664}{750}$$

$$\Rightarrow \quad V_3 = 0.413\ \text{m}^3$$

$$W_{1-2} = \frac{P_1V_1 - P_2V_2}{n-1} = \frac{mR(T_2 - T_1)}{1-n}$$

$$\mathbf{W_{1-2} = -476.82\ kJ}$$

$$W_{2-3} = P_2(V_3 - V_2) = 750(0.413 - 0.54)$$

$$= -95.25\ \text{kJ}$$

$$W_{3-1} = P_1V_1 \ln\left(\frac{P_3}{P_1}\right)$$

$$= 150\times2.0664 \ln\frac{750}{150}$$

$$\mathbf{W_{3-1} = 498.66\ kJ}$$

$$\text{Net work done} = W_{1-2} + W_{2-3} + W_{3-1}$$

$$= -73.21\,\text{kJ}$$

$$Q_{1-2} = \left(\frac{\gamma - n}{\gamma - 1}\right) \times \text{Work done}$$

$$= \frac{1.4 - 1.2}{1.4 - 1.0} \times W_{1-2}$$

$$\mathbf{Q_{1-2} = -238.41\ kJ}$$

$$Q_{2-3} = \text{change in enthalpy} = mC_P\,\Delta t = mC_P\left(T_3 - T_2\right)$$

$$= \frac{m\gamma R}{\gamma - 1}\left(T_3 - T_2\right)$$

$$= \frac{3 \times 1.4 \times 0.287}{1.4 - 1}\left(360 - 470.76\right)$$

$$\mathbf{Q_{2-3} = -333.775\ kJ}$$

$$Q_{3-1} = \text{work done} = 498.86\,\text{kJ}$$

$$\therefore \quad \text{Net heat} = Q_{1-2} + Q_{2-3} + Q_{3-1}$$

$$\mathbf{Net\ heat = -73.21\ kJ}$$

$$\therefore \quad \text{Net heat} = \text{Net work done}$$

Thus the I-law for the cycle satisfies.

Example 2.19 A certain oxygen cylinder has a capacity 250 litres and contains oxygen at a pressure of 3 MPa and a temperature of 20°C. The stop valve is opened and some oxygen is used. If the pressure and temperature of oxygen left in the cylinder falls to 1.8 MPa and 16°C respectively, determine the mass of oxygen used.

If after the stop valve is closed the oxygen remaining in the cylinder attains its original temperature of 20°C, determine the amount of heat, transferred through the cylinder walls from the atmosphere.

Solution

Capacity of the cylinder $= V = 250$ litre $= 0.250\ \text{m}^3$ as 1000 litre $= 1\ \text{m}^3$

Density of O_2 at 0°C, 0.1013 MPa $= 1.43\ \text{kg/m}^3$

$\therefore$ Mass of given O_2 at 0°C and 0.1013 MPa $= 1.43 \times 0.25$

$$\therefore \quad m_1 = 0.3575\ \text{kg at STA}$$

$$P_2 = 3\ \text{MPA};$$

$$P_3 = 1.8\ \text{MPa}$$

$$T_2 = 20°C = 293 \text{ K}$$

$$T_3 = 16°C = 289 \text{ K}$$

$$PV = mRT$$

Since V is constant, as volume of the cylinder does not change, we have,

$$\frac{P}{T.m} = \frac{R}{V} = \text{Constant}$$

$$\therefore \quad \frac{P_1}{T_1 m_1} = \frac{P_2}{T_2 m_2}$$

$$\therefore \quad \frac{0.1013}{273 \times 0.3575} = \frac{3}{293 \times m_2}$$

$$\therefore \quad m_2 = 9.865 \text{ kg}$$

Similarly, $$\frac{P_1}{T_1 m_1} = \frac{P_3}{m_3 T_3}$$

$$= \frac{0.1013}{273 \times 0.3575} \times \frac{1.8}{289 \times m_3}$$

$$\therefore \quad m_3 = 6.001 \text{ kg}$$

$$\Delta m = m_2 - m_3$$

$$= 9.865 - 6.001$$

$$\mathbf{\Delta m = 3.864 \text{ kg}}$$

Also we have, $$\frac{PV}{mT} = R$$

$$\therefore \quad R = \frac{3000 \times 0.25}{9.865 \times 293} = 0.26 \text{ kJ/kg-K}$$

Heat transfer, $Q = m\, C_V\, (T_4 - T_3)$

{where, $T_4 = T_2$, the original temperature}

$$C_V = \frac{R}{\gamma - 1}$$

$$= \frac{R}{\left(\dfrac{C_P}{C_V} - 1\right)}$$

$$= \frac{0.26}{1.4-1} = 0.65 \text{ kJ/kg-K}$$

$$\therefore \quad Q = 6.001 \times 0.65 (293-289)$$

$$\mathbf{Q = 15.603 \text{ kJ}}$$

Example 2.20 A gas initially at pressure 510 kPa and volume 142 litres undergoes: a process and final pressure of 170 kPa and volume of 275 litres during which enthalpy decreased by 65 kJ. Take C_V = 0.718 kJ/kg-K. Determine:

(i) Change in internal energy.

(ii) Specific heat at constant pressure.

(iii) Specific gas constant.

Data: Intial pressure $P_1 = 510 \text{ kPa} = 510 \text{ kN/m}^2$

Intial volume $V_1 = 142\, l = 142 \times 10^{-3} \text{ m}^3$

Final pressure $P_2 = 170 \text{ kPa} = 170 \text{ kN/m}^2$

Final volume $V_2 = 275\, l = 275 \times 10^{-3} \text{ m}^3$

Change in enthalpy $= -65 \text{ kJ} \left[\because \text{ Enthalpy is being decreased}\right]$

$C_V = 0.718$ kJ/kg-K

Solution

(i) We have, from first law of thermodynamics,

$$\Delta H = \Delta U + \Delta(PV)$$

$$\Delta U = \Delta H - \Delta(PV)$$

$$\Delta U = -65 \times 10^3 - (170 \times 275 - 510 \times 142)$$

$$= -65 \times 10^3 - (46750 - 72420)$$

$$= -65000 + 25670 = -39330$$

$$\therefore \quad \mathbf{\Delta U = -39.33 \text{ kJ}}$$

(ii) Also we know that,

$$\frac{\Delta H}{\Delta U} = \frac{C_P}{C_V} = \gamma$$

$$\therefore \quad C_P = C_V \times \frac{\Delta H}{\Delta U} = 0.718 \times \frac{-65}{-39.33}$$

$$\mathbf{C_P = 1.186 \text{ kJ/kg-K}}$$

(iii) Also, $C_P - C_V = R$

$$R = 1.1866 - 0.718$$

$$\mathbf{R = 0.4686\ kJ/kg\text{-}K}$$

Example 2.21 An engine has a piston area of 0.1 m^2 contains gas at a pressure of 1.5 MPa. The gas expands according to a process which is represented by a straight line on *P-V* plane. The final pressure is 0.15 MPa. Calculate work done by gas on piston if the stroke is 0.3 m.

Data: Area of the piston = 0.1 m^2

P_1 = 1.5 MPa $\quad$ P_2 = 0.15 MPa

Stroke = 0.3 m $\quad$ W = ?

Solution

According to the formula,

Change in volume = Area of the piston × Stroke

$= 0.1 \times 0.3 = 0.03$ m

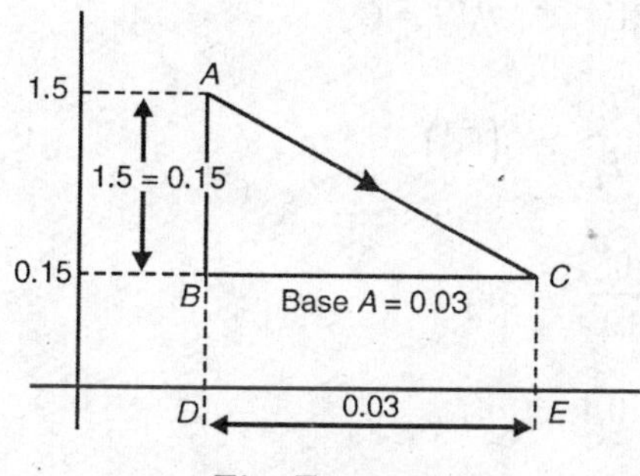

Fig. Ex. 2.21

Also we know that,

Work done by gas = Area under the curve (in this case a straight line) on the *P-V* diagram

∴ Net work done = Area of the rectangle *BCED* + Area of triangle *ABC*

$$W = \left[\{0.15 \times 10^6 \times 0.3\} + \left\{\frac{1}{2}(1.5 - 0.15) \times 10^6 \times 0.03\right\}\right]$$

$$\mathbf{W = 24.75\ kJ}$$

Example 2.22 In a piston cylinder arrangement the pressure varies inversely proportional to the square of volume during a process. The initial pressure in the cylinder is 20 bar and the initial volume is 0.1 m^3. The final pressure after the expansion is 200 kPa. Estimate the work done.

Data: $P \propto \frac{1}{V^2} \Rightarrow P = \frac{C}{V^2}$ and $PV^2 = C$

$$P_1 = 20 \text{ bar} = 20\times10^5 \ \frac{\text{N}}{\text{m}^2}$$

$$V_1 = 0.1 \text{ m}^3$$

$$P_2 = 200 \text{ kPa} = 200\times10^3 \ \frac{\text{N}}{\text{m}^2}$$

$$W = ?,\ V_2 = ?$$

Solution

We know that for a Polytropic process,

$$\text{Work done} = \frac{P_1 V_1 - P_2 V_2}{n-1} \qquad (1)$$

Since V_2 is not given,

$$P_1 V_1^2 = P_2 V_2^2 = C$$

$$V_1^2 \times \frac{P_1}{P_2} = \left(V_2^2\right)$$

$$0.1^2 \times \frac{20\times10^5}{200\times10^3} = V_2^2$$

$$0.1 = V_2^2$$

$$0.3162 \text{ m}^3 = V_2 \qquad (2)$$

Now from Eq. (1),

$$W = \frac{20\times10^5 \times 0.1 - 200\times10^3 \times 0.3162}{2-1}$$

$$W = \frac{200000 - 63245.55}{1}$$

W = 136754.45 kJ

or **W = 136.75445 kJ**

Second Method :

We know that work done in a closed system,

$$= \int P \cdot dV$$

But, Law for the process is,

$$PV^2 = C$$

or

$$P = \frac{C}{V^2}$$

∴ From Eq. (3),

$$W = \int P \cdot dV - \int_{V_1}^{V_2} \frac{C}{V^2} \cdot dV$$

$$= C\int_{V_1}^{V_2} V^{-2} \cdot dV = C\left[\frac{V^{-2+1}}{-2+1}\right]_{V_1}^{V_2} = C\left[-\frac{1}{V}\right]_{V_1}^{V_2}$$

$$W = C\left[-\frac{1}{V_2} + \frac{1}{V_1}\right] \quad ...(4)$$

Now,

$$P_1 V_1^2 = C$$

∴

$$20 \times 10^5 \times 0.1^2 = C = 20{,}000$$

Now, from Eqs (2), (4), (5),

$$W = 20000\left[-\frac{1}{0.316} + \frac{1}{0.1}\right] = 20000[-3.162 + 10]$$

$$= 20000[6.838]$$

W = 136754.45

W = 136754.45 kJ

Example 2.23. A cylinder contains 0.12 cubic m of air at 1 bar and 90°C. It is compressed to 0.03 cubic m, the final pressure being 6 bar. Find the index of compression, increase in internal energy and heat transferred. Take:

Characteristic gas constant = 0.287 kJ/kg-K; and

Specific heat at constant volume = 0.7176 kJ/kg-K.

Data:

$$V_1 = 0.12 \text{ m}^3 \qquad P_1 = 1 \text{ bar} = 1 \times 10^5 \frac{\text{N}}{\text{m}^2}$$

$$T_1 = 90 + 273 = 363 \text{ K} \qquad V_2 = 0.03 \text{ m}^3$$

$$P_2 = 6 \times 10^5 \frac{\text{N}}{\text{m}^2} \qquad n = ?$$

$$\Delta U = ? \qquad Q = ?$$

$$R = 0.287 \frac{\text{kJ}}{\text{kg-K}} \qquad C_v = 0.7176 \frac{\text{kJ}}{\text{kg-K}}$$

Solution

(i) We know that for the Polytropic process,

$$P_1V_1^n = P_2V_2^n = C$$

$$\frac{P_2}{P_1} = \left(\frac{V_1}{V_2}\right)^n$$

$$\left(\frac{6\times10^5}{1\times10^5}\right) = \left(\frac{0.12}{0.03}\right)^n$$

$$6 = (4)^n$$

$$\ln 6 = n \ln 4$$

$$\frac{\ln 6}{\ln 4} = n$$

$$\frac{1.79175}{1.38629} = n$$

$$\mathbf{n = 1.29248 = 1.3}$$

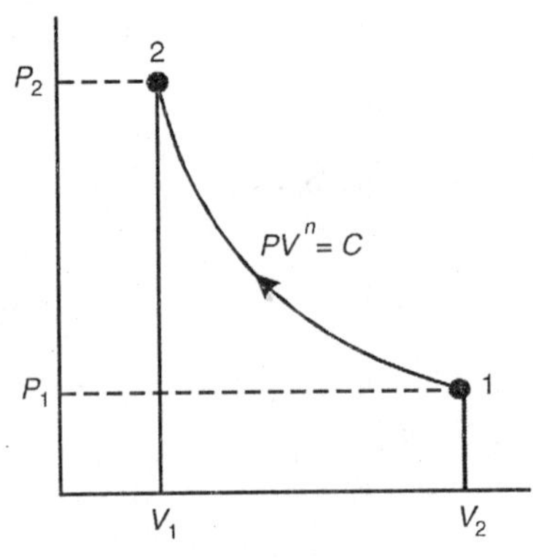

Fig. Ex. 2.23

(ii) Now change in IE $= \Delta u = C_V \cdot (T_2 - T_1)$

To find T_2, since it is a polytropic process,

$$\frac{T_2}{T_1} = \left(\frac{V_1}{V_2}\right)^{n-1}$$

$$T_2 = T_1\left(\frac{V_1}{V_2}\right)^{n-1}$$

$$T_2 = 363\left(\frac{0.12}{0.03}\right)^{1.3-1}$$

$$= 363\times4^{0.3} = 363\times1.5157$$

$$\mathbf{T_2 = 550.199\ K}$$

$$\therefore \quad \Delta u = C_V\,(T_2 - T_1)\,\frac{\text{kJ}}{\text{kg}}$$

$$= 0.7176(550.199 - 363)$$

$$\mathbf{\Delta u = 134.334\ kJ/kg}$$

(iii) We know from the I-law,

$$q - w = \Delta u$$

$$q = \Delta u + w$$

And now to find $W, W_{1-2} = \dfrac{P_2V_2 - P_1V_1}{1-n} = \dfrac{mR(T_2 - T_1)}{1-n}$

For unit mass, $W_{1-2} = \dfrac{R(T_2 - T_1)}{1-n}$

$$= \frac{0.287(550.199 - 363)}{1 - 1.3}$$

$$= \frac{53.70}{-0.3}$$

$$\mathbf{W_{1-2} = -179.01\ kJ/kg}$$

Now heat transfer, from Eq. (1) will be,

$$q = \Delta u + W$$

$$= 134.334 + (-179.01)$$

$$\mathbf{q = -44.6765\ \frac{kJ}{kg}}$$

Example 2.24 Process 1: Air initially at 100 kPa and 50° C undergoes reversible adiabatic compression such that its volume is reduced to 1/5th of initial volume.

Process 2: Then 940 kJ/kg of heat is added to this air at constant volume.

Process 3: Process 2 is followed by reversible adiabatic expansion up to initial volume.

Process 4: Finally heat is rejected at constant volume so as to reach the initial condition.

Draw the four processes on one PV-diagram. Determine the maximum temperature, and heat rejected per kg of air. Assume adiabatic index of compression and expansion of 1.4 and constant volume specific heat as 0.717 kJ/kg-K.

Determine : (i) T_{max} = ?, (ii) Heat rejected/kg of air ?

Solution

Process (1 – 2) : $P_1 = 100$ kPa

$$PV^{\gamma} = C$$

$$V_2 = \frac{1}{5} V_1 = 0.2\, V_1$$

$$T_1 = 50 + 273 = 523 \text{ K}$$

Process (2 –3) : $q_s = 940 \frac{\text{kJ}}{\text{kg}}$

$$= C_V \cdot dT$$

$$\text{at } V = C$$

Process (3 –4) : $PV^\gamma = C$

$$V_1 = V_4$$

Process (4 – 1) : $V = C$: $\gamma = 1.4,\ C_V = 0.717 \frac{\text{kJ}}{\text{kg-K}}$

For the adiabatic 1 – 2 :

$$\frac{T_2}{T_1} = \left(\frac{V_1}{V_2}\right)^{\gamma-1}$$

$$T_2 = \left(\frac{V_1}{V_2}\right)^{\gamma-1} \times T_1$$

$$= \left(\frac{1}{0.2}\right)^{1.4-1} \times 323 \text{ as } \frac{V_2}{V_1} = 0.2$$

$$T_2 = \left(\frac{1}{0.2}\right)^{0.4} \times 323 = (5)^{0.4} \times 323$$

$$= 0.90 \times 323$$

$$\mathbf{T_2 = 614.88\ K}$$

For Process (2 – 3) : $q_s = C_V\,(T_3 - T_2)$

$$940 = 0.717(T_3 - 614.88)$$

$$\frac{940}{0.717} = T_3 - 614.88$$

$$1311.0181 = T_3 - 614.88$$

$$\mathbf{1925.898\ K = T_3}$$

This is the maximum temperature.

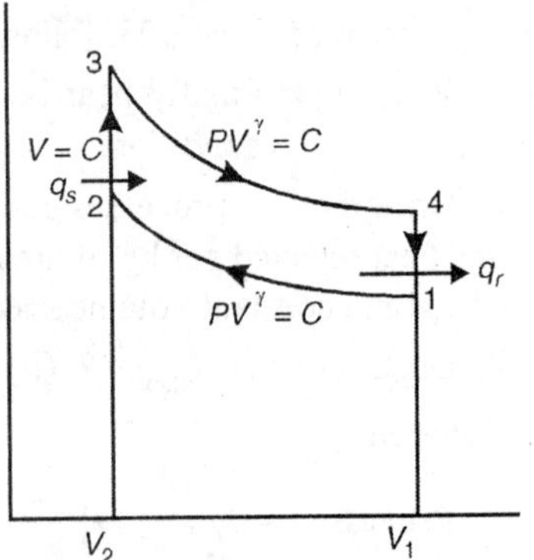

Fig. Ex. 2.24

Now for the adiabatic 3 – 4:

$$\frac{T_4}{T_3} = \left(\frac{V_3}{V_4}\right)^{\gamma-1}$$

$$T_4 = \left(\frac{V_2}{V_1}\right)^{\gamma-1} \times T_3$$

$$= (0.2)^{1.4-1} \times 1925.9$$

$$= 0.2^{0.4} \times 1925.9$$

$$\mathbf{T_4} = \mathbf{1011.686\ K}$$

Now heat rejected, $q_r = C_V(T_1 - T_4) = 0.717(323 - 1011.69)$

$$\mathbf{q_r} = \mathbf{-493.7878\ \frac{kJ}{kg}}$$

– ve sign implies heat rejection.

Example 2.25 Two vessels *A* and *B*, both containing Nitrogen are connected by a valve which is opened to allow the contents to mix and achieve an equilibrium temperature of 27°C. Before mixing, the following information is known about the gases in the two vessels.

Vessel A: Pressure = 1.5 MPa,
Temperature = 50°C
and Contents = 0.5 kg mol.

Vessel B: Pressure = 0.6 MPa,
Temperature = 20°C
and Contents = 2.5 kg

Calculate the final equilibrium pressure, and the heat transferred to the surroundings. If the vessels had been perfectly insulated, calculate the final temperature and pressure which would have been reached. Take $\frac{C_p}{C_v} = 1.4$, molecular weight = 28 kg/kg mol and $R = 0.297$ kJ/kg for Nitrogen.

Solution

For the gas in vessel *A*

$$m_A = n_A \cdot M = 0.5 \times 28 = 14 \text{ kg}$$

$$P_A V_A = m_A R T_A$$

$$\therefore \quad V_A = \frac{14 \times 0.297 \times 323}{1.5 \times 10^3}$$

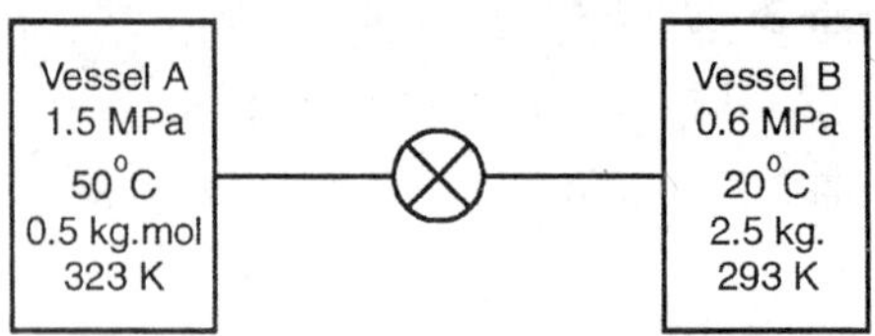

Fig. Ex. 2.25

$$V_A = 0.895 \text{ m}^3$$

For vessel B, $$V_B = \frac{m_B \, R \, T_B}{P_B} = \frac{2.5 \times 0.297 \times 293}{0.6 \times 10^3} = 0.362 \text{ m}^3$$

$\therefore$ Total volume of A and $B = V_A + V_B = 1.258 \text{ m}^3$

Also, Total mass $= 16.5 \text{ kg} = 0.297 \text{m}$

Final temperature after mixing $= 300$ K

For the final condition after mixing,

$$P.V. = mRT$$

$$P = \frac{16.5 \times 0.297 \times 300}{1.258} = 1168.64 \text{ kPa}$$

Since there is no work transfer,

$$\text{Heat transfer} = \text{Change in internal energy}$$
$$= U_2 - U_1$$

Taking $U = 0 \text{ kJ}$ at $T = 0$ K

Initial internal energy, $U_1 = m_A \, C_V T_A + m_B \, C_V T_B$

where $$C_V = \frac{R}{\gamma - 1} = \frac{0.297}{0.4}$$

$$\mathbf{C_V = 0.743 \text{ kJ/kg-K}}$$

$$\therefore \quad U_1 = [(14 \times 323) + (2.5 \times 293)] \times 0.743$$

$$\mathbf{U_1 = 3904.09 \text{ kJ}}$$

Final Internal Energy, $U_2 = m \cdot C_V \cdot T = 16.5 \times 0.743 \times 300$

$$\mathbf{U_2 = 3677.85 \text{ kJ}}$$

$$\therefore \quad \text{Heat transfer} = 3677.85 - 3904.05$$

Heat transfer = –226.24 kJ

If the vessels had been perfectly insulated,

$$U_1 = U_2$$

$$m_A C_V T_A + m_B C_V T_B = m C_V T$$

$$\therefore \quad T = \frac{m_A T_A + m_B T_B}{m}$$

$$= \frac{(14 \times 323) + (2.5 \times 293)}{16.5}$$

T = 318.45 K = 45.45°C

The final pressure $P = \dfrac{mRT}{V}$

$$= \frac{16.5 \times 0.297 \times 318.45}{1.258}$$

$\therefore$ **P = 1240.51 kPa**

THEORY QUESTIONS

1. Define an Ideal Gas.
2. Explain the concept of an Ideal Gas.
3. Define an Ideal gas, give the approximate conditions for the gaseouss phase of a pure substance to behave as an ideal gas.
4. What are the ideal gas laws ?
5. Write a note on Equation of State or Characteristic Gas Equation.
6. Write a short note on Universal Gas Constant.
7. Define universal gas constant. What are its units ?
8. What is the difference between specific gas constant and universal gas constant? How are these two related ?
9. With the help of Joule's experiment prove that I.E. is a function of temperature.
10. With the help of Joule's experiment, prove that I.E. is a function of temperature.
11. Explain the constant internal enemy experiment of an ideal gas. Also explain upon what property is the internal energy and enthalpy of ideal gas depends ?
12. Prove that $C_p - C_v = R$. Also prove $C_v = \dfrac{R}{\gamma - 1}; C_p = \dfrac{\gamma . R}{\gamma - 1}$.
13. Prove that,

    (a) $C_p - C_v = R$

    (b) $C_v = \dfrac{R}{\gamma - 1}$

(c) $C_p = \frac{\gamma . R}{\gamma - 1}$

14. Derive the relation between C_p, C_v and R.

15. Derive the relation between C_v, γ, and R and also C_p, γ and R.

16. Prove that $\gamma = \frac{C_p}{C_v}$.

17. Explain Constant Volume Process with *P-V* and *T-S* diagrams. Also write the equation for work done and derive the equation for heat transferred and change of I.E.

18. Explain constant pressure process with *P-V* and *T-S* diagrams. Write the equation for work done and derive the equations for heat transferred.

19. What is an Isothermal process ? Represent this on *P-V* and *T-S* diagrams. Write the equation for work done, change in I.E. and heat transferred.

20. What is a polytropic process, show it on *P-V* diagram. Write the equation for work done and derive the equation for heat supplied.

21. Define polytropic specifc heat. What is the significance of negative sign?

22. Show that the polytropic sp. heat is given by

$$C_n = C_v\left(\frac{k-n}{n-1}\right)$$

where k is specific heat ratio.

23. What is an adiabatic process? Write the equations for work done, heat supplied.

PROBLEMS FOR PRACTICE

1. 2 kg of air at 247°C expands isothermally from a volume of 1 m³ to a volume of 3 m³. Determine the work done during the process. Take:

R = 287 Nm/kg-K

(**Ans.** 328 kJ)

2. A gas at 10 bar, 327°C occupies a volume of 0.02 m³. This gas expands according to $PV^{14} = C$ upto a pressure of 1 bar. Find the temperature and work done during the process.

(**Ans.** T_2 = 321.16 K; 24.1 kJ)

3. 3 litre of air at 1200 kN/m² and 87°C expand polytropically according to $PV^{1.3} = C$ to a pressure of 120 kN/m². If for air $\gamma = 1.4$, $C_v = 0.718$ find,

(i) Polytropic specific heat.

(ii) Heat transfer per kg of air during the process and state its direction.

(iii) Change of specific IE of air.

(**Ans.** (i) – 0.239 kJ/kg-K ; (ii) 35.51 kJ/kg ; (iii) –106.27 kJ/kg)

4. 0.22 kg of a gas having a volume of 0.14 m³ and a pressure of 1.4 bar is compressed to a pressure of 14 bar according to the, law $PV^{1.25} = C$. Estimate the amount of heat transferred during the process by calculating the work done and change in I.E. Take $C_p = 1.041$ kJ/kg-K and $C_v = 0.743$ kJ/kg-K.

5. A cylinder has a capacity of 300 litre and contains oxygen at a pressure of 3.1 MN/m² and temperature 18°C. The stop valve is opened and some gas is used. If the pressure and temperature of the gas left in the cylinder falls to 1.7 MN/m² and 15°C respectively, determine the mass of the oxygen used. If after the stop valve is closed, the oxygen remaining in the cylinder gradually attains its initial temperature of 18°C, determine the amount of heat transferred through the cylinder walls from atmosphere.

   The density of oxygen at 0°C and 0.1013 MN/m² be taken as 1.43 kg/m³ and

   $$\frac{C_p}{C_v} = 1.4.$$

6. 0.2 m³ of mixture of fuel and air at 1.2 bar and 60°C is compressed till its pressure becomes 12 bar and temperature 270°C. Then it is ignited suddenly at constant volume and its pressure becomes twice the pressure at the end of compression. Find the maximum temperature reached and change in I.E. Also find the heat transfer during the compression process consider the mixture as perfect gas. Take

   $C_p = 1.005$ kJ/kg-K,

   and $\gamma = 1.4$

7. A gas initially at a pressure of 510 kPa and volume 142 litres undergoes a process and final pressure of 170 kPa and volume of 275 litres during which enthalpy decreased by 65 kJ. Take $C_v = 0.718$ kJ/kg-K, determine,
   (a) Change in I.E.
   (b) C_p
   (c) Specific gas constant

8. 10 kg of air is heated at constant pressure from a temperature of 100°C to 200°C. Calculate the heat added during the process and also change in I.E. Take $R = 0.287$ kJ/kg-K, $\gamma = 1.4$.

9. 50 litre of air at 1.013 bar and 100°C temperature is compressed to 28 bar. Volume of air at the end of polytropic compression is found to be 4 litres. Air is now heated at constant volume till pressure rises to 56 bar.

   Assuming $C_p = 1.00$ kJ/kg-K and $C_v = 0.71$ kJ/kg-K, determine
   (a) Polytropic index of compression
   (b) Entropy change in each process.

   Sketch the process on *P-V* and *T-S* diagram

10. 3 kg of an ideal gas expanded from a pressure of 7 bar and volume 1.5 m³ to a pressure of 1.4 bar and volume 4.5 m³. If the change in I.E. during the process is 520 kJ. Determine,
    (a) Specific gas constant

(b) Initial and final temperature

(c) Change in entropy is heat exchange. Take $C_p = 1.05$ kJ/kg-K

11. 0.01 kg of certain gas occupies a volume of 0.003 m³ at a pressure of 7 bar and a temperature of 131°C. Calculate the molar mass of the gas. When the gas is allowed to expand until the pressure is 1 bar and final volume is 0.02 m³, calculate final temperature.

12. 1 kg of perfect gas is compressed from 1.1 bar and 27°C, according to the law, $PV^{1.3} = C$, until the pressure is 6.6 bar. Calculate the heat flow to or from the cylinder watts when the gas is argon having molecular mass of 40 kg/mol and C_p = 0.52 kJ/kg.

13. A certain quantity of gas has a volume of 0.11 m³ at 0.15 MN/m² and 300 K.It is compressed to a pressure of 1.5 MN/m² according to the law $PV^{1.2}$ = constant. If the molecular mass of the gas is 27.9 and $C_p = 29.04$ kJ/kmol K determine heat transfer and its direction.

14. An ideal gas having $\gamma = 1.4$ and volume 1.5 m³ at 1 bar and 27°C is compressed according to the law $PV^{1.25} = C$, to a pressure of 6 bar. Calculate volume, temperature at the end of the process. Evaluate heat transfer, work done and the change of entropy during the process.

15. Following pressure – volume data is available for a process:

P (bar)	20	17.5	15	12.5	10	7.5	5	2.5
V (m³)	1.00	1.1	1.2	1.3	1.4	1.5	1.6	1.7

Estimate the work done during the process.

16. Air with initial condition as 27°C, 1 bar and 1.5 m³ is compressed according to the law $PV^{1.3} = C$, to a pressure of 7.5 bar. Calculate final volume, temperature at the end of the process. Evaluate heat transfer, W.D. and change of entropy during the process. Assume $\gamma = 1.4$, $R = 0.287$ kJ/kg-K.

ꕤ ꕤ

3

First Law of Thermodynamics

CHAPTER OBJECTIVES

After reading this chapter you will be able to learn the following

- Joule's Experiment to study the law of Conservative of Energy.
- I-law law for a process, Perpetual Motion Machine of the first kind (PMM-I)
- Work done for Closed System and Open Systems.
- Meaning of $-\int vdp$ and $\int Pdv$ for the processes.
- Continuity equation.

3.1 INTRODUCTION

Energy is inherent in all matters. Energy may appear in many different forms. Conversion can be made from one form of energy to another. We are unable to define the general term energy in a simple way, but we can define with precision the various forms in which it appears.

One of the consequences of Einstein's theory of Relativity is that mass may be converted into energy and energy into mass, the relation being given by the famous equation,

$$\text{Energy } (E) = m \text{ (mass)} \times C^2 \qquad \text{(where } C - \text{velocity of light)}$$

Energy is a scalar quantity, not a vector quantity.

Except the nuclear reaction, (where mass is converted into energy) total energy of the universe is constant. For this matter, First Law Law of Thermodynamics can be expressed as following.

Energy can neither be created nor destroyed (except in nuclear reactions. This is the law of conservation of energy.

3.2 JOULE'S EXPERIMENT

Joule carried out experiments in 1843, which led to the formulation of the I Law of Thermodynamics. He took a known quantity of water in a rigid vessel, which was insulated adiabatically from the surroundings. The vessel was fitted with a paddle wheel and a thermometer as shown in Fig. 3.1.

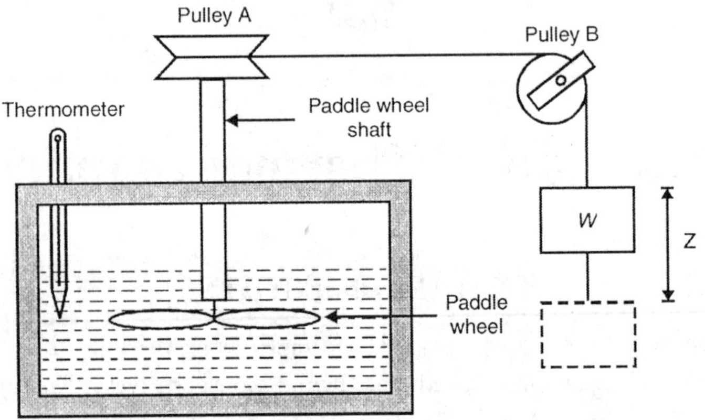

Fig. 3.1. *Joules experiment.*

Now let a certain amount of work W_{1-2} be done upon the closed system by the paddle wheel. The quantity of work can be measured by the fall of weight W which drives the paddle wheel.

Let t_1 be the initial temperature of water before work transfer and after work transfer let the temperature rise to t_2. So, the system is changing its state from state (1) to state (2) and the process 1–2 undergone by the system is shown in Fig. 3.2.

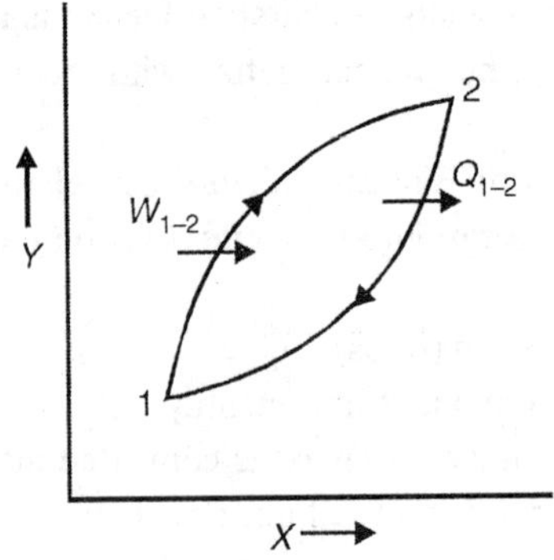

Fig. 3.2

Now let the insulation be removed. Then the system will interact by heat transfer (i.e. heat will be dissipated from the system to the surroundings) till the system comes to original temperature t_1. The amount of heat transfer Q_{2-1} from the system during the process 2–1 can be estimated: ($Q_{2-1} = mC_p\,\Delta T$).

The system thus undergoes a cycle, which consists of a definite amount of work input W_{1-2} to the system, followed by a heat transfer Q_{2-1} from the system.

Joule, repeated this experiment for different weights and different heights and in each case he found that, net work transfer during a cycle is proportional to net heat transfer,

i.e. $$\Sigma\delta W \propto \Sigma\delta Q$$

Or in other words "*when a closed system undergoes any cyclic process, the cyclic integral of work is proportional to the cyclic integral of heat*". This is known as the first law of thermodynamics for a closed system undergoing a cycle.

i.e. $$\oint \delta W \propto \oint \delta Q$$

$$\oint \delta W = J\oint \delta Q \qquad \text{Since in SI units J} = 1$$

$$\oint \delta W = \oint \delta Q$$

Thus, the I law states the general principle of conservation of energy i.e. "*Energy can neither be created nor be destroyed, but energy can be converted from one form to the other.*"

3.3 FIRST LAW FOR A CLOSED SYSTEM UNDERGOING A PROCESS

If a closed system undergoes a change of state or a process and during which, both work transfer and heat transfer are involved, then the net energy transfer will be stored within the system. If Q is the amount of heat transferred to the system and W is the amount of work transferred from the system, during the process as shown in Fig. 3.3, then the net energy transfer ($Q - W$) will be stored in the system.

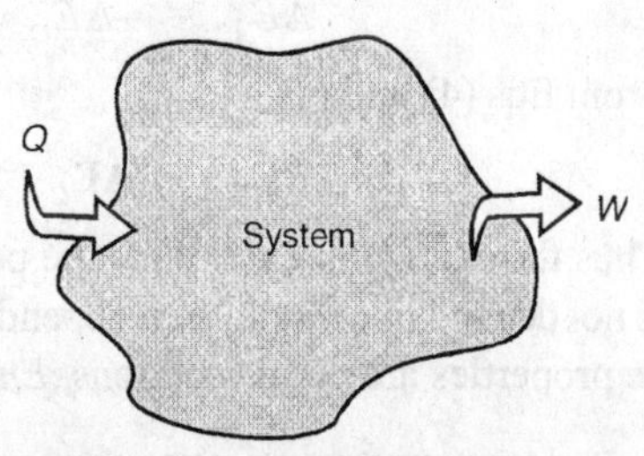

Fig. 3.3

Energy in storage is neither heat nor work, but is called as Internal energy or simply energy of the system.

$$\therefore \qquad Q - W = \Delta E$$

or $$Q = \Delta E + W$$

where ΔE is the change in energy. Here Q, W and ΔE all are expressed in joules.

3.4 ENERGY — A PROPERTY OF THE SYSTEM

Consider a system which changes its state from state (1) to state (2) by following the path A, and returns from state (2) to state (1) by following the path B as shown in Fig. 3.4. So the system undergoes a cycle. Now writing the I-law for the path A,

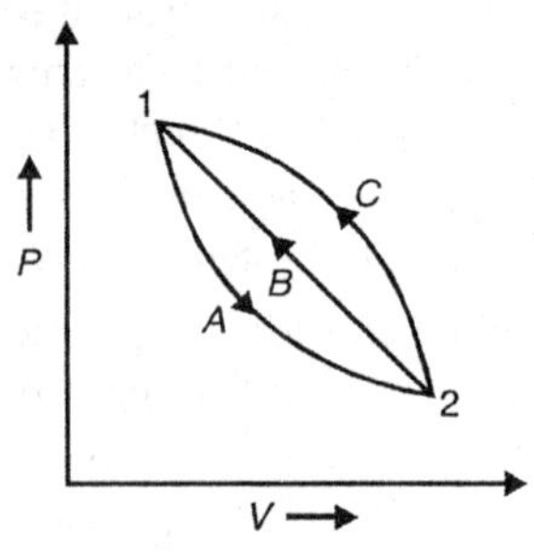

Fig. 3.4

$$Q_A - W_A = \Delta E_A \qquad (1)$$

and for path B.

$$Q_B - W_B = \Delta E_B \qquad (2)$$

The processes 1–A–2 and 2–B–1 together will constitute a cycle for which,

$$\oint \delta W = \oint \delta Q$$

i.e. Total work transfer during cycle = Total heat transfer during the cycle.

$$W_A + W_B = Q_A + Q_B$$

Rewriting $\quad W_B - Q_B = Q_A - W_A \qquad (3)$

$$\therefore \quad -(Q_B - W_B) = Q_A - W_A \qquad (4)$$

$$-\Delta E_B = \Delta E_A$$

i.e. $\quad \Delta E_A = -\Delta E_B \qquad (5)$

Similarly if the system returns from point (2) to state point (1) by following the path C instead of path B then,

$$\Delta E_A = -\Delta E_C \qquad (6)$$

From Eqs (4) and (5),

$$\mathbf{\Delta E_B = \Delta E_C}$$

Thus the change in energy for the path B and C are same. Hence change in energy does not depend upon path, so it depends on end states. Hence, it is a point function and since properties are point functions, *Energy is a property of the system.*

3.5 DIFFERENT FORMS OF STORED ENERGIES

The total energy E is made up of Kinetic Energy (K.E.), Potential Energy (P.E.) and Internal Energy (I.E.).

$$\therefore \quad E = \text{K.E.} + \text{P.E.} + \text{I.E.}$$

Kinetic Energy is due to the motion of the fluid in the system. Potential Energy is the energy due to the gravitational force.

A part of the total energy which is stored in the molecular and atomic structure is known as *Internal energy* and is denoted by U.

In the absence of Kinetic Energy and Potential Energy i.e. when

$$\text{KE} = 0 \text{ and } \text{PE} = 0$$

$$E = U, \text{ and the I law for process,}$$

$$Q = \Delta E + W$$

becomes

$$Q = \Delta U + W$$

In the differential form, Eqs (1) and (2) become

$$\delta Q = dE + \delta W$$

i.e.

$$\delta Q = dE + PdV \left(\text{as } \delta W = PdV\right)$$

Also,

$$\delta Q = dU + \delta W \left(\text{when } E = U\right)$$

i.e.

$$\delta Q = dU + PdV$$

When KE and PE are considered, then the I law for a process will be

$$\delta Q - \delta W = dU + d(\text{KE}) + d(\text{PE})$$

In the integrated form, Eq. (5) becomes

$$Q_{1\text{-}2} - W_{1\text{-}2} = U_2 - U_1 + \frac{m\left(C_2^2 - C_1^2\right)}{2} + mg\left(Z_2 - Z_1\right)$$

where C is the velocity and Z is the height.

For unit mass, Eq. (6) will be,

$$q_{1\text{-}2} - w_{1\text{-}2} = u_2 - u_1 + \frac{\left(C_2^2 - C_1^2\right)}{2} + g\left(Z_2 - Z_1\right)$$

3.6 INTERNAL ENERGY (U)

"It is the energy stored in the molecular structure due to heat and work interactions".

or

In case of gases we know that the gas is made up of a number of molecules, which are moving continuously. Internal Energy is the energy which arises from the motion of these molecules.

If the temperature of the gas is increased, the molecular activity increases, therefore the I.E. also increases. Thus I.E. is a function of temperature and its value can be increased or decreased by adding or removing heat to or from the system.

Due to practical difficulties it is very very difficult to determine the absolute value of Internal energy. Fortunately, in most of the thermodynamic applications change in I.E. is used, since changes in the states of a system is considered. Change in I.E. is denoted by ΔU and $\Delta U = U_2 - U_1$.

For unit mass, $\Delta u = u_2 - u_1$

3.7 ENTHALPY (H)

It is an extensive property, since its value depends on mass. Enthalpy of a substance is given by the sum of Internal energy and Pressure–Volume product.

i.e. $\mathbf{H = U + PV}$ and specific enthalpy, $\mathbf{h = u + Pv}$

3.8 SPECIFIC HEAT AT CONSTANT PRESSURE AND SPECIFIC HEAT AT CONSTANT VOLUME

"*Specific heat of a gas is defined as the amount of heat required to rise the temperature of unit mass of a gas through one degree*".

Let C = General specific heat

Q = Heat transfer (J)

ΔT = Change in temperature K

m = mass (kg)

Then from the definition, $C = \dfrac{Q}{m \cdot \Delta T}\left[\text{units will be } \dfrac{\text{J}}{\text{kg-K}}\right]$ (1)

or $Q = m \cdot C \cdot \Delta T$ (2)

In the differential form, $\delta Q = m \cdot C \cdot dT$ (3)

And for unit mass, $\delta q = C \cdot dT$ (4)

or $\mathbf{C = \dfrac{\delta q}{dT}}$ (5)

There are 2-types of specific heats,

(1) Specific heat at constant volume C_v

(2) Specific heat at constant pressure C_p

1. Specific Heat at Constant Volume C_v

"*It is the amount of heat required to rise the temperature of unit mass of a gas through one degree when the volume is constant*".

If a unit mass of a gas is taken in a closed vessel and is heated, the volume of the gas remains constant, but the temperature increases. As the volume remains constant, there is no external work done by the gas and as temperature of the gas increases, there is increase in internal energy of the gas.

Therefore heat supplied to the gas is completely utilised in increasing the I.E. of the gas,

And $C_V = \left(\dfrac{\delta q}{dT}\right)_V = \left(\dfrac{du}{dT}\right)_V$

Therefore C_v is also defined as, "*the rate of change of specific internal energy with respect to temperature when the volume is kept constant.*"

2. Specific Heat at Constant Pressure C_p

"It is the amount of heat required to rise the temperature of unit mass of a gas through one degree when the pressure is kept constant."

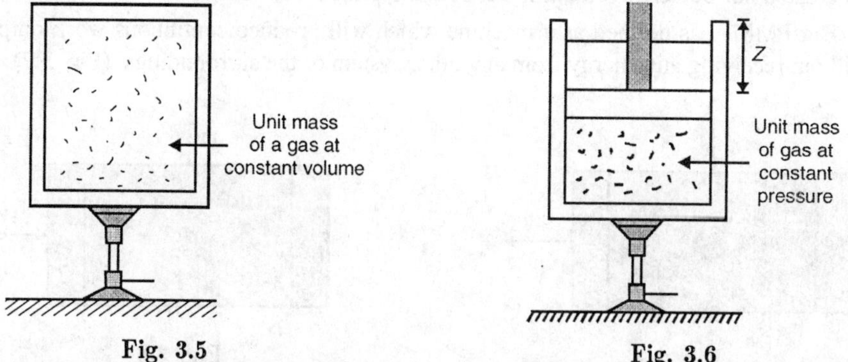

Fig. 3.5 Fig. 3.6

Consider a unit mass of a gas in a cylinder fitted with a frictionless piston as shown in the Fig. 3.6. When the gas is heated, the piston moves up, maintaining the same pressure. But the volume and temperature of the gas increases during heating. As there is increase in volume, there is external work done by the gas and as there is increase in temperature, there is increase in I.E.

Thus heat supplied when the pressure is constant, is utilised for two purposes,

(a) To do some external work.

(b) To increase the I.E. of the gas.

Whereas in case of constant volume heating, the heat supplied is completely utilised for increasing the I.E.

∴ Specific heat at constant pressure is greater than specific heat at constant volume.

i.e. $C_P > C_V$ and, $C_P = \left[\frac{\delta q}{dT}\right]_P$

Also $$C_P = \left[\frac{dh}{dT}\right]_P$$

Therefore, C_p is also defined as, *"the rate of change of specific enthalpy with respect to temperature when the pressure is kept constant."*

3.9 ADIABATIC INDEX

It is the ratio of specific heat at constant pressure to the specific heat at constant volume and is given by,

$$K \text{ or } \gamma = \frac{C_p}{C_v}$$

Note: For air, $C_p = 1.005 \frac{\text{kJ}}{\text{kg - K}}$ and $C_v = 0.718 \frac{\text{kJ}}{\text{kg - K}}$ and $\gamma = 1.4$

3.10 PERPETUAL MOTION MACHINE OF FIRST KIND [PMM – 1]

First law states the general principle of conservation of energy, i.e. energy can neither be created nor be destroyed but it can be transformed from one form to the other.

But PMM–1 is defined as a machine which will produce continuous work output without receiving any energy from any other system or the surroundings. (Fig. 3.7)

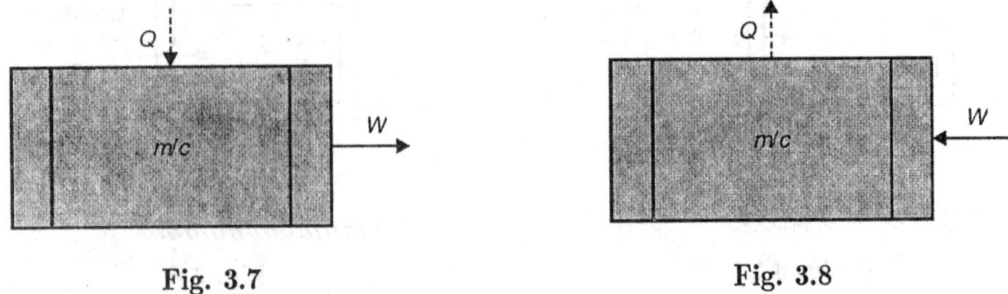

Fig. 3.7 Fig. 3.8

It will create energy and thus violates I law of thermodynamics. All the attempts made so far to make such a machine have failed, thus they show the validity of I–law. Thus PMM–1 is just a conceptual machine.

Converse of PMM–1 is also true, i.e. there can be no machine which would continuously consume work, without producing some other form of energy. (Fig. 3.8)

3.11 FLOW PROCESS, CONTROL VOLUME, CONTROL SURFACE

3.11.1 Flow Process

Processes performed in the open systems are called as Flow-processes. Figure 3.9 shows a portion of steam power plant. High pressure steam enters the steam turbines at Section (1), expands and produces work output and then it leaves the turbine at section (2).

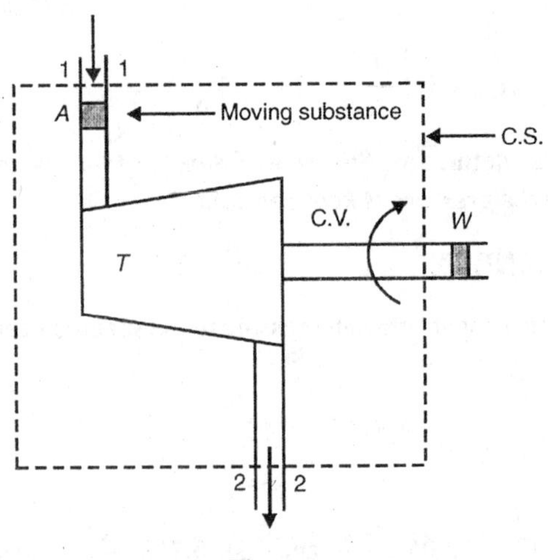

Fig. 3.9

For analysing the expansion process, 2-methods are used.

In the first method certain mass of the fluid Marked *A* is considered and this is suppose to flow through the turbine.

In the second method, certain fixed volume of the system known as *control volume* (CV) is considered.

The moving substance flows through this control volume. The surface of the control volume is known as *control surface* (*CS*).

3.12 FLOW WORK OR FLOW ENERGY

In case of flow processes certain amount of work or energy is required to push the fluid into and out of the system. This work energy is known as *flow work* or *flow energy.*

Consider a flow process as shown in Fig. 3.10.

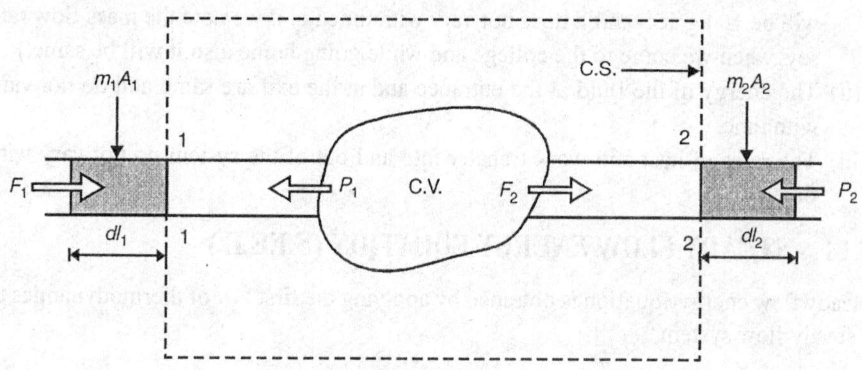

Fig. 3.10

Let F_1 be the force which forces the mass m_1 of cross sectional area A_1 into the system against the pressure P_1 of the system. Now consider a small amount of work done on the system in causing the displacement dl_1 for the mass m_1.

Thus the small amount of work done on the system in causing the displacement dl_1,

$$\delta W = -F_1 \times dl_1$$

Note. –ve sign, since the work is done on the system.

But $$F_1 = P_1 \times A_1$$

$\therefore$ $$\delta W = -P_1 \times A_1 \times dl_1$$

Since $$A_1 \times dl_1 = dV_1 = \text{small amount of displacement volume.}$$

$\therefore$ $$\delta W = -P_1 \times dV_1$$

$\therefore$ Total flow work at Sec. (1) $= -P_1 V_1$

Similarly work done by the system to force the fluid out of the system at Sec. (2) $= P_2V_2$

$\therefore$ **Net flow work** $= P_2V_2 - P_1V_1$

and **for unit mass** $= P_2v_2 - P_1v_1$

Note. Since the flow work is entirely expressed in terms of properties of the system, the net flow work depends on the end states and it is a *Property.* Whereas the other forms of work are path functions so in problems involving flow processes, flow work is to be calculated separately.

3.13 CONDITIONS OF STEADY FLOW SYSTEM

A steady flow process should satisfy the following conditions:

(i) The mass flow rate into and out of the system are equal and do not vary with time i.e. mass in the system does not change.
(If the mass flow rate at the inlet is 10 kg/sec, then the mass flow rate at the exit will be 10 kg/sec. and it does not vary with time i.e. if we note the mass flow rate say, when we come to the college and while going home also it will be same.)

(ii) The energy of the fluid at the entrance and at the exit are same and do not vary with time.

(iii) The rates of heat and work transfer into and out of the system do not vary with time.

3.14 STEADY FLOW ENERGY EQUATION (S.F.E.E.)

Steady flow energy equation is obtained by applying the first law of thermodynamics to a steady flow system.

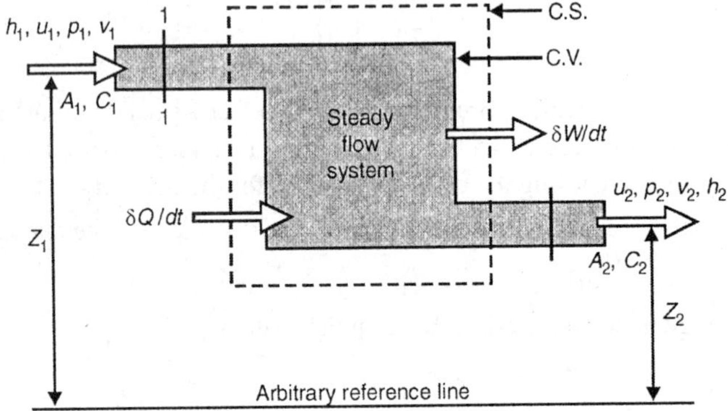

Fig. 3.11

Let,

$\dot{m}$ = Mass flow rate in kg/sec

$A_1 A_2$ = Cross sectional areas at sec 1–1 and 2–2 respectively in m^2

$C_1 C_2$ = Velocities in m/sec

$P_1\ P_2$ = Absolute pressures in N/m^2
$v_1\ v_2$ = Specific volumes in m^3/kg
$u_1\ u_2$ = Specific internal energies in J/kg
$h_1\ h_2$ = Specific enthalpies in J/kg
$Z_1\ Z_2$ = Elevations above arbitrary reference datum line in m
Q = Net heat transfer in Joules

$\frac{\delta Q}{dt}$ = Net rate of heat transfer in J/sec

$\frac{\delta Q}{dm}$ = q = Net rate of heat transfer in J/kg

W = Net work transfer in joules

$\frac{\delta W}{dt}$ = Net rate of work transfer in J/sec

$\frac{\delta W}{dm}$ = w = Net rate of work transfer in J/kg

t = time in seconds

Figure 3.11 shows a steady flow system. Fluid entes at sec. 1–1 and leaves at sec. 2–2.

Since it is a steady flow system. According to 2nd condition.

Total energy at the entrance = Total energy at the exit

∴ We can write,

$$\dot{m}\begin{bmatrix}\text{Energy carried}\\ \text{into the system}\end{bmatrix} + \dot{m}\begin{bmatrix}\text{Specific enthalpy}\\ \text{of entering fluid}\end{bmatrix} + \begin{bmatrix}\text{Net rate of heat}\\ \text{transfer into}\\ \text{system}\end{bmatrix}$$

$$=$$

$$\dot{m}\begin{bmatrix}\text{Energy carried}\\ \text{out of the system}\end{bmatrix} + \dot{m}\begin{bmatrix}\text{Specific enthalpy}\\ \text{of the fluid}\\ \text{at the exit}\end{bmatrix} + \begin{bmatrix}\text{Net rate of work}\\ \text{transfer by the}\\ \text{system}\end{bmatrix}$$

i.e. $$\dot{m}[\text{KE}_1 + \text{PE}_1] + \dot{m}(h_1) + \frac{\delta Q}{dt} = \dot{m}[\text{KE}_2 + \text{PE}_2] + \dot{m}(h_2) + \frac{\delta W}{dt}$$

i.e. $$\dot{m}\left[\frac{C_1^2}{2} + g\,Z_1\right] + \dot{m}(h_1) + \frac{\delta Q}{dt} = \dot{m}\left[\frac{C_2^2}{2} + g\,Z_2\right] + \dot{m}(h_2) + \frac{\delta W}{dt}$$

Note. m of KE and PE will be unity, since we are deriving the SFEE for unit mass.

$$\therefore \dot{m}\left[\frac{C_1^2}{2}+g\,Z_1+h_1\right]+\frac{\delta Q}{dt} = \dot{m}\left[\frac{C_2^2}{2}+g\,Z_2+h_2\right]+\frac{\delta W}{dt} \quad (1)$$

This is steady flow energy equation on *time basis.*

Also, since $h_1 = u_1 + P_1v_1$ and $h_2 = u_2 + P_2v_2$

$\therefore$ Eq. (1) becomes,

$$\therefore \dot{m}\left[\frac{C_1^2}{2}+gZ_1+u_1+P_1v_1\right]+\frac{\delta Q}{dt} = \dot{m}\left[\frac{C_2^2}{2}+gZ_2+u_2+P_2v_2\right]+\frac{\delta W}{dt}$$

or $$\dot{m}\left[KE_1+PE_1+IE_1+FE_1\right]+\frac{\delta W}{dt} = \dot{m}\left[KE_2+PE_2+IE_2+FE_2\right]+\frac{\delta W}{dt} \quad (2)$$

3.15 STEADY FLOW ENERGY EQUATION ON MASS BASIS

For deriving this, we have to consider $\dot{m}$ = 1 kg/sec and all other quantities will be for per kg mass such as $\delta W/dm$ and $\delta Q/dm$.

$\therefore$ Eq. (1) becomes,

$$\left[\frac{C_1^2}{2}+gZ_1+h_1\right]+\frac{\delta Q}{dm} = \left[\frac{C_2^2}{2}+gZ_2+h_2\right]+\frac{\delta W}{dm}$$

$$\therefore \left[\frac{C_1^2}{2}+gZ_1+h_1\right]+q = \left[\frac{C_2^2}{2}+gZ_2+h_2\right]+w$$

This is the SFEE on mass basis.

3.16 IN A STEADY FLOW SYSTEM W.D.= $-\int v.dP$ WHEN CHANGES IN K.E. AND P.E. ARE NEGLECTED

We know that, SFEE on mass basis. From Eq. (3)

$$\left[\frac{C_1^2}{2}+gZ_1+h_1\right]+q = \left[\frac{C_2^2}{2}+gZ_2+h_2\right]+w \quad (1)$$

i.e. $$\left[\frac{C_1^2}{2}+gZ_1+u_1+P_1v_1\right]+q = \left[\frac{C_2^2}{2}+gZ_2+u_2+P_2v_2\right]+w$$

$$\therefore \quad q = u_2-u_1+P_2v_2-P_1v_1+\frac{C_2^2-C_1^2}{2}+g\left(Z_2-Z_1\right)+w$$

$$\therefore \qquad q = \Delta u + \Delta(Pv) + \Delta(\text{KE}) + d(\text{PE}) + w$$

In differential form,

$$\delta q = du + d(Pv) + d(\text{KE}) + d(\text{PE}) + \delta w \qquad (2)$$

But we also know that I law for a closed system undergoing a process is,

$$Q - W = \Delta E = \Delta U$$

For unit mass, $q - w = \Delta u$

i.e. $$q = \Delta u + w$$

In the differential form, $\delta q = du + \delta w$

i.e. $$\delta q = du + Pdv \qquad (3)$$

Assuming that δq is same for closed system and for the steady flow system, equating Eqs (2) and (3)

$$du + Pdv = du + d(Pv) + d(\text{KE}) + d(\text{PE}) + \delta w \qquad (4)$$

i.e. $$du + Pdv = du + Pdv + vdP + d(\text{KE}) + d(\text{PE}) + \delta w$$

$\therefore$ We can write, $-vdP = \delta w + d(\text{KE}) + d(\text{PE})$

$\therefore$ For total mass, $$-\int vdP = w + \Delta(\text{KE}) + \Delta(\text{PE}) \qquad ...(5)$$

When change in KE and PE are neglected,

$$-\int \mathbf{vdP} = \mathbf{w}$$

This work done is given by the area behind the curve 1–2 as shown in Fig. 3.12.

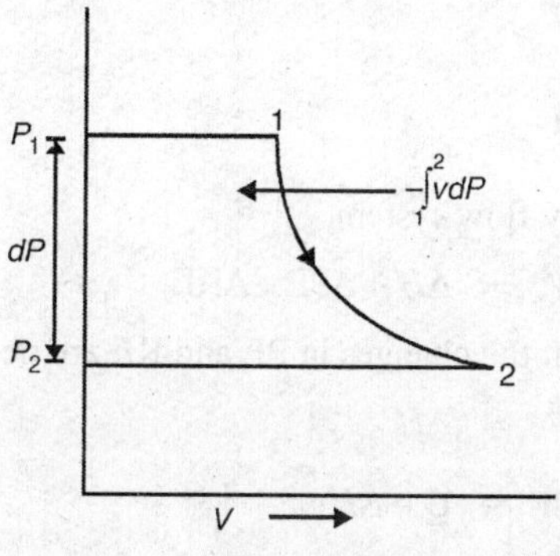

Fig. 3.12

3.17 SIGNIFICANCE OF $\int Pdv$ IN CASE OF STEADY FLOW PROCESS AND NON-FLOW PROCESS

Upto Eq. (4) derivation is same as in the above section.

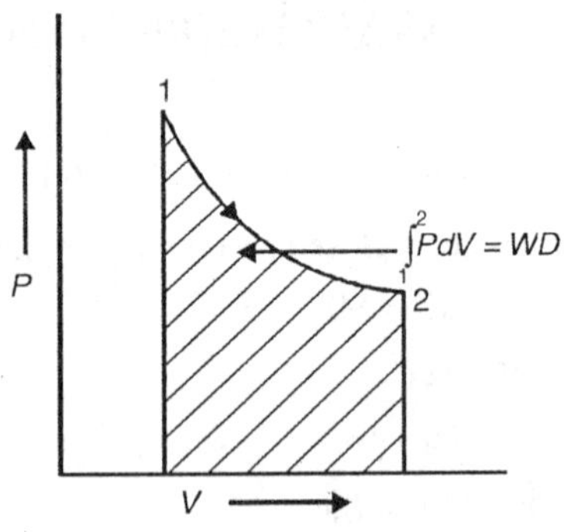

Fig. 3.13

From Eq. (4),

$$du + PdV = du + d(Pv) + d(\text{KE}) + d(\text{PE}) + \delta w$$

$$Pdv = d(Pv) + d(\text{KE}) + d(\text{PE}) + \delta w$$

∴ For total mass, $$\int_1^2 \mathbf{Pdv} = \mathbf{\Delta(Pv) + \Delta(KE) + \Delta(PE) + w}$$

For a non-flow process (i.e. for a closed system),

$\int Pdv$ = work done and is given by the area under the curve 1–2 as shown in Fig. 3.13.

3.18 PROOFS

3.18.1 $-\int \mathbf{VdP = Q - \Delta H}$

We know that, for a steady flow system,

$$Q - W = \Delta H + \Delta \text{PE} + \Delta \text{KE} \tag{1}$$

Consider the case when the changes in PE and KE are neglected,

$$Q - W = \Delta H$$

∴ $$W = Q - \Delta H \tag{2}$$

Also we know that, in a reversible steady flow process,

Work done $= -\int vdP$ when changes in KE and PE are neglected. (3)

Hence from (2) and (3) we can write,

∴ $$-\int \mathbf{vdP = Q - \Delta H}$$

3.18.2 $\int P dv = Q - \Delta U$

We know that, for a non-flow process,

$$Q - W = \Delta U$$

$$W = Q - \Delta U \quad (1)$$

And for a Non-flow process,

$$W = \int P dv \quad (2)$$

$\therefore$ From (1) and (2) $\quad \int \mathbf{P dv} = \mathbf{Q - \Delta U}$

3.18.3 $-\int vdP - \int Pdv = -\Delta Pv$

We know that,

$$-\int vdP = w + \Delta KE + \Delta PE$$

Also we know that, significance of $\int Pdv$ in case of steady flow process and non-flow process,

$$\int Pdv = w + \Delta(PE) + \Delta(KE) + \Delta Pv \quad (3)$$

$\therefore$ Substracting Eq. (2) from (1) we get,

$$-\int vdP - \int Pdv = -\Delta Pv$$

3.19 APPLICATIONS OF ENERGY EQUATIONS

(a) Boiler. It is a steam generator.

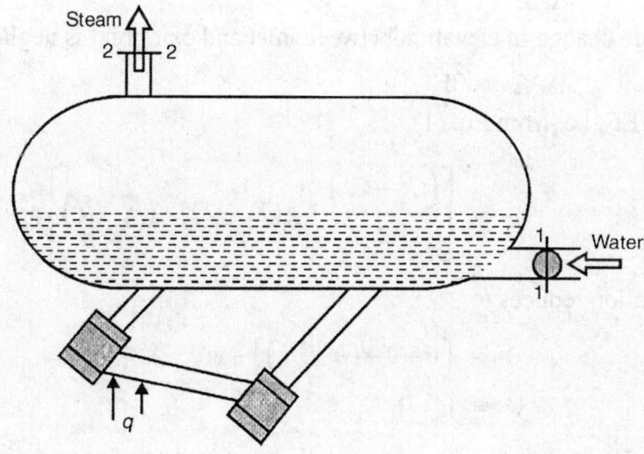

Fig. 3.14

Characteristics of the boiler are:

(1) Change in velocity of the fluid at the entrance and exit is very small, and hence ΔKE can be neglected, i.e. $\Delta KE = 0$.

(2) Change in elevation between the inlet and exit point is very small so change in PE may be neglected, i.e. $\Delta PE = 0$.

(3) Since no work is done in a boiler $w = 0$.

$\therefore$ From SFEE on mass basis,

$$\left[\frac{C_1^2}{2} + gZ_1 + h_1\right] + q = \left[\frac{C_2^2}{2} + gZ_2 + h_2\right] + w$$

i.e.

$$q = \left[\frac{\left(C_2^2 - C_1^2\right)}{2} + g\left(Z_2 - Z_1\right) + \left(h_2 - h_1\right)\right] + w \qquad (1)$$

On substituting the above conditions, Eq. (1) becomes,

$$q = \left[0 + 0 + \left(h_2 - h_1\right)\right] + 0$$

i.e.

$$\mathbf{q = h_2 - h_1}$$

(b) Turbine (Engine)/Compressor (or Pump)

The turbines and engines are power producing devices whereas Compressors, Pumps, Blowers etc. are power consuming devices.

Charcteristics of a turbine are:

(i) Since it is insulated, negligible heat transfer.

$$\therefore \quad q = 0$$

(ii) Change in velocity of the fluid at the entrance and at the exit is negligible.

$$\therefore \quad \Delta KE = 0$$

(iii) Since change in elevation between inlet and exit point is negligible,

$$\Delta PE = 0$$

$\therefore$ From SFEE, i.e. from Eq. (1),

$$q = \left[\frac{\left(C_2^2 - C_1^2\right)}{2} + g\left(Z_2 - Z_1\right) + \left(h_2 - h_1\right)\right] + w$$

$\therefore$ The equation reduces to,

$$0 = \left[0 + 0 + \left(h_2 - h_1\right)\right] + w$$

or

$$\mathbf{w = h_1 - h_2}$$

Here work is produced at the expense of enthalpy. Similarly for an Adiabatic Pump/ Compressor, work is done on the system, so w is – ve.

∴ SFEE becomes,

$$w = h_2 - h_1 \quad \text{(since w is – ve)}$$

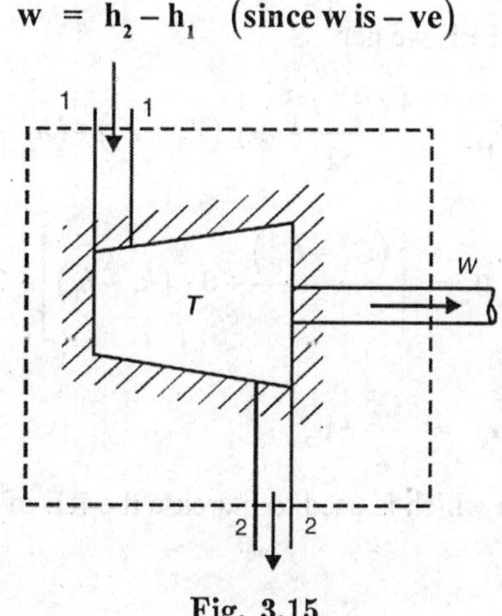

Fig. 3.15

(c) Throttling Process:

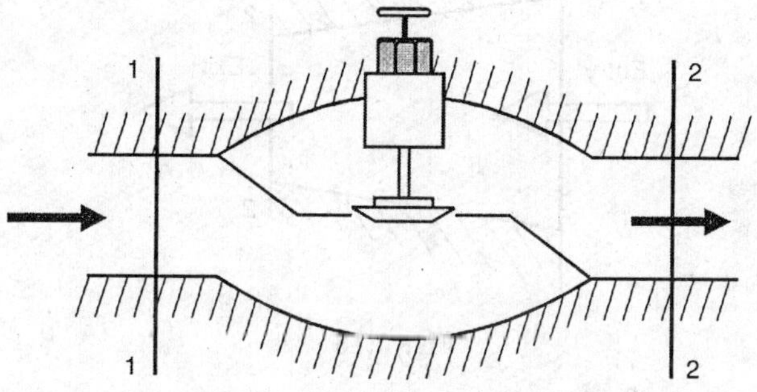

Fig. 3.16

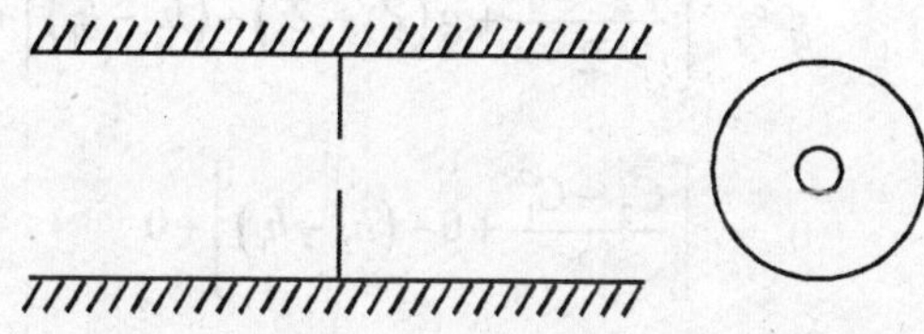

Fig. 3.17

When a fluid flows through a partially opened valve or an orifice then there is an appreciable pressure drop and the flow is said to be throttled.

Characteristics of throttling process are,

(i) $q = 0$

(ii) $w = 0$

(iii) $\Delta PE = 0$

Substituting these in SFEE we get

$$q = \left[\frac{C_2^2 - C_1^2}{2} + g(Z_2 - Z_1) + (h_2 - h_1)\right] + w$$

$$0 = \left[\frac{(C_2^2 - C_1^2)}{2} + 0 + (h_2 - h_1)\right] + 0$$

i.e.
$$\frac{C_1^2}{2} + h_1 = \frac{C_2^2}{2} + h_2$$

(d) Nozzle. It is a device which is used to increase the KE of flowing fluids

(i) $q = 0$

(ii) $w = 0$

(iii) $\Delta PE = 0$

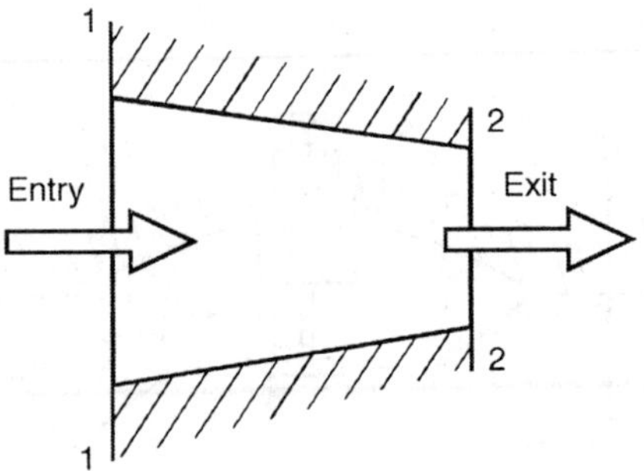

Fig. 3.18

∴ Substituting these in SFEE we get,

$$q = \left[\frac{C_2^2 - C_1^2}{2} + g(Z_2 - Z_1) + (h_2 - h_1)\right] + w$$

$$0 = \left[\frac{C_2^2 - C_1^2}{2} + 0 + (h_2 - h_1)\right] + 0$$

i.e.
$$\frac{C_1^2}{2} + h_1 = \frac{C_2^2}{2} + h_2$$

(e) Condenser or to show in a steady flow process with multiple streams of fluid without mixing.

$$\frac{\dot{m}_1}{\dot{m}_3} = \frac{h_4 - h_3}{h_1 - h_2}$$

Figure 3.19 shows a surface condenser. In this case, cooling water enters at Section 1– 1 passes through the tubes and leaves the condenser at Section 2 – 2.

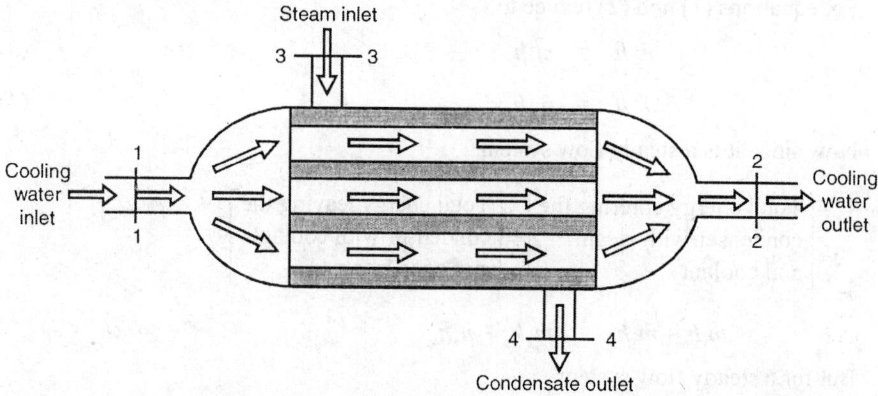

Fig. 3.19

Steam to be condensed enters the condenser at Section 3–3 and when it comes in contact with the cold surface of water tubes, it gets condensed and the condensate leaves the condenser at Section 4–4.

So, in this case working fluids are water and steam i.e. two fluids flow through the system and since they do not mix with each other, it may be considered as a steady flow system with multiple streams of fluid without mixing.

Let,

$\dot{m}_1$ $\dot{m}_2$ — Mass flow rates of cooling water at Sec. 1–1 and 2–2 in kg/sec.

$\dot{m}_3$ $\dot{m}_4$ — Mass flow rates of steam and condensate Sec. 3–3 and 4–4 respectively in kg/sec

h_1 h_2 — Specific enthalpies of cooling water at Sec. 1–1 and 2–2 respectively in J/kg.

h_3 h_4 — Specific enthalpies of steam and condensate at Sec. 3–3 and 4–4 respectively.

Since it is steady flow system, writing SFEE for the coolant.

$$\dot{m}_1\left[\frac{C_1^2}{2} + gZ_1 + h_1\right] + \frac{\delta Q}{dt} = \dot{m}_2\left[\frac{C_2^2}{2} + gZ_2 + h_2\right] + \frac{\delta W}{dt} \quad (1)$$

and for steam,

$$\dot{m}_3\left[\frac{C_3^2}{2} + gZ_3 + h_3\right] + \frac{\delta Q}{dt} = \dot{m}_4\left[\frac{C_4^2}{2} + gZ_4 + h_4\right] + \frac{\delta W}{dt} \quad (2)$$

Neglecting the changes in KE and PE, since they are very small and as the condenser is insulated $\delta Q / dt = 0$ and as no work is done in the condenser $\delta W / dt = 0$.

So, equations (1) and (2) reduce to,

$$\dot{m}_1 h_1 = \dot{m}_2 h_2 \qquad (3)$$

$$\dot{m}_3 h_3 = \dot{m}_4 h_4 \qquad (4)$$

Now since it is a steady flow system,

$$\begin{bmatrix} \text{Total energy entering the} \\ \text{condenser with steam} \\ \text{and coolant} \end{bmatrix} = \begin{bmatrix} \text{Total energy leaving the} \\ \text{condenser with coolant} \\ \text{and condensate} \end{bmatrix}$$

$$\therefore \qquad \dot{m}_1 h_1 + \dot{m}_3 h_3 = \dot{m}_2 h_2 + \dot{m}_4 h_4$$

But for a steady flow system

$$\dot{m}_1 = \dot{m}_2 \text{ and } \dot{m}_3 = \dot{m}_4$$

$$\therefore \qquad \dot{m}_1 (h_1 - h_2) = \dot{m}_3 (h_4 - h_3)$$

$$\therefore \qquad \frac{\dot{\mathbf{m}}_1}{\dot{\mathbf{m}}_3} = \frac{\mathbf{h}_4 - \mathbf{h}_3}{\mathbf{h}_1 - \mathbf{h}_2} \qquad \text{Hence proved.}$$

3.20 CONTINUITY EQUATION OR LAW OF CONSERVATION OF MASS

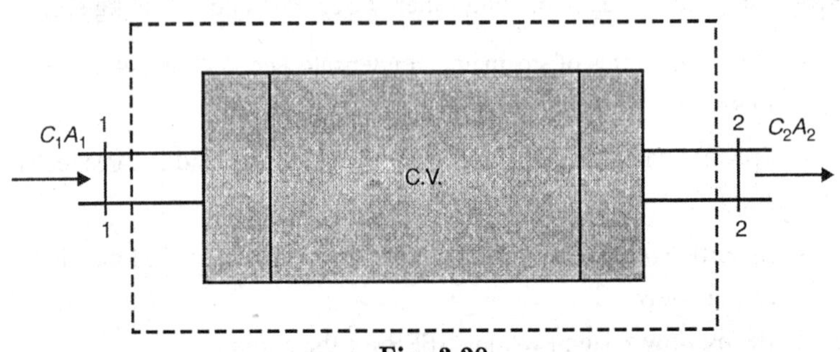

Fig. 3.20

Continuity equation is based on the principle of conservation of mass

Let $\rho_1 \rho_2$ — Densities of the fluid at sec 1–1 and 2–2 respectively in kg/m^3

$A_1 A_2$ — Cross sectional areas at sec 1–1 and 2–2 respectively in m^2

$C_1 C_2$ — Velocities of the fluid at sec 1–1 and 2–2 respectively in m/sec

Then the volume of the fluid entering at sec 1–1 per unit time $= A_1 \times C_1$

$\left(\because A_1 \text{ is in } m^2 \text{ and } C_1 \text{ is in m/sec then resultant unit will be } m^3/\text{sec}\right)$

$\therefore$ Mass of the fluid entering at sec 1–1 per unit time,

$$= \text{Density} \times \text{Volume}$$

$$= \rho_1 \times A_1 \times C_1$$

Simlarly mass of fluid leaving at sec 2–2 per unit $= \rho_2 A_2 C_2$

Since it is steady flow system,

Mass flow rate at sec $1-1 =$ Mass flow rate at sec $2-2$

$$\therefore \quad \rho_1 A_1 C_1 = \rho_2 A_2 C_2$$

Since $$\rho = \frac{1}{v}$$

$$\therefore \quad \frac{A_1 C_1}{v_1} = \frac{A_2 C_2}{v_2}$$

or $$\rho AC = \text{constant}$$

$$\frac{AC}{v} = \text{constant}$$

These are the continuity equations.

3.21 LIMITATIONS OF FIRST LAW OF THERMODYNAMICS

We have seen that first law of thermodynamics is simply a law of conservation of energy and has its own limitations. These limitations are:

1. The law considers all forms of energies equivalent i.e. the first law of thermodynamics is a law of energy equivalence. This law is necessary condition so far as the account of energy balance is concerned with the possibility of transformation of one kind of energy to another.
2. The first law does not consider the direction of energy transformation.
3. The first law of thermodynamics does not consider the grade of the energy or energy reservoirs. It, assumes that all energy reservoirs are identical.

The second law of thermodynamics relates to the direction of energy exchange processes.

One of the statements of the second law of thermodynamics is Clausius statement and states that no cyclic process is prossible whose result is the flow of heat out of a heat reservoir at one temperature and the flow of an equal quantity of heat into a second reservoir at a higher temperature, without external work. Or, heat can not flow from one body at a lower temperature to the another body at higher temperature without any external work.

This principle is used in a refrigerator. For refrigerator external power is supplied to the compressor.

Without proof or discussion we here define the third law of thermodynamics.

The third law states that the entropy of a pure substance is zero at absolute zero temperature.

SOLVED EXAMPLES

Example 3.1 During a cycle consisting of 4-processes, the heat transfers are 60 kJ, – 8kJ, – 34 kJ and 6 kJ. Determine the net work for the cycle.

Solution

We know that from the first law for a closed system undergoing a cycle,

Total heat transfer in a cycle = Total work transfer in a cycle

i.e. $$\oint \delta Q = \oint \delta W$$

As given in the problem, the cycle consists of 4-processes and heat transfers for the processes are given.

$$\therefore \quad \oint \delta Q = 60\text{ kJ} - 8\text{ kJ} - 34\text{ kJ} + 6\text{ kJ}$$

$$\oint \delta Q = 24\text{ kJ} = \oint \delta W$$

$\therefore$ **Net work transfer in a cycle = 24 kJ**

Example 3.2 In a non-flow reversible process 300 kJ of heat leaves the system consisting of a gas. The Internal Energy of the gas remains the same. Calculate the work done.

Data: $Q = -300$ kJ (Since heat flows out of the system)

U = Constant $\therefore \quad \Delta U = 0$

$W = ?$

Solution

We know that first law for a closed system undergoing a process $Q - W = \Delta U$. As given in problem $Q - W = 0$, since $\Delta U = 0$.

$$\mathbf{Q = W = -300\ kJ}$$

i.e. 300 kJ of work is done on the gas.

Example 3.3 In a non-flow reversible process, the pressure and volume are related by

$$P = \left[\frac{3.5}{V} + 3\right] \text{bar}$$

where P is in bar and V is in m^3. During the process the volume changes from 1.5 m^3 to 4.5 m^3 and heat added is 2000 kJ. Determine change in Internal Energy (I.E.)?

Data:

$$P = \left[\frac{3.5}{V}+3\right]\times 10^2 \text{ kPa}$$

$$V_1 = 1.5 \text{ m}^3$$

$$V_2 = 4.5 \text{ m}^3$$

$$Q = 2000 \text{ kJ}$$

$$\Delta U = ?$$

Solution

We know that first law for a closed system undergoing a process,

$$Q - W = \Delta U$$

Q is given, ΔU is to be found

∴ To find W,

Also we know that,

$$\text{Work done} = \int P dV \text{ in closed system}$$

Substituting

$$P = \left[\frac{3.5}{V}+3\right]\times 10^2 \text{ kPa}$$

We get

$$\text{W.D.} = \int\left[\frac{3.5}{V}+3\right]\times 10^2 \, dV$$

$$= 100\int_{1.5}^{4.5}\left(\frac{3.5}{V}+3\right) dV \text{ kJ}$$

$$= 100\left[(3.5 \ln V + 3V)\right]_{1.5}^{4.5}$$

$$= 100\left[3.5 \ln\frac{4.5}{1.5}+3(4.5-1.5)\right]$$

$$= 100\left[3.5 \ln 3 + 3\times(3)\right]$$

$$= 100[3.845+9] = 100[12.845]$$

$$\text{Work done} = 1284.5143 \text{ kJ}$$

$$\therefore \quad Q - W = \Delta U$$

$$2000 - 1284.5143 = \Delta U$$

$$\therefore \quad \mathbf{\Delta U = 715.4857}$$

Example 3.4 In a boiler water enters with an enthalpy of 35 kJ/kg and steam leaves with the enthalpy of 705 kJ/kg. Find the heat transfer per kg of steam. The change in KE and PE may be neglected.

Solution

We know that, SFEE on mass basis,

$$\left[\frac{C_1^2}{2} + gZ_1 + h_1\right] + q = \left[\frac{C_2^2}{2} + gZ_2 + h_2\right] + w$$

$$(0+0+h_1)+q = (0+0+h_2)+w$$

$$h_1 + q = h_2 + w$$

But in a boiler since no work is done $w = 0$

$$\therefore \quad q = h_2 - h_1$$

$$q = 705 - 35$$

$$\therefore \quad \mathbf{q = 670\ kJ/kg}$$

Example 3.5 A piston and cylinder machine contains a fluid system which passes through a complete cycle of 4-processes. During a cycle the sum of all heat transfers is – 170 kJ. The system completes 100 cycles/min. Complete the following table showing the method for each item and calculate the net rate of work output in kW.

Process	Q (kJ/min)	W (kJ/min)	4E (kJ/min)
a–b	0	2170	—
b–c	21,000	0	—
c–d	– 2100	—	– 36600
d–a	—	—	—

Solution

Process a–b

$$Q = \Delta E + W$$

$$0 = \Delta E + 2170$$

$$\Delta \mathrm{E} = \mathbf{-2170\ kJ/min}$$

Process b–c

$$Q = \Delta E + W$$

$$21000 = \Delta E + 0$$

$$\therefore \quad \Delta \mathrm{E} = \mathbf{21000\ kJ/min}$$

Process c–d

$$Q = \Delta E + W$$

$$-2100 = -36600 + W$$

$$\mathrm{W} = \mathbf{34500\ kJ/min}$$

Process d–a

It is given that, sum of all heat transfers/cycle

i.e. $\oint Q = -170 \text{ kJ}$

Secondly, the system completes 100 cycles/min

$\therefore$ Total heat transfer/min $= -17000$ kJ/min

$$Q_{a-b} + Q_{b-c} + Q_{c-d} + Q_{d-a} = -17000 \text{ kJ/min}$$

$$0 + 21000 - 2100 + Q_{d-a} = -1700$$

$$\therefore \quad \mathbf{Q_{d-a} = -35900 \text{ kJ/min}}$$

While studying energy a property of the system, we have studied that change in energy for the process $1 - a - 2 = \Delta E_a$

And change in energy for the process,

$$2 - b - 1 = \Delta E_b\,(-\text{ve})$$

$\therefore$ Net change in energy for the cycle

$$= \Delta E_a - \Delta E_b = 0$$

$\therefore$ We can write cyclic integral of any property is zero or $\oint dE = 0$

So, $\Delta E_{a-b} + \Delta E_{b-c} + \Delta E_{c-d} + \Delta E_{d-a} = 0$

$$\therefore \quad -2170 + 21000 - 36600 + \Delta E_{d-a} = 0$$

$$\therefore \quad \mathbf{\Delta E_{d-a} = 17770 \text{ kJ/min}}$$

We know that, $Q_{d-a} = \Delta E_{d-a} + W_{d-a}$

$$W_{d-a} = Q_{d-a} - \Delta E_{d-a} = -35900 - 17770$$

$$\mathbf{W_{d-a} = -53670 \text{ kJ/min}}$$

Lastly rate of work output since

$$\oint \delta Q = \oint \delta W$$

$$= -17000 \text{ kJ/min}$$

$$\oint \delta \mathbf{W} = \textbf{Net rate of work} = \mathbf{-283.33 \text{ kW or kJ/sec}}$$

Example 3.6 A certain cycle consists of 3 processes. The energy transfer in each process are tabulated below.

Process	Q (kJ/kg)	w (kJ/kg)	Δu (kJ/kg)
1–2	50	—	20
2–3	– 30	– 40	—
3–1	—	—	– 30

If the network done l kg of fluid is + 30 kJ/kg complete the table. Also find the power developed, if the mass of fluid in the cycle is 0.1 kg and the system completes 10 such cycles per sec.

Solution

Since all the equalities are given for unit mass,

For process 1–2

$$q = \Delta u + w$$
$$50 = 20 + w$$
$$\therefore \quad \mathbf{w = 30\ kJ/kg}$$

For process 2–3

$$q = \Delta u + w$$
$$-30 = \Delta u - 40$$
$$\mathbf{\Delta u = 10\ kJ/kg}$$

For process 3–1

It is given that, net W.D./kg $= -30$ kJ/kg

$$\therefore \quad W_{1-2} + w_{2-3} + w_{3-1} = 30 \text{ kJ/kg}$$
$$30 - 40 + w_{3-1} = 30$$
$$30 - 40 + w_{3-1} - 30 = 0$$
$$\therefore \quad \mathbf{W_{3\text{-}1} = 40\ kJ/kg}$$
$$\therefore \quad q = \Delta u + w$$
$$q = -30 + 40$$
$$\mathbf{q = 10\ kJ/kg}$$

It is given that W.D/kg = 30 kJ/kg and mass of the fluid flowing = 0.1 kg/cycle and system completes 10 cycle per sec.

$$\therefore \quad \dot{m} = 0.1 \text{ kg/cycle} \times 10 \text{ cycle/sec}$$
$$\therefore \quad \dot{m} = 1 \text{ kg/sec}$$

$\therefore$ We know that W.D/kg $= 30$ kJ/kg

$\therefore$ Rate of W.D. or power $= 30 \text{ kJ/kg} \times 1 \text{ kg/sec} = 30 \text{ kJ/sec} = 30 \text{ kW}$

Remember: $$\frac{\delta w}{dt} = \dot{m}\frac{\delta w}{dm}$$

Unit: $$\frac{kJ}{sec} = \frac{kg}{sec} \times \frac{kJ}{kg}$$

$$\frac{kJ}{sec} = \frac{kJ}{sec}$$

Example 3.7 The I.E. of a certain substance is given by the following equation,

$$u = 3.56\,Pv + 84,$$

where u is in kJ/kg, P is in kPa, v is in m^3/kg.

A system composed of 3 kg of this substance expands from an initial pressure of 500 kPa and a volume of 0.22 m^3 to a final pressure of 100 kPa in a process in which pressure and volume are related by $PV^{1.2} = C$.

(i) If the expansion is quasi-static, find Q, ΔU and W for the process.

(ii) In another process the same system expands according to the same pressure-volume relation as in case (i) and from the same initial state and same final state as in case (i), but the heat transfer for this case is 30 kJ. Find the work transfer for this process.

(iii) Explain the difference in work transfer in case (i) and (ii).

Data : u is in kJ/kg

P is in kPa

$$v = \text{m}^3/\text{kg}$$

$$m = 3\text{ kg}$$

$$P_1 = 500\text{ kPa}$$

$$V_1 = 0.22\text{ m}^3$$

$$P_2 = 100\text{ kPa}$$

$$PV^{1.2} = C$$

$$Q = ? \quad \Delta U = ? \quad W = ?$$

Solution

It is given that $\quad u = 3.56\,Pv + 84$

Change in IE for unit mass

$$\Delta u = u_2 - u_1$$

$$= 3.56\,P_2V_2 + 84 - 3.56P_1V_1 - 84$$

$$\Delta u = 3.56(P_2V_2 - P_1V_1)$$

$\therefore$ Total change in IE $= \Delta U = 3.56(P_2V_1 - P_1V_1)$

Note: V is the total volume for 3 kg mass amd it is an extensive property.

$\therefore$ It depends on mass and P is an intensive property.

Since in Eq. (1) V_2 is not given,

$\therefore$ from $\quad P_1\,V_1^{1.2} = P_2\,V_2^{1.2}$

$$\therefore \quad V_2 = V_1\left(\frac{P_1}{P_2}\right)^{\frac{1}{1.2}} = 0.22\left(\frac{500}{100}\right)^{\frac{1}{1.2}}$$

$$V_2 = 0.22 \times 3.83 = 0.845 \text{ m}^3$$

Now from (1) $\Delta U = 3.56[100 \times 0.845 - 500 \times 0.22]$ kJ

$$= 356[1 \times 0.845 - 5 \times 0.22] \text{ kJ}$$

$$\mathbf{\Delta U = -91 \text{ kJ}}$$

For a quasi-static process, W.D. = $\int PdV$

Since it is a polytropic process,

$$W_{1-2} = \frac{P_2V_2 - P_1V_1}{1-n}$$

$$W_{1-2} = \frac{100 \times 0.845 - 500 \times 0.222}{1 - 1.2}$$

$$\mathbf{W_{1-2} = 127.5 \text{ kJ}}$$

$$\therefore \quad Q = \Delta U + W$$

$$Q = -91 + 127.5$$

$$\mathbf{Q = 36.5 \text{ kJ}}$$

Case (ii) Here $Q = 30$ kJ

Since the end states are same as in case (i)

$$W = Q - \Delta U = 30 - (-91)$$

$$\mathbf{W = 121 \text{ kJ}}$$

(iii) The work in case (ii) is not equal to $\int PdV$. So the process is not quasi-static.

Example 3.8 The I.E. of a certain substance is given by the equation $U = 3.5\ PV + 80$. Where U is in kJ, pressure P is in kPa, and volume V is in m^3. A system composed of 5 kg of this substance expands from an initial pressure of 5 bar and volume of 0.22 m^3 to a final pressure of 1 bar in a process in which pressure and volume are related by $PV^{1.2} = C$.

Find for quasi-static process,

(1) Heat transfer, (2) Total change in I.E. (3) Non-flow work

Data: U in kJ ; P in kPa

Solution

V in m^3; $m = 5$ kg

$$P_1 = 5 \text{ bar} = 500 \text{ kPa}$$

$$V_1 = 0.22 \text{ m}^3; \quad P_2 = 1 \text{ bar} = 100 \text{ kPa}$$

$$P_1V_1^{1.2} = P_2V_2^{1.2} = C$$

$$\therefore \quad V_2 = V_1 \times \left(\frac{P_1}{P_2}\right)^{\frac{1}{1.2}}$$

$$V_2 = 0.22 \times (5)^{1/1.2} = 0.84 \text{ m}^3$$

Now ΔU i.e. change in I.E. will be,

$$\Delta U = U_2 - U_1 = 3.5(P_2V_2 - P_1V_1)$$

$$= 3.5[100 \times 0.84 - 500 \times 0.22]$$

$$\mathbf{\Delta U = -91 \text{ kJ}}$$

$$W_{1-2} = \frac{P_1V_1 - P_2V_2}{n-1}$$

$$= \frac{500 \times 0.22 - 100 \times 0.84}{0.2}$$

$$\mathbf{W_{1-2} = 130 \text{ kJ}}$$

$$Q = \Delta U + W = -91 + 130$$

$$\mathbf{Q = 39 \text{ kJ}}$$

Example 3.9 A fluid is confined in a cylinder by a spring loaded frictionless piston, so that the pressure in the fluid is a linear function of the volume, $P = a + bV$. The internal energy of the fluid is given by the following equation $U = 34 + 3.15\ PV$.

Where U is in kJ, P in kPa, V in m³. If the fluid changes from an initial state of 170 kPa, 0.03 m³ to a final state of 400 kPa, 0.06 m³ with no work other than that done on the piston. Find the direction and magnitude of the work and heat transfer.

Data: $P = a + bV$; $U = 34 + 3.15\ PV$

$P_1 = 170$ kPa ; $V_1 = 0.03$ m³

$P_2 = 400$ kPa ; $V_2 = 0.06$ m³

$W = ?$ $Q = ?$

Solution

The change in I.E. of the fluid during the process is given as,

$$U_2 - U_1 = 3.15(P_2V_2 - P_1V_1)$$

Note : Units $\frac{\text{kN}}{\text{m}^2} \times \text{m}^3 = \text{kNm} = \text{kJ}$

$$U_2 - U_1 = 3.15[400 \times 0.06 - 1.70 \times 0.03]$$

$$= 100\times3.15[4\times0.06-1.70\times0.03]$$

$$U_2-U_1 = 315\times0.189$$

$$\mathbf{U_2-U_1 = 59.5\ kJ}$$

Now $$P = a+bV$$

$$170 = a+b\times0.03$$

$$400 = a+b\times0.06$$

From these two equations on solving we get,

$$a = -60\ \text{kN/m}^2$$

$$b = 7667\ \text{kN/m}^5$$

We know that, W.D. during the process $W_{1-2} = \int PdV$

$$= \int_{V_1}^{V_2}(a+bV)dV$$

$$= a(V_2-V_1)+b\frac{(V_2^2-V_1^2)}{2}$$

$$= (V_2-V_1)\left[a+\frac{b}{2}(V_1+V_2)\right]$$

$$= 0.03\ \text{m}^3\left[-60\frac{\text{kN}}{\text{m}^2}+\frac{7667}{2}\frac{\text{kN}}{\text{m}^5}\times0.09\ \text{m}^3\right]$$

$$\mathbf{W_{1-2} = 8.55\ kJ}$$

Since W_{1-2} is positive, work is done by the system,

∴ Heat transfer involved is given by

$$Q = \Delta U+W$$

$$Q_{1-2} = U_2-U_1+W_{1-2} = 59.5+8.55$$

$$Q_{1-2} = 68.05\ \text{kJ}$$

Since Q_{1-2} is positive, heat flows into the system during the process.

Example 3.10 The expression for the I.E. of a certain closed system is given by $U = M + N(PV)$ where P is the pressure, V is the volume and M and N are constants. The system now undergoes a reversible process where $Q = 0$. Show that the relation between P and V is given by, $PV^K = C$ where $K = \frac{N+1}{N}$.

Solution

As given in the problem,

$$U = M + N(P.V) \qquad (1)$$

differentiating we get,

$$dU = O + N.P.\,dV + N.V.\,dP$$

We know that from first law for the process,

$$Q - W = \Delta U$$

$$Q = \Delta U + W$$

or $$\delta Q = dU + \delta W$$

$$\delta Q = dU + P.dV$$

Since $$Q = 0,\ 0 = dU + P.dV$$

Given, $$dU = -P.\,dV \qquad (2)$$

Equating (1) and (2)

$$-P.\,dV = N.\,P.dV + N.V\ dP$$

$$O = N.\,P.dV + N.V\ d\,P + P.\,dV$$

$$= (N+1)\,P.\,dV + N.V.\,dP$$

Dividing throughout by N, we get

$$O = \frac{N+1}{N}\,P.dV + V.dP$$

i.e. $$-V.\,dP = \frac{N+1}{N}\,P.dV$$

$$-VdP = K.\,P.\,dV \text{ as given } \frac{N+1}{N} = K$$

$$-\frac{d\,P}{P} = K.\frac{dV}{V}$$

Integrating, $$\ln\frac{1}{P} = \ln V^K$$

or $$\mathbf{PV^K = C}$$

where $$K = \frac{N+1}{N}$$

Example 3.11 Following data refers to a steady flow process.

	At entrance	**At exit**
Enthalpy (kJ/kg)	4000	4100
Velocity (m/sec)	50	20
Height (m)	50	10

Mass flow rate $\dot{m}$ = 1 kg/sec

Heat transfer rate = 200 kJ/sec to the system.

Determine the power capacity of the system and state whether it is a power producing system or otherwise.

Solution

$$\dot{m} = 1\,\text{kg/sec}$$

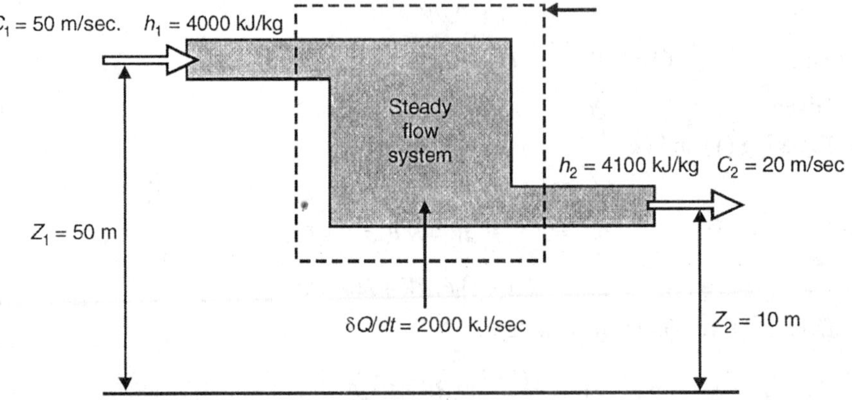

Fig. Ex. 3.11

Since heat transfer is given in kJ/sec, we will have to use SFEE on time basis.

$$\dot{m}\left[\frac{C_1^2}{2}+gZ_1+h_1\right]+\frac{\delta Q}{dt} = \dot{m}\left[\frac{C_2^2}{2}+gZ_2+h_2\right]+\frac{\delta W}{dt}$$

Since power is to found,

$$\frac{\delta W}{dt} = \dot{m}\left[\frac{C_1^2-C_2^2}{2}+g(Z_1-Z_2)+(h_1-h_2)\right]+\frac{\delta Q}{dt}$$

$$\frac{\delta W}{dt} = \frac{1\,\text{kg}}{\text{sec}}\left[\frac{50^2-20^2}{2\times1000}\frac{\text{kJ}}{\text{kg}}+\frac{9.81(40)}{1000}\frac{\text{kJ}}{\text{kg}}+(4000-4100)\right]+200\frac{\text{kJ}}{\text{sec}}$$

$$\frac{\delta W}{dt} = \frac{1\,\text{kg}}{\text{sec}}\left[1.05\frac{\text{kJ}}{\text{kg}}+0.3924\frac{\text{kJ}}{\text{kg}}+-100\frac{\text{kJ}}{\text{kg}}\right]+200\frac{\text{kJ}}{\text{sec}}$$

$$\frac{\delta W}{dt} = 1.05\frac{\text{kJ}}{\text{sec}}+0.3924\frac{\text{kJ}}{\text{sec}}-100\frac{\text{kJ}}{\text{sec}}+200\frac{\text{kJ}}{\text{sec}}$$

$$\frac{\delta \mathbf{W}}{\mathbf{dt}} = \mathbf{101.4484\ kJ/sec\ or\ kW}$$

Since $\frac{\delta W}{dt}$ is + ve, it is a power producing system.

Example 3.12 In a steady flow system 135 kJ/kg of work is done by the system. The specific volume of the fluid, pressure and velocity at the inlet are 0.37 m³/kg, 600 kPa, 16 m/sec. The inlet is 32 m above the floor level. The discharge conditions are 0.62 m³/kg, 100 kPa and 270 m/sec. The total heat loss between inlet and discharge is 9 kJ/kg. In flowing through this system, does the specific I.E. increase or decrease and by how much? The outlet is at flow level.

Solution

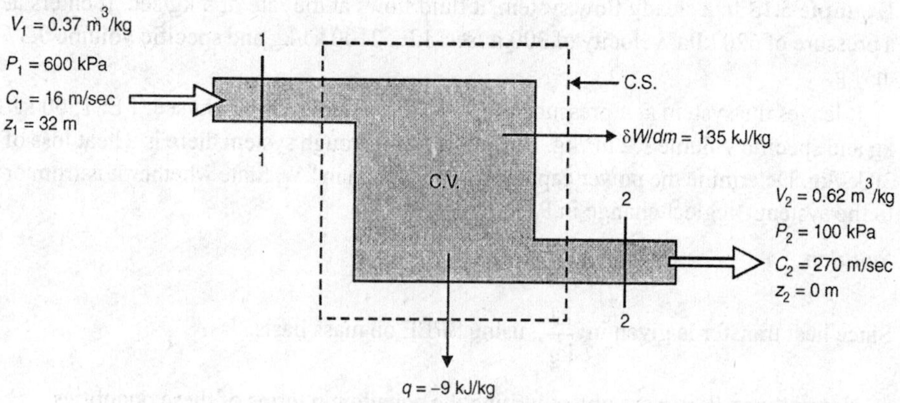

Fig. Ex. 3.12

Since work and heat transfer are given in $\frac{\text{kJ}}{\text{kg}}$

Using SFEE on mass basis,

$$\left[\frac{C_1^2}{2} + gZ_1 + h_1\right] + q = \left[\frac{C_2^2}{2} + gZ_2 + h_2\right] + w$$

Since pressure, specific volumes are given and change in Sp. I.E., is to be found, substitute $h_1 = u_1 + P_1 v_1$ and $h_2 = u_2 + P_2 v_2$ we get,

$$\left[\frac{C_1^2}{2} + gZ_1 + u_1 + P_1 v_1\right] + q = \left[\frac{C_2^2}{2} + gZ_2 + u_2 + P_2 v_2\right] + w$$

$$\therefore \quad u_1 - u_2 = \frac{C_2^2 - C_1^2}{2} + g(Z_2 - Z_1) + (P_2 v_2 - P_1 v_1) + w - q$$

$$u_1 - u_2 = \frac{270^2 - 16^2}{2 \times 1000} \frac{\text{kJ}}{\text{kg}} + \frac{9.81(-32)}{1000} \frac{\text{kJ}}{\text{kg}}$$

$$+ (100 \times .62 - 600 \times 0.37) \frac{\text{kJ}}{\text{kg}} + 135 \frac{\text{kJ}}{\text{kg}} - \left(-9 \frac{\text{kJ}}{\text{kg}}\right)$$

$$u_1 - u_2 = 36.45 - 0.314 - 160 + 135 + 9$$

$$u_1 - u_2 = 36.45 - 20.136 \ \text{kJ/kg}$$

$$u_1 - u_2 = 20.136 \ \text{kJ/kg}$$

$$\therefore \quad \mathbf{u_2 - u_1 = -20.136 \ kJ/kg}$$

∴ Specific I.E. decreases by 20.236 kJ/kg

Example 3.13 In a steady flow system, a fluid flows at the rate of 5 kg/sec. It enters at a pressure of 620 kPa, velocity of 300 m/sec. I.E., 2100 kJ/kg and specific volume 0.37 m^3/kg.

It leaves the system at a pressure of 130 kPa, a velocity of 150 m/sec; I.E. 1500 kJ/kg and specific volume 1.2 m^3/kg. During its flow through system there is a heat loss of 30 kJ/kg. Determine the power capacity of the system in kW. State whether it is from or to the system. Neglect change in P.E.

Solution

Since heat transfer is given in $\frac{\text{kJ}}{\text{kg}}$, using SFEE on mass basis.

Note : Since *P*, *u*, *v* are given, writing the equation in terms of these quantities.

$$\left[\frac{C_1^2}{2} + gZ_1 + u_1 + P_1 v_1\right] + \frac{\delta Q}{dm} = \left[\frac{C_2^2}{2} + gZ_2 + u_2 + P_2 v_2\right] + \frac{\delta W}{dm}$$

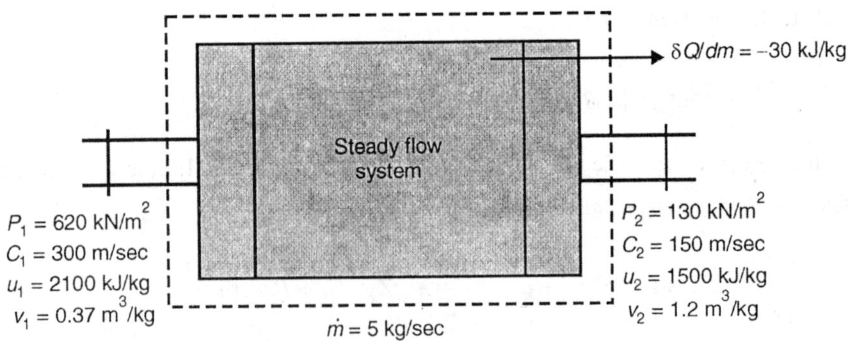

Fig. Ex. 3.13

Since power capacity is to be found out,

$$\frac{\delta W}{dm} = \left[\frac{C_1^2 - C_2^2}{2} + g(Z_1 - Z_2) + (u_1 - u_2) + (P_1 v_1 - P_2 v_2)\right] + \frac{\delta Q}{dm}$$

Neglecting ΔPE as given in the problem,

$$\frac{\delta W}{dm} = \left[\frac{C_1^2 - C_2^2}{2} + (u_1 - u_2) + (P_1 v_1 - P_2 v_2)\right] + \frac{\delta Q}{dm}$$

$$\frac{\delta W}{dm} = \frac{300^2 - 150^2}{2 \times 1000}\frac{\text{kJ}}{\text{kg}} + (2100 - 1500) + (620 \times 0.37 - 130 \times 1.2)$$

$$-30\frac{\text{kJ}}{\text{kg}}$$

$$\frac{\delta W}{dm} = \left[\frac{67500}{2000} + 600 + 73\right]\frac{\text{kJ}}{\text{kg}} - 30\frac{\text{kJ}}{\text{kg}}$$

$$= 33.75 + 600 + 73 - 30$$

$$\frac{\delta \mathbf{W}}{\mathbf{dm}} = \mathbf{676.75\ kJ/kg}$$

Since power is required to be expressed in kW,

Using $\dfrac{\delta W}{dt} = \dot{m}\dfrac{\delta W}{dm}$

Note: Mass flow rate = 5 kg/sec is given

$$\therefore \quad \text{Power} = 676.75\frac{\text{kJ}}{\text{kg}} \times 5\frac{\text{kg}}{\text{sec}}$$

$$\textbf{Power} = \mathbf{3383.75\ kW} = \frac{\delta \mathbf{W}}{\mathbf{dt}}$$

Example 3.14. The following data is obtained from tests on an air compressor.

	At Inlet	At Exit
Pressure (kPa)	100	500
Sp. volume (m^3/kg)	0.6	0.15
Sp. Internal Energy (kJ/kg)	50	125
Velocity of air (m/sec)	8	4

Rate of flow of air is 5 kg/sec.

Heat rejected to cooling water = 45 kW

Find the power required to drive the compressor in kW.

Solution

Since heat is rejected is given in kJ/sec., using SFEE on time basis,

$$\dot{m}\left[\frac{C_1^2}{2}+gZ_1+h_1\right]+\frac{\delta Q}{dt} = \dot{m}\left[\frac{C_2^2}{2}+gZ_2+h_2\right]+\frac{\delta W}{dt}$$

$$\therefore \quad \frac{\delta Q}{dt}-\frac{\delta W}{dt} = \dot{m}\left[\frac{C_2^2-C_1^2}{2}+g(Z_2-Z_1)+(P_2v_2-P_1v_1)+(u_2-u_1)\right]$$

$$-45-\frac{\delta W}{dt} = 5\left[\frac{4^2-8^2}{2\times1000}+0+(500\times0.15-100\times0.6)+(125-50)\right]$$

$$-45-\frac{\delta W}{dt} = 5\left[\frac{16-64}{2000}+(75-60)+75\right]$$

$$-45-\frac{\delta W}{dt} = 5[-0.024+15+75]$$

$$\Rightarrow \quad \frac{\delta \mathbf{W}}{\mathbf{dm}} = \mathbf{-494.88\ kW}$$

∴ Power required to drive the compressor = 494.88 kW.

Example 3.15. Steam enters a steam turbine with a velocity of 40 m/sec., enthalpy 2500 kJ/kg and leaves with a velocity of 90 m/sec and enthalpy 2030 kJ/kg. Heat losses from the turbine insulation to the surroundings are 250 kJ/min, and the steam flow rate is 5000 kg/hr. Neglect the change of P.E. Find the power developed by the turbine in kW.

Solution

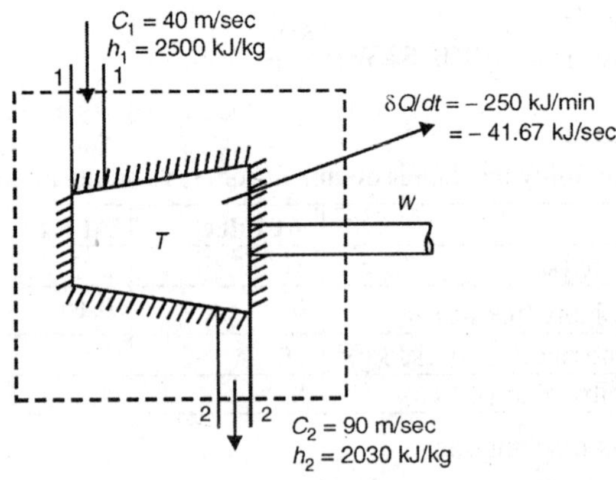

Fig. Ex. 3.15

$$\text{Steam flow rate} = 5000 \text{ kg/hr} = \frac{5000}{3600} \text{ kg/sec} = 1.389 \text{ kg/sec}$$

$$\dot{m}\left[\frac{C_1^2}{2} + gZ_1 + h_1\right] + \frac{\delta Q}{dt} = \dot{m}\left[\frac{C_2^2}{2} + gZ_2 + h_2\right] + \frac{\delta W}{dt}$$

$$\dot{m}\left[\frac{C_1^2 - C_2^2}{2} + (h_1 - h_2)\right] + \frac{\delta Q}{dt} = \frac{\delta W}{dt}$$

$$1.389 \frac{\text{kg}}{\text{sec}}\left[\frac{40^2 - 90^2}{2 \times 1000} \frac{\text{kJ}}{\text{kg}} + (2500 - 2030)\frac{\text{kJ}}{\text{kg}}\right] - 4.167 \frac{\text{kJ}}{\text{sec}} = \frac{\delta W}{dt}$$

$$1.389\left[-3.25 \frac{\text{kJ}}{\text{kg}} + (470)\frac{\text{kJ}}{\text{kg}}\right] - 4.167 \frac{\text{kJ}}{\text{sec}} = \frac{\delta W}{dt}$$

$$1.389 \frac{\text{kg}}{\text{sec}}\left[466.75 \frac{\text{kJ}}{\text{kg}}\right] - 4.167 \frac{\text{kJ}}{\text{sec}} = \frac{\delta W}{dt}$$

$$\frac{\delta \mathbf{W}}{\mathbf{dt}} = \mathbf{644.149\ kW}$$

Example 3.16 Steam enters a steam turbine with a velocity of 16 m/sec and specific enthalpy 2990 kJ/kg. The steam leaves the turbine with a velocity of 37 m/sec and specific enthalpy of 2530 kJ/kg. The heat lost to the surroudings is 25 kJ/kg. The steam flow rate is 3,60,000 kg/hr. Determine the work output from the turbine in kW.

Solution

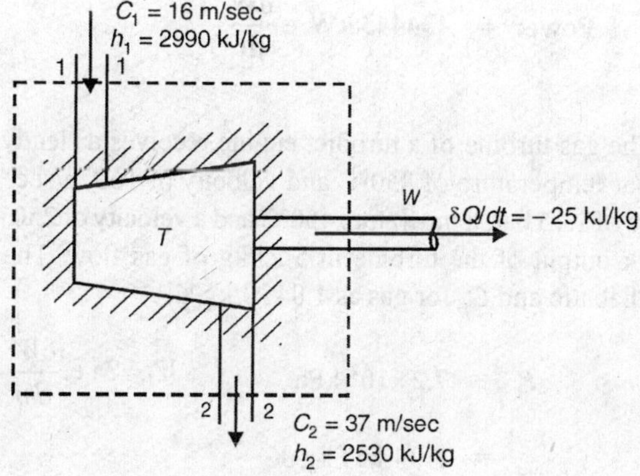

Fig. Ex. 3.16

Since heat lost is given in kJ/kg, using SFEE on mass basis,

$$\left[\frac{C_1^2}{2}+gZ_1+h_1\right]+\frac{\delta Q}{dm} = \left[\frac{C_2^2}{2}+gZ_2+h_2\right]+\frac{\delta W}{dm}$$

Since nothing is mentioned about elevation, neglecting ΔPE.

$$\therefore \quad \frac{C_1^2}{2}+h_1+\frac{\delta Q}{dm} = \frac{C_2^2}{2}+h_2+\frac{\delta W}{dm}$$

$$\therefore \quad \frac{\delta W}{dm} = \frac{\left(C_1^2-C_2^2\right)}{2}+\left(h_1-h_2\right)+\frac{\delta Q}{dm}$$

$$\frac{\delta W}{dm} = \frac{16^2-37^2}{2\times 1000}\frac{\text{kJ}}{\text{kg}}+\left(2990-2530\right)\frac{\text{kJ}}{\text{kg}}-25\frac{\text{kJ}}{\text{kg}}$$

$$\frac{\delta W}{dm} = \frac{-1113}{2000}+460-25$$

$$\mathbf{\frac{\delta W}{dm} = 434.443\ \frac{kJ}{kg}}$$

To express power in kW, using

$$\frac{\delta W}{dt} = \dot{m}\frac{\delta W}{dm}$$

It is given that mass flow rate =360000 kg/hr = 100 kg/sec.

$$\therefore \quad \text{Power} = 434.443\frac{\text{kJ}}{\text{kg}}\times 100\frac{\text{kg}}{\text{sec}}$$

$$\mathbf{Power = 434443\ kW = \frac{\delta W}{dt}}$$

Example 3.17 The gas turbine of a turbojet engine receives a steady flow of gases at pressure of 7.2 bar temperature of 850°C and velocity of 160 m/sec. It discharges the gases at a pressure of 1.15 bar, temperature 450°C and a velocity of 250 m/sec. Determine the external work output of the turbine in 5 kJ/kg of gas flow. The process may be assumed to be adiabatic and C_p for gas as 1.04 kJ/kg K.

Data: $P_1 = 7.2\times 10^2$ kPa $\qquad W = ?$ i.e. $\frac{\delta W}{dm} = ?$

$$C_{p\ \text{gases}} = 1.04\ \text{kJ/kg-K}$$

$T_1 = 850°\text{C}$ $\qquad C_1 = 160$ m/sec

Solution

Insulation (Adiabtic Turbine) $\therefore Q = 0$

$$P_2 = 1.15 \times 10^2 \text{ kPa}$$

$$T_2 = 450°\text{C}$$

$$C_2 = 250 \text{ m/sec}$$

We know that steady Flow Energy equation on mass basis is given by,

$$\left[\frac{C_1^2}{2} + gZ_1 + h_1\right] + \frac{\delta Q}{dm} = \left[\frac{C_2^2}{2} + gZ_2 + h_2\right] + \frac{\delta W}{dm} \quad (1)$$

Since Adiabtic Expansion process $\frac{\delta Q}{dm} = 0$

Since they have not mentioned anything about elevation $gZ_1 = 0, gZ_2 = 0$. Then Eq. (1) becomes,

$$\left[\frac{C_1^2}{2} + h_1\right] = \left[\frac{C_2^2}{2} + h_2\right] + \frac{\delta W}{dm}$$

i.e.
$$\frac{\left(C_1^2 - C_2^2\right)}{2} + (h_1 - h_2) = \frac{\delta W}{dm} \quad (2)$$

Since
$$h = C_P T$$

$\therefore$
$$\frac{\delta W}{dm} = (h_1 - h_2) + \frac{\left(C_1^2 - C_2^2\right)}{2}$$

$$= C_P(850 - 450) + \frac{160^2 - 250^2}{2 \times 1000}$$

$$= 1.04(850 - 450) + \frac{160^2 - 250^2}{2 \times 1000} \quad 416 + (-18.45)$$

$$\frac{\delta \text{W}}{\text{dm}} = \textbf{397.55 kJ/kg}$$

Example 3.18 The power output of an adiabatic steam turbine is 5 MW, and the inlet and exit conditions are as under

	Inlet	A Exit
Pressure	2 MPa	0.15 bar
Temperature (°C)	400	Dryness 0.9
Velocity (m/sec)	50	180
Elevation m	10	6

Determine the work done per unit of mass of the steam flowing through turbine and mass flow rate of the steam.

Solution

We know that, since work done per unit of mass of steam is to be found, using SFEE on mass basis.

$$\left[\frac{C_1^2}{2}+gZ_1+h_1\right]+q = \left[\frac{C_2^2}{2}+gZ_2+h_2\right]+w$$

or
$$w = (h_1-h_2)+\frac{(C_1^2-C_2^2)}{2}+g(Z_1-Z_2)$$

Since $h_1 = 3248.7 \text{ kJ/kg}$

Note. From steam tables coresponding to and 20 bar (2 MPa).

$$h_2 =$$

$h_f + x \cdot h_{fg}$. Since the steam is wet at 0.15 bar from steam tables

and
$$h_2 = 226+0.9\times2373.2 = 2361.88 \text{ kJ/kg}$$

$$w = (3248.7-2361.88)+\frac{50^2-180^2}{2000}+\frac{4\times9.81}{1000}$$

$$w = 871.91 \text{ kJ/kg} = \frac{\delta W}{dm}$$

Also, we know that, $\frac{\delta W}{dt} = \dot{m}\frac{\delta W}{dm}$

$$5000\frac{\text{kJ}}{\text{sec}} = \dot{m}\times 871.91 \text{ kJ/kg}$$

$\therefore$ $$\dot{\mathbf{m}} = \mathbf{5.734 \text{ kg/sec}}$$

Example 3.19 A turbine operating on air has inlet conditions as10 bar, 750 K and 200 m/s. While exit conditions are 1.25 bar and 40 m/sec. The mass flow rate of air is 1000 kg/hr. The flow of air is assumed to be reversible adiabatic (isentropic). Calculate,

(i) The temperature of air at exit.

(ii) The power output of turbine.

Assume C_p = 1.053 kJ/kg-K and k = 1.375 adiabatic index

Solution

$$\dot{m} = 1000 \text{ kg/hr} = 0.2777 \text{ kg/sec}$$

To find T_2, since the expansion process is adiabatic,

We know that, $$\frac{T_2}{T_1} = \left(\frac{P_2}{P_1}\right)^{\frac{k-1}{k}}$$

$$\Rightarrow \quad T_2 = T_1 \times \left(\frac{P_2}{P_1}\right)^{\frac{k-1}{k}} = 750 \times \left(\frac{1.25 \times 100}{10 \times 100}\right)^{\frac{1.375-1}{1.375}}$$

$$\mathbf{T_2 = 425.367\ K}$$

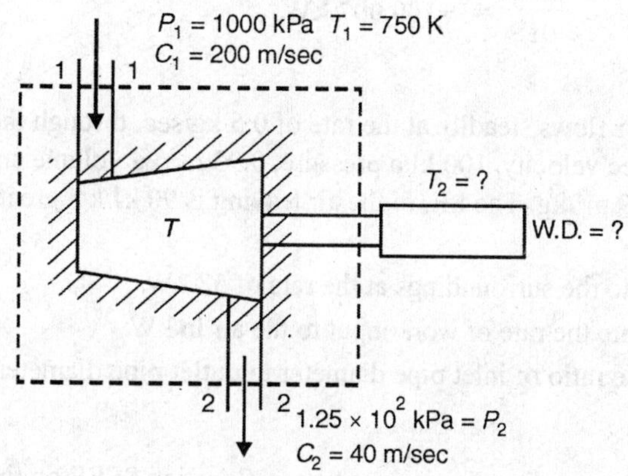

Fig. Ex. 3.19

We also know that, the SFEE is,

$$\dot{m}\left[\frac{C_1^2}{2} + gZ_1 + h_1\right] + \frac{\delta Q}{dt} = \dot{m}\left[\frac{C_2^2}{2} + gZ_2 + h_2\right] + \frac{\delta W}{dt}$$

Neglecting the ΔPE, and $\frac{\delta Q}{dt} = 0$ since the turbine is adiabatic.

$$\dot{m}\left[\frac{C_1^2}{2} + h_1\right] = \dot{m}\left[\frac{C_2^2}{2} + h_2\right] + \frac{\delta W}{dt}$$

$$-\frac{\delta W}{dt} = \dot{m}\left[\frac{(C_1^2 - C_2^2)}{2} + h_2 - h_1\right] \quad (1)$$

$$h_2 - h_1 = \Delta h = m \times C_p \,.\, \Delta T$$

$$= 1 \times 1.053 \times (425.367 - 750) \quad [\because m = 1 \text{ for unit mass}]$$

$$(h_2 - h_1) = -341.83855 \text{ kJ} \tag{2}$$

Substituting (2) in (1) we get,

$$-\frac{\delta W}{dt} = 0.277 \frac{\text{kg}}{\text{sec}} \left[\frac{40^2 - 200^2}{2 \times 1000} + (-341.838) \right]$$

$$-\frac{\delta W}{dt} = 0.277[-19.2 - 341.838] = 0.277[-361.038]$$

$$= -100.007 \text{ kW}$$

$$\frac{\delta \mathbf{W}}{\mathbf{dt}} = \mathbf{+100.007\ kW.}$$

Example 3.20 Air flows steadily at the rate of 0.5 kg/sec, through the air compressor, entering at 7 m/sec velocity, 100 kPa pressure, 0.95 m^3/kg volume and leaving at 5 m/sec, 700 kPa, 0.19 m^3/kg. The I.E. of the air leaving is 90 kJ/kg greater than that of air entering.

Heat rejected to the surroundings at the rate of 58 kW.

(a) Calculate the rate of work input to the air in kW.

(b) Find the ratio of inlet pipe diameter to outlet pipe diameter.

Solution

Heat lost to the surroundings is given in kJ/sec. So using SFEE on time basis.

$$\dot{m}\left[\frac{C_1^2}{2} + gZ_1 + h_1\right] + \frac{\delta Q}{dt} = \dot{m}\left[\frac{C_2^2}{2} + gZ_2 + h_2\right] + \frac{\delta W}{dt}$$

Since, P, u, and v are given.

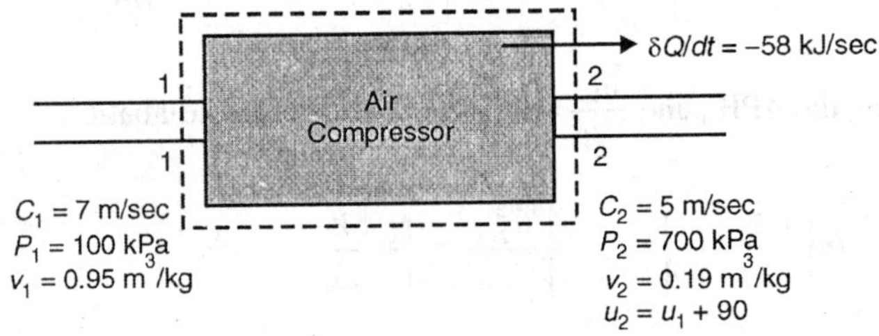

Fig. Ex. 3.20

$$\dot{m}\left[\frac{C_1^2}{2} + gZ_1 + u_1 + P_1 v_1\right] + \frac{\delta Q}{dt} = \dot{m}\left[\frac{C_2^2}{2} + gZ_2 + u_2 + P_2 v_2\right] + \frac{\delta W}{dt}$$

Since power is to be found,

$$\frac{\delta W}{dt} = \dot{m}\left[\frac{C_2^2 - C_1^2}{2} + g(Z_2 - Z_1) + (P_2 v_2 - P_1 v_1) + (u_2 - u_1)\right] + \frac{\delta Q}{dt}$$

$$= 0.5\frac{\text{kg}}{\text{sec}}\left[\frac{5^2 - 7^2}{2\times 1000} + 0 + (700\times 0.19 - 100\times 0.95)\frac{\text{kJ}}{\text{kg}}\right.$$

$$\left. + 90\frac{\text{kJ}}{\text{kg}}\right] - 58\frac{\text{kJ}}{\text{sec}}$$

$$= 0.5\frac{\text{kg}}{\text{sec}}\left[(-0.012 + 38 + 90)\frac{\text{kJ}}{\text{kg}}\right] - 58\frac{\text{kJ}}{\text{sec}}$$

$$= 0.5\frac{\text{kg}}{\text{sec}}\left[127.988\frac{\text{kJ}}{\text{kg}}\right] - 58\frac{\text{kJ}}{\text{sec}}$$

$$\frac{\delta W}{dt} = -63.994\frac{\text{kJ}}{\text{sec}} - 58\frac{\text{kJ}}{\text{sec}}$$

$$\frac{\delta W}{dt} = \mathbf{-121.994\ kW}$$

(b) From continuity equation,

$$\dot{m} = \frac{A_1 C_1}{v_1} = \frac{A_2 C_2}{v_2}$$

$$\therefore \quad \frac{A_1}{A_2} = \frac{v_1}{v_2}\cdot\frac{C_2}{C_1} = \frac{0.95}{0.19}\times\frac{5}{7}$$

$$\mathbf{\frac{A_1}{A_2} = 3.57}$$

Also we know that, $$\frac{A_1}{A_2} = \frac{\frac{\pi}{4}\cdot d_1^2}{\frac{\pi}{4}\cdot d_2^2} = 3.57$$

$$\therefore \quad \frac{d_1}{d_2} = \sqrt{3.57}$$

$$\mathbf{\frac{d_1}{d_2} = 1.89}$$

Example 3.21 At the inlet to a certain nozzle, the enthalpy of the fluid passing is 3000 kJ/kg and the velocity is 60 m/sec. At the discharge end the enthalpy is 2757 kJ/kg. The nozzle is horizontal. The heat loss during the flow is negligible.

(1) Find the velocity at the exit.

(2) If the inlet area is 0.1 m^2 and specific volume is 0.187 m^3/kg, find the mass flow rate.

(3) If the specific volume at the outlet is 0.498 m^3/kg.

Find the area at the exit of the nozzle.

Solution

Since the nozzle is horizontal, $\Delta PE = 0$

Since the heat loss during the flow is negligible, $q = 0$

Since no work is done in the nozle, $w = 0$

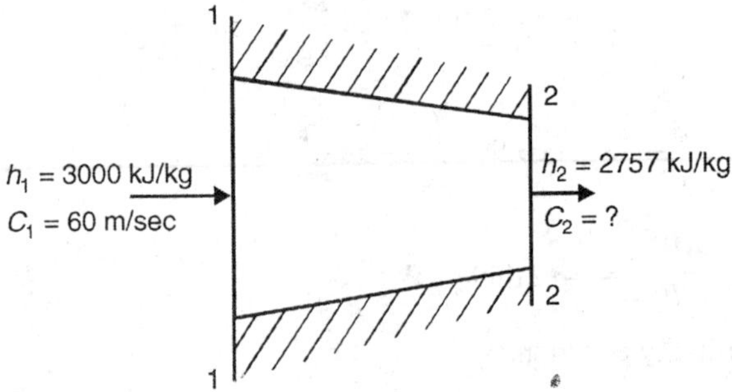

Fig. Ex. 3.21

We know that SFEE on mass basis is,

$$\left[\frac{C_1^2}{2} + gZ_1 + h_1\right] + q = \left[\frac{C_2^2}{2} + gZ_2 + h_2\right] + w \qquad (1)$$

or

$$q - w = \frac{C_2^2 - C_1^2}{2} + g(Z_2 - Z_1) + h_2 - h_1$$

On substituting the given characteristics, Eq. (1) reduces to,

$$h_1 + \frac{C_1^2}{2} = h_2 + \frac{C_2^2}{2}$$

$$3000\frac{\text{kJ}}{\text{kg}} + \frac{60^2}{2 \times 1000} = 2757\frac{\text{kJ}}{\text{kg}} + \frac{C_2^2}{2 \times 1000}$$

$$3000 + 1.8 = 2757 + \frac{C_2^2}{2000}$$

$$244.8 = \frac{C_2^2}{2000}$$

$$\therefore \quad C_2^2 = 489600$$

$$\therefore \quad \mathbf{C_2 = 699.71\ m/sec}$$

(ii) Mass flow rate $\dot{m} = \frac{A_1\,C_1}{v_1} = \frac{A_2\,C_2}{v_2}$

$$= \frac{0.1\,\text{m}^2 \times 60\,\text{m/sec}}{0.187}$$

$$\mathbf{\dot{m} = 32.08\ kg/sec}$$

(iii) Area at the exit. $\dot{m} = \frac{A_1 C_1}{v_1} = \frac{A_2 C_2}{v_2}$

$$A_2 = \frac{m v_2}{C_2} = \frac{32.08 \times 0.498}{699.71}$$

$$\therefore \quad \mathbf{A_2 = 0.0228\ m^3}$$

Example 3.22 A nozzle is a device used to increase velocity of steam flowing through it. At inlet the specific enthalpy of steam is 3000 kJ/kg. Specific volume 0.187 m^3/kg and area 0.1 m^2 and at outlet sp. enthalpy 2762 kJ/kg, specific volume 0.498 m^3/kg. There is no loss of heat, nozzle is horizontal and stream enters with the velocity of 60 m/sec.

Find: (i) Velocity of steam of inlet, (ii) Mass flow rate of steam in kg/hr, (iii) Exit area of nozzle.

Solution

For nozzle since no work is done $w = 0$.

We know that SFEE on mass basis is,

$$\left[\frac{C_1^2}{2} + gz_1 + h_1\right] + q = \left[\frac{C_2^2}{2} + gz_2 + h_2\right] + w$$

On substituting characteristic, we get,

$$h_1 + \frac{C_1^2}{2} = h_2 + \frac{C_2^2}{2}$$

$$3000 + \frac{60^2}{2\times1000} = 2762 + \frac{C_2^2}{2\times100}$$

$$\Rightarrow \quad \mathbf{C_2 = 692.53\ m/sec}$$

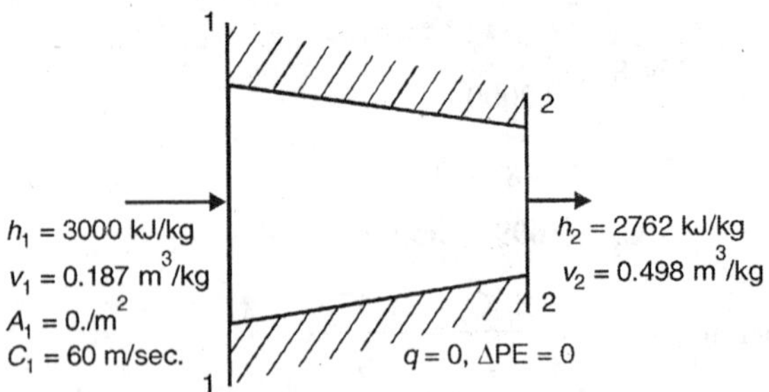

Fig. Ex. 3.22

(ii) Mass flow rate

$$\dot{m} = \frac{A_1 C_1}{v_1} = \frac{A_2 C_2}{v_2}$$

$$= \frac{0.1 \times 60}{0.187}$$

$$\dot{\mathbf{m}} = \mathbf{32.856\ kg/sec}$$

(iii) Area at the exit,

$$\dot{m} = \frac{A_1 C_1}{v_1} = \frac{A_2 C_2}{v_2}$$

$$A_2 = \frac{m \,.\, v_2}{C_2} = \frac{32.856 \times 0.4948}{692.53}$$

$$\mathbf{A_2 = 0.0231\ m^2}$$

Example 3.23 Air enters a nozzle with a velocity of 40 m/sec. The decrease in the enthalpy in the nozzle is 180000 J/kg. Determine the exit velocity. Assume nozzle to be adiabatic.

Solution

We know that, SFEE on mass basis is,

$$\left[\frac{C_1^2}{2} + gZ_1 + h_1\right] + q = \left[\frac{C_2^2}{2} + gZ_2 + h_2\right] + w \qquad (1)$$

Since for a nozzle characteristics are,

$$\Delta PE = 0,\ w = 0,\ q = 0 \text{ since adiabatic}$$

Equation (1) reduces to,

$$\frac{C_1^2}{2} + h_1 = \frac{C_2^2}{2} + h_2$$

$$\frac{C_1^2}{2}+(h_1-h_2) = \frac{C_2^2}{2}$$

$$\frac{(40)^2}{2\times1000}+180000 = \frac{C_2^2}{2\times1000}$$

$$\therefore \qquad \mathbf{C_2 = 601.33\ m/sec}$$

Example 3.24 Air is compressed continuously by an air compressor according to the law $PV^{1.3} = C$. The pressure at the inlet is 1 bar, and at the outlet is 5 bar. The volume of the air changes from 3 m³/kg to 0.8 m³/kg and the velocity changes 25 m/sec at inlet to 130 m/sec at the delivery. The delivery connection is 12 mt above the inlet. What is the shaft power of the compressor. Is it a power absorbing or power producing system.

Solution

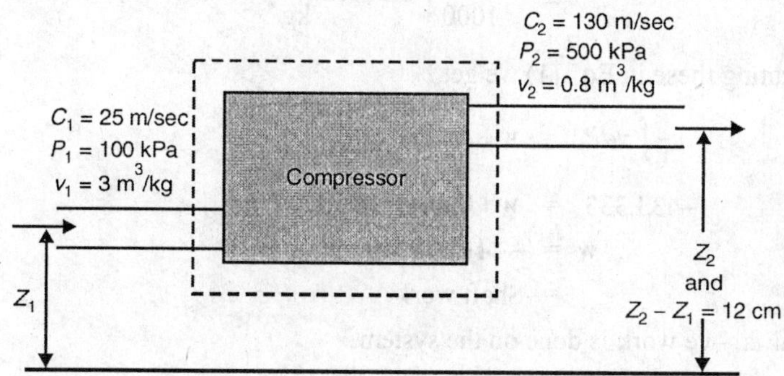

Fig. Ex. 3.24

Note. If the steady flow process or system follows. $PV^n = C$ then follow this method to solve the problem.

i.e. use $$-\int v\,dP = w+\Delta KE+\Delta PE$$

We know that,

	Isothermal ($Pv = C$) (For unit mass)	Polytropic ($Pv^n = C$)
(i) W.D. in closed system = $\int Pdv$	$P_1v_1 \ln_e \frac{v_2}{v_1}$	$\frac{P_2v_2 - P_1v_1}{1-n}$
(ii) W.D. for open system = $-\int v\,dP$	Steady flow system, $P_1v_1 \ln_e \frac{v_2}{v_1}$ or $\frac{P_1}{P_2}$	$\frac{n}{n-1}\times P_2v_2 - P_1v_1$

For Polytropic Process,

$$-\int_1^2 vdP = \frac{n}{n-1}\times P_1v_1 - P_2v_2$$

$$= \frac{1.3}{1.3-1} \times (100 \times 3 - 500 \times 0.8) \frac{\text{kJ}}{\text{kg}}$$

$$= -433.333 \text{ kJ/kg}$$

and $$\Delta\text{KE} = \frac{1}{2}\left(\frac{C_2^2 - C_1^2}{1000}\right) = \frac{130^2 - 25^2}{2 \times 1000}$$

and $$\Delta\text{KE} = 8.1375 \frac{\text{kJ}}{\text{kg}}$$

$$\Delta\text{PE} = g(Z_2 - Z_1)$$

$$= \frac{9.81 \times 12}{1000} 0.11772 \frac{\text{kJ}}{\text{kg}}$$

Substituting these in Eq. (1) we get,

$$\therefore \quad -\int v dP = w + \Delta\text{KE} + \Delta\text{PE}$$

$$-433.333 = w + 8.1375 + 0.11772$$

$$\therefore \quad \mathbf{w = -441.588 \text{ kJ/kg}}$$

$$= \text{Shaft work}$$

Since w is –ve work is done on the system.

∴ It is a power absorbing system.

Example 3.25 A fluid at the rate of 10 kg/sec is compressed adiabatically from 5 bar to 50 bar in a steady flow process. Calculate the power required assuming that the specific volume of fluid being as 0.001 m³/kg which remains almost constant.

Solution

Since the compression process follows an adiabatic process, we have to use,

$$\text{W.D.} = -\int v dP = v.(P_2 - P_1)$$

$$\therefore \quad \text{Power required} = \dot{m} \times v \times (P_2 - P_1)$$

$$= 10 \text{ kg/sec} \times 0.001 \text{ m}^3/\text{kg}\,(50-5) \times 100$$

$$\textbf{Power required} = \mathbf{45 \text{ kW}}$$

Example 3.26 A blower handles 2 kg/sec of air at 20°C and consumes a power of 30 kW. The inlet and outlet velocities are 100 in/sec and 150 m/sec respectively. Estimate the exit air temperature assuming adiabatic conditions.

Take R for air $= 0.287$ kJ/kg-K

$$\gamma = \frac{C_p}{C_v} = 1.4$$

and $C_p = 1.005$ kJ/kg-K

Solution

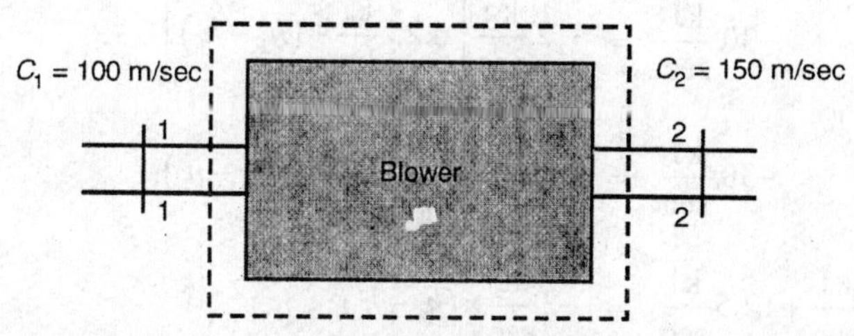

Fig. Ex. 3.26

$$\dot{m} = 2 \text{ kg/sec}$$

$$\frac{\delta W}{dt} = -30 \text{ kW} = -30 \text{ kJ/sec.}$$

$$t_2 = ?\quad t_1 = 20°C$$

Since $\frac{\delta W}{dt}$ is given using the SFEE on time basis.

$$\dot{m}\left[\frac{C_1^2}{2} + gZ_1 + h_1\right] + \frac{\delta Q}{dt} = \dot{m}\left[\frac{C_2^2}{2} + gZ_2 + h_2\right] + \frac{\delta W}{dt}$$

$$\frac{\delta W}{dt} = -\dot{m}\left[\frac{C_2^2 - C_1^2}{2} + g(Z_2 - Z_1) + (h_2 - h_1)\right] + \frac{\delta Q}{dt}$$

Since adiabatic conditions are given, $\frac{\delta Q}{dt} = 0$

Neglecting the change in PE, we have,

$$\frac{\delta W}{dt} = -\dot{m}\left[\frac{C_2^2 - C_1^2}{2} + (h_2 - h_1)\right] \quad (1)$$

$$\frac{\delta W}{dt} = -\dot{m}\left[\frac{C_2^2 - C_1^2}{2} + (h_2 - h_1)\right]$$

$$-30\frac{\text{kJ}}{\text{sec}} = -2\frac{\text{kg}}{\text{sec}}\left[\frac{150^2-100^2}{2\times1000}+(h_2-h_1)\right]$$

$$-30\frac{\text{kJ}}{\text{sec}} = -2\frac{\text{kg}}{\text{sec}}\left[\frac{12500}{2000}+(h_2-h_1)\right]$$

$$-30\frac{\text{kJ}}{\text{sec}} = -2\frac{\text{kg}}{\text{sec}}\left[6.25\frac{\text{kJ}}{\text{kg}}+(h_2-h_1)\right]$$

$$-30\frac{\text{kJ}}{\text{sec}} = -12.5\frac{\text{kJ}}{\text{sec}}-2\frac{\text{kg}}{\text{sec}}\times(h_2-h_1)$$

$$-30\frac{\text{kJ}}{\text{sec}}+12.5\frac{\text{kJ}}{\text{sec}} = -2\frac{\text{kg}}{\text{sec}}\times(h_2-h_1)$$

$$\frac{-17.5\ \text{kJ/sec}}{2\ \text{kg/sec}} = (h_2-h_1)$$

$$\mathbf{8.75\frac{kJ}{kg} = (h_2-h_1)}$$

We also know from the definition of C_P,

$$\left(\frac{dh}{dT}\right)_P = C_P$$

or $$dh = C_P\,.\,dT \qquad (m=1\text{ for unit mass})$$

or $$\text{Change enthalpy } \Delta h = C_P\,\Delta T$$

$$h_2-h_1 = C_P(t_2-t_1)$$

$$8.75 = 1.005\,(t_2-20)$$

$$\mathbf{t_2 = 28.75°C}$$

Example 3.27 In a water cooled compressor 0.6 kg of air is compressed/sec. Power required to run the compressor is 40 kW. Heat lost to the cooling water is 30% of input and 10% of input is lost in bearings and other frictional effects. Air enters the compressor of 1 bar and 30°C. If the changes in PE and KE are neglected, estimate the exit air temperature. Take $C_p = 1.005$ kJ/kg-K.

Data : $\dot{m} = 0.6\ \text{kg/sec.}$ $\quad \frac{\delta W}{dt} = 40\ \text{kW}$

$P_1 = 1\ \text{bar} = 10^5\ \text{N/m}^2 \quad T_1 = 30°\text{C}$

$$\Delta \text{ PE} = 0, \ \Delta \text{ KE} = 0. \qquad T_2 = ?$$

Solution

We know that, $\Delta H = m\,C_P\left(T_2 - T_1\right)$

$$= 0.6 \times 1.005\left(T_2 - 30\right)$$

$$\Delta H = 0.603\,(T_2 - 30) \qquad (1)$$

Net heat lossses, $= 40\%$ of input

$$= \frac{40}{100} \times 40\,\text{kW}$$

$$\frac{\delta Q}{dt} = 0.4 \times 40 = 16 \text{ kW}.$$

SFEE is given by,

$$\dot{m}\left[\frac{C_1^2}{2} + gZ_1 + h_1\right] + \frac{\delta Q}{dt} = \dot{m}\left[\frac{C_2^2}{2} + gZ_2 + h_2\right] + \frac{\delta W}{dt}$$

Since $\Delta\text{KE} = 0$ and $\Delta\text{PE} = 0$

$$\dot{m}h_1 + \frac{\delta Q}{dt} = \dot{m}h_2 + \frac{\delta W}{dt}$$

$$\therefore \quad \frac{\delta W}{dt} - \frac{\delta Q}{dt} = \dot{m}\left(h_1 - h_2\right) = \Delta H \qquad (2)$$

From (1) and (2) $40 - 16 = 0.603\left(T_2 - 30\right)$

$$\mathbf{T_2 = 70^\circ\ C}$$

Example 3.28 A room is fitted with 2 fans each consuming 0.2 kW power. There are three lamps in the room which are provided to heat the room each consuming 200 W. Ventilation air enters the room with the enthalpy of 85 kJ/kg and leaves the room with the enthalpy of 60 kJ/kg. The rate of air flow is 100 kg/hr. There are five persons in the room and heat generated by each person is 600 kJ/hr. Determine the rate at which the heat is to be removed by a room cooler, so that steady state is maintained in the room.

Solution

Air flow rate = 100 kg/hr

Heat generated by each person = 600 kJ/hr

= 0.1666 kJ/sec

∴ For 5 persons = 0.8333 kJ/sec

Since work input is given in kW using SFEE on time basis.

$$\dot{m}\left[\frac{C_1^2}{2}+gZ_1+h_1\right]+\frac{\delta Q}{dt} = \dot{m}\left[\frac{C_1^2}{2}+gZ_2+h_2\right]+\frac{\delta W}{dt}$$

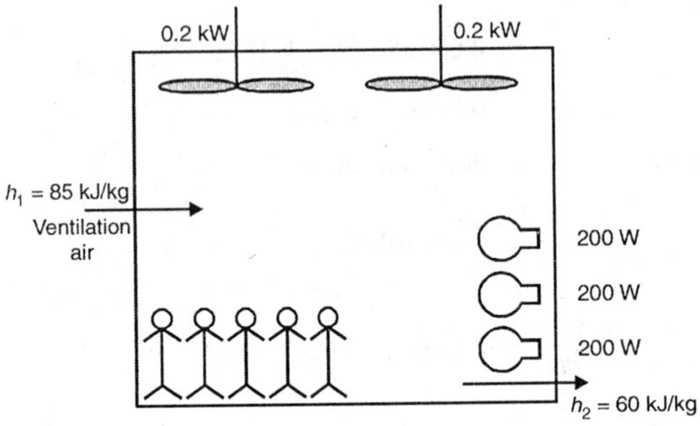

0.2 kW
0.2 kW
h_1 = 85 kJ/kg
Ventilation air
200 W
200 W
200 W
h_2 = 60 kJ/kg

Fig. Ex. 3.28

Neglecting ΔKE and ΔPE.

$$\dot{m}\,[h_1-h_2]+\frac{\delta Q}{dt} = \frac{\delta W}{dt}$$

$$\text{i.e. } \dot{m}\,[h_1-h_2]+\left[\begin{pmatrix}\text{heat added due}\\ \text{to lamps}\end{pmatrix}+\begin{pmatrix}\text{heat generated}\\ \text{by 5 persons}\end{pmatrix}-\begin{pmatrix}\text{heat to be removed}\\ \text{by the cooler}\end{pmatrix}\right]$$

$$= \text{Work input to the fans (which should be } -\text{ve)}$$

$$\therefore \frac{100}{3600}\frac{\text{kJ}}{\text{sec}}\left[85\frac{\text{kJ}}{\text{kg}}-60\frac{\text{kJ}}{\text{kg}}\right]+\left[0.6\frac{\text{kJ}}{\text{sec}}+0.833\frac{\text{kJ}}{\text{sec}}-X\right] = -0.4\frac{\text{kJ}}{\text{sec}}$$

$$0.6944+[0.6+0.833-X] = -0.4$$

$$0.6944+0.6+0.833+0.4$$

$\therefore$ X = **2.574 kJ/sec.** **Heat to be removed by Cooler**

Example 3.29 In a steady flow device, the inlet and outlet conditions are given below. Determine the heat loss/gain by the system.

Property	**Inlet**	**Outlet**
Pressure (bar)	10	0.15
Specific volume (m^3/kg)	0.206	8.93
Specific enthalpy (kJ/kg)	2827	2341
Velocity (m/s)	20	120
Elevation m	3.2	0.5

The fluid flow rate through the device is 2.1 kg/s. The work output of the device i s 750 kW.

Data:
$$Q = ?$$
$$\frac{\delta W}{dt} = +750 \text{ kW}$$

Solution

Since work output is given in $\frac{\delta W}{dt}$, using SFEE on time basis,

$$\dot{m}\left[\frac{C_1^2}{2} + gZ_1 + h_1\right] + \frac{\delta Q}{dt} = \dot{m}\left[\frac{C_2^2}{2} + gZ_2 + h_2\right] + \frac{\delta W}{dt}$$

$$= \dot{m}\left[\frac{C_2^2 - C_1^2}{2} + g(Z_2 - Z_1) + (h_2 - h_1)\right] + \frac{\delta W}{dt}$$

$$= 2.1\frac{\text{kg}}{\text{sec}}\left[\frac{120^2 - 20^2}{2\times1000} + \frac{9.81(0.5-3.2)}{1000} + (2341-2827)\right] + 750$$

$$= 2.1[7 + (-0.026487) + (-486)] + 750$$

$$= 2.1[7 - 0.026487 - 486] + 750$$

$$= 2.1[-479.026] + 750$$

$$= -1005.9556 + 750$$

$$\frac{\delta Q}{dt} = \mathbf{-255.95562 \text{ kW}}$$

Note. –ve sign implies that heat will flow out of system.

Example 3.30 Air at a temperature of 15°C passes through a heat exchanger at a velocity of 30 m/c where its temperature is raised to 800°C. It then enters a turbine with the same velocity of 30 m/s and expands until the temperature falls to 650°C. On leaving the turbine, the air is taken at a velocity of 60 m/s to a nozzle where it expands until the temperature has fallen to 500°C. If the air flow rate is 2 kg/s calculate,

(a) Rate of heat transfer to the air in the heat exchanger.

(b) Power output from the turbine, assuming no heat loss, and

(c) Velocity of air at exit from nozzle, assuming no heat loss, take the specific enthalpy of air as

$h = C_p T$, where C_p = 1.005 kJ/kg K, and T = Temperature

Solution

For heat exchanger:

$$q_{1-2} + \dot{m}\left(h_1 + \frac{C_1^2}{2} + gz_1\right) = w_{1-2} + \dot{m}\left(h_2 + \frac{C_2^2}{2} + gz_2\right)$$

Then

$$q_{1-2} = \dot{m}\left(h_2 - h_1\right) \qquad (\because C_1 = C_2)$$

$$= \dot{m}\, C_P \left(T_2 - T_1\right)$$

$$q_{1-2} = 2 \times 1.005 \times (800 - 15)$$

$$\mathbf{q_{1-2} = 1577.85\ kJ/s}$$

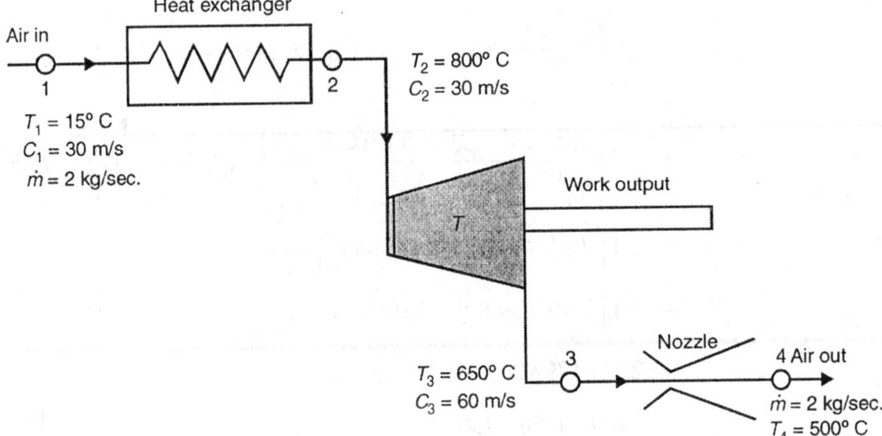

Fig. Ex. 3.30

For the turbine:

$$q_{2-3} + \dot{m}\left(h_2 + \frac{C_2^2}{2} + gz_2\right) = w_{2-3} + \dot{m}\left(h_3 + \frac{C_3^2}{2} + gz_3\right)$$

$$\therefore \quad w_{2-3} = \dot{m}\left[\left(h_2 - h_3\right) + \frac{C_2^2 - C_3^2}{2}\right] \qquad (\because q_{2-3} = 0)$$

$$= \dot{m}\left[C_P\left(T_2 - T_3\right) + \frac{C_2^2 - C_3^2}{2}\right]$$

$$= 2\left[1.005(800 - 650) + \left(\frac{30^2 - 60^2}{2 \times 1000}\right)\right]$$

$$\mathbf{W_{2\text{-}3} = 298.8\ kW}$$

For nozzle:

$$q_{3-4} + \dot{m}\left[h_3 + \frac{C_3^2}{2} + gz_3\right] = w_{3-4} + \dot{m}\left[h_4 + \frac{C_4^2}{2} + gz_4\right]$$

$$C_4^2 - C_3^2 = 2\times1000\times C_P(T_3 - T_4)$$

$(\because q_{3-4} = 0$ and $w_{3-4} = 0)$

$$\therefore \quad C_4^2 - C_3^2 = 2\times1000\times C_P(T_3 - T_4) = 2\times1000\times1.005(650-500)$$

$$= 301500$$

$$\therefore \quad \mathbf{C_4 = 552.36\ m/s}$$

Example 3.31 Helium gas is expanded polytropically in a turbine, from 4 bar, 300°C to 1 bar such that final volume is 2.5 times the volume at inlet. Velocity of gas at exit is 50 m/s. What is the mass flow rate of gas required to produce 1 MW turbine output? How much is the heat transfer during the process ? Also determine exit area of turbine.

Assume specific heat of helium = 5.193 kJ/kg-K at constant pressure.

Solution

We know that $R = \frac{G}{M} = \frac{8.314}{4} = 2.0785$

$$v_1 = \frac{RT_1}{P_1} = \frac{2.0785\times573}{400} = 2.9774\ \frac{\text{m}^3}{\text{kg}}$$

$$v_2 = 2.5\ v_1 = 7.4436\ \text{m}^3/\text{kg}$$

For Polytropic Process, $P_1v_1^n = P_2v_2^n$

$$\therefore \quad n = \frac{l_n\left(\frac{P_2}{P_1}\right)}{l_n\left(\frac{v_1}{v_2}\right)} = 1.513$$

and
$$\frac{T_2}{T_1} = \left(\frac{P_2}{P_1}\right)^{\frac{n-1}{n}}$$

$$\therefore \quad T_2 = 358.11\ \text{K}$$

$$\text{Specific work} = -\int_1^2 v\,dP = \frac{n}{n-1}RT_1\left[1-\frac{P_2v_2}{P_1v_1}\right]$$

$$= \frac{1.513}{0.513}\times 2.0785\times 573\times\left[1-\frac{1\times 2.5}{4}\right]$$

$$= 1317.2 \text{ kJ/kg}$$

$$\Delta\text{KE} = \frac{1}{2}C_2^2 = 1250\frac{\text{J}}{\text{kg}} \text{ or } 1.25 \text{ kJ/kg}$$

$$q - w_s = \Delta h + \Delta\text{KE}$$

and $$-\int v\,dP = w_s + \Delta\text{KE} \quad \text{as} \quad \Delta\text{PE} = 0$$

∴ $$\mathbf{W_s = 1317.2 - 1.25 = 1315.95 \text{ kJ/kg}}$$

and $$q = 1315.95 + C_P(T_2 - T_1) + 1.25$$

$$= 1315.95 + 5.193(358.11 - 573) + 1.25$$

$$\mathbf{q = 201.28 \text{ kJ/kg}}$$

Then $$\mathbf{\dot{m} = \frac{100}{w_s} \text{ kJ/s} = 0.7599 \text{ kg/s}}$$

∴ $$\mathbf{\text{Exit area } A_2 = \frac{\dot{m}\,v_2}{C_2} = 0.1131 \text{ m}^2}$$

THEORY QUESTIONS

1. Write a note on 'Joule's Experiment'?
2. State and explain the first law of thermodynamics for a closed system undergoing a cyclic change.
3. Derive an equation for energy transfer for a closed system.
4. Explain the First Law of Thermodynamics and hence show their cyclic integral of δQ and cyclic integral of δW are equal for a closed system.
5. State and explain the first law of thermodynamics for a closed system undergoing a process.
6. Prove that 'Internal Energy' is a property of a system.
7. Write a note on different forms of stored energies.

8. Define the term 'Enthalpy' (H).
9. Write a note on 'Internal Energy' (U).
10. Distinguish between heat and internal energy.
11. Define specific heat at constant pressure and at constant volume for a perfect gas.
12. Define Adiabatic Index (g).
13. What is Perpetual Motion Machine of first kind (PMM-1) ?
14. Define a flow process and explain the terms Control Volume and control surface.
15. Explain the term flow work and discuss how it differs from the work done by a closed system.
16. Write a note on Flow work or Flow energy.
17. How is 'Steady flow system characterized'?
18. What are the conditions for a steady flow process?
19. State the conditions of First Law of Thermodynamics for a Steady Flow Processes.
20. Explain the essentials of non-flow and flow systems. Give suitable examples of each. What are advantages of steady flow system over non-flow system.
21. Derive the Steady Flow Energy Equation (SFEE).
22. Apply first law of thermodynamics to steady flow system and derive the equation of energy.
23. Derive Steady Flow Energy Equation on mass basis.
24. Explain the significance of $-\int v\,dP$
25. Explain the significance of $\int P dv$ in case of steady flow process and non-flow process.
26. Prove that, $-\int vdP = Q - \Delta H$.
27. Prove that, $\int Pdv = Q - \Delta U$.
28. Obtain the energy equations for the following open systems:
    - (a) Boiler
    - (b) Turbine
    - (c) Compressor
    - (d) Throttling process
    - (e) Nozzles
    - (f) Condenser
29. Derive Continuity equation.

PROBLEMS FOR PRACTICE

1. A centrifugal air compressor delivers 15 kg of air/min. Air enters at I bar, I0 m/sec, 0.5 m^3/kg. It is discharged at 7 bar, 80 m/sec and 0.15 m^3/kg. During the process, its enthalpy increases by I60 kJ/kg and the air rejects 720 kJ/min of heat to the surroundings. Find,
   (i) Power required to run the compressor.
   (ii) Ratio of inlet to outlet pipe diameter. Neglect ΔPE.
2. For a steady flow system, following particulars are noted,

	Inlet	Exit
1. Pressure (bar)	1	5
2. Vol (m^3/kg)	2	0.58
3. Velocity (m/sec)	20	30
4. Elevation (m)	0	2

   If the mass flow rate of the working fluid is 1 kg/sec., find the power of the system. Is it a power absorbing or power producing system?
   Assume that the steady flow system follows $Pv^{1.3} = C$.
3. One kg of working substance undergoes a reversible constant pressure process at 1.2 bar during which its volume changes from 1 m^3 to 1.8 m^3 and temperature changes from 50°C to 37°C. The sp. heat of a substance at constant pressure is given by $C_p = \left(1.1 + \dfrac{40}{t+30}\right)$ kg/kg°C where t is in °C. Find out
   (a) Heat supplied (b) Work done
   (c) Change in I.E. (d) Change in enthalpy
4. A fluid is contained in a cylinder piston arrangement. The piston is spring loaded and frictionless such that the pressure of the fluid varies with its volume according to the law $P = a + bV$. The internal energy of the fluid is given by $V = 30 + 3.2\,PV$ where U is in kJ. P in kPa, V in m^3. The fluid undergoes a change of state from 150 kPa and 0.025 m^3 to 450 kPa and 0.05 m^3. Assuming no work other than that on the piston determine the direction and magnitude of work and heat transfer.
5. What at the rate of 10 kg/sec is compressed adiabatically from 5 bar to 50 bar in a steady flow process. Calculate the power required assuming that sp. volume of water being as 0.00I m^3/kg which remains almost constant.
6. Steam enters a turbine steadily at 10 mPa and 550°C with a velocity of 60 m/s and leaves at 25 kPa with a dryness fraction of 0.95. A heat loss of 30 kJ/kg occurs during the process. The inlet area of the turbine is 150 cm^2 and exit area is 1400 cm^2. Determine : (1) mass flow rate of steam. (2) exit velocity (3) power output.
7. A closed system receives 200 kg of heat at constant volume. The system then undergoes a constant pressure process during which 150 kg of heat is rejected

and 50 kJ of work is done on the system. Calculate how much of work is required to be done if the system is to be restored to its original state by means of an adiabatic process. If the initial internal energy is 50 kJ, determine the values of internal energy at the end of the constant volume and constant pressure processes.

8. In a reversible steady flow process 225 kW of heat is rejected per kg of the fluid passing through the system. During the process, pressure and volume change from 6 bar and 0.075 m³/kg to 2 bar and 0.2 m³/kg. Assuming that flow process follows the law PV = constant and ignoring changes in K.E. and P.E., determine the change in enthalpy of the fluid during the process.

9. Air flow steadily at the rate of 0.4 kg/sec through an air compressor, entering at 6 m/sec with pressure of 1 bar and sp. volume of 0.85 m³/kg and leaving at 4.5 m/sec with a pressure of 6.9 bar and sp. volume of 0.16 m³/kg. The sp. internal energy of the air leaving is 88 kJ/kg greater than that of the air entering, cooling water in the jacket surrounding the cylinder absorbs heat from the air at the rate of 59 kg/sec. Calculate the power required to drive the compressor and the inlet and outlet pipe areas.

10. In the turbine of a gas turbine unit the gases flow through the turbine at 17 kg/sec and the power developed by the turbine is 14 MW. The sp. enthalpies of the gases at inlet and outlet are 1200 kJ/kg and 360 kJ/kg respectively. Calculate the rate at which heat is rejected from the turbine. Find also the area of the inlet pipe given that sp. volume of the gases at inlet is 0.5 m³/kg.

11. A steam turbine operates under steady flow conditions, receiving steam and leaving the steam at the following rate.

	Inlet	Outlet
Pressure	15 bar	130 kPa
I.E. (kJ/kg)	2500	1500
Velocity (m/sec)	300	200
Elevation (m)	3	0
Volume (m³)	0.15	1.2 m³ /kg

12. A closed system undergoes a cycle 1–2–3–1. If $Q_{12} = 30$, $Q_{23} = 10$, $W_{12} = 5$, $W_{31} = 25$ and $\Delta E_{31} = 15$, determine W_{23} and ΔE_{23}.

13. The following data is available for a test on air compressor for an air flow rate of 6 kg/sec.

	Inlet	Outlet
Pressure	95 kPa	7.5 bar
Sp. Internal Energy (kJ/kg)	40	150
Sp. Volume (m³/kg)	0.95	0.162
Velocity of air (m/sec)	6	1

Estimate: (1) Power required to drive the compressor in kW. (2) Ratio of inlet pipe diameter to outlet pipe diameter.

14. A non-flow system undergoes a frictionless process according to the law $P = \frac{4.5}{v} + 2$ where P is in bar, V is in m³/kg. During this process volume changes from 0.12 m³/kg to 0.04 m³/kg and temperature increases by 133°C. The change in I.E. of fluid is given by $du = C_v \, dT$ where $C_V = 0.71$ kJ/kg K and dT is temperature change. Find out (I) Heat transfer, (2) Change in enthalpy. Assume fluid mass of 5 kg.

ꕤ

4

Second Law of Thermodynamics

CHAPTER OBJECTIVES

After reading this chapter you will be able to learn the following

- Limitations of I-law of Thermodynamics.
- Concpets of Heat and Work Reservoirs.
- Refrigerator, Heat Pump, Statements of II-law of Thermodynamics and their equivalence.
- Reversibility and Irreversibility.
- Carnot's Engine, Carnot's Cycle and Carnot Theorem.
- Continuity equation.

4.1 INTRODUCTION

The first law of thermodynamics deals with the conversion of energy. However it does not specify the direction of the process. The second law of thermodynamics is about the direction, quality and quantity.

4.2 LIMITATIONS OF FIRST LAW OF THERMODYNAMICS

First-law states that, when a closed system undergoes any cyclic process, then the cyclic integral of work is equal to the cyclic integral of heat

i.e. $$\oint \delta W = \oint \delta Q$$

Thus the law merely states that work transfer during a cycle is equal to the heat transfer and does not place any restriction on the direction of heat and the work transfer. It does not specify whether the process is possible in a particular direction or not at all.

According to this law it can be assumed that the energy transfer can take place in either direction, since it does not specify the direction of energy transfer. But by practical experience it is observed that, even if a proposed cycle satisfies the first law, it doesn't ensure that the cycle will occur actually. This resulted in the formulation of *Second Law of Thermodynamics*. Thus, a cycle will proceed only if both I and II laws of thermodynamics are satisfied.

Thus, the second law involves the fact that processes will proceed in a certain direction but not in the opposite direction, for example,

(i) Consider the Joule's experiment in which the fall of weight *W* rotates the paddle wheel and increases the temperature of water. Here work is converted into heat. However reverse of this process is not possible i.e. by heating water, paddle wheel will not rotate and lift the weight. This shows that the system can operate in a cyclic process only when work is done on the system and heat is rejected out of the system. But it cannot work in a cyclic process when both work and heat transfer are positive even though such a process will not violate the first law of thermodynamics.

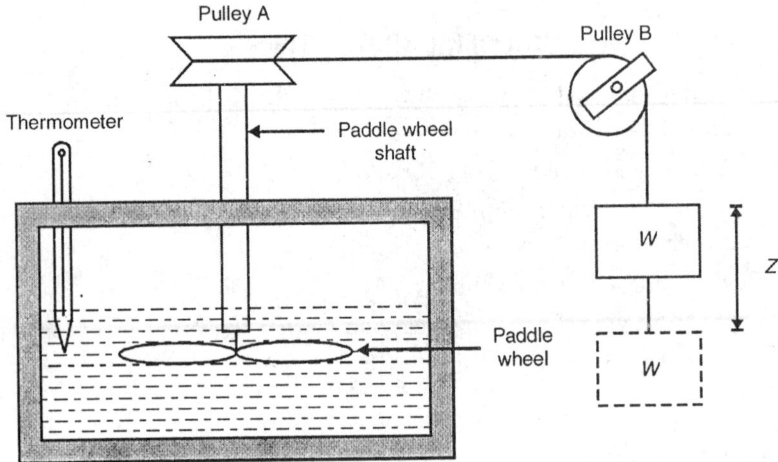

Fig. 4.1. *Joule's experiments.*

(ii) Consider hot tea in a cup, here heat flows from tea to the cooler surroundings. Once it is cooled, it can never be heated by the addition of heat from the cooler surroundings, without doing any work on it.

(iii) A cup of ice cream when kept in atmosphere absorbs heat and melts. But it will not solidify giving back heat without doing any work on it.

(iv) When a vehicle is stopped by applying mechanical brakes, the brakes get hot. The I.E. of the brake increases by an amount equal to the decrease in K.E. of the vehicle. According to the first law, the hot brake on cooling, should give back its increase in I.E. to the wheels of the vehicle, causing it to regain its speed. This never happens in practice.

From the above examples it is obvious that, even if the first law is satisfied the processes will proceed in a particular direction while they are impossible in the opposite direction. Thus the first law of thermodynamics is necessary but not sufficient condition for the processes to take place.

It has also been practically found that all forms of energies are not equally convertible into work and the first law is silent about the extent of conversion of energy. It is necessary to study *Second Law of Thermodynamics* in this regard.

Points to Remember

- First Law is silent about the direction.
- It gives no indication about extent of conversion of energy.

4.3 HEAT ENGINE

Heat engine converts heat energy into mechanical energy. Heat engine works on thermodynamic cycle like Otto, Diesel, Rankine in which there is a net heat transfer to the system and a net work transfer from the system.

As we have studied, heat engines are broadly classified into two types.

(i) External combustion engines; (ii) Internal combustion engines.

Figure 4.2 shows a simple steam power plant. In this power plant an amount of heat Q_1 is supplied from the furnace (Also called as high temperature reservoir) to the water in the boiler drum. An amount of heat Q_2 is rejected to the low temperature reservoir, like a coolant in the condenser and in doing so, an amount of work W will be produced.

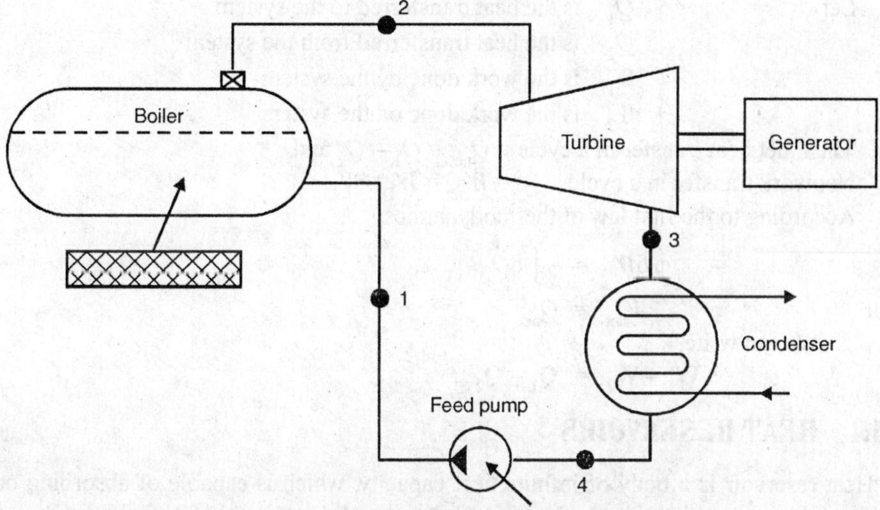

Fig. 4.2. *A simple steam power plant.*

The thermal efficiency η_{Th} of the heat engine is defined as,

$$\eta_{Th} = \frac{\text{Energy output}}{\text{Energy input}} \quad \text{or} \quad \eta_{Th} = \frac{\text{Energy sought for}}{\text{Energy that costs}}$$

$$\eta_{Th} = \frac{\text{Work output}}{\text{Heat supplied}} \quad \text{or} \quad \eta_{Th} = \frac{W}{Q_1} = \frac{Q_1 - Q_2}{Q_2}$$

$$\eta_{Th} = 1 - \frac{Q_2}{Q_1}$$

In case of IC engines combustion of air and fuel takes place inside the engine cylinder. The products of combustion will be directly acting on the pistons of the IC engines for producing the power:

Now consider an IC engine as shown in Fig. 4.3.

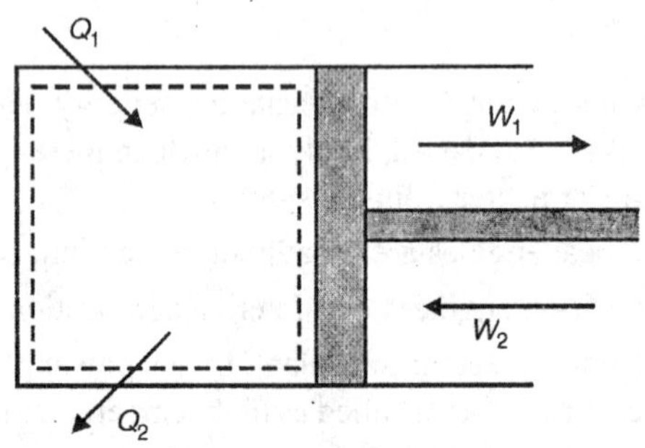

Fig. 4.3

Let Q_1 is the heat transferred to the system
Q_2 is the heat transferred from the system
W_1 is the work done by the system
W_2 is the work done on the system

Then, net heat transfer in a cycle $= Q_{net} = Q_1 - Q_2$ and
Net work transfer in a cycle $= W_{net} = W_1 - W_2$.
According to the first law of thermodynamics,

$$\oint \delta W = \oint \delta Q$$

or $$W_{net} = Q_{net}$$

$\therefore$ We can write,

$$\mathbf{W_1 - W_2 = Q_1 - Q_2}$$

4.4 HEAT RESERVOIRS

"Heat reservoir is a body of infinite heat capacity, which is capable of absorbing or rejecting any quantity of heat without suffering any change in any of its thermodynamic properties".

A heat reservoir which supplies an amount of heat Q_1 to the heat engine operating in heat engine cycle is called the High Temperature Reservoir (HTR) or Source.

The heat reservoir to which an amount of heat Q_2 is rejected by the heat engine is called the Low Temperature Reservoir (LTR) or Sink.

4.5 REFRIGERATOR

"It is a device, which produces and maintains the temperature below atmospheric temperature" or

"It is a device which operating in cycle, maintains an enclosed space at a temperature lower than the temperature of the surroundings".

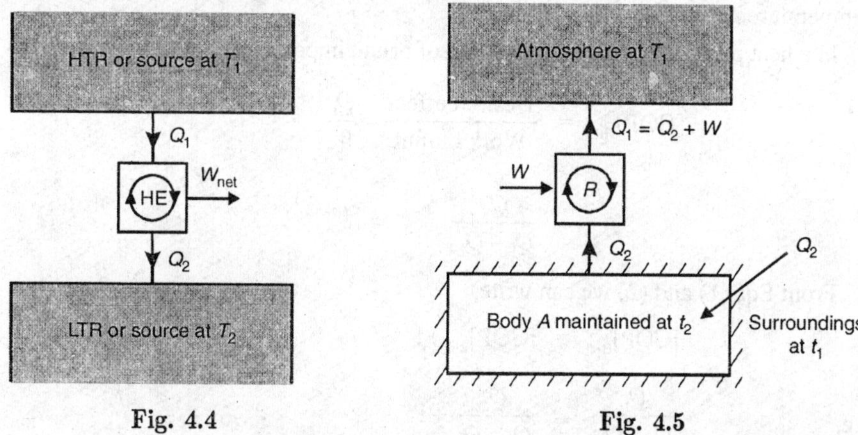

Fig. 4.4 Fig. 4.5

Let a body A is to be maintained at a temperature t_2, which is lower than the atmospheric temperature t_1. Even though the body, A is insulated, there will be a heat leakage Q_2 into the body because of the temperature difference between the body and the surroundings. In order to maintain the body A at constant temperature t_2, same amount of heat is to be removed at the same rate, at which it is leaking into the body.

A refrigerator is a device which operating in a cycle, removes heat Q_2 from the body t_2 and transfers this heat to the atmosphere at t_1 by consuming an amount of work W.

In the refrigeration cycle, body A, Q_2 and W are of prime importance. Just like the efficiency of heat engine cycle, here a parameter called *co-efficient of performance* is important. The copies defined by,

$$[\text{COP}]_{\text{ref.}} = \frac{\text{Desired effect}}{\text{Work output}} = \frac{Q_2}{W} = \frac{Q_2}{Q_1 - Q_2} \quad (1)$$

4.6 HEAT PUMP

It is just opposite to. the refrigerator, i.e. "It is a device operating in a cycle, maintains a body say B at a temperature higher than the temperature of the surroundings".

Because of the temperature difference between the body and the surroundings there will be a heat leakage Q_1 from the body. The body will be maintained at constant temperature t_1, if heat is supplied into the body at the same rate at which heat leaking out of the body.

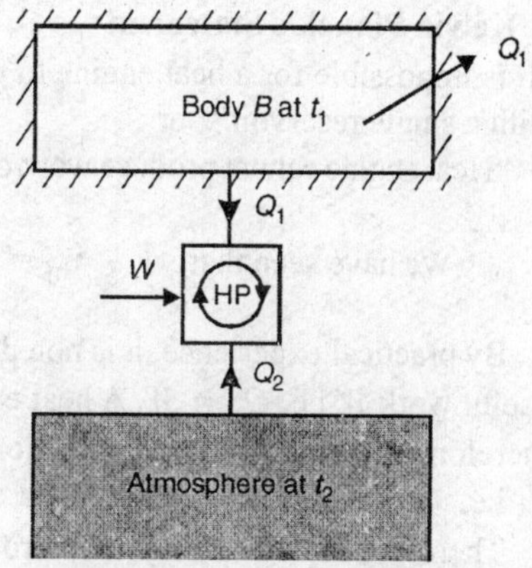

Fig. 4.6. *Heat Pump.*

A heat pump is a device which operating in a cycle extracts heat from the atmosphere and transfers an amount of heat Q_1 to the body B and maintains the body B at constant temperature t_1.

In a heat pump, body B, Q_1 and W are of prime importance.

$$[\text{COP}]_{\text{HP}} = \frac{\text{Desired effect}}{\text{Work output}} = \frac{Q_1}{W}$$

$$= \frac{Q_1}{Q_1 - Q_2}$$

From Eqs (1) and (2) we can write,

$$[\text{COP}]_{\text{HP}} = [\text{COP}]_{\text{Ref}} + 1$$

i.e.

$$\frac{Q_1}{Q_1 - Q_2} = \frac{Q_2}{Q_1 - Q_2} + 1$$

$$\frac{Q_1}{Q_1 - Q_2} = \frac{Q_2 + Q_1 - Q_2}{Q_1 - Q_2}$$

$$\frac{Q_1}{Q_1 - Q_2} = \frac{Q_1}{Q_1 - Q_2}$$

$$\text{LHS} = \text{RHS}$$

$$\mathbf{[COP]_{HP} = [COP]_{ref} + 1}$$

4.7 STATEMENTS OF SECOND LAW OF THERMODYNAMICS

There are two important statements of second law. They are,

1. Kelvin Planck Statement,
2. Clausius Statement.

1. Kelvin Planck's Statement

"It is impossible for a heat engine to produce net work in a cycle, if it exchanges heat with a single reservoir". or

"Heat engine cannot produce work output, if it exchanges heat with a single reservoir".

We have seen that, $\eta_{\text{Th}} = \dfrac{\text{Energy output}}{\text{Energy input}} = \dfrac{W}{Q_1}$

By practical experience, it is noted that total heat supplied cannot be converted into useful work W i.e. $Q_1 \neq W$. A heat engine can never be 100% efficient. But $W < Q_1$, therefore there has to be heat rejection

i.e. $Q_2 > 0$.

Therefore for a heat engine to produce network in a cycle, it has to exchange heat with two reservoirs. (Ref. Fig. 4.7)

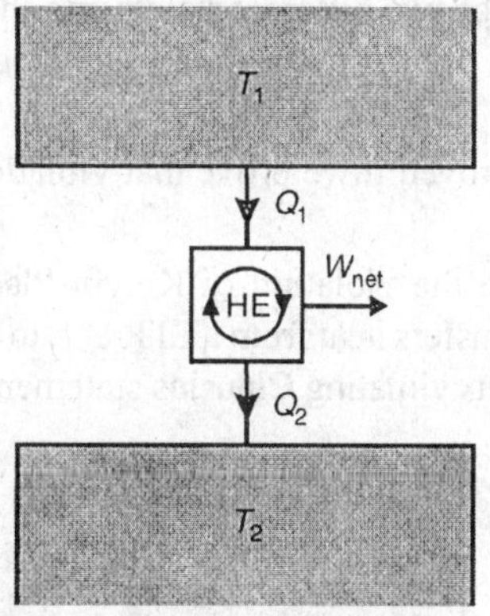

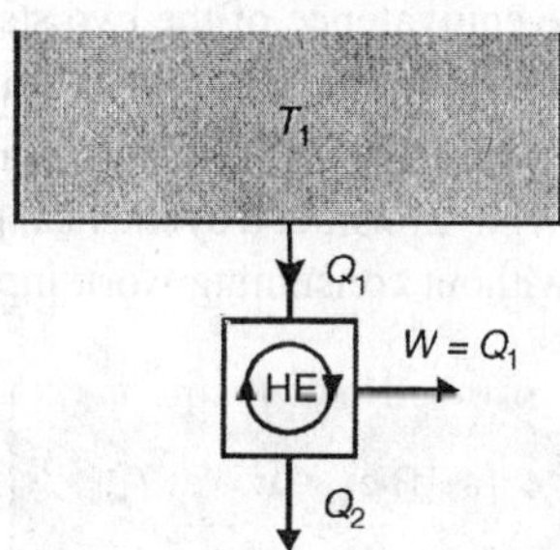

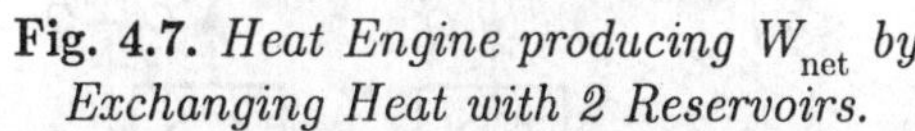
Fig. 4.7. *Heat Engine producing* W_{net} *by Exchanging Heat with 2 Reservoirs.*

Fig. 4.8. *PMM-2.*

But if, $Q_2 = 0$, then $Q_1 = W$ or $\eta_{Th} = 100\%$, the heat engine will produce network in a complete cycle by exchanging heat with only one reservoir, thus it violates Kelvin Planck's statement. Such a heat engine is called a *Perpetual Motion Machine of the second kind. (PMM2)* Ref. Fig. 4.8. A PMM-2 is impossible and it is just a conceptual engine. All the attempts made so far to make such a machine have failed. Thus they show the validity of Kelvin Planck's, statement.

2. Clausius Statement

We know that, heat always flow from a hot body to a cold body. The reverse process never occurs by itself.

Clausius statement is given as, "It is impossible to construct a device, which operating in a cycle will produce no effect other than transfer of heat from a low temperature body to a high temperature body". Or

"Heat cannot flow by itself from a cold body to a hot body, in order to achieve this some work must be expended".

For example, Refrigerator—in this case heat is removed from the cold body *A* and transferred to the atmosphere but by consuming work input *W*.

4.8 EQUIVALENCE OF KELVIN PLANCK AND CLAUSIUS STATEMENTS

We have studied, the Kelvin Plank and Clausius statement as following.

1. Kelvin Planck's Statement. "For a heat engine to produce W_{net} it has to exchange heat with two reservoirs".

2. Clausius Statement. "Heat cannot flow by itself from a cold body to a hot body. In order to achieve this some work must be expended".

At first sight these two statements will appear to be two different statements and are unconnected. But it can be easily shown that, these are the two parallel statements of second law and are equivalent in all respects.

The equivalence of the two statements will be proved if we prove that violation of one statement results into the violation of the other.

(a) Violation of Clausius statement leads into the violation of Kelvin Planck's statement. Consider a cyclic heat pump P, which transfers heat from a LTR at t_2 to HTR at t_1, without consuming work input (i.e. $W = 0$), thus violating Clausius statement.

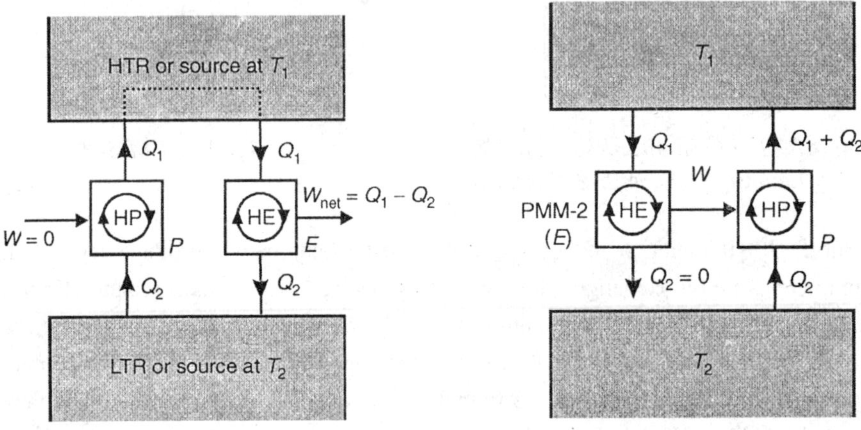

Fig. 4.9. *Violation of Clausius Statement.*

Fig. 4.10. *Violation of Kelvin–Planck Statement.*

Now, let us assume a cyclic heat engine E which also operates between the same two reservoirs at t_1 and t_2 respectively. The rate of working of the heat engine is such that it draws an amount of heat Q_1 from HTR equal to that discharged by the heat pump. Then the HTR may be eliminated and heat Q_1 discharged by the heat pump may be directly fed to the heat engine. So, the heat pump P and heat engine E acting together will form a heat engine, operating in cycles and produce net work by exchanging heat with one reservoir. This violates the Kelvin Planck's statement.

(b) Violation of Kelvin Planck statement leads into the violation of Clausius statement: Let us consider a PMM-2 (E), which produces W_{net} in a cycle by exchanging heat with only one reservoir at t_1 and thus violates Kelvin Planck's statement (Fig. 4.10).

Now, let us assume a cycle heat pump P extracting heat Q_2 from LTR at t_2 and supplying heat to HTR at t_1 by consuming work W equal to that PMM-2 supplies in a complete cycle. So, E and P together will form a heat pump and producing the complete effect of transferring heat from LTR to HTR, without any external aid.

Points to Remember

- At first sight the two statements appear to be different but these two are the two parallel statements and they are equivalent in all respects.

4.9 REVERSIBILITY AND IRREVERSIBILITY

Quasi-means nearly or almost. So, quasi-static process means nearly static process or a process which proceeds with extreme slowness. A quasi-static process proceeds from one equilibrium state to another equilibrium state till the end of process. All the states passed through by the system, during the process are all equilibrium states.

A quasi-static process is represented on the *PV*-diagram by a continuous curve and if we carry out the process in the reverse direction, then it is possible to retrace the same path. Hence it is known as a *Reversible process.*

Whereas in case of *Irreversible process* only end states are equilibrium states and all the intermediate states are non-equilibrium states and is represented by means of a dotted curve as shown in Fig. 4.11.

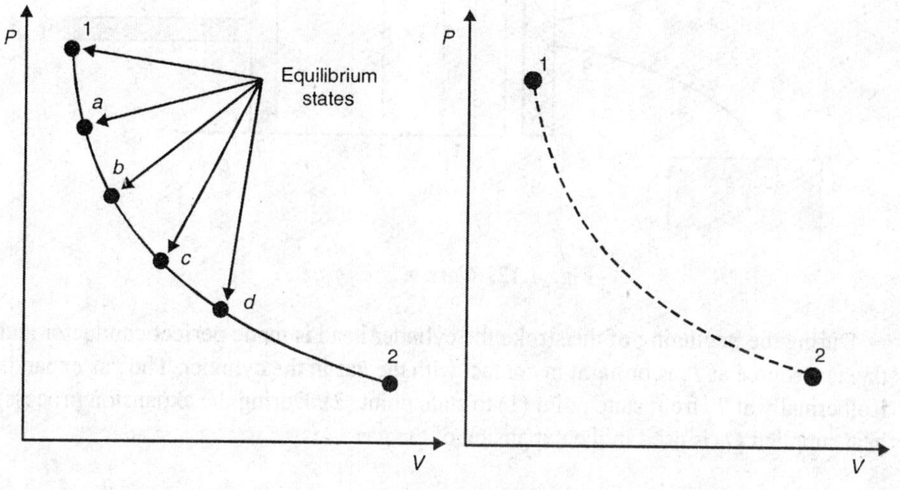

Fig. 4.11

Conditions of Reversibility or Factors Making the Process Irreversible

(i) A process should be quasi-static

(ii) There should be no friction

(iii) Both the system and the surroundings should restore their initial state after the process is reversed.

4.10 CARNOT ENGINE AND CARNOT CYCLE

Carnot, a French Engineer, was the first to introduce the concept of reversible cycle. Carnot devised an engine, which is named after him as Carnot engine. Carnot engine works between high temperature and low temperature reservoirs as shown in Fig. 4.12.

Carnot engine works on carnot cycle. It consists of an alternate series of two reversible isothermal and two reversible adiabatic processes. Since each process is reversible, the Carnot cycle as a whole is reversible.

Carnot cycle (Figs 4.12 and 4.13) is independent of nature of working fluid. It can work with any working substance like gas, vapour or any other working substance. Let us assume that the working fluid of the engine is a gas.

Figure 4.12 shows the proposed reversible engine working between source at T_1 and sink at T_2. It consists of a cylinder, which has a piston working in it without friction. The walls of the cylinder and piston are assumed to be perfect insulators of heat. The cylinder head is so arranged that it can be a perfect conductor and perfect insulator alternatively.

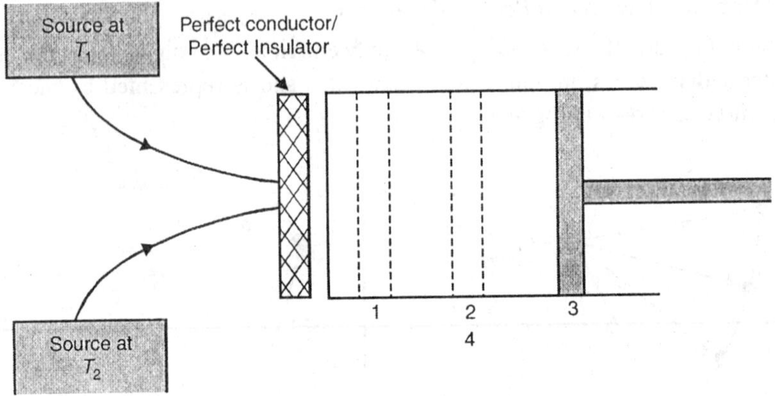

Fig. 4.12. *Carnot's engine.*

During the beginning of the stroke the cylinder head is made perfect conductor and the heat source at T_1 is brought in contact with the gas in the cylinder. The gas expands isothermally at T_1 from state point (1) to state point (2). During the expansion process, heat supplied Q_1 is used in the expansion of the gas.

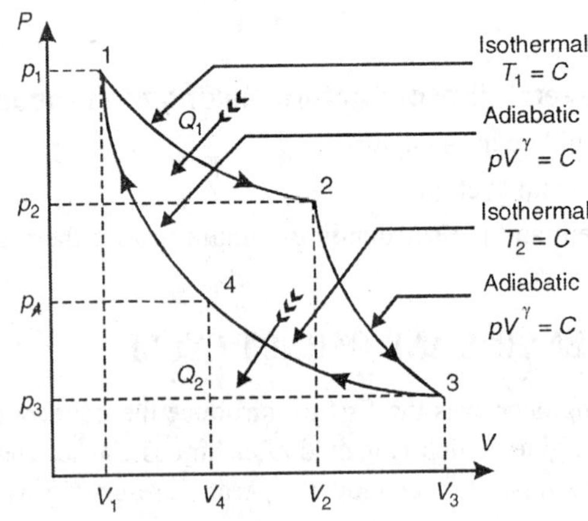

Fig. 4.13

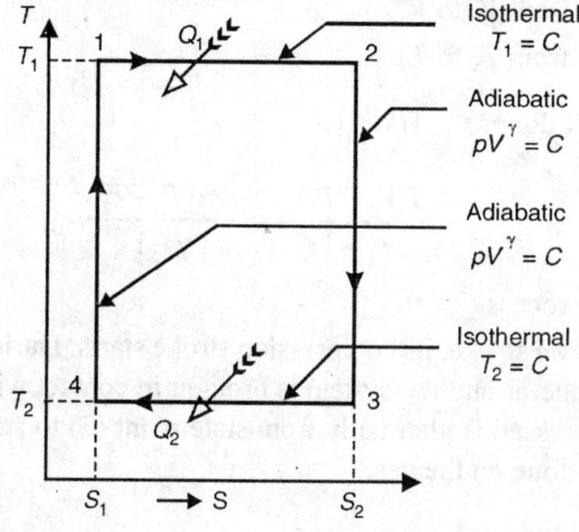

Fig. 4.14

Process 1–2

It is a Isothermal expansion process.

Law for the process is $PV = C$

Pressure falls from P_1 to P_2

Volume increases from V_1 to V_2 and, $T = C$.

$$\text{Work done} = \int_1^2 PdV$$

$$W_{1-2} = P_1V_1 \ln \frac{V_2}{V_1}$$

Since, $P_1V_1 = mRT_1$

$$W_{1-2} = mRT_1 \ln \frac{V_2}{V_1}$$

and For an Isothermal process,

$$Q_1 = W_{1-2} = mRT_1 \ln \frac{V_2}{V_1} \quad (1)$$

(**Note.** For derivation refer Ideal gases work done during a Isothermal process).

When the state point (2) is reached, the cylinder head is made perfect insulator and the gas in the cylinder is allowed to expand till the end of the stroke by following $PV^{\gamma} = C$.

Process 2–3

It is an adiabatic expansion process.

Pressure falls from P_2 to P_3

Volume increases from V_2 to V_3

Temperature falls from T_1 to T_2

$$\text{Work done} = \int P dV$$

$$W_{2-3} = \frac{P_2V_2 - P_3V_3}{\gamma - 1} = \frac{mR(T_1 - T_2)}{\gamma - 1}$$

For an adiabatic process $Q = 0$.

Now after expansion stroke, its compression stroke starts. During this cylinder head is made perfect conductor and the system is brought in contact with the sink at T_2 and the gas is now compressed isothermally from state point (3) to state point (4). During this process work is done on the gas.

Process 3–4

It is an Isothermal compression process.

Pressure increases from P_3 to P_4

Volume decreases from V_3 to V_4 and $T_2 = C$

$$\text{W.D.} = W_{3-4} = P_3V_3 \ln \frac{V_3}{V_4}$$

i.e.

$$= -P_3V_3 \ln \frac{V_3}{V_4}$$

$$W_{3-4} = -mRT_2 \ln \frac{V_3}{V_4}$$

– ve sign implies work is done on the gas during compression.

For isothermal process $Q = W$

$$\therefore \quad Q_{3-4} = Q_2 = -mRT_2 \ln \frac{V_3}{V_4} \tag{2}$$

Again the cylinder head is made perfect insulator and the gas is compressed adiabatically to state point (1).

Process 4–1

It is an adiabatic compression process.

Pressure increases from P_4 to P_1.

Volume decreases from V_4 to V_1

Temperature increases from T_2 to T_1

$$\text{W.D.} = \frac{(P_4V_4 - P_1V_1)}{\gamma - 1} = \frac{-(P_1V_1 - P_4V_4)}{\gamma - 1}$$

and $Q = 0$ for an adiabatic process.

We know that, $\eta_{th} = \frac{\text{Work done}}{\text{Heat applied}} = \frac{W}{Q_1} = \frac{Q_1 - Q_2}{Q_1}$

$\therefore$ From (1) and (2) $\eta_{th} = \dfrac{mRT_1 \ln \frac{V_2}{V_1} - mRT_2 \ln \frac{V_3}{V_4}}{mRT_1 \ln \frac{V_2}{V_1}}$

Note that the ratio $\frac{V_2}{V_1} = \frac{V_3}{V_4}$

$\therefore$ $\eta_{Th} = \dfrac{mR \ln \frac{V_2}{V_1}(T_1 - T_2)}{mR \ln \frac{V_2}{V_1} \cdot T_1}$

$$\eta_{th} \text{ or } \eta_{Carnot} = \frac{T_1 - T_2}{T_1}$$

Note: Carnot cycle is a reversible cycle. If the processes are carried out in anticlockwise direction then it will work as heat pump.

4.11 CARNOT THEOREM

Two cyclic heat engines E_A and E_B operating between the same source and sink out of which E_B is reversible (Fig. 4.15).

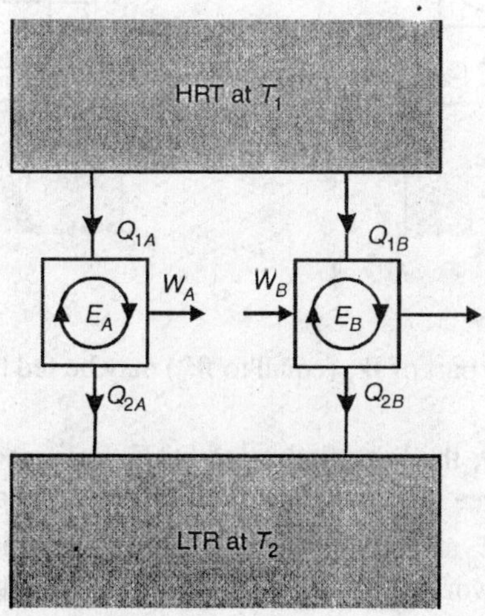

Fig. 4.15

Carnot's theorem states that "*No engine operating between two thermal reservoirs can have an efficiency more than that of a Carnot's engine operating between the same two reservoirs*".

i.e. "*Carnot's engine will be more efficient than any other engine operating between the same temperature reservoirs*".

Let two heat engines E_A and E_B operate between the source at temperature T_1 and the sink at temperature T_2 as shown in Fig. 4.15.

Let E_A be any irreversible engine and E_B be any reversible heat engine. We have to prove that the efficiency of E_B is more than that of E_A. Now, let us assume that this is not true and η_A is greater than η_B. Let the rates of heat transfer are such that,

$$Q_{1A} = Q_{1B} = Q_1$$

and since

$$\eta_A > \eta_B$$

$$\frac{W_A}{Q_{1A}} > \frac{W_B}{Q_{1B}}$$

$$\therefore \quad W_A > W_B$$

Now, let E_B be reversed, since E_B is a reversible heat engine.

Note. Here the magnitudes of heat and work quantities will remain the same, but their directions will be reversed as shown in Fig. 4.16.

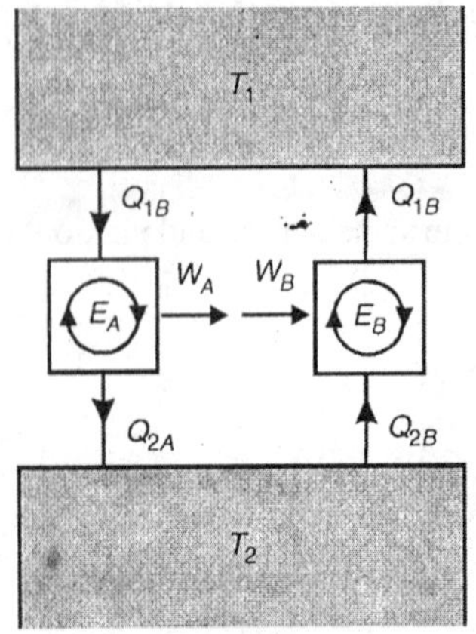

Fig. 4.16

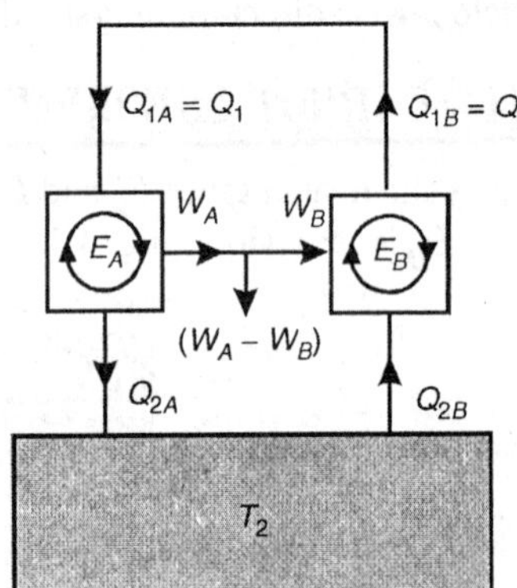

Fig. 4.17

Since $W_A > W_B$, some part of W_A (equal to W_B) may be fed to drive the reversed heat engine E_B.

Since $Q_{1A} = Q_{1B} = Q_1$ the heat discharged by E_B (reversed heat engine B) may be supplied to E_A. The source may therefore be eliminated as shown in Fig. 4.17.

The net result is that E_A and E_B together will constitute a heat engine, which operating in a cycle, produces, network $(W_A - W_B)$ while exchanging heat with a single reservoir at T_2.

Thus violates the Kelvin Planck's statement of the second law. Hence the assumption than $h_A > h_B$ is wrong.

$$\therefore \quad \eta_B \geq \eta_A$$

4.12 REVERSIBLE CYCLE

If a process can occur in a reverse order, and if the initial state all energies transferred or transformed during the process can be completely restored in both system and environment, the process is called *reversible.* If a process really reversible, then no after effects or changes are evident in the system or in the environment when the process occurs in the forward and then in the reverse direction. An irreversible process that is not reversible. A quasi-static process implies an infinitely slow process since all potential differences acting on the system are infinitesimally small. Such a process may be thought of as an infinite succession of equilibrium states. It may be stopped at any time and made to proceed in the opposite direction, thereby reversing the original process in every detail and restoring the system and environment to its original state.

We know that the thermodynamic cycle is defined as a cyclic process end states of which are same. If all the processes in a cycle are reversible process, then the cycle is called a *reversible cycle.*

Reversible process and consequently reversible cycle is an ideal one. There are no losses in the form of friction, temperature differences for heat transfer to take place. Presence of fraction makes the process mechanically irreversible and the temperature difference for heat transfer during the process makes the process thermally irreversible.

Therefore reversible cycle is an ideal cycle.

Engines working on reversible cycle are called reversible engines and we have seen that:

(a) No heat engine operating in cycles between two reservoirs at different temperatures can have a greater efficiency than a reversible engine operating between the same two reservoirs,

(b) All reversible engines operating between two reservoirs at given temperatures have the same efficiency.

Carnot cycle is a reversible cycle and its efficiency depends only on the temperature.

In reversible engines, heat is received while the working substance is at the same constant temperature as the source and heat is rejected while the working substance is at the same constant temperature as the sink.

Points to Remember

$$\eta_{Carnot} = \frac{T_1 - T_2}{T_1}$$

- From this equation we can see that, efficiency can be increased either by increasing T_1 or by decreasing T_2. But in case of IC engines since T_2 is the temperature of exhaust gases, hence it cannot be decreased less than a certain value when high temperatures are used, then the metals used in the construction of engine cylinder, cylinder head etc. will not withstand high temperature stresses and will lose some of the properties. So if we have to increase T_1, then the metals which can withstand high temperature stresses are to be used in the construction of the engine even though they are costly.

(a) It is very very difficult to construct operate and to measure the efficiency of carrot engine. It is not possible to construct such an engine. Since during the same stroke of the piston, isothermal and adiabatic processes are to be achieved.

- For isothermal process the engine has to run very slowly so that the temperature during the process is maintained constant. Such a slow speed is not desirable as the size of the engine operating at very low speed is very large. The adiabatic process needs complete insulation and the engine has to run very fast.
- As in this cycle isothermal and adiabatic processes take place during the same stroke, the piston has to move very slowly for the first half stroke and very fast for the second half stroke, but this is not possible to design.

(b) The cylinder head is perfect conductor for half part of the stroke and is a perfect insulator during the remaining stroke. This is also very difficult to obtain in actual practice.

(c) It is not possible to perform frictionless reversible process.

FORMULAE

1. Thermal efficiency of the heat engine (any engine)

$$\eta_{Th} = \frac{\text{Energy output}}{\text{Energy input}}$$

$$= \frac{\text{Workdone}}{\text{Heat supplied}}$$

$$= \frac{W}{Q_1} = \frac{Q_1 - Q_2}{Q_1}$$

2. $$[\text{COP}]_{ref} = \frac{Q_2}{W} = \frac{Q_2}{Q_1 - Q_2} = \frac{T_2}{T_1 - T_2}$$

3. $$[\text{COP}]_{HP} = \frac{Q_1}{W} = \frac{Q_1}{Q_1 - Q_2} = \frac{T_1}{T_2 T_2}$$

$$\text{Also, } [\text{COP}]_{HP} = [\text{COP}]_{Ref} + 1$$

4. $$\eta_{Carnot} = \frac{T_1 - T_2}{T_1} = 1 - \frac{T_2}{T_1} \quad \text{(a)}$$

$$\text{Also } \eta_{\text{any engine}} = \frac{Q_1 - Q_2}{Q_1} = 1 - \frac{Q_2}{Q_1} \quad \text{(b)}$$

For maximum efficiency

$$\eta_{\text{any engine}} = \eta_{\text{Carnot}}$$

$$\therefore \quad \frac{Q_2}{Q_1} = \frac{T_2}{T_1}$$

5. $$\eta_{\text{Carnot}} > \eta_{\text{any engine}}$$

SOLVED EXAMPLES

Example 4.1 What will be the maximum efficiency of a heat engine operating between 227° C and 27° C?

Data:

$$T_1 = 227 + 273 = 500 \text{ K}$$
$$T_2 = 27 + 273 = 300 \text{ K}$$
$$\eta_{\text{Carnot}} = ?$$

Solution

Maximum efficiency or Carnot engine is given by,

$$\eta_{\text{Carnot}} = \frac{T_1 - T_2}{T_1} = \frac{500 - 300}{500} = 0.4$$

$$\therefore \quad \eta_{\text{Carnot}} = \mathbf{40\%}$$

Example 4.2 What would be the COP of the Carnot refrigerator and heat pump working between temperature limits of 27° C and 600 K?

Data: $T_1 = 600$ K, $T_2 = 273 + 27 = 300$ K

Find: COP of the carnot refrigerator and heat pump.

Solution

We know that,

$$[\text{COP}]_{\text{ref}} = \frac{Q_2}{Q_1 - Q_2} = \frac{T_2}{T_1 - T_2} = \frac{300}{600 - 300}$$

$$\therefore \quad [\mathbf{COP}]_{\mathbf{ref}} = \mathbf{1}$$

$$[\text{COP}]_{\text{HP}} = \frac{Q_2}{Q_1 - Q_1} = \frac{T_2}{T_1 - T_2} = \frac{600}{600 - 300}$$

$$\therefore \quad [\mathbf{COP}]_{\mathbf{HP}} = \mathbf{2}$$

Alternatively

$$[\text{COP}]_{\text{HP}} = [\text{COP}]_{\text{Ref}} + 1 = 1 + 1 = 2$$

Example 4.3 A heat engine operates on Carnot cycle between source and sink temperatures 227°C and 27°C respectively. If the heat engine receives 400 kJ from the source, find the net work done, heat rejected to the sink and efficiency of the heat engine.

Data: $T_1 = 227°C + 273 = 500$ K, $T_2 = 27°C + 273 = 300$ K, $Q_1 = 400$ kJ

Find: $W_{net} = ?$, $Q_2 = ?$, $\eta_{Carnot} = ?$

Solution

(i) $$\eta_{Carnot} = \frac{T_1 - T_2}{T_1}$$

$$= \frac{500 - 300}{500}$$

$$\eta_{Carnot} = \mathbf{0.4 = 40\%}$$

(ii) $$\eta = \frac{\text{Work done}}{\text{Heat supplied}} = \frac{W_{net}}{Q_1}$$

$\therefore$ $$W_{net} = \eta_{Carnot} \times Q_1$$

$$= 0.4 \times 400 \text{ kJ}$$

$$\mathbf{W_{net} = 160 \text{ kJ}}$$

(iii) $$W_{net} = Q_1 - Q_2$$

$$160 \text{ kJ} = 400 - Q_2$$

$\therefore$ $$\mathbf{Q_2 = 240 \text{ kJ}}$$

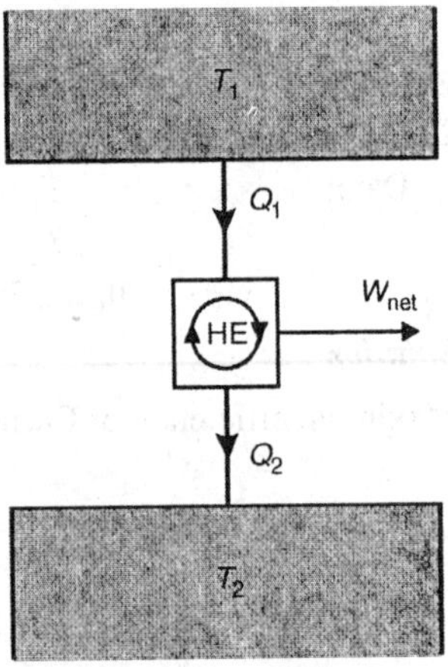

Fig. Ex. 4.3

Example 4.4 A carnot cycle heat engine operates between a source and sink temperatures of 227°C and 27°C. If the heat engine receives 25 kJ of heat from the source, calculate:

1. Work done. 2. Heat rejected. 3. η_{Carnot}

What will be the COP if it operates as a heat pump and when it works as a refrigerator?

Data: $T = 227 + 273 = 500$ K, $T_2 = 27 + 273 = 300$ K, $Q_1 = 425$ kJ

Find: $W_{net} = ?$, $Q_2 = ?$, $\eta_{Carnot} = ?$, $[COP]_{HP} = ?$, $[COP]_{ref} = ?$

Solution

(i) $$\eta_{Carnot} = \frac{T_1 - T_2}{T_1} = \frac{500 - 300}{500}$$

$$\eta_{Carnot} = \mathbf{0.4 = 40\%}$$

(ii) $$\eta = \frac{W_{net}}{Q}$$

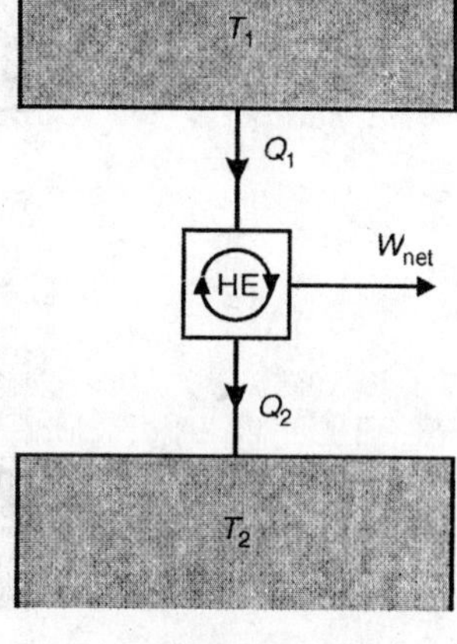

Fig. Ex. 4.4

$$0.4 = \frac{W_{net}}{425}$$

$$\therefore \quad \mathbf{W_{net} = 170\ kJ}$$

(iii)

$$W_{net} = Q_1 - Q_2$$

$$170 = 425 - Q_2$$

$$\therefore \quad \mathbf{Q_2 = 255\ kJ}$$

(iv) When it operates as a heat pump:

$$[COP]_{HP} = \frac{Q_1}{W} = \frac{Q_1}{Q_1 - Q_2}$$

$$= \frac{T_1}{T_1 - T_2} = \frac{500}{500 - 300}$$

$$[COP]_{HP} = \mathbf{2.5}$$

(v) When its operates as a refrigerator :

$$[COP]_{Ref} = \frac{Q_2}{W} = \frac{Q_2}{Q_1 - Q_2} = \frac{T_1}{T_1 - T_2}$$

$$= \frac{300}{500 - 300}$$

$$[COP]_{Ref} = \mathbf{1.5}$$

Example 4.5 A domestic refrigerator maintains a temperature of – 10°C. The ambient (atmospheric) temperature is 35°C. If heat leaks into the freezer at the continuous rate of 1.75 kJ/sec. What is the least power necessary to pump this heat out continuously?

Data: Freezer temperature $T_2 = -10 + 273 = 263$ K, $T_1 = 35 + 273 = 308$ K.

Solution

For minimum power requirements or for maximum COP, we have,

$$\frac{Q_2}{Q_1} = \frac{T_2}{T_1}$$

$$\therefore \quad Q_1 = \frac{T_1}{T_2} \times Q_2 = \frac{308}{263} \times 1.75 \text{ kJ/sec.}$$

$$Q_1 = 2.049 \text{ kJ/sec}$$

$$\therefore \quad W = Q_1 - Q_2 = 2.049 - 1.75$$

$$\mathbf{W = 0.2994\ kJ/sec}$$

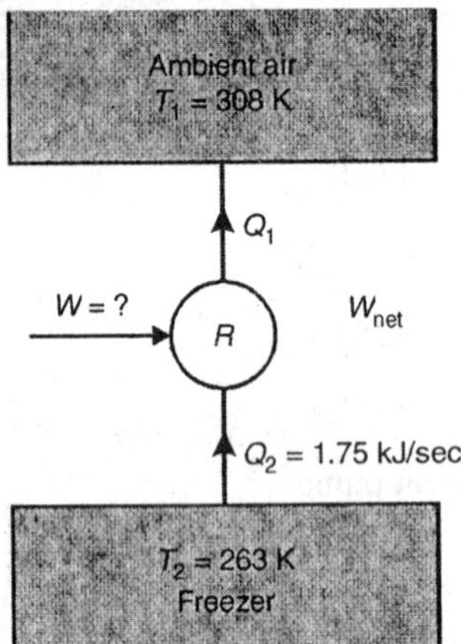

Fig. Ex. 4.5

Example 4.6 A heat pump is used to maintain an auditorium hall at 25°C, when the atmospheric temperature is 10°C. The heat leakage from the hall is 1500 kJ/min. Calculate the power required to run the actual heat pump, if the COP of the actual heat pump is 30% of the COP of the Carnot heat pump working between the same temperature limits.

Data: $T_1 = 25 + 273 = 298$ K, $T_2 = 10 + 273 = 283$ K, heat leakage $Q_1 = 1500$ kJ/min = 25 kJ/sec.

Solution

It is given that,

$$[\text{COP}]_{\text{actual}} = 0.3 \times [\text{COP}]_{\text{Carnot HP}}$$

We know that,

$$[\text{COP}]_{\text{Carnot HP}} = \frac{T_1}{T_1 - T_2} = \frac{298}{298 - 283}$$

$$= 19.867$$

$$\therefore \quad [\text{COP}]_{\text{actual HP}} = 0.3 \times 19.867 = 5.96$$

Also,

$$[\text{COP}]_{\text{HP}} = \frac{Q_1}{W}$$

$$5.96 = \frac{25 \text{ kJ/sec}}{\text{W.D.}}$$

Work done or power required,

$$W = \frac{25}{5.96}$$

$$\mathbf{W = 4.195 \text{ kJ/sec}}$$

Fig. Ex. 4.6

Example 4.7 The C.O.P. of a Carnot refrigerator is 6, when it maintains the temperature of –3°C in the evaporator. Determine the condenser temperature and refrigerating effect if the power required to run the refrigerator is 7.5 kW.

Data: $T_2 = 273 - 3 = 270$ K, $[\text{COP}]_{\text{Ref.}} = 6$, $T_1 = ?$

$$W = 7.5 \text{ kW}$$

Solution

$$\text{Now } [\text{COP}]_{\text{Ref}} = \frac{T_2}{T_1 - T_2} = \frac{\text{Evaporator temprature}}{\text{Condenser temperature} - \text{Evaporator temperature}}$$

$$6 = \frac{270}{T_1 - 270}$$

$$6(T_1 - 270) = 270$$

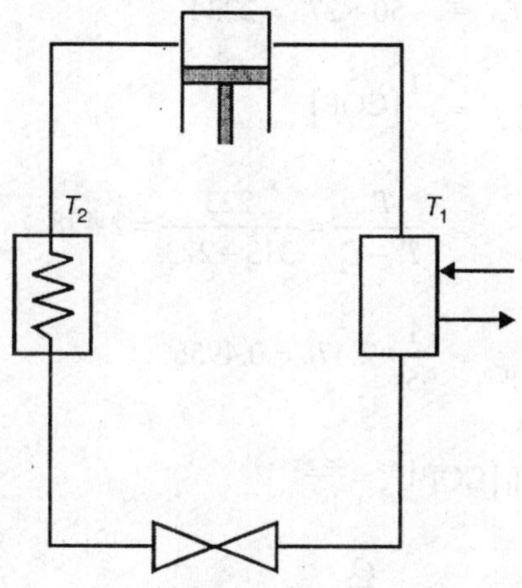

Fig. Ex. 4.7

$$6T_1 = 270 + 6 \times 270$$

$$6T_1 = 270 + 1620$$

$$\therefore \quad T_1 = 315 \text{ K}$$

It is also given that the power required is,
= 7.5 kW = 7500 watts = 7500 J/sec.
We also know that,

$$[\text{COP}]_{\text{Ref}} = \frac{Q_2}{W} = \frac{\text{Desired effect}}{\text{Work input}}$$

$$= \frac{\text{Refrigerating effect/sec}}{\text{Work supplied/sec}}$$

$$6 = \frac{\text{Refrigerating effect}}{7500}$$

$$\therefore \quad \text{Refrigerating effect} = 45000 \ \text{J/sec.}$$

Example 4.8 A fish freezing plant of 100 tonnes capacity is to be maintained at – 50°C when the outside atmospheric temperature is 40°C. The actual C.O.P. of the refrigeration system used is 1/5 of the theoretical Carnot refrigerator working between the same temperature limits. Calculate the power required to run the plant.

Solution

Take 1 tonne of refrigeration in SI units = 3.516 kJ/sec of heat to be removed (Q_2).

$$\therefore \quad 100 \text{ ton of capacity} = 100 \times 3.516 = 351.6 \text{ kJ/sec} = Q_2$$

$$T_1 = 40 + 273 = 313 \text{ K}$$

$$T_2 = -50 + 273 = 223 \text{ K}$$

$$\text{Since} \quad [\text{COP}]_{\text{Ref}} = \frac{1}{5}[\text{COP}]_{\text{Carnot ref.}}$$

$$\therefore \quad [\text{COP}]_{\text{Carnot}} = \frac{T_2}{T_1 - T_2} = \frac{223}{313 - 223} = 2.478$$

$$\text{But} \quad [\text{COP}]_{\text{actual ref.}} = \frac{1}{5} \times 2.478 = 0.4956$$

$$\text{Also we know that, } [\text{COP}]_{\text{Ref}} = \frac{Q_2}{W}$$

$$W = \frac{Q_2}{[\text{COP}]_{\text{Ref}}}$$

$$\mathbf{W} = \frac{\mathbf{351.6}}{\mathbf{0.4956}} = \mathbf{709.44 \text{ kJ/sec.}}$$

Example 4.9 The efficiency of a Carnot engine rejecting heat to a cooling pond at 28°C is 30%. If the cooling pond receives 1050 kJ/min. What is the power developed by the cycle in kW. Also find the temperature of the source.

Data: $T_2 = 28 + 273 = 301$ K, $Q_0 = 1050$ kJ/min = 17.5 kJ/sec., $\eta_{\text{Carnot}} = 0.3$

Find: $W = ?$, $T_1 = ?$

Solution

$$\text{We have,} \quad \eta_{\text{Carnot}} = \frac{T_1 - T_2}{T_1}$$

$$0.3 = \frac{T_1 - 301}{T_1} = 1 - \frac{301}{T_1}$$

$$\frac{301}{T_1} = 1 - 0.3$$

$$\frac{301}{0.7} = T_1$$

$$\therefore \quad \mathbf{T_1 = 430\ K}$$

We also know that, $\eta = \dfrac{Q_1 - Q_2}{Q_1}$

$$= 1 - \frac{Q_2}{Q_1}$$

$$0.3 = 1 - \frac{17.5}{Q_1}$$

$$\frac{17.5}{Q_1} = 1 - 0.3$$

$$\frac{17.5}{0.7} = Q_1 = 25 \text{ kJ/sec} = 25 \text{ kW}$$

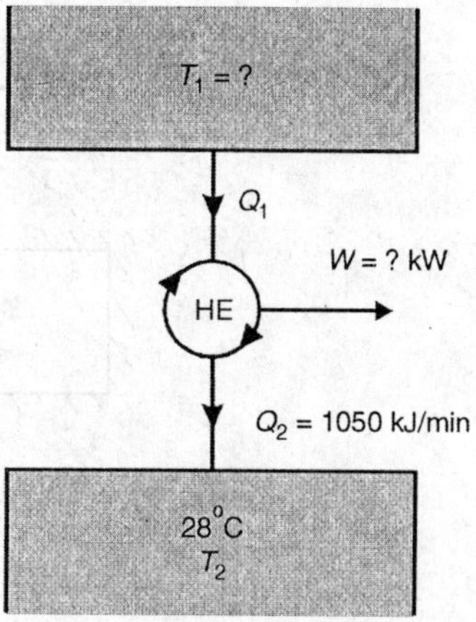

$\therefore$ Power = Rate of doing work,

$$W = Q_1 - Q_2$$

$$\mathbf{W = 25 - 17.5 = 7.5\ kW}$$

Fig. Ex. 4.9

Example 4.10 300 kg of fish at 5° C is to be frozen at – 2°C. Specific heat above freezing point is 3.182 kJ/kg-K and latent heat of fusion is 234.5 (kJ/kg Freezing paint is – 2°C. A refrigerator is to reject heat to surrounding at, 40°C. A refrigerator has actual COP 60% of Carnot refrigerator operating in the same temperature limits. How much power must be supplied to remove the heat in 10 hours.

Solution

We know that,

$$[\text{COP}]_{\text{Ref.}} = \frac{\text{Desired effect}}{\text{Work input}} = \frac{Q_2}{W}$$

$$= \frac{Q_2}{Q_1 - Q_2} = \frac{T_2}{T_1 - T_2} \quad (1)$$

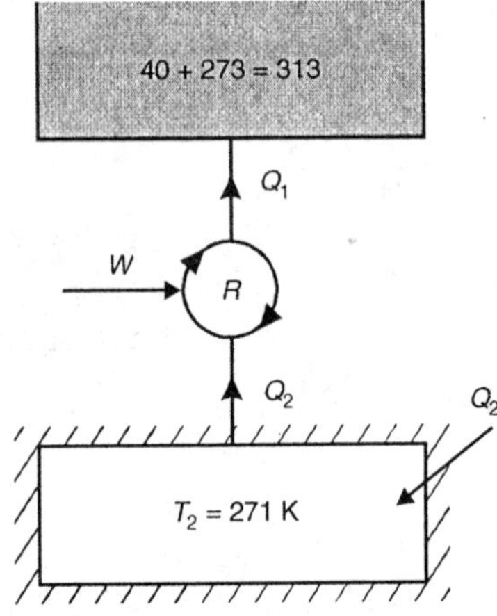

Fig. Ex. 4.10

Also

$$Q_2 = m[C_P \cdot \Delta t + \text{Latent heat of fusion}]$$

$$= m[C_P\, \Delta t + L_f]$$

$$= 300[3.187 \times [5 - (-2)] + 234.5]$$

$$= 300[3.187 \times 7 + 234.5]$$

$$Q_2 = 77032.2 \text{ kJ}$$

$$\eta_{\text{Carnot ref}} = \frac{T_2}{T_1 - T_2} = \frac{271}{313 - 271}$$

$$= 6.452$$

$$\therefore \quad [\text{COP}]_{\text{actual ref.}} = 0.6 \times 6.452 = 3.871$$

Also $$[\text{COP}]_{\text{actual}} = \frac{Q_2}{W} = \frac{77032.2}{W} = 3.871$$

$$\therefore \quad W = 19897.616 \text{ kJ}$$

$$\therefore \text{Rate of work output} = \frac{19897.616}{10 \times 3600}$$

Rate of work input = 0.552 kW

Example 4.11 An engineer claims his engine to develop 3.68 kW, on testing the engine consumes 0.44 kg of fuel in one hour. It has the calorific value (C.V.) of 41,870 kJ/kg. The maximum temperature recorded in the cycle is 1400°C and minimum is 350° C. Is the engineer justified in his claim ?

Solution

$$W = 3.68 \text{ kW} = 3.68 \text{ kJ/sec}$$

$$T_1 = 1400 + 273 = 1673 \text{ K}$$

$$T_2 = 350 + 273 = 623 \text{ K}$$

We know that, $\eta_{Carnot} = \dfrac{T_1 - T_2}{T_1}$

$$= \frac{1673 - 623}{1673} = 0.6276 = 62.76\%$$

For actual engine, heat supplied/hr = C.V. × Mass of fuel consumption

= 41870 kJ/kg × 0.44 kg/hr = 18422.8 kJ/hr.

Work development = 3.68 kW = 3.68 kJ/sec

Work development = 13248 kJ/hr

$\therefore$ Efficiency of actual engine $= \dfrac{\text{W.D.}}{\text{Heat supplied}} = \dfrac{13248}{18422.8}$

Efficiency of actual engine = 71.91%

The actual efficiency of the engine is greater than the efficiency of carnot engine operating between the same temperature limits. So the claim of the engineer is not justified.

Example 4.12 An inventor claims that his newly developed heat engine will produce 0.184 kW power, when the heat is supplied to engine at the rate of 70 kJ/min. The cycle operates between the maximum temperature of 1092°C and minimum temperature of 149°C. Make complete analysis of the inventors claim and clearly state whether it is worth investing any amount of money on such a project. Give reasons for your suggestions.

Solution

$$W = 0.184 \text{ kW power}$$

$$Q = 70 \text{ kJ/min} = \frac{70}{60} \text{ kJ/sec} = 1.17 \text{ kW}$$

$$T_1 = 1092°\text{C} + 273 = 1365 \text{ K}$$

$$T_2 = 149° + 273 = 422 \text{ K}$$

$$\therefore \quad \eta_{Carnot} = \frac{T_1 - T_2}{T_1}$$

$$= \frac{1365 - 422}{1365} = 0.6908 = 69.08\%$$

$$\therefore \quad \eta_{actual} = \frac{\text{W.D.}}{\text{Heat supplied}} \quad \frac{0.184}{1.17}$$

$$= 0.1577$$

$$\eta_{actual} = \mathbf{15.77\%}$$

Since the efficiency of the actual engine is too low there is no point in investing any amount in such a project.

Example 4.13 An inventor claims to have invented refrigeration machine operating between – 23°C and 27°C. It consumes 1 kW electrical power and gives 20000 kJ of refrigerating effect in one hour. Comment on his claim.

Data: $T_1 = 300$ K $\qquad T_2 = 250$ K

$$W_{input} = 1 \text{ kW} \qquad Q_2 = \text{Refrigerating} = \frac{20000}{3600} \text{kJ/sec}$$

$$Q_2 = 5.55 \text{ kJ/sec}$$

Solution

We know that,

$$[\text{COP}]_{\text{Carnot ref}} = \frac{T_2}{T_1 - T_2} = \frac{250}{300 - 250}$$

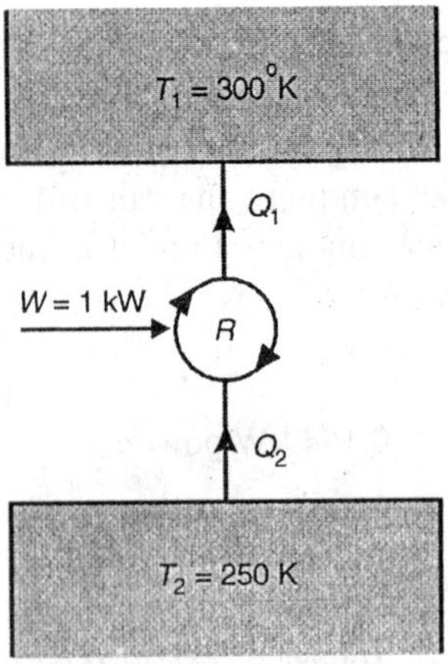

Fig. Ex. 4.13

$$[\text{COP}]_{\text{Carnot ref}} = 5 \quad (1)$$

$$\therefore \quad [\text{COP}]_{\text{actual ref}} = \frac{\text{Refrigerating effect}}{\text{Work input}} = \frac{5.55}{1}$$

$$= 5.55 \quad (2)$$

$$[\text{COP}]_{\text{actual ref.}} > [\text{COP}]_{\text{Carnot ref.}}$$

∴ Claim of the inventor is not valid.

Example 4.14 A reversible engine working in a cycle takes 4800 kJ of heat from a source at 800 K per minute and develops 20 kW power. The engine rejects heat to two reservoirs at 300 K and 360 K. Determine the heat rejected to each sink in kJ/min.

Solution

It is given that,

W.D. by the reversible engine =20 kW = 20 kJ/sec = 20 × 60 kJ/min = 1200 kJ/min.

(Since we have to find heat rejected in kJ/min)

$$Q_1 = 4800 \text{ kJ/min} \qquad \therefore W = Q_1 - Q_2$$

$$Q_2 = Q_1 - W$$

$$Q_2 = 4800 - 1200 = 3600 \text{ kJ/min}$$

Since this heat is rejected to two sinks (a) and (b)

$$\left.\begin{aligned} Q_2 &= Q_{2a} + Q_{2b} \\ \therefore \quad Q_{2b} &= Q_2 - Q_{2a} \\ Q_{2b} &= 3600 - Q_{2a} \end{aligned}\right\} \quad (1)$$

We also know that, $\dfrac{Q_1}{T_1} = \dfrac{Q_2}{T_2}$

$$\therefore \quad \frac{Q_1}{T_1} - \frac{Q_2}{T_2} = 0$$

$$\therefore \quad \frac{Q_1}{T_1} - \left(\frac{Q_{2a}}{T_{2a}} + \frac{Q_{2b}}{T_{2b}}\right) = 0$$

Let Q_{2a} be the heat rejected to sink a

Q_{2b} be the heat rejected to sink b

$$\frac{4800}{800} - \left(\frac{Q_{2a}}{300} + \frac{Q_{2b}}{360}\right) = 0$$

$$\frac{4800}{800}-\left(\frac{Q_{2a}}{300}+\frac{3600-Q_{2a}}{360}\right) = 0 \text{ (From Eq. 1)}$$

$$6-\left[\frac{Q_{2a}}{300}+\frac{3600}{360}-\frac{Q_{2a}}{360}\right] = 0$$

i.e. $$6-\frac{Q_{2a}}{300}-10+\frac{Q_{2a}}{360} = 0$$

i.e. $$-4+\left[\frac{Q_{2a}}{360}-\frac{Q_{2a}}{300}\right] = 0$$

i.e. $$Q_{2a}\left[\frac{1}{360}-\frac{1}{300}\right] = 4$$

$$Q_{2a}\times\left[\frac{300-360}{360\times 300}\right] = 4$$

$$Q_{2a}\times\left[\frac{-60}{108000}\right] = 4$$

$$Q_{2a}[-0.000555] = 4$$

$$\mathbf{Q_2 = -7200.00\ kJ/min}$$

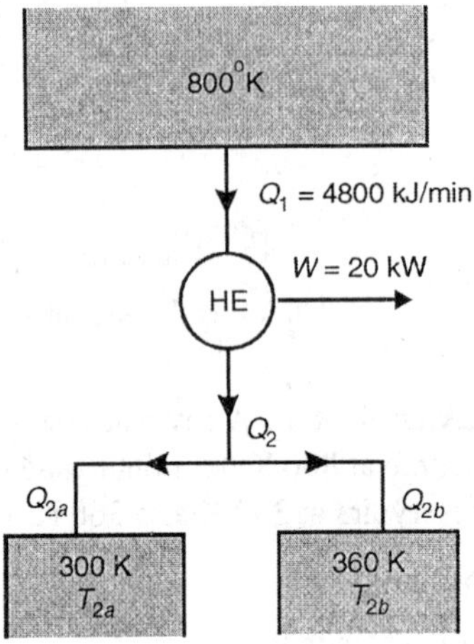

∴ Since $$Q_2 = Q_{2a}+Q_{2b}$$

$$3600 = -7200 + Q_{2b}$$

∴ $$Q_{2b} = 3600+7200$$

$$\mathbf{Q_{2b} = 10800\ kJ/min}$$

Fig. Ex. 4.14

Example 4.15 A reversible engine working on reversible cycle is supplied with 5000 kJ/min of heat from a source at 527°C while it develops 25 kW of power. This engine rejects heat to two reservoirs at two different temperatures equal to 27°C and 97°C respectively. Determine the efficiency of the engine and heat rejected to each reservoir.

Solution

$$T_1 = 527+273$$

$$Q_1 = 5000 \text{ kJ/min}$$

$$W = 25 \text{ kW} = 25 \text{ kJ/sec}$$

$$= 25\times 60 = 1500 \text{ kJ/min}$$

and $$Q_2 = \text{Heat rejected/min}$$

We know that, $W = Q_1 - Q_2$

$\therefore$ $Q_2 = Q_1 - W$

$$Q_2 = 5000 - 1500 = 3500 \text{ kJ/min}$$

$$Q_2 = Q_{2a} + Q_{2b}$$

$$\therefore \left.\begin{aligned} Q_{2a} &= Q_2 - Q_{2b} \\ Q_{2a} &= 3500 - Q_{2b} \end{aligned}\right\} \quad (1)$$

$$\text{Efficiency of the engine} = \frac{\text{W.D.}}{\text{Heat supplied}}$$

$$= \frac{W}{Q} = \frac{1500}{5000} \times 100 = 30\%$$

For reversible engines, we have,

$$\frac{Q_1}{T_1} = \frac{Q_2}{T_2}$$

i.e. $$\frac{Q_1}{T_1} - \frac{Q_2}{T_2} = 0$$

i.e. $$\frac{Q_1}{T_1} - \left[\frac{Q_{2a}}{T_{2a}} + \frac{Q_{2b}}{T_{2b}}\right] = 0$$

$$\frac{5000}{800} - \left[\frac{3500 - Q_{2b}}{370} + \frac{Q_{2b}}{300}\right] = 0$$

$\therefore$ We get, $\mathbf{Q_{2b} = -5090.14}$ **kJ/min**

$$\therefore \quad Q_{2a} = 3500 - Q_{2b} = 3500 + 5090.14$$

$\mathbf{Q_{2a} = 8590.14}$ **kJ/min**

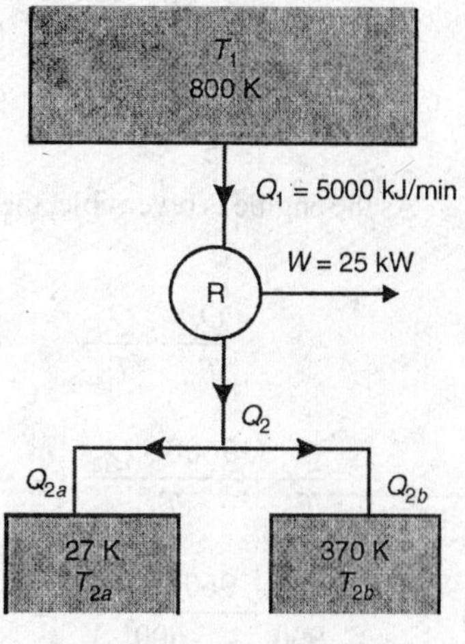

Fig. Ex. 4.15

Example 4.16 A reversible engine is supplied with heat from two constant temperature sources at 900 K and 600 K and rejects heat to a constant temperature sink at 300 K. If the engine executes a number of complete cycles while developing 100 kW, and rejecting 3600 kJ of heat per min. Determine the heat supplied by each source per minute and efficiency of the engine.

Solution

Now the work developed by the engine

$$= 100 \text{ kW}$$
$$= 100 \text{ kJ/sec} = 100 \times 60 \text{ kJ/min}$$
$$= 6000 \text{ kJ/min}$$

We know that, $W = Q_1 - Q_2$

Given $Q_2 = 3600 \text{ kJ/min}$

$\therefore$ $W + Q_2 = Q_1$

$$600 + 3600 = Q_1$$

$\therefore$ $Q_1 = 9600 \text{ kJ/min}$

Assume Q_{1a} is the heat supplied by the source (*a*)

and Q_{1b} is the heat supplied by the source (*b*)

and therefore $Q_1 = Q_{1a} + Q_{1b}$

$\therefore$ $Q_{1b} = Q_1 - Q_{1a}$

i.e. $Q_{1b} = 9600 - Q_{1a}$ (1)

As the engine is reversible, the required condition for the engine to work is, $\dfrac{Q_1}{T_1} = \dfrac{Q_2}{T_2}$

$$\frac{Q_{1a}}{T_{1a}} + \frac{Q_{1b}}{T_{1b}} = \frac{Q_2}{T_2} \quad (2)$$

$$\frac{Q_{1a}}{T_{1a}} + \frac{9600 - Q_{1a}}{T_{1b}} = \frac{Q_2}{T_2}$$

$$\frac{Q_{1a}}{900} + \frac{9600 - Q_{1a}}{600} = \frac{3600}{300}$$

$$\frac{Q_{1a}}{900} + \frac{9600}{600} - \frac{Q_{1a}}{600} = 12$$

$$\frac{Q_{1a}}{900} - \frac{Q_{1a}}{600} + 16 = 12$$

$$Q_{1a}\left[\frac{1}{900} - \frac{1}{600}\right] = -4$$

$$Q_{1a}\left[\frac{600 - 900}{900 \times 600}\right] = -4$$

$$Q_{1a}\left[\frac{-300}{900 \times 600}\right] = -4$$

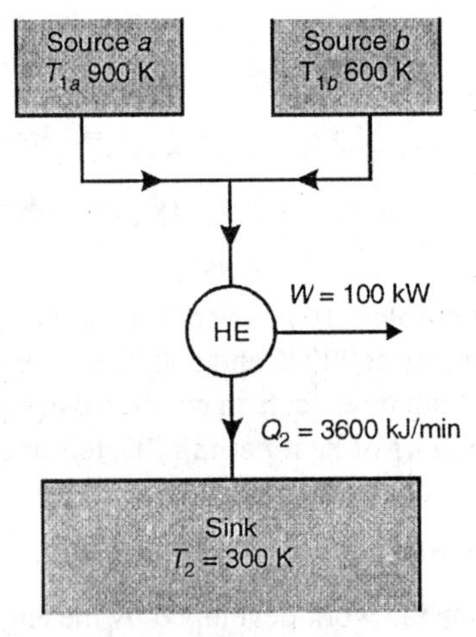

Fig. Ex. 4.16

$$Q_{1a}[-0.000555] = -4$$

$$Q_{1a} = 7200 \text{ kJ/min}$$

$$\therefore \quad Q_{1b} = 9600 - 7200$$

$$Q_{1b} = 2400 \text{ kJ/min}$$

i.e. Heat supplied by the source b

$$\text{Now, The Engine efficiency} = \frac{\text{W.D.}}{\text{Heat supplied}}$$

$$= \frac{6000}{Q_1}$$

$$= \frac{6000}{9600} = 0.625$$

i.e. $\eta = 62.5\%$

Example 4.17 Two reversible engines R_1 and R_2 are connected in series between a heat source S and cold body C. R_1 receives 900 kJ from S, at 540 K and R_2 rejects heat to C at 210 K. Assuming equal thermal efficiencies from R_1 and R_2, determine,

(i) The temperature at which heat is rejected by R_1 and received by R_2.

(ii) The work output W_1 and W_2 of the two engines.

(iii) The heat rejected by R_2 to the cold body C.

Data: $Q_1 = 900$ kJ , $T_1 = 540$ K, $T_2 = 210$ K

Solution

Thermal efficiencies of R_1 and R_2 are equal to

$$\therefore \quad \text{(i)} \quad \eta_{R_1} = \frac{T_1 - T_3}{T_1}$$

$$\text{and} \quad \eta_{R_2} = \frac{T_3 - T_2}{T_3}$$

$$\text{Since} \quad \eta_{R_1} = \eta_{R_2}$$

$$\therefore \quad \frac{T_1 - T_3}{T_1} = \frac{T_3 - T_2}{T_3}$$

$$T_3T_1 - T_3^2 = T_1 T_3 - T_1 T_2$$

$$T_1 T_3 - T_1 T_3 + T_1 T_2 = T_3^2$$

$$\therefore \quad T_3^2 = T_1 T_2 = 540 \times 210$$

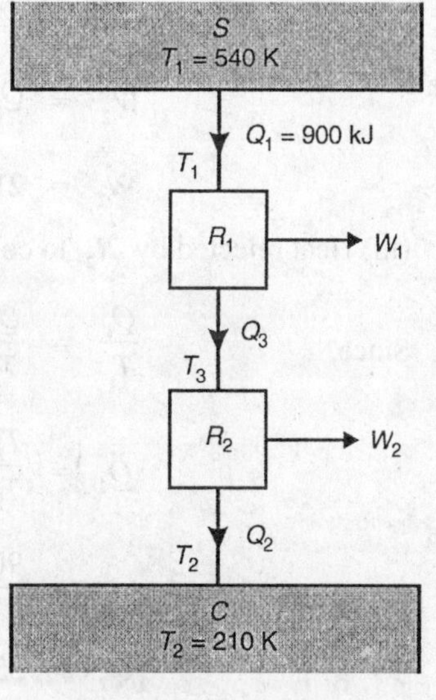

Fig. Ex. 4.17

$$\therefore \quad T_3^2 = 113400$$

$$\therefore \quad T_3 = 336.75 \text{ K}$$

(ii) We know that, $W_1 = Q_1 - Q_3$ (1)

We also know that, $\frac{Q_1}{T_1} = \frac{Q_2}{T_2} = \frac{Q_3}{T_3}$

$$\therefore \quad Q_3 = \frac{T_3}{T_1} \times Q_1$$

Substituting this in Eq. (1)
We get,

$$\therefore \quad W_1 = Q_1 - Q_1 \frac{T_3}{T_1}$$

$$= 900 - 900 \times \frac{336.75}{540}$$

$$\mathbf{W_1 = 338.75\ kJ}$$

Similarly $W_2 = Q_3 - Q_2 = Q_1 \times \frac{T_3}{T_1} - Q_2$

From $\frac{Q_1}{T_1} = \frac{Q_2}{T_2}$

$$Q_2 = \frac{T_2}{T_1} \times Q_1 \text{ (i.e. } Q_2 \text{ in terms of } Q_1\text{)}$$

$$W_2 = Q_1 \times \frac{T_3}{T_1} - Q_1 \times \frac{T_2}{T_1} = 900 \times \frac{336.75}{540} - 900 \times \frac{210}{540}$$

$$\therefore \quad \mathbf{W_2 = 211.25\ kJ}$$

(iii) Heat rejected by R_2 to cold body (Q_2)

Since, $\frac{Q_1}{T_1} = \frac{Q_2}{T_2}$

$$Q_2 = \frac{T_2}{T_1} \times Q_1$$

$$= \frac{900 \times 210}{540}$$

$$\mathbf{Q_2 = 350\ kJ}$$

Example 4.18 A reversible heat engine operates between two reservoirs at temperatures of 600°C and 40°C. The engine drives a reversible refrigerator, which operates between reservoirs at temperature 40°C and – 20°C. The heat transfer to the heat engine is 2000 kJ and the net work output of the combined engine refrigerator plant is 360 kJ.

(a) Evaluate the heat transfer to the refrigerator and the net heat transfer to the reservoir at 40°C.

(b) Reconsider (a) given that the efficiency of the heat engine and the C.O.P. of the refrigerator are each 40% of their maximum possible values.

Solution

(a) Maximum efficiency of the heat engine

$$\eta_{max} = \frac{\text{W.D.}}{\text{Heat supplied}}$$

$$= \frac{W_1}{Q_1} = \frac{T_1 - T_2}{T_1}$$

i.e. $$\eta_{Carnot} \text{ or } \eta_{max} = 1 - \frac{T_2}{T_1} = 1 - \frac{313}{873} = 0.642$$

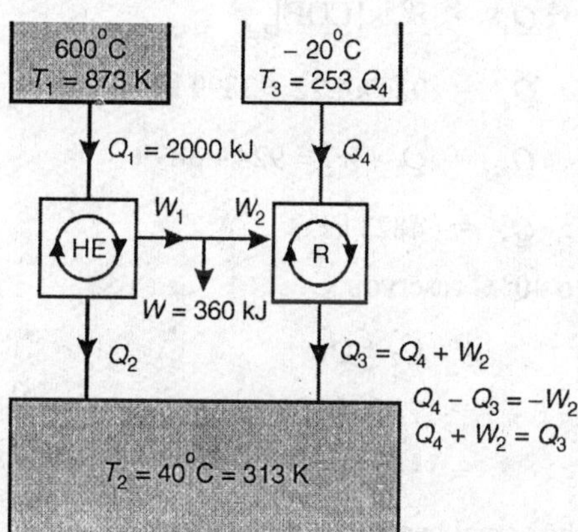

Fig. Ex. 4.18

$\therefore$ From $$\eta_{max} = \frac{W_1}{Q_1} = 0.642$$

$$\therefore \quad 0.642 \times 2000 = W_1$$

$$\therefore \quad 1284 \text{ kJ} = W_1$$

Since $$Q_1 - Q_2 = W_1$$

$$Q_2 = Q_1 - W_1$$

$$Q_2 = 2000 - 1284$$
$$= 716 \text{ kJ}$$

Since $W_1 - W_2 = W = 360 \text{ kJ}$

$$\therefore \quad W_2 = W_1 - W$$
$$= 1284 - 360$$
$$W_2 = 924 \text{ kJ}$$

Now we know that,

$$[\text{COP}]_{\text{Ref}} = \frac{\text{Desired effect}}{\text{Work supplied}} = \frac{Q_2}{W} = \frac{Q_2}{Q_1 - Q_2} = \frac{T_2}{T_1 - T_2}$$

But from Fig. Ex. 18 $[\text{COP}]_{\text{Ref}} = \dfrac{T_3}{T_2 - T_3} = \dfrac{253}{313 - 253} = 4.22$

Also $[\text{COP}]_{\text{Ref}} = \dfrac{Q_4}{W_2}$ (Ref. Fig. Ex. 3.18) $= 4.22$

$$\therefore \quad Q_4 = W_2 \times [\text{COP}]_{\text{Ref}}$$
$$Q_4 = 924 \times 4.22 = 3899 \text{ kJ}$$
$$\therefore \quad Q_3 = Q_4 + W_2 = 924 + 3899$$
$$Q_3 = 4823 \text{ kJ}$$

$\therefore$ Heat rejected to 40° C reservoir

$$= Q_2 + Q_3$$
$$= 716 + 4823$$
$$= 5539 \text{ kJ}$$

(b) Efficiency of the actual heat engine cycle,

$$\eta_{\text{actual}} = 0.4\,\eta_{\text{max}}$$
$$= 0.4 \times 0.642$$
$$= 0.2568$$

Also since $\eta_{\text{actual}} = \dfrac{W_1}{Q_1}$

$$\therefore \quad W_1 = 0.2568 \times 2000$$
$$= 513.6 \text{ kJ}$$

Since $\quad W_1 - W_2 = W$

$\therefore \quad W_2 = W_1 - W$

$$W_2 = 513.6 - 360$$

$$= 153.6 \text{ kJ}$$

and since $\quad Q_1 - Q_2 = W_1$

$\therefore \quad Q_2 = Q_1 - W_1$

$$= 2000 - 513.6$$

$$Q_2 = 1486.4 \text{ kJ}$$

$[\text{COP}]$ of actual refrigerator cycle,

$$[\text{COP}]_{\text{actual ref.}} = \frac{Q_4}{W_2}$$

$$= 0.4 \times 4.22$$

$$= 1.69$$

$\therefore \quad Q_4 = 153.6 \times 1.69$

$$Q_4 = 259.6 \text{ kJ}$$

$\therefore \quad Q_2 = Q_4 + W_2$

$$Q_2 = 159.6 + 153.6$$

$$Q_2 = 413.2 \text{ kJ}$$

$\therefore$ Heat rejected to 40°C reservoir

$$= Q_2 + Q_3$$

$$= 413.2 + 1486.4 = 1899.6 \text{ kJ.}$$

Example 4.19 A heat pump working on the Carnot cycle takes in heat from a reservoir at 5°C and delivers heat to a reservoir at 60° C. The heat pump is driven by a reversible heat engine which takes heat from a reservoir at 840° C and rejects heat at 60°C. The reversible heat engine also drives a machine that absorbs 36 kW. If the heat pum extracts 17 kJ/sec from 5°C reservoir, determine,

(i) The rate of heat supply from 840°C source.

(ii) The, rate of heat rejection to 60°C sink.

Data: $T_1 = 60 + 273 = 333$ K, $T_2 = 5 + 273 = 278$ K, $T_3 = 840 + 273 = 1113$ K.

Solution

$$\text{We know that, } [\text{COP}]_{\text{HP}} = \frac{Q_1}{Q_1 - Q_2} = \frac{T_1}{T_1 - T_2}$$

$$\therefore \quad [\text{COP}]_{\text{HP}} = \frac{T_1}{T_1 - T_2} = \frac{333}{333 - 278} = 6.05$$

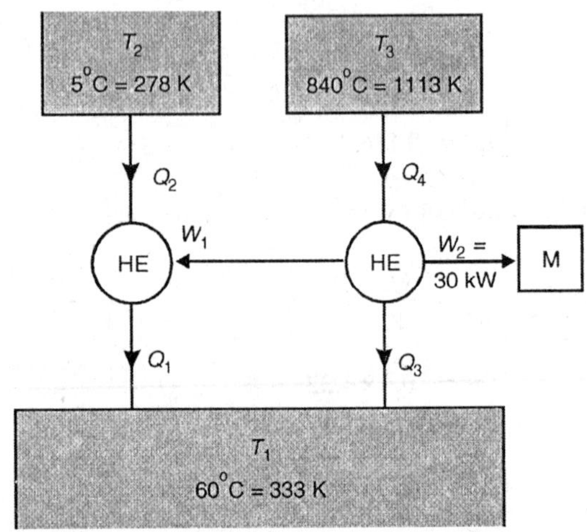

Fig. Ex. 3.19

Also, $$6.05 = \frac{Q_1}{Q_1 - Q_2}$$

$$6.05 = \frac{Q_1}{Q_1 - 17}$$

$$\Rightarrow \quad Q_1 = 20.37\text{kW or kJ / sec.}$$

For the engine $$\eta_{\text{Carnot}} = \frac{T_3 - T_1}{T_3} = \frac{1113 - 333}{1113} = 0.7$$

Also, $$0.7 = \frac{W_1 + 30}{Q_4} \qquad \text{(a)}$$

We know that, $$W_1 = Q_1 - Q_2 = 20.37 - 17$$

$$W_1 = 3.37$$

From (a) and (b) we have,

$$Q_4 = 47.67 \text{ kW}$$

and $$W_E = W_1 + W_2 = 3.37 + 30 = 33.37$$

Also $$W_E = Q_4 - Q_3$$

$$\Rightarrow \quad Q_3 = Q_4 - W_E = 47.67 - 33.37$$

$$\mathbf{Q_3 = 14.3\ kW\ or\ kJ/sec}$$

$\therefore$ Heat rejected to 333 K sink $= Q_1 + Q_3 = 20.37 + 14.3 = 34.67$ kJ

Example 4.20 A heat pump is used to heat a bunglow in the winter and then reversed to cool the bunglow in the summer. The interior temperature is to be maintained at 25°C. Heat transfer through the walls and roof is estimated to be 2800 kJ/hr°C temperature difference between the inside and outside. Estimate:

(i) If the outside temperature in winter is 5°C, what is the minimum power required to drive the heat pump ?

(ii) If the power input is the same as in part (i), what is the maximum outside temperature for which the inside temperature can be maintained at 25°C ?

Solution

(i) In winter : $$T_1 = 25 + 273 = 298\ \text{K}$$

$$T_2 = 5 + 273 = 278\ \text{K}$$

Now heat loss through walls $= 2800$ kJ/hr°C

$$\therefore \quad \text{Net heat loss} = 2800 \times (25 - 5)$$

$$Q_1 = 56000\ \text{kJ/hr.}$$

For minimum power requirements of for maximum efficiency $[\text{COP}]$ of heat pump should be maximum,

$\therefore$ Heat Pump work on reverse Carnot cycle

$$[\text{COP}]_{HP} = \frac{T_1}{T_1 - T_2} = \frac{298}{298 - 278} = 14.9$$

$\therefore$ Also, $$14.9 = \frac{Q_1}{W}$$

$$W = \frac{Q_1}{14.9} = \frac{56000}{14.9} = 3758.45\ \text{kJ/hr.}$$

$$\therefore \quad \text{Minimum Power} = \frac{3758.4}{3600}$$

$$\textbf{Minimum Power} = \mathbf{1.045\ kW}$$

(ii) For summer $$T_2 = 25 + 273 = 298\ \text{K}$$

$$T_1 = \text{outside temprature} > T_2$$

$$Q_2 = \text{Heat to be removed/hr.} = \text{Heat loss/hr}$$

$$Q_2 = 2800\,(T_1 - 298) \text{ kJ/hr.}$$

As the heat pump is reversed to work as refrigerator in the summer,

$$[\text{COP}] = \frac{T_2}{T_1 - T_2} = \frac{Q_2}{W}$$

$$\frac{298}{T_1 - 298} = \frac{2800(T_1 - 298)}{3758.45}$$

$$(T_1 - 298)^2 = 460$$

$$\mathbf{T_1 = 318\ K = 45°C}$$

Example 4.21. Three Carnot engines R_1, R_2, R_3 operate in series between two heat reservoirs which are at temperatures of 1000 K and 300 K as shown.

Calculate intermediate temperatures if amount of work produced by these engines is in the proportions of 5 : 4 : 3.

Solution

From Fig. Ex. 3.21 by applying Carnot's Theorem we can write,

$$\frac{Q_1 - Q_2}{Q_1} = \frac{W_1}{Q_1} = \frac{T_1 - T_2}{T_1} \quad (1)$$

and

$$\frac{Q_2 - Q_3}{Q_2} = \frac{W_2}{Q_2} = \frac{T_2 - T_3}{T_2} \quad (2)$$

From (1) and (2) we get

$$\frac{W_1}{W_2} = T_1 - T_2 \times \frac{Q_1}{T_1} \times \frac{T_2}{Q_2} \times \frac{1}{T_2 - T_3}$$

$$= \frac{T_1 - T_2}{T_2 - T_3}\left(\text{as}\,\frac{Q_1}{T_1} = \frac{Q_2}{T_2}\right)$$

Therefore for Carnot engines operating in series, the temperature difference between the reservoirs, is proportional to the work ratios.

$$\therefore 1000 - T_1 : T_1 - T_2 : T_2 - 300 = 5:4:3$$

$$\therefore \quad \frac{1000 - T_1}{T_1 - T_2} = \frac{5}{4}$$

and $$\frac{T_1 - T_2}{T_2 - 300} = \frac{4}{3}$$

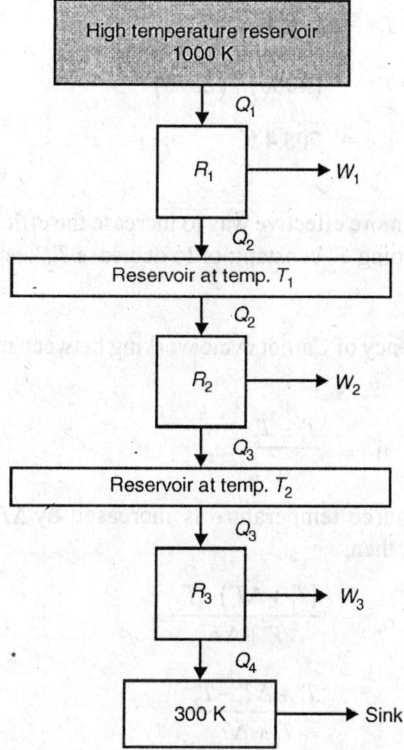

Fig. Ex. 4.21

From Eq. (1), we get,

$$4000 - 4\,T_1 = 5\,T_1 - 5\,T_2$$

or $$4000 - 9\,T_1 = -5\,T_2 \quad (3)$$

Similarly from Eq. (2), we get,

$$3\,T_1 - 3\,T_2 = 4\,T_2 - 1200$$

or $$3\,T_1 = 7\,T_2 - 1200$$

Multiplying by 3 throughout we get,

or $$9\,T_1 = 21\,T_2 - 3600 \quad (4)$$

Adding Eqs (3) and (4), we get,

$$4000 - 9\,T_1 + 9\,T = -5\,T_1 + 21\,T_2 - 3600$$

$\therefore$ $$16\,T_2 = 7600$$

$$\therefore \quad \mathbf{T_2} = \mathbf{475\ K}$$

Substitute $T_2 = 475$ K in Eq. (3) we get,

$$\therefore \quad 4000 - 9\,T_1 = (-5)(475)$$

$$\therefore \quad 9\,T = (4000) + (2375)$$

$$\therefore \quad \mathbf{T_1} = \mathbf{708.4\ K}$$

Example 4.22 Which is more effective way to increase the efficiency of Carnot engine, either to increase T_1 keeping T_2 constant; or to decrease T_2 keeping T_1 constant.

Solution

We know that, the efficiency of Carnot cycle working between the temperature limits T_1 and T_2 is given by,

$$\eta = \frac{T_1 - T_2}{T_1}$$

Consider that the source temperature is increased by ΔT maintaining the sink temperature T_2 constant, then,

$$\eta_1 = \frac{(T_1 + \Delta T) - T_2}{T_1 + \Delta T}$$

$$= \frac{T_1 + \Delta T - T_2}{T_1 + \Delta T} \qquad \text{(a)}$$

Now consider that the sink temperature is decreased by ΔT maintaining source temperature T_1 constant, then,

$$\eta_2 = \frac{T_1 - (T_2 - \Delta T)}{T_1}$$

$$= \frac{T_1 + \Delta T - T_2}{T_1} \qquad \text{(b)}$$

Comparing Eqs. (a) and (b) it is obvious that the numerators of both the equations are same and denominator of Eq. (b) is less than the denominator of Eq. (a) as $T_1 < (T_1 + \Delta T)$

$$\therefore \quad \eta_2 > \eta_1$$

In order to increase the efficiency of Carnot cycle, decreasing to T_2 will be more effective.

Example 4.23 A reversible engine operates between T_1 and T ($T_1 > T$). The energy rejected by this engine is received by a second reversible engine at the same temperature T. The second engine rejects the energy at temperature T_2 ($T_2 < T$). Show that

(i) temperature T is the arithmetic mean of temperature T_1 and T_2 if the engine produces same amount of work output and

(ii) the temperature T is the geometric mean of T_1 and T_2 if the engines have the same cycle efficiency.

Solution

(i) $$\frac{Q_1}{T_1} = \frac{Q_1 - W}{T} = \frac{Q_1 - 2W}{T_2}$$

$$= C \text{ some constant}$$

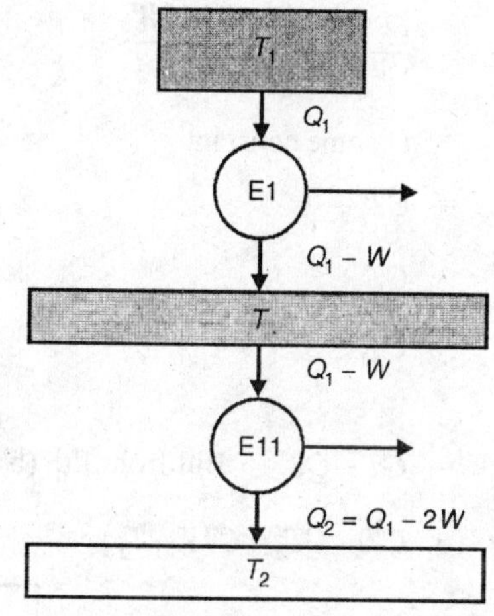

Fig. Ex. 4.23

$$Q_1 = CT_1, Q_1 - W = CT,$$

$$Q_{1-2}\, W = CT_2$$

$\therefore$ $$T = \frac{T_1 + T_2}{2} = \text{Arithmetic mean}$$

(ii) $$\eta = \frac{T_1 - T}{T_1} = \frac{T - T_2}{T}$$

$\therefore$ $$T = \sqrt{T\,T_2} = \text{Geometric mean}$$

Example 4.24 Two heat pumps are connected in series between two heat reservoirs at T_1 and T_2. Heat pump A pumps heat from a reservoir at T_2 and rejects heat to a reservoir at T, while the heat pump B pumps heat from reservoir at T to n reservoir at T_1. If $T_1 > T_2$, show that (a) intermediate temperature T is the arithmetic mean of temperatures T_1 and T_2 if the work input to both the pumps is equal and (b) the temperature is the geometric mean of temperature T_1 and T_2 if the heat pumps have equal C.O.P.

Solution

(a) For equal work inputs, we can write,

$$\frac{Q_2}{T_2} = \frac{Q}{T} = \frac{Q_1}{T_1} \qquad \text{(a)}$$

But $$Q = Q_2 + W$$

and $$Q_1 = Q + W = Q_2 + 2W$$

$$\therefore \quad \frac{Q_2}{T_2} = \frac{Q_2 + W}{T} = \frac{Q_2 + 2W}{T_1}$$

$$= C \text{ some constant}$$

$$\therefore \quad Q_2 = CT_2 \qquad \text{(b)}$$

$$Q_2 + W = CT \qquad \text{(c)}$$

$$Q_2 + 2W = CT_1 \qquad \text{(d)}$$

Now from Eq. (c),

$$W = CT - Q_2 \qquad \text{But from Eq. (b) } Q_2 = CT_2$$

$$\therefore \quad W = CT - CT_2 = C(T - T_2) \qquad \text{(e)}$$

From Eq. (d)

$$2W = CT_1 - CT_2 \qquad \text{(f)}$$

$$W = C\frac{(T_1 - T_2)}{2}$$

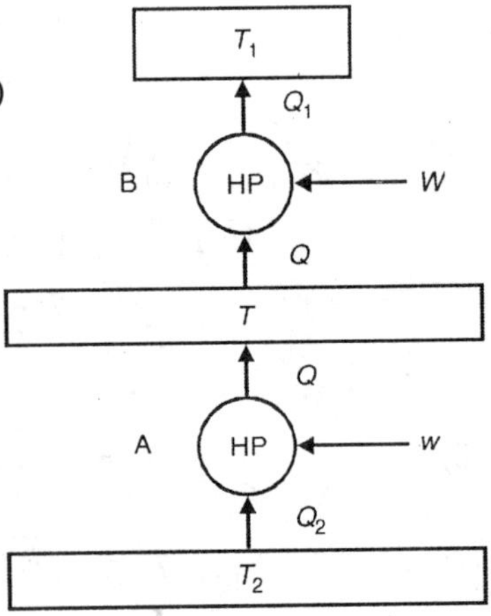

From Eqs (e) and (f)

$$C(T - T_2) = C\frac{(T_1 - T_2)}{2}$$

or $$(T - T_2) = \frac{T_1 - T_2}{2}$$

$$\therefore \quad 2T - 2T_2 = T_1 - T_2$$

Fig. Ex. 3.24

or $$T = \frac{T_1 + T_2}{2} \quad \text{which is arithmetic mean}$$

(b) For equal C.O.P.

$$[\text{C.O.P.}]_{\text{HP(A)}} = \frac{T_1}{T_1 - T}$$

and $$[\text{C.O.P.}]_{\text{HP(B)}} = \frac{T}{T - T_2}$$ and for equal C.O.P.

$$\therefore \quad \frac{T_1}{T_1 - T} = \frac{T}{T - T_2}$$

or $$T = \sqrt{T_1 T_2}$$ which is the geometric mean.

Example 4.25. A reversible engine works between three thermal reservoirs A, B, C. The engine absorbs an equal amount of heat from the thermal reservoirs. A and B kept at temperatures T_A and T_B respectively, and rejects heat to the thermal reservoir C kept at temperature T_C. The efficiency of the engine is μ times the efficiency of the reversible engine, which works between the two reservoirs A and C. Prove that

$$\frac{T_A}{T_B} = (2\mu - 1) + 2(1 - 2\mu)\frac{T_A}{T_C}.$$

Solution

Given that, efficiency of the Engine =
$\mu \times \eta$ of engine working between T_A
and T_C

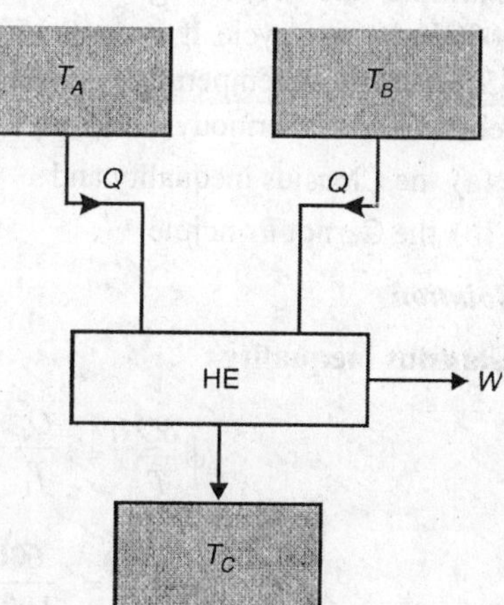

Fig. Ex. 4.25

$$\therefore \text{ Efficiency of engine } A = \frac{T_A - T_C}{T_A}$$

$$\text{and Output of Engine } A = Q \times \frac{T_A - T_C}{T_A}$$

Similarly,

$$\text{Efficiency of Engine } B = \frac{T_B - T_C}{T_B}$$

$$\text{And Output of Engine } B = Q \times \frac{T_B - T_C}{T_B}$$

$$\therefore \quad \text{Net work output} = Q \times \left(\frac{T_A - T_C}{T_A}\right) + Q \times \left(\frac{T_B - T_C}{T_B}\right)$$

and Net input $= Q + Q = 2Q$

$\therefore$ Efficiency $= \dfrac{\text{Net work output}}{\text{Net input}}$

$$= \frac{Q\left(\dfrac{T_A - T_C}{T_A}\right) + Q\left(\dfrac{T_B - T_C}{T_B}\right)}{2Q}$$

$$= \frac{1}{2}\left(1 - \frac{T_C}{T_A} + 1 - \frac{T_C}{T_B}\right)$$

But since $\eta = \mu \times \eta$ of Engine working between T_A and T_C

$$\therefore \quad \frac{1}{2}\left(1 - \frac{T_C}{T_A} + 1 - \frac{T_C}{T_B}\right) = \mu\left(1 - \frac{T_C}{T_A}\right)$$

Multiplying both sides by $\dfrac{T_A}{T_C}$ and rearranging terms we get the required result

$$\frac{T_A}{T_B} = (2\mu - 1) + 2(1 - 2\mu)\frac{T_A}{T_C}.$$

Example 4.26 A heat engine receives 600 kJ of heat from a high temperature source at 1000 K during a cycle. It converts 150 kJ of this heat to net work and rejects the remaining 450 kJ to a low temperature sink at 300 K. Determine if this heat engine violates the second law of thermodynamics on the basis of

(a) the Clausius inequality and

(b) the Carnot Principle

Solution

Clausius inequality :

$$\oint \frac{\delta Q}{T} = \frac{Q_1}{T_1} - \frac{Q_2}{T_2}$$

$$= \frac{600}{1000} - \frac{450}{300}$$

$$= -0.9 \text{ kJ/K}$$

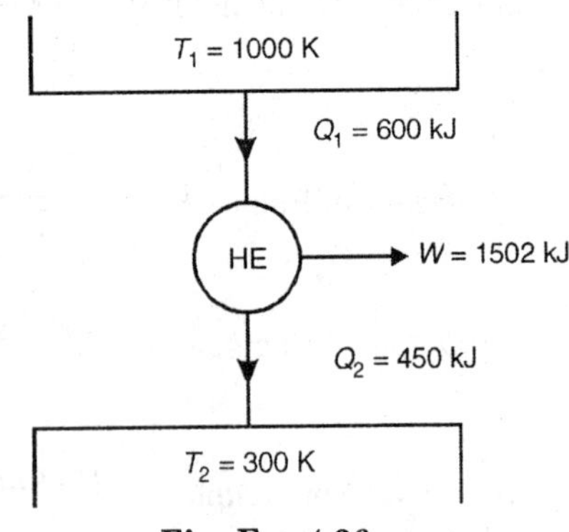

Fig. Ex. 4.26

Since $\oint \dfrac{\delta Q}{T}$ is negative, the heat engine does not violate second law of thermodynamics.

Carnot Principle :

$$\eta_{\text{th (HE)}} = \frac{Q_1 - Q_2}{Q_1} = \frac{150}{600} = 25\%$$

$$\eta_{\text{th (Carnot HE)}} = \frac{T_1 - T_2}{2T_1} = \frac{1000 - 300}{1000} = 70\%$$

Since η_{th} of actual engine working between the same temperature limits is less than theat of Carnot engine, the heat engine does not violate second law of thermodynamics.

Example 4.27. A heat pump is used to maintain a house at a constant temperature of 23°C. The house is losing heat to the outside air through the walls and the windows at a rate of 60,000 kJ/h while the energy generated within the house from people, lights and appliances amounts to 4000 kJ/h. For a COP of 2.5, determine the required power input to the heat pump.

Solution

$$[\text{COP}]_{\text{HP}} = \frac{Q_1}{W}$$

Now,

$$Q_1 = \begin{pmatrix}\text{Heat energy lost through} \\ \text{walls and windows}\end{pmatrix} - \begin{pmatrix}\text{Energy generated} \\ \text{from people, lights etc}\end{pmatrix}$$

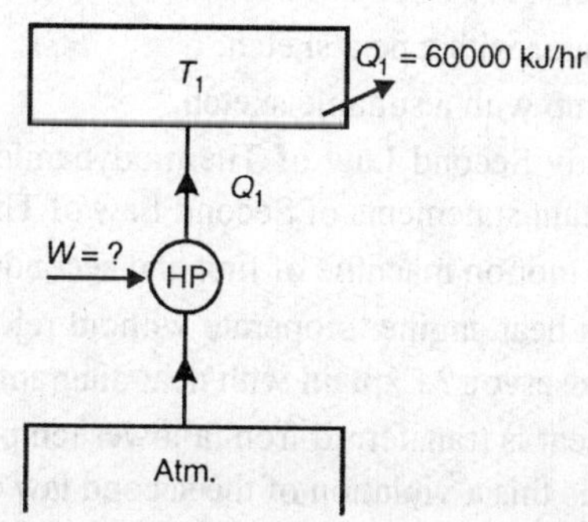

Fig. Ex. 4.27

$$\therefore \quad Q_1 = 60{,}000 - 4{,}000$$

$$= 56{,}0000 \text{ kJ/h}$$

$$= 15.556 \text{ kJ/s}$$

$$\therefore \quad W = \frac{15.556}{2.5} = 6.22 \text{ kW}$$

Example 4.28 A heat pump is used to maintain an auditorium hall at 24°C when the atmospheric temperature is 10°C. The heat lost from the hall is 1500 kJ/min. Calculate the power required to run the heat pump if its COP is 30 percent of Carnot machine, working between the same temperature limits.

Solution

$$COP_{HP(Carnot)} = \frac{T_2}{T_2 - T_1} = \frac{297}{297 - 283}$$

$$= 21.21$$

$$COP_{HP\,(Actual)} = 0.3 \times COP_{HP(Carnot)}$$

$$= 0.3 \times 21.21 = 6.36$$

$$COP = \frac{Q}{W}$$

$$\therefore \quad \mathbf{W} = \frac{\mathbf{1500/60}}{\mathbf{6.36}} = \mathbf{3.928\ kW}$$

1500 kJ/min
T_2 = 297 K
HP
W
T_1 = 283 K

Fig. Ex. 3.28

THEORY QUESTIONS

1. Explain the Limitation of First Law of Thermodynamics.
2. Explain Heat engine and Heat engine cycle.
3. What is a heat reservoir ? Explain its types.
4. Explain a refrigerator with a neat sketch.
5. Explain a heat pump with a suitable sketch.
6. Why we shall study Second Law of Thermodynamics? Give an example and explain the important statements of Second Law of Thermodynamics.
7. Explain perpetual motion machine of first and second kind.
8. Is it possible for a heat engine to operate without rejecting any waste heat to a low temperature reservoir? Explain with neat diagram.
9. In a refrigerator heat is transferred from a lower temperature medium to higher temperature one. Is this a violation of the second law of thermodynamics?
10. Prove that Kelvin Planck and Clausius statements are equivalent in all respects.
11. State and explain Kelvin–Planck constant and Clausius statements of Second Law of Thermodynamics. Prove that violation of Kelvin–Planck statement leads to violation of Clausius statement.
12. What is a Quasi-static process ? Explain reversibility and irreversibility. Enumerate the factors making the process irreversible.
13. Define thermodynamic reversibility. Under what conditions a process is said to be reversible?

14. Explain Carnot Engine and Carnot Cycle. Prove, that $\eta_{\text{Carnot}} = \dfrac{T_1 - T_2}{1}$.
15. State and explain Carnot's theorem.
16. Explain why Carnot cycle cannot be used in practical engine?
17. How the efficiency of IC engines is increased ?

PROBLEMS FOR PRACTICE

1. Two reversible heat engines *A* and *B* are operating between the same temperature limits. Engine *A* develops 37.5 kW with 30% efficiency. The engine *B* receives 7535 kJ/min of heat from source. Determine the power developed, by engine *B* and heat rejected by each engine.

   (Ans. W_B = 37.675 kW, Q_{2A} = 5250 kJ/min Q_{2B} = 5274.5 kJ/min)
2. An inventor of a heat engine claims that his engine is 75% efficient which receives heat at 100°C and rejects at 20°C. How do you rate his claim ? What is the maximum efficiency you would thihk to be realistic ?

   (Ans. Claim is wrong, 21.4%)
3. A heat engine is used to drive a heat pump. The heat transfers from the heat engine and from the heat pump are used to heat the water circulating through the radiators of a building. The efficiency of the heat engine is 27% and the C.O.P. of the heat pump is 4. Evaluate the Patio of heat transfer to the circulating water to the heat transfer to the engine.

   (Ans. 1.81)
4. The C.O.P. of a refrigerator is 6, when it maintains the temperature of 3°C in the evaporator: Determine the condenser temperature and refrigerating effect if the power required to run the refrigerator is 7.5 kW.
5. An engineer claims his engine to develop 3.68 kW on testing, the engine consumes 0.44 kg of fuel in 1 hour. It has calorific value of 41870 kJ/kg. The maximum temperature recorded in the cycle is 1400°C and minimum is 350°C. Is the engineer justified in his claim ?
6. An inventor claims that his newly developed heat engine will produce 0.184 kW power, when the heat is supplied to engine at the rate of 70 kJ/min. The cycle operates between maximum temperature of 1092°C and minimum temperature of 149°C. Make complete analysis of the inventors claim and clearly state whether it is worth investing any amount of money on such a project. Give reasons for your suggestions.
7. A heat engine is used to drive a heat pump. The heat transfer from the heat engine and from the heat pump are used to heat the water circulating through the radiators of a building. The efficiency of the heat engine is 27% and C.O.P. of the heat pump is 4. Evaluate the ratio of heat transfer to the circulating water to the heat transfer to the engine.

8. A reversible heat engine operates between two reservoirs at temperature of 650°C and 35°C. The engine drives a reversible refrigerator. Which operates between temperature of 35°C and – 20°C. The heat transfer to the engine is 1000 kJ/min. The work output (net) of the combined device is 180 kJ/min.

9. Estimate the refrigerating effect in kW and net heat transfer to the reservoir at 35°C. What is the power developed by the engine in kW.

10. A reversible heat engine used for a satellite, operates between a hot reservoir at T_1 and radiating panel- at T_2. The radiation from the panel is proportional to its area and T_2^4. For a given work output and fixed T_1, show that the area of the panel will be minimum when

$$\frac{T_1}{T_2} = \frac{4}{3}$$

11. An inventor claims to have developed a refrigerator machine, which operates between – 20°C and 30°C and consumes and kW of power. The machine gives refrigerating effect of 21.6 MJ/hr. Comment on the claim of the inventor.

12. A heat pump is used to heat at room in winter and to cool in summer. The interior temperature is to be maintained at 20°C. The heat transfer through the wall and roof is 2400 kJ/hr per degree temperature difference between inside and outside.
    (i) If the outside temperature in winter is 0°C, what is the maximum power required to drive the heat pump ?
    (ii) If the power input is the same as that in part (i), what is the maximum outside temperature for which inside can be maintained at 20°C.

    A Carnot engine between 537°C and 37°C develops 5 kW of power. Determine,
    (i) The thermal efficiency.
    (ii) Heat supplied per sec.
    (iii) The change of entropy per second during heat rejection.

13. A heat pump working on Carnot cycle takes in, heat from a reservoir at 5°C and delivers heat to a reservoir at 60°C. The heat pump is driven by a reversible heat engine which takes in heat from a reservoir at 840°C and rejects heat at 60°C. The reversible heat eng 1e also drives a machine that absorbs 30 kW. If pump extracts 17 kJ/sec from 5°C reservoir determine.
    (i) The rate of heat supply from 840°C source.
    (ii) The rate of heat rejection to 60°C sink

14. A Carnot cycle receives heat at 527°C causing increase in entropy equal to 5 kJ/kg K. The engine delivers 2000 kJ/kg of work. Determine the efficiency of the cycle and lowest temperature in the cycle.

15. A small metallic object 4 kg in mass of sp. heat $C_p = 0.5$ kJ/kg-K at a temperature of 227°C is thrown into a lake at temperature of 27°C. Calculate the change in entropy of the universe.

16. An inventor claims that he has designed an engine which will produce a power output of 170 kW while consuming 0.5 kg/min of fuel having C.V. of 42000 kJ/kg. He states that the upper and lower temperature of the engine are 670 K and 230 K respectively. Do you believe his claim? Make your comments:

17. A Carnot engine operates between the temperature limits of 180°C and 60°C and produces 200 kN m/sec of work. Determine
    (i) Heat supplied in kW
    (ii) Change of entropy during heat rejection
    (iii) Thermal efficiency of the engine.

18. Consider two Carnot heat engines operating in series. The first engine receives heat from a reservoir at 1000 K and rejects the waste heat to another reservoir at temperature *T*. The second engine receives this energy rejected by the first one, converts some of it to work and rejects the rest to a reservoir at 300 K. If the thermal efficiencies of both engines are the same, determine the temperature *T.*

19. A house is to be maintained at a temperature of 15°C by means of a heat pump by pumping heat from the atmosphere. Heat losses through the walls are estimated to be 0.75 kW per unit temperature difference between the inside of the house and the outside atmosphere. If the atmosphere temperature is –10°C. Calculate the power required to drive the heat pump if C.O.P. of the actual heat pump is 25% of C.O.P. of Carnot's heat pump working between the same temperature limits.

20. A heat engine receives half of its heat at a temperature of 1000 K and the rest at 5000 K while rejecting heat to a sink at 300 K. What is the maximum possible thermal efficiency of this heat engine ?

21. Source *A* can supply energy at the rate of 12000 kJ /min at 320°C. A second source *B* can supply energy at the rate of 12000 kJ/min at 70°C. Which source *A* or *B* would you choose, to supply energy to an ideal reversible heat engine that is to produce large amount of flow. If the temperature of surroundings is 35°C.

22. A domestic refrigerator maintains a temperature of –12°C. The ambient air temperature is 35°C. If heat leaks into the freezer at the continuous rate of 2 kJ/sec, determine the least power necessary to min the refrigerator.

23. A reversible engine takes 1190 kJ/min from a reservoir at 700 K and develops 190 kJ of work/min when executing complete cycle. The engine rejects heat to two reservoirs at 600 K and 500 K. Find the heat rejected to each sink.

24. A system undergoes a reversible cycle during which it exchanges heat with three thermal reservoirs *A, B, C* at T_1, T_2 and T_3. The amount of heat exchanged is Q_1, Q_2 and Q_3.

Show that

$$\frac{Q_1}{\frac{1}{T_2}-\frac{1}{T_3}} = \frac{Q_1}{\frac{1}{T_3}-\frac{1}{T_1}} = \frac{Q_1}{\frac{1}{T_1}-\frac{1}{T_2}}.$$

25. Two Carnot heat pumps are arranged in series. Pump 1 takes 150 kJ/cycle from heat sink at 300 K. The heat output from this heat pump serves as heat input to pump 2, which delivers its output to a heat reservoir at 1200 K. If the heat pumps have same C.O.P., how much energy (kJ/cycle) must be delieverd to pump 1 and to pump 2.

࿇ ࿈

5

Availability

CHAPTER OBJECTIVES

After reading this chapter you will be able to learn the following

- Grades of Energy, Dead State, Availability of various Systems, Irreversibility and Effectiveness.
- Second Law Efficiency and Helmontz and Gibbs Functions.
- Entropy Change.
- Availability of Heat Reservoirs.

5.1 INTRODUCTION

Energy exists in many forms like kinetic energy, potential energy, electrical energy, heat energy etc. While it is possible to convert most of the energies like potential energy, kinetic energy, electricity etc. into work, only some part of heat energy can be converted into work. That part of energy that can be converted into work is *Available energy* and the remainder is *Unavailable portion.*

Availability is about work potential or quality of the energy. When thermal energy is used to do work entire heat energy cannot be converted to useful work as per the second law of thermodynamics. The part that can be converted to useful work is referred to as the *Available energy* or *Availability* or *Energy*. The quality of energy is measured by Availability. As energy is used in a process it loses quality, its availability or Energy decreases.

5.2 HIGH/LOW GRADE OF ENERGY

Energies like electrical energy or potential energy which can be converted almost entirely to do work are high-grade energy. Heat on the other hand is low-grade energy since only a part of it can be converted into work. Heat is of the lowest grade in which energy exists in a very disordered state. The lower the temperature at which the heat energy exists the lower would be the grade of energy.

However low grade energy is used to produce high-grade energy. For example electricity a high-grade energy is produced in thermal or nuclear power plants using heat a low-grade energy source.

High Grade Energy Sources	**Low Grade Energy Sources**
Mechanical work	Heat
Electricity	Heat derived from combustion
Kinetic energy	Kinetic energy
Tidal power	Heat from nuclear fission

5.3 AVAILABILITY

Available energy is that of energy which is available for doing work. It is either work or that portion of energy that can be converted wholly into work. Concept of availability comes from the Second Law of Thermodynamics where some portion of heat has to be rejected to a sink to produce work. The work has to be with reference to some datum. Generally earth's atmosphere is taken as reference and is called the dead state designated by '0'. Dead state implies the temperature and pressure existing in the surroundings and are designated by T_0 and P_0.

Availability or Exergy is the maximum portion of energy that can be converted into work by ideal process that reduces the system to dead state.

Dead State

Any system that has temperature T and pressure P can do useful work till the temperature and pressure are reduced to T_0 and P_0. When the temperature and pressure are equal to that of the earth or dead state all transfer of energy, stops although the system contains internal energy which would be unavailable. When a system is in equilibrium with the surroundings its potential to do any work ceases. Thus when the pressure P of the system reduces to the atmospheric pressure P_0, temperature T becomes equal to T_0 and likewise the KE and PE become equal to that of the surroundings no more work can be obtained.

The properties of the system in dead state are denoted by subscript 0 i.e. as P_0, T_0, H_0, S_0, U_0, C_0 etc. Generally dead state temperature is taken at 25°C (298 K) and P_0 as 1.01325 bar unless otherwise indicated.

5.4 AVAILABILITY OF VARIOUS SYSTEMS

We can study the availability of

1. Work Reservoir
2. Heat Reservoir
3. Closed System
4. Steady Flow System

(a) Work Reservoir. A work reservoir is a source of infinite work and it would not come to equilibrium with the surroundings. However the amount of W withdrawn from the work reservoir can be converted fully into useful work in the absence of any dissipative effects. Therefore Availability of work reservoir is $A = \Delta W$.

(b) Heat Reservoir.

(i) Infinite Heat Reservoir (at constant temperature).

(ii) Finite Heat Reservoir (where temperature changes).

5.5 IRREVERSIBILITY

Most of the real processes in nature are irreversible due to mixing, friction and heat transfers with finite temperature difference. In a reversible process there is no net increase in the entropy of the universe. In a reversible process the change in the entropy of a system equals the change in the entropy of the universe. However in the case of an irreversible process the reduction in the entropy of a system is less than the increase of entropy of the heat receiving system. This leads to loss of availability or an equivalent increase in unavailability—which however are equal. In any process, aim is always to get maximum work. In an expansion process we try to get the maximum work output. In compression process aim is to have minimum work input. Thus both in expansion as well as in compression processes aim is to maximize the work as per sign convention of work.

Maximum work (W_{rev}) during a process between two states can be obtained by reversible process. However in actual practice the actual work (W_{act}) between these two states is always less than the W_{rev}. The difference between the two is called Irreversibility.

$$I = W_{rev} - W_{act}$$

In irreversibility (I)/Exergy destruction = wasted work potential in a process. It is lost opportunity to do work. The greater the irreversibility the greater is the loss of work that could have been performed. Thus irreversibility gives the quality of the process.

Let there be two processes, one reversible and another irreversible between two states and let heat δQ_{rev} and δQ be added giving work output of W_{rev} and W_u respectively.

Thus $$\delta Q_{rev} = dE + \delta W_{rev} \qquad (1)$$

and $$\delta Q = dE + \delta W \qquad (2)$$

$$\delta I = (1) - (2)$$

$$\delta I = \delta Q_{rev} - \delta Q = TdS - \delta Qs$$

$$= \delta W_{rev} - \delta W$$

Integrating $$I = W_{rev} - W_{act} = \text{Lost work}$$

5.6 EFFECTIVENESS

The effectiveness of a system is the ratio of the useful or actual work done to the maximum or reversible work.

Thus, $$\text{Effectiveness} = \delta W / \delta W_{rev}$$

5.7 SECOND LAW EFFICIENCY

So far we have been using the efficiencies and COP based on the First Law of Thermodynamics. Based on the increasing use of availability which gives indication about possibility of maximum work output or minimum work input if the process is carried out reversibly — concept of Second Law efficiency is being used these days. Second law efficiency is a measure of reversible operation.

Let us assume that an engine operating between say 1200 K and 300 K has – η_{Th} of 30% (by measuring brake power and dividing by the heat input—based on First Law of Thermodynamics).

However, had the work been done in a reversible manner i.e. by Carnot Cycle.

$$\eta_{rev} = \left(1 - \frac{T_0}{T}\right)$$

$$= (1 - 300/200) = 0.75 = 75\%$$

The η_{Th} is far less than the η_{rev}. We can infer that η_I is not realistic measure of the performance. To overcome we define η_{II} for various heat-work machines

$$\eta_{II} = \eta_{th}/\eta_{rev} \text{ (For heat engines)}$$

$$\eta_{II} = W/W_{rev} \text{ (For power or work producing devices)}$$

$$\eta_{II} = W_{rev}/W \text{ (For power absorbing devices)}$$

where W = useful work and

W_{rev} = maximum or reversible work

$$\eta_{II} = \text{COP}/\text{COP}_{rev} \text{ (For refrigerators and heat pumps)}$$

5.8 IMPORTANCE OF AVAILABILITY/IRREVERSIBILITY/ EFFECTIVENESS/η_{II}

These concepts are based on the second law of thermodynamics and are applicable to both cycle and process. These indicate the departure of an actual process from an ideal or reversible process.

5.9 HELMONTZ AND GIBBS FUNCTIONS

Both these functions are properties (based on combination of properties just like $h = u + pv$.

Helmontz Function $F = U - TS$

or $f = u + Ts$ (per unit mass)

This function is useful for closed systems that undergo reversible, isothermal process. Decrease in the value of F equals the max work that can be done by system undergoing an isothermal process at temperature equal to that of the surroundings.

5.10 GIBBS FUNCTION

$$G = H - Ts$$

or $g = h - Ts$ (For unit mass)

5.11 SUMMARY

Availability or Exergy is the maximum useful work that can be obtained from a system under ideal conditions by reducing it to the dead state. Availability A of various systems are

(a) Heat Reservoir $A = \Delta Q - T_0 \ \Delta S$

(b) Closed System $A = A - A_0 = (U - T_0 S + P_0 V) - (U_0 - T_0 S_0 - P_0 + P_0 V_0)$

(c) Steady Flow System $A =$

$$A - A_0 = m\left\{\left(h - T_0 S + \frac{C^2}{2} + gZ\right) - (h_0 - T_0 S_0 + \frac{C_0^2}{2} + gZ_0)\right\}$$

Irreversibilty = $W_{rev} - W = T_0 - T_0 \Delta S$

Effectivness = W/W_{rev}

5.12 ENTROPY CHANGE ΔS

The unavailable energy $UA = T_0 \Delta S$. Thus change in entropy during a process is required to find out avilability and unavailability.

From Ist Law

$$\delta Q = \delta u + \delta w$$

$$\delta Q = Au + Pdv$$

Divide by T

$$\frac{\delta Q}{T} = \frac{du}{T} + Pdv$$

$$= C_v \frac{dt}{T} + R\frac{dv}{V} \qquad \left(\text{Since} \frac{P}{T} = \frac{R}{V}\right)$$

But

$$\frac{\delta Q_{rev}}{T} = ds$$

$\therefore$

$$ds = C_V \frac{dT}{T} + R\frac{dv}{V}$$

Integrating between 1 and 2

$$s_2 - s_1 = C_V \ln\frac{T_2}{T_1} + R\frac{V_2}{V_1}$$

$$\Delta S = C_V \ln\frac{P_2 V_2}{P_1 V_1} + R\ln\frac{V_2}{V_1} \qquad (1)$$

$$= C_V\left(\ln\frac{P_2}{P_1} + \ln\frac{V_2}{V_1}\right) + R\ln\frac{V_2}{V_1}$$

$$= C_V\left(\ln\frac{P_2}{P_1}\right) + \ln\frac{v_2}{v_1} + (R + C_V)$$

But $$C_V + R = C_p$$

$$\therefore \quad \Delta S = C_V \ln\frac{P_2}{P_1} + C_p \ln\frac{V_2}{V_1} \quad (2)$$

We can also put $\dfrac{v_2}{v_1} = \dfrac{P_1T_2}{P_2T_1}$ in Eq. (1)

$$\therefore \quad \Delta S = C_V \ln\frac{T_2}{T_1} + R\ln\frac{P_1T_2}{P_2T_1}$$

$$= C_V \ln\frac{T_2}{T_1} + R\left(\ln\frac{P_1}{P_2} + \ln\frac{T_2}{T_1}\right)$$

$$= (C_V + R)\ln\frac{T_2}{T_1} + R\ln\frac{P_1}{P_2}$$

$$\Delta S = C_p \ln\frac{T_2}{T_1} + R\ln\frac{P_1}{P_2} \quad (3)$$

Entropy Changes for Various Processes are

(1) Constant volume $\quad s_2 - s_1 = C_V \ln\dfrac{T_2}{T_1}$

(2) Constant pressure $\quad s_2 - s_1 = C_P \ln\dfrac{T_2}{T_1}$

(3) Iso thermal $\quad s_2 - s_1 = R\ln\dfrac{V_2}{V_1}$

(4) Adiabatic $\quad s_2 - s_1 = 0$

(5) Polytropic $\quad s_2 - s_1 = C_V\left(\dfrac{n-r}{n-1}\right)\ln\dfrac{T_2}{T_1}$

5.13 AVAILABILITY OF HEAT RESERVOIR

(a) Infinite Heat Reservoir. In such a reservoir temperature remains constant when small amount of heat δQ is withdrawn from it.

As shown in Fig. 5.1, let a small quantity of heat δQ be withdrawn from the heat reservoir which is at average temperature T. For getting maximum work, let there be a Carnot engine operating between temperatures T and T_0 (sink temperature being of the surroundings).

$$\frac{\delta W}{\delta Q} = \left(1-\frac{T_0}{T}\right)$$

or
$$\delta W = \left(1-\frac{T_0}{T}\right)(-\delta Q) = T_0\frac{\delta Q_{rev}}{T}$$

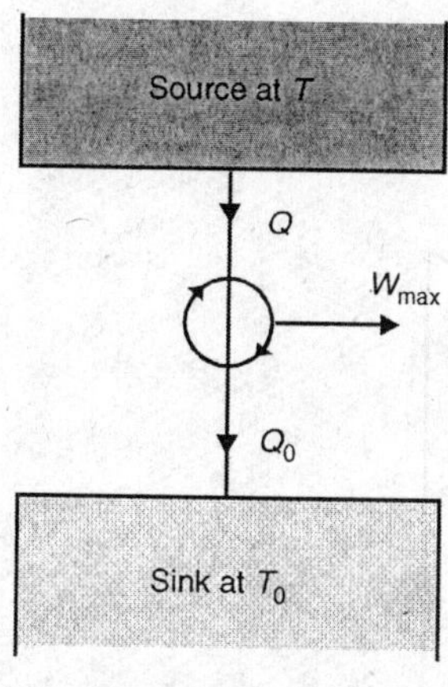

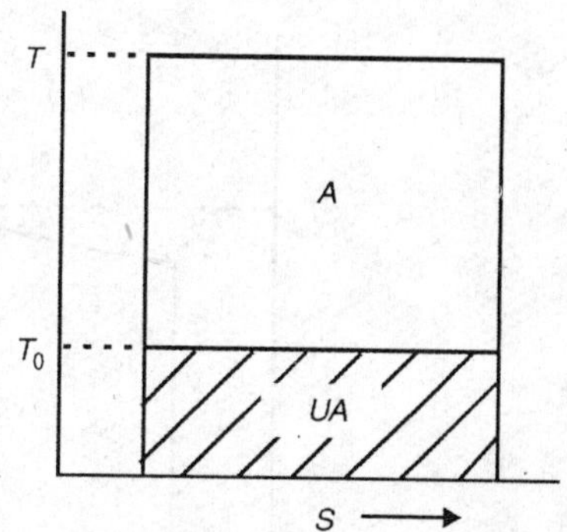

Fig. 5.1

Integrating,

$$W_{max} = \int_{T_0}^{0} = A = -\delta Q - \int_{0}^{1} T_0\,dS$$

$$= Q + T_0\,(s_0 - s_1)$$

$$= Q - T_0\,(S_1 - S_0)$$

Thus
$$W_{max} = Q - T_0\,\Delta S$$

$$= \text{Heat Added} - T_0\,(\text{change in entropy})$$

(b) Finite Heat Reservoir. In this case when heat is added or withdrawn from the reservoir which has finite capacity the temperature does not remain constant.

Consider a small quantity δ heat being withdrawn from the reservoir (1) at temperature T and producing entropy change dS as shown. Using a Carnot engine to derive maximum work between T and T_0; we have

$$\delta W'_{Carnot} = \left(1-\frac{T_0}{T}\right)(-\delta Q)$$

Integrating between 0 and 1 we would get

$$W_{max} = A = \int_0^1 (-\delta Q) + \int_0^1 T_0 \frac{\delta Q_{rev}}{T}$$

$$= \int_0^1 (-\delta Q) + \int_0^1 T_0 dS$$

$$= Q - T_0 (s_1 - s_0)$$

$$= \text{Heat Added or withdrawn} - T_0 \Delta S$$

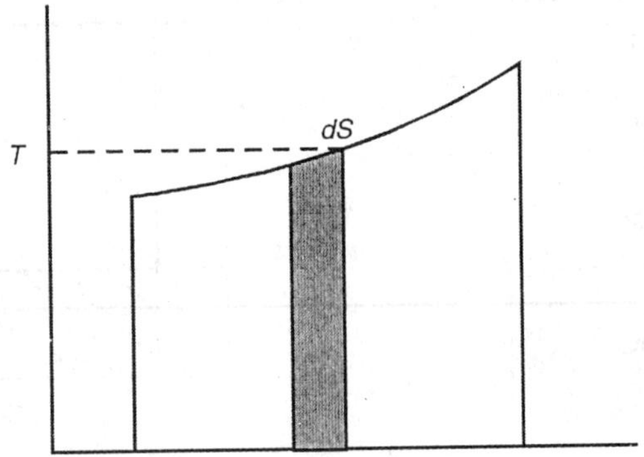

Fig. 5.2

$$A = Q - T_0 \Delta S$$

$$UA = T_0 (s_1 - s_0) = T_0 \Delta S$$

For the heat reservoir where temperature does not remain constant we would use appropriate equation for entropy change as applicable.

SOLVED EXAMPLES

Example 5.1 0.5 kg of air at 500 K is heated reversibly at constant pressure to 2000 K. Find the available and unavailable energies and the heat added. Take $T_0 = 300$ K and C_P = 1.005 kJ/kg-K.

Solution

$$\Delta Q = m C_P (T_2 - T_1)$$

$$= 0.5 \times 1.005 (2000 - 500)$$

$$= 753.75 \text{ kJ}$$

Entropy changes at constant pressure

$$\Delta S = mC_P \ln\frac{T_2}{T_1}$$

$$= 0.5\times1.005\ln\frac{2000}{500}$$

$$= 0.6966 \text{ kJ/kg-K}$$

$$UA = T_0\times\Delta S = 300\times0.6966$$

$$= 208.98 \text{ kJ}$$

$$\mathbf{A} = \mathbf{753.75 - 208.98 = 544.77 \text{ kJ}}$$

Example 5.2 One kg of air is heated from 70°C to 270°C at constant pressure. Determine (a) Heat supplied (b) Entropy change (c) Available energy for sink temperature of 20°C and – 20°C. Assume C_P = 1.005 kJ/kg-K for air.

Solution

$$T_1 = 273+70 = 343 \text{ K}$$

$$T_2 = 543 \text{ K}$$

(a) Heat supplied $= mC_P\Delta T$

$$= 1\times1.005\times(270.70) = 201 \text{ kJ}$$

(b) ΔS at const. pressure $= m\,Cp\ln\frac{T_2}{T}$

$$= 1\times1.005\ln\frac{543}{343}$$

$$= 0.46167 \text{ kJ/kg-K}$$

(c) (i) Available energy for sink temperature and $20+273 = 293$ K

$$UA = T_0\Delta S$$

$$= 293\times0.808 = 135.27 \text{ kJ}$$

$$A = \Delta Q - UA = 201-135.27 = 65.73 \text{ kJ}$$

(ii) Available and unavailable energy at sink temperature of

$$273-20 = 253 \text{ K}$$

$$T_0 = 253 \text{ K}$$

$$UA = T_0\Delta S = 116.80$$

$$\mathbf{A} = \mathbf{201-116.80 = 84.2 \text{ kJ}}$$

Example 5.3 100 kg of water at 100°C is mixed with 50 kg of water at 50°C, while the surrounding temperature is 27°C. Determine the decrease in availability due to mixing.

Data: $T_0 = 273 + 27 = 300$ K; $m_1 = 100$ kg

$T_1 = 273 + 100 = 373$ K; $m_2 = 50$ kg

$T_2 = 273 + 50 = 323$ K

Solution

Temperature after mixing will be

$$(100+50)\,t_m = 100\,t_1 + 50\,t_2$$

$$= \frac{100\times100+50\times50}{150} = \frac{2\times100+50}{3}$$

$$t_m = \frac{250}{3} = 83.33°\text{C}$$

$$T_m = 273+83.33 = 356.33\text{ K}$$

(c) Heat given by hot water at 100°C(373 K)

$$Q = m_1\times C_P\times(100-83.33)$$

$$= 100\times4.187\times16.67 = 6979.73\text{ kJ}$$

$$\text{Entropy decrease} = m_1 C_P \ln\frac{T_1}{T_m}$$

$$= 100\times4.187\times\ln\frac{373}{356.33}$$

$$\Delta S = 19.1434\text{ kJ/K}$$

$$UA_{100} = T_o\,\Delta S = 300\times19.1434 = 5743.0\text{ kJ}$$

$$\therefore \text{Availability}\quad A_{100} = \Delta\,Q_{100} - UA_{100}$$

$$= 6979.73 - 5753.6 = 1226.13$$

(d) Heat received by cold water at 50°C

$$Q_{50} = m_2 C_P \Delta T$$

$$= 50\times4.187\times(83.33-50) = 6977.64\text{ kJ}$$

$$\text{Entropy gain by cold water} = m_2 C_P \ln\frac{T_m}{T_2}$$

$$= 50\times4.187\ln\frac{356.33}{323} = 20.56\text{ kJ/K}$$

$$UA_{50} = T_o\,\Delta S = 300 \times 20.56$$

$$= 6168 \text{ kg}$$

$$A_{50} = \Delta Q_{50}^{\cdot} - UA_{50}$$

$$= 6977.64 - 6168 = 809.64 \text{ kJ}$$

(e) Decrease in Availability = (1) – (2)

$$= 1226.13 - 809.64$$

Decrease in Availability = 416.49 kJ

Example 5.4 Exhaust gas leave an IC engine at 827° and 1.0 atmosphere after doing 1000 kJ of work per kg of gas in the engine (C_{Pg} = 1.1). The surrounding temperature is 25°C.

(a) How much available energy is cost by throwing away exhaust gases.

(b) What is the ratio δ the lost available energy to the engine work.

Solution

$$T_1 = 827 + 273 = 1100$$

$$T_0 = 25 + 273 = 298$$

Entropy change

$$\Delta S = ?$$

$$= m \cdot C_{Pg} \cdot \ln \frac{T_1}{T_0}$$

$$= 1 \times 1.1 \times \ln \frac{1100}{298}$$

$$= 1.4365 \text{ kJ/kg-K}$$

(a)

$$UA = T_0 \times \Delta S$$

$$= 298 \times 1.4365$$

$$= 428.1 \text{ kJ/kg-K}$$

Thus lost available energy = 428 kJ/kg.

(b) Ratio of lost available energy/useful work

$$= \frac{428.1}{1000} = 0.428$$

Example 5.5 A hot spring produce water at a temperature δ 57°C, at the rate of 0.1 m³/ minute. The water into a large lake with minimum temperature of 15°C. What is the rate of working of an ideal heat engine which use all the available energy.

Solution

Now $$T_1 = 273 + 57 = 330 \text{ K}$$

$$T_0 = 273 + 15 = 288 \text{ K}$$

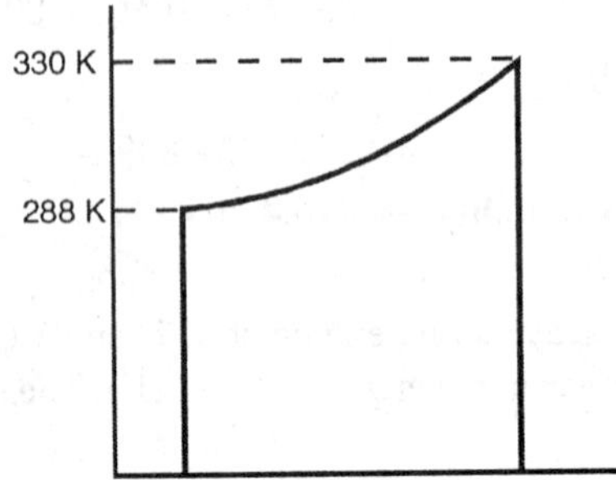

Fig. Ex. 5.5

$$\dot{m} = 0.1 \text{ m}^3\text{/min}$$

$$= \frac{0.1 \times 1000}{60} \text{kg/sec} = 1.667 \text{ kg/s}$$

(a) Heat given by the cooling water/s

$$\Delta Q = m\, C_P (T_1 - T_0)$$

$$= 1.6667 \times 4.187 \times (57 - 15)$$

$$= 293.09 \text{ kJ/s}$$

(b) Change of entropy of spring water

$$= \dot{m}\, C_P \ln \frac{T_1}{T_0} = 1.667 \times 4.187 \times \ln \frac{330}{288}$$

$$\Delta S = 0.94999 \text{ kJ/K}$$

$$UA = T_0 \Delta S = 288 \times \Delta S = 273.6 \text{ kJ}$$

(c) Available energy $= Q - UA = 293.09 - 273$

$$\cong 19.5 \text{ kJ/s} = 19.5 \text{ kW}$$

Thus rate of working of an ideal heat engine which are all energy is 19.5 kW.

Example 5.6 1 kg of air of pressure p_1 and temperature 900 K is mixed with 1 kg of air at the same pressure and 500 K. Determine the loss in availability if surrounding temperature is 27°C. Take C_P for air = 1.0 J/kg-K.

Solution

$$T_0 = 27 + 273 = 300 \text{ K}$$

Temprature of air after mixing one kg each will be (900 + 500) / 2 = 700 K

Change in entropy.

(a) For air at 900 K

$$\Delta S = \dot{m}C_P \ln\frac{T_2}{T_1}$$

$$= 1\times1\times\ln\frac{700}{900} = -0.2513 \qquad (1)$$

(b) For air at 500 K

$$\Delta S = \dot{m}C_P \ln\frac{700}{500} = 1\times1\times\ln\frac{700}{500}$$

$$= 0.3365$$

(c) Net change in entropy after mixing is

$$(1)+(2) = 0.0852$$

$$\text{Loss in availability} = m\times T_0\left(\text{Net change in entropy}\right)$$

$$= (1+1)300\times0.0852$$

$$= 2\times300\times0.0852$$

$$\textbf{Loss in Availability} = \textbf{51.12 K}$$

Example 5.7 500 kJ of heat from a finite temperature source at 1000 K is supplied to 2 kg of gas initially at 2 bar and 350 K in a closed tank. Find the loss in available energy due to the above heat transfer. Take C_V of gas on 0.8 kJ/kg-K and temperature of surrounding as 27°C.

Solution

With the addition of 500 kJ heat, the temperature of the gas will increase.

Thus $\quad$ Heat added $= \dot{m}_{\text{gas}} \times C_V\left(T_2 - 350\right)$

where 350 K is the initial temperature

$$500 = 2\times0.8\times(T_2-350)$$

This gives $\quad (T_2 - 350) = \dfrac{500}{2\times0.8} = 312.5$

or $\quad T_2 = 350+312.5 = 662.5\text{ K}$

Available *UA* Energy with Source

Now $\quad UA = T_0\Delta S$

$$T_0 = 27+273 = 300\text{ K}$$

Change in entropy of the source at temperature of 1000 while 500 kJ is taken out is

$$\Delta S = \frac{\Delta Q}{T} = \frac{500}{1000} = 0.5 \text{ kJ/kg-K}$$

$$UA = T_0 \Delta S = 300 \times 0.5$$

$$= 150 \text{ kJ}$$

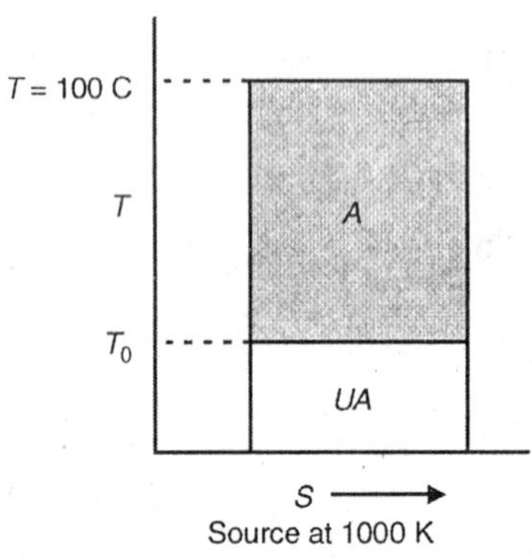

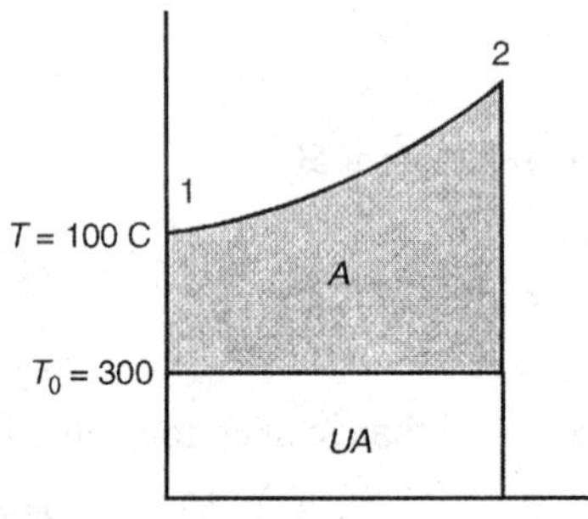

Fig. Ex. 5.7

$$A = Q - UA = 500 - 150$$

$$= 350 \text{ kJ}$$

An *UA* of the gas

Change in entropy of the gas at constant volume is given by

$$\Delta S = m_g\, C_V \ln \frac{T_2}{T_1}$$

$$= 2 \times 0.8 \times \ln \frac{662.5}{350}$$

$$= 1.0209398$$

$$UA = T_0 \times \Delta S = 300 \times \Delta S = 306.28 \text{ K}$$

$$A = Q - UA$$

$$= 500 - 306.28 = 193.7 \text{ kJ}$$

Loss in Availability due to heat transfer is

$$(1) + (2) = 350 - 193.7$$

$$\mathbf{A = 156.3 \text{ kJ}}$$

Example 5.8 Calculate the availability of a system which absorbs 1000 kJ of heat from a source at following constant temperature, while the environment temperature all three cases remain 27°C. (a) 600 K ; (b)1000 K ; (c) 1500 K.

Solution

Now $$A = Q - T_0\,\Delta S$$

where $$Q = 1\text{ H added} = 1000\text{ kJ}$$

$$\Delta S = \text{Change in entropy, it will be different in all three cases}$$

Entropy change at constant temperature is.

$$\Delta S = \int \frac{TQ}{T} = \frac{\Delta Q}{T} = \frac{1000}{T}$$

Thus $$A = Q - T_0\,\Delta S$$

$$= 1000 - 300 \times \frac{1000}{T} = 1000\left(1 - \frac{300}{T}\right)$$

At 600 K

$$A_{600} = 1000\left(1 - \frac{300}{600}\right) = 1000(1 - 0.5) = 500\text{ kJ}$$

At 1000 K

$$A_{1000} = 1000\left(1 - \frac{300}{1000}\right) = 700\text{ kJ}$$

At 1500 K

$$A_{1500} = 1000\left(1 - \frac{300}{1500}\right) = 800\text{ kJ}$$

Thus A increaseas temperature increases.

Example 5.9 800 kJ of heat is added to a system during vaporisation process occurring at 150°C while pressure remain constant. Determine the available and unavailable energy of the surrounding temperature is 25°C.

Solution

$$UA = \text{Unavailable Energy} = T_0\,\Delta S$$

$$A = \text{Available Energy} = Q - UA$$

Now $$Q = 800\text{ kJ}$$

$$T_0 = 273 + 25 = 298\text{ K}$$

$$T = 273 + 150 = 423$$

$$\Delta S = \int \frac{\delta Q_{\text{rev}}}{T}$$

$$= \frac{Q}{T}\text{(since temperature remain constant)}$$

$$UA = T_0 \times \Delta S$$

$$= 298 \times \Delta S = 563.59 \text{ kg}$$

$$A = Q - UA = 800 - 563.59$$

$$\mathbf{A = 236.40 \text{ kg}}$$

Example 5.10 20 kg of water at 90°C is mixed with 32 kg of water at 30°C or constant pressure. Calculate the decrease in availability of surrounding air at 27°C.

Solution

$$\text{Decrease in Availability} = A)_{20} + A)_{32} - A)_{52}$$

i.e. Sum of the original availability of the unmixed 20 kg and 32 kg water less the availbility of the mixture.

Temperature the mixture t_3

$$m_1 C_P t_1)_{20} + m_2 C_P t_2 = m_3 C_P t_3)_{52}$$

or $$20 \times 90 + 32 \times 30 = 52 \times t_3$$

$$\Rightarrow \quad t_3 = 53°\text{C (Approximately)}$$

Now $T_0 = 300$ K; $T_1 = 273 + 90 = 363$ K, $T_2 = 303$ and $T_3 = 273 + 53 = 326$ K

We have to find net change in entropy. Resulting due to 20 kg water at 363 K after mixing getting lower temperature 326

$$\Delta S_{20} = m_{20} C_P \ln\frac{363}{326} = 20 \times 4.187 \times \ln\frac{363}{326}$$

$$= 9.0025 \quad (1)$$

Similarly for 32 kg work at 303

$$\Delta S_{32} = m_{32} C_P \ln\frac{303}{326} = -9.8028 \quad (2)$$

$$\text{Net change in entropy} = (1) + (2)$$

$$= -0.80037$$

$$\text{Loss in Availability} = T_0\ \Delta S = 300 \times 0.80037$$

$$\textbf{Loss in Availability} = \mathbf{240.112 \text{ kJ}}$$

On mixing the water at $90°\text{C}(363\text{ K})$

Example 5.11 From a fire box of a boiler 1000 kg of heat leave hot gases at 1400°C and is given to steam at 250°C. Atmospheric temperature is 20°C. Divide the energy into available and unavailable portions (a) As it leave the hot gas (b) As it reaches the steam (c) Also show that the equation that

Increase in unavailability = $T_0\,\Delta S$ holds good.

Solution

Now $$T_0 = 273+20 = 293\text{ K}$$

$$T_2 = 273+250 = 523\text{ K}$$

$$T_1 = 273+1400 = 1673\text{ K}$$

Case (a) Hot Gas at 1673 K

$$UA = T_0\,\Delta S$$

Since heat is taken out at constant temperature

$$\Delta S_{\text{Gas}} = \frac{Q}{T} = \frac{1000}{1673}\text{ kJ/kg-K}$$

$$UA = 293\times\frac{1000}{1673} = 175.13\text{ kJ}$$

$$A = -1000-(-UA) = -175.13$$

$$= -824.87\text{ kg}$$

(Heat is leaving the system)

Case (b) Heat reaching steam at 523 K

$$\Delta S_{\text{steam}} = \frac{Q}{T} = \frac{1000}{523}$$

$$UA = T_0\,\Delta S = 293\times\frac{1000}{523} = 560.23\text{ kJ}$$

$$A = Q - T_0\,\Delta S = 1000-560.23 = 439.77\text{ kJ}$$

(c) To prove that increase in $UA = T_0\,\Delta S$, we have to find

$$\Delta S_{\text{total}} = \Delta S_{\text{gas}} + S_{\text{steam}}$$

$$= -\frac{1000}{1673}+\frac{1000}{523} = 1.3143$$

$$T_0\,\Delta S = 293\times 1.3143 = 385.1$$

Increase in unavailability

$$= (1)+(2)$$

$$= 175.13+560.23$$

$$= 385.1$$

Since $(3) = (4)$, hence proved.

Example 5.12 900 kJ of heat is added to a system during vaporisation process occurring at 200°C and at a constant pressure. Determine the available energy if the surrounding are at 17°C.

Solution

Available $A = Q - T_0\,\Delta S$

where $Q =$ Heat added $= 900$ kJ

$$T_0 = 17+273 = 290 \text{ K}$$

$\Delta S =$ Entropy changes

$= \dfrac{\Delta Q}{T}$ since temperature is constant

$$= \frac{900}{273+200} = \frac{900}{473} \text{ kJ/kg-K}$$

$$\therefore \quad A = 900 - 290 \times \frac{900}{473}$$

$$= 900 - 551.797$$

$$\mathbf{A = 348.203 \text{ kJ}}$$

Example 5.13 An amount of 90 kJ of heat is added to a touch of air that is originally at 26°C. The touch contains 2.8 kg of air. The surrounding temperature is 15°C. Find available and unavailable energy.

Solution

Here the heat is added at constant volume. C_V for air $= 0.718$ kJ/kg-K with the addition of 90 kJ heat, the temperature of air in the touch will increase by ΔT

Thus

$$90 = mC_V\Delta T$$

or

$$90 = 2.8 \times 0.718\,\Delta T$$

$$\Delta T = 44.76$$

T_2 i.e. the temperature of air after heat addition

$$T_2 = 26 + 44.76 = 70.76$$
$$= 70.76 + 273 = 343.76$$
$$\cong 444 \text{ K}$$

Now change of entropy ΔS for constant volume increase

$$\Delta S = C_V \ln \frac{T_2}{T_0}$$

$$= 0.718 \ln \frac{444}{273+15}$$

$$\mathbf{\Delta S = -0.31079}$$

Example 5.14 A temperature of 2000°C is obtained in a furnace by burning fuel in air at atmosphere pressure and ambient tempeature of 27°C. The gases can be assumed to be perfect gas with C_p = 1.0 kJ/kg-K. Determine the availability of heat in the products of combustion.

Solution

$$T_0 = 27 + 273 = 300 \text{ K}$$

For unit mass of gas

(a) Heat addition $Q = mC_P \Delta T$

$$= 1 \times 1.00 \times (2000 - 27)$$

$$= 1 \times 1983 = 1983 \text{ kJ/kg}$$

(b) We have now to find change is entropy during the process

$$\Delta S = mC_P \ln \frac{T_1}{T_2}$$

$$= 1 \times 1.0 \ln \frac{2000+273}{27+273}$$

$$= \ln \frac{2773}{300} = 2.025 \text{ kJ/kg-K}$$

(c) $UA = T_0 \Delta S = 300$

$$= 300 \times 2.025 = 607.522 \text{ kJ/kg}$$

(d) $A = Q - UA = 1983 - 607.522$

$$= 1375.478 \text{ kJ/kg}$$

THEORY QUESTIONS

1. Explain
   (a) Dead state
   (b) Availability
   (c) Unavailable energy
2. Write short notes on:
   (a) Irreversibility
   (b) Effectiveness
   (c) Second law efficiency
3. Comment on 'There is no energy crisis but available energy crisis'.
4. Derive expressions'for available energies for heat source
   (a) At constant temperature
   (b) At variable temperature

PROBLEMS FOR PRACTICE

1. Determine available and unavailable apart of energy of a system which absorbs 15 MJ of heat from a source at 227°C. The surroundings are at 17°C.

   (**Ans.** $UA = 8.7$ MJ, $A = 6.3$ MJ)

2. A source at 227°C receives 7200 kJ/min from a source at a temperature of 727°C. The atmospheric temperature is 27°C. Assuming that the temperature of the system and that of the source remain constant, find the decrease in availability after the heat transfer.

   (**Ans.** 2160 kJ)

3. A steel ball of 800 kg at 1250 K is cooled to 500 K. Using steel ball as source of energy, compute available and unavailable energies. Assume specific heat of steel to be 0.5 kJ/kg-K and ambient temperature to be 27°C.

   (**Ans.** $UA = 110$ MJ, $A = 190$ MJ)

4. Air 0.2 kg initially at a temperature of 575 K receives 300 kJ heat reversibly at constant pressure. Find available and unavailable portions of the heat added. Take C_p for air to be 1.005 kJ/kg-K and ambient temperature to be 27°C.

   (**Ans.** $UA = 77.16$ kJ, $A = 222.54$ kJ)

5. During a vaporization process at constant pressure taking place at 180 C, 850 kJ of heat is added while ambient temperature is 299 K. Find available and unavailable energies.

   (**Ans.** $UA = 55.53$, $A = 296.46$ kJ)

6. A steam stable condenses at 1000° C rejecting 3000 kJ of heat while another sample of heat condenses at 500 C rejecting the same quantity of heat. Determine

the difference in the available energy between the two processes. Take sink temperature to be 27°C.

(Ans. A = 2840 kJ, UA = 1160 kJ)

7. A constant temperature source is maintained at 1000 K and the surroundings are at 290 K. If 4000 kJ heat is transferred from the source reversibly, determine available and unavailable portions of heat.

(Ans. A = 2840 kJ, UA = 1160 kJ)

8. From a heat reservoir 1000 kJ of heat is withdrawn at 600 K. Find the availability and unavailability of heat if surroundings are at 290 K.

(Ans. A = 516.7 kJ, UA = 483.3 kJ)

9. 500 kJ of heat was removed at constant temperature at 527°C and supplied a system at constant temperature at 427°C. Find the loss in available energy as a result of this irreversible heat transfer. Take ambient temperature as 27°C.

(Ans. 26.79 kJ)

ﻌ ﻍ

6

Entropy

CHAPTER OBJECTIVES

After reading this chapter you will be able to learn the following

- Concept of Entropy, Clausius Theorem—for a process and cycle.
- Entropy as a Property.
- Clausius Inequality.
- Entropy changes in Irreversible process, Isolated system, Universe and Entropy and disorder.
- Property Relations from Energy equations and change of entropy equations for various processes.
- Thrid Law of Thermodynamics.

6.1 INTRODUCTION

We have studied the I–law of thermodynamics. Analysis of I–law leads to the definition of a derived property known as Internal energy. Analysis of second law will lead to the definition of another derived property known as Entropy.

Clausius discovered that, when a small amount of heat δQ is supplied to a system, which is at a absolute temperature T, then it will undergo a process, and the ratio $\frac{\delta Q}{T}$ is same for all reversible processes. He assigned the value $\frac{\delta Q}{T} = dS$ and called S as entropy.

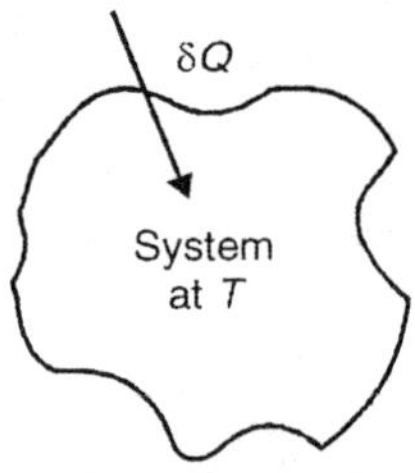

Fig. 6.1

The term entropy is taken, from Greek work 'tropee' meaning transformation. Thus, when a small amount of heat δQ is transformed to a system, the entropy changes by an amount dS. This change in entropy is regarded as the transformation content of the system.

Note that entropy is a thermodynamics property, it increases with the addition of heat and it decreases with the removal of heat.

6.2 CLAUSIUS THEOREM FOR A PROCESS AND FOR A CYCLE

Let a system be taken from an equilibrium state (1) to another equilibrium state (2) following the reversible path 1–2. Let (a) and (b) be two reversible adiabatic, which pass through the points (1) and (2) respectively. A reversible isotherm (c) is drawn, such that the area under 1–3–4–2 is equal to the area under the curve 1–2.

From the first law,

$$Q = \Delta U + W$$

For process 1–2,

$$Q_{1-2} = U_2 - U_1 + W_{1-2} \text{ and for the process } 1-3-4-2$$

$$Q_{1-3-4-2} = U_2 - U_1 + W_{1-3-4-2}$$

As we have assumed that the area under $1-3-4-2$ is equal to the area under 1–2.

i.e. $$W_{1-2} = W_{1-3-4-2}$$

$\therefore$ $$Q_{1-2} = Q_{1-3-4-2}$$

$$= Q_{1-3} + Q_{3-4} + Q_{4-2}$$

But $$Q_{1-3} = 0$$

and $$Q_{4-2} = 0 \; (\because \text{ adiabatic})$$

$\therefore$ $$Q_{1-2} = Q_{3-4}$$

Thus any reversible path can be replaced by a zig-zag path between the same end states, consisting of a reversible adiabatic process followed by a reversible isothermal

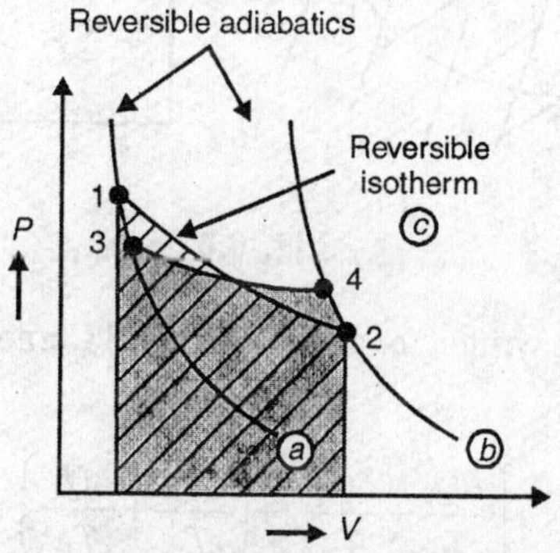

Fig. 6.2

process, which is again followed by a reversible adiabatic process, such that the heat transferred during the isothermal process is the same as that transferred during the original process.

Consider a reversible cycle as shown in Fig. 6·2. It is divided into a large number of strips by means of reversible adiabatic. Each strip may be closed at the top and bottom by means of reversible isotherms. The original closed cycle is thus replaced by a zig-zag closed path, consisting of alternate adiabatic and isothermal processes, such that the heat transferred during all the isothermal processes is equal to the heat transferred in the original cycle.

Now, for cycle 1 – 2 – 3 – 4, δQ_1 is heat absorbed reversibly at T_1 and δQ_2 is heat rejected reversibly at T_2.

Then, $$\frac{\delta Q_1}{T_1} = \frac{\delta Q_2}{T_2}$$

If heat supplied is taken as positive and heat rejected as negative, then

$$\frac{\delta Q_1}{T_1} + \frac{\delta Q_2}{T_2} = 0$$

Similarly for another elemental cycle 5 – 6 – 7 –8

$$\frac{\delta Q_3}{T_3} + \frac{\delta Q_4}{T_4} = 0$$

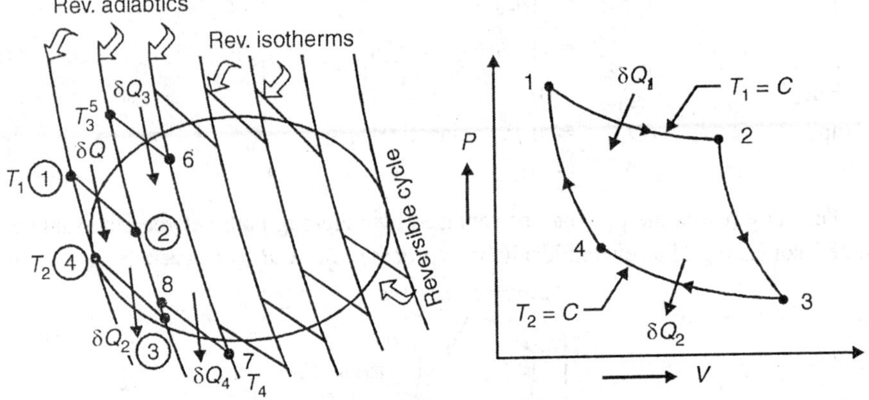

Fig. 6.3

If similar equation are written for all the elemental Carnot cycles, then for the whole (complete) original cycle.

$$\left(\frac{\delta Q_1}{T_1} + \frac{\delta Q_2}{T_2}\right) + \left(\frac{\delta Q_3}{T_3} + \frac{\delta Q_4}{T_4}\right) + \ldots = 0$$

or $$\oint \frac{dQ}{T} = 0$$

Thus the cyclic integral of $\delta Q/T$ for a reversible cycle is equal to zero. This is known as **Clausious Theorem.**

6.3 ENTROPY A PROPERTY

To prove this, we have to prove that the change of entropy does not depend upon path but it depends upon end states. Then, we will able to say that, entropy is a property of the system.

Consider a system which changes its state from state point (1) to state point (2) by following the reversible path a and returns from state point (2) to state point (1) by following the reversible path b.

Then the two paths $1-a-2$ and $2-b-1$ together will form a cycle.

Now from Clausius theorem,

$$\oint \frac{\delta Q}{T} = 0$$

(1–a–2–b–1)

Note. May be read as integral for the reversible path $1-a-2-b-1$.

The above integral may be replaced as the sum of two integrals one for the path a and other for the path b.

$$a\int_{R1}^{2} \frac{\delta Q}{T} + b\int_{R2}^{1} \frac{\delta Q}{T} = 0$$

Note. May be read as integral $\delta Q/T$ for the reversible path 1–a–2.

$$\therefore \quad a\int_{R1}^{2} \frac{\delta Q}{T} = -b\int_{R1}^{2} \frac{\delta Q}{T}$$

Since path b is reversible

$$a\int_{R1}^{2} \frac{\delta Q}{T} = b\int_{R2}^{1} \frac{\delta Q}{T}$$

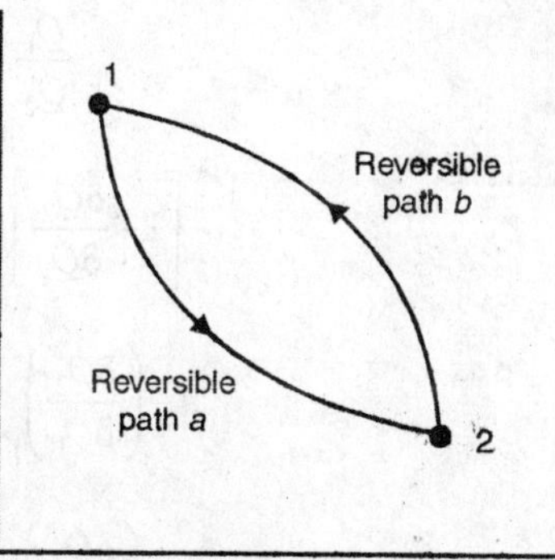

Fig. 6.4

The magnitude of $\dfrac{\delta Q}{T}$ (i.e. dS) is same for the paths a and b and it does not depend upon the end states, hence it is point function and we know that properties are point–functions, hence it is a property of the system.

6.4 CLAUSIUS INEQUALITY

According to Clausius Theorem,

$$\oint_R \frac{\delta Q}{T} = 0$$

In the II law, during Carnot theorem, we have proved that the efficiency of the Reversible engine is more than that of an Irreversible engine.

$$\eta_R > \eta_I$$

or $$\eta_I < \eta_R \quad (1)$$

We also know that,

$$\eta \text{ of any engine} = \frac{\text{Heat supplied} - \text{Heat rejected}}{\text{Heat supplied}} = \frac{Q_1 - Q_2}{Q_1}$$

or For small amount of heat δQ,

$$\eta = \frac{\delta Q_1 - \delta Q_2}{\delta Q_1} = 1 - \frac{\delta Q_2}{\delta Q_1}$$

Hence the Eq. (1) will therefore be,

$$\left[1 - \frac{\delta Q_2}{\delta Q_1}\right]_I < \left[1 - \frac{\delta Q_2}{\delta Q_1}\right]_R$$

But for reversible engine,

$$\frac{Q_1}{Q_2} = \frac{T_1}{T_2}$$

$$\left[1 - \frac{\delta Q_2}{\delta Q_1}\right] = 1 - \frac{T_2}{T_1}$$

$$\therefore \quad 1 - \left(\frac{\delta Q_2}{\delta Q_1}\right)_I < 1 - \frac{T_2}{T_1}$$

$$\therefore \quad \left(\frac{\delta Q_2}{\delta Q_1}\right)_I > \frac{T_2}{T_1}$$

or opposite $$\left(\frac{\delta Q_1}{\delta Q_2}\right) < \frac{T_1}{T_2}$$

or $$\left(\frac{\delta Q_1}{T_1}\right)_I - \left(\frac{\delta Q_2}{T_2}\right)_I < 0$$

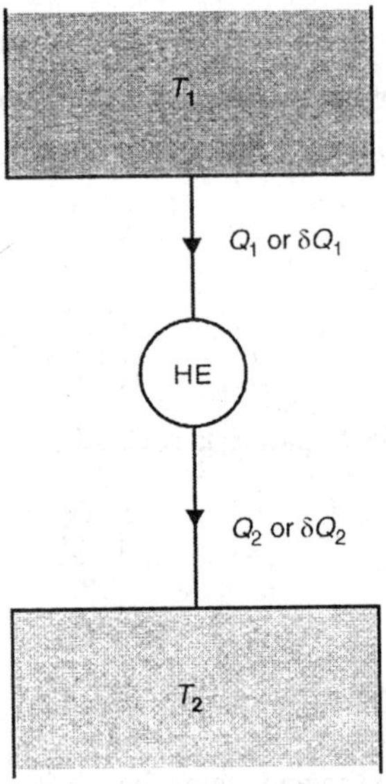

Fig. 6.5

We know that, heat added is positive. Therefore δQ_1 is positive and heat rejected is negative. Therefore δQ_2 is negative.

$$\therefore \qquad \left(\frac{\delta Q_1}{T_1}\right)_I + \left(\frac{\delta Q_2}{T_2}\right)_I < 0$$

i.e. The algebraic sum of $\delta Q/T$, for an irreversible cycle is always less than zero.

$$\therefore \qquad \oint_R \frac{\delta Q}{T} < 0 \text{ for an Irreversible cycle.}$$

And we know that from Clausius theorem $\oint \frac{\delta Q}{T} = 0$ for a reversible cycle. Combining results for reversible and irreversible cycle, we may write,

$$\oint \frac{\delta Q}{T} \leq 0$$

Thus expression is known as Clausius Inequality. It implies whether any cyclic process is reversible or irreversible or impossible.

Note: According to Clausius Inequality

(i) $\oint \frac{dQ}{T} = 0$ for a reversible cycle.

(ii) $\oint \frac{dQ}{T} < 0$ for a irreversible cycle.

(iii) $\oint \frac{dQ}{T} > 0$ for a impossible cycle.

(iv) All temperatures must be in K.

(v) Heat received by the engine is positive.

(vi) Heat rejected by the engine is negative.

6.5 CHANGE OF ENTROPY IN AN IRREVERSIBLE PROCESS

We know that, change in entropy for a reversible process is given by,

$$\left(\frac{\delta Q}{T}\right)_R = (dS)_R \qquad (1)$$

Now to find the value of change in entropy in an Irreversible process.

Consider a system, which change its state from state point (1) to state point (2) by following the reversible path a and returns from state point (2) to state point (1) by following the irreversible path b as shown in Fig. 6.6.

Since cyclic integral of any property is zero and entropy is a property we can write,

$$\oint dS = a\int_1^2 (dS)_R + b\int_2^1 (dS)_I = 0 \qquad (2)$$

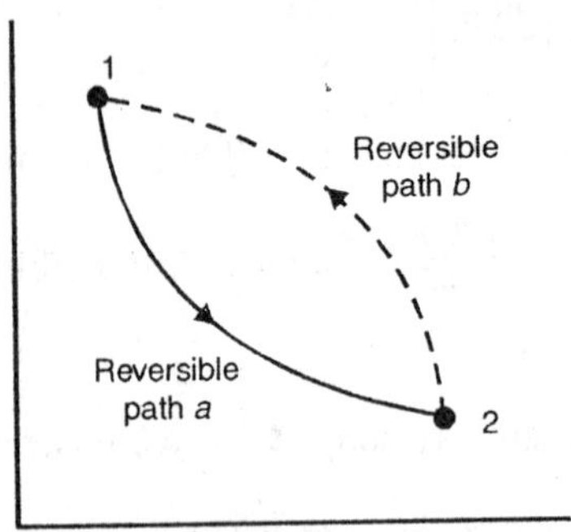

Fig. 6.6

Now, from Eq. (1), for a reversible process we have,

$$a\int_1^2 (dS)_R = a\int_1^2 \left(\frac{\delta Q}{T}\right)_R \tag{3}$$

Substituting this in Eq. (2) we get,

$$a\int_1^2 \left(\frac{\delta Q}{T}\right)_R + b\int_2^1 (dS)_I = 0 \tag{4}$$

Since the processes 1 – a – 2 and 2 – b – 1 together will form Irreversible cycle. Applying Clausius inequality,

$$\oint \frac{\delta Q}{T} < 0$$

We get from Eq. (4),

$$a\int_1^2 \left(\frac{\delta Q}{T}\right)_R + b\int_2^1 \left(\frac{\delta Q}{T}\right)_I < 0 \tag{5}$$

Subtracting Eq. (5) from Eq. (4) we get,

$$b\int_1^2 (dS)_I > a\int_2^1 \left(\frac{\delta Q}{T}\right)_I$$

In general, $(dS)_I > \left(\frac{\delta Q}{T}\right)_I$ (6)

Combining Eqs (1) and (6), we can write,

$$dS \geq \frac{\delta Q}{T}$$

Where equality sign is for reversible process and inequality sign is for an Irreversible process (from Eq. 6).

Note. The effect of irreversibility is always to increase the entropy of the system.

6.6 CHANGE OF ENTROPY FOR AN ISOLATED SYSTEM

We know that, in an isolated system, matter, work or heat cannot cross the boundary of the system. Hence according to the first law of thermodynamics, the IE of the system will remain constant.

Since for an isolated system $\delta Q = 0$, from equation,

$$dS = \frac{\delta Q}{T} \tag{1}$$

We get, (dS) isolated ≥ 0 (2)

Equation (2) states, that entropy of an isolated system either increases or remains constant and never decreases. This is known as the *Principle of increase of entropy.*

6.7 ENTROPY OF UNIVERSE

We know that all natural process are irreversible processes and during irreversible processes entropy increases and hence entropy of the universe always increases.

6.8 ENTROPY AND DISCORDER

Consider an isolated system comprising of two gases O_2 and H_2 in a separated box as shown in Fig. 6.7. When the partition or barrier is removed, the molecules of both gages

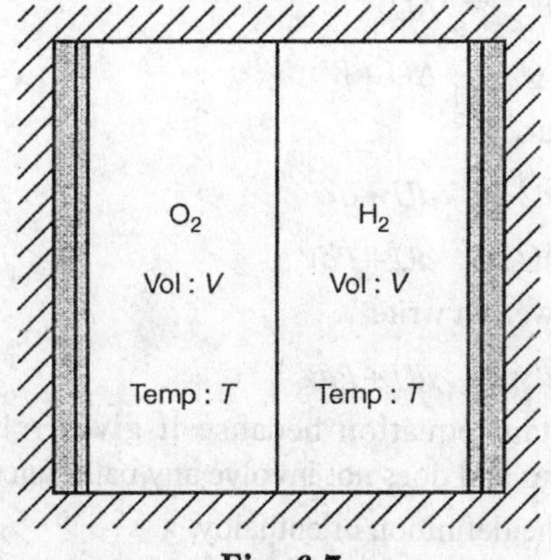

Fig. 6.7

get more space to move randomly and therefore collision take place between the mole molecules of the same gas as well as of both the gases, and after some time, equilibrium will be established.

If we assume that initially both gases have the same volume and temperature then after removing the partition each gas occupies double the previous volume while the temperature and the I.E. of the system remains the same, but we find that the entropy increase.

The reason is obvious, as the mixing process is irreversible and irreversible process is always associated with the increase in entropy.

Thus an irreversible process always tends to take the system (isolated) to state of greater disorder. It is a tendency on the part of nature to proceed to a state of greatest disorder. And an isolate system always tends to a state of greater entropy. So there is a close link between entropy and disorder.

It may be stated roughly that "**The entropy of a system is a measure of the degree of molecular disorder existing in the system**". When heat is supplied to the system, the disorderly motion of molecules increases and so the entropy of the system increases. The reverse occurs when heat is removed from the system.

6.9 PROPERTY RELATIONS FROM ENERGY EQUATIONS

For deriving the equation for change of entropy it is essential to know equation based on I and II law of thermodynamics.

We know that, change in entropy for a reversible process is given by,

$$(dS)_R = \left(\frac{\delta Q}{T}\right)_R$$

i.e.
$$(\delta Q)_R = T \cdot dS \qquad (1)$$

Also we know that, the first–law applied to a closed system undergoing a process gives,

$$Q - W = \Delta U$$

or
$$Q = \Delta U + W$$

In the differential form,

$$\delta Q = dU + \delta W$$

i.e.
$$\delta Q = dU + PdV \qquad (2)$$

$\therefore$ From (1) and (2) we can write,

$$T \cdot dS = dU + PdV \qquad (3)$$

This is very important equation because it gives relationship between all thermodynamic properties and does not involve any path functions.

We also know from the definition of enthalpy,

$$H = U + PV$$

or
$$dH = dU + PdV + VdP$$

$$dH = T.dS + V.dP \text{ from Eq. (3)}$$

or
$$T.dS = dH - V.dP \qquad (4)$$

From Eq. (3) i.e.

$$T.dS = dU + P.dV$$

$$\therefore \quad dS = \frac{dU}{T} + \frac{PdV}{T}$$

$$\therefore \quad S_2 - S_1 = \int_1^2 \frac{dU}{T} - \int_1^2 \frac{PdV}{T} \qquad (5)$$

From Eq. (4) i.e.

$$T.dS = dH - V.dP$$

$$dS = \frac{dH}{T} + \frac{V.dP}{T}$$

$$\therefore \quad S_2 - S_1 = \int_1^2 \frac{dH}{T} - \int_1^2 \frac{V \times dP}{T} \qquad (6)$$

6.10 GENERAL EQUATIONS FOR CHANGE IN ENTROPY

We know that, the firlst law applied to a closed system undergoing a process gives,

$$Q - W = \Delta U$$

Now consider a unit mass of a gas which changes its states from pressure P_1, specific volume v_1, and temperature T_1 to a new state P_2, V_2, T_2.

For unit mass $q = \Delta u + w$

In the differential form

$$\delta q = du - \delta w \qquad (1)$$

Now, from the definition of specific heat at constant volume i.e.

$$C_v = \left(\frac{\delta q}{dT}\right)_v = \left(\frac{du}{dT}\right)_v$$

From this $\quad du = C_v \cdot dT$

$\therefore$ Eq. (1) becomes,

$$\delta q = C_v \cdot dT + P \cdot dv \qquad (2)$$

Dividing throughout by T,

$$\frac{\delta q}{T} = C_v \frac{dT}{T} + \frac{P}{T} dv \qquad (3)$$

But $\quad \dfrac{\delta q}{T} = ds$

$$\therefore \quad \frac{\delta q}{T} = ds = C_v \frac{dT}{T} + \frac{P}{T} dv \tag{4}$$

From perfect gases, for unit mass, $Pv = RT$ or $\frac{P}{T} = \frac{R}{v}$

Substituting this in Eq. (4)

$$ds = C_v \frac{dT}{T} + R \frac{dv}{v} \tag{5}$$

$$\therefore \quad \int_1^2 ds = \int_1^2 C_v \frac{dT}{T} + \int_1^2 R. \frac{dv}{v}$$

$$\mathbf{s_2 - s_1 = C_v \ln \frac{T_2}{T_1} + R \ln \frac{v_2}{v_1}} \tag{6}$$

This equation gives change in entropy for unit mass, in terms of temperature ratio and volume ratio.

From perfect gases, $C_p - C_v = R$ and susbtituting this in Eq. (6) we get,

$$s_2 - s_1 = C_v \ln \frac{T_2}{T_1} + C_p \ln \frac{v_2}{v_1} - C_v \ln \frac{v_2}{v_1}$$

$$= C_p \ln \frac{v_2}{v_1} + C_v \left[\ln \frac{T_2}{T_1} - \ln \frac{v_2}{v_1} \right]$$

$$= C_p \ln \frac{v_2}{v_1} + C_v \left[\ln \frac{T_2}{T_1} \cdot \frac{v_1}{v_2} \right] \tag{7}$$

From characteristic equation

$$P.v = RT$$

$$\text{or} \quad \frac{P_1 v_1}{T_1} = \frac{P_2 v_2}{T_2} \qquad \therefore \quad \frac{P_2}{P_1} = \frac{T_2 v_1}{T_1 v_2} \tag{8}$$

Substituting Eq. (8) in Eq. (7) we get,

$$\mathbf{s_2 - s_1 = C_p \ln \frac{v_2}{v_1} - C_p \ln \frac{P_2}{P_1}} \tag{9}$$

Again substituting $C_p - C_v = R$ or $C_p - R = C_v$ in Eq. (6) we get,

$$s_2 - s_1 = C_p \cdot \ln \frac{T_2}{T_1} - R \cdot \ln \frac{T_2}{T_1} + R \cdot \ln \frac{T_2}{T_1}$$

$$= C_p.\ln\frac{T_2}{T_1} - R.\left(\ln\frac{T_2}{T_1} - \ln\frac{v_2}{v_1}\right)$$

$$= C_p \cdot \ln\frac{T_2}{T_1} - \left(\ln\frac{T_2}{T_1} \cdot \frac{v_1}{v_2}\right)$$

From Eq. (8) $$\mathbf{s_2 - s_1 = C_p.\ln\frac{T_2}{T_1} - R\ln\frac{P_2}{P_1}} \quad (10)$$

It is to be noted that for a given change, Eqs (6), (9) and (10) will give the same result. The choice of equation is a matter of convenience.

6.11 CHANGE IN ENTROPY DURING VARIOUS PROCESSES

We have studied the equations for work done, heat transferred change in I.E. in the ideal gas chapter. Now to derive the equations for change of entropy during (i) constant volume process (ii) constant pressure process (iii) isothermal process (iv) polytropic process.

(i) Constant Volume Process

Change of entropy:

Here $$v_1 = v_2 \text{ hence } \ln\frac{v_2}{v_1} = \ln 1 = 0$$

Hence from Eq. (6) (of last article)

$$s_2 - s_1 = C_v \ln\frac{T_2}{T_1} + R\ln\frac{v_2}{v_1}$$

$$\mathbf{s_2 - s_1 = C_v\ln\frac{T_2}{T_1}}$$

and from Eq. (9) i.e.

$$s_2 - s_1 = C_p \ln\frac{v_2}{v_1} - C_v \ln\frac{P_2}{P_1}$$

$$\mathbf{s_2 - s_1 = C_v\ln\frac{P_2}{P_1}}$$

(ii) Constant Pressure Process

Change in Entropy:

Here $$P_1 = P_2$$

$$\therefore \quad \ln\frac{P_1}{P_2} = \ln 1 = 0$$

Hence from equation which are in terms or pressure ratio (From equations derived in last article)
i.e. from Eq. (9),

$$s_2 - s_1 = C_p \ln \frac{v_2}{v_1}$$

And from Eq. (10),

$$s_2 - s_1 = C_p \ln \frac{T_2}{T_1}$$

(iii) Isothermal Process
Change in Entropy,

Here $$T_1 = T_2$$

$$\therefore \quad \ln \frac{T_1}{T_2} = \ln 1 = 0$$

Hence from equations which are in terms of temperature ratio.
i.e. from Eq. (6)

$$s_2 - s_1 = R \ln \frac{v_2}{v_1}$$

And from Eq. (10)

$$s_2 - s_1 = -R \ln \frac{P_2}{P_1}$$

$$s_2 - s_1 = R \ln \frac{P_2}{P_1}$$

(iv) Polytropic Process:
(a) We know that heat transfer in a polytropic process,

$$q = \left[\frac{\gamma - n}{1 - n}\right] C_v (T_2 - T_1)$$

In the differential form,

$$\delta q = \left[\frac{\gamma - n}{1 - n}\right] C_v \cdot dT$$

Since $$\frac{\delta q}{T} = ds$$

$$\therefore \quad ds = \frac{\delta q}{T} = \left[\frac{\gamma - n}{1 - n}\right] C_v \cdot \frac{dT}{T}$$

or
$$\int_1^2 ds = \int_1^2 \left[\frac{\gamma - n}{1-n}\right] \cdot C_v \frac{dT}{T}$$

$\therefore$
$$\mathbf{s_2 - s_1 = \left[\frac{\gamma - n}{1-n}\right] . C_v \ln \frac{T_2}{T_1}} \quad (1)$$

This is the change in entropy, for unit mass in a polytropic process in terms of temperature ratio.

(b) Change of entropy: In terms of volume ratio

We know that from (1),

$$s_2 - s_1 = \left[\frac{\gamma - n}{1-n}\right] C_v \ln \frac{T_2}{T_1}$$

But, for a polytropic process

$$\frac{T_2}{T_1} = \left(\frac{V_1}{V_2}\right)^{n-1}$$

$\therefore$ Substituting this value of $\frac{T_2}{T_1}$ in Eq. (1), we get

$$s_2 - s_1 = \left[\frac{\gamma - n}{1-n}\right] C_v \ln \left(\frac{V_1}{V_2}\right)^{n-1}$$

$$= \frac{\gamma - n}{1-n} C_v (1-n) \ln \frac{V_2}{V_1}$$

$$= (\gamma - n) C_v \ln \frac{V_2}{V_1}$$

$$\mathbf{s_2 - s_1 = \frac{\gamma - n}{1-n} R \ln \frac{V_2}{V_1}} \qquad \left(\text{As } C_v = \frac{R}{\gamma - 1}\right)$$

(c) Change of entropy polytropic process: In terms of pressure ratio:

We know that from (1)

$$s_2 - s_1 = \frac{\gamma - n}{1-n} C_v \ln \frac{T_2}{T_1}$$

But, for a polytropic process

$$\frac{T_2}{T_1} = \left(\frac{P_2}{P_1}\right)^{\frac{n-1}{n}}$$

Substituting this value of $\frac{T_2}{T_1}$ in Eq. (1), we get

$$s_2 - s_1 = \frac{\gamma - n}{1-n} C_v \ln\left(\frac{P_2}{P_1}\right)^{\frac{n-1}{n}}$$

$$s_2 - s_1 = \frac{\gamma - n}{1-n} \cdot \frac{R}{\gamma - 1} \cdot \frac{1-n}{n} \ln \frac{P_2}{P_1}$$

$$= \frac{\gamma - n}{1-n} \cdot \frac{R}{n} \cdot \ln \frac{P_2}{P_1} 2$$

$$s_2 - s_1 = \frac{\gamma - n}{n} C_v \ln \frac{P_2}{P_1} \quad (3)$$

(d) Adiabatic Process: change of entropy for a reversible adiabatic process.

As we have studied during a reversible adiabatic process, entropy reamin constant ($S = C$) and the process is called as Isentropic process. Since $S = C$, $dS = 0$ i.e. change in entropy is zero.

6.12 REPRESENTATION OF VARIOUS PROCESS ON T–S CHART

Various processes are shown on T–S diagram in Fig. 6.8. Note that constant volume lines on T–S diagram have steep slope compared to constant pressure lines.

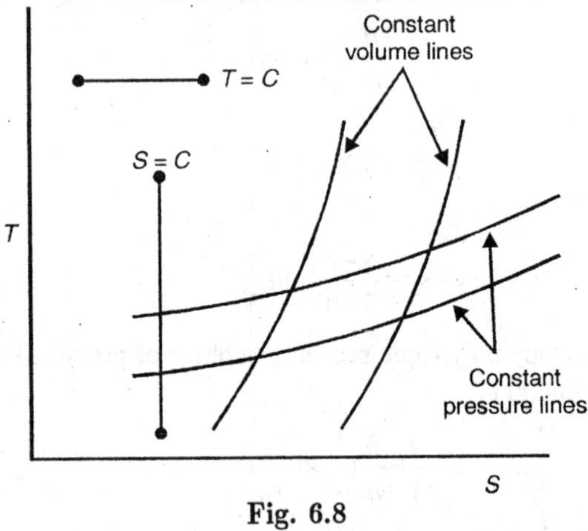

Fig. 6.8

6.13 PURE SUBSTANCE AND CHANGE OF ENTROPY

We define a pure substance as a system which is homogeneous in composition and chemical aggregation that remains invariable with time.

In many cases we have to deal with system of fixed chemical composition with changing phases, as for example, in case of vapour power generation and referigeration cycles. In the former system water is commonly used whereas in the latter system ammonia, freon etc. are used. In these applications we find the working substance as liquid in one part of the cycle and as vapour or mixture of vapour and liquid in another part of the cycle and as vapour or mixture of vapour and liquid in another part of the cycle. The change in phase of the substance in the cycle is of physical nature only as its chemical composition remains unchanged. These fall in the category of pure substance.

We define a pure substance as a system which is homogeneous in composition and chemical aggregation than remain invariable with time.

A pure substance can exist in three different states of aggregation viz. solid, liquid and gas. It may exist in a single phase like solid, liquid, gas or more than one phase which are in equilibrium with each other. Thus a mixture of water and ice, water and steam, water and ice are all examples of pure substance.

Third law of thermodynamics states "*The entropy of a pure substance in complete thermodynamic equilibrium becomes zero at the absolute zero of temperature*".

This law enables the absolute entropies of pure substances to be calculated from the fundamental definition of entropy, with S_0 set equal to zero.

$$\therefore \quad S = \int_0^T \frac{\delta Q_{rev}}{T} \text{ pure substance in equilibrium}$$

Therefore, the change of entropy of a pure substance when it undergoes a reversible process, can be calculated as in case of ideal gases.

Thus, ΔS_w = Change in entropy of 1 kg water when heated at constant pressure from 0° C to saturation temperature.

$$= \int_{273}^{T_{sat}} \frac{\delta Q_{rev}}{T}$$

$$= C_{p_w} \int_{273}^{T_{sat}} \frac{dT}{T}$$

$$= C_{p_w} \log_e \frac{T_{sat}}{273} = C_{p_w} \ln \frac{T_{sat}}{273}$$

During evaporation at constant saturation temperature, change of entropy is given by

$$\Delta S_{fg} = \frac{\text{Latent heat}}{\text{Saturation temperature}} = \frac{h_{fg}}{T_{sat}}$$

Again for superheating process, the change of entropy (temperature increases from T_{sat} to T_{sup}).

$$\Delta S_{sup} = C_{p_v} \log_e \frac{T_{sup}}{T_{sat}} = C_{p_v} \ln \frac{T_{sup}}{T_{sat}}$$

6.14 THIRD LAW OF THERMODYNAMICS

Statistical analysis suggests that the entropy of a substance tends to zero as absolute zero temperature is reached.

This point has been so intensly investigated, that it is probably safe to say that at absolute zero temperature, **the entropy of a pure substance in some perfect crystalline form becomes zero, a generalisation known as the Third law.**

It is found that the specific heats approach zero as temperature tends to zero and the difference $C_p - C_v$ also approaches zero, becuase the co-efficient of thermal expansion approaches zero.

FORMULAE

1. I–law for a closed system undergoing a cycle,

$$\oint \delta W = \oint \delta Q$$

2. I–law for a closed system undergoing a process,

$$Q - W = \Delta U$$

where Q = Heat transfer in J

W = Work transfer in J

ΔU = Change of Internal Energy in J.

3. Specific heat at constant pressure,

$$C_P = \left(\frac{\delta q}{dT}\right)_P = \left(\frac{dh}{dT}\right)_P$$

$$\Rightarrow \quad \delta q = C_P \cdot dT \text{ for unit mass}$$

And for total mass $\delta Q = m \cdot C_P \cdot dT$

4. Specific heat at constant volume

$$C_v = \left(\frac{\delta q}{dT}\right)_V = \left(\frac{du}{dT}\right)_V$$

$$\delta q = C_V \cdot dT \text{ for unit mass}$$

And for total mass $\delta Q = m \cdot C_V \cdot dT$

Contd.

5. Adiabatic index k or $\gamma = \dfrac{C_P}{C_V}$

Note for air $C_P = 1.005 \dfrac{\text{kJ}}{\text{kg-K}}$; $C_V = 0.718 \dfrac{\text{kJ}}{\text{kg-K}}$

and $\gamma = 1.4$

6. Continuity equation or law of conservation of mass

$$\rho_1 A_1 C_1 = \rho_2 A_2 C_2$$

or $$\frac{A_1 C_1}{v_1} = \frac{A_2 C_2}{v_2} \text{ as } \rho = \frac{1}{v}$$

where ρ = Density (kg/m^3)

A = Cross sectional area (m^2)

C = Velocity of fluid in (m/sec)

7. Equation of state $PT = mRT$

and R for air $= 0.287 \text{kJ/kg-K}$

8. Relations between C_P, C_V and R, γ

(i) $C_P - C_V = R$

(ii) $C_V = \dfrac{R}{\gamma - 1}$

(iii) $C_P = \dfrac{\gamma \cdot R}{\gamma - 1}$

9. For constant volume process ($V = C$)

(i) Work done in a closed system,

$$W_{1-2} = \int P.dV = 0 \text{ as } V = C \text{ and } dV = 0$$

(ii) From the I–law $Q - W = \Delta U$

$$Q - 0 = \Delta U = m\ C_V \cdot dT$$

$$\Rightarrow \quad Q = \Delta U = m\ C_V \cdot dT$$

10. For constant pressure process ($P = C$)

(i) Work done in a closed system,

Contd.

$$W_{1-2} = \int_1^2 P.dV$$

$$W_{1-2} = P(V_2 - V_1)$$

(ii) Change in Internal energy

$$\Delta U = U_2 - U_1 = m \cdot C_V \cdot dT$$

(iii) Heat transfer $\delta Q = dH = m \cdot C_P \cdot dT$

11. For Isothermal process ($PV = C$)

(i) Work done $W_{1-2} = \int P.dV = P_1V_1 \ln \frac{V_2}{V_1}$

or $$W_{1-2} = P_1V_1 \ln \frac{V_2}{V_1}$$

(ii) Change in I.E. $= \Delta U = m \cdot C_V \cdot dT = 0$ as

$$T = C$$

$$\therefore \quad dT = 0$$

(iii) From I–law $Q - W = \Delta U$ as $\Delta U = 0$

$$Q = W = P_1V_1 \ln \frac{V_2}{V_1}$$

12. Polytropic process $(PV^{\gamma}) = C$

(i) Work done $W_{1-2} = \int P.dV$

$$W_{1-2} = \frac{P_2V_2 - P_1V_1}{1-n}$$

$$W_{1-2} = \frac{mR(T_2 - T_1)}{1-n}$$

As $PV = mRT$

(ii) Heat transfer during polytropic process,

$$q = C_V \left[\frac{\gamma - n}{1-n}\right](T_2 - T_1)$$

$$q = C_n(T_2 - T_1)$$

where C_n = polytropic specific heat $= C_V \left[\frac{\gamma - n}{1-n}\right]$

Contd.

Also $$Q = \frac{\gamma - n}{\gamma - 1} \times \text{work done in a polytropic process}$$

13. For adiabatic process $PV^{\gamma} = C$

(i) $$W_{1-2} = \int P.dV = \frac{P_2V_2 - P_1V_1}{1-\gamma}$$

(ii) Heat transfer $Q = 0$

14. For,

(i) Constant pressure process,

(ii) Constant volume process,

(iii) Constant temperature process.

To find the property at second state use :

$$\frac{P_1V_1}{T_1} = \frac{P_2V_2}{T_2}$$

15. For polytropic process to find the properties at the second state use :

$$\frac{T_2}{T_1} = \left(\frac{P_2}{P_1}\right)^{\frac{n-1}{n}} = \left(\frac{V_1}{V_2}\right)^{n-1}$$

and for adiabatic process, in place of n, use γ.

SOLVED EXAMPLES

Example 6.1 A heat engine is supplied with 278 kJ/sec of heat at constant fixed temperature of 283° C and rejection takes place at 5°C. The following results were reported,

(i) 208 kJ/sec are rejected.

(ii) 139 kJ/sec are rejected.

(iii) 70 kJ/sec are rejected.

Classify which of the results report a reversible cycle or irreversible cycle or impossible cycle.

Solution

(i)
$$\oint \frac{\delta Q}{T} = \left(\frac{Q_1}{T_1}\right) + \left(\frac{Q_2}{T_2}\right)$$

$$= \left(\frac{278}{283+273}\right) + \left(\frac{-208}{5+273}\right)$$

$$= 0.5 - 0.748 = -2.48 < 0$$

∴ The cycle is Irreversible.

(ii)
$$\oint \frac{\delta Q}{T} = \left(\frac{278}{556}\right)+\left(-\frac{139}{278}\right)$$
$$= \frac{278}{556}-\frac{139}{278} = 0.5-0.5 = 0$$

∴ Cycle is reversible.

(iii)
$$\oint \frac{\delta Q}{T} = \left(\frac{278}{556}\right)+\left(-\frac{70}{278}\right)$$
$$= 0.5-0.252 = 0.248 > 0$$

This cycle is impossible according to Clausius inequality.

Example 6.2 2 kg of a gas is heat at constant volume in a non flow reversible process, till its pressure is doubled. The initial temperature of gas is 50°C. Determine change in entropy.

Take $C_v = 0.652$ kJ/kg-K and $C_p = 0.91$ kJ/kg-K.

Data: $m = 2$ kg, $P_2 = 2P_1$, or $\frac{P_2}{P_1} = 2$ (given), $T_1 = 50°C + 273$, $S_2 - S_1 = ?$, $C_v =$ 0.652 kJ/kg-K, $C_p = 0.91$.

Solution

We know that change of entropy for a constant volume process is,

$$S_2 - S_1 = m \cdot C_V \cdot \ln\frac{P_2}{P_1}$$

$$\left[\text{Units} : \text{kg}\times\frac{\text{kJ}}{\text{kg-K}}\times\text{ratio}=\frac{\text{kJ}}{\text{K}}\right]$$

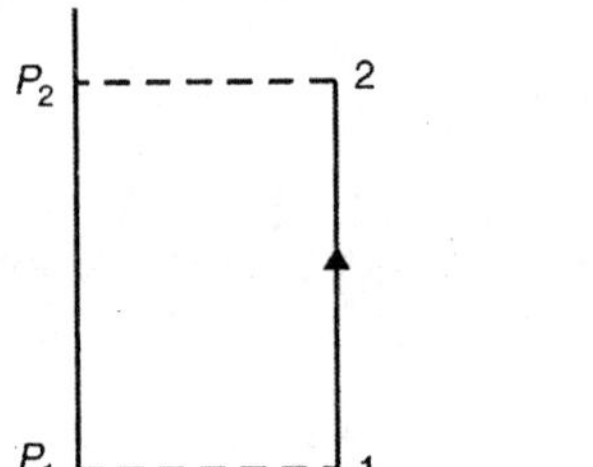

Fig. Ex. 6.2

$$= 2\times 0.652\times \ln = 2\times 0.652\times 0.6931$$

$$\mathbf{S_2 - S_1 = 0.904\ kJ/K}$$

Example 6.3 A quantity of gas has a pressure of 8 bar and it occupies a volume of 0.015 m³ at 160°C. The gas is expand isothermally to a volume of 0.085 m³. Determine change of entropy.

Data: $P_1 = 8\text{ bar} = 800\text{ kN/m}^2$ $V_1 = 0.015\text{ m}^3$

$T_1 = 160 + 273 = T_2$ $V_2 = 0.085\text{ m}^3, S_2 - S_1 = ?$

Solution

We know that change in entropy during isothermal process is given by,

$$S_2 - S_1 = mR.\ln\frac{v_2}{v_1}$$

$$S_2 - S_1 = mR.\ln\frac{V_2}{V_1}$$

Note. $\frac{v_2}{v_1}$ is a ratio, we can write ratio of specific volume is equal to the ratio of total volume

i.e. $$\frac{v_2}{v_1} = \frac{V_2}{V_1}$$

We know that $$P_1V_1 = mRT_1$$

or $$\frac{P_1V_1}{T_1} = mR$$

Substituting this in above equation we can write,

$$S_2 - S_1 = \frac{P_1V_1}{T_1}\ln\frac{V_2}{V_1}$$

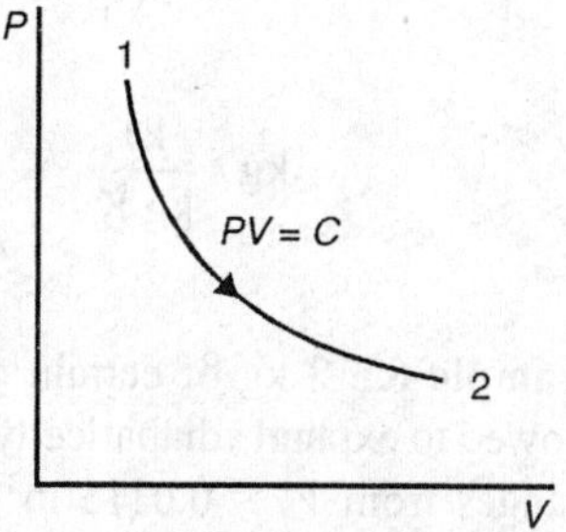

Fig. Ex. 6.3

Note. $$\frac{\text{kN/m}^2 \times \text{m}^3}{\text{K}} = \frac{\text{kNm}}{\text{K}}$$

$$= \frac{800 \times 0.015}{433} \times \ln\frac{0.0585}{0.015} = \frac{\text{kJ}}{\text{K}}$$

$$\mathbf{S_2 - S_1 = 0.048\ kJ/K}$$

Example 6.4 3 kg of air at a pressure of 5 bar and 250°C expands reversibly in a polytropic process. The final pressure is 1.5 bar and index of expansion, is 1.25. Find the change in entropy during the process. Take $C_v = 0.652$ kJ/kg-K and $\gamma = 1.4$.

Data: $m = 3\text{ kg}$ $P_1 = 15\text{ bar} = 1500\text{ kN/m}^2$

$T_1 = 250 + 273$ $PV^n = C$

$P_2 = 1.5\text{ bar} = 150\text{ kN/m}^2$

$n = 125$ $S_2 - S_1 = ?$

Solution

Change in entropy in a polytropic process is given by (equation in terms of pressure ratio).

$$S_2 - S_1 = m\left[\frac{\gamma - n}{n} \cdot C_V \cdot \ln\frac{P_1}{P_2}\right]$$

$$= 3 \times \left[\frac{1.4 - 1.25}{1.25} \times 0.652 \times \ln\frac{15}{1.5}\right]$$

$$= 3[0.12 \times 0.652 \times 2.303]$$

$$S_2 - S_1 = 3[0.18015]$$

$$S_2 - S_1 = 0.5405 \text{ kJ/K}$$

Note. Unit of m is kg and

$$C_V = \frac{\text{kJ}}{\text{kg-K}}$$

$$\therefore \quad \text{kg} \times \frac{\text{kJ}}{\text{kg K}} = \frac{\text{kJ}}{\text{K}}$$

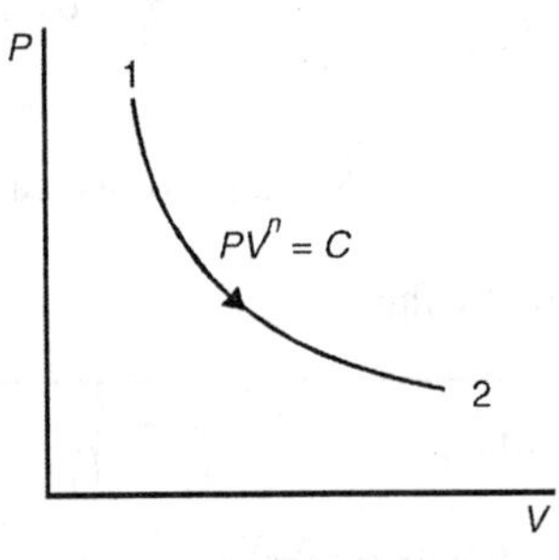

Fig. Ex. 6.4

Example 6.5 2 kg of certain gas with C_p = 0.85 kJ/kg-K and C_v = 0.70 kJ/kg-K is allowed to expand adiabatically through a partly opened valve, where by the its volume changes from V_1 = 0.0113 m³ to some higher value and during the process entropy increases by 0.8 kJ/K. Find the volume after expansion of gas.

Data:

$$m = 0.85 \qquad C_p = 0.85$$

$$C_v = 0.70 \qquad PV = C$$

$$V_1 = 0.113 \text{ m}^3 \qquad V_2 = ?$$

$$S_2 - S_1 = 0.8 \text{ kJ/K}$$

Solution

For free expansion process, $W = 0$ and given the process is adiabatic

$$Q = 0$$

$\therefore$ From I–law,

$$Q = \Delta U + W$$

$$0 = \Delta U + 0 = \Delta U = 0$$

$$\therefore \quad \text{Initial} = T_{\text{final}}$$

i.e.

$$T_1 = T_2$$

Using the general equation for change in entropy,

$$S_2 - S_1 = m\left[C_V \ln\frac{T_2}{T_1} + R\ln\frac{v_2}{v_1}\right]$$

Since $\quad T_1 = T_2$

$$S_2 - S_1 = mR\ln\frac{v_2}{v_1}$$

$$0.8 = 2\times[0.85 - 0.70]\ln\frac{v_2}{v_1}$$

$$\frac{0.8}{2\times 0.15} = \ln\frac{v_2}{v_1}$$

$$\ln\frac{v_2}{v_1} = 2.667$$

$$\therefore \quad \frac{v_2}{v_1} = 14.397 = \frac{V_2}{V_1}$$

Since $\quad \dfrac{v_2}{v_1} = \dfrac{V_2}{V_1}$ ratio

$$\therefore \quad V_2 = 13.397\times 0.0113\ \text{m}^3$$

$$\mathbf{V_2 = 0.1627\ m^3}$$

Fig. Ex. 6.5

Example 6.6 A quantity of gas has a pressure of 300 kN/m² when its volume is 0.025 m² and its temperature is 37°C. If the value of is 0.287 kJ/kg-K determine the mass of the gas present.

If the pressure of gas is now increased to 0.9 MN/m² while the volume remains constant what will be the new temperature of the gas.

Solution

(1) By characteristic gas equation

$$PV = mRT$$

$$\therefore \quad m = \frac{PV}{RT} = \frac{300\times 10^3 \times 0.025}{0.287\times 10^3 \times 310} = 0.084\ \text{kg}$$

(2) We have

$$\frac{P_1V_1}{T_1} = \frac{P_2V_2}{T_2} \text{ and in this case } V_1 = V_2$$

$$\therefore \quad T_2 = \frac{P_2 T_1}{P_1} = 310 \times \frac{0.9 \times 10^6}{300 \times 10^3} = 930 \text{ K}$$

$$\therefore \quad T_2 = 930 - 273 = 657°\text{C}$$

Example 6.7 10 kg of air is heated at constant pressure, from a temperature of 100°C to 200°C. Calculate the heat added during the process and also the change in I.E. Take gas constant $R = 0.287$ kJ/kg-K and $\gamma = 1.4$ for air.

Data: $m = 10$ kg, $t_1 = 100$ °C

$t_2 = 200$°C, $T_1 = 373$ K

$T_2 = 473$ K, $R = 0.287$ kJ/kg-K

$\gamma = 1.4$

Solution

We know that, heat added during constant pressure process.

$$= mC_P (T_2 - T_1)$$

Also, $$C_V = \frac{R}{\gamma - 1} = \frac{0.287}{1.4 - 1} = 0.7175$$

and $$C_P - C_V = R$$

$$\Rightarrow \quad C_P = R + C_V$$

$$C_P = 0.287 + 0.7175$$

$$\mathbf{C_p = 1.0045 \text{ kJ/kg-K}}$$

$\therefore$ Substituting (2) and (3) in (1) we get,

$$\text{Heat added} = mC_P (T_2 - T_1)$$

$$= 10 \times 1.0045 (200 - 100) = 1004.5 \text{ kJ}$$

$$\text{Charge in I.E.} = \Delta U = mC_V (T_2 - T_1) = 10 \times 0.7175 (200 - 100)$$

$$\mathbf{\Delta U = 717.5 \text{ kJ}}$$

Example 6.8 A mass of 0.8 kg of air at 1 bar and 25°C is contained in a gas tight frictionless piston cylinder device. The air is now compressed to a final pressure of 5 bar. During the process heat is transferred from the air such that the temperature inside the cylinder remains constant. Calculate the heat transfer and work done during the process and direction of each in the process.

Data: $m = 0.8$ kg, $P_1 = 1$ bar $= 10^5$ N/m²

$T_1 = 298$ K, $P_2 = 5$ bar $= 5 \times 10^5$ N/m²

$T = C$, Isothermal process.

Solution

$\therefore$ We know that, for an isothermal process,

$$Q = W = P_1V_1 \ln\frac{P_1}{P_2} = mRT_1 \ln\frac{P_1}{P_2} = -110.11 \text{ kJ}$$

As Q is –ve $\therefore$ W is also –ve.

Example 6.9 1 kg of air at an initial condition of 3 bar and 175°C expands adiabetically without friction to a final condition of 1 bar. Calculate:

(a) External work done during expansion.

(b) Change in Internal Energy.

Take $\gamma = 1.4$ and $R = 0.287$ kJ/kg-K

Data: $m = 1$ kg $\quad P_1 = 3 \times 105$ N/m^2 $= 3 \times 10^2$ kN/m^2

$T_1 = 175 + 273 = 448$ K $\quad P_2 = 1 \times 10^2$ kN/m^2

$\gamma = 1.4$ $\quad R = 0.287$ kJ/kg-K

Solution

(a) We know that the work done in an adiabatic expansion process is given by,

$$W_{1-2} = \frac{P_1V_1 - P_2V_2}{\gamma - 1}$$

Here V_1 and V_2 are not given,

Since it is an adiabatic process,

$$P_1V_1^\gamma = P_2V_2^\gamma$$

$$\frac{P_1}{P_2} = \left(\frac{V_2}{V_1}\right)^\gamma$$

$$\left(\frac{P_1}{P_2}\right)^{1/\gamma} = \frac{V_2}{V_1}$$

$$\therefore \quad V_2 = V_1 \times \left(\frac{P_1}{P_2}\right)^{1/\gamma}$$

To find V_1 $\quad P_1V_1 = mRT_1$

$$V_1 = \frac{mRT_1}{P_1} = \frac{1 \times 0.287 \times 448}{3 \times 10^2}$$

$$\text{Note units : }\left[\frac{\text{kg}\dfrac{\text{kJ}}{\text{kg-K}}\times\text{K}}{\text{kN/m}^2}=\frac{\text{kJ}}{\text{kN/m}^2}=\frac{\text{kNm}}{\text{kN}}\times\text{m}^2=\text{m}^3\right]$$

∴ From Eq. (2),

$$V_2 = 0.429\times\left(\frac{300}{100}\right)^{1/1.4} = 0.429\times(3)^{1/1.4}$$

$$\mathbf{V_2 = 0.940\ m^3}$$

∴ Now from Eq. (1), work done,

$$W_{1-2} = \frac{P_1V_1 - P_2V_2}{\gamma - 1}$$

$$= \frac{300\times0.429 - 100\times0.940}{1.4-1} = \frac{34.7}{0.4}$$

$$\mathbf{W_{1-2} = 86.75\ kJ}$$

(b) We also know from the I–law, that for a process, $Q = \Delta U + W$. Since it is adiabatic process $Q = 0$

∴ $W = -\Delta U = 86.75\ \text{kJ} \Rightarrow \mathbf{\Delta U = -86.75\ kJ}$

Therefore internal energy decreases.

Example 6.10 Find the change of Internal energy and enthalpy of 3 kg of air, when temperature changes from 25°C to 100°C.

Take $C_p = 1$ kJ/kg-K and $C_v = 0.71$ kJ/kg-K.

Solution

In the problem, the type of process is not given, but since temperature change, it is not an isothermal process.

∴
$$\Delta U = m\cdot C_V\cdot(T_2 - T_1)$$

$$= 3\ \text{kg}\times0.71\ \text{kJ/kg-K}\times(373-298)\ \text{K}$$

$$= 3\times0.71\times75$$

$$\mathbf{\Delta U = 159.75\ kJ}$$

As change of enthalpy $= \Delta H = mC_P\cdot(T_2 - T_1)$

$$= 3\times1\times(75)$$

$$\mathbf{\Delta H = 225\ kJ}$$

Also from the first law, $Q - W = \Delta U$

We know that heat supplied is utilised for,

(i) Increasing I.E.

(ii) For doing external work

And heat transfer,

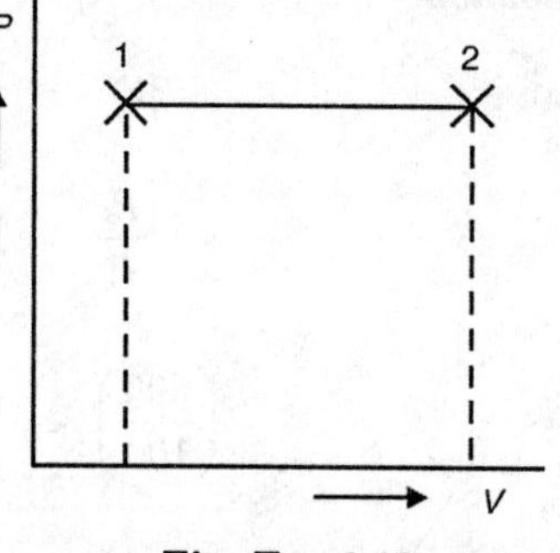

Fig. Ex. 6.12

$$\delta q = dh$$

or $$\delta Q = dH = mC_p\Delta T$$

Also, $$\Delta U = mC_p\Delta T$$

$$\frac{\Delta U}{m \cdot C_V} = \Delta T$$

$$\frac{-80 \text{ kJ}}{2 \text{ kg} \times 0.8 \text{ kJ/kg-K}} = \Delta T \quad (3)$$

$$= \frac{-80}{1.6} = \Delta T$$

$$-50\text{K} = \Delta T$$

$\therefore$ From (3) $$\delta Q = m.C_P.\Delta T = 1 \times 1.12 \times (-50)$$

$$\delta Q \text{ or } Q = -112 \text{ kJ}$$

$\therefore$ From Eq. (2),

$$Q - W = \Delta U$$

$$Q - \Delta U = W$$

$$-112 + 80 = W$$

$\therefore$ $$\mathbf{W = -32 \text{ kJ}}$$

Example 6.13 A cylinder contains 0.12 m³ of air at 1 bar and 100°C. The air is compressed to 0.03 m³. The final being 6 bar.

Determine,

(i) The value of index n

(ii) Mass of air in the cylinder.

(iii) Increase in internal energy .

Take $\gamma = 1.4$ $R = 0.287$ kJ/kg-K and $C_v = 0.72$ kJ/kg-K

Data: $V_1 = 0.12 \text{ m}^3$ $\qquad P_1 = 100 \text{ kPa}$

$T_1 - 100 + 273 = 373 \text{ K}$ $\qquad V_2 = 0.03 \text{ m}^3$

$P_2 = 600 \text{ kPa}$

Solution

So
$$P_1V_1^n = P_2V_2^n$$

$$\frac{P_2}{P_1} = \left(\frac{V_1}{V_2}\right)^n$$

$$\left(\frac{600}{100}\right) = \left(\frac{0.12}{0.03}\right)^n$$

$$6 = (4)^n$$

$$\log_e 6 = n\log_e (4)$$

$$\frac{\log_e 6}{\log_e 4} = n$$

$$\frac{1.79175}{1.38629} = n$$

$$n = 1.29248$$
$$n = 1.3$$

(ii) To find mass of air in the cylinder

We know that, $P_1V_1 = mRT_1$

$$100\times 0.12 = m\times 0.287\times 383$$

$\therefore$ $\mathbf{m = 0.112096\ kg}$

To find the increase in internal energy

(iii) We have, $\Delta U = m\cdot C_V(T_2 - T_1)$

Here T_2 is not known, so to find T_2, since it is a polytropic process,

$$\therefore \quad \frac{T_2}{T_1} = \left(\frac{V_1}{V_2}\right)^{n-1} = \left(\frac{P_2}{P_1}\right)^{\frac{n-1}{n}}$$

$$T_2 = T_1\times\left(\frac{V_1}{V_2}\right)^{n-1} = 373\times\left(\frac{0.12}{0.03}\right)^{1.3-1}$$

$$= 373\times(4)^{0.3}$$

$$= 373\times 1.516$$

$$\mathbf{T_2 = 565.36\ K}$$

$$\therefore \quad \Delta U = m\cdot C_V(T_2 - T_1)$$

$$= 0.112\ \text{kg} \times 0.72(565-373)$$

$$\Delta \mathbf{U} = \mathbf{15.51\ kJ\ (Increase)}$$

Example 6.14 Calculate the change of,

(i) Internal energy and

(ii) Enthalpy of 2 kg of air when the temperature changes from 20°C to 90°C. Take $C_p = 1$ kJ/kg-K and $C_v = 0.71$ kJ/kg-K

Solution

(i)
$$\Delta U = m \cdot C_V \cdot (T_2 - T_1) = 2 \times 0.71(90-20)$$
$$\Delta U = 99.4\ \text{kJ}$$

(ii)
$$\Delta H = m.C_P.(T_2 - T_1) = 2 \times 1 \times (90-20)$$
$$\boldsymbol{\Delta H = 140\ \text{kJ}}$$

Example 6.15. 10 kg of air are heated in a right vessel from 20°C to 100°C. If the ratio of specific heats is 1.4, estimate the values of C_p and C_v the changes in internal energy and enthalpy.

Solution

We have,
$$T_1 = 20°,\ T_2 = 100°\text{C}$$

$$C_V = \frac{R}{\gamma - 1}$$

and
$$\gamma = 1.4 \text{ and } R = 0.287$$

$$\therefore \quad C_V = \frac{0.287}{0.4} = 0.7175\ \text{kJ/kg-K}$$

$$C_P = R + C_V = 0.287 + 0.7175 = 1.0045\ \text{kJ/kg°K}$$

We know that,

Change in internal energy

$$dU = mC_V(T_2 - T_1)$$
$$= (10)0.7175(100-20)$$
$$\mathbf{dU = 574\ kJ}$$

And change in enthlpy

$$dH = mC_P(T_2 - T_1) = (10)(1.0045)(100-80) = \mathbf{803.6kJ}$$

Example 6.16 A piston and cylinder machine contains 1 kg of air, initially $v = 0.8$ m^3 and $T = 290$ K. The air is then compressed in a slow frictionless process specific volume 0.2 m^3/kg and a temperature 580 K. The law for compression is $PV^{1.5} = 0.75$ with P in bar and V in m^3/kg. Determine work done and heat transferred during the process. Assume for air $C_p = 1.0$ kJ/kg-K, $C_v = 0.743$ kJ/kg-K and $R = 0.287$ kJ/kg-K.

Data: $m = 1$ kg, $V_1 = 0.8$ m^3/kg

$T_1 = 290$ K, $V_1 = 0.2$ m^3/kg

$T_2 = 580$ K, W.D. = ? Q = ?

Solution

Law for the process $= PV^{1.5} = C = 0.75$

We know that, for a polytropic process,

$$\text{W.D.} = \frac{P_2V_2 - P_1V_1}{1-n} = \frac{mR(T_2 - T_1)}{1-n}$$

$$= \frac{1 \times 0.287(580 - 290)}{1 - 1.5} = \frac{83.23}{-0.5}\text{ kJ}$$

$$= -166.46 \text{ kJ}$$

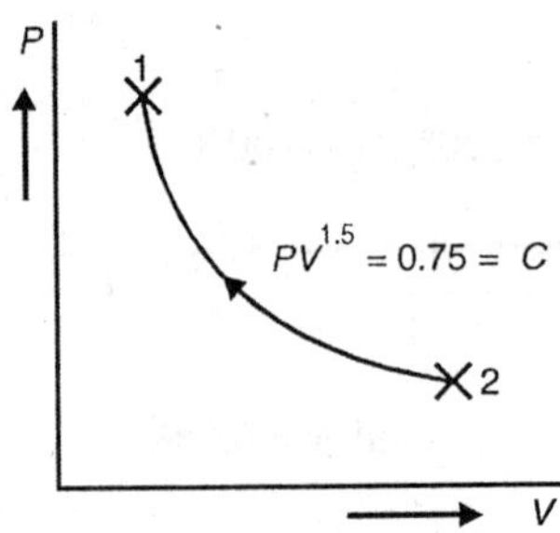

Fig. Ex. 6.16

The sign implies work is done on the air.

Also,

$$\Delta U = m.C_v dT = 1 \times 0.743 \times (580 - 290)$$

$$= 206.77 \text{ kJ}$$

From I-law, for a process,

$$Q - W = \Delta U$$

$$Q = \Delta U + W$$

$$= 206.77 + (-166.46)$$

$$\mathbf{Q = 40.31\ kJ}$$

Example 6.17 One kg of N_2 at a temperature of 250° C occupies a volume of 0.2 m^3. The gas undergoes a constant pressure expansion without friction to a final volume of 0.36 m^3. Find final pressure, final temperature, work done, change in internal energy, heat transfer and change of entropy.

Take $C_v = 0.743$ kJ/kg-K and $R =$ A297 kJ/kg-K and also represent the process on P–V and T–S diagrams.

Data:

$$m = 1\text{ kg} \qquad T_1 = 250 + 273 = 523\text{ K}$$

$$V_2 = 0.36\text{ m}^3 \qquad V_1 = 0.2\text{ m}^3.$$

$$P_2 = ? \qquad P = C \qquad T_2 = ?$$

$$W = ? \qquad \Delta U = ? \qquad Q = ?\ \Delta S = ?$$

Solution

We know that,

$$\frac{P_1V_1}{T_1} = \frac{P_2V_2}{T_2}$$

Since $P = C,\ P_1 = P_2 = C$

$$\therefore \quad \frac{V_1}{T_1} = \frac{V_2}{T_2}$$

$$\therefore \quad T_2 = T_1 \times \frac{V_2}{V_1} = 523 \times \frac{0.36}{0.20}$$

$$T_2 = 941.4\text{ K}$$

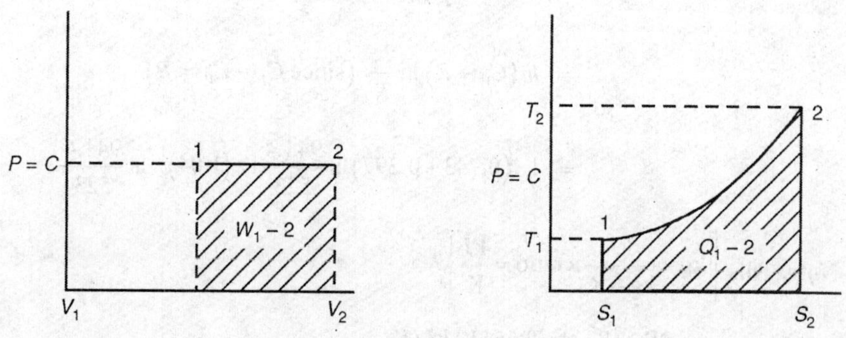

Fig. Ex. 6.17

Since $P_1V_1 = mRT_1$

$$P_1 = \frac{mRT_1}{V_1} = \frac{1 \times 0.297 \times 523}{0.2}$$

Note. Units : $\left[\dfrac{\text{kg}\times\dfrac{\text{kJ}}{\text{kg}-\text{K}}\times\text{K}}{\text{m}^3}=\dfrac{\text{kJ}}{\text{m}^3}=\dfrac{\text{kNm}}{\text{m}^2}=\text{kPa}\right]$

$$P_1 = 776.655 \text{ kPa} = P_2$$

(iii) Work done in a non-flow process = $\int PdV$

(since it is a constant pressure process).

$$= P(V_2 - V_1)$$

$$= 776.655(0.36 - 0.20)$$

$$W_{1-2} = 124.26 \text{ kJ}$$

(iv)
$$\Delta U = m.C_V.(T_2 - T_1) = 1\times 0.743\times(941.4 - 523)$$

$$= 310.87 \text{ kJ} (+\text{ve so I.E. increases})$$

Note. This $\left[\text{kg}\times\dfrac{\text{kJ}}{\text{kg}-\text{K}}\times\text{K}=\text{kJ}\right]$

(v) We also know from the I-law, for a closed system undergoing a process,

$$Q = \Delta U + W = 310.87 + 124.26$$

$$Q = 435.13 \text{ kJ}$$

(vi) Change in entropy $= m\cdot C_P\cdot \ln\dfrac{T_2}{T_1}$

$$= m(C_V + R)\ln\frac{T_2}{T_1} \quad (\text{since } C_P - C_V = R)$$

$$= 1\times(0.743 + 0.297)\ln\frac{941.4}{523} 1\times(1.04)\times\ln\frac{941.4}{523}$$

Note. This $\left[\text{kg}\times\dfrac{\text{kJ}}{\text{kg}-\text{K}}\times\text{ratio}=\dfrac{\text{kJ}}{\text{K}}\right]$

$$S_2 - S_1 = 0.66112 \text{ kJ/K}$$

Example 6.18 1 m^3 of gas is filled in a closed tank. The initial condition of the gas is 3 bar and 50°C. The gas is heated until the pressure becomes 5 bar. Find the change in internal energy, work done, heat supplied, change in entropy. Take $R = 0.287$ kJ/kg-K and $C_v = 0.711$ kJ/kg-K.

Solution

Since it is a constant volume heating process,

$$V_1 = V_2 = C = 1\text{ m}^3$$

$$P_1 = 3\text{ bar} = 3\times10^5\text{ N/m}^2$$

$$T_1 = 50+273 = 323\text{ K}$$

$$P_2 = 5\text{ bar} = 5\times10^5\text{ N/m}^2$$

$$R = 0.287\text{ kJ/kg-K} = 287\text{ J/kg-K}$$

$$C_V = 0.711\text{ kJ/kg-K} = 711\text{ J/kg-K}$$

We know that change in IE in a constant volume process is given by,

$$\Delta U = m\cdot C_V\left(T_2 - T_1\right) \qquad (1)$$

Here m and T_2 are not given

$\therefore$ To find T_2

Since $$\frac{P_1V_1}{T_1} = \frac{P_2V_2}{T_2}$$

as $$V_1 = V_2 = C$$

$$T_2 = \frac{P_2}{P_1}\times T_1$$

$$T_2 = \frac{5\times10^5}{3\times10^5}\times 323$$

$$T_2 = 538.33\text{ K}$$

And since $$P_1V_1 = mRT_1$$

$$m = \frac{R_1V_1}{RT_1} = \frac{3\times10^5\times1}{287\times323}$$

Note. Units: $\left[\dfrac{\dfrac{\text{N}}{\text{m}^2}\times\text{m}^3}{\dfrac{\text{J}}{\text{kg-K}}\times\text{K}} = \dfrac{\text{Nm}}{\text{J/kg}} = \dfrac{\text{J}}{\text{J/kg}} = \text{kg}\right]$

$$m = 32362\text{ kg}$$

$\therefore$ From Eq. (1),

$$\Delta U = m \cdot C_V (T_2 - T_1)$$

$$= 3.236 \times 711 \times (538.33 - 323)$$

Note this $\left[\text{kg} \times \frac{\text{J}}{\text{kg-K}} \times \text{K} = \text{J} \right]$

$$\Delta U = 595462.7 \text{ J} = 495.4627 \text{ kJ}$$

(ii) Work done. In a constant volume heating process, heat supplied is completely utilised for increasing the I.E. of the gas and no external work is done since volume remains constant.

$$W_{1-2} = \int P dv = 0$$

As $V = C,\ dV = 0$

(iii) From first law

$$Q - W = \Delta U$$

$$Q = \Delta U + W$$

$$Q = \Delta U \text{ as } W = 0$$

$$Q = \Delta U = 495.4627 \text{ kJ}$$

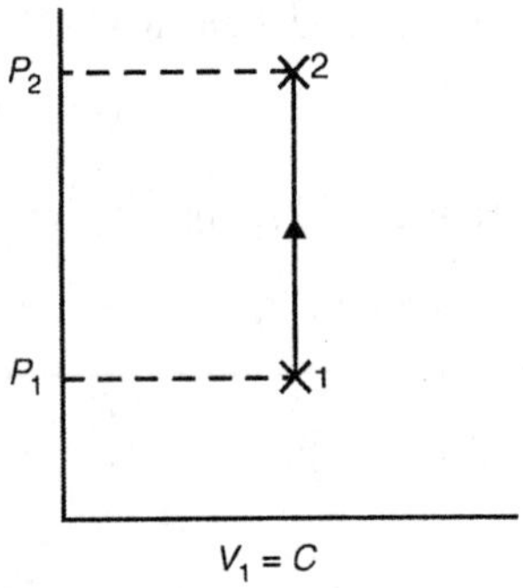

(iv) $$S_2 - S_1 = mC_V \ln \frac{P_2}{P_1}$$

$$= 3.23 \times 0.711 \times \ln \frac{5 \times 10^5}{3 \times 10^5}$$

Note this $\left[\text{kg} \times \frac{\text{J}}{\text{kg-K}} \times \text{ratio} = \frac{\text{kJ}}{\text{K}} \right]$

$$S_2 - S_1 = 1.173 \text{ kJ/K}$$

Fig. Ex. 6.18

Example 6.19 2 kg of air at 150°C and 3 bar expands according to $PV^{1.2} = C$ to a final pressure of 1 bar. Find W, Q, and ΔU.

Take $R = 0.287$ kJ/kg-K and $\gamma = 1.4$

Data: $m = 2$ kg $\quad T_1 = 150 + 273 = 423$ K

$P_1 = 3 \times 10^3$ N/m² $\quad n = 1.2 \quad P_2 = 1 \times 10^5$ N/m²

Find: (i) m = ? (ii) ΔU = ? Q = ? $R = 0.287$ kJ/kg-K, $\gamma = 1.4$

Solution

(i) We know that, work done in a polytropic process is given by,

$$W_{1-2} = \int PdV = \frac{P_1V_1 - P_2V_2}{n-1}$$

$$= \frac{mR(T_1 - T_2)}{n-1}$$

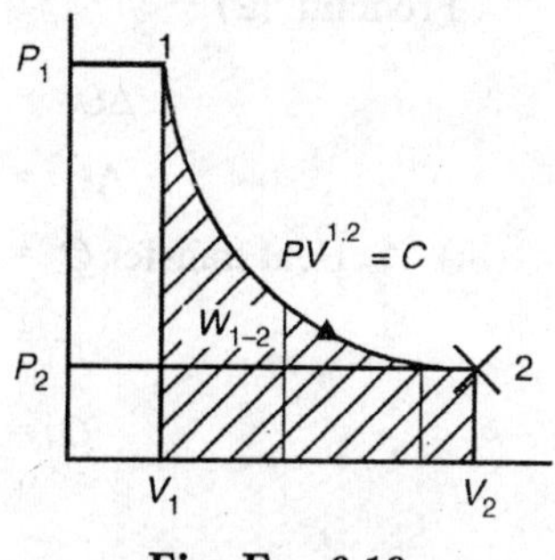

Fig. Ex. 6.19

Since T_2 is not given, to find T_2, use the relation

$$\frac{T_2}{T_1} = \left(\frac{P_2}{P_1}\right)^{\frac{n-1}{n}}\left[\frac{V_1}{V_2}\right]^{n-1}$$

Since it is a polytrope process.

[Note that if it is $P = C; V = C; T = C$ process then use $\frac{P_1V_1}{T_1} = \frac{P_2V_2}{T_2}$ to find the unknown]

$$\therefore \quad \frac{T_2}{T_1} = \left(\frac{P_2}{P_1}\right)^{\frac{n-1}{n}}$$

$$T_2 = T_1 \times \left(\frac{P_2}{P_1}\right)^{\frac{n-1}{n}} = 423 \times \left[\frac{1\times10^5}{3\times10^5}\right]^{\frac{1.2-1}{1.2}}$$

$$= 423 \times \left[\frac{1}{2}\right]^{\frac{0.1}{1.2}}$$

$$\mathbf{P_2 = 325.25\ K}$$

W_{1-2} from Eq. (1)

$$= \frac{mR(T_1 - T_2)}{n-1} = \frac{2\times0.287(423 - 352.25)}{1.2 - 0.2}$$

$$W_{1-2} = 203\ \text{kJ}$$

(ii) Change in IE,

$$\Delta U = m \cdot C_V (T_2 - T_1)$$

Now, $$\frac{C_P}{C_V} = \gamma = 1.4$$

$$C_P - C_V = T = 0.287$$

And $$C_V = \frac{R}{\gamma - 1} = \frac{0.287}{1.4-1} = 0.7175\ \text{kJ/kg-K}$$

∴ From Eq. (2)

$$\therefore \qquad \Delta U = 2\times 0.7175\times(352.25-423)$$

$$\mathbf{\Delta U = -101.52\ kJ\ (Decreases)}$$

(iii) ∴ Heat transfer $Q = \Delta U + W$

$$= -102.52+203\ \text{kJ}$$

$$Q = 101.48\ \text{kJ}$$

(iv) $$S_2 - S_1 = m\left[\frac{\gamma-n}{n}\times C_V\times \ln\frac{P_2}{P_1}\right]$$

$$= 2\left[\frac{1.4-1.2}{1.2}\times 0.7175\times \ln\frac{3}{1}\right]$$

$$\mathbf{S_2 - S_1 = 0.26916\ kJ/K}$$

For selecting the equation to find the change of entropy, note that it is a polytropic process and pressure ratio is given, also we can select the equation in which temperature ratio is given for this problem, since temperature are given. Both the equations will give the same answer.

Example 6.20 One kg of Nitrogen at a temperature of 150°C occupies a volume of 0.2 m³. The gas undergoes a fully restricted constant pressure expansion without friction to a final volume of – 0.36 m³. Evaluate the final pressure, final Take C_p = 743 J/kg-K, R = 0.297 kJ/kg-K.

Solution

(a) We know that,

$$P_1V_1 = mRT_1,\ m = 1$$

$$\therefore \qquad P_1 = \frac{RT_1}{V_1} = \frac{0.297\times 10^3\times 423}{0.2}$$

$$= 628.155\ \text{kPa}$$

$$= P_2 = \text{final pressure.}$$

(b) We know that for a constant pressure process,

$$\frac{P_1V_1}{T_1} = \frac{P_2V_2}{T_2}\ (P_1 = P_2)$$

$$\therefore \qquad \frac{V_1}{T_1} = \frac{V_2}{T_2}$$

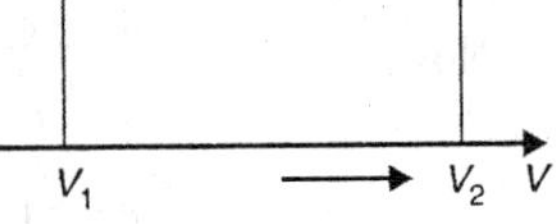

Fig. Ex. 6.20

$$\therefore \quad T_2 = \frac{V_2}{V_1} \cdot T_1 = \frac{0.36}{0.2} \times 423$$

$$= 761.4 \text{ K or } 488.4°\text{C}$$

(c) Work done in constant pressure process is given by

$$\delta Q = dH$$

$$\therefore \text{ For unit mass,} \quad \delta q = dh = C_P dT = C_P (T_2 - T_1)$$

$$= 0.743 \times (761 - 423) = 251.43 \text{ kJ/kg}$$

(d) Change in entropy

$$\therefore \quad s_2 - s_1 = C_P \ln \frac{T_2}{T_1}$$

$$= 0.743 \times \ln \frac{761.4}{423}$$

$$\mathbf{s_2 - s_1 = 0.437 \text{ kJ/kg-K}}$$

Example 6.21 1 kg of ice at –5°C is exposed to the surrounding atmosphere which is at 20°C. The ice melts and comes into thermal equilibrium with the atmosphere which is at 20°C. The ice melts and comes into thermal equilibrium with the atmosphere. Determine the entropy increase of the universe. Show the melting process on *T–S* diagram. Take, C_p of ice = 2.093 kJ/kg K, Latent heat of fusion of ice = 333.3 kJ/kg.

Solution

The melting is shown on *T–S* diagram $T_1 = 26$ K, $\quad T_2 = T_3 = 273$ K, $T_4 = 293$ K

With reference to *T–S* diagram we have

$$S_2 - S_1 = C_P \log_e \frac{T_2}{T_1}$$

$$= 2.093 \log_e \frac{273}{268}$$

$$= 2.093 \times \log_e 1.0186$$

$$= 0.0387 \text{ kJ/kg-K}$$

$$S_3 - S_2 = h_{fg} / T_3$$

$$= \frac{33.3}{273} = 1.221 \text{ kJ/kg-K}$$

$$S_4 - S_3 = C_{Pw} \log_e \frac{T_4}{T_3}$$

$$= 4.187 \log_e \frac{293}{273}$$

$$= 0.296 \text{ kJ/kg-K}$$

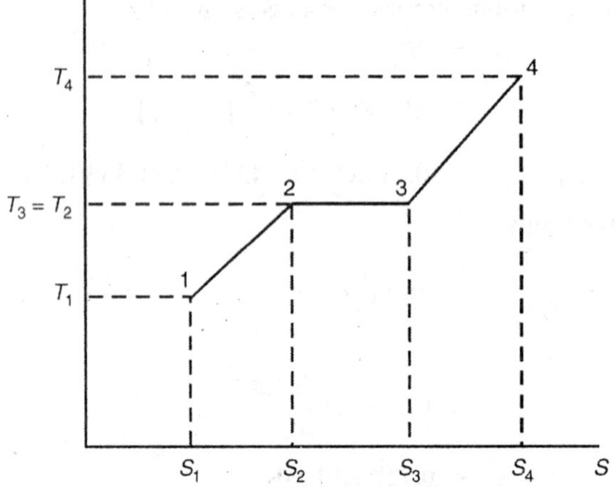

Fig. Ex. 6.21

∴ Total change in entropy of ice/water system

$$= S_4 - S_1 = 0.296 + 0.0387 + 1.221$$

$$= 1.556 \text{ kJ/kg-K}$$

Heat which is taken from surroundings

$$= 2.093(5) + 33.3 + 4.187(20)$$

$$= 10.465 + 333.3 + 83.74$$

$$= 427.505 \text{ kJ}$$

This heat is given by surrounding at constant temperature 293 K

∴ Change in entropy of the surrounding

$$= \frac{427.505}{293} = 1.4591 \text{ kJ/kg-K}$$

∴ Increase of entropy of universe

$$= \Delta S_p = 1.5560 - 1.4591$$

Increase of entropy of universe = 0.0969 kJ/kg-K

THEORY QUESTIONS

1. State and prove the principle of increase of entropy of the universe. Give some practical examples.
2. Prove that entropy is a property.
3. State and explain Clausius Theorem.
4. Explain Clausius inequality.
5. Write a short note on: Entropy and disorder.
6. Prove for polytropic process:

$$S_2 - S_1 = \frac{\gamma - n}{1 - n} C_v \ln\left(\frac{T_2}{T_1}\right)$$

7. Explain the principle of increase of entropy. What is its significance ?
8. Give the criteria of reversibility, irreversibility and impossibility of a thermodynamic cycle.
9. State and prove the principle of increase of entropy of the universe. Give some practical examples.
10. Derive the express for Clausius Inequality.
11. Straight change of entropy during a polytropic process for an ideal gas is given by,

$$\Delta S = C_v[k - n]\ln\frac{V_2}{V_1}$$

where k = superheat ratio, n = Polytropic index.

12. What is the principle of increase of entropy? Explain with two practical examples.
13. P.T. for a poly thermo process:

$$\Delta S = \frac{\zeta - n}{\zeta - 1} . R . \ln\frac{V_2}{V_1}$$

14. What do you understand by the term reversibility, when need in the thermodynamic source?

List to the best of your knowledge and explain all the possible irreverbilities that may be present in a system.

PROBLEMS FOR PRACTICE

1. 0.5 kg of air is compressed reversibly and adiabatically from 80 kPa 60°C to 0.4 MPa and is then expanded at constant pressure to the original volume. Sketch the process on *P–V* and *T–S* diagram, compute work transfer heat transfer and change in entropy for whole path.

   Take R = 0.287 kJ/kg-K and γ = 1.4. **(Ans.** 93.5 kJ, 0.578 kJ /K)

2. A certain mass of air initially at a pressure of 480 kPa and temperature 190° C is expanded adiabatically to a pressure of 94 kPa. It is then heated at constant volume until it attains its initial temperature when the pressure is found. to be 150 kPa. State the type of compressor necessary to bring the system back to its original pressure and volume. Determine (1) the index of adiabatic expansion (2) The work done per kg of air and (3) the change in specific entropy of air.

   R for air = 0.29 kJ/kg K. **(Ans.** 1.402, –31.59 kJ/kg)

3. An ideal gas having γ or K = 1.35 and volume is 1.5 m³ at 100 kPa and 77°C. The gas is compressed according tot he law $PV^{1.25} = C$ to a pressure 3 MPa. Calculate the volume and temperature at the end of compression. Evaluate the heat transfer, work done, ΔU and change of entropy during the process.

   Take C_p = 0.98 kJ/kg-K. **(Ans.** –167 kJ, –584.5 kJ, 417.5 kJ, –0.322 kJ/K)

4. Show that when 1 kg of perfect gas is expanded through a volume ratio r the law of expansion being PV^n = constant, the change of entropy is given $(C_p - nC_v)$ Inr.

5. 50 litres of air at 1.013 bar and 100°C temperature is compressed to 28 bar. Volume of air at the end of polytropic compression is found to be 4 litres. Air is now heated at constant volume till pressure rises to 56 bar. Assuming C_p = 1.00 kJ/kg-K and C_v = 0.71 kJ/kg-K over range of working determine.

   (i) Polytropic index of compression.

   (ii) Entropy change in each process. Sketch the process on $P–V$ and $T–S$ diagram.

   (Ans. (i) 1.31, (ii) –2.987, 10.5 kJ/K)

6. 1 kg of Nitrogen at a temperature of 150°C occupies a volume of 0.2 m³. The gas undergoes a fully restricted constant pressure expansion without friction to a volume of 0.36 m³. The gas is then expanded isothermally to a volume of 0.5 m³. Represent the process on $P–V$ and $T – S$ diagrams and determine the change in entropy for each process and overall change in entropy.

   Take C_v = 743 J/kg-K and R = 0.297 kJ/kg-K. **(Ans.** 0.6089, 0.0962, 0.7051)

7. 4 kg of air are compressed from 40°C and 125° kPa to 250°C and 875 kPa. It is then throttled to 57 kPa. Finally it is cooled to a pressure of 125 kPa and 180°C. Calculate the overall change in entropy and also for each process.

   Take C_p = 1.055 kJ/kg-K, C_v = 0.717 kJ/kg-K. **(Ans.** 1.4859)

8. A solar thermal power plant flat plate collects or parabolic concentrators of sufficient area on which solar energy is incident are used. Flat plate collectors can give highest temperature in cycle as 90°C while parabolic concentrators give a temperature of 250°C. If the ambient temperature is 27°C. Find the minimum theoretical, area required of the solar system developing 10 kW power.

(i) By using flat plate collector having 60% collection efficient.

(ii) By using parabolic concentrator having 50% collection efficiency.

Assume incident plate solar energy collector, we can write.

(**Ans.** (i) 96 m (ii) 46.7 m^2)

9. A novel reversible heat engine plot in (T–S) diagram as a circle. The maximum and minimum temperatures are 1100 K and 200 K respectively and the maximum entropy change in the cycle is 2 kJ/K. Calculate the heat added to the cycle, heat rejected, the net work output and thermal efficiency of the cycle.

(**Ans.** 2006.86 kJ, 593.119, 1413.7Z kJ, 70.45)

10. Certain gas at a pressure of 1.4 MN/m^2 and 360° C is expanded adiabatically to a pressure of 100 kJ/m^2. The gas is then heated at constant volume until it attains 300°C when the pressure is found to be 220 kN/m^2 and finally it is compressed isothermally to the original pressure of 1.4 MN/m^2. Sketch the process on P–V and T–S diagrams. For 0.23 of gas and evaluate.

(i) ΔU during adiabatic expansion.

(ii) ΔS for each process.

Assume: C_v = 0.705 kJ/kg-K. (**Ans.** (i) 55.987 kJ (ii) 0, 0.1278 kJ /K)

11. 60 litres of hydrogen at 20°C and 1 bar is compressed adiabatically to 9.8 bar. It is then cooled at constant volume to a pressure P_3 such as to have temperature of 20°C and further expanded isothermally so as to reach initial state: Find,

(i) Work done during each process.

(ii) Change in internal energy during each process.

Take C_p = 14.4 kJ/kg-K

C_v = 10.28 kJ/kg-K

for hydrogen and show the process on P–V and T–S plane.

(**Ans.** (i) –13.788 kJ (ii) 13.788 kJ)

12. 3 kg of air at a pressure of 150 kPa and temperature 360 K is compressed polytropically to 750 kPa according to law $PV^{1.2}$ = constant. The air is then cooled to initial temperature at constant pressure. The air is then heated at constant temperature till it reaches original pressure of 150 kPa. Draw the cycle on PV diagram and determine net work and heat.

(**Ans.** –73.21 kJ, –73.21 kJ)

13. Process 1: Air initially at 100 kPa and 50°C undergoes reversible adiabatic compression such that its volume is reduced to 1/5th of initial volume:

Process 2: Then 940 kJ/kg of heat is added to this air at constant volume.

Process 3: Process 2 is followed by reversible adiabatic expansion up to initial volume.

Process 4: Finally heat is rejected at constant volume so as to reach the initial condition.

Draw the four processes on one *PV* diagram. Determine the maximum temperature, and heat rejected per kg of air. Assume adiabatic index of compression and expansion of 1.4 and constant volume specific heat as 0:717 kJ/kg-K.

(**Ans.** 1926 K, – 493.8 kJ/kg)

ꗃ ꗄ

7

Reciprocating Air Compressors

CHAPTER OBJECTIVES

After reading this chapter you will be able to learn the following

- Classification of Compressors, PV diagrams with and without clearance, work Input, Isothermal efficiency and Isentropic Efficiency.
- Methods to Improve Isothermal Efficiency.
- Actual indicator diagram, Free Air Delivered.
- Multi Stage Compressors, work done, saving in work done etc.
- Capacity Control of Compressors.

7.1 INTRODUCTION

Air compressor is a power absorbing machine which provides high pressure air. It takes the air from the atmosphere, compresses it to a high pressure and high pressure air will be stored in a storage vessel (reservoir) from where it can be taken out for use.

There are many uses of high pressure air in industry. The main uses of compressed air are,

1. For inflating automobile tyres.
2. To clean workshop machines, generators etc.
3. To operate air operated drills, hammers (Pneumatic tools).
4. To inject fuel in the diesel engine cylinder.
5. In spray painting.
6. To operate air brakes in automobiles.
7. To operate compressed air engines/air motors in mines.

(In mines I.C. engines and electricity are not used because of fire risks. So high pressure air operated machines are used).

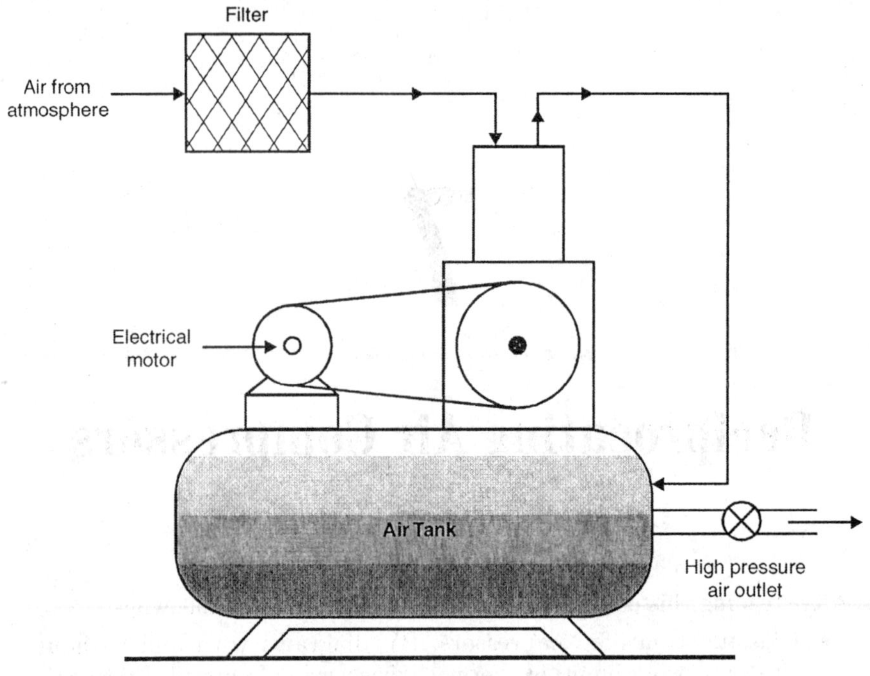

Fig. 7.1

7.2 CLASSIFICATION

Air compressors may be broadly classified as given below: (i) Reciprocating and (ii) Rotary.

The principle parts of a reciprocating air compressor are same as that of a reciprocating I.C. engine. The air compressor may also be as follows.

(i) Single Acting or Double Acting. The air is admitted to one side of the piston or air is admitted to either side of the piston alternately.

(ii) Single Stage/Multistage. In case of single stage air is compressors, air is compressed in a single cylinder completely, whereas in case of multistage air compressor, air where air is compressed in several stages. And each stage in pressure of air increases.

7.3 SINGLE STAGE/SINGLE ACTING RECIPROCATING AIR COMPRESSOR

It can be seen that the construction is just similar to reciprocating I.C. engine. In this case crank of the compressor is driven by means of electrical motor or engine output shaft.

During suction stroke, piston will be moving downwards and the pressure in the cylinder falls below the atmospheric pressure. As a result of which inlet valve opens and the air is drawn from the atmosphere. Then during compression stroke air will be compressed by means of upward moving piston and delivered through the delivery valve, when the pressure in the cylinder increases beyond the prespecified value of delivry valve.

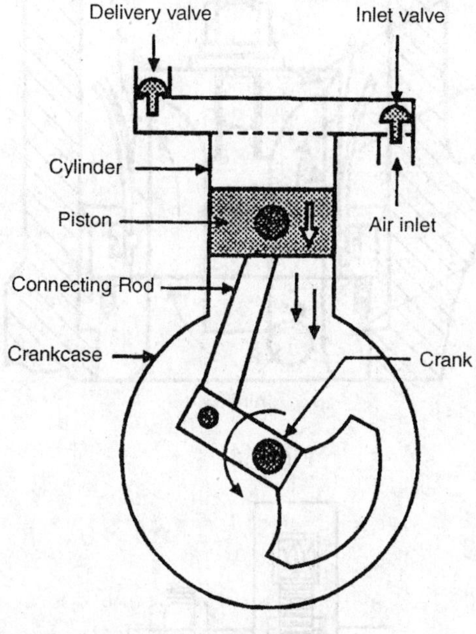

Fig. 7.2

7.3.1 Compressor Valves and Types

Compressor valves are automatic i.e. there is no valve operating mechanism provided, but they are operated because of the pressure difference.

Types of Compressor valves:

(i) Reed types

(ii) Plate type

(iii) Poppet type

(iv) Thimble valves

(i) Reed Type Valves. Figure 7.3 shows the reed type valve during compression the pressure of air increases and the pressure devleoped at (1) increases beyond the delivery chamber pressure (2), then the valve plate (which has 6–8 reeds) lifts and the air is delivered.

(ii) Plate Type Valves. As shown in Fig. 7.4 valve plate is held between valve seat and valve gaurd. In between valve plate and valve guard spring is provided. When the air pressure increases at (1), then the valve plate is lifted against then compression of spring and high pressure air is delivered to side (2).

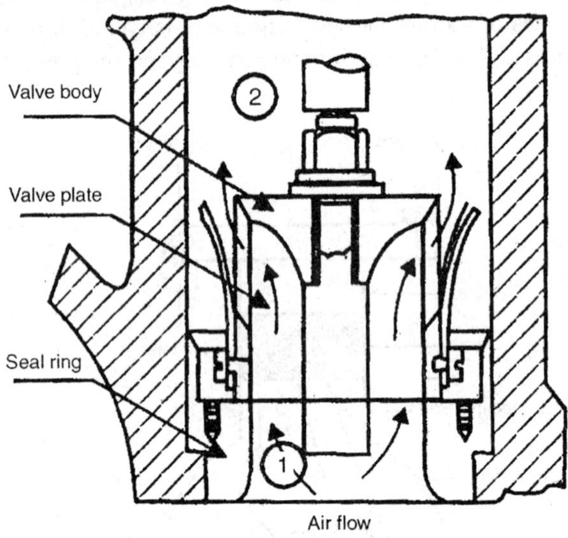

Fig. 7.3

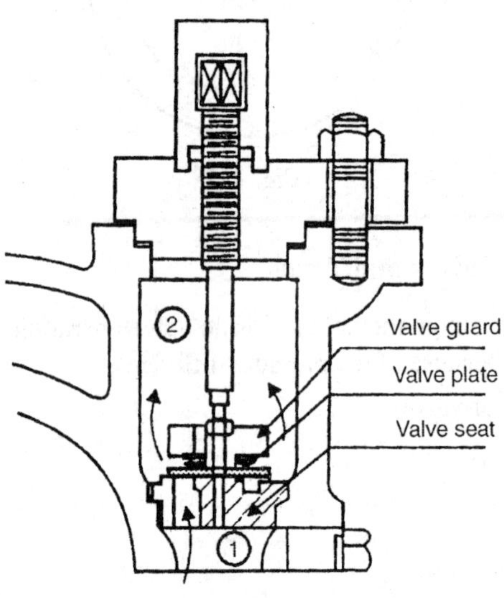

Fig. 7.4

7.4 INDICATOR DIAGRAM

The events described for the single stage air compressor can be represented on P–V diagram as shown in Fig. 7.5. The diagram is drawn for the compressor without clearance volume.

Clearance Volume. It is the volume of space provided between piston head and cylinder head when the piston is at TDC. This is provided in order to avoid striking of piston heat with cylinder head. Also in this clearance volume, during the delivery stroke some volume of air will be trapped, when the piston comes to TDC. This volume of air during the downward stroke expands and produces the necessary vacuum.

During suction stroke, the charge of air is drawn along 4–1 at constant pressure P_1, which is slightly below P_{atm}. At point 1 piston completes suction stroke and starts compression stroke. During starting both the valves being closed air will be compressed along 1–2. At point 2 pressure P_2 is reached which is slightly higher than the pressure in the receiver.

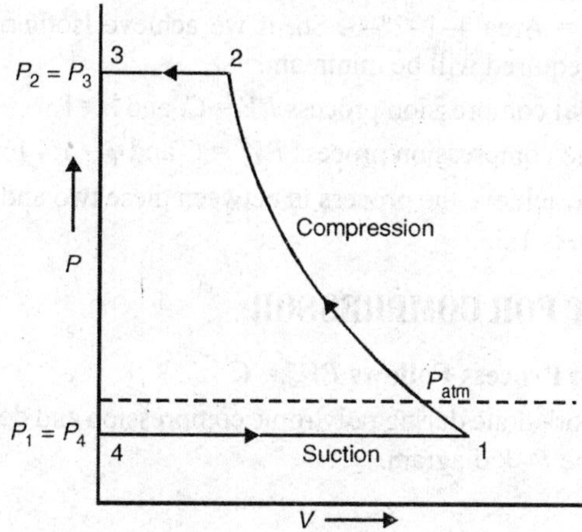

Fig. 7.5

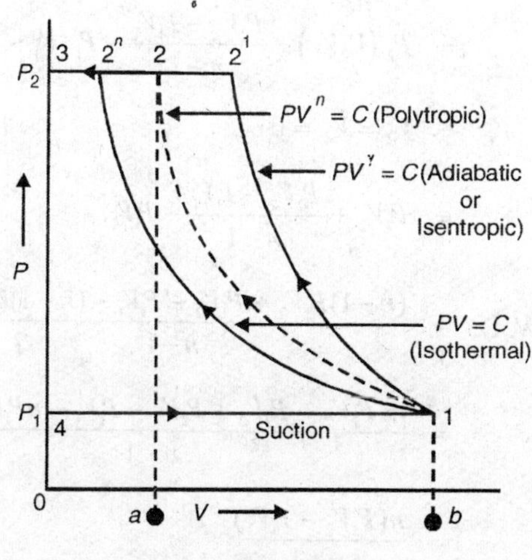

Fig. 7.6

The delivery valve at this point opens and the air is delivered along 2–3 into the receiver.

The net work input required for compression and delivery of air per cycle is given by the area 1–2–3–4 on the $P–V$ diagram.

The amount of work done on air depends upon the nature of compression curve. If the compression occurs very rapidly in a non-conducting cylinder so that heat transfer is zero, then the compression will be practically isentropic.

If it is carried out very slowly and heat of compression is extracted from air by jacket cooling water, then the compression will approach isothermal.

However in actual practice neither of these conditions can be fulfilled and the actual compression will be in between these two and is called as the polytropic compression as shown in Fig. 7.6.

From Fig. 7.6 we can see that the isentropic work done = Area 4–1–2'–3 and Isothermal work done = Area 4–1–2"–3. So, if we achieve isothermal compression curve, the work input required will be minimum.

Note: For isothermal compression process $PV = C$, and $n = 1$

For isentropic compression process $PV^{\gamma} = C$ and $\gamma = 1.4$ for air.

In actual practice we achieve the process in between these two and for water cooled cylinders the value of n is 1.3.

7.5 WORK INPUT FOR COMPRESSOR

(i) If the Compression Process Follows $PV^n = C$

As shown in Fig. 7.4 work done during polytropic compression and delivery is given by the area 1–2–3–4 on the $P–V$ diagram.

Now

Area $1-2-3-4$ = [Area $2-3-0-a$] + [Area $1-2-a-b$] – [Area $4-1-0-b$]

$$= P_2\left(V_2 V_3\right) + \frac{P_2V_2 - P_1V_1}{n-1} - P_1\left(V_1 - V_4\right)$$

As $$V_C = V_3 = V_4 = 0$$

$$= P_2V_2 + \frac{P_2V_2 - P_1V_1}{n-1} - P_1V_1$$

$$\text{W. D.} = \frac{(n-1)P_2V_2 + P_2V_2 - P_1V_1 - (n-1)(P_1V_1)}{n-1}$$

$$= \frac{n.P_2V_2 - P_2V_2 + P_2V_2 - P_1V_1 - nP_1V_1 + P_1V_1}{n-1}$$

$$= \frac{n(P_2V_2 - P_1V_1)}{n-1}$$

$$= \frac{n.P_1V_1\left\{\dfrac{P_2V_2}{P_1V_1} - 1\right\}}{n-1}$$

But as $PV^n = C$; $= \dfrac{V_2}{V_1} = \left(\dfrac{P_1}{P_2}\right)^{1/n} = \left(\dfrac{P_2}{P_1}\right)^{-1/n}$

$$\therefore \quad \text{W.D.} = \frac{n}{n-1} P_1 V_1 \left\{ \frac{P_2}{P_1} \left(\frac{P_2}{P_1}\right)^{-1/n} - 1 \right\}$$

Work Done or Indicated Power Required to Drive Compressors

$$\textbf{W.D or I.P.} = \frac{\mathbf{n}}{\mathbf{n-1}} \mathbf{P_1 . V_1} \left\{ \left(\frac{\mathbf{P_2}}{\mathbf{P_1}}\right)^{\frac{n-1}{n}} - 1 \right\}$$

Note generally,

W = in kW $\qquad n$ = index of compression

no units

P_1 = kN/m² $\qquad V_1$ = m³/sec

P_2 = kN/m² (delivery pressure)

Also

$$\textbf{W.D.} = \frac{\mathbf{n}}{\mathbf{n-1}} \mathbf{mRT_1} \left\{ \left(\frac{\mathbf{P_2}}{\mathbf{P_1}}\right)^{\frac{n-1}{n}} - 1 \right\}$$

As $P_1V_1 = mRT_1$ and as air is the ideal gas.

(ii) If the Compression Process follows $PV^\gamma = C$ (i.e. Adiabatic Process). The replace n by γ and the derivation is same as above and

$$\text{W.D.} = \frac{\gamma}{\gamma-1} P_1 V_1 \left[\left(\frac{P_2}{P_1}\right)^{\frac{\gamma-1}{\gamma}} - 1 \right] \text{kW}$$

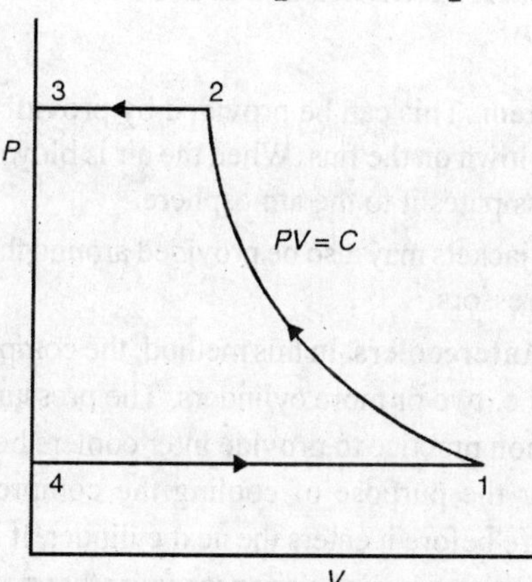

Fig. 7.7

(iii) If the Compression is Isothermal i.e. it follows PV = const. Then W.D = Area 1 – 2 – 3 – 4 = Area under 2 – 3 + Area under 1 – 2 – Area under 4 – 1

$$= P_2V_2 + P_1V_1 \ln\frac{V_1}{V_2} - P_1V_1$$

$$\textbf{W.D.} = \mathbf{P_1V_1 \ln\frac{V_1}{V_2}}$$

or

$$\textbf{W.D.} = \mathbf{P_1V_1 \ln\frac{P_2}{P_1}}$$

As $\quad P_1V_1 = P_2V_2 = C$

Note: All the expressions are arranged to give positive value, since we are interested in the magnitude of work input.

7.6 ISOTHERMAL EFFICIENCY

It is defined as the ratio of,

$$\eta_{iso} = \frac{\text{Isothermal work of compressor}}{\text{Actual work input}}$$

And Isothermal work input $= P_1V_1 \ln\dfrac{P_2}{P_1}$ or $= mRT_1 \ln\dfrac{P_2}{P_1}$ as we have studied in earlier article.

Note: η_{iso} gives an indication of closeliness of the compression process to isothermal compression i.e. lower the value of n higher is η_{iso}.

7.7 MEASURES OR METHODS TO IMPROVE ISOTHERMAL EFFICIENCY

(i) Effective Cooling System. This can be provided by providing fins on the cylinder and cylinder head and air blown on the fins. When the air is blown, it removes heat from the compressed air and dissipates it to the atmosphere.

Secondly cooling water jackets may also be provided around the cylinder and cylinder head for large sized compressors.

(ii) Multistaging with Intercoolers. In this method, the compression of air is carried out in two or more stages i.e. two or more cylinders. The pressure of air is increased in each cylinder. It is a common practice to provide inter coolers between the cylinders of multistage compressor for the purpose of cooling the compressed air to intake (or atmosphere) air temperature before it enters the next cylinder. It is this cooling between the cylinders that keeps the compression very near to isothermal.

(iii) Spray Injection. In this method a small amount of water is injected or sprayed into the compressed air. It immediately vapourises by absorbing heat from air and thereby keeps the temperature of air low.

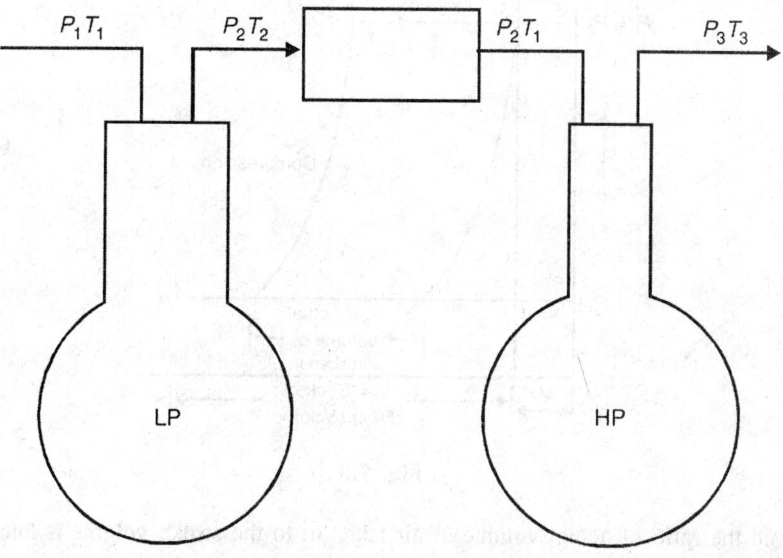

Fig. 7.8

(iv) Suitable Cylinder Dimensions. To have effective heat transfer, surface to volume ratio should be higher i.e. a cylinder with large dia. and short stroke should be selected. But this has the limitation since increased dia increases clearance volume and hence reduces volumetric efficiency.

7.8 EFFECT OF CLEARANCE ON VOLUMETRIC EFFICIENCY

The volume of space provided between piston heat and cylinder head when the piston is at TDC is known as clearance volume. In actual compressors it is provided to safeguard the piston from striking the cylinder head.

The events taking place in a compressor with clearance are same as those taking place in a compressor without clearance.

Figure 7.9 shows the *P–V* diagram for a single stage/single acting reciprocating air compressor with clearances volume. In this case both compression and expansion curves are assumed to be polytropic following the law $PV^n = C$.

As shown in Fig. 7.9, 1– 2 is polytropic compression, 2 – 3 is delivery stroke. At point 3, piston reaches TDC and delivery of high pressure air stops and some air is trapped in the clearance volume at the delivery pressure P_2.

As the piston moves downwards this air which is trapped in the clearance volume expands according to $PV^n = C$ and produces the necessary vacuum. At point 4 the inlet valve opens and actual volume of air taken is along 4–1.

Hence $V_1 - V_4 = V_a$ = Actual volume of air taken in and $V_1 - V_3$ = stroke volume = Swept volume.

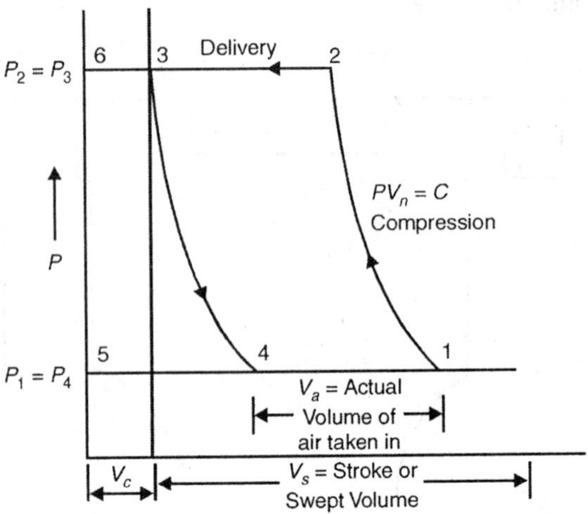

Fig. 7.9

And the ratio of actual volume of air taken in to the stroke volume is known as volumetric efficiency i.e.

$$\eta_{vol.} = \frac{\text{Actual volume of air taken in}}{\text{Stroke volume}}$$

$$= \frac{V_1 - V_4}{V_1 - V_3} = \frac{V_a}{V_s}$$

i.e.
$$\eta_{vol.} = \frac{\mathbf{V_a}}{\mathbf{V_s}} \quad (1)$$

For the process 3–4

$$P_3V_3^n = P_4V_4^n$$

$$\left(\frac{P_3}{P_4}\right)^{1/n} = \frac{V_4}{V_3}$$

$$V_3 \times \left(\frac{P_3}{P_4}\right)^{1/n} = V_4$$

$$V_c\left(\frac{P_2}{P_1}\right)^{1/n} = V_4 \; [\because V_3 = V_c \text{ and } P_3 = P_2 \text{ and } P_4 = P_1] \quad (2)$$

As we know actual volume of air taken in,

$$V_a = V_1 - V_4$$

$$= V_1 - V_c\left(\frac{P_3}{P_4}\right)^{1/n} = (V_c + V_s) - V_c\left(\frac{P_3}{P_4}\right)^{1/n} \quad \text{as } V_1 = V_c + V_s$$

or

$$= V_s + V_c - V_c\left(\frac{P_3}{P_4}\right)^{1/n} = V_s - V_c\left(\frac{P_3}{P_4}\right)^{1/n} + V_c$$

$$V_a = V_s - V_c\left\{\left(\frac{P_2}{P_1}\right)^{1/n} - 1\right\}$$

i.e.

$$V_a = V_s - V_c\left\{\left(\frac{P_2}{P_1}\right)^{1/n} - 1\right\} \quad ...(3)$$

$\therefore$ From (1),

$$\eta_{vol} = \frac{V_a}{V_s} = \frac{V_s - V_c\left\{\left(\frac{P_2}{P_1}\right)^{1/n} - 1\right\}}{V_s} = 1 - \frac{V_c}{V_s}\left\{\left(\frac{P_2}{P_1}\right)^{1/n} - 1\right\}$$

$$\eta_{vol} = \mathbf{1 - C\left\{\left(\frac{P_2}{P_1}\right)^{1/n} - 1\right\}}$$

or

$$\eta_{vol} = \mathbf{1 - C \times \left(\frac{P_2}{P_1}\right)^{1/n} + C}$$

or

$$\eta_{vol} = \mathbf{1 + C - C\left(\frac{P_2}{P_1}\right)^{1/n}}$$

Let the ratio $\frac{V_c}{V_s} = C =$ clearance ratio and $\frac{P_2}{P_1} =$ pressure ratio r.

The variation of volumetric efficiency with clearance volume and polytropic index is shown in Fig. 7.10. We note from this figure that

(i) As r increases η_{vol} decreases.
(ii) As n increases η_{vol} increases.
(iii) As c increases η_{vol} decreases.

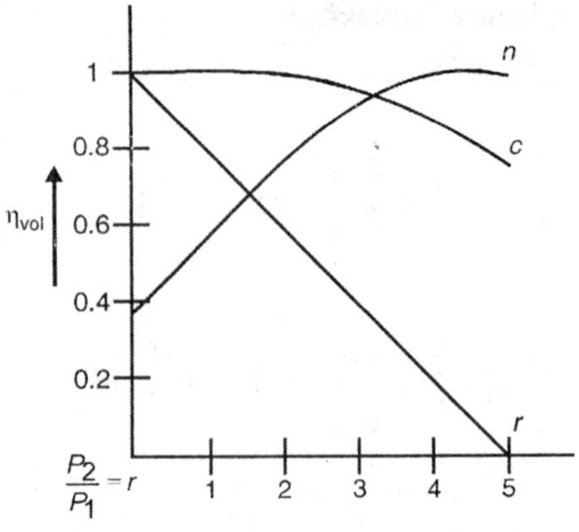

Fig. 7.10

Note.

(i) η_{vol} ranges from 60 to 85% and

(ii) Clearance ratio ranges from 4 to 10%

(iii) *n* ranges from 1.25 to 1.35 for air.

7.9 EXPRESSION FOR WORK INPUT WHEN CLEARANCE IS CONSIDERED

The work input required to compress the air with clearance is given by the area 1 – 2 – 3 – 4. (Ref. Fig. 7.9)

i.e. W_{input} = Area 1 – 2 – 3 – 4

= [Area 1 – 2 – 6 – 5] – [Area 3 – 4 – 5 – 6]

Assuming index of compression to expansion to be same as *n*.

$$W_{input} = \frac{n}{n-1} P_1V_1 \left[\left(\frac{P_2}{P_1} \right)^{\frac{n-1}{n}} - 1 \right] - \frac{n}{n-1} P_4V_4 \left\{ \left(\frac{P_3}{P_4} \right)^{\frac{n-1}{n}} - 1 \right\}$$

But $P_4 = P_1,\ P_3 = P_2$

$$\therefore \quad W_{input} = \frac{n}{n-1} P_1V_1 \left\{ \left(\frac{P_2}{P_1} \right)^{\frac{n-1}{n}} - 1 \right\} - \frac{n}{n-1} P_1V_4 \left\{ \left(\frac{P_2}{P_1} \right)^{\frac{n-1}{n}} - 1 \right\}$$

$$\mathbf{W_{input} = \frac{n}{n-1} P_1 (V_1 - V_4) \left\{ \left(\frac{P_2}{P_1} \right)^{\frac{n-1}{n}} - 1 \right\}}$$

Therefore, it can be seen that from above equation work will depend upon volume of air taken/cycle and will not depend upon clearance. If clearance is neglected then $V_1 - V_4 = V_1$ and we get the expression for work without clearance as,

$$W_{input} = \frac{n}{n-1} P_1 V_1 \left[\left(\frac{P_2}{P_1} \right)^{\frac{n-1}{n}} - 1 \right]$$

$$\therefore \quad IP = \text{Indicated work/cycle} \times \frac{N}{60}$$

If V_1 is in m^3, then resultant unit will be kJ. Then multiply by $N/60$ to get power in kW.

7.10 ACTUAL INDICATOR DIAGRAM

In actual air compressor, some air is present in the clearance volume. This air expands during suction stroke.

At point 4, the inlet valve will not open in actual practice because of valve inertia and pressure differential required to open the valve. Thus the pressure drops away until the valve is forced to open. Some valve bounce will take place and slowly intake will become nearly steady. This negative pressure, is known as *Intake depression.*

A similar situation occurs at point 2. i.e. at point 2 pressure has to increase otherwise valve will not open.

$\therefore$ Actual work required to compress and deliver the air will be greater than theoretical work.

and,
$$\eta_{mech} = \frac{\text{Indicator work}}{\text{Actual work}} = \frac{\text{Indicator work}}{\text{Brake work}} = \frac{IP}{BP}$$

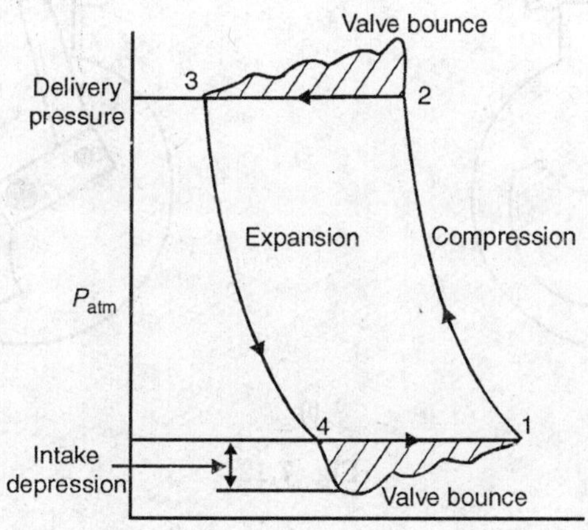

Fig. 7.11

7.11 FREE AIR DELIVERY (FAD)

It is the actual volume of air delivered by an air compressor reduced to either NTP or STP conditions or intake conditions. We know that from continuity equation.

Mass flow rate at the inlet of compressor = Mass flow rate at outlet of compressor.

Since $$PV = \dot{m} RT$$

$$\frac{PV}{RT} = \dot{m}$$

$$\therefore \quad \frac{P_1V_1}{RT_1} = \frac{P_2V_2}{RT_2} = \frac{P_fV_f}{T_f} \text{ neglecting clearance}$$

where P_fV_f and T_f are free air conditions and V_f will be FAD. For convenience take $P_f = 101.325$ kPa, $T_f = 288$ K if not given and V_f can be calculated.

Note. $$\frac{P_fV_f}{T_f} = \frac{P_1(V_1 - V_4)}{T_1} = \frac{P_2(V_2 - V_3)}{T_2} \text{ if clearance is considered.}$$

7.12 MULTISTAGE AIR COMPRESSOR

A 2 stage air compressor with water jacketed cylinder and inter cooler is shown in Fig. 7.12 (i).

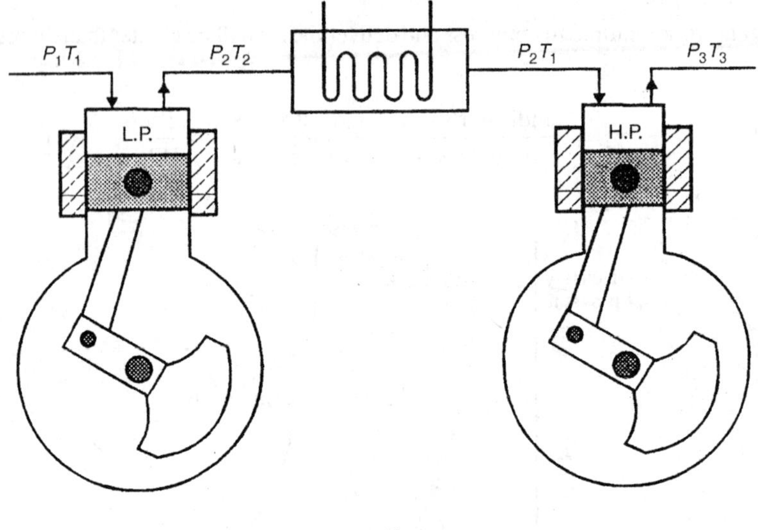

(i)

Fig. 7.12

The suction in LP cylinder ends at 1 and the air is compressed polytropically to 2'. The LP cylinder then delivers the air along $2' - P_2$ to the intercooler. Where air is cooled to initial intake temperature T_1 at constant pressure P_2 by circulating cold water in the inter cooler. When the air is cooled in the intercooler to initial temperature T_1, the cooling is perfect. The air when it is cooled at constant pressure, suffers reduction in volume from $2' - P_2$ to $P_2 - 2$. The cooled air is drawn into the HP cylinder along P_2 -2 for II-stage compression, where it is compressed polytropically to final pressure P_3 along $2 - 3$ and then delivered at constant pressure P_3 along $3 - P_3$. [Ref. Fig. 7.12 (iii)].

Figure 7.12 (iv) shows the combined indicator diagram for LP and HP cylinder of a single acting 2-stage air compressor with perfect intercooling.

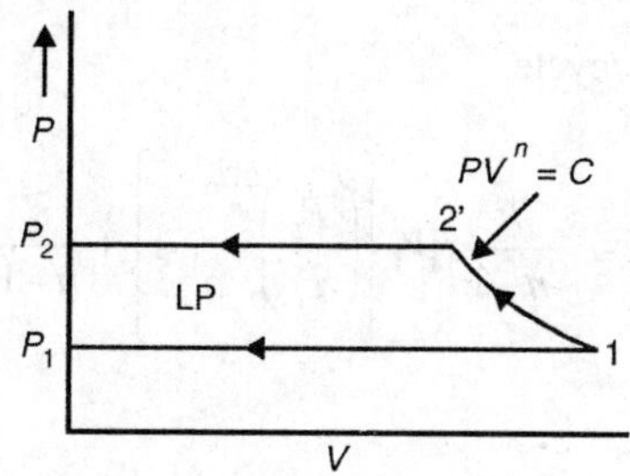

(ii) Ind. diagram for LP cylinder

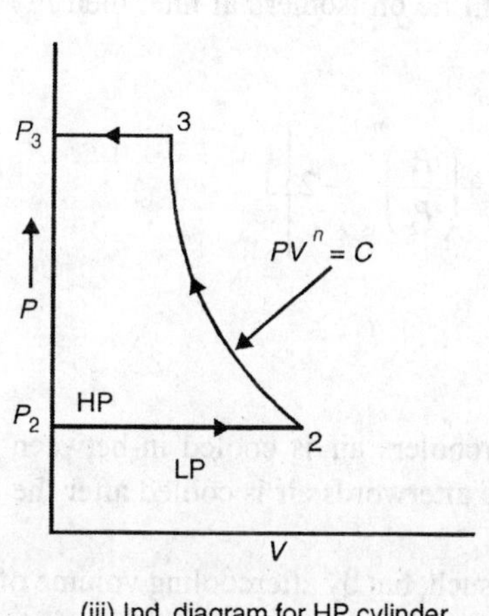

(iii) Ind. diagram for HP cylinder

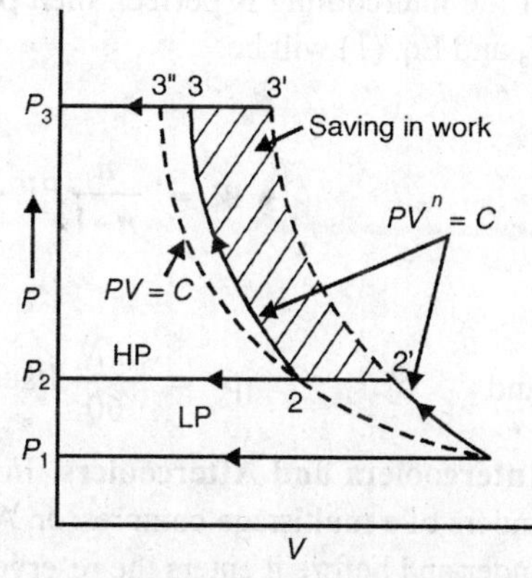

(iv) Combined Ind. diagram of 2 stage/comp. with perfect intercooling

Fig. 7.12 *Contd.*

LP diagram is shown as $P_1 - 1 - 2' - P_2$ and HP diagram is shown as $P_2 - 2 - 3 - P_3$. The reduction of work required due to intercooling is shown by shaded are $2 - 3 - 3' - 2'$, when the intercooling is perfect, the point 2 lies on the isothermal line $1 - 3''$.

Work required in LP cylinder/cycle.

$$= \frac{n}{n-1} P_1V_1 \left\{ \left(\frac{P_2}{P_1} \right)^{\frac{n-1}{n}} - 1 \right\}$$

And work required in HP cylinder/cycle

$$= \frac{n}{n-1} P_2V_2 \left\{ \left(\frac{P_3}{P_2} \right)^{\frac{n-1}{n}} - 1 \right\}$$

∴ Total work required /cycle,

$$W = \frac{n}{n-1} P_1V_1 \left\{ \left(\frac{P_2}{P_1} \right)^{\frac{n-1}{n}} - 1 \right\} + \frac{n}{n-1} P_2V_2 \left\{ \left(\frac{P_3}{P_2} \right)^{\frac{n-1}{n}} - 1 \right\} \quad (1)$$

and

$$\text{IP of compressor} = \frac{WN}{60} \text{ J/sec or W.}$$

If the intercooling is perfect, then point 2 will lie on isothermal line, then $P_1V_1 = P_2V_2$ and Eq. (1) will be,

$$W = \frac{n}{n-1} P_1V_1 \left\{ \left(\frac{P_2}{P_1} \right)^{\frac{n-1}{n}} + \left(\frac{P_3}{P_2} \right)^{\frac{n-1}{n}} - 2 \right\} \text{J} \quad (2)$$

and

$$\text{IP} = \frac{WN}{60} \text{ J/sec or W} \quad (3)$$

Intercoolers and Aftercoolers. In the intercoolers air is cooled in-between the cylinders of a multistage compressor. And in the afterwords air is cooled after the last cylinder and before it enters the reservoir.

By aftercooling there is no saving in work as such, but by aftercooling volume of air decreases. So more amount of air can be stored in the reservoir. Hence capacity (mass of air to be stored) of air increases.

Ideal Intercooler Pressure. It may be noted that, saving in work increases as intercooling is increased. When the intercooling is perfect, saving of work is maximum and work input to compressor is given by Eq. (2).

It may further be noted that saving in work also varies with the chosen intercooler pressure P_2. When P_1 and P_3 are fixed, the best value, of intercooler pressure P_2 shall be fixed to give minimum work.

This value of P_2 can be found by differentiating the Eq. (2) w.r.t. P_2 and equating it to zero.

$$\therefore \quad W = \frac{n}{n-1}P_1V_1\left[\left(\frac{P_2}{P_1}\right)^{\frac{n-1}{n}} + \left(\frac{P_3}{P_2}\right)^{\frac{n-1}{n}} - 2\right] \tag{3}$$

$$W = \frac{n}{n-1}P_1V_1\left[\left(\frac{P_2}{P_1}\right)^{a} + \left(\frac{P_3}{P_2}\right)^{a} - 2\right] \text{ where } a = \frac{n-1}{n}$$

$$W = \text{Constant} \times \left\{\left(\frac{P_2}{P_1}\right)^{a} + \left(\frac{P_3}{P_2}\right)^{a} - 2\right\}$$

Differentiating w.r.t. P_2 and equating it to zero for minimum work,

$$\frac{dW}{dP_2} = \frac{aP_2^{a-1}}{P_1^a} - \frac{aP_3^a}{P_2^{a+1}} = 0$$

Dividing by a throughout,

$$\frac{P_2^{a-1}}{P_1^a} = \frac{P_3^a}{P_2^{a+1}}$$

$$\therefore \quad P_2^{2a} = (P_1P_3)^a$$

Taking ath root throughout,

$$P_2^2 = P_1P_3 \qquad \Rightarrow P_2\sqrt{P_1P_3} \text{ or } \frac{P_2}{P_1} = \frac{P_3}{P_2}$$

This shows that for minimum work required or maximum efficiency, the intercooler pressure is the geometric mean of initial and final pressure, i.e. pressure ratio in each stage is same.

$\therefore$ Equation (3) will be,

$$W = \frac{n}{n-1}P_1V_1\left\{\left(\frac{P_2}{P_1}\right)^{\frac{n-1}{n}} + \left(\frac{P_3}{P_2}\right)^{\frac{n-1}{n}} - 2\right\}$$

$$W = \frac{2n}{n-1}P_1V_1\left\{\left(\frac{P_2}{P_1}\right)^{\frac{n-1}{n}} - 1\right\} \text{since} \frac{P_2}{P_1} = \frac{P_3}{P_2}$$

$$= \frac{2n}{n-1} P_1 V_1 \left\{ \left(\frac{(P_1 P_3)^{1/2}}{P_1} \right)^{\frac{n-1}{n}} - 1 \right\}$$

$$= \frac{2n}{n-1} P_1 V_1 \left\{ \left[\left(\frac{P_3}{P_1} \right)^{1/2} \right]^{\frac{n-1}{n}} - 1 \right\}$$

$$\mathbf{W} = \frac{\mathbf{2n}}{\mathbf{n-1}} \mathbf{P_1 V_1} \left\{ \left(\frac{\mathbf{P_3}}{\mathbf{P_1}} \right)^{\frac{\mathbf{n-1}}{\mathbf{2n}}} - \mathbf{1} \right\}$$

and $\min IP = \frac{WN}{60}$ J/sec

Similarly for 3 stage compressor

$$\mathrm{W} = 3 \frac{\mathrm{n}}{\mathrm{n}-1} \left\{ \left(\frac{\mathrm{P_4}}{\mathrm{P_1}} \right)^{\frac{\mathrm{n}-1}{3\mathrm{n}}} - 1 \right\}$$

And similarly for x-stage compress

$$W = \frac{xn}{n-1} P_1 V_1 \left\{ \left(\frac{P_{x+1}}{P_1} \right)^{\frac{n-1}{xn}} - 1 \right\}$$

Now in general,

$$\frac{P_2}{P_1} = \frac{P_3}{P_2} = \frac{P_4}{P_3} = \frac{P_{x+1}}{P_1} = k$$

where k is the pressure ratio is each stage.

From this,

$$P_2 = k P_1$$

$$P_3 = kP_2 = k\, kP_1 = k^2 P_1$$

$$P_4 = kP_3 = k.k^2 P_1 = k^3 P_1$$

$$P_{x+1} = kP_x = k^x P_1$$

$$\therefore \quad k^x = \left(\frac{P_{x+1}}{P_1} \right) \text{ or } k = \left(\frac{P_{x+1}}{P_1} \right)^{\frac{1}{x}}$$

i.e. $k = (\text{Pressure Range})^{1/x}$

7.13 CAPACITY CONTROL OF COMPRESSORS

Generally compressors are not operated at full capacity. Depending upon the demand for high pressure air, they are operated.

Various types of controls for compressors are:

(i) Throttle control

(ii) Blowing air to atmosphere

(i) Throttle Control. When the demand for high pressure air is more and when less amount of air is present is the reservoir tank, then less pressure from the reservoir will act on the piston (1). Hence because of less pressure piston moves upwards and more amount of air is taken during suction stroke opposite operation will take place when the demand for high per air is less.

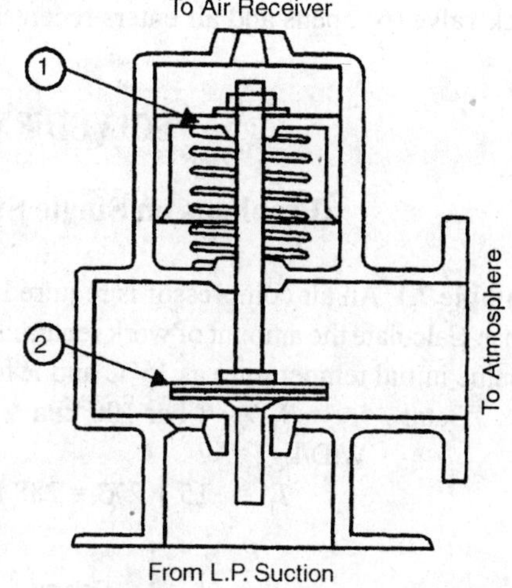

Fig. 7.13. *Throttle control.*

(ii) Blowing Air to Atmosphere. When the pressure in the receiver increases beyond the required limit, then because of high pressure air piston

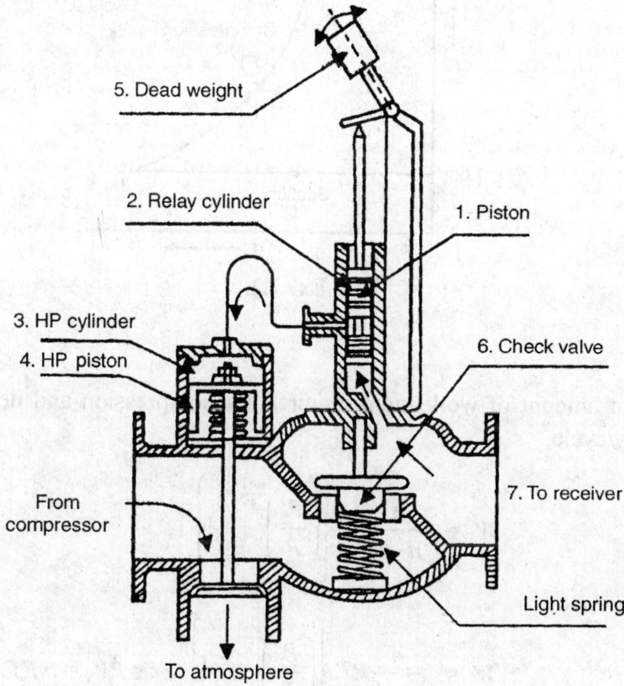

Fig. 7.14. *Blowing air to atmosphere.*

(1) of relay cylinder (2) in lifted and high pressure air enters cylinder (3) and presses the piston (4) downwards and high pressure air from the compressor is directly blown off to atmosphere.

And when the pressure comes below normal in the receiver, then piston (1) of relay cylinder moves down because of dead weights (5). Because of compressor air pressure check valve (6) opens and air enters receiver (7).

SOLVED EXAMPLES

(Problems on Single Stage Air Compressor)

Example 7.1 An air compressor is required to compress air from a pressure of 1 bar to 10 bar. Calculate the amount of work required per kg of air, when $n = 1.2$ for compression. Assume initial temperature as 15°C and R for air is 0.287 kJ/kg-K.

Data: P_1 = 1 bar 100 kPa P_2 = 10 bar = 1000 kPa

W.D/kg = ? n = 1.2

T_1 = 15 + 273 = 288 K R = 0.287 kJ/kg-K.

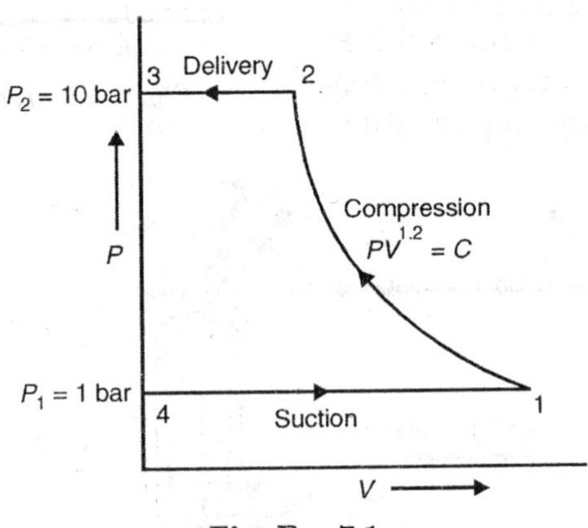

Fig. Ex. 7.1

Solution

We know that amount of work input required for compression and delivery of air polytropically/cycle,

$$W = \frac{n}{n-1} P_1 V \left\{ \left(\frac{P_2}{P_1} \right)^{\frac{n-1}{n}} - 1 \right\}$$

$$W = \frac{n}{n-1} RT_1 \left\{ \left(\frac{P_2}{P_1} \right)^{\frac{n-1}{n}} - 1 \right\} \quad \text{as } PV_1 = mRT_1$$

$$= 4.712\times10^{-4}\left(\frac{550}{97}\right)^{1/1.3}$$

$$V_4 = 1.79\times10^{-3}\,\text{m}^3$$

Then $$\frac{P_1(V_1-V_4)}{T_1} = \frac{P_f V_f}{T_f}$$

$$\mathbf{V_f = 7.625\times10^{-3}\,m^3}$$

(i) $$\text{FAD} = V_f\times N = 3.814\,\text{m}^3/\text{min}$$

(ii) $$\eta_{vol} = \frac{V_1-V_4}{V_1-V_3} = 0.8656 \Rightarrow 86\%$$

(iii) $$\frac{T_1}{T_2} = \left(\frac{P_1}{T_2}\right)^{\frac{n-1}{n}} = \left(\frac{V_2}{V_1}\right)^{n-1}$$

$$\mathbf{T_2 = 437.29\,K}$$

(iv) $$\text{Cycle power} = \frac{n}{n-1}P_1(V_1-V_4)\left[\left(\frac{P_2}{P_1}\right)^{\frac{n-1}{n}}-1\right]\times\frac{N}{60}$$

$$= \frac{1.3}{0.3}\times97\,(8.106\times10^{-3})\left[\left(\frac{550}{97}\right)^{\frac{0.3}{1.3}}-1\right]\times\frac{500}{60}$$

$$\textbf{I.P.} = \mathbf{13.983\,kW}$$

Example 7.5 A single stage, single acting air compressor works between 1 bar and 16 bar compression follows $PV^{1.3} = C$. The piston speed is 200 m/min. It runs at 350 rpm. It has an IP consumption of 30 kW and η_{vol} is 85%. Estimate the cylinder bore and stroke.

Data: $P_1V_1^{1.3} = P_2V_2^{1.3}$ Piston speed = 2 LN = 200 m/min

N = 350 rpm IP = 30 kW

η_{vol} = 85% D = ?

L = ?

Solution

Since Piston speed = 2 LN =200 m/min

= 2 × L × 350 = 200 m /min

$$= \frac{1.2}{1.2-1} \times 100 \times \frac{3}{60}\left[\left(\frac{500}{100}\right)^{\frac{1.2-1}{1.2}} - 1\right]$$

$$= \frac{kN}{m^2} \times \frac{m^3}{sec}[\text{ratio} - 1] = \frac{kJ}{sec} = kW$$

W.D. or I.P = **9.229 kW**

Also $\eta_{iso} = \frac{\text{Isothermal work of compression}}{\text{IP}}$

$$= \frac{P_1 V_1 \ln \frac{P_2}{P_1}}{9.229}$$

$$= \frac{100 \times \frac{3}{60} \times \ln \frac{500}{100}}{9.229} = \frac{8.05}{9.229}$$

η_{iso} = **87.2 %**

Example 7.3. A single stage single acting/compressor with bore 15 cm and stroke 20 cm run at 200 rpm. Air is drawn in at 1 bar and 25°C. Air is delivered at 6 bar, law of compression is $PV^{1.3} = C$. Determine FAD and indicated power required to drive the compressor in kW.

Data:
D = 0.5 m L = 0.2 m
N = 200 rpm P_1 = 100 kPa
$T_1 = 25 + 273 = 298$ K
FAD = ? IP = ?

Solution

We know that,

$$\frac{P_1 V_1}{T_1} = \frac{P_2 V_2}{T_2} = \frac{P_f V_f}{T_f}$$

Now V_1 = volume of air handled/min

$$= \frac{\pi}{4} D^2 LN$$

$$= \frac{\pi}{4} \times 0.15^2 \times 0.2 \times 200$$

$$= 0.707 \, m^3/min$$

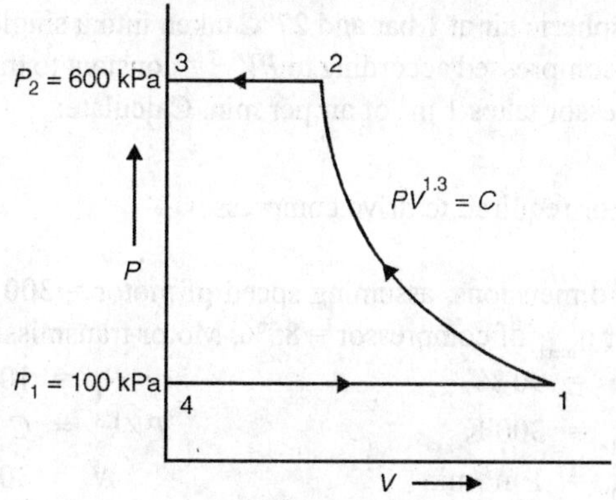

Fig. Ex. 7.3

To find FAD i.e. V_f

$$\frac{P_1 V_1}{T_1} = \frac{P_f V_f}{T_f}$$

Since nothing is mentioned take,

$$P_f = 101325\,\text{kPa}$$

$$T_f = 15+273 = 288\,\text{K}$$

$$\therefore \quad \frac{100\times 0.7087}{283} = \frac{101.325\times V_f}{288}$$

$$\mathbf{0.674\ m^3/min = V_f}$$

$$\text{IP} = \frac{n}{n-1}P_1V_1\left[\left(\frac{P_2}{P_1}\right)^{\frac{n-1}{n}}-1\right]$$

$$= \frac{1.3}{1.3-1}\times 100\times\frac{0.707}{60}\left[\left(\frac{600}{100}\right)^{\frac{0.3}{1.3}}-1\right]$$

$$\textbf{IP} = \textbf{2.615 kW}$$

Example 7.4 A single stage, single acting reciprocating air compressor has a bore of 200 mm and a stroke of 300 mm. It runs at a speed of 500 rpm. The clearance volume is 5% of swept volume and polytropic index is 1.3 throughout. Intake pressure and temperature are 97 kPa, 20°C and the compression pressure is 550 kPa. Detertermine:

(i) FAD in m³/min

(ii) Air delivery temperature

Example 7.7 Atmospheric air at 1 bar and 27°C taken into a single stage single acting air compressor. It is compressed according to $PV^{1.3}$ = constant to the discharge pressure of 7 bar. The compressor takes 1 m^3 of air per min. Calculate:

(1) IP
(2) Power of motor required to drive compressor.
(3) η_{iso}, and
(4) The cylinder dimensions, assuming speed of motor = 300 rpm, stroke to bore ratio = 1.5: 1; η_{mech} of compressor = 85%, Motor transmission efficiency 90%.

Data:

$$\eta = 90\,\% \qquad P_1 = 100 \text{ kPa}$$

$$T_1 = 300 \text{ K} \qquad PV^{1.3} = C$$

$$V_1 = 1 \text{ m}^3/\text{min} \qquad N = 300 \text{ rpm}$$

$$\frac{L}{D} = 1.5 \quad \Rightarrow \quad L = 1.5\text{ D}$$

$$P_2 = 700 \text{ kPa}$$

Solution

Neglecting clearance

$$\text{IP} = \frac{n}{n-1}P_1V_1\left[\left(\frac{P_2}{P_1}\right)^{\frac{n-1}{n}} - 1\right]$$

$$= \frac{1.3}{0.3}\times 100\times\frac{1}{60}\times\frac{1}{60}\left\{\left(\frac{7}{1}\right)^{\frac{0.3}{1.3}} - 1\right\}$$

$$\textbf{IP} = \textbf{4.09 kW}$$

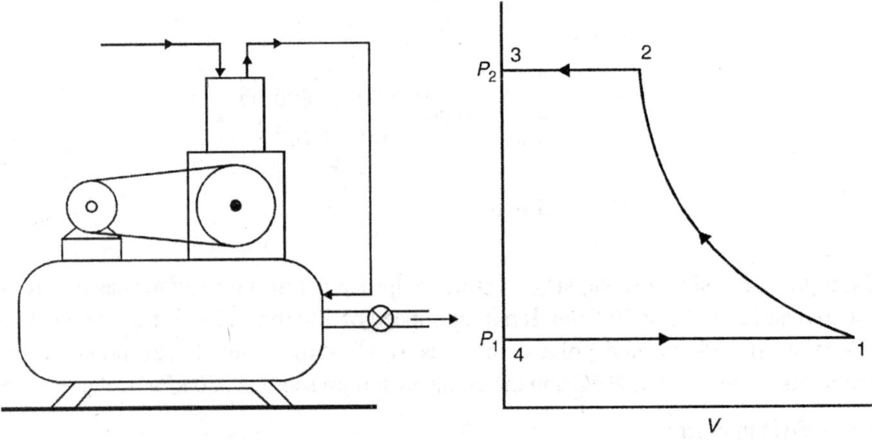

Fig. Ex. 7.7

Note. m = 1 kg as work input /kg is required

$$W = \frac{1.2}{1.2-1} \times 0.287 \times 288 \left\{ \left(\frac{1000}{100} \right)^{\frac{1.2-1}{1.2}} - 1 \right\}$$

$$= \frac{1.2}{0.2} \times 0.287 \times 288\{(10)^{0.2/1.2} - 1\}$$

$$= 495.936\ \{1.46779 - 1\}$$

$$= 495.936\ \{0.46779\}$$

$$\mathbf{W = 231.9985\ kJ/kg}$$

Example 7.2 A single stage, single acting air compressor delivers air at 5 bar. The suction temperature is 20°C and suction pressure is 1 bar volume of air entering the compressor is 3 m³/min. Index of compression is 1.2. Calculate isothermal efficiency and power required to drive the compressor. Neglect clearance volume.

Data: $T_1 = 20 + 273 = 293$ K $\qquad V_1 = 3\ \text{m}^3/\text{min}$

$n = 1.2 \qquad 7_{iso} = ?$

W.D. = ?

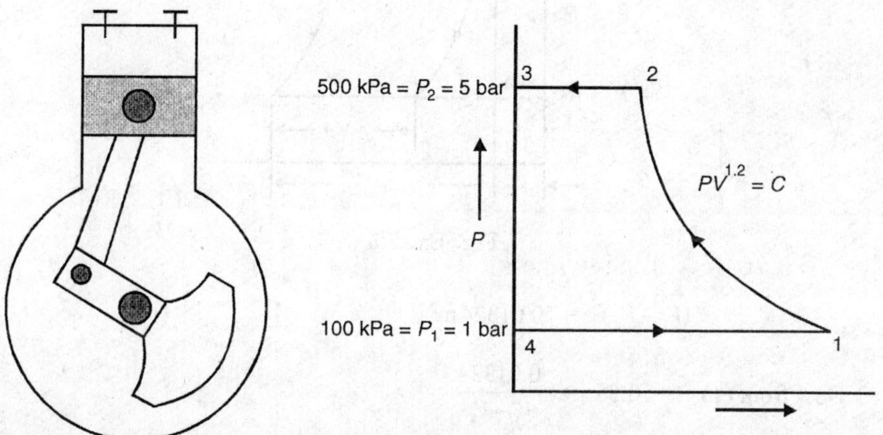

Fig. Ex. 7.2

Solution

We know that,

$$\text{W.D. or I.P} = \frac{n}{n-1} P_1 V_1 \left\{ \left(\frac{P_2}{P_1} \right)^{\frac{n-1}{n}} - 1 \right\}$$

$$L = 0.286 \text{ m}$$

$\therefore$ $$\text{Stroke } L = 286 \text{ mm}$$

$$\eta_{vol} = \frac{V_a}{V_s} = \frac{V_1 - V_4}{V_1 - V_3}$$

$$\text{IP} = \frac{n}{n-1} P_1 (V_1 - V_4) \left[\left(\frac{P_2}{P_1} \right)^{\frac{n-1}{n}} - 1 \right] \times \frac{N}{60} \frac{\text{kJ}}{\text{sec}}$$

$$30\,\text{kW} = \frac{1.3}{0.3} \times 100\ (V_1 - V_4) \left[\left(\frac{1600}{100} \right)^{\frac{0.3}{1.3}} - 1 \right] \times \frac{350}{60}$$

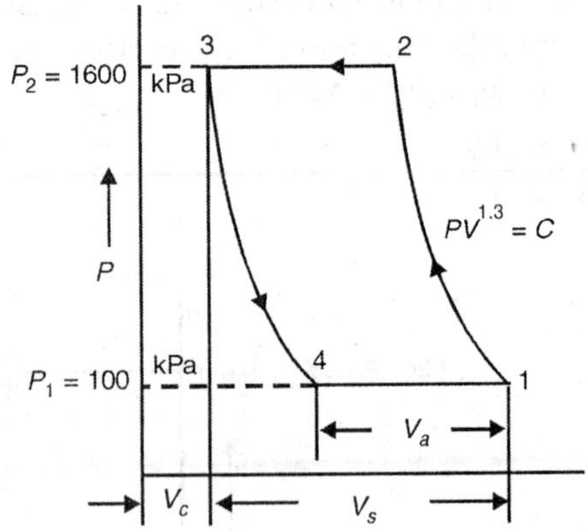

Fig. Ex. 7.5

$$(V_1 - V_4) = 0.01324\,\text{m}^3$$

Now from (1) $$0.85 = \frac{0.01324}{V_1 - V_3}$$

$$(V_1 - V_3) = 0.01558\,\text{m}^3 = V_s$$

$\therefore$ $$V_s = 0.01558\,\text{m}^3 = \frac{\pi}{4} D^2 L$$

$\therefore$ $$D = 0.263 \text{ m}$$

$$\mathbf{D = 263\,mm}$$

Example 7.6 Actual volume of air taken in at 0.98 bar and 27°C by a single stage, acting reciprocating air compressor is 8.5 m³/min. The delivery conditions are 6.5 bar.

The clearances volume is 5% of stroke volume compression and expansion follows $PV^{1.3} = C$. If the speed is 300 rpm. L/D ratio is 1:1. Find the cylinder dimensions by finding the volumetric efficiency required to run the compressor. Also find I.P.

Data: $V_a = 8.5\ \text{m}^3/\text{min}$ $P_2 = 6.5 \times 10^2\ \text{kPa}$

$T_1 = 27 + 273\text{K}$ $P_1 = 0.98 \times 10^2\ \text{kPa}$

$\frac{V_c}{V_s} = 0.05$ $n = 1.3, \text{N} = 300\ \text{rpm}$

Solution

We know that,

$$\eta_{vol} = 1 - C\left\{\left(\frac{P_2}{P_1}\right)^{1/n} - 1\right\}$$

$$= 1 - 0.05\left\{\left(\frac{6.5 \times 10^2}{0.98 \times 10^2}\right)^{1/1.3} - 1\right\}$$

$$\boldsymbol{\eta_{vol} = 0.8352 = 83.57\%}$$

Also volumetric efficiency,

$$\eta_{vol} = \frac{V_a}{V_s}$$

$$\therefore \quad 0.8356 = \frac{8.5}{V_s}$$

$$V_s = 1.7712\ \text{m}^3/\text{min} = \frac{\pi}{4} D^2 LN$$

$$\mathbf{D = L = 0.3508\ m}$$

and

$$\text{IP} = \frac{n}{n-1} P_1(V_1 - V_4)\left[\left(\frac{P_2}{P_1}\right)^{\frac{n-1}{n}} - 1\right]$$

$$= \frac{n}{n-1} P_1 V_a = \left\{\left(\frac{P_2}{P_1}\right)^{\frac{n-1}{n}} - 1\right\}$$

$$= \frac{1.3}{1.3-1} \times \frac{0.98 \times 10^2 \times 8.5}{60}\left[\left(\frac{6.5}{0.98}\right)^{\frac{0.3}{1.3}} - 1\right]$$

$$\mathbf{IP = 33.6\ kW}$$

(iii) Cycle power

(iv) η_{iso} neglecting clearance volume.

Solution

$$\frac{V_c}{V_s} = 0.05 = \frac{V_3}{V_1 - V_3} \quad (1)$$

$$V_s = \text{Stroke volume}$$

$$= \frac{\pi}{4} 0.2^2 \times 0.3$$

$$V_s = V_1 - V_3 = 9.425 \times 10^{-3} \text{m}^3$$

$$\therefore \quad 0.05 = \frac{V_3}{V_1 - V_3} = \frac{V_3}{9.425 \times 10^{-3}}$$

$$\Rightarrow \quad V_3 = 4.712 \times 10^{-4} \text{m}^3$$

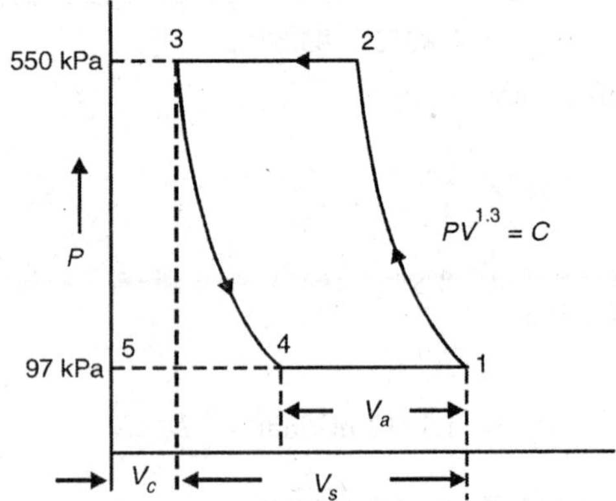

Fig. Ex. 7.4

Now
$$V_s = V_1 - V_3$$

$$9.524 \times 10^{-3} = V_1 - 4.712 \times 10^{-4}$$

$$\therefore \quad V_1 = 9.896 \times 10^{-3} \text{m}^3$$

$$p_3 v^n{}_3 = P_4 V_4^n$$

$$\therefore \quad V_4 = V_3 \left(\frac{P_3}{P_4}\right)^{1/n} = V_3 \left(\frac{P_2}{P_1}\right)^{1/n}$$

Now $$\eta_{vol} = \frac{V_a}{V_s} = \frac{24.472539}{38.8249} = 0.7235062$$

$$\eta_{vol} = \mathbf{72.35\ \%}$$

Example 7.10 A single stage single acting reciprocating air compressor delivers 10 m^3 of free air per minute and it is running at 200 rpm. It takes air at a pressure of 1 bar and 303 K and delivers at 6 bar. Take $n = 1.25$ and clearance volume is 6% of stroke volume.

Determine

(i) Delivery temperature

(ii) Volumetric efficiency

(iii) Stroke volume

(iv) Power required and

(v) Dimensions of cylinder if $L = 1.2$ D

Data: FAD = 10 m^3/min = V_f

N = 200 rpm $\quad V_c = 0.06\ V_s$

T_2 = ? $\quad \eta_{vol}$ = ?

V_s = ? $\quad$ IP = ?

L = ? $\quad$ and D = ?

Solution

(i) We know that,

$$\frac{T_2}{T_1} = \left(\frac{P_2}{T_1}\right)^{\frac{n-1}{n}}$$

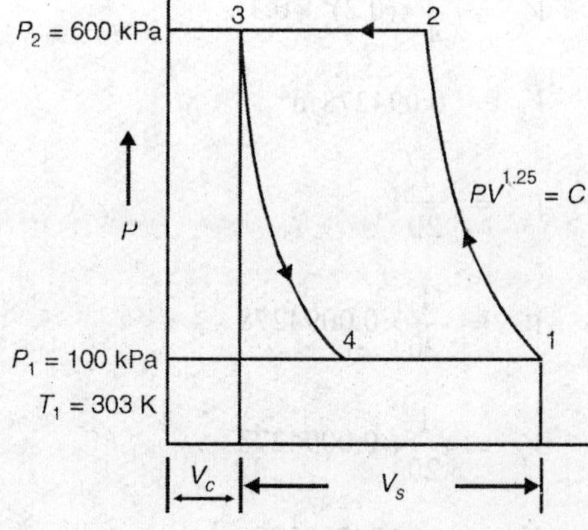

Fig. Ex. 7.10

Solution

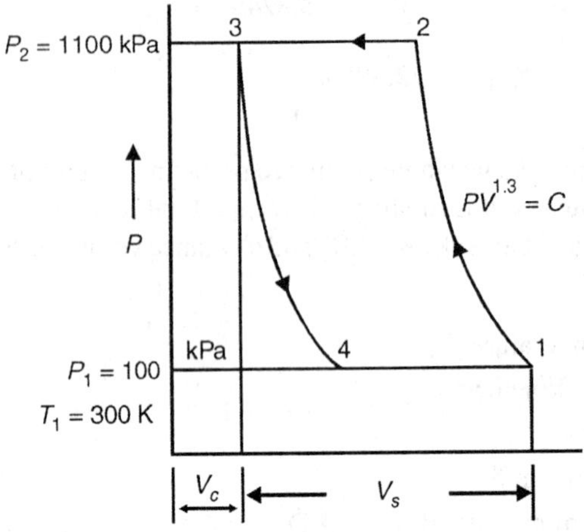

Fig. Ex. 7.8

We know that, work input to a compressor or I.P. of a compressor with clearance,

$$= \frac{n}{n-1} P_1 (V_1 - V_4) \left\{ \left(\frac{P_2}{P_1} \right)^{\frac{n-1}{n}} - 1 \right\} \qquad (1)$$

Now find V_1 and V_4. We also know that,

$$\text{Stroke volume} = V_s = V_1 - V_3 = \frac{\pi}{4} d^2 L$$

$$V_s = \frac{\pi}{4} \times (0.2)^2 \times 0.3$$

$$V_s = 0.094278 \, \text{m}^3$$

As given $\quad V_c = \frac{1}{20} V_s$

$\therefore \quad V_c = \frac{1}{20} \times 0.0094278$

$\therefore \quad V_c = \frac{1}{20} \times 0.0094278$

$\therefore \quad V_c = 0.0004712389 \, \text{m}^3$

From Fig. 7.8 $\quad V_1 = V_c + V_s = 0.0004712389 + 0.009427$

$$\mathbf{V_1 = 0.0098990\,m^3}$$

Now to find V_4 consider polytropic 3 – 4 for which

$$P_3 V_3^n = P_4 V_4^n$$

$$V_4 = V_3 \times \left(\frac{P_3}{P_4}\right)^{1/n}$$

$$V_4 = V_c \left\{\frac{P_2}{P_4}\right\}^{1/n}$$

As $\quad P_3 = P_2$ and $P_4 = P_1$

$$\therefore \quad V_4 = 0.0004712389 \left\{\frac{1100}{100}\right\}^{1/1.3}$$

$$\mathbf{V_4 = 0.0029894139\ m^3}$$

Now from Eq. (1) ,

$$\text{IP} = \frac{n}{n-1}((V_1 - V_4)\left\{\left(\frac{P_2}{P_1}\right)^{\frac{n-1}{n}} - 1\right\}$$

$$= \frac{1.3}{1.3-1} \times 100 \times (0.009890 - 0.0029804139) \times \left\{\left(\frac{1100}{100}\right)^{\frac{1.3-1}{1.3}} - 1\right\}$$

$$= 4.33 \times 100 \times (0.069186) \times \{1.739 - 1\}$$

$$\mathbf{IP = 2.2141\ kJ\ /cycle}$$

Since $\quad \text{N} = 650\ \text{rpm}$

$$\text{IP} = 2.2141 \times \frac{650}{60}$$

$$\mathbf{IP = 23.9835\ kJ\ /\ sec\ or\ kW.}$$

Since Mechanical efficiency is given

$$\text{Actual power input} = \frac{\text{IP}}{\eta_{\text{mech}}} = \frac{23.98365}{0.85} = 28.2158\,\text{kW}$$

Also we know that, volumetric efficiency.

$$\eta_{\text{vol}} = \frac{V_a}{V_s} = \frac{V_1 - V_4}{V_1 - V_3}$$

Solution

(1) We know that, for a polytropic process,

$$\frac{T_1}{T_2} = \left(\frac{P_2}{P_1}\right)^{\frac{n-1}{n}}$$

$$\therefore \quad T_2 = T_1 \times \left(\frac{P_2}{P_1}\right)^{\frac{n-1}{n}} = 288 \times \left(\frac{7}{1}\right)^{\frac{1.25-1}{1.25}}$$

$$= 288 \times (7)\frac{0.25}{1.25} = 288 \times 1.475$$

$$\mathbf{T_2 = 425.022\ K}$$

(2) For volumetric efficiency, we know that,

$$\eta_{vol} = 1 + C - C\left(\frac{P_2}{P_1}\right)^{\frac{1}{n}} = 1 + 0.06 - 0.06(7)^{\frac{1}{1.25}}$$

$$= 1 + 0.06 - 0.06 \times 4.73 = 1 + 0.06 - 0.2859$$

$$= 0.7754034$$

$$\therefore \quad \mathbf{\eta_{vol} = 77.54\%}$$

(3) For volume of air delievered we know that,

$$\frac{T_2}{T_1} = \left(\frac{P_2}{P_1}\right)^{\frac{n-1}{n}} = \left(\frac{V_2}{V_1}\right)^{n-1}$$

$$\left(\frac{V_2}{V_1}\right)^{n-1} = \left(\frac{P_1}{P_2}\right)^{\frac{n-1}{n}}$$

or

$$\frac{V_2}{V_1} = \left(\frac{P_1}{P_2}\right)^{\frac{1}{n}}$$

$$V_2 = V_1 \times \left(\frac{P_1}{P_2}\right)^{\frac{1}{n}} = 6\,\text{m}^3/\text{min} \times \left(\frac{1}{7}\right)^{1/1.25}$$

$$= 6 \times 0.2108 = 1.2649\ \text{m}^3/\text{min}$$

$$= 1.2649\,\text{m}^3/\text{min} = \frac{1.2649}{200}\,\text{m}^3/\text{stroke}$$

$$\mathbf{V_2 = 0.006324742\ m^3/stroke}$$

Note. Actual work IP required to drive the compressor is more than IP, due to work required to overcome frictional lossess in compressor.

$\therefore$ Shaft power = IP + FP for compressor

and

$$\text{Compressor } \eta_{mech} = \frac{\text{IP}}{\text{shaft power}}$$

$$\therefore \quad \text{Shaft Power} = \frac{\text{IP}}{\eta_{mech} \text{ of compressor}} = \frac{4.09}{0.85}$$

$$= 4.81 \text{ kW}$$

$$\text{And Power input to motor} = \frac{\text{Shaft power}}{\text{Tranmission } \eta \text{ (inclusive of motor and drive } \eta)}$$

$$= \frac{4.81}{0.9} = 5.35 \text{ kW}$$

$$\text{Also Iso } \eta, \ \eta_{iso} = \frac{\text{Iso. work}}{\text{Indi. work}}$$

$$\text{Iso work} = P_1 V_1 \ln \frac{P_2}{P_1} = 150 \times \frac{1}{60} \times \ln \frac{7}{1} = 3.24 \text{ kW}$$

$$\therefore \quad \eta_{iso} = \frac{3.24}{4.09} = 0.792 \text{ i.e. } 79.2\%$$

Volume of air taken in/min

$$= \frac{\pi}{4} \cdot D^2 LN = \frac{\pi}{4} D^2 (1.5D) \times 300$$

$$\mathbf{D = 0.1414 \text{ m}}$$

$$\mathbf{L = 1.5 \times 0.1414 = 0.2121 \text{ m}}$$

Example 7.8 A single stage single acting reciprocating air compressor has a cylinder dia 20 cm and piston stroke 30 cm and running at 650 rpm. Air taken from atmosphere which is at 1 bar and 27°C and is delivered at 11 bar. Assuming polytropic compression and expansion with index $n = 1.3$, find the power required to drive the compressor when its mechanical efficiency is 85%. The compressor has a clearance which is 1/20 of the stroke volume. Find volumetric efficiency of the compressor.

Data: d = 20 cm = 0.2 m $\quad$ L = 30 cm = 0.3 m

N = 650 rpm $\quad$ W/sec = ?

η_{mech} = 85% $\quad$ $V_c = \frac{1}{20} V_c$

η_{vol} = ?

$$T_2 = T_1\left(\frac{P_2}{P_1}\right)^{\frac{n-1}{n}} = 303\times\left(\frac{600}{100}\right)^{\frac{1.25-1}{1.25}}$$

$$T_2 = 303\times\left(\frac{6}{1}\right)^{\frac{0.25}{1.25}}$$

$$\mathbf{T_2 = 433.584\ K}$$

(ii) We also know that, volumetric efficiency,

$$\eta_{vol} = 1+C-C\times\left(\frac{P_2}{P_1}\right)^{1/n} = 1+0.06-0.06\times\left(\frac{6}{1}\right)^{\frac{1}{1.25}}$$

$$= 1 + 0.06 - 0.06 \times 4.19 = 0.8084$$

$$\boldsymbol{\eta_{vol} = 80.84\ \%}$$

(iii) To find stroke volume, we know that

$$\frac{P_f V_f}{T_f} = \frac{P_1(V_1 - V_4)}{T_1}$$

$$\frac{P_f V_f}{T_f}\times\frac{T_1}{P_1} = (V_1 - V_4)$$

$$\frac{101.325\times 10\,\text{m}^3/\text{min}}{288}\times\frac{303}{100} = (V_1 - V_4)$$

$$\mathbf{(V_1 - V_4) = 10.86\ m^3/min = V_a}$$

Since the compressor is running at 200 rpm.

$$(V_1 - V_4) = V_a = \frac{10.66}{200}\,\text{m}^3$$

$$V_a = 0.0533\ \text{m}^3$$

Also since $\quad \eta_{vol} = \dfrac{V_a}{V_s}$

$$0.8084 = \frac{0.533}{V_s}$$

$$\mathbf{V_s = 0.0659327\,m^3}$$

(iv) $\quad$ I.P. required $= \dfrac{n}{n-1}P_1(V_1 - V_4)\left[\left(\dfrac{P_2}{P_1}\right)^{\frac{n-1}{n}} - 1\right]\dfrac{\text{rpm}}{60}$

$$= \frac{1.25}{1.25-1} \times 100 \times 0.533 \left\{ \left(\frac{6}{1} \right)^{\frac{0.25}{1.25}} - 1 \right\} \times \frac{200}{60}$$

$$= 5 \times 100 \times 0.0533(1.43 - 1) \times \frac{200}{60}$$

$$\textbf{I.P.} = \textbf{38.284 kW}$$

(v) To find cylinder dimensions :

We know that, stroke volume,

$$V_s = 0.0659327 = \frac{\pi}{4} D^2 \times L = \frac{\pi}{4} D^2 \times 1.2\, D$$

$$0.0659327 = D^3 \times 0.9425$$

$$\frac{0.0659327}{0.9425} = D^3$$

$$0.0699567 = D^3 \quad \therefore \quad \mathbf{0.4120\ m = D}$$

As $L = 1.2\text{ D} \quad \therefore \quad L = 1.2 \times 0.4120$

$$\mathbf{L = 0.49445\ m}$$

Example 7.11 The following data were recorded for a single acting single stage reciprocating air compressor; speed : 200 rpm.

Free air delivered = 6 m³/min

Volume of intake air = 6 m³/min

$n = 1.25$

Clearance volume = 6% of stroke volume

Suction pressure = 1 bar

Delivery pressure = 7 bar

Temperature of air at inlet = 288 K

Determine:

(i) Temperature of air at delivery

(ii) Volumetric efficiency

(iii) Volume of air delivered at the outlet stroke

(iv) Compressor indicated power in kW neglecting clearance.

(v) Isothermal efficiency of the compressor.

Data: $N = 200$ rpm, FAD $= V_f = 6\ \text{m}^3/\text{min}$

$V_1 = 6\ \text{m}^3/\text{min}$, $PV^{1.25} = C$

$V_c = 0.06\ V_s$, $P_1 = 100$ kPa

$P_2 = 700$ kPa, $T_1 = 288$ K

Find: $T_2 = ?$, $\eta_{vol} = ?$

$V_a = ?$, $\eta_{iso} = ?$

$$= \frac{0.0098990 - 0.0029804139}{0.0098990 - 0.000471239} = \frac{0.0069186}{0.0094278}$$

$$\eta_{vol} = \mathbf{0.73385 \text{ or } 73.385\%}$$

Example 7.9 Calculate the volumetric efficiency of the compressor having a cylinder diameter 410 mm and stroke 610 mm. Compressor makes 420 rpm and delivers 30 kg/min of air at 1.01325 bar and 15° C.

Data: $\eta_{vol} = ?$ $\qquad d = 410 \text{ mm} = 0.41 \text{ m}$

$L = 610 \text{ mm} = 0.61 \text{ m}$

$N = 420 \text{ rpm}$ $\qquad m = 30 \text{ kg/min}$

$P = 1.01325 \text{ bar} = 101.325 \text{ kPa}$

$T = 15 + 273 = 288 \text{ K}$

Solution

We know that volumetric efficiency,

$$\eta_{vol} = \frac{V_a}{V_s}$$

and $V_s = \text{Stroke volume} = \frac{\pi}{4} d^2 \times L \times N$

Note. $m^2 \times m \times rmp = \text{m}^3/\text{min}$

$$= \frac{\pi}{4} \times 0.41^2 \times 0.61 \times 420$$

$$V_s = 33.824 \text{ m}^3/\text{min}$$

As 30 kg of air is delivered at 101.325 kPa of 288 K,

As $PV = mRT$

Actual volume of air handled

$$V_a = \frac{mRT}{P} = \frac{30 \text{ kg/min} \times 0.287 \times 288}{101.325}$$

Note.
$$V = \frac{\frac{\text{kg}}{\text{min}} \times \frac{\text{kJ}}{\text{kg-K}} \times \text{K}}{\text{K}\frac{\text{N}}{\text{m}^2}} = \frac{\text{K-Nm}}{\text{min}} \times \frac{\text{m}^2}{\text{kN}}$$

$$V = \text{m}^3/\text{min}$$

$$V_a = 24.472539 \text{ m}^3/\text{min}$$

(4) We know that, indicated power,

$$\text{I.P.} = \frac{n}{n-1} P_1 V_1 \left\{ \left(\frac{P_2}{P_1} \right)^{\frac{n-1}{n}} - 1 \right\}$$

$$= \frac{1.25}{1.25-1} \times 100 \times \frac{6}{10} \text{m}^3/\text{sec} \left\{ (7)^{0.25/1.25} - 1 \right\}$$

$$= 5 \times 100 \times 0.1\{1.4757 - 1\}$$

Note.

$$\frac{\text{kN}}{\text{m}^2} \times \frac{\text{m}^3}{\text{sec}} = \frac{\text{kNm}}{\text{sec}} = \frac{\text{kJ}}{\text{sec}} = \text{kW}$$

$$\textbf{IP} = \textbf{23.785 kW}$$

(5) For isothermal efficiency, we know that,

$$\eta_{iso} = \frac{\text{Isothermal work}}{\text{Actual work (or IP)}}$$

$$\text{Isothermal work} = P_1 V_1 \ln \frac{P_2}{P_1}$$

$$= 100 \times \frac{6}{60} \times \ln \frac{7}{1}$$

$$\text{Isothermal work} = 19.459 \text{ kW}$$

Then Isothermal efficiency from Eqs (1) and (2)

$$\eta_{isothermal} = \frac{19.459}{23.785} = 0.8181294$$

$$\boldsymbol{\eta_{iso}} = \textbf{81.8125 \%}$$

Example 7.12. A single stage single acting air compressor is working between 1 bar and 7 bar, initial intake temperature is 27°C. Take FAD as 7.5 m³/min and clearance ratio as 0.04, $L : D = 1.3$ and N = 300 rpm and n = 1.3.

Determine:

(a) Volumetric efficiency
(b) Bore and stroke dimensions
(c) Indicated power
(d) Isothermal efficiency

Data:

$V_f = 7.5$ m min $\quad C = 0.04$

$L/D = 1.3 \quad N = 300$ rpm

$\eta_{vol} = ? \quad D = ?$

$L = ? \quad \text{IP} = ?$

$\eta_{iso} = ?$

Solution

(a) We know that, volumetric efficiency of an air compressor is given by,

$$\eta_{vol} = 1+C-C\left\{\frac{P_2}{P_1}\right\}^{1/n} = 1+0.04-0.04\left\{\frac{700}{100}\right\}^{1/1.3}$$

$$= 1+0.04-0.04\times 4.467$$

$$\eta_{vol} = 0..8612955$$

$$\therefore \quad \boldsymbol{\eta_{vol} = 86.13\ \%}$$

(b) We also know that,

$$\frac{P_f V_f}{T_f} = \frac{P_1(V_1 - V_4)}{T_1}$$

$$\frac{P_f V_f}{T_f}\times\frac{T_1}{p_1} = V_1 - V_4 = V_a$$

$$\frac{101.325\times 7.5\ \text{m}^3/\text{min}}{288}\times\frac{300}{100} = (V_1 - V_4) = V_a$$

$$\therefore \quad \mathbf{V_a = 7.9160\,m^3/min}$$

Since the compressor is running at 300 rpm

$$V_a = \frac{7.9160}{300}\text{m}^3$$

$$\mathbf{V_a = 0.0263866\ m^3}$$

Also $$\eta_{vol} = \frac{V_a}{V_s} = 0.8613$$

$$\therefore \quad 0.8613 = \frac{0.263867}{V_s}$$

$$\therefore \quad \mathbf{V_s = 0.0306358\ m^3}$$

and $$= 0.0306358 = \frac{\pi}{8}D^2 L = \frac{\pi}{4}D^2\times 1.3D$$

$$0.0306358 = 1.0210\ D^3$$

$$0.0300051 = D^3$$

$$\mathbf{0.310741\ m = D}$$

and $$\mathbf{L = 1.3\ D = 0.1039634\ m}$$

(c) Now to find induced power,

$$\text{W or IP} = \frac{n}{n-1} P_1 V_a \left\{ \left(\frac{P_2}{P_1} \right)^{\frac{n-1}{n}} - 1 \right\}$$

$$= \frac{1.3}{1.3-1} \times 100 \times 0.26386 \left\{ \left(\frac{7}{1} \right)^{\frac{0.3}{1.3}} - 1 \right\}$$

Note: Unit $\text{kN/m}^2 \times \text{m}^3 = \text{kJ}$

$$= 4.33 \times 100 \times 0.26386 \{1.5668 - 1\}$$

$$\text{IP} = 6.476289 \text{ kJ}$$

As the compressor is running at 300 rpm.

$$\text{Power} = \frac{\text{WN}}{60} = 6.476289 \times \frac{300}{60}$$

$$\textbf{Power} = \textbf{32.38145 kW}$$

(d) Finally to find Isothermal efficiency

$$\eta_{vol} = \frac{\text{Isothermal work}}{\text{Actual work}} = \frac{P_1 V_1 \ln \frac{P_2}{P_1}}{\frac{n}{n-1} P_1 V_1 \left\{ \left(\frac{P_2}{P_1} \right)^{\frac{n-1}{n}} - 1 \right\}}$$

$$= \frac{\ln \frac{P_2}{P_1}}{\frac{n}{n-1} \left\{ \left(\frac{P_2}{P_1} \right)^{\frac{n-1}{n}} - 1 \right\}}$$

$$= \frac{\ln \left(\frac{7}{1} \right)}{\frac{1.3}{1.3-1} \left\{ \left(\frac{7}{1} \right)^{0.3-1.3} - 1 \right\}}$$

$$= \frac{1.9459}{2.456275} = 0.7922$$

$$\eta_{vol} = \textbf{79.22\%}$$

Fig. Ex. 7.12

Example 7.13 For a single stage single acting air compressor, actual volume of air taken in is 10 m³/min, initial intake pressure 1.013 bar, initial temperature 27°C. Final pressure 900 kN/m². Clearance is 6% of stroke volume, $L : D = 1.25$ and compressor is running at 400 rpm, take $n = 1.3$.

Determine:

(i) Volumetric efficiency

(ii) Cylinder dimensions

(iii) Indicated power

Data:

$V_a = 10$ m³/mn		$P_1 = 1.013$ bar $= 101.3$ kN/m²
$T_1 = 27 + 273 = 300$ K		$P_2 = 900$ kN/m².
$V_c = 0.06\ V_s$		$N = 400$ rpm

Solution

(i) $h_{vol} = ?$ (ii) $L = ?\ D = ?$ (iii) $IP = ?$

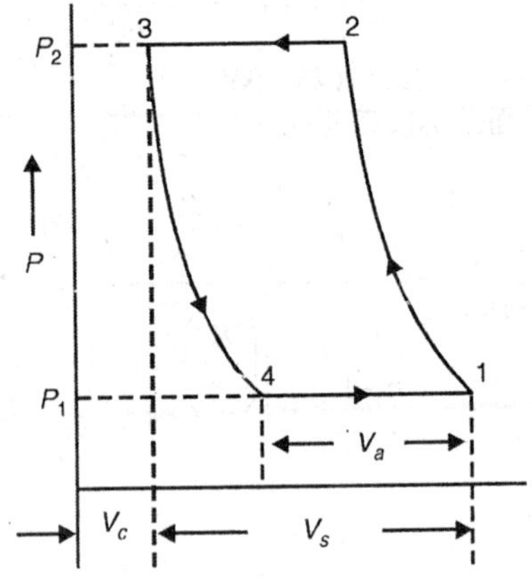

Fig. Ex. 7.13

(i) e know that the volumetric efficiency of a compressor is given by,

$$\eta_{vol} = 1 + C - C \times \left(\frac{P_2}{P_1}\right)^{1/n}$$

$$= 1 + 0.06 - 0.06 \times \left(\frac{900}{101.3}\right)^{1/1.3}$$

$$= 1 + 0.06 - 0.06 \times 5.366$$

$$\eta_{vol} = 0.73799 \cong 0.7380$$

$$\eta_{vol} = \mathbf{73.80\%}$$

(ii) We also know that,

$$\eta_{vol} = \frac{\text{Actual volume of air taken in}}{\text{Stroke volume}}$$

$$\therefore \quad 0.7380 = \frac{10\ \text{m}^3/\text{min}}{V_s}$$

$$V_s = \frac{10\ \text{m}^3/\text{min}}{0.7380}$$

$$V_s = 13.550\ \text{m}^3\ /\text{min}$$

Since $N = 400$ rpm

$$V_s = \frac{13.550}{400}$$

$$V_s = 0.03387 = \frac{\pi}{4}D^2 \times L = \frac{\pi}{4}D^2 \times 1.25D$$

$$0.03387 = 0.9817 \times D^3$$

$$D^3 = 0.0344996$$

$$\frac{0.3387}{0.9817} = D^3$$

$$(0.0344996)^{1/3} = D$$

$$\mathbf{D = 0.3255\ m}$$

and $\mathbf{L = 0.4069\ m}$

(iii) We also know that, IP of compressor is given by

$$= \frac{n}{n-1}P_1V_a\left\{\left(\frac{P_2}{P_1}\right)^{\frac{n-1}{n}} - 1\right\}$$

$$= \frac{1.3}{1.3-1}\times 101.3 \times \frac{10}{60}\left\{\left(\frac{900}{101.3}\right)^{\frac{1.3-1}{1.3}} - 1\right\}$$

Note. $\left[\frac{\text{kN}}{\text{m}^2}\times\frac{\text{m}^3}{\text{sec}} = \frac{\text{kNm}}{\text{sec}} = \frac{\text{kJ}}{\text{sec}} = \text{kW}\right]$

$$= 4.33 \times 101.3 \times 0.166 \times \left\{(8.88)^{0.3/1.3} - 1\right\}$$

$$= 4.33 \times 101.3 \times 0.166 \times \{1.655 - 1\}$$

$$\mathbf{IP = 47.71\ kW}$$

Example 7.14 A single stage **double acting** compressor running at 120 rpm and power input = 75 kW, piston speed = 200 m/min. Suction pressure 1 bar and delivery pressure 10 bar, η_{vol} = 85%. Assuming $PV^{1.25} = C$ for expansion and compression find cylinder bore and V_c as a percent of V_s.

Data: Double acting; $N = 120$ rpm $\quad P = 75$ kW

Piston speed $= 2LN$ m/min = 200 m/min $\quad \eta_{vol} = 0.85$

$D = ?$ $\quad \dfrac{V_c}{V_s} = ?$

Solution

Note. N = Number of complete cycles/min

= RPM for single acting

= 2 × rpm for double acting

We know that,

$$\text{Piston speed} = 2LN = 200$$

$$L = \frac{200}{2 \times N} = \frac{200}{2 \times 120}$$

$$L = 0.8333 \text{ m}$$

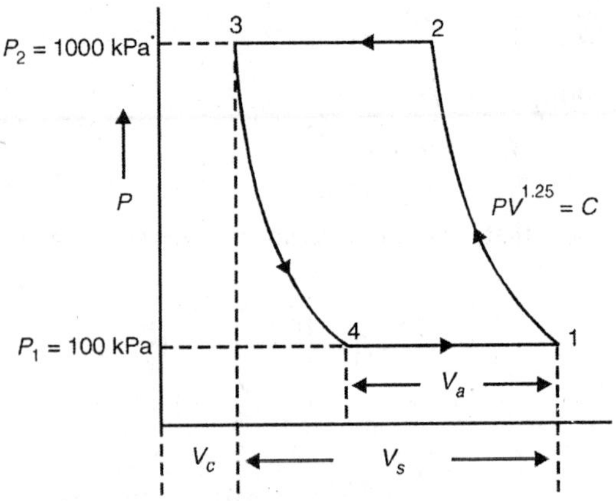

Fig. Ex. 7.14

Also

$$\text{power} = \frac{\text{WN}}{60}$$

$$75 = \frac{\text{W} \times (2 \times 120)}{60}$$

$$\mathbf{W = 18.75\ kJ/cycle}$$

Also work done per cycle is given by

$$W = \frac{n}{n-1} P_1 V_a \left\{ \left(\frac{P_2}{P_1} \right)^{\frac{n-1}{n}} - 1 \right\}$$

$$18.75 = \frac{1.25}{1.25-1} \times 100 \times V_a \left\{ \left(\frac{10}{1} \right)^{\frac{0.25}{1.25}} - 1 \right\}$$

$$\Rightarrow \quad \mathbf{V_a = 0.064\,m^3 = Actual\ vol.}$$

Then, we know that volumetric efficiency is given by,

$$\eta_{vol} = \frac{V_a}{V_s}$$

$$0.85 = \frac{0.064}{V_s}$$

$$V_s = 0.075\,\text{m}^3 = \frac{\pi}{4} D^2 \times L$$

$$0.075 = \frac{\pi}{4} \times D^2 \times 0.833$$

$$0.075 = D^2 \times 0.65447$$

$$0.11459 = D^2$$

$$\mathbf{0.3385\ m = D}$$

Now to find C, we know that,

$$\eta_{vol} = 1 + C - C \left\{ \frac{P_2}{P_1} \right\}^{1/n}$$

or

$$= 1 - C \left\{ \left(\frac{P_2}{P_1} \right)^{1/n} - 1 \right\}$$

$$\therefore \quad 0.85 = 1 - C\left\{ (10)^{1/0.25} - 1 \right\}$$

$$0.85 = 1 - C\{5.31\}$$

$$0.85 = 1 - 5.31\,C$$

$$5.31\,C = 1 - 0.85$$

$$C = \frac{0.15}{5.31} = 0.0282$$

or
$$C = \frac{V_c}{V_s} = 0.0282$$

$$\mathbf{V_c = 2.82\ \%\ V_s}$$

Example 7.15. A single stage, **double acting** air compressor has a cylinder diameter of 400 mm and a stroke of 300 mm. The clearance volume is 6% of V_s. The intake pressure is 0.95 bar and temperature is 25°C. If the delivery pressure is, 5 bar and speed is 250 rev/min, calculate the mass rate of air compressed. The index of compression and expansion is 1.3 and assume $R = 0.287$ kJ/kg-K.

Data: **Double Acting**, so use $N = 2 \times$ rpm

$D = 0.4$ m $L = 0.3$ m

$V_c = 0.06\ V_s$ $P_1 = 95$ kPa

$T_1 = 25 + 273 = 298$ K

$P_2 = 500$ kPa $N = 250$ rpm

$m = ?$ kg/sec $n = 1.3$

and $R = 0.287$ kJ/kg-K for air.

Solution

We know that, $P_1V_1 = mRT_1$

$$\dot{m} = \frac{P_1V}{RT_1} \qquad (1)$$

Also, we know that, stroke volume

$$V_s = \frac{\pi}{4}D^2 \times L = \frac{\pi}{4} \times 0.4^2 \times 0.3$$

$$\mathbf{V_s = 0.037699\,m^3}$$

As $V_c = 0.06 V_s$

$\therefore$ $V_c = 0.06 \times 0.0377$

$$\mathbf{V_c = 0.002262\,m^3}$$

$\therefore$ $V_1 = V_c + V_s$

$$V_1 = 0.002262 + 0.0377$$

$$V_1 = 0.0399619\ m^3$$

$$V_1 = 0.002262 + 0.0377$$

$$\mathbf{V_1 = 0.0399619\ m^3}$$

Now from Eq. (1)

$$\dot{m} = \frac{P_1 V_1}{RT_1} = \frac{95 \times 0.0399619}{0.287 \times 298}$$

$$\dot{\mathbf{m}} = \mathbf{0.0443887\,kg}$$

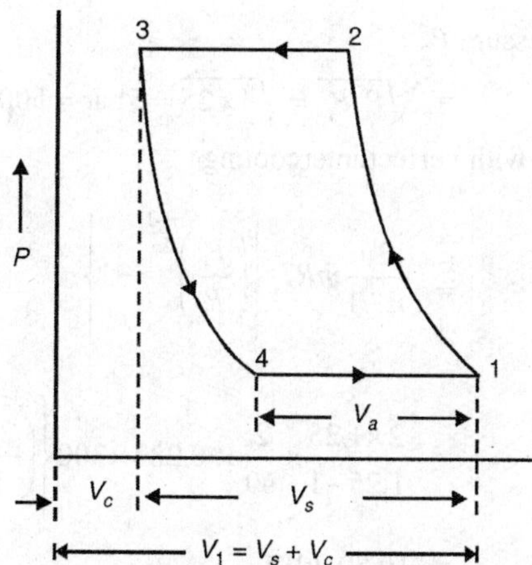

Fig. Ex. 7.15

Now to find mass of air handled in kg/sec

$$\dot{m} = 0.0443887 \times \frac{2N}{60}$$

(Double acting)

$$= 0.0443887 \times \frac{2 \times 250}{60}$$

$$\dot{\mathbf{m}} = \mathbf{0.3699\,kg\,/\,sec}$$

(Problems on Multistage Air Compressors)

Example 7.16 A 2-stage reciprocating air compressor handles 2 kg/min of air corresponding to condition of 1 bar and 27°C. It is delivered of 25 bar. The polytropic index of compression is 1.25 and the intercooling is perfect. Neglect in clearance find,

(1) Intermediate pressure for maximum η.
(2) Power required in kW.
(3) $\eta_{isothermal}$
(4) Saving in power required compared with that for single stage compression.

Data: $m = 2$ kg/min $\quad P_1 = 1$ bar = 100 kPa

$T_1 = 300$ K $\quad P_3 = 25$ bar = 2500 kPa

$n = 1.25$

Solution

Perfect intercooling

(I) Intermediate pressure P_2

$$= \sqrt{P_1 P_3} = \sqrt{1 \times 25} = 5 \text{ bar} = 500 \text{ kPa}$$

(2) Power required with perfect intercooling

$$= \frac{2n}{n-1} \dot{m} R T_1 \left\{ \left(\frac{P_3}{P_1} \right)^{\frac{n-1}{2n}} - 1 \right\}$$

$$= \frac{2 \times 1.25}{1.25 - 1} \times \frac{2}{60} \times 0.287 \times 300 \left\{ \left(\frac{2500}{100} \right)^{\frac{0.25}{2.5}} - 1 \right\}$$

$$= 10.89 \text{ kW}$$

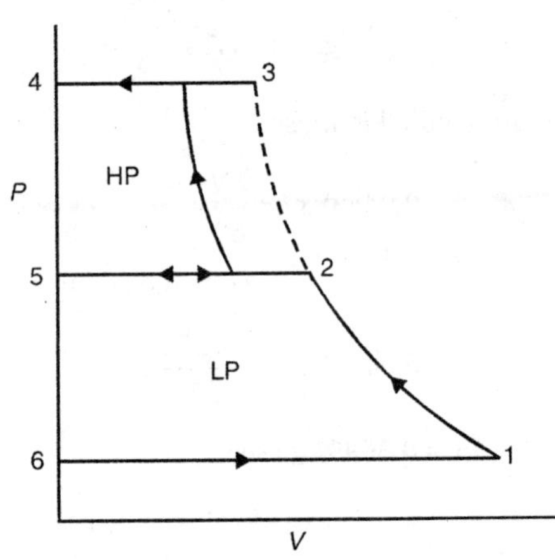

Fig. Ex. 7.16

(3) Isothermal Power input

$$= P_1 V_1 \ln \frac{P_3}{P_1} = \dot{m} R T_1 \ln \frac{P_3}{P_1}$$

$$= \frac{2}{60} \times 0.287 \times 300 \ln \frac{2500}{100} = 9.24 \text{ kW}$$

$$\therefore \quad \eta_{iso} = \frac{\text{Isothermal power input}}{\text{Actual work input}} = \frac{9.24}{10.89}$$

$$= 0.848$$

$$\boldsymbol{\eta_{iso} = 84.8\%}$$

(4) Power input (single stage)

$$= \frac{n}{n-1} \times \dot{m}RT_1 \left\{ \left(\frac{P_3}{P_1}\right)^{\frac{n-1}{n}} - 1 \right\}$$

$$= \frac{1.25}{0.25} \times \frac{2}{60} \times 0.287 \times 300 \left\{ \left(\frac{2500}{100}\right)^{\frac{0.25}{1.25}} - 1 \right\}$$

$$= 12.96 \text{ kW}$$

$$\text{Saving in power} = 12.96 - 10.89$$

Saving in power = 2.07 kW

Example 7.17 A 2 stage single acting reciprocating air compressor draws in air at 1 bar and 300 K. The delivery pressure is 12 bar and intermediate pressure is ideal for minimum work and the intercooling is perfect. The compression follows the law $PV^{1.3} = C$. Flow rate of air through the compressor is 0.15 kg/sec.

Determine:

(1) Power required to drive the compressor.

(2) Saving in power required to drive the compressor compared with single stage working.

(3) Isothermal η.

Data: $P_1 = 1 \text{ bar} = 100 \text{ kPa}$ $\quad T_1 = 300 \text{ K}$

$P_3 = 12 \text{ bar} = 1200 \text{ kPa}$ $\quad PV^{1.3} = C$

$m = 0.15 \text{ kg/sec}$

Solution

Since ideal intermediate pressure is there,

$$P_2 = \sqrt{P_1 P_3} = 3.46 \text{ bar} = 346 \text{ kPa}$$

With ideal intermediate pressure and perfect intercooling,

$$\text{IP (2 stage)} = \frac{2n}{n-1} mRT_1 \left\{ \left(\frac{P_3}{P_1}\right)^{\frac{n-1}{2n}} - 1 \right\}$$

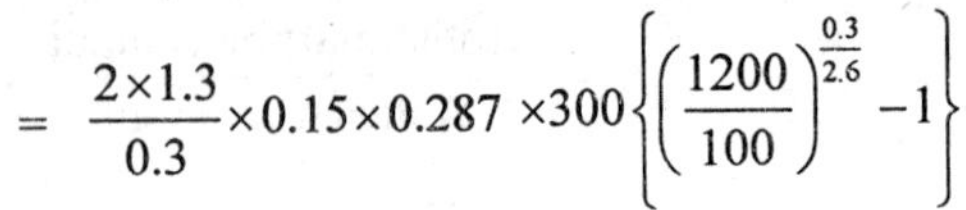

$$= \frac{2 \times 1.3}{0.3} \times 0.15 \times 0.287 \times 300 \left\{ \left(\frac{1200}{100} \right)^{\frac{0.3}{2.6}} - 1 \right\}$$

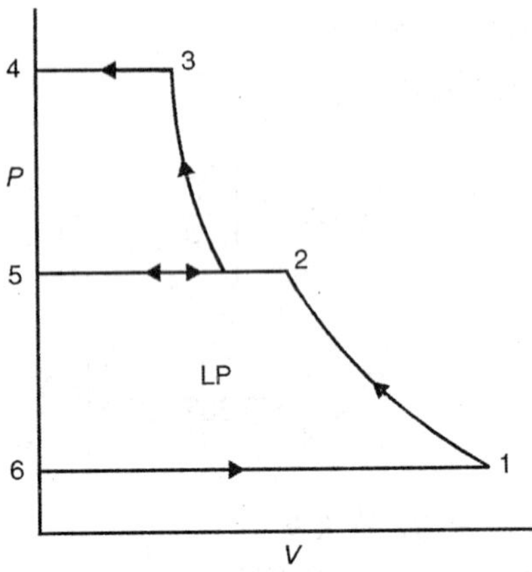

Fig. Ex. 7.17

IP (2 stage) = 37.16 kW

$$\text{IP (single stage)} = \frac{n}{n-1} mRT \left\{ \left(\frac{P_3}{P_1} \right)^{\frac{n-1}{n}} - 1 \right\}$$

$$= \frac{1.3}{0.3} \times 0.15 \times 0.287 \times 300 \left\{ \left(\frac{1200}{100} \right)^{\frac{0.3}{1.3}} - 1 \right\}$$

IP (single stage) = 43.33 kW

Saving in power = 43.33 – 37.16

Saving in Power = 6.17 kW

and Isothermal work $= \dot{m}RT_1 \ln \frac{P_3}{P_1}$

$$= 0.15 \times 0.287 \times 300 \ln \frac{1200}{100} = 32.09 \text{ kW}$$

$$\therefore \quad \eta_{iso} = \frac{\text{Isothermal work}}{\text{IP}} = \frac{32.09}{37.16}$$

$$\boldsymbol{\eta_{iso} = 0.863 = 86.3\% \text{ (for multistage)}}$$

Example 7.18 A 2 stage single acting air compressor takes in air at 1 bar and 300 K. Air is discharged at 10 bar. The intermediate pressure is ideal and intercooling is perfect. The law of compression is $PV^{1.3} = C$. The rate of discharge is 0.1 kg/sec. Find.

1. Power required to drive compressor.
2. Saving in work compared with single stage.
3. $\eta_{isothermal}$ for multistage and single stage.
4. Heat rejected in intercooler.

Take: $R = 0.287$ and $C_p = 1$ kJ/kg-K

Solution

For ideal conditions,

$$P_2 = \sqrt{P_1P_3} = \sqrt{1\times 10} = 3.162 \text{ bar}$$

and

$$\text{IP (2 stage)} = \frac{2n}{n-1}\dot{m}RT_1\left\{\left(\frac{P_3}{P_1}\right)^{\frac{n-1}{2n}} - 1\right\}$$

$$= \frac{2\times 1.3}{1.3-1}\times 0.1\times 0.287\times 300\left\{\left(\frac{10}{1}\right)^{\frac{1.3\times 1}{2\times 1.3}} - 1\right\} = 22.62 \text{ kW}$$

$$\text{IP single in work} = \frac{n}{n-1}\dot{m}RT_1\left\{\left(\frac{P_3}{P_1}\right)^{\frac{n-1}{n}} - 1\right\}$$

$$= \frac{1.3}{0.3}\times 0.1\times 0.28\times 300\left\{\left(\frac{10}{1}\right)^{\frac{0.3}{1.3}} - 1\right\} = 26.05 \text{ kW}$$

$$\text{Saving in work} = 26.05 - 22.62$$

Saving in work = 3.43 kW

(iii) $$\text{Isothermal work} = m.RT_1 \ln\frac{P_3}{P_1} = 0.1\times 0.287\times 300\times \ln\frac{10}{1}$$

$$= 19.82 \text{ kW}$$

$$\eta_{iso} = \frac{19.82}{22.62} = 0.8765$$

η_{iso} **= 87.65% with multistage**

$$\eta_{iso} = \frac{19.82}{26.05} = 0.7617$$

$$\eta_{iso} = \mathbf{76.17\%} \textbf{ with single stage.}$$

(4) Heat rejected in intercooler.
Temperature rise after I stage

$$T_2 = T_1\left(\frac{P_2}{P_1}\right)^{\frac{n-1}{n}} = 300(3.162)^{0.3/1.3} = 390.95\,\text{K}$$

The heat lost in intercooler.

$$= \dot{m}C_p(T_2 - T_1) = 0.1 \times 1 \times (390.95 - 300)$$

$$= 9.095 \text{ kJ/sec}$$

Example 7.19 A single acting two stage reciprocating air compressor with complete intercoooling delivers 10 kg/min of air at 16 bar pressure. The suction occurs at 1 bar and 15°C. The compression and expansion processes are reversible polytropic with polytropic index $n = 1.25$. Calculate

(i) The power required
(ii) The isothermal efficiency
(iii) The free air delivered
(iv) Heat transferred in intercooler

Data: m = 10 kg/min, P_3 = 16 bar,
P_1 = 16 bar, T_1 = 15 + 273,
n = 1.25.

Solution

For ideal condition

$$P_2 = \sqrt{P_1P_3} = \sqrt{1\times16} = 4\,\text{bar}$$

(i) Power require

$$= \frac{2n}{n-1}\dot{m}RT_1\left\{\left(\frac{P_2}{P_1}\right)^{\frac{n-1}{2n}} - 1\right\}$$

$$= \frac{2\times1.25}{1.25-1}\times10\times0.287(15+273)\left\{(4)^{\frac{1.25-1}{2\times1.25}} - 1\right\}$$

Power = 44 kW

$$\text{Isothermal work} = mRT_1 \ln\frac{P_3}{P_1}$$

$$= \frac{10}{60}\times0.287\times(15+273)\ln\left(\frac{16}{1}\right)$$

$$= 38.2 \text{ kW}$$

(ii) Isothermal efficiency $= \dfrac{\text{Isothermal work}}{\text{Actual work}}$

$$\eta_{iso} = \frac{38.2}{44} = 0.868 = 86.8\%$$

(iii) $$\text{FAD} = \frac{mRT_1}{P_1} = \frac{10 \times 0.287 \times 288 \times 1000}{1 \times 10^5}$$

$$= 8.266 \text{ m}^3/\text{min}$$

(iv) Temperature after end of compression upto intermediate pressure

$$\Rightarrow \quad T_2 = T_1 \times \left(\frac{P_2}{P_1}\right)^{\frac{n-1}{n}}$$

$$T_2 = 288 \times (4)^{\frac{0.25}{1.25}}$$

$$T_2 = 380 \text{ K}$$

Heat transferred in the intercooler $Q = m C_p (T_2 - T_1)$

$$= \frac{10}{60} \times 1.005 \times 1.005(380 - 288)$$

$$\mathbf{Q = 15.14 kW}$$

Example 7.20 In a 3 stage compressor air is compressed from 98 kPa to 500 kPa. Calculate for 1m^3 of air per sec.

(1) Work under ideal condition for $n = 1.3$,

(2) Isothermal work

(3) Saving in work due to multistaging

(4) Isothermal efficiency.

Data: $P_1 = 98$ kPa $\quad P_4 = 500$ kPa

$V = 1 \text{ m}^3/\text{sec}$

Solution

Work input or IP for 3-stage

$$= \frac{3n}{n-1} P_1 V_1 \left\{ \left(\frac{P_4}{P_1}\right)^{\frac{n-1}{n}} - 1 \right\}$$

$$= \frac{3\times 1.3}{1.3-1}\times 98\times 1\left\{\left(\frac{500}{98}\right)^{\frac{0.3}{3\times 1.3}}-1\right\} = \mathbf{170\,kW}$$

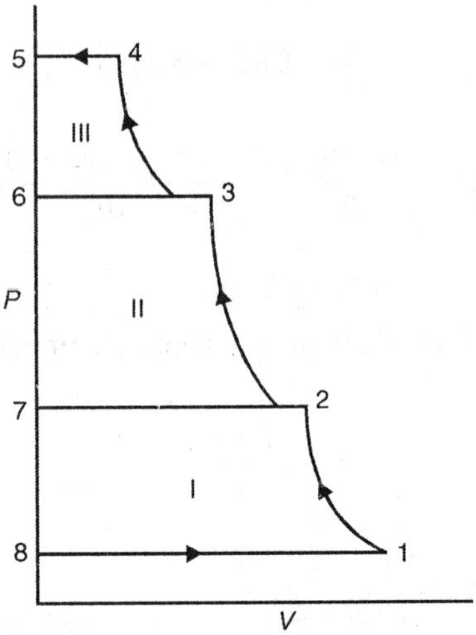

Fig. Ex. 7.20

For single stage

$$\text{IP} = \frac{n}{n-1}P_1V_1\left\{\left(\frac{P_4}{P_1}\right)^{\frac{n-1}{n}}-1\right\}$$

$$= \frac{1.3}{1.3-1}\times 98\times 1\left\{\left(\frac{500}{98}\right)^{\frac{0.3}{1.3}}-1\right\}$$

$$= \mathbf{193.1}$$

$\therefore$ **Saving in work 193.1 – 170 = 23.1 kW**

$$\text{Isothermal work} = P_1\,V_1\ln\frac{P_{\text{final}}}{P_1} = 90\times 1\times \ln\frac{500}{98} = 159.7\,\text{kW}$$

$\therefore$ Isothermal efficiency for multistage,

$$\eta_{\text{iso}} = \frac{159.7}{170} = 93.89\%$$

$$\eta_{\text{iso}} \text{ for single stage} = \frac{159.7}{193} = 82.7\%$$

Example 7.21 A 3 stage reciprocating air compressor compresses air from 1 bar and 17°C to 35 bar. The law for compression is $PV^{1.25} = C$ for all stages. Assuming perfect intercooling and neglecting clearance find the minimum work required to compress 15 m³/min of free air. Also find intermediate pressures.

Data: 3-stage

$$P_1 = 100 \text{ kN/m}^2 \qquad T_1 = 290 \text{ K}$$
$$P_4 = 3500 \text{ kN/m}^2 \qquad n = 1.25$$
$$V_1 = 15 \text{ m}^3/\text{min}$$

Solution

$$\text{IP (3 stage)} = \frac{3n}{n-1} P_1 V_1 \left\{ \left(\frac{P_4}{P_1} \right)^{\frac{n-1}{3n}} - 1 \right\}$$

$$= \frac{3 \times 1.25}{1.25 - 1} \times 100 \times \frac{15}{60} \left\{ \left(\frac{3500}{100} \right)^{\frac{0.25}{3 \times 1.25}} - 1 \right\}$$

$$= 100.29 \text{ kJ/sec or kW}$$

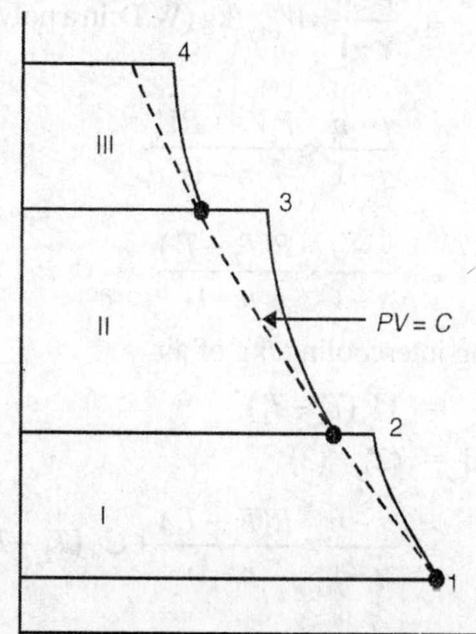

Fig. Ex. 7.21

Now
$$\frac{P_2}{P_1} = \frac{P_3}{P_2} = \frac{P_n}{P_3} = k$$

and
$$k = \left(\frac{P_{x+1}}{P_1} \right)^{1/x}$$

$$= \text{(Pressure Range)}^{1/x} = \left(\frac{3500}{100}\right)^{\frac{1}{3}} = 3.27$$

$\therefore$ $$\mathbf{P_2 = kP_1 = 3.27\,bar}$$

$$\mathbf{P_3 = P_1k^2 = 10.674\,bar}$$

Example 7.22 Prove that heat rejected per stage per kg of air in a reciprocating air compressor with perfect intercooling is given by $\left[C_p + C_v\left(\frac{\gamma - n}{n-1}\right)\right]$ $(T_2 - T_1)$ where $(T_2 - T_2)$ is the temperature rise during each stage.

Solution

Total heat rejected in a reciprocating compressor

= Heat rejected during compression + Heat rejected during intercooling (1)

Now heat rejected during compression per kg,

$$Q\text{ (poly).} = \frac{\gamma - n}{\gamma - 1} \times W_{\text{poly}}/\text{kg (W.D. in a poly compression process)}$$

$$= \frac{\gamma - n}{\gamma - 1} \times \frac{P_2V_2 - P_1V_1}{n-1}$$

$$= \frac{\gamma - n}{\gamma - 1} \times \frac{R(T_2 - T_1)}{n-1} \quad (2)$$

And heat rejected during intercooling /kg of air

$$= C_p(T_2 - T_1)$$

$$\text{Total heat rejected} = (2) + (3)$$

$$= \frac{\gamma - n}{\gamma - 1} \times \frac{R(T_2 - T_1)}{n-1} + C_p(T_2 - T_1)$$

But $$R = C_p - C_v \text{ and } \gamma = \frac{C_p}{C_v}$$

$\therefore$ $$R = \gamma C_v - C_v$$

$$R = C_v(\gamma - 1)$$

$\therefore$ $$C_v = \frac{R}{\gamma - 1}$$

∴ Total heat rejected /kg/stage

$$= \frac{\gamma - n}{n-1} \times C_v(T_2 - T_1) + C_p(T_2 - T_1)$$

$$= \left\{C_p + C_v\left(\frac{\gamma - n}{n-1}\right)\right\}(T_2 - T_1)$$

Example 7.23 A 4 stage compressor works between limits of 1 bar and 115 bar. The index of compression in each stage is 1.29. The temperature at the start of compression in each stage is 35°C and the intermediate pressure are so chosen that the work is divided equal among the stages (∴ pressure ratio same). Neglecting clearance find,

(1) Pressures P_2, P_3 and P_4
(2) Delivery temperature in each stage
(3) Individual work/kg and
(4) Isothermal work and, $\eta_{isothermal}$

Data: $P_1 = 1$ bar $\quad P_5 = 115$ bar
$n = 1.29 \quad T_1 = 35°C = 308$ K

Solution

For ideal intermediate pressure

$$\frac{P_2}{P_1} = \frac{P_3}{P_2} = \frac{P_4}{P_3} = \frac{P_5}{P_4} = k$$

$$\therefore \quad P_2 = kP_1$$

$$P_3 = kP_2 = k^2 P_1$$

$$P_{x+1} = k^x P_1$$

i.e. $$P_5 = k^4 P_1$$

$$\frac{P_5}{P_1} = k^4$$

$$\left(\frac{P_5}{P_1}\right)^{1/4} = k; \left(\frac{115}{1}\right)0.25 \Rightarrow k = 3.27$$

$$P_2 = 327.47 \text{ kPa}$$

$$P_3 = 1072.38 \text{ kPa}$$

$$P_4 = 3511.74 \text{ kPa}$$

Since compression is polytropic,

$$\frac{T_2}{T_1} = \left(\frac{P_2}{P_1}\right)^{\frac{n-1}{n}}$$

$$T_2 = 308(3.27)^{0.29/1.29}$$

$$= 401.39 = 402\,\text{K}$$

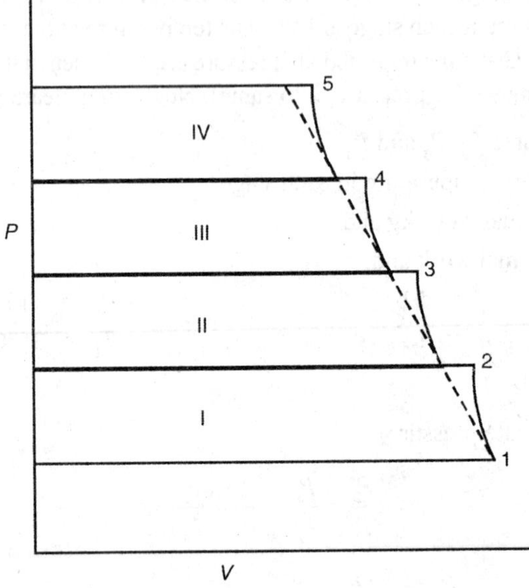

Fig. Ex. 7.23

Delivery temperature from each stage.

$$\text{IP (4 stage)} = \frac{4n}{n-1} 1.RT_1\left[\left(\frac{P_5}{P_1}\right)^{\frac{n-1}{4n}} - 1\right]$$

$$= \frac{4\times1.29}{0.29}\times0.287\times308\{(115)^{0.29/5.26} - 1\}$$

$$= 480.7\ \text{kJ/kg}$$

$$\text{Isothermal work} = mRT_1 \ln\frac{P_5}{P_1} \text{ and } m = 1\,\text{kg (for kJ/kg)}$$

$$= 0.287\times308\ln\left(\frac{115}{1}\right)$$

$$= 419.43 \text{ kJ/kg}$$

$$\therefore \text{ Isothermal efficiency} = \eta_{iso} = \frac{419.45}{480.7}$$

$$\eta_{iso} = 0.872 = 87.2\%$$

Example 7.24 A single acting, two stage RAC running at 280 rpm delivers air at a pressure of 18 bar, and the conditions at the commencement of compression being 0.98 bar and 305 K.

The intermediate pressure is 4 bar and the clearance volume of LP cylinder is 5% of stroke volume (swept volume). Equation for compression and expansion for each cylinder is $PV^{1.25} = C$. The compressor has a free air capacity of 2.25 m³/min measured under free conditions of 1 bar and 290 K. Take mass of air handled as 2.7 kg/min.

Determine

(i) η_{vol} of L.P. cylinder.

(ii) Theoretical work spent in driving the compressor.

(iii) Dimension of L.P. cylinder if $L = D$

The given compressor has a perfect intercooling.

Data: $V_c = 0.05\ V_s$ for L.P. cylinder

$V_f = 2.25$ m³/min $\qquad P_f = 100$ kPa

$T_f = 290$ K $\qquad \dot{m} = 2.7$ kg/min

Solution

(i) Now volumetric efficiency of L.P. cylinder

$$\eta_{vol}(\text{L.P.}) = 1 - C\left\{\left(\frac{P_2}{P_1}\right)^{\frac{1}{1.25}} - 1\right\}$$

$$= 1 - 0.05\left\{\left(\frac{4}{0.98}\right)^{\frac{1}{1.25}} - 1\right\} = 0.896 \text{ and } 89.6\%$$

(ii) Theoretical work input, for 2-stage air compressor, we know that,

$$\text{IP (2 stage)} = \frac{2n}{n-1}\dot{m}RT_1\left\{\left(\frac{P_3}{P_1}\right)^{\frac{n-1}{2n}} - 1\right\}$$

$$= \frac{2 \times 1.25}{1.25 - 1} \times \frac{2.7}{60} \times 0.287 \times 305\left\{\left(\frac{18}{0.98}\right)^{\frac{1.25-1}{2\times 1.25}} - 1\right\}$$

$$= 13.307 \text{ kW}$$

(iii) We also know that,

$$\frac{P_f V_f}{T_f} = \frac{P_1(V_1 - V_4)}{T_1}$$

$$\frac{P_f V_f}{T_f} \times \frac{T_1}{P_1} = (V_1 - V_4)$$

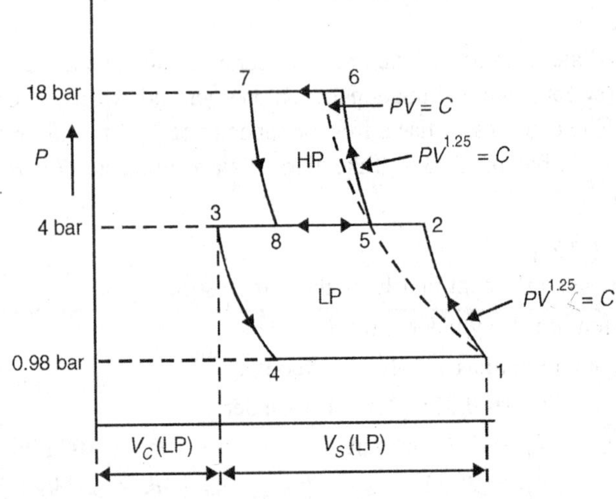

Fig. Ex. 7.24

$$\frac{100 \times 2.25 \text{m}^3/\text{min}}{210} \times \frac{305}{0.98 \times 100} = (V_1 - V_4)$$

$$2.41 \text{m}^3/\text{min} = (V_1 - V_4)$$

$$\eta_{\text{vol LP}} = \frac{(V_1 - V_4)}{V_s} = 0.896$$

$$\frac{2.41}{2.896} = V_s = 2.6897 \text{m}^3/\text{min} = \frac{\pi}{4} D^2 \times L \times N$$

$$2.6897 = \frac{\pi}{4} D^3 \times 280$$

D = 0.23040 m = L

Example 7.25 A 2 stage air compressor delivers air at a rate of 1.35 kg/sec. The suction pressure is 1 bar, intermediate pressure is 7 bar and delivery pressure is 42 bar. Air enters L.P. cylinder at 290 K is intercooled to 305 K before inlet to H.P. and delivered at 415 K. The clearance volumes of L.P. and H.P. cylinders are 6% and 8% of stroke volumes. Assuming the law of compression and expansion be same, find,

(i) Index of compression and expansion

(ii) Volumetric efficiency of L.P. and H.P. cylinders

(iii) Total power

Data: $m = 1.35$ kg/sec $\qquad V_c = 0.06\, V_s$ (L.P)

$V_c = 0.08\, V_s$ (H.P) $\qquad V_{s\,LP} = V_1 - V_3$

$V_{s\,HP} = V_5 - V_7$

Solution

For the process 5–6

$$\frac{T_6}{T_5} = \left(\frac{P_6}{P_5}\right)^{\frac{n-1}{n}}$$

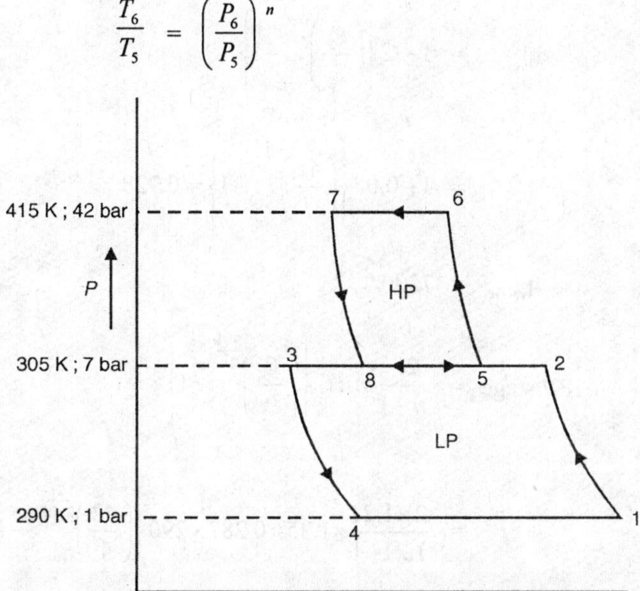

Fig. Ex. 7.25

$$\frac{415}{305} = \left(\frac{42}{7}\right)^{\frac{n-1}{n}}$$

$$1.36 = (6)^{n-1/n}$$

$$\ln 1.36 = \frac{n-1}{n} \ln 6$$

$$0.171 = \frac{n-1}{n}$$

$\therefore$ **n = 1.21** (We can get this value by considering process 1–2 also)

$$\eta_{vol}(\text{L.P.}) = 1 - C\left\{\left(\frac{P_2}{P_1}\right)^{1/n} - 1\right\}$$

$$= 1 - 0.06\left\{\left(\frac{7}{1}\right)^{1/n} - 1\right\}$$

$$= 1 - 0.06\left\{\left(\frac{7}{1}\right)^{1/n} - 1\right\} = 0.7603 \Rightarrow 76.03\%$$

$$\eta_{vol\,HP} = 1 - C\left\{\left(\frac{P_3}{P_2}\right)^{1/n} - 1\right\}$$

$$= 1 - 0.08\left\{\left(\frac{42}{7}\right)^{\frac{1}{1.21}} - 1\right\} = 0.728$$

$$\boldsymbol{\eta_{vol\,HP} = 72.8\%}$$

$$\text{IP 2 stage} = \frac{2n}{n-1}\dot{m}RT_1\left\{\left(\frac{P_3}{P_2}\right)^{\frac{n-1}{2n}} - 1\right\}$$

$$= \frac{2\times1.21}{1.21-1}\times1.35\times0.287\times290\left\{\left(\frac{42}{1}\right)^{\frac{1.21-1}{2\times1.21}} - 1\right\}$$

$$= 1.294.831\times\left\{\left(\frac{42}{1}\right)^{\frac{0.21}{2.42}} - 1\right\}$$

IP 2 stage = 496.073 kW

THEORY QUESTIONS

1. State the uses of compressed air.
2. Explain classification of air compressor.
3. Explain: (i) Volumetric efficiency, (ii) Isothermal efficiency, (iii) FAD.
4. Obtain an expression for volumetric efficiency of an air compressor in terms of clearance ratio index ofcompression and pressure ratio.
5. Explain the reasons for low volumetric efficiency of a reciprocating air compressor.
6. Discuss the reasons for multistage compression. Also explain the additional benefits of multistage compression, if any.
7. Bring out the factors that influence the volumetric efficiency of a reciprocating. air compressor.
8. What are the advantages of multistage compression ?
9. Explain the working of a two stage reciprocating air compressor with intercooler.
10. Prove that the volumetric efficiency of air compressor is given by,

$$\eta_{vol} = 1 + C - C\left[\frac{P_d}{P_s}\right]^{1/n}$$

where C = Clearance ratio, P_d = Discharge pressure, P_s = Suction pressure
11. Explain the effect of clearance on volumetric efficiency.
12. Explain the methods to improve the isothermal efficiency of air compressor.
13. Obtain an expression for IP of single stage air compressor.
14. Derive the expression for the volumetric efficiency of a reciprocating air comp. in terms of clearance ratio; pressure ratio; and index of compression.
15. Explain why staging is done in reciprocating compression.
16. Explain clearly how the actual indicator design for a single stage reciprocating compression in different than theoretical indicator design.
17. What are the advantage of multistage compressor ?
18. Explain the reasons for low volumetric efficiency of reciprocating air compressor.
19. Obtain an express for how volumetric efficiency of reciprocating air compressor.
20. State difference uses of compress air.
21. Why are multistage lamp are preferred in single stage ? Explain with?
22. When is multistage compressor used for air ? What are its advantage ?
23. With the help of *PV* design, explain how work is seemed by a 2 stage compressor in the same pressure limit. Write expansion for amount of work seemed because of 2-state opm. State ending for major maximum work saving.
24. Explain clearly how actual *P–V* diagram for a single stage air compressor deviates from the theoretical diagram.

PROBLEMS FOR PRACTICE

1. A 2 stage single acting air compressor takes in air at 1 bar and 300 K. Air is discharged at 10 bar. The intermediate pressure is ideal and intercooling is perfect. The law of compression is $PV^{1.3} = C$. The rate of discharge is 0.1 kg/sec. Find

   (i) Power required to drive the compressor.

   (**Ans.** 22.62 kW)

   (ii) Saving in work compared with single stage.

   (**Ans.** Saving in work = 3.43 kW)

   (iii) Isothermal efficiency for single and multistage.

   (**Ans.** Iso η_{single} = 76.17%, Iso η_{mult} = 87.65%)

2. A single acting single stage compressor is belt driven from an electric motor at 400 rpm. The cylinder diameter is 15 cm and stroke is 17.5 cm. The air is compressed from 1 bar to 7 bar and law of compression is $PV^{1.3} = C$. Find power of motor, if transmission efficiency is 97%, and η_{mech} = 90%. Neglect clearance effects.

ଛ୦ ଓଃ

8

Rotary Air Compressors

CHAPTER OBJECTIVES

After reading this chapter you will be able to learn the following

- Classification of Compressors, Stagnation Properties of Flowing Fluids.
- Roots blower, Vane type, Screw type Centrifugal Compressors, and Axial Compressors.
- Work (Euler's work) for C.I. Compressors.
- Slip and Slip Factor of Centrifugal Compressors.
- Pre-Whirl and Losses in C.F. Compressors.

8.1 INTRODUCTION

Compressor is a device which compresses air/gases or vapours from low pressure to high pressure. It needs external energy input in the form of work. Out of the total work input to the compressor, some work is utilised to compress the fluid while the remaining is lost in overcoming friction, some work is lost to cooling medium, etc.

Compressors are mainly grouped into two categories:

(i) Positive displacement compressor; and

(ii) Dynamic compressors.

In a positive displacement compressor, the pressure of the gas is increased by reduction of its volume i.e. by positive displacement of the gas to the delivery side.

While in a dynamic compressor, the kinetic energy imparted to the gas by the rotation of the rotor (impeller) is converted into pressure energy partly in the rotor and the remaining in the diffuser. Thus, the pressure rise is developed due to the dynamic action of gas.

In the subsequent article, various types of compressors are listed. Compressors are extensively used for a variety of applications (Refer Sec. 8.15).

8.2 CLASSIFICATION OF COMPRESSORS

Compressors are classified on the basis of several criteria as follows.

(a) On the basis of design and principle of operation. Based on the design and principle of operation, compressors are classified into two main groups and further sub-divided into sub-groups as shown in Flow Chart.

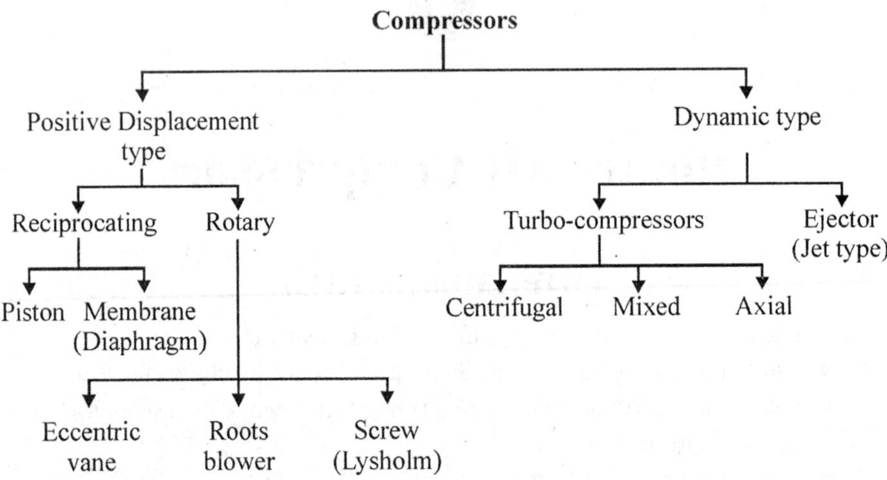

(b) According to delivery pressure. Low pressure (upto 10 bar), medium pressure (10 – 80 bar), high pressure (80 – 1000 bar).

Hyper-compressors are multistage reciprocating compressors with delivery pressure upto 1000 bar.

(c) According to the pressure ratio (as per ASME code). Fans — Pressure ratio upto 1.1

Blowers — Pressure ratio between 1.1 to 2.3

Compressor — Pressure ratio above 2.3

(d) According to free air delivered (capacity). Small capacity — Upto 9 m^3/min

Medium capacity — Between 9 – 3000 m^3/min

Large capacity — Above 3000 m^3/min

(e) According to number of stages adopted. Single stage, Multi-stage

(f) According to the drive (prime mover). Electric motor driven, I.C. engine driven.

The present chapter deals with Rotary positive displacement compressors and dynamic compressors.

8.2.1 Stagnation Properties of a Flowing Fluid

When a flowing fluid having same velocity is brought to rest, it is said to have attained a stagnation state. The final stagnation state is governed by the manner in which it is attained. A reversible adiabetic or isentropic process has a great significance.

For an isentropic stagnation process, the steady flow energy equation simplifies to

$$h + \frac{V^2}{2} = h_o, \text{ kJ/kg}$$

where h_o is stagnation enthalpy and h is its initial enthalpy when fluid is flowing with a velocity of V m/s.

During the stagnation process, the kinetic energy of the fluid is converted to enthalpy, thereby pressure and temperature of the fluid increases as shown in Fig. 8.1.

The properties of the fluid at the stagnation state are referred as stagnation properties e.g. stagnation pressure, stagnation temperature, stagnation density, etc. The stagnation state and stagnation properties are denoted by lower suffix o.

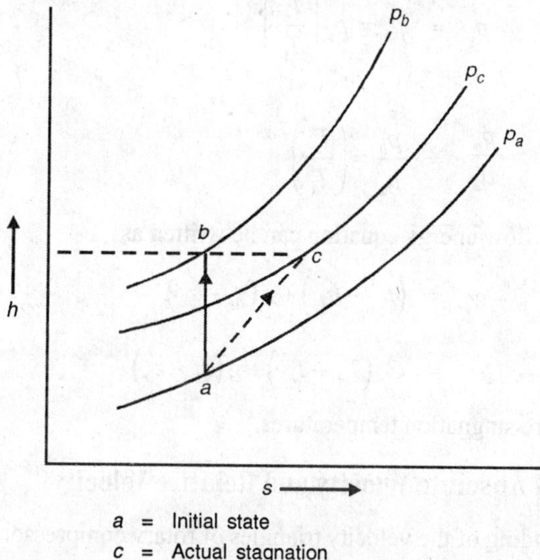

Fig. 8.1

An isentropic stagnation state is the state, fluid attains when the stagnation is reversible adiabetic or isentropic, shown by process $a - b$ on the h–s diagram. Actual irreversible process with friction or heat transfer is shown by process $a - c$. It may be noted that stagnation enthalpy h_b and h_c is same in both the processes. However actual stagnation pressure p_c is lower than isentropic stagnation process p_b, due to entropy increase in the actual process on account of friction. Now, for an ideal gas undergoing isentropic stagnation process,

$$h_o = h + \frac{V^2}{2}$$

$$\therefore \quad C_p T_o = C_p T + \frac{V^2}{2} \qquad h = f(T) \text{ for ideal gas}$$

$$\therefore \quad T_o = T + \frac{V^2}{2C_p}$$

where T_o = Stagnation temperature

$$\frac{V^2}{2C_p} = \text{Dynamic temperature}$$

Now, stagnation pressure p_o can be found as

$$p_o = p_b = p_a \left(\frac{T_o}{T_1}\right)^{\frac{\gamma-1}{\gamma}}$$

Also, we have,
$$\frac{p_o}{p_a} = \frac{p_b}{p_a} = \left(\frac{T_o}{T_1}\right)^{\frac{1}{\gamma-1}}$$

Similarly, steady flow energy equation can be written as

$$q - w_s = \left(h_{o_b} - h_{o_a}\right) + g\left(z_b - z_a\right)$$

$$= C_p\left(T_{o_b} - T_{o_a}\right) + g\left(z_b - z_a\right)$$

where T_{o_b} and T_{o_a} are stagnation temperatures.

8.2.2 Concept of Absolute Velocity and Relative Velocity

For better understanding of the velocity triangles of rotary compressors, the concept of absolute velocity and relative velocity is necessary.

8.2.2.1 Absolute velocity and relative velocity

Absolute velocity. It is defined as the velocity of a moving object as sensed by a stationary observer. In a true sense, no observer can be stationary on the earth as the earth is constantly moving slowly. However, its slow motion can be ignored.

Relative velocity. It is defined as the velocity of a moving object as sensed by an observer who is moving with its own velocity.

To illustrate this concept further, consider an object moving with a velocity V m/s. When a stationary observer looks at this object, the observer can get a correct feel about the magnitude and direction of the moving object. However, if the observer also has its own velocity, V m/s, then the observer gets only an apparant feel about the magnitude and direction of the moving object. This apparant feel of velocity is the relative velocity. Let us consider the following examples.

Here, the relative velocity V_r is the vector difference of the two absolute velocity vectors. Figure 8.2 (b) shows the two moving objects moving in opposite direction. The relative velocity V_r is the vector sum of the two absolute velocity vectors. In general, the procedure to obtain the relative velocity between the two objects moving in their own directions can be stated as follows.

From a common starting point, set both absolute velocity vectors in their magnitude and direction. The line joining the ends of the two absolute velocity vectors represents the relative velocity which is the vector difference of the known absolute velocity vectors, see Fig. 8.2 (c).

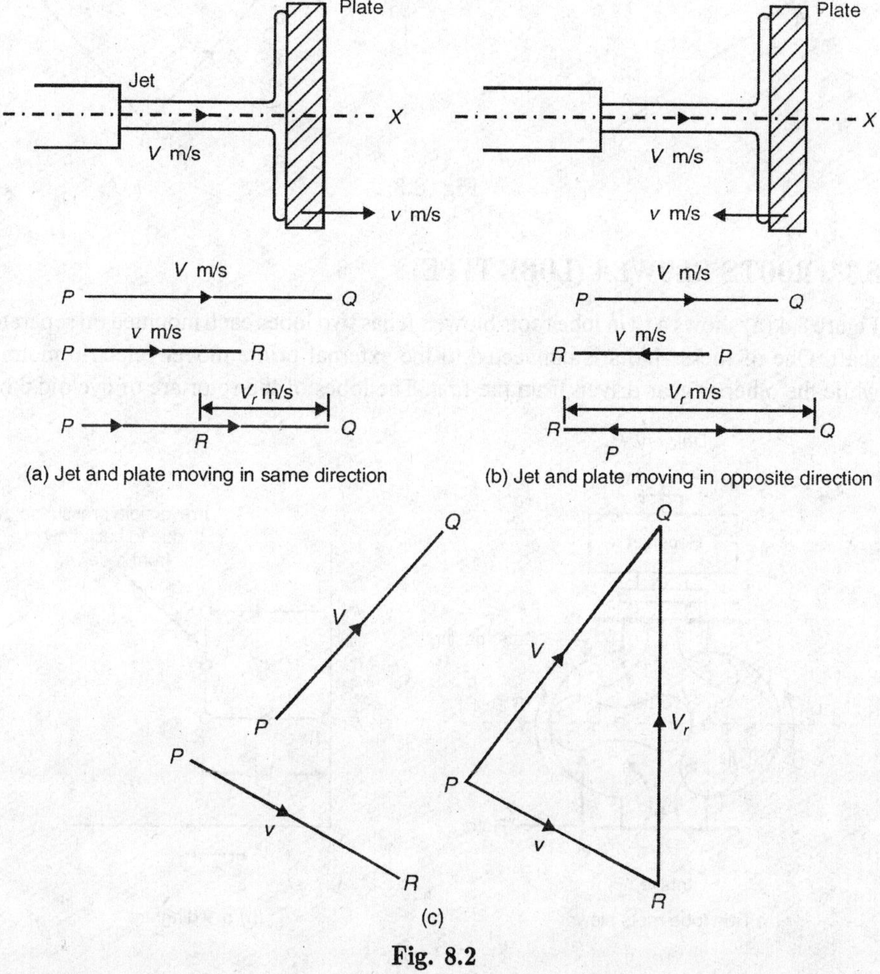

Fig. 8.2

Similarly, when a relative velocity and one of the absolute velocities are known, the unknown absolute velocity is given by the vector sum of the known relative velocity and known absolute velocity. The graphical procedure is as follows.

Draw the known relative velocity vector to some scale, in its magnitude and direction. In succession i.e. starting from the end of the relative velocity vector draw to same scale, the known absolute velocity vector in its magnitude and direction. Then the line joining the starting point to end point, representing vector sum, gives the unknown absolute velocity, as shown in Fig. 8.3.

$$A = \text{Known relative velocity, } V_r\text{, m/s}$$

$$B = \text{Known relative velocity, } u\text{, m/s}$$

Then C = Vector sum $(V_r + u)$ = unknown absolute velocity vector

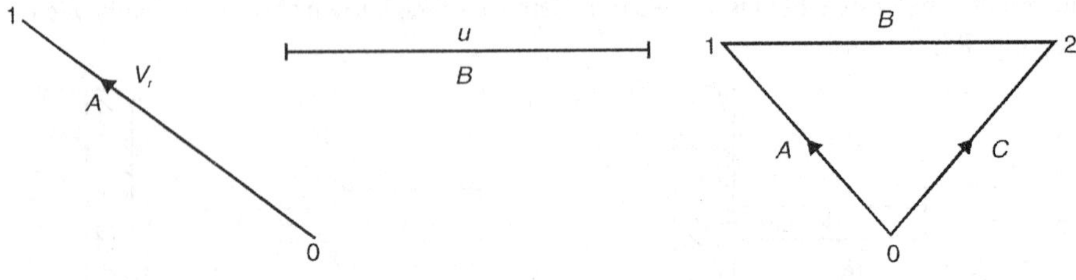

Fig. 8.3

8.3 ROOTS BLOWER (LOBE TYPE)

Figure 8.4 (a) shows a twin lobe roots blower. It has two lobes each mounted on separate shaft. One of these shafts is connected to the external prime mover (electric motor) while the other is gear driven from the first. The lobes of the rotor are of cycloidal or

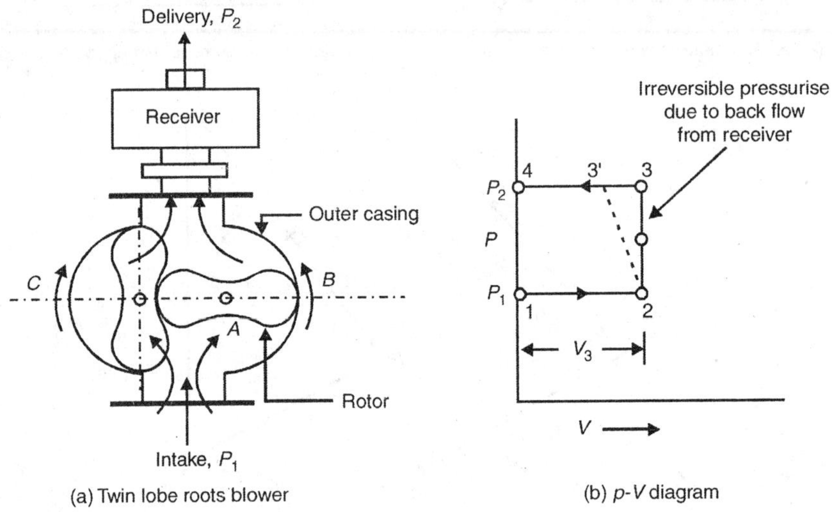

(a) Twin lobe roots blower

(b) p-V diagram

Fig. 8.4

envolute profile. Throughout all angular positions, the high pressure delivery side remains sealed from the low pressure suction side by the closely mating lobes. The wear between the lobes is avoided by a small clearance of 0.1 – 0.2 mm. The clearance results in leakage of air thereby reducing volumetric efficiency of the blower.

Suction takes place through the intake port. The entrapped air between the lobes and casing is carried forward during the rotation and is finally discharged to the delivery port. There is no change in the flow area and no reduction of volume of air. However, when the delivery port is open, blower discharges air into the high pressure reservoir causing irreversible pressure rise as shown in the *p-V* diagram [Fig. 8.4 (b)]. The dotted line in *p-V* diagram shows compression process of a reciprocating compressor. The area represents excess work required due to irreversible pressure rise.

Let, V = Volume of entrapped air in space A i.e. lobe face and casing, m^3

p_1 = Suction pressure, kPa

p_2 = Delivery pressure, kPa

N = Speed, r.p.m.

Volume of Free Air delivered/revolution = $4V$

$$\therefore \quad \text{Work required/rev.} = (p_2 - p_1) \times 4V \text{ kJ/rev.} \quad (1)$$

$$\therefore \quad \text{Power required} = (p_2 - p_1) \times 4V \times \frac{N}{60} \text{ kW}$$

$$\text{Further isentropic work reqd./s} = \frac{\gamma}{\gamma - 1} p_1 \times 4V \left[\left(\frac{p_2}{p_1} \right)^{\frac{\gamma-1}{\gamma}} - 1 \right] \times \frac{N}{60} \text{ kW} \quad (2)$$

Roots efficiency η_{root} is defined as the ratio of isentropic work required to the actual work required.

$$\therefore \quad \eta_{roots} = \frac{\frac{\gamma}{\gamma - 1} \times p_1 4V \left[\left(\frac{p_2}{p_1} \right)^{\frac{\gamma-1}{\gamma}} - 1 \right]}{(p_2 - p_1) \times 4V}$$

$$= \frac{\frac{\gamma}{\gamma - 1} \times p_1 \times 4V \left[\left(\frac{p_2}{p_1} \right)^{\frac{\gamma-1}{\gamma}} - 1 \right]}{p_1 (r_p - 1) \times 4V} = \frac{\frac{r}{r-1} \left[r_p^{\frac{\gamma-1}{\gamma}} - 1 \right]}{r_p - 1}$$

where r_p = pressure ratio = $\frac{p_2}{p_1}$

$$\therefore \qquad \eta_{roots} = \frac{C_p}{R} \frac{\left[r_p^{\frac{\gamma-1}{\gamma}} - 1 \right]}{r_p - 1} \qquad (3)$$

From Eq. (3), it is seen that with increase of pressure ratio r_p, roots efficiency η_{roots} decreases.

Roots blower have capacity ranging between 0.14 m³/min to 1400 m³/min and can develop pressure ratio of 2 in a single stage and of 3 in two stages.

The roots (lobes) are made of steel and are finished with close precision. Extremely fine clearances are required to be maintained during their manufacture and assembly.

8.4 ROTARY VANE COMPRESSOR

This compressor is also called as 'sliding vane compressor'. Figure 8.5 shows a multi-vane rotary compressor. It consists of a rotor eccentrically housed in the casing. Rotor has several radial slots in it, each housing a spring loaded vane. These vanes are made of steel or synthetic fibrous material. Larger the number of vanes, internal leakage of air decreases due to small pressure difference prevailing between the adjacent spaces around the rotor.

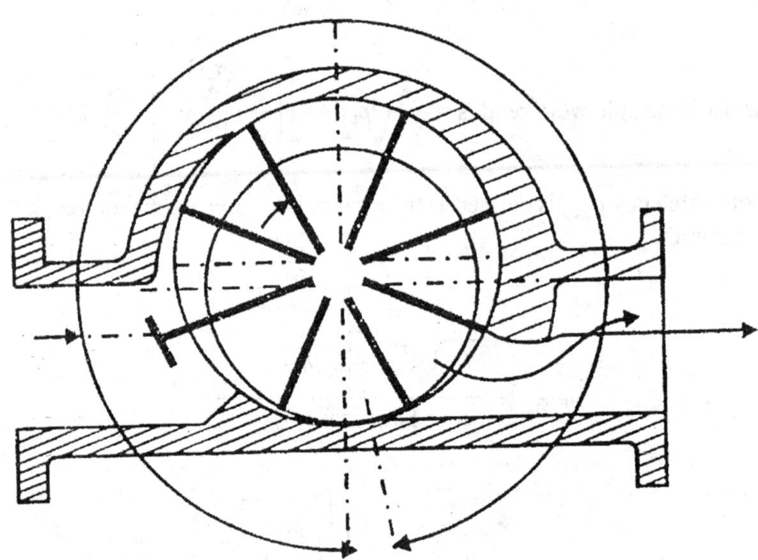

Fig. 8.5. *Vanes compressors.*

High pressure ratio requires large number of vanes (20 – 30). The casing has intake and delivery openings. These compressors are often used for capacities upto 150 m^3/min and for pressure ratios upto 8.5. For a given pressure ratio and FAD, vane compressor requires less work input than that for roots blower.

When the rotor rotates, vanes are driven out of the rotor towards the casing due to centrifugal force. The space between the two adjacent vanes, rotor and the casing increases creating vacuum. Thus, the gas is drawn in, from the suction opening. When the rotor crosses the point just opposite to its eccentricity, suction starts. As the rotor continues to rotate, the entrapped gas is compressed due to reduction in volume. The high pressure gas is then discharged through delivery opening. The *p-V* diagram of vane compressor is shown in the Fig. 8.6. Usually, half of the total pressure rise is developed

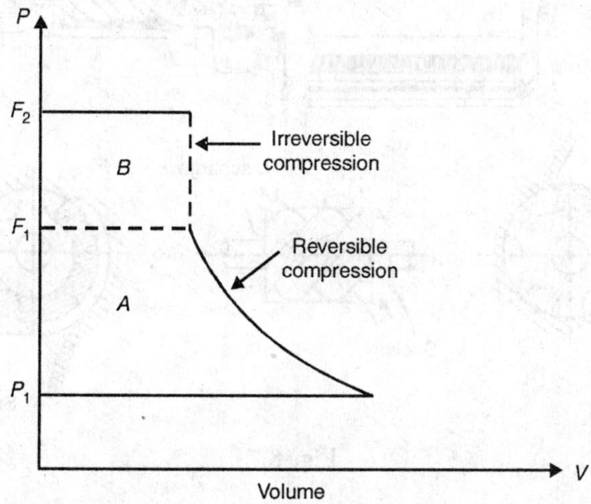

Fig. 8.6. *p-V diagram.*

during the internal reversible compression and the remaining pressure rise occurs irreversibly when the entrapped gas is released to the delivery side, due to back flow of high pressure air from the receiver.

Due to uniform torque, the compressor does not require any flywheel and is light weight.

Let, ε = eccentricity of rotor, m

L = axial length of rotor, m

t = thickness of each vane, m

D = diameter of rotor, m

n = number of vanes

$\therefore$ Swept volume/revolution $= 2L \times \varepsilon\,(\pi D - nt)$ m^3/rev. (4)

8.5 SCREW (LYSHOLM) COMPRESSOR

Figure 8.7 shows a screw (Lysholm) compressor. It consists of two rotors meshing with each other closely. Each rotor is helical with large pitch. The male rotor has three or four lobes while the female rotor has four, five or six recesses. The entrapped air is continuously compressed as it flow axially, due to narrowing of passages formed by the rotors. The inlet and discharge openings are somewhat oblique as shown in Fig. 8.7.

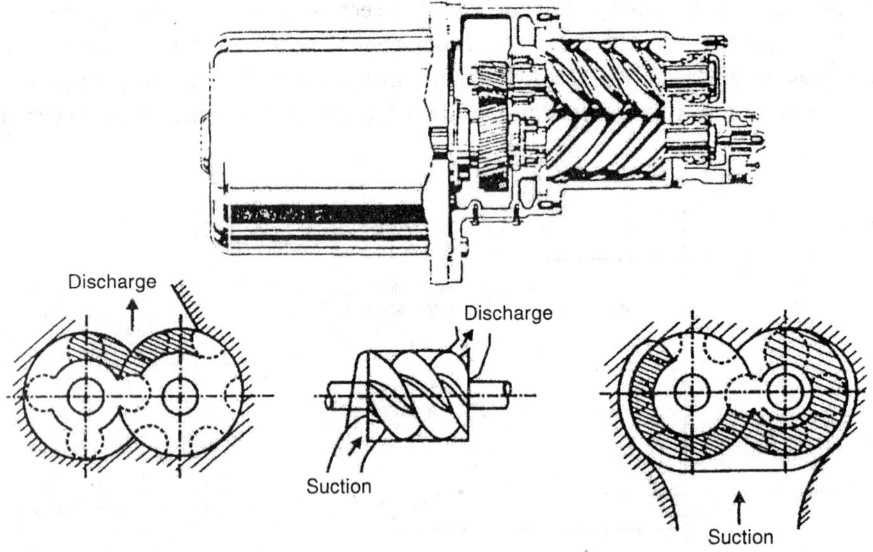

Fig. 8.7

The cycle of working of screw compressor consists of four processes namely:

(i) Suction. When the meshing rotors begins to disengage near the suction opening, gas is drawn in due to low pressure created. This is continuous so long as rotors unmesh.

(ii) Transportation. As the rotation of the rotor continues, the space filled with gas isolates from suction opening. This space is then gradually transported around the outer periphery at constant pressure until the male lobe starts meshing with the space being transported.

(iii) Compression. When rotors remesh from suction end gradually, the entrapped gas is compressed.

(iv) Delivery. Soon after the space filled with entrapped gas is exposed to discharge opening, the gas is driven out into the delivery line due to continuous penetration of male rotor into the space.

The compressor is run at very high speeds, 3000 r.p.m. for large compressors and upto 32000 r.p.m. for small units. Rotors have peripheral velocity of 50 m/s to 100 m/s. The rotors made of Carbon steel need precision machining. The variation of volumetric efficiency and adiabatic efficiency versus pressure ratio is shown in Fig. 8.8. High pressure ratio increases internal leakage of gas back to suction opening and reduces volumetric efficiency.

These compressors have capacity ranging between 3 m^3/min to 1000 m^3/min and pressure ratios of 4 (in single stage) and 8 – 11 (in double stage). The compressor does not need any lubrication, except for bearings. Thus, there is no contamination of gas with lubricating oil. These compressors are widely used in food and chemical industries.

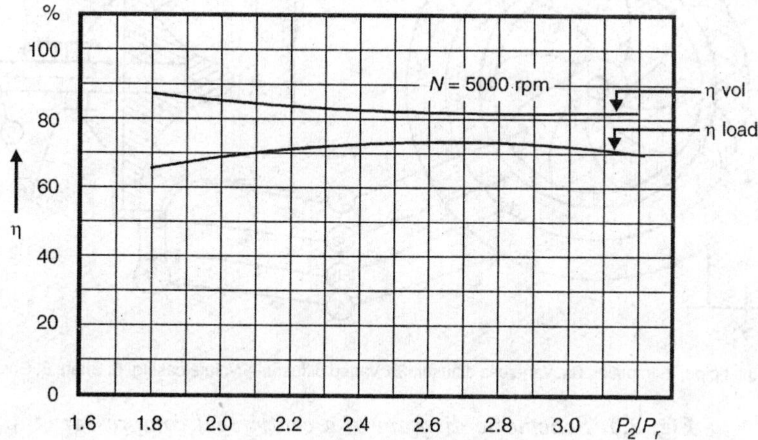

Fig. 8.8. *Variation of efficiencies.*

The rotors are dynamically balanced, hence, it runs vibration free and does not need heavy foundation. Its main drawback is operating noise of high frequency.

Let, A_A = Cross-sectional area between two lobes of male rotor, m^2

A_B = Cross-sectional area between two lobes of female rotor, m^2

n_A = Number of lobes in the male rotor

L = Length of rotor, m

N_A = Speed of male rotor, r.p.m.

η_{vol} = Volumetric efficiency

$\therefore$ Volume of gas sucked/rev. = $(A_A + A_B)\, n_A\, L\, N_A \times \eta_{vol}$, m^3/min. (5)

8.6 CENTRIFUGAL COMPRESSORS

This is a dynamic compressor. The compression and the pressure rise of air is achieved by dynamic action. The compressor is often directly coupled to a prime mover and is driven at high speed.

8.6.1 Construction and Principle of Operation

Figure 8.9 shows a schematic diagram of a centrifugal compressor. It consists of inlet (suction) pipe (1), an impeller (rotor) with blades or vanes (2), a vaned or vanless diffuser (3), a volute or scroll casing (4), an outlet or delivery pipe (5), a shaft (6) and shaft seal (7). The impeller is keyed to the shaft and the rotating assembly is dynamically balanced.

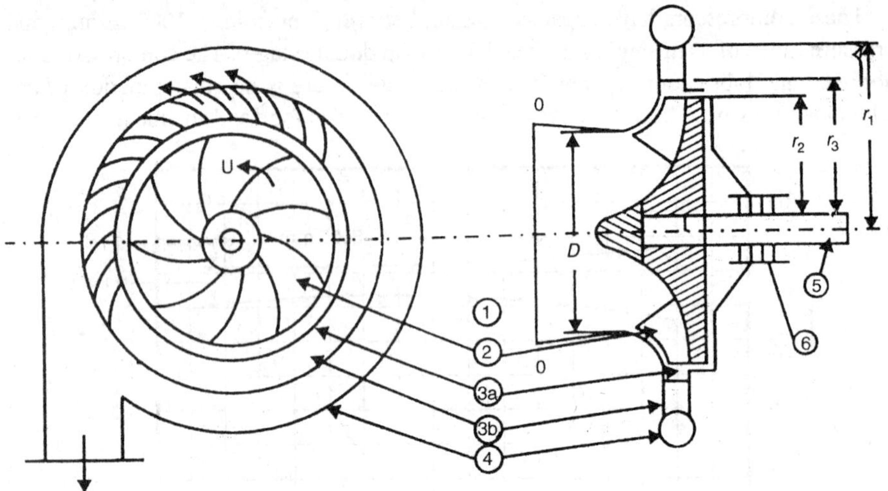

1. Inlet pipe, 2. Impeller, 3a. Vaneless diffuser, 3b. Vaned diffuser, 4. Volute casing, 5. Shaft, 6. Sealing.

Fig. 8.9. *Schematic diagram of a centrifugal compressor.*

The thermodynamic process of compression and pressure rise over the rotor and in the diffuser is shown on h–S diagram in Fig. 8.10.

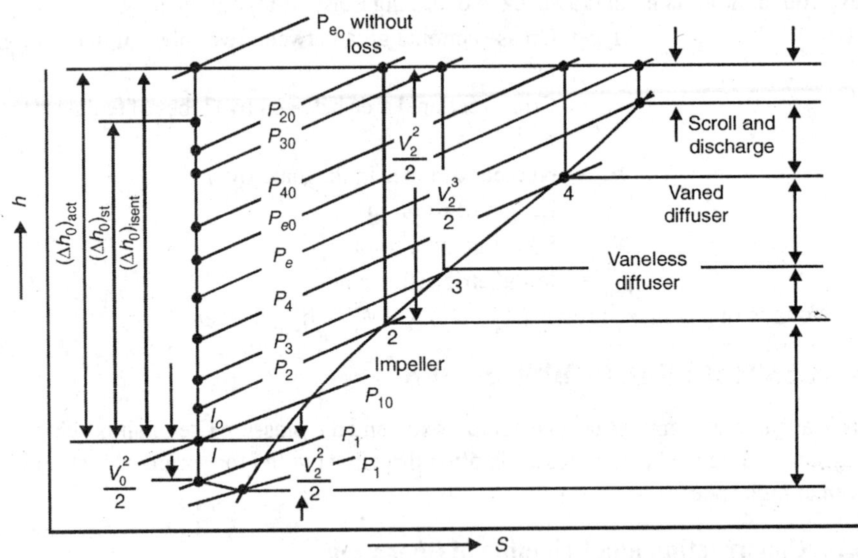

Fig. 8.10. *Centrifugal compressor process on h–S chart.*

Due to the rotation of impeller, vacuum is created at the eye of the impeller. Thus, air is sucked in from the surrounding ambient, at initial stagnation pressure p_{I_0} and stagnation temperature T_{I_0}, represented by point a on h–S diagram.

The static ambient condition is represented by point o. The air is somewhat accelerated in the inlet pipe. The inlet pipe admits or directs air for smooth and shockless entry into the eye of the impeller. In the inlet pipe, the direction and value of absolute velocity of air changes. Due to the acceleration of flow in the inlet pipe, velocity increases from V_o to V_1 thereby decreasing pressure and temperature of air. The acceleration process is non-isentropic due to friction and is shown by process 0 – 1 on h–S diagram. Therefore, at inlet to the impeller, the pressure and temperature are p_1 and T_1 respectively.

The work input to the shaft sets the impeller rotating. Air continuously accelerates as it flows over the impeller, thus, gaining kinetic energy. The vanes or blades are so mounted on the impeller as to provide gradually increasing cross-section area between two successive vanes to provide diffuser action. The reduction of kinetic energy $\left(\frac{V_{r_2}^2 - V_{r_1}^2}{2}\right)$ gets converted into pressure rise. Further, air is impressed with centrifugal head $\left(\frac{u_2^2 - u_1^2}{2}\right)$. The diffuser action and the centrifugal action thus converts a part of kinetic energy imparted to air by the impeller into pressure energy. The compression process is shown by line 1 – 2 on h–S diagram.

The vanes or blades of the impeller may have radial, backward or forward curvature.

Air leaves the impeller outlet at very high absolute velocity V_2 which is further converted into pressure in a vaneless or vaned diffuser and volute (scroll) casing. In the vaneless diffuser the absolute velocity of air decreases from V_2 to V_3, process 2–3 on h–S diagram. It also stabilises the flow leaving the vaned impeller tips for shockless entry into the vaned diffuser. The velocity of air further decreases from V_3 to V_4 in the vaned diffuser to increase pressure further. This process is represented by line 3–4 on h–S diagram.

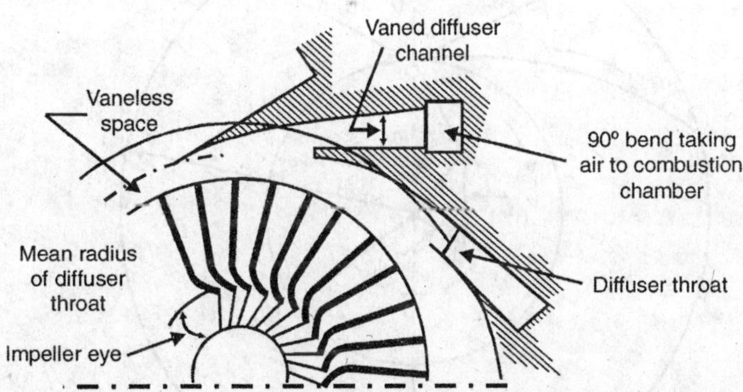

Fig. 8.11. *Radial centrifugal compressor coupled with gas turbines.*

The vaned diffuser discharges air into the volute (scroll) casing and finally air leaves from the outlet pipe. Process 4 – e on h–S diagram represents the process in the volute casing and outlet pipe. Centrifugal compressors used in gas turbine plants have a 90° bend in place of volute casing, to direct the air towards the combustion chamber as shown in Fig. 8.11. Finally, air has static and stagnation pressure of p_e and p_{e_o} at the exit of the compressor. If there would be no frictional losses whatsoever, the final pressure would have been together, as represented by $p_{e_{o(wl)}}$.

A typical multistage centrifugal compressor is shown in Fig. 8.12.

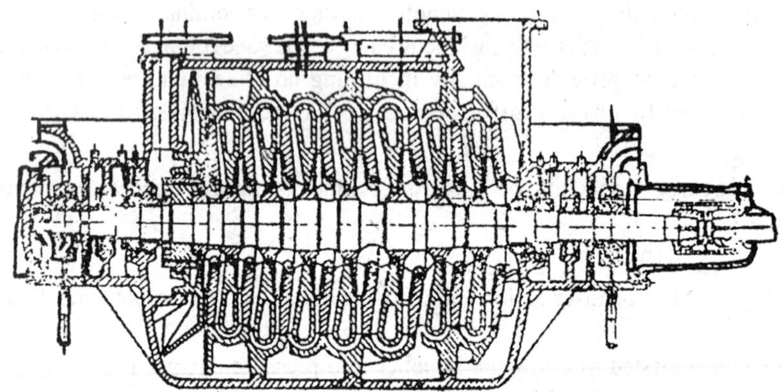

Fig. 8.12. *Multistage centrifugal compressor.*

8.6.2 Velocity Diagrams of a Centrifugal Compressor

Figure 8.13 shows velocity diagrams at the inlet and outlet of impeller and at inlet and outlet of diffusers.

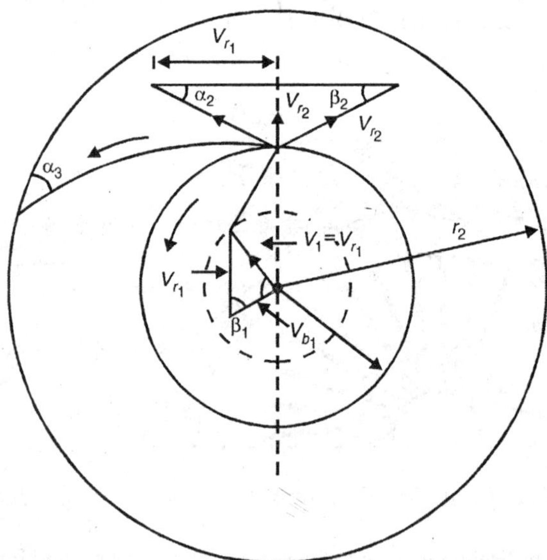

Fig. 8.13. *Velocity at inlet and outlet.*

Nomenclature. The following notations shall be adopted hereafter for energy analysis of the centrifugal compressor. Suffix '1' denotes parameters at inlet and '2' at outlet.

β_1 = Angle of the rotor blade at inlet
β_2 = Angle of the rotor blade at outlet
α_1 = Angle made by entering air or exit angle of guide blade
α_2 = Angle made by the outgoing air from rotor blade
V_1 and V_2 = Absolute velocity of air at inlet and outlet of rotor, m/s
V_{r_1} and V_{r_2} = Relative velocity of air at inlet and outlet of rotor, m/s
V_{f_1} and V_{f_2} = Velocity of flow at inlet and outlet of rotor, m/s
V_{w_1} and V_{w_2} = Velocity of whirl at inlet and outlet of rotor, m/s
u_1 and u_2 = Mean peripheral velocity of blade tip at inlet and outlet, m/s
r_1 and r_2 = Inner and outer radii of rotor, m
m = Mass flow rate of air, kg/s
α_3 = Vaned diffuser inlet angle or vaneless diffuser outlet angle

At the inlet to rotor, air enters with absolute velocity V_1 making an angle α_1 to the direction of motion of blade (usually $\alpha_1 = 90°$), without any shock and its whirl component $V_{w_1} = 0$.

Inlet triangle of velocities is now drawn to scale setting V_1 in the radial direction, u_1 in the tangential direction to inlet periphery. The vector joining end points of V_1 and u_1 represents relative velocity V_{r_1} at inlet. The blade tip at inlet has angle β_1 i.e. its curvature at inlet lies in the direction of V_{r_1}.

At the outlet from rotor, air leaves with a relative velocity V_{r_2} at an angle β_2 with the direction of motion. Now, as u_2 is known, V_2 can be found by vectorial addition.

In a vaneless diffuser, the flow is assumed to be logarithmic spiral and free-vortex. Air enters the vaned diffuser with velocity V_3 at an angle α_3 and leaves the diffuser with a velocity.

8.6.3 Work Requirement (Euler's work) for a Centrifugal Compressor

The work required/kg of air in a stage of a centrifugal compressor can be found by applying the moment of momentum theorem.

As per the Newtonian equation, force is given by rate of change of momentum.

Similarly, rate of change of moment about the centre of rotation gives torque.

Consider 1 kg of air flowing through the impeller (rotor). Theoretical torque supplied by impeller,

$$T = \left(V_{w_2} r_2 - V_{w_1} r_1\right) \text{ N-m/kg}$$

Now, for an angular velocity of impeller w rad/s.

Theoretical work required/kg of air = $T \times w$ w/kg

$$\left(V_{w_2}r_2 - V_{w_1}r_1\right) \ w/\text{J/kg} \tag{6}$$

Equation (6) is known as Euler's equation and gives Euler's work.

Further, if air enters radially i.e. when $\alpha_1 = 90°$ and $V_{w1} = 0$,

Theoretical work reqd./kg = $V_{w2}\,u_2$ J/kg (7)

The theoretical work reqd./kg also represents theoretical or virtual head developed in the absence of all losses. Thus, virtual or theoretical head; $H_{\text{virt.}}$ is given by

$$H_{\text{virt.}} = \frac{V_{w_2}u_2}{g} \tag{8}$$

From inlet and outlet velocity triangles, we have

$$V_{r_1}^2 = V_1^2 + u_1^2 - 2\ V_{w_1}, u_1 \text{ vectorially}$$

and

$$V_{r_1}^2 = V_2^2 + u_2^2 - 2\ V_{w_2}, u_2 \text{ vectorially}$$

Substituting the value of $V_{w1}\ u_1$ and $V_{w2}\ u_2$ from the above equations in Eq. (6), we get theoretical work required/kg.

$$= \frac{V_{r_2}^2 - V_1^2}{2} + \frac{V_{r_1}^2 - V_{r_2}^2}{2} + \frac{u_2^2 - u_1^2}{2} \ w/\text{kg} \tag{9}$$

In Eq. (9), first term $\left(\frac{V_2^2 - V_1^2}{2}\right)$ represents the increase in kinetic energy per kg of fluid in the impeller which has to be converted into pressure energy in the diffuser.

The last two terms signify the pressure rise in the impeller due to diffusion action and centrifugal action respectively. The second term is known as diffusion action as the relative velocity of fluid decreases from inlet to outlet of impeller. The last term represents centrifugal action due to increase of peripheral velocity of the vane tips from inlet to outlet.

Therefore, out of kinetic energy imparted by the impeller to the fluid, the part of the kinetic energy converted into pressure energy in impeller is

$$\int_1^2 \frac{dp}{p} = \left(\frac{V_{r_1}^2 - V_{r_2}^2}{2}\right) + \left(\frac{u_2^2 - u_1^2}{2}\right) \tag{10}$$

Similarly, if velocity at outlet of the diffuser is V_4 then,

Decrease of kinetic energy in the diffuser

$$\frac{V_2^2 - V_4^2}{2} = \int_2^4 \frac{dp}{p} + \Delta h_{\text{diff-loss}}$$

Thus, decrease of the kinetic energy of fluid in the diffuser is partly converted into pressure and the remaining is lost irreversibly as heat losses in the diffuser.

8.6.4 Dimensionless Parameters of Centrifugal Compressors

For the design and performance analysis of centrifugal compressor, several dimensionless parameters are very useful. These are defined and explained in that follows.

(a) Flow coefficient (ϕ). It is defined as the actual mass flow rate to the mass flow rate referred to the tip of the impeller, corresponding to tip peripheral velocity.

$$\therefore \quad \text{Flow coefficient, } \phi = \frac{\dot{m}}{\rho_2 A_2 u_2} \tag{11}$$

where, $\dot{m}$ = actual mass flow rate, kg/s

ρ_2 = density of fluid at the outlet tip of impeller, kg/m^3

A_2 = flow area at the exit tip of impeller, m^2

Applying continuity equation,

$$\text{Flow cofficient, } \phi = \frac{\rho_2 \pi D_2 b_2 V_{f_2}}{\rho_2 \pi D_2 b_2 u_2} = \frac{V_{f_2}}{u_2} \tag{12}$$

Value of flow coefficient often lies between 0.28 and 0.32, its optimum value being 0.3.

(b) Head coefficient 'λ'. It is defined as ratio of enthalpy increase in the stage to the kinetic energy corresponding to the tip peripheral velocity.

Thus,

$$\text{Head coefficient, } \lambda = \frac{\Delta h}{u_2^2/2} \tag{13}$$

where Δh = Enthalpy increase in the stage, per kg of fluid. The enthalpy increase includes the work supplied /kg of fluid compressed, as given by the Euler's Eq. (6) plus half of the heat corresponding to rotor friction losses. So we get,

$$\text{Head coefficient, } \lambda = \frac{2\left[\left(V_{w_2} u_2 - V_{w_1} u_1\right) + \frac{1}{2}\alpha u_2^2\right]}{u_2^2} \tag{14}$$

$$\lambda = 2\left[\mu \bar{V}_{w_{2\infty}} - \left(\frac{D_1}{D_2}\right)^2 \bar{V}_{w_1}\right] + \alpha$$

where, μ = slip factor

$$\bar{V}_{w_{2\infty}} = \frac{\bar{V}_{w_{2\infty}}}{u_2} \text{ and } \bar{V}_{w_1} = \frac{\bar{V}_{w_1}}{u_1}$$

For inlet velocity in axial direction,

$$V_1 = V_{f_1} \text{ and } V_{w_1} = 0$$

$$\therefore \quad \lambda = 2\mu \bar{V}_{w_{2\infty}} + \alpha \tag{15}$$

Now, if $\alpha = 0$, head coefficient for infinite number of blades is given by

$$\lambda = 2\mu \bar{V}_{w_{2\infty}} \tag{16}$$

and head coefficient with finite number of blades is

$$\lambda = 2\mu \bar{V}_{w_{2\infty}} \tag{17}$$

(c) Pressure coefficient (Ψ). It is defined as the ratio of isentropic enthalpy rise to the kinetic energy corresponding to tip peripheral velocity. Thus,

$$\text{Pressure coefficient, } \psi = \frac{\Delta h_{\text{isen}}}{u_2^2/2}$$

Considedring isentropic efficiency, η_{isent}, we get

$$\psi = \frac{\eta_{\text{isent}} \times \Delta h}{u_2^2/2} = \lambda \times \eta_{\text{isent}} \tag{18}$$

where $$\eta_{\text{isent}} = \frac{\Delta h_{\text{isent}}}{\Delta h}$$

(d) Reaction (Ω). It is defined as the ratio of temperature rise in the rotor to the temperature rise in the stage.

Thus,

$$\text{Reaction } \Omega = \frac{\Delta T_R}{\Delta T_{\text{stage}}} = 1 - \frac{C_p \Delta T_{\text{stator}}}{C_p \Delta T_{\text{stage}}}$$

$$= 1 - \frac{V_2^2 - V_1^2}{2C_p \Delta T_{\text{stage}}}$$

where it is assumed that $V_1 = V_4$ i.e. initial and final velocity is same. Now, for axial entry at inlet, $V_{w1} = 0$.

$$\text{Reaction } \Omega = 1 - \frac{V_{w_2}^2 + V_{f_2}^2 - V_{f_1}^2}{2u_2 V_{w_2}} \tag{19}$$

8.6.5 Slip Factor of a Centrifugal Compressors

Under ideal conditions, fluid particles follow exactly the same path of blade profile such that relative velocity at impeller outlet tip is inclined with the tangential direction at blade tip angle β_2 only, irrespective of mass flow rate, speed etc.

Such an ideal flow is possible when impeller has infinite number of blades of no thickness. Figure 8.14 shows velocity triangle for radial blades. Dotted lined diagram represents ideal conditions while full lined diagram represents actual conditions.

In actual practice, when impeller has finite number of blades, fluid is trapped between the impeller vanes due to its inertia and the fluid is reluctant to flow over the impeller.

This causes a pressure difference across the blades. There is a high pressure at the leading face and low pressure at the trailing face. This pressure difference generates a relative velocity gradient and formation of eddies.

The fluid is thus discharged at a certain average angle β'_2 which is less than β_2.

Therefore fluid is said to have slipped with respect to impeller during its flow across it.

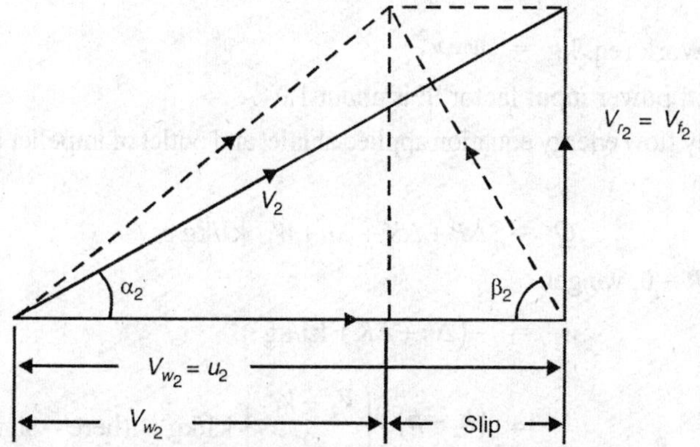

Fig. 8.14. *Slip over blackward curved bladed impller.*

The slip factor μ is defined as the *ratio of actual velocity of whirl to the ideal velocity of whirl.*

$$\text{Slip factor, } \mu = \frac{\text{Actual velocity of whirl}}{\text{Ideal velocity of whirl}}$$

$$\mu = \frac{V_{wa}}{V_{wi}} = \frac{V_{w_{2a}}}{u_2} \text{ (for radial blades)} \tag{20}$$

Slip factor depends upon number of blades and is usually 0.9. It does not reduces the efficiency but only reduces the head developed.

8.6.6 Pressure Ratio of Compression in a Centrifugal Compressor

Slip factor is also defined as the ratio of work required under actual conditions to the work required under ideal conditions.

Thus,
$$\mu = \frac{V_{w_2} u_2 - V_{w_1} u_1}{V_{w_{2i}} u_2 - V_{w_{1i}} u_1}$$

But with radial entry, $V_{w_1} = V_{w_{1i}} = 0$

$$\mu = \frac{V_{w_2}}{V_{w_{2i}}}$$

$$\text{Work} = \mu \times V_{w_2} \times u_2$$

$$= \mu u_2^2 \quad (21)$$

Actual work required is greater than theoretical work given by Eq. (21) due to friction, windage etc.

Power input factor Ψ is introduced now such that

$$\text{Actual Work req./kg} = \psi \mu u_2^2 \quad (22)$$

The value of power input factor Ψ is about 1.04.

As per steady flow energy equation applied at inlet and outlet of impeller and assuming no heat transfer,

$$Q = \Delta P + \Delta K + \Delta h + W_s \text{ kJ/kg}$$

With $Q = 0$, $\Delta P = 0$, we get

$$w_s = -(\Delta h + \Delta K) \text{ kJ/kg}$$

$$= (h_2 - h) + \left(\frac{V_2^2 - V_1^2}{2}\right) \text{ kJ/kg}$$ (here – sign indicates work required)

$$= C_p T_{o_2} - C_p T_{o_1} \quad (23)$$

But, work required/kg $= \psi \mu u_2^2$

$$\therefore \quad T_{o_2} - T_{o_1} = \frac{\psi \mu u_2^2}{C_p} \quad (24)$$

Actual stagnature temperature T_{o_1} is given by

$$T'_{o_2} - T_{o_1} = \frac{\psi \mu u_2^2}{C_p} \times \eta_{\text{isent}}$$ where η_{isent} = isentropic efficiency

Therfore, pressure ratio

$$\frac{p_{o_2}}{p_{o_1}} = \left(\frac{T_{o_2}}{T_{o_1}}\right)^{\frac{\gamma}{\gamma-1}}$$

$$= \left[1 + \frac{T'_{o_2} - T_{o_1}}{T_{o_1}}\right]^{\frac{\gamma}{\gamma-1}}$$

$$= \left[1 + \frac{\eta_{isent}\ \psi \mu u_2^2}{C_p T_{o_1}}\right]^{\frac{\gamma}{\gamma-1}} \quad (25)$$

If all factors are unity as in case of isentropic compression, then,

$$\frac{p_{o_2}}{p_{o_1}} = \left[1 + \frac{u_2^2}{C_p T_{o_1}}\right]^{\frac{\gamma}{\gamma-1}} \quad (26)$$

The slip factor μ and the power input factor Ψ are different and independent from each other. Whereas the power input factor Ψ signifies the increase in the work input required which is lost in overcoming friction. This energy lost gets converted into heat and causes temperature rise. The temperature rise of compressed air is often not a loss as in gas turbine plants, the hot compressed air elevates the temperature in the combustion chamber without increase of fuel supply. Yet the power input factor should be low and compression should ideally be isentropic.

But the slip factor limits the work capacity of the compressor even with isentropic working. The slip factor should be as high as possible. High slip factor increases V_{w_2}. When $V_{w_2} = u_2$, work input is maximum which can be gainfully utilised.

Slip factor increases with greater number of vanes which in turn decreases area of flow passage for air. A compromise is often necessary. The present practice is to adopt about 20 vanes to obtain a slip factor of 0.9.

An impirical formula proposed by Stanitz is

$$\eta = 1 - \frac{0.63\pi}{n}$$

where, n = number of blades

Higher tip speed increases the pressure ratio.

Maximum tip speed is restricted to 460 m/s for the common materials.

Also reducing the inlet temperature T_{o_1} raises the pressure ratio for a given work input.

8.6.7 Influence of Impeller Blade Geometry

The blades of the impeller have various shapes which can be described based on the blade angle at outlet β_2. Depending on the value of β_2, the blades have different curvatures compared with the forward direction of rotation of the impeller. The various shapes of blades are:

(i) Backward curved blades; $\beta_2 < 90°$

(ii) Radial blades; $\beta_2 = 90°$

(iii) Forward curved blades; $\beta_2 > 90°$

Figure 8.15 shows the velocity triangle for each case of the above.

In all cases, blade tip velocity at exit u_2 is kept the same.

It is seen from Fig. 8.15 that for the backward curved blades, the tangential component V_{w2} is least compared with those is other cases. Therefore, for a given speed work input required for the impeller is low, as given by Euler's work (Eq. 6).

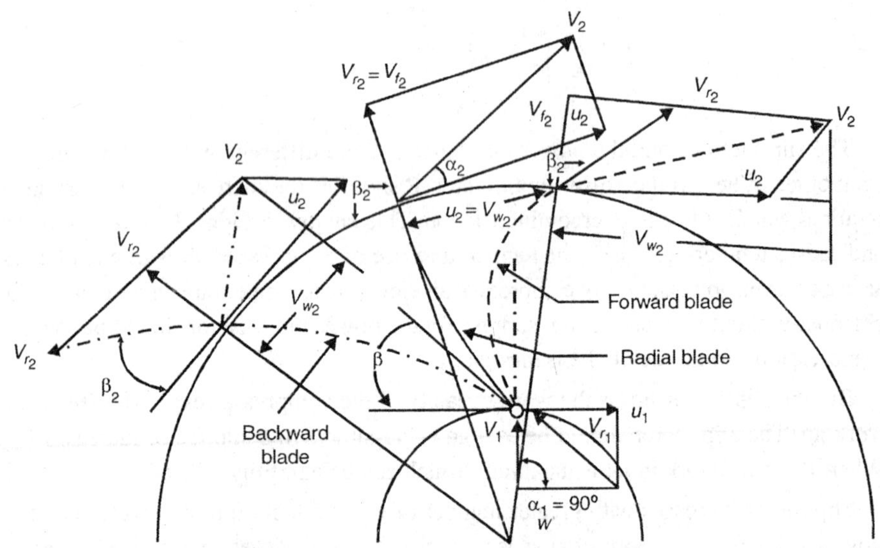

Fig. 8.15. *Velocity diagram for each blade shape.*

In the case of forward curved blades, V_{w2} is maximum, thus the impeller needs maximum work input. Too, the value of V_2 is highest.

And in case of radial blades, V_{w2} lies in between, that is, backward and forward curved blades.

Therefore, backward curved blades are best giving high efficiency.

The influence of blade geometry can also be explained by dimensionless parameters as follows.

The head coefficient is one dimensionless parameter which is used to express the work required per stage of centrifugal compressor.

For a working fluid treated as ideal gas, isentropic efficiency $\eta_{isent} = 1$. As such head coefficient λ and pressure coefficient Ψ are same, the work input to the fluid is then given by,

$$\psi_i = 2\bar{V}_w = 2\frac{V_{w_{2i}}}{u_2} \tag{27}$$

But,

$$\beta_2 = \frac{V_{f_2}}{u_2 - V_{w_{2i}}} = \frac{V_{f_2}/u_2}{1 - \dfrac{V_{w_{2i}}}{u_2}} \tag{28}$$

where, $\frac{V_{f_2}}{u_2} = \phi =$ flow coefficient

Therefore, $$\psi_i = 2\left(1 - \frac{\phi}{\tan \beta_2}\right) \tag{29}$$

It is seen from Eq. (29) that for a given value of flow coefficient ϕ, the work input required per stage increases as blade angle at outlet β_2 increases.

For backward curved blades, $\beta_2 < 90°$ and with $\tan \beta_2 = \phi$, the $\Psi_i = 0$.

For radial blades, $\beta_2 = 90°$, $\Psi_i = 2$ and is independent of flow coefficient ϕ.

For forward curved blades, $\beta_2 > 90°$, $\tan \beta_2$ is negative and when $\phi = -\tan \beta_2$, $\Psi_i = 4$.

Further blade angle at outlet β_2 influences the reaction Ω of the stage. For an impeller with infinite number of blades,

$$\text{Reaction } \Omega = 1 - \frac{V_{w_2}}{u_2}$$

$$= 1 - \frac{\psi_i}{4}$$

Thus, as pressure coefficient Ψ increases, the reaction of the stage decreases.

8.6.8 Influence of Compressor Geometry on the Performance of Centrifugal Compressor

Figure 8.16 (a) and (b) illustrates variation of rection of the stage Ω with pressure coefficient Ψ.

8.6.9 Pre-Whirl

The gas turbines and jet engines use centrifugal compressors running at high speed. There is often a possibility of development of a shock wave in the flow passage. This can be avoided if the Mach Number at any point in the flow passage is restricted below unity.

The maximum value of Mach Number referred to relative velocity is found at inlet.

$$\therefore \text{Mach Number at inlet} = \frac{V_{r_1}}{\text{Velocity of sound}}$$

$$M_1 = \frac{V_{r_1}}{\sqrt{\gamma R T_1}} \tag{30}$$

where, $T_1 =$ Static temperature at inlet, K

For a given flow rate, if V_{r_1} and u_1 is high for shockless operation of the compressor, the fluid is given an initial pre-rotation or pre-whirl with the help of fixed guide blades attached to the casing at inlet.

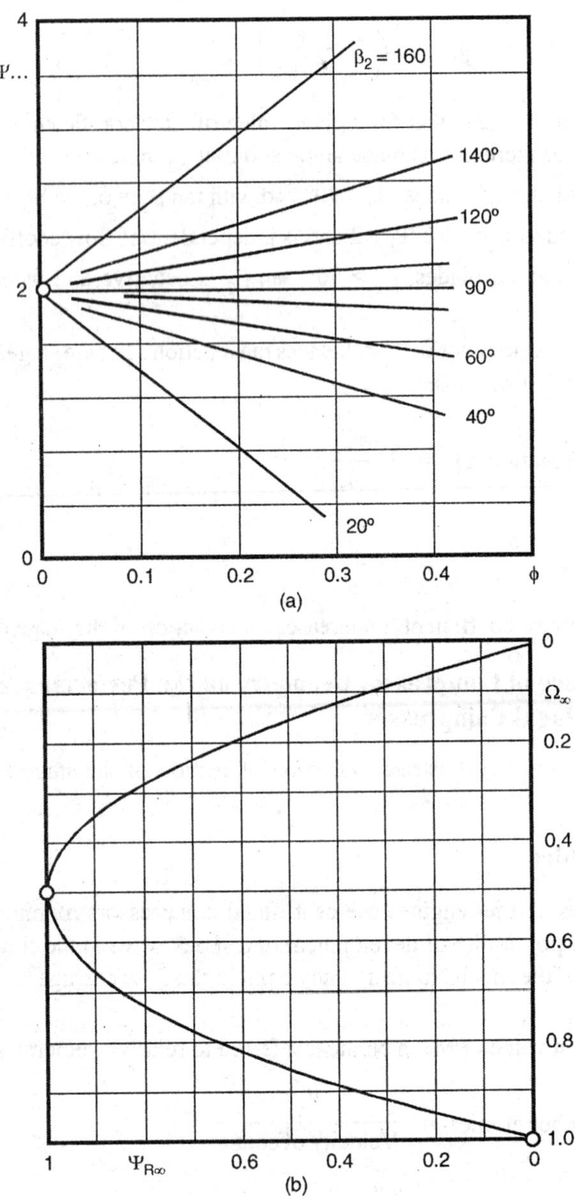

Fig. 8.16. *Influence of blade shape.*

The entering air passing through the fixed guide blades is given pre-whirl. Thus V_{r_1} is reduced without affecting V_{f_1} or mass flow rate.

8.6.10 Losses in a Centrifugal Compressor

The frictional losses in a centrifugal compressor occur mainly at two places viz. over the impeller and in the diffuser.

I. Losses Over the Impeller

The fluid friction losses over the impeller are given by

$$\Delta h_{imp} = \Delta h_1 + \Delta h_2 + \Delta h_3 + \Delta h_4 + \Delta h_d$$

where Δh_1 = Loss in the inducer = $\xi_1 \dfrac{V_{r1}^2}{2}$ J/kg where, = ξ_1 = 0.2 to 0.3

Δh_2 = Loss due to change of direction from axial to radial at the impeller inlet

$= \xi_2 \dfrac{V_1^2}{2}$ J/kg where, ξ_2 = loss coefficient = 0.1 to 0.2

Δh_3 = Loss due to friction in the blade passages

$= \xi_3 \dfrac{Vr_1^2}{2}$ J/kg where, ξ_3 = loss coefficient 0.2 to 0.4

Δh_4 = Loss due to design deviations

$= \xi_4 \dfrac{(\Delta Vr_1)^2}{2}$ J/kg

where, ΔV_{r_1} = Change in relative velocity due to design deviation

= 0.54 to 0.60

Δh_d = Defriction loss in the impller

$= \sigma \dfrac{u_2^2}{2}$ J/kg

where, σ = 0.003 to 0.008

II. Losses in the Diffuser

The losses in the diffuser (vaned or vanless) are due to irreversible inversion of friction force in boundary layer on the surface of diffuser.

All these losses tend to reduce the pressure ratio and increase the work input required thereby decreasing the efficiency of the compressor.

8.7 AXIAL FLOW COMPRESSORS

In these compressors, flow of air is in the axial direction. Figure 8.17 shows the construction of an axial flow compressor. It consists of alternately placed rows of stator blades and rotor blades. The rotor blades are mounted on the rotor while the stator blades are attached to the casing.

The conversion of kinetic energy into pressure head i.e. the diffusion action takes place partly over the rotor blades and partly over the stator blades. The stator blades also direct air over the successive rotor blades for a shockless entry. The flow passages

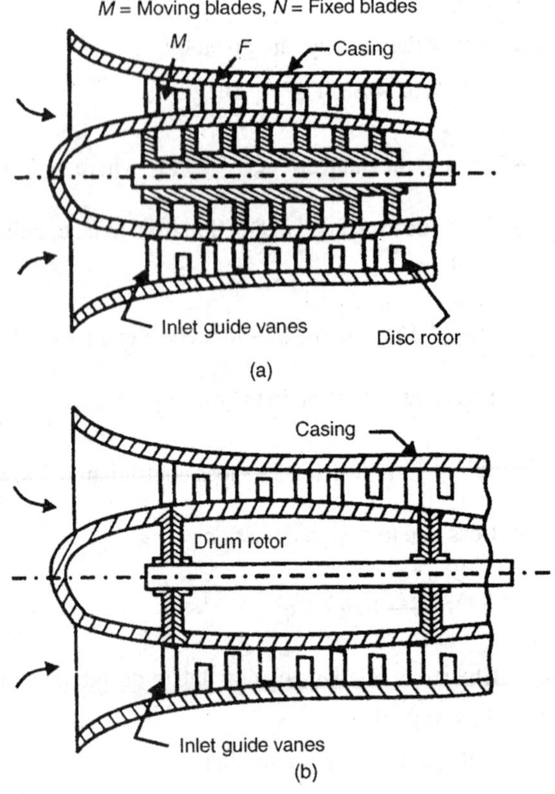

Fig. 8.17. *Axial flow compressor.*

between rotor blades as well as stator blades provide diverging area. One row of rotor blades preceded by a row of stator blades is known as a stage of axial flow compressor. A single stage compressor can achieve a pressure ratio of 1.12 to 1.2 only. Therefore, for higher pressure ratio, often several stages of axial flow compressor are necessary.

8.7.1 Construction and Principle of Operation

Rotors are of two types viz. drum type and disc type (shown in Fig. 8.17). The disc type rotors being light weight are preferred for aircraft applications. The drum type rotors are used for static industrial applications. (Volume of casing converges from low pressure suction end to high pressure discharge end). It helps to maintain more or less a constant velocity throughout the length of compressor, though the density of fluid increases gradually. Constant axial velocity throughout the stage is an important design consideration of axial flow compressor. Contraction of the flow annulus is achieved by contraction of casing and diverging the rotor.

The materials used for various components of the compressor are listed below:

(i) **Rotor blades.** Fibrous composites, Aluminium, Titanium, Steel, Nickel alloy.

(ii) **Rotor.** Steel for shaft, disc. In aircrafts, first stage uses Titanium while the later stages use Nickel steel.

(iii) **Stator blades.** These are made of same materials as those for rotor blades.

(iv) **Casing.** It is either a casting of magnesium, aluminium, steel, iron or fabricated from Titanium or steel.

The blades of compressor are of aerofoil section. The axial flow compressors can be of impulse type or reaction type. The blade profile and variation of pressure and velocity in the stage are shown in the Fig. 8.17 (a). While in Fig. 8.17 (b), the difference in the shape of Axial flow turbines and Axial flow compressor is illustrated.

8.7.2 Velocity Diagrams of Axial Flow Compressors

As explained earlier, in the axial flow compressor, fluid is imparted with kinetic energy from the rotor. This kinetic energy of the fluid is converted into pressure energy by diffuser action. The process occurs over several stages.

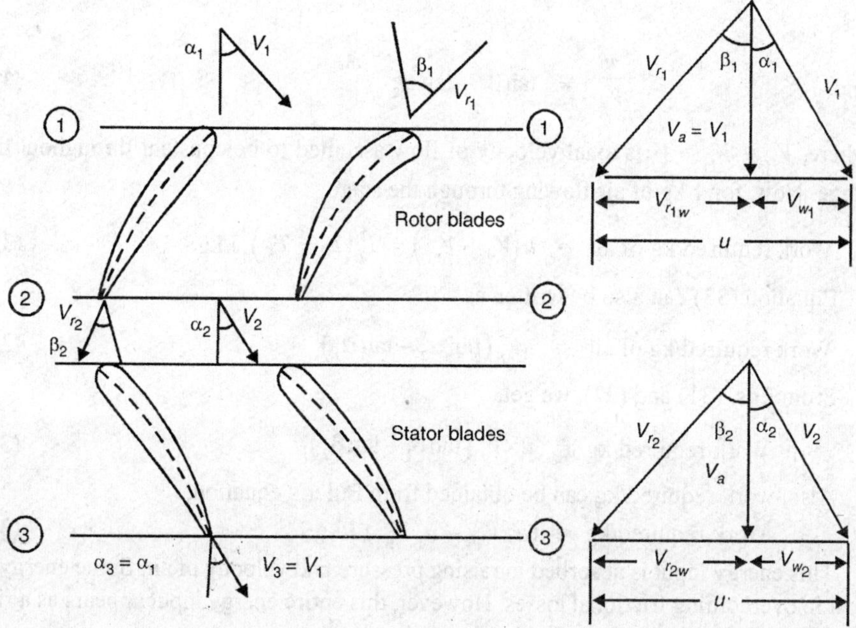

Fig. 8.18. *Velocity traingle for compressor stages.*

The energy analysis is considered for one stage of the compressor. The flow is assumed to occur tangential to the mean blade speed of u.

Figure 8.18 shows velocity triangles for one stage of the compressor. Air approaches rotor blade with absolute velocity V_1 at an angle α_1 to the axial direction. Combining V_1

with blade velocity u_1 gives relative velocity V_{r_1} at inlet, inclined at angle β_1 with the direction of motion.

While air flows through the diverging passage between two successive rotor blades, its absolute velocity increases and air leaves the rotor with relative humidity V_{r_2} at an angle β_2. The reduction in the relative velocity from V_{r_1} at inlet to V_{r_2} at exit causes some pressure rise over the rotor. The absolute velocity at exit V_2 can be found by vector difference between V_{r_2} and u_2. The absolute velocity at exit V_2 makes an angle α_2 with the direction of motion. The air then enters the next stage via stator blades. Over the stator blades, the pressure of air further increases due to diffusion action. Air leaves the stator blades with absolute velocity V_3 (less than V_2) making an angle α_3 with the axial direction. The stator blades are so designed that $V_3 = V_1$ and $\alpha_3 = \alpha_1$ for a shockless entry over the next stage.

It may particularly be noted that unlike in centrifugal compressors, in this case all angles are measured with axial direction and not from tangential direction.

Referring to the velocity triangles shown in Fig. 8.18 we have,

$$\frac{u}{V_f} = \tan\alpha_1 + \tan\beta_1 \tag{31}$$

and

$$\frac{u}{V_f} = \tan\beta_2 + \tan\alpha_2 \tag{32}$$

where, $V_{f1} = V_{f2} = V_f$ is axial velocity of flow assumed to be constant throughout the stage. Now, for 1 kg of air flowing through the stage

$$\text{Work required/kg of air} = u\left(V_{w_2} - V_{w_1}\right) = C_p\left(T_{o_2} - T_{o_1}\right) \text{ J/kg} \tag{33}$$

Equation (33) can also be written as

$$\text{Work required/kg of air} = uV_f\left(\tan\alpha_2 - \tan\alpha_1\right) \tag{34}$$

From Eqs. (31) and (32), we get,

$$\text{Work required/kg} = u \times V_f\left(\tan\beta_1 - \tan\beta_2\right) \tag{35}$$

Also, work required/kg can be obtained from Euler's equation,

$$\text{Work required/kg} = (u_2\, V_{w_2} - u_1\, V_{w_1}) \text{ J/kg} \tag{36}$$

This energy input is absorbed in raising pressure and velocity of air. Some energy is lost in overcoming frictional losses. However, this entire energy input appears as a rise in "stagnation" temperature of air, ΔT_{os}. If $V_3 = V_1$, then

$$\Delta T_{o_s} = \Delta T_{\text{static}} = \frac{uV_f}{C_p}\left(\tan\beta_1 - \tan\beta_2\right) \tag{37}$$

Considering work done factor λ, the actual stage temperature rise is

$$\text{Actual } \Delta T_{o_s} = \Delta T_{\text{static act}} = \frac{\lambda}{C_p} \times u \times V_f\left(\tan\beta_1 - \tan\beta_2\right)$$

And the pressure ratio R_s is given by,

$$R_s = \left[1+\frac{\eta_s \Delta T_{o_s}}{T_{o_i}}\right]^{\frac{\gamma}{\gamma-1}} \tag{38}$$

where, T_{oi} = inlet stagnation temperature and η_s is isentropic efficiency of the stage

$$\eta_s = \frac{\text{Isentropic temperature rise}}{\text{Actual temperature rise}}$$

Using velocity triangles and cosine rule, Eq. (38) can be written as

$$\text{Work required/kg} = \frac{V_{r1}^2 - V_{r2}^2}{2} + \frac{V_2^2 - V_3^2}{2} \tag{39}$$

where, $V_3 = V_1$

Equation (39) is represented graphically on h–S diagram in Fig. 8.19.

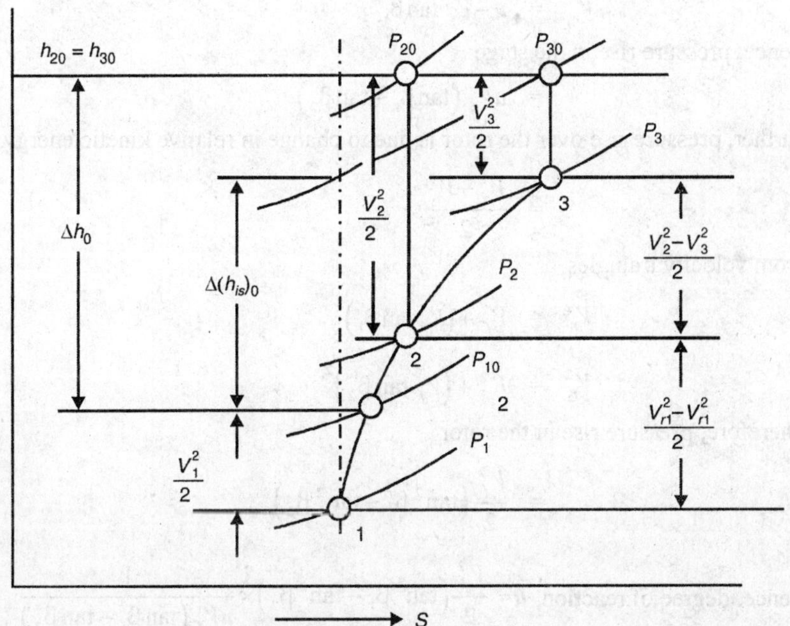

Fig. 8.19. *Process in axial flow compressor without GV on h-S diagram.*

The first term on the right hand side of Eq. (39) represents the increase of kinetic energy in the rotor which is converted into pressure due to diffusion action in the rotor itself while the second term represents increase of kinetic energy over the rotor that has to be converted into pressure energy in stator blades.

Comparing the Eq. (39) of work required/kg with Eq. (39) applicable for centrifugal compressor, it is seen that the term centrifugal head $\left(\frac{u_2^2 - u_2^2}{2}\right)$ is missing for axial flow compressor. Therefore, pressure ratio per stage in axial flow compressor is much less than that of centrifugal compressor.

8.7.3 Degree of Reaction of Axial Flow Compressor

The degree of reaction is defined as the ratio of pressure rise in the rotor to the pressure rise in the stage.

$$\text{Degree of reaction, } R = \frac{\text{Pressure rise in the door}}{\text{Pressure rise in the stage}} \quad (40)$$

Total pressure rise in the stage of a compressor is given by work input per stage

$$= u\,(V_{w_2} - V_{w_1}) \quad (41)$$

From velocity triangles,

$$V_{w_2} = u - V_f \tan\beta_2$$

and

$$V_{w_1} = u - V_f \tan\beta_1$$

Hence, pressure rise in the stage

$$= uV_f\left(\tan\beta_1 - \tan\beta_2\right) \quad (42)$$

Further, pressure rise over the rotor is due to change in relative kinetic energy.

$$= \frac{V_{r_1}^2 - V_{r_2}^2}{2} \quad (43)$$

From velocity traingles,

$$V_{r_1}^2 = V_f^2 + \left(V_f \tan\beta_1\right)^2$$

and

$$V_{r_2}^2 = V_f^2 + \left(V_f \tan\beta_2\right)^2$$

Therefore, pressure rise in the rotor

$$= \frac{V_f^2}{2}\left(\tan^2\beta_1 - \tan^2\beta_2\right) \quad (44)$$

$$\text{Hence, degree of reaction, } R = \frac{V_f^2}{2}\left(\tan^2\beta_1 - \tan^2\beta_2\right) \times \frac{1}{uV_f\left(\tan\beta_1 - \tan\beta_2\right)}$$

$$= \frac{V_f}{2u}\left(\tan\beta_1 - \tan\beta_2\right) \quad (45)$$

Axial flow compressors are designed for degree of reaction of 50% i.e. half of the total pressure rise occurs over the rotor and the remaining half over the state blades.

$$\therefore \quad R = 0.5 = \frac{V_f}{2u}(\tan\beta_1 - \tan\beta_2)$$

$$\therefore \quad \frac{u}{V_f} = (\tan\beta_1 - \tan\beta_2) \quad (46)$$

Also, from velocity traingles,

$$\frac{u}{V_f} = (\tan\alpha_1 - \tan\beta_1) \quad (47)$$

and

$$\frac{u}{V_f} = (\tan\alpha_2 - \tan\beta_2) \quad (49)$$

From Eqs (47), (48) and (49), it follows that

$$\alpha_1 = \beta_2 \text{ and } \alpha_2 = \beta_1$$

Also constant velocity of flow through the stage,

$$V_f = V_1\cos\alpha_1 = V_3\cos\alpha_3$$

But,

$$V_3 = V_1 \text{ (assumed)}$$

$$\therefore \quad \alpha_1 = \alpha_3$$

As $\alpha_1 = \beta_2 = \alpha_3$ and $\alpha_2 = \beta_1$, rotor blade and stator bade are symmetrical. In the above deductions, work done factor is assumed to be unity.

For work done factor lower than unity, the degree of reaction differs slightly.

8.7.4 Cascade or Diffuser Efficiency

It is defined as the ratio of actual pressure rise to the isentropic pressure rise. Thus,

$$\eta_c = \frac{P_2 - P_1}{P_{2_{is}} - P_1} \quad \ldots(50)$$

Friction in rotor and stator blade passages, shock losses, etc. reduce the efficiency of the cascade (stage).

8.7.5 Polytropic Efficiency

Consider the compression of air in a multistage axial flow rotary compressor shown on *T–S* diagram in Fig. 8.20 given on next page.

It is customary to assume that all stages have equal isentropic efficiency where isentropic efficiency is given by

$$\eta_{is} = \text{stage isentropic efficiency} = \frac{\text{Isentropic temp. rise in the stage}}{\text{Actual temp. rise in the stage}}$$

Now, overall isentropic efficiency

$$\text{Overall} \qquad \eta_{is} = \frac{T_{o_{2'}} - T_{o_1}}{T_{o_2} - T_{o_1}}$$

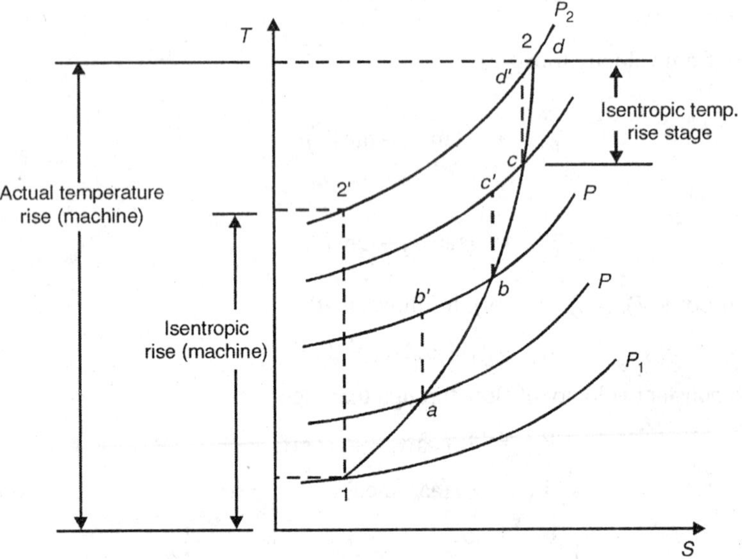

Fig. 8.20. *Multistage compression.*

$$\text{and,} \quad \text{Polytropic efficiency} = \frac{\Sigma(\text{Isentropic temp. rise in the stage})}{T_{o_2} - T_{o_1}}$$

$$= \frac{\left(T_{o_{a'}} - T_{o_1}\right) + \left(T_{o_{b'}} - T_{o_a}\right) + \left(T_{o_{c'}} - T_{o_c}\right) + \left(T_{o_{d'}} - T_{o_d}\right)}{T_{o_2} - T_{o_1}}$$

8.7.6 Polytropic Efficiency η Terms of *n*

Consider an actual compression process in a stage of an axial flow compressor, as shown in Fig. 8.21.

Actual compression process 1 – 2 follows the polytropic law $pV^n \cong C$.

$$\therefore \qquad \text{Polytropic efficiency, } \eta = \frac{T_{o_{2'}} - T_{o_1}}{T_{o_2} - T_{o_1}}$$

$$= \frac{\dfrac{T_{o_{2'}}}{T_{o_1}} - 1}{\dfrac{T_{o_2}}{T_{o_1}} - 1}$$

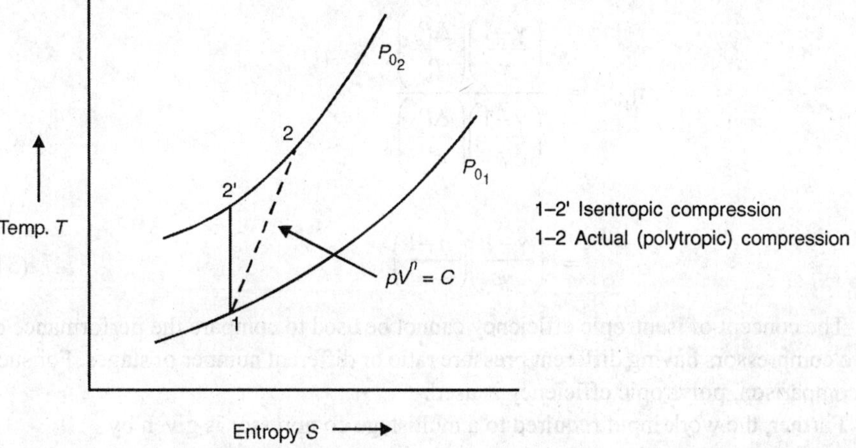

Fig. 8.21. *Actual compression process in a stage.*

$$= \frac{\left(\dfrac{P_{o_2}}{P_{o_1}}\right)^{\frac{\gamma-1}{\gamma}} - 1}{\left(\dfrac{P_{o_2}}{P_{o_1}}\right)^{\frac{n-1}{n}} - 1}$$

$$= \frac{\left(\dfrac{P_{o_1} + \Delta P_o}{P_{o_1}}\right)^{\frac{\gamma-1}{\gamma}} - 1}{\left(\dfrac{P_{o_1} + \Delta P_o}{P_{o_1}}\right)^{\frac{n-1}{n}} - 1}$$

$$= \frac{\left(1 + \dfrac{\Delta P_o}{P_{o_1}}\right)^{\frac{\gamma-1}{\gamma}} - 1}{\left(1 + \dfrac{\Delta P_o}{P_{o_1}}\right)^{\frac{n-1}{n}} - 1}$$

Expanding the above equation binomially and ignoring higher order terms, we may write,

$$\eta_{poly} = \frac{\left(\frac{\gamma-1}{\gamma}\right)\left(\frac{\Delta P_o}{P_{o_1}}\right)}{\left(\frac{\gamma-1}{\gamma}\right)\left(\frac{\Delta P_o}{P_{o_1}}\right)}$$

$$= \left(\frac{\gamma-1}{\gamma}\right)\left(\frac{n-1}{n}\right) \quad (51)$$

The concept of isentropic efficiency cannot be used to compare the performance of two compressors having different pressure ratio or different number of stages. For such a comparison, polytropic efficiency is used.

Further, the work input required to a multistage compressor is given by

$$\text{Work input required/kg of air} = C_p\left(T_{o_2} - T_{o_1}\right)$$

$$= C_p \times \left(T_{o_2} - T_{o_1}\right) \times \eta_{o_{isent}}$$

$$= C_p \times \eta_{o_{isent}} \times T_{o_1}\left[\left(\frac{P_{o_2}}{P_{o_1}}\right)^{\frac{\gamma-1}{\gamma}} - 1\right] \quad (52)$$

8.7.7 Work Done Factor

It is rather difficult to draw the actual velocity triangles at the root, mean and tip of the blades as accurate estimation of contraction of main stream cannot be done.

The axial velocity at the mean section is somewhat greater than the average axial velocity.

The actual temperature rise is somewhat lower than that estimated from the velocity triangles.

An empirical relation suggested by Hotwell, for stagnation enthalpy rise across a stage is

$$C_p \Delta T_o = \lambda u\left(V_{f_2} \tan\alpha_2 - V_{f_1} \tan\alpha_1\right)$$

where λ is work done factor which is approximately 0.85. Hotwell further suggested that the work done factor depends upon the number of stages. Greater the number of stages, the work done factor reduces.

8.7.8 Losses in Axial Flow Compressor Stage

During the flow of fluid through a stage of a compressor, various losses occur. The pressure loss is due to three factors, namely

(a) profile losses on the blade surface; (b) skin friction on the annulus walls and (c) secondary flow losses.

Figure 8.22 illustrates various losses influencing the stage efficiency.

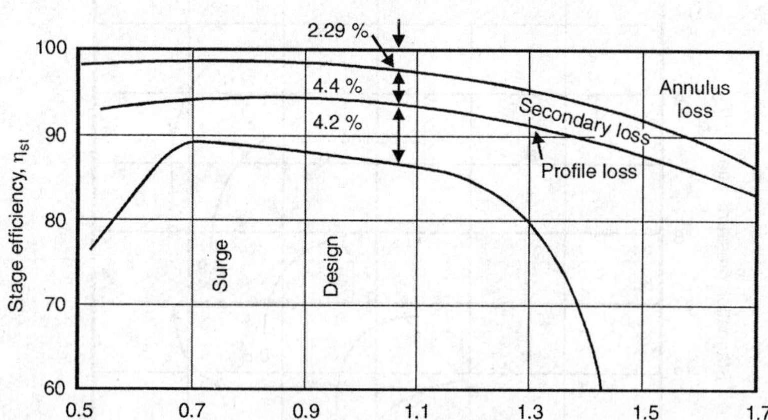

Fig. 8.22. *Losses in compressor stage.*

(a) Profile losses. This is a pressure loss of two dimensional cascade arising from the skin friction on the blade surface and due to mixing of fluid particles after the blade.

(b) Skin friction losses on the annulus walls. The total pressure loss arise from the skin friction on the annulus walls and some secondary loss.

(c) Secondary flow losses. These losses occur due to combined effects of curvature and boundary layer.

8.7.9 Performance of Axial Flow Compressors

Figure 8.23 shows the performance characteristics of an axial flow compressor using dimensionless parameters namely stagnation pressure ratio $\frac{P_{o_2}}{P_{o_1}}$, dimensionless mass flow $\frac{m\sqrt{T_{o_1}}}{P_{o_1}}$ for dimensionless constant speeds $\frac{N}{\sqrt{T_{o_1}}}$ relative to design value.

It is seen that a fixed value of $\frac{N}{\sqrt{T_{o_1}}}$ covers only a narrow range of mass flow as compared with that in centrifugal compressors. As speed increases, the curves become steeper. Surging and choking impose limit at the upper end of $\frac{N}{\sqrt{T_{o_1}}}$.

The range of stable operation of axial flow compressor is narrow. Due care has to be taken to match the components for avoiding instability during operation when these compressors are used in gas turbine.

The variation of power, pressure ratio and efficiency of an axial flow compressor against flow rate is depicted in the Fig. 8.24.

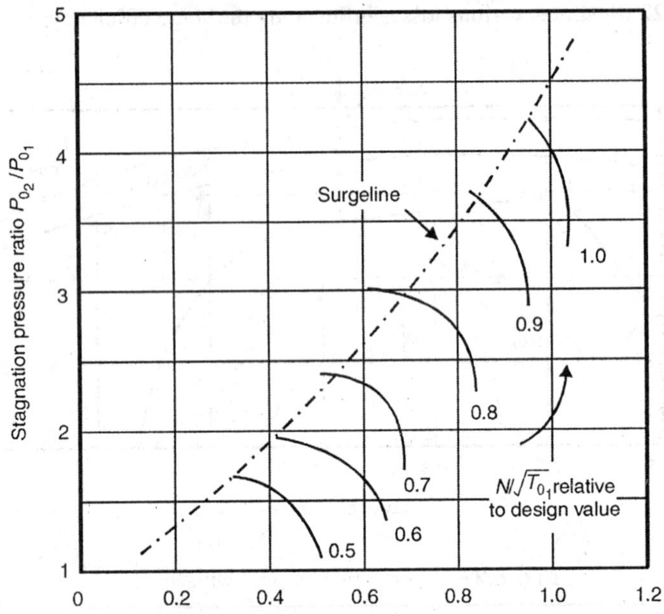

Fig. 8.23. *Axial compressor characteristics.*

Figure 8.24 shows variation of pressure ratio versus volume flow rate for axial flow compressor.

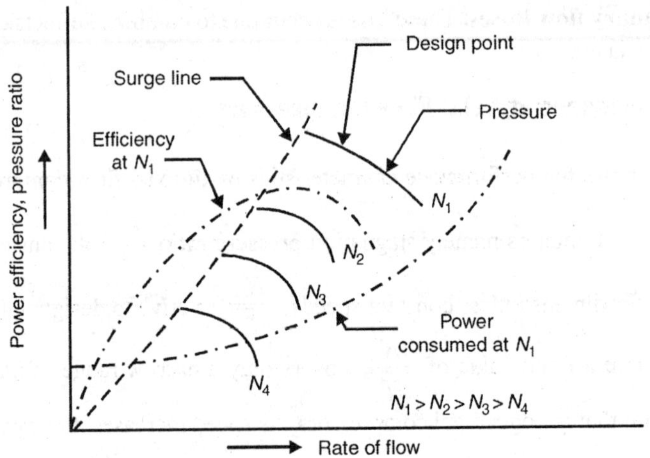

Fig. 8.24. *Performance of axial flow compressor.*

8.8 SURGING

In the operation of centrifugal and axial flow compressor, there is an instability known as surging. It is caused due to unsteady, periodic and reversal of flow through the compressor when it is operated at less mass flow rate than that corresponding to maximum pressure.

Surging may lead to mechanical failure. The rotor is subjected to alternating stresses during this irregular working and may result in damage to the bearings, rotor blades and seals. In the extreme case, the rotor shaft may bend.

Consider the pressure ratio (head) versus mass flow rate curve of a centrifugal or axial flow compressor as shown in the Fig. 8.25. The curve consists of part *AB* having a positive slope and part *BC* having a negative slope. Point *A* represents fully closed discharge valve while *B* represents full open discharge valve.

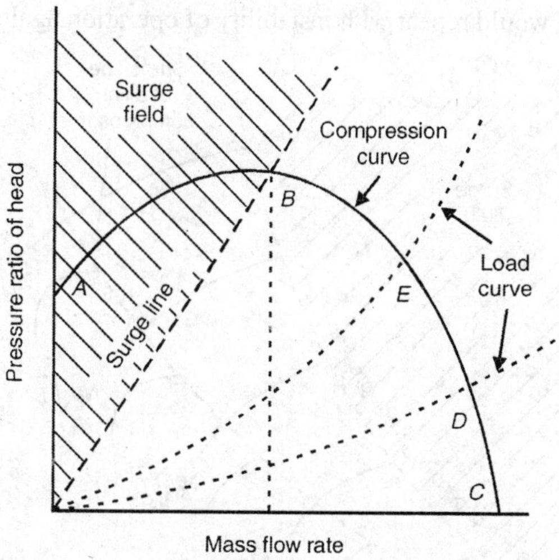

Fig. 8.25. *Surging.*

At point *A*, the flow rate is zero and pressure developed is called shut-off head. And at point *B*, pressure head developed is zero when the mass flow rate is maximum. Thus at point *B*, efficiency is zero.

When the discharge valve is opened gradually from its fully closed condition, the discharge of fluid commences and static pressure gradually builts-up due diffuser adding to the pressure rise. Further opening of discharge valve increases the pressure till point *B* is reached. At this point, efficiency is maximum for a given speed, inlet pressure and temperature.

Now, when delivery valve is opened beyond point *B*, the mass flow rate increases but pressure rise and efficiency decreases. This trend continues till point *C* when the pressure ratio approaches unity while the mass flow rate is maximum but the efficiency, is zero.

The opening and closing of a discharge valve acts as an external load on the compressor. The compressor operates in response to the external load. The intersection of the compressor curve and the load curve represents the operating point, say as at point *D* in Fig. 8.25.

When the discharge valve is closed further thus increasing the external load, the back pressure in the discharge line increases. As a result, new operating point shifts to *E*. The new operating point is possible and stable as the compressor develops greater pressure head to offset the increased back pressure in the discharge line.

With further closing of the discharge valve, external load increases, the operating point shifts in the region *A–B*. During this operation, mass flow rate is less than the estimated value, say corresponding to point *B*. The compressor develops less head than that existing in the discharge line. As such, no discharge of fluid is possible. This leads to momentary flow reversal. A short while after, the fluid from the discharge line leaves and the back pressure would reduce. The delivery of fluid from the compressor would restart and the cycle would repeat with instability of operation again.

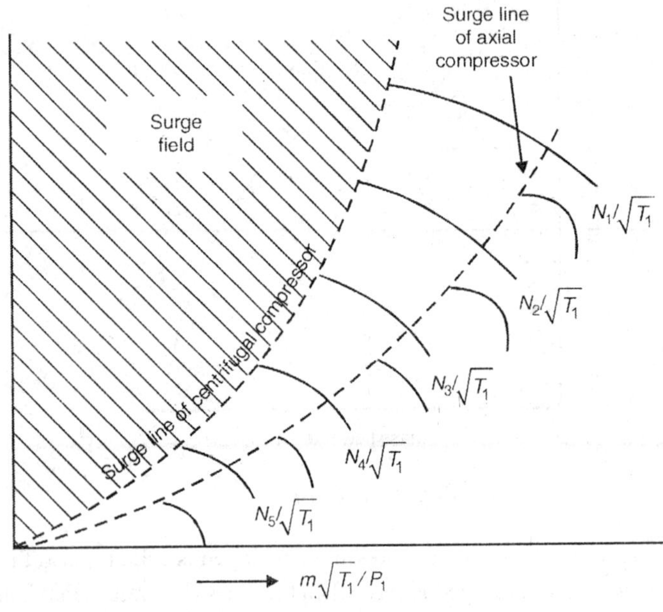

Fig. 8.26. *Surge line.*

Therefore, when the flow rate from the compressor is less than the estimated value, a surge or pulsation commences. Air surges back and forth through the whole compressor rather than giving a unidirection steady stream of flow. The unstable operation of the compressor prevails in the region of positive slope of the curve shown in Fig. 8.25.

Various methods of avoiding and correcting surging conditions are listed below. It can be controlled,

(a) by speed control.

(b) by throttling inlet.

(c) by provision of flow control system.

(c–1) by venting discharge to the atmosphere.

(c–2) by passing fluid from discharge line to suction line.

(c–3) by providing integral expander on the by-pass line.

(d) by cut-off capacity control.

(e) by inlet guide vanes or adjustable diffuser vanes.

Figure 8.26 shows the surge fields of centrifugal compressor and axial flow compressor on pressure ratio $\frac{P_{o_2}}{P_{o_1}} - \frac{m\sqrt{T_{o_1}}}{P_{o_1}}$ plane. It is seen that the surge field of the centrifugal compressor is narrow as compressed with that of the axial flow compressor.

8.9 CHOKING

Referring to the Fig. 8.25, when the mass flow rate is increased beyond B, the pressure ratio decreases and efficiency falls, as the air angle differs significantly from the vane angle causing the break-away of the air-stream.

This continues till point C where the pressure ratio becomes unity and efficiency zero. Entire power input is lost in overcoming internal friction. The point D on the curve BC, represents the maximum mass flow rate, known as 'choking mass flow'. The phenomena of choking limits the maximum mass flow rate.

As pressure ratio falls to unity, the theoretical mass flow rate is maximum. This occurs when the Mach number corresponding to inlet relative velocity becomes sonic.

8.10 STALLING

Stalling of a stage of axial flow compressor is defined as the *aerodynamic stall or breakaway of the flow stream from the suction side of the blade aerofoil*. It may be due to less mass flow rate than the designed value or due to non-uniformity in the blade profile.

The phenomena of stall preceds surging. A multistage compressor may operate in the stable unsurged region with some of its stages stalling. Thus, stalling is a local phenomena while surging is a complete system phenomena. The phenomena of stalling has been extensively researched by Smith and Fletcher.

A rotating stall is illustrated in Fig. 8.27.

The non-uniformity of the flow or geometry of blade profile causes blade B to stall. The air now flows on to the blade A at an increased incidence due to blockage of channel AB while blade C receives air at reduced incidence. As a result blade A stalls while blade C may unstall. In this way, stall 'cell' shifts along the stage in the direction of lift of the blades. Thus, stall can be stated as a reduction in the lift force at higher angles of incidence.

A rotating stall may rotate in the opposite direction to that of the rotor at about half the rotational speed. It may lead to aerodynamically induced vibrations resulting in fatigue failure of the attached components. Axial flow compressors are more prone to stalling.

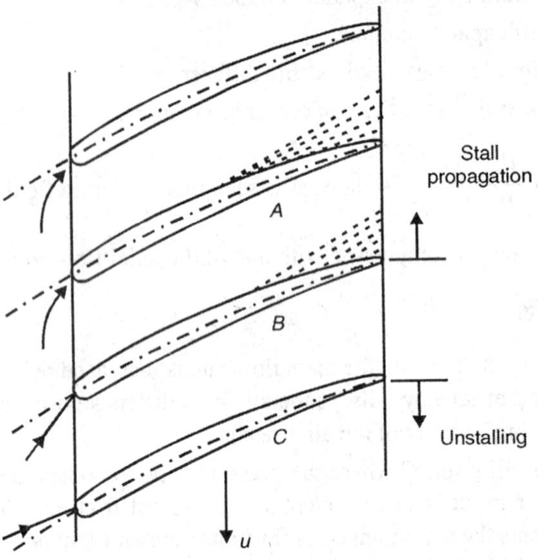

Fig. 8.27. *Mechanism of stall propagations.*

8.11 PERFORMANCE CHARACTERISTICS OF CENTRIFUGAL FLOW COMPRESSOR

The performance of centrifugal compressors is specified by illustrating the variation of delivery pressure and temperature with mass flow rate for various speeds. However, these characteristics are further influenced by other variables like inlet pressure and temperature.

To understand the interdependence of these variables, dimensionless parameters are often useful, such as

$$\frac{P_{o_2}}{P_{o_1}}, \frac{T_{o_2}}{T_{o_1}}, \frac{m\sqrt{T_{o_1}}}{P_{o_1}}, \frac{ND}{\sqrt{RT_{o_1}}}$$

Even isentropic efficiency is also expressed in a similar way as

$$\eta_{\text{isent}} = \frac{T_{o_2'} - T_{o_1}}{T_{o_2} - T_{o_1}} = \frac{\left(\frac{P_{o_2}}{P_{o_1}}\right)^{\frac{\gamma-1}{\gamma}} - 1}{\left(\frac{T_{o_2}}{T_{o_1}} - 1\right)}$$

Figure 8.28 shows the plot of such parameters.

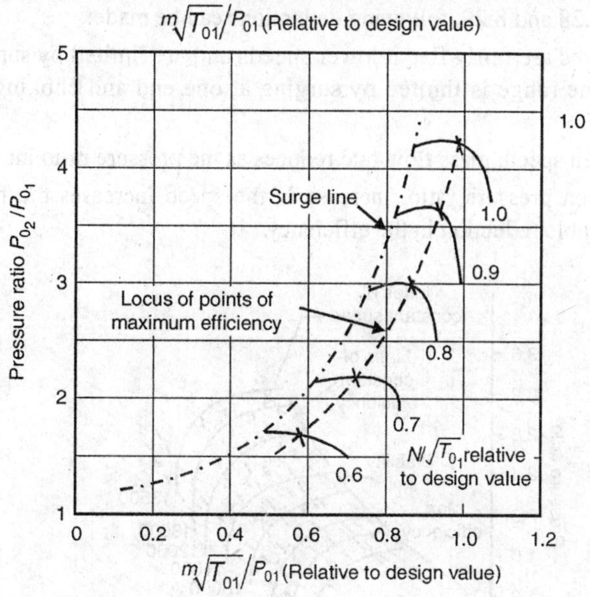

Fig. 8.28. *Centrifugal compressor charcateristics.*

Points of low mass flow at the extremities of constant speed curves can be joined to obtain surge line. While the curve obtained by joining extremities on the right hand side of constant speed curves represents choking limit occurring at low pressure ratio. The compressor is operated only between these extreme limits.

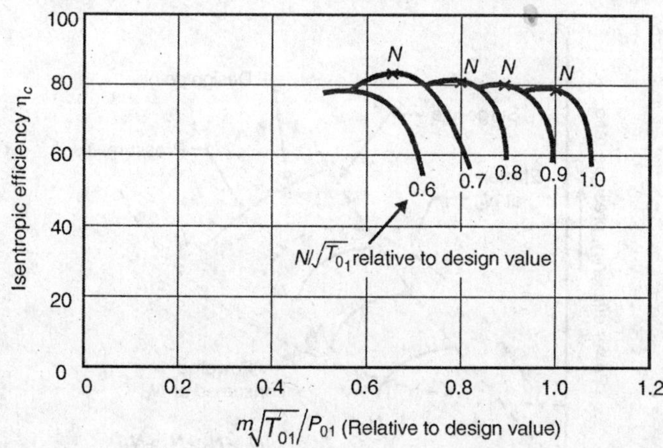

Fig. 8.29. *Efficiency and mass flow for centrifugal characteristics.*

Figure 8.29 shows the variation of isentropic efficiency with $\frac{m\sqrt{T_{o_1}}}{P_{o_1}}$ for various values of $\frac{N}{\sqrt{T_{o_1}}}$ relative to the design value. The maximum efficiency is almost same at all speeds.

From Figs 8.28 and 8.29, following deductions can be made:

1. The curves are fairly flat at lower speeds and are limited by surging. At high speed, the range is limited by surging at one end and choking at the other end.
2. At a given speed, mass flow rate reduces as the pressure ratio increases.
3. At a given pressure ratio, increase in the speed increases the flow rate with considerable reduction in the efficiency.

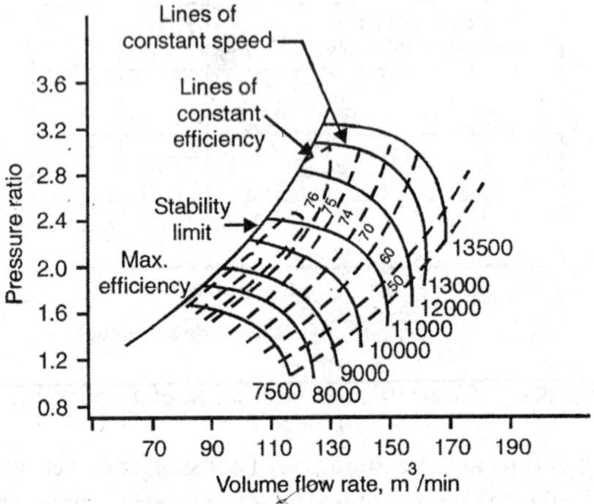

Fig. 8.30. *Performance charcateristics of centrifugal flow compressors.*

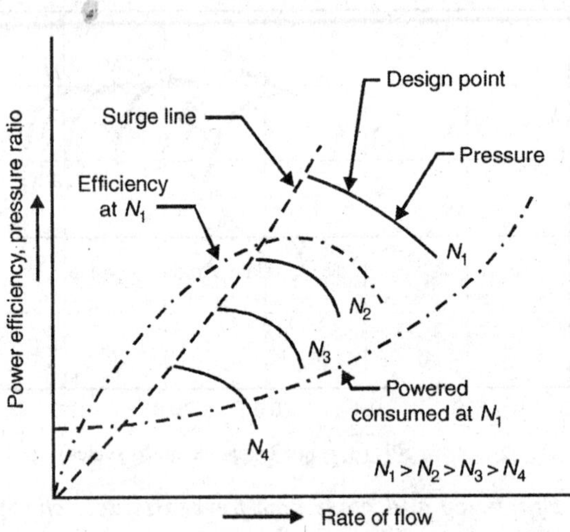

Fig. 8.31. *Performance curves of an centrifugal flow compressors.*

Similar performance characteristics using absolute parameters are shown in Figs 8.30 and 8.31.

8.12 PERFORMANCE CHARACTERISTICS OF AXIAL FLOW COMPRESSORS

The variation of pressure ratio $\frac{P_{o_2}}{P_{o_1}}$ aganist $\frac{m\sqrt{T_{o_1}}}{P_{o_1}}$ relative to design value for various values of $\frac{N}{\sqrt{T_{o_1}}}$ relative to design value is shown in the Fig. 8.32.

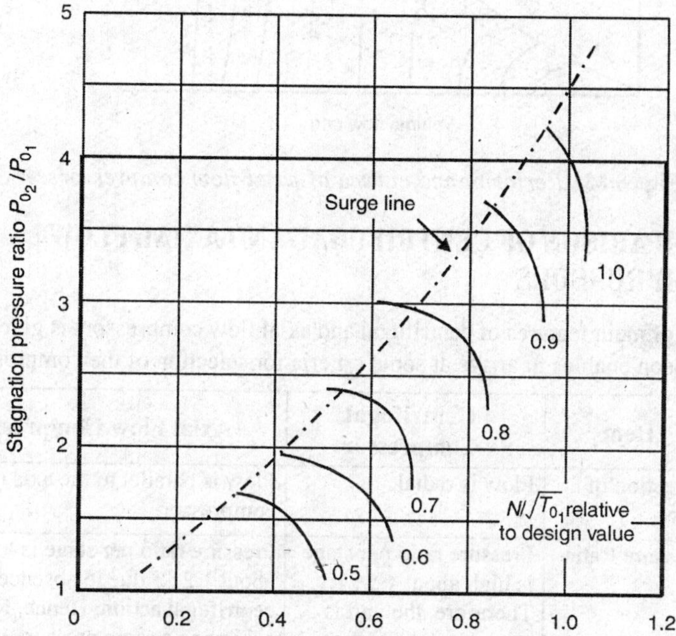

Fig. 8.32. *Axial compressor characteristics.*

It is seen that, for a given value of $\frac{N}{\sqrt{T_{o_1}}}$, these curves cover a much narrow range of mass flow rate as compared with those for centrifugal compressors. Also at higher rotational speeds, the curves become steep, nearly vertical. Therefore, the range of stable operation of axial flow compressor is much narrower. Hence, great care is necessary to match the components of a gas turbine plant for avoiding instability of operation. The phenomena of surging and stalling can hardly be distinguished as occurrence of one may lead to occurrence of the other. Stalling of these compressors results in severe vibrations of blade.

The performance characteristics using absolute parameters are also shown in Fig. 8.33.

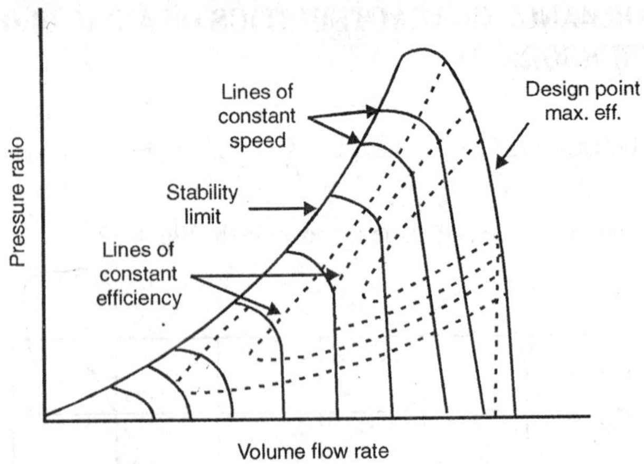

Fig. 8.33. *Performance curves of axial flow compressors.*

8.13 COMPARISON OF CENTRIFUGAL AND AXIAL FLOW COMPRESSORS

Comparison of main features of centrifugal and axial flow compressors is given below. The comparison enables to arrive at some criteria for selection of the compressor.

S. No.	Item	Centrifugal Compressor	Axial Flow Compressor
1.	Direction of flow	Flow is radial	Flow is parallel to the axis of the compressor.
2.	Pressure Ratio	Pressure ratio per stage is high about 4.5 : 1. Therefore, the unit is compact. Supersonic compressors use pressure ratio of 10 per stage with poor efficiency. Operation is simple and safe.	Pressure ratio per stage is low, about 1.2 : 1 due to absence of centrifugal action. Hence, for achieving a same pressure per stage of a centrifugal compressor, several stages are required. Thus, the unit becomes bulky and less rugged. The supersonic compressor has pressure ratio of 10 per stage with efficiency dropping rapidly. The compression is achieved due to shock wave but the operation is risky.
3.	Isentropic efficiency	It is about 80% to 82%.	It is about 86% to 88%.
4.	Operating range	It has a wide range of operation between surging and choking. The head-discharge curve is flat.	It has narrow operating range between surging and choking. The head-discharge curve is steep.

Contd.

S. No.	Item	Centrifugal Compressor	Axial Flow Compressor
5.	Part load performance	It is better.	It is poor.
6.	Frontal area	It has a large frontal area.	It has smaller frontal area for a given mass floor rate and pressure than that of centrifugal compressor. Therefore, it is more suitable for jet engines.
7.	Effect of deposits on the blade surfaces	Accumulated deposits on the blade surfaces due to the flow of contaminated fluid has no adverse effect on the performance.	Performance is adversely affected due to accumulated deposits on the blade surfaces when contaminated fluid flows.
8.	Starting torque required	Need low starting torque.	Need high starting torque.
9.	Construction	Simple and rigid	Complicated construction.
10.	Cost	Low cost.	High cost.
11.	High altitude operation	Less sensitive to icing trouble at high altitude.	Highly sensitive to icing trouble at high altitude.
12.	Multistaging	Difficult for multistaging. Maximum delivery pressure of 400 bar possible.	Most suitable for multistaging.
13.	Field of application	For blowing engines in steel mills, low pressure refrigeration, central air conditioning plants, fertiliser industry, supercharging of I.C. engine, gas pumping for transportation, petrochemical industries.	Jet engines, gas turbine plants, steel mills, etc.

8.14 COMPARISON OF CENTRIFUGAL AND RECIPROCATING COMPRESSORS

The distinguishing features of these compressors are tabulated below.

Item	Reciprocating Compressor	Centrifugal Compressor
1. Mechanical balance	Poor mechanical balance and greater vibrations due	Absence of reciprocating masses provides better mechanical

Contd.

Item	Reciprocating Compressor	Centrifugal Compressor
	to presence of reciprocating masses.	balance and vibration free operation.
2. Pressure ratio per stage	High upto 5–8	Comparatively low upto 4.5.
3. Multistaging capability	Suitable for high delivery pressures with multistaging, delivery pressure upto 5000 bar possible.	Suitable for medium delivery pressure. Delivery pressure of 400 bar possible with multistaging.
4. Volume flow rate	Capable of delivering small volume flow rates.	Capable of delivering large volume of flow rates.
5. Operational flexibility	Greater flexibility in pressure range and capacity.	No flexibility in pressure range and capacity.
6. Maintenance cost	High.	Low.
7. Effect of pressure ratio	For pressure ratios above 2, efficiency is high.	High efficiency for pressure ratio below 2.
8. Drive	Suitable for slow speed drive only.	Suitable for direct coupling to high speed drives like turbines.
9. Attention required	Greater attention required.	Less attention is adequate.
10. Contamination with lubricating oil.	Greater chance of mixing working fluid with lubricating oil.	No chance of contamination of working fluid with lubricating oil.
11. Suitability	Suitable for low and medium and high delivery pressures with low/medium volume flow rates.	Suitable for low and medium pressures with large volume flow rates.

8.15 APPLICATIONS OF COMPRESSED AIR/GASES

The various applications of compressed air/gases are

1. For industrial furnaces for supporting combustion.
2. For infloating the vehicle tubes.
3. For actuating pneumatic tools such as hammers, chiesels, drills, screw drivers, etc.
4. For industrial and household cleaning of floors, equipments, machine tools, components etc.
5. For pneumatic conveyor belts.
6. For pneumatic brakes of automobiles.
7. For air conditioning plants.

8. For ventilation of mines.
9. For production of aerated soft drinks.
10. For the production of synthetic ammonia, hydrogenation of brown coal, polymerisation of ethylene.
11. For air lift pumps, glass blowing.
12. For supercharging of IC engines.
13. For long distance transportation of industrial gases.
14. In gas turbine plants and jet engines.
15. For cabin pressurisation of air crafts.

SOLVED EXAMPLES

Example 8.1 A centrifugal compressor has a total pressure head ratio 4:1 and discharges 18.2 kg of air per second. It runs at 15000 rpm. This inlet stagnation temperature is 15° C. The slip factor is 0.9 and power input factor is 1.4. It has isentropic efficiency of 80%.

Determine the outer diameter of the impeller.

Solution

Let suffix *o* denote the stagnation temperature.

$$\frac{T_{o_2}}{T_{o_1}} = \left(\frac{P_{O_2}}{P_{O_1}}\right)^{\frac{\gamma-1}{\gamma}} = 4^{\frac{0.1}{1.4}}$$

$$\therefore \quad T_{o_2} = 288 \times 4^{\frac{0.1}{1.4}}$$

$$= 427.97 \text{ K}$$

$\therefore$ Now, isentropic efficiency $\eta_{is} = \dfrac{T_{o_2} - T_{o_1}}{T_{o_{2'}} - T_{o_1}}$

$\therefore$ Actual temperature rise $\left(T_{o_2} - T_{o_1}\right) = \dfrac{427.97 - 288}{0.8}$

$$= 174.96 \text{ K}$$

Work required/kg of air $= \dfrac{\sigma \mu u_2^2}{1000} = C_p\left(T_{o_2} - T_{o_1}\right)$

$$\therefore \text{ Blade speed at outlet } u_2 = \sqrt{\frac{1.005 \times 1000 \times 174.96}{1.04 \times 0.9}}$$

$$= 433.42 \text{ m/s}$$

But $$u_2 = \frac{\pi D_2 N}{60}$$

$$D_2 = \frac{60 \times 433.42}{\pi \times 15000} = 0.5518$$

$$= 55.18 \text{ cm}$$

Example 8.2 A rotary vane compressor has pressure ratio of 1.4 and the intake pressure of 1.0 bar. The compressor discharges 00.6 m³/rev. of FDA.

Calculate the work input required per revolution to drive the compressor when

(1) there is no internal compression (Roots blower)

(2) there is 50% pressure rise due to internal isentropic compression before back flow occurs (vane compressor).

Also find the efficiency of compressor.

Data: $V_1 = 0.6 \text{ m}^3/\text{rev}$ $p_1 = 1.0$ bar $p_2 = 1.4$ bar

Solution

(i) When there is no internal compression (Roots blower):

The p–V diagram is shown below.

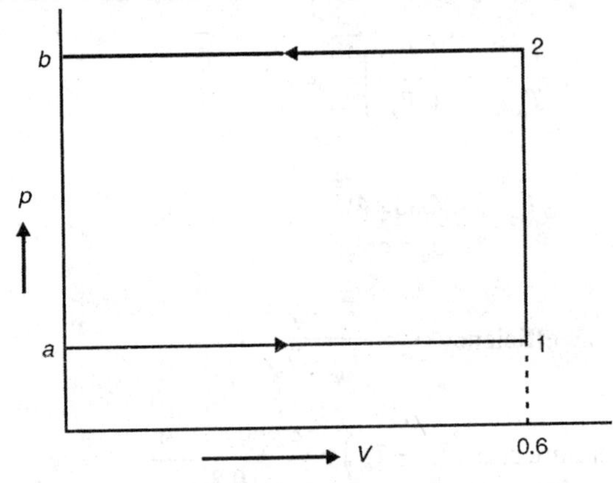

Fig. Ex. 8.1 (a)

$$\text{Work req./rev.} = (P_2 - P_1) \times V_1$$

$$= (1.4 - 1.0) \times 10^2 \times 0.06$$

$$= \mathbf{2.4 \text{ kJ/rev.}}$$

(ii) When half the pressure rise occurs due to internal compression (vane compressor).

The p–V diagram is shown below.

Data: $p_1 = 1$ bar

$p_3 = ?$

The intermediate pressure $P_2 = P_1 + \frac{(P_2 - P_1)}{2}$

$$= 1 + \frac{1.4 - 1.0}{2}$$

$$= 1.2 \text{ bar}$$

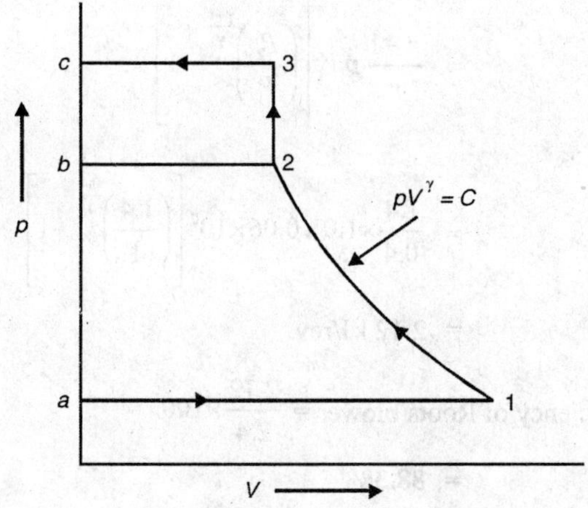

Fig. Ex. 8.1 (b)

Now, work required/rev. for isentropic compression 1 – 2

$$\text{Area } a-1-2-b = \frac{\gamma}{\gamma - 1} p_1 V_1 \left[\left(\frac{P_2}{P_1} \right)^{\frac{\gamma-1}{\gamma}} - 1 \right]$$

$$= \frac{1.4}{0.4} \times 1.0 \times 10^2 \times 0.06 \left[\left(\frac{1.2}{1.0} \right)^{\frac{r-1}{r}} - 1 \right]$$

$$= 1.24 \text{ kJ/rev.}$$

and $$V_2 = V_1 \left(\frac{P_2}{P_1} \right)^{\frac{\gamma-1}{\gamma}} = \left(\frac{1.2}{1} \right)^{\frac{0.4}{1.4}} \times 0.06 = 0.05266 \text{ m}^3$$

Work required/rev. for irreversible pressure rise

= area 2–3–c–b–2

$$= (P_3 - P_2) \times 10^2 \times V_2$$

$$= (1.4 - 1.2)10^2 \times 0.05266$$

$$= 1.0532 \text{ kJ/rev.}$$

$\therefore$ Total work required/rev. $= 1.24 + 1.0536$

$$= \mathbf{2.1772 \text{ kJ/rev.}}$$

Ideal (minimum) work required/rev. is when the entire compression is isentropic throughout.

Thus, the ideal (minimum) work required/rev.

$$= \frac{r-1}{r} p_1 V_1 \left[\left(\frac{P_2}{P_1} \right)^{\frac{\gamma-1}{\gamma}} - 1 \right]$$

$$= \frac{1.4}{0.4} \times 1.0 \times 0.06 \times 10^2 \left[\left(\frac{1.4}{1} \right)^{\frac{0.4}{1}} - 1 \right]$$

$$= 2.12 \text{ kJ/rev.}$$

Therefore, efficiency of Roots blower $= \dfrac{2.12}{2.4} \times 100$

$$= \mathbf{88.3\%}$$

Similarly, efficiency of vane compressor $= \dfrac{2.12}{2.1772} \times 100$

$$= \mathbf{97.3\%}$$

Example 8.3 A centrifugal compressor delivers 10 m/s free air while running at 9000 r.p.m. The air is compressed from 1 bar and 20°C to 4 bar with an isentropic efficiency of 0.82.

Blades are radial at outlet of the impeller and the flow velocity is 62 m/s which is constant throughout. The ratio of outer to inner radii of the impeller is 2 and the slip factor is 0.9. The blade area coefficient at inlet is 0.9.

Determine:

(1) temperature of air discharged
(2) theoretical power required
(3) diameters of the impeller at inlet and outlet
(4) the breadth of impeller at inlet
(5) impeller blade angle at inlet
(6) diffuser blade angle at inlet

Data:

$$N = 9000 \text{ r. p. m.}, \quad \text{F. A. D.} = 10 \text{ m}^3/\text{s}$$

$$P_1 = 1.0 \text{ bar}, \quad T_{o_2} = 293 \text{ K}$$

$$P_2 = 4.0 \text{ bar}, \quad \eta_{isent} = 0.82$$

Solution

Slip factor, $\mu = 0.9$

$$V_{f_1} = V_{f_2} = 62 \text{ m/s} \quad \frac{R_2}{R_1} = 2.0$$

$$k_1 = 0.9$$

$$T_{o_2} = T_{o_1}\left(\frac{P_{o_2}}{P_{o_1}}\right)^{\frac{\gamma-1}{\gamma}} = 293\,(4)^{\frac{0.4}{1.4}}$$

$$= \mathbf{433.56\ K}$$

$$\text{Actual temp. rise} = \left(T_{o_2} - T_{o_1}\right) = \frac{T_{o_2} - T_{o_1}}{\eta_{isent}}$$

$$= \frac{433.56 - 293}{0.82}$$

$$= 173.85$$

$$\therefore \quad T_{o_2} = 293 + 173.85 = 466.85 \text{ K}$$

$$\text{Mass flow rate } \dot{m} = \frac{p_1 V}{RT_{o_1}} = \frac{1\times10^2\times10\times60}{0.287\times293}$$

$$= 713.5 \text{ kg/min.}$$

Now, work required/s = Theoretical power input required.

$$= \dot{m}\, C_P\left(T_{o_2} - T_{o_1}\right)$$

$$= \frac{713.5}{60}\times1.005\times173.85$$

$$= 2077.7 \text{ kW}$$

For radial blades, the work input required is given by

$$\text{Work input required/kg} = \frac{\sigma\mu u_2^2}{1000} = C_p\left(T'_{o_2} - T_{o_1}\right)$$

$$\therefore \quad u_2 = \sqrt{\frac{1000\times1.005\times173.85}{0.9}}$$

$$= 440.6 \text{ m/s} = \frac{\pi D_2 N}{60}$$

$$\therefore \qquad D_2 = \frac{60 \times 440.6}{\pi \times 9000} = 0.9349 \text{ cm}$$

$$= \mathbf{93.49\ cm}$$

and
$$D_1 = \frac{D_2}{2} = \mathbf{46.745\ cm}$$

where k = blade area coefficient at inlet

$$b_1 = \frac{600}{60 \times \pi \times 0.46745 \times 62 \times 0.9}$$

$$= 0.122 \text{ m} = 12.2 \text{ cm}$$

Referring to inlet velocity triangle (Fig. 8.18)

$$\tan \beta_1 = \frac{V_{f_1}}{u_1} = \frac{62 \times 2}{440.0} \qquad \left[\because u_1 = \frac{1}{2} u_2\right]$$

$$= 0.2814$$

$$\therefore \qquad \beta_1 = 15.7°$$

Also,
$$\tan \alpha_2 = \frac{V_{f_2}}{\mu u_2} = \frac{62}{0.9 \times 440.6}$$

$$= 0.1563$$

$$\alpha_2 = 8.9°.$$

Example 8.4 A centrifugal air compressor delivers air with a stagnation pressure ratio of 4 :1. Its speed is 15000 r.p.m. The stagnation temperature at inlet is 17°C and the power input factor is 1.04. It has overall diameter 0.5538 m and isentropic efficiency of 0.8.

Determine the slip factor.

Data: N = 15000 r.p.m., $\eta_{isent} = 0.8$

$\psi = 1.04$, $T_{o_1} = 280$ K

Solution

$$\frac{P_{o_2}}{P_{o_1}} = 4, \qquad D_2 = 0.5538 \text{ m}$$

$$\therefore \qquad T_{o_2} = T_{o_1}\left(\frac{P_{o_2}}{P_{o_1}}\right)^{\frac{\gamma-1}{\gamma}} = 290 \times 4^{\frac{0.4}{1.4}}$$

$$= 431 \text{ K}$$

Actual stagnation temperature after compression T'_{o_2} can be calculated as

$$T'_{o_2} - T_{o_1} = \frac{T_{o_2} - T_{o_1}}{\eta_{isent}} = \frac{431-290}{0.8}$$

$$= 176.25 \text{ K}$$

Now, work required/kg $\frac{\Psi\mu u_2^2}{1000} = C_p\left(T'_{o_2} - T_{o_1}\right)$

But, $u_2 = \frac{\pi D_2 N}{60} = \frac{\pi \times 0.5538 \times 15000}{60}$

$$= 435 \text{ m/s}$$

Ship factor $\mu = \frac{1.005(176.25)\times 1000}{1.04 \times 435^2}$

$$= \mathbf{0.9}$$

Example 8.5 An aeroengine uses centrifugal compressor for supercharging. The compressor delivers 3 kg/s of air. Air enters the compressor at 1 bar, 280 K and 90 m/s after compression in the impeller, air has pressure 1.5 bar temperature 335 K and 230 m/s.

Determine:

(1) isentropic efficiency

(2) power required to drive the compressor

(3) overall efficiency.

Assume that all the kinetic energy of air gained in the impeller is entirely converted into pressure in the diffuser.

Solution

$$\dot{m} = 3 \text{ kg/s}$$

$p_1 = 1$ bar $\quad T_1 = 280$ K, $\quad V_1 = 90$ m/s

$p_2 = 1.5$ bar $\quad T_2 = 335$ K, $\quad V_2 = 230$ m/s

Referring to the T–S diagram,

Actual work required/kg of air in the impeller

$$= C_p(T_2 - T_1) + \left(\frac{V_2^2 - V_1^2}{2\times 1000}\right)$$

$$= 1.005(335-280) + \frac{230^2 - 90^2}{2\times 1000}$$

$$= 81.71 \text{ kJ/kg}$$

$$\frac{T_2}{T_1} = \left(\frac{P_2}{P_1}\right)^{\frac{\gamma-1}{\gamma}} \qquad \therefore \quad T_2 = 280 \times 1.5^{\frac{0.4}{1.4}} = 314.4 \text{ K}$$

Isentropic work required/kg of air

$$= C_p(T_2 - T_1) + \frac{V_2^2 - V_1^2}{2\times1000}$$

$$= 1.005(314.4 - 280) + \frac{230^2 - 90^2}{2\times1000}$$

$$= 56.97 \text{ kJ/kg}$$

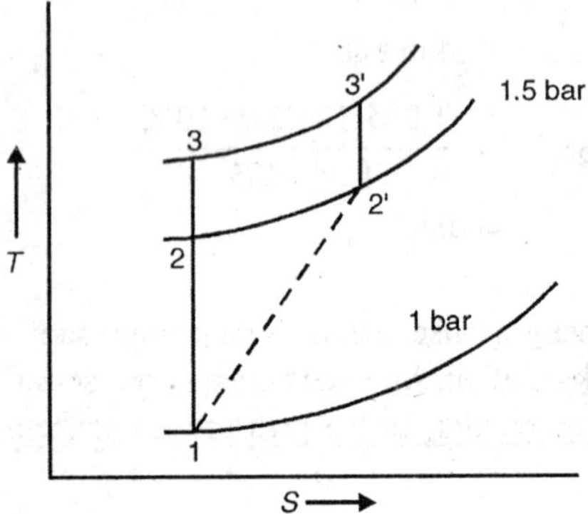

Fig. Ex. 8.5

$$\therefore \text{ Isentropic efficiency} = \frac{\text{Isentropic work reqd./kg}}{\text{Actual work reqd./kg}}$$

$$= \frac{56.97}{87.71}$$

$$= 69.72\%$$

Power required = Actual work required/kg × m

= 81.71 × 3 = **245.13 kW**

As kinetic energy gained in the impeller is converted into pressure, therefore,

$$C_p(T_3' - T_2') = \frac{V_2^2 - V_1^2}{2\times1000}$$

$$1.005\ (T_3' - 335) = \frac{230^2 - 90^2}{2000}$$

$$\mathbf{T_3' = 357.278\ K}$$

Now, the pressure of air after leaving the diffuser

$$= P_3 = P_2\left(\frac{T_3'}{T_2'}\right)^{\frac{\gamma-1}{\gamma}}$$

$$= 1.5\left(\frac{357.28}{335}\right)^{\frac{1.4}{0.4}}$$

$$= 1.879 \text{ bar}$$

In case of isentropic compression the discharged temperature from diffuser is

$$T_2' = T_1\left(\frac{P_2}{P_1}\right)^{\frac{r-1}{r}} = 280\left(\frac{1.879}{1}\right)^{\frac{0.4}{1.4}}$$

$$= 335.35 \text{ K}$$

$$\text{Overall isentropic efficiency} = \frac{335.35-280}{357.28-280}$$

$$= \mathbf{71.62\%}$$

Example 8.6 A single suction centrifugal compressor delivers 6 kg/s of air. The ambient air conditions are 1 bar and 20°C. It runs at 22000 r.p.m. with an isentropic efficiency 82%. Air is compressed from 1 bar to 4.2 bar total (stagnation) pressure. Air enters the eye of the impeller with a velocity of 150 m/s without any pre-whirl. Assuming the ratio of whirl speed to tip speed as 0/9, calculate

(a) rise in stagnation temperature during compression if the change in kinetic energy is negligible

(b) tip diameter of the impeller

(c) power required

(d) diameter of the eye of the impeller if the hub diameter is 10 cm.

Data: $T_1 = 293$ K, $V_1 = 150$ m/s

$N = 22000$ r. p. m., $\eta_{isent} = 82\%$

Solution

$$\frac{V_{w_2}}{u_2} = 0.9, \qquad P_{o_2} = 4.2 \text{ bar}$$

$$p_1 = 1.0 \text{ bar} \qquad \dot{m} = 6 \text{ kg/s}$$

$$V_{w_1} = 0$$

$$T_{o_1} = T_1 + \frac{V_1^2}{2C_P} = 293 + \frac{150^2}{2\times1.005\times1000}$$

$$= 304.19 \text{ K}$$

$$P_{o_1} = P_1\left(\frac{T_{o_1}}{T_1}\right)^{\frac{\gamma-1}{\gamma}}$$

and $$T_{o_2} = T_{o_1}\left(\frac{P_{o_2}}{P_{o_1}}\right)^{\frac{\gamma-1}{\gamma}} = T_{o_1}\left[\frac{P_{o_2}}{P_1\left(\frac{T_{o_1}}{T_1}\right)^{\frac{\gamma-1}{\gamma}}}\right]^{\frac{\gamma-1}{\gamma}}$$

$$= T_{o_1}\left(\frac{p_{o_2}}{p_{o_1}}\right)^{\frac{\gamma-1}{\gamma}}$$

$$= 293\left(\frac{4.2}{1}\right)^{\frac{0.4}{1.4}} = 441.68 \text{ K}$$

Referring to the Fig. Ex. 8.6 given below.

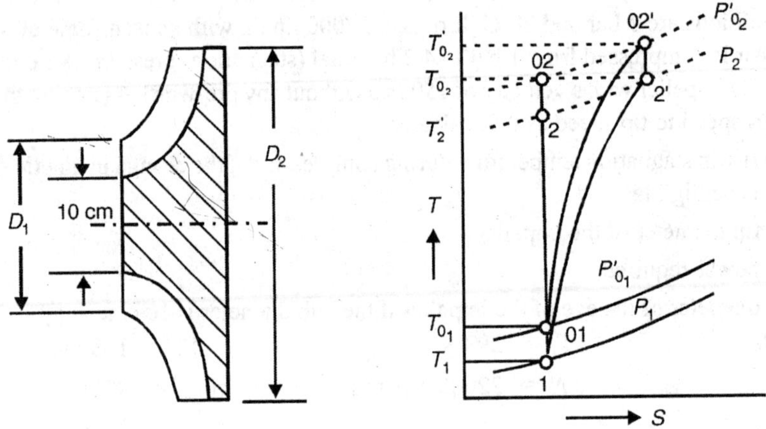

Fig. Ex. 8.6

But, $$\frac{V_{w_r}}{u_2} = 0.9$$

$$\therefore \quad 168.5 = \frac{u_2^2 \times 0.9}{1000}$$

$$\therefore \quad u_2 = 432.69 \text{ m/s}$$

and $$u_2 = 432.69 = \frac{\pi D_2 N}{60}$$

$$\therefore \quad D_2 = 0.3756 \text{ m}$$

$$= \mathbf{37.56\ cm}$$

Similarly, power required= Work required/kg × $\dot{m}$

$$= 168.5 \times 6$$
$$= \mathbf{1011.00\ kW}$$

Now, using continuity equation,

$$\dot{m} = \frac{\pi}{4}\left(D^2{}_1 - D^2{}_h\right)V_1 \times \rho_1$$

where, $$\rho_1 = \frac{P_1}{RT_1} = \frac{1\times10^5}{0.287\times293}$$
$$= 1.189\ \text{kg/m}^3$$

$\therefore$ $$\rho = \frac{\pi}{4}\left(D_1^2 - 0.1^2\right)\times 90\times 1.189$$

$\therefore$ $$D_1 = 0.2298\ \text{m}$$
$$= \mathbf{22.98\ cm}$$

Example 8.7 A centrifugal compressor is used as a supercharger for an aircraft engine developing 750 kW power and having specific fuel consumption of 0.273 kg/kW-h. The supercharger supplies fuel : air mixture at 1.25 bar. The air : fuel ratio is 17. Air enters the supercharger at 0.55 bar and 0°C if the adiabetic efficiency of the supercharger is 85%, find the volume of mixture to be supplied and the power required to drive the supercharger.

Assume $C_p = 1.0$ kJ/kg-K and $C_v = 0.277$ kJ/kg-K.

Solution

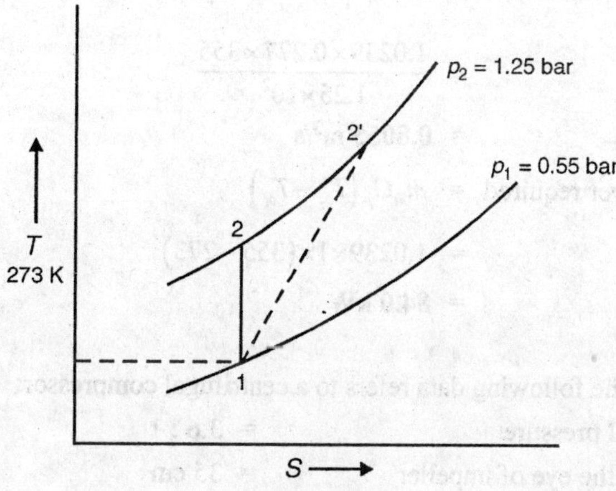

Fig. Ex. 8.7

Fuel consumption of the jet engine = Specific fuel comsumption × Power developed

$$= \frac{750\times0.273}{3600}\ \text{kg/s}$$

$$\dot{m}_f = 0.05688 \text{ kg/s}$$

$$\text{Air supplied/s} = \dot{m}_f \times \text{air : fuel ratio}$$

$$= 0.5688 \times 17$$

$$\dot{m}_a = 0.967 \text{ kg/s}$$

Hence, mass of mixture $= \dot{m}_f + \dot{m}_a$

$$m_m = 1.0239 \text{ kg/s}$$

Now,

$$r = \frac{C_p}{C_v} = \frac{C_p}{C_p + R} = \frac{1.0}{1.0 - 0.277} = 1.383$$

$$T_{o_2} = T_{o_1}\left\{\frac{P_{o_2}}{P_{o_1}}\right\}^{\frac{\gamma-1}{\gamma}} = 273\left\{\frac{1.25}{0.55}\right\}^{\frac{0.4}{1.4}}$$

$$= 342.7 \text{ K}$$

$$\text{Isentropic efficiency, } \eta_{isent} = \frac{T'_{o_2} - T_{o_1}}{T_{o_2} - T_{o_1}}$$

$$0.85 = \frac{342.7 - 273}{T'_{o_2} - 273}$$

$$\therefore \quad T_{o_2} = 355 \text{ K}$$

$\therefore$ Volume of mixture to be supplied, $V_m = \dfrac{\dot{m}_m RT'_{o_2}}{P_o}$

$$= \frac{1.0239 \times 0.277 \times 355}{1.25 \times 10^2}$$

$$= \mathbf{0.8055\ m^3/s}$$

$$\text{And, Power required} = \dot{m}_m C_p \left(T'_{o_2} - T_{o_1}\right)$$

$$= 1.0239 \times 1 \times (355 - 273)$$

$$= \mathbf{84.0\ kW}$$

Example 8.8 The following data refers to a centrifugal compressor:

Ratio of total pressure	= 3.6 : 1
Diameter of the eye of impeller	= 35 cm
Axial velocity at inlet	= 140 m/s
Mass flow rate of air	= 12 m kg/s
Velocity in the delivery duct	= 120 m/s
Tip speed of impeller	= 460 m/s
Speed	= 16000 r.p.m

Total head isentropic efficiency = 80%

Pressure coefficient = 0.73

Ambient conditions = 1.013 bar and 15°C

Calculate:

(i) Static pressure ratio

(ii) Static pressure and temperature at inlet and outlet of compressor

(iii) Work reqd./kg of air

(iv) Theoretical power required.

Data: $r_p = 3.6$

$T_{o_1} = 288$ K, $\eta_{isent} = 0.80$

$\dot{m} = 12$ kg/s $\psi = 0.73$

Solution

$$T_{o_2} = T_{o_1}(r_p)h^{\frac{\gamma-1}{\gamma}} = 288(3.66)^{\frac{0.4}{1.4}} \text{ K}$$

$$\therefore \quad T'_{o_2} - T_{o_1} = \frac{T_{o_2} - T_{o_1}}{\eta_{isent}} = \frac{288\left[3.6^{\frac{0.4}{1.4}} - 1\right]}{0.80} = 159.7°\text{C}$$

$$T_{o_2} = 288 + 159.7 = \mathbf{447.1\ K}$$

Now static temperature at exit $= T'_{o_2} - \dfrac{120°}{2C_p}$

$$= 447.1 - \frac{120°}{2 \times 1.005 \times 10^3}$$

$$T_2 = 440 \text{ K}$$

Also static pressure at exit

$$P_2 = P_{o_2} - \frac{\rho_2 \times 120^2}{2}$$

where $= 3.65 - \dfrac{P_2 \times 120^2}{0.287 \times 440 \times 2 \times 10^3}$ $\qquad \rho_2 = \dfrac{P_2}{RT_2}$

$= 3.45$ bar and $P_{o_2} = 1.013 \times 3.6$

$= 3.65$ bar

Now static condition at inlet

$$T_{o_1} = T_1 + \frac{140^2}{2 \times C_p} = T_1 + \frac{140^2}{2 \times 1.005 \times 10^3} = T_1 + 9.75$$

$$T_1 = 288 - 9.5$$
$$= 278.25 \text{ K}$$

Also, $$P_1 = P_{o_1} - \frac{140^2 \times P_1}{2 \times 0.287 \times 278.25 \times 10^3}$$
$$= 1.013 - 0.123\, P_1$$

$\therefore$ $$\mathbf{P_1 = 0.9 \text{ bar}}$$

$\therefore$ **Static presure ratio** $= \mathbf{\dfrac{P_2}{P_1} = \dfrac{3.45}{0.9} = 3.83}$

Furthur, work required/kg of air $= C_p\left(T'_{o_2}, T_{o_1}\right)$
$$= 1.005\,(159 - 1)$$
$$= \mathbf{159.89 \text{ kJ/kg}}$$
$$= \dot{m} C_p\left(T'_{o_2}, T_{o_1}\right)$$
$$= 10 \times 159.89 \text{ kW}$$
$$= \mathbf{1598.9 \text{ kW}}$$

Example 8.9 A centrifugal compressor running at 18000 r.p.m. admits at 300 K. The other data is as follows:

Isentropic total head efficiency = 0.76

Outer diameter of blade tip = 550 mm

Slip factor = 0.82

Calculate:

(i) Temperature rise of air

(ii) Static pressure ratio.

Assume that absolute velocity of air at inlet and exit to be same.

Take $C_p = 1.005$ kJ/kg-K

Solution

$$T_1 = 300 \text{ K}, \qquad N = 18000 \text{ r. p. m.}$$
$$D_2 = 0.55 \text{ m} \qquad \eta_{isent} = 0.76$$
$$\mu = 0.82$$

$$u_2 = \frac{\pi D_2 N}{60} = \frac{\pi \times 0.55 \times 18000}{60}$$
$$= 518.36 \text{ m/s}$$

Work required/kg of air $= \left(V_{w_2} u_2 - V_{w_1} u_1\right)$ N-m/kg

But, $$V_{w_1} = 0 \text{ and } \mu = \frac{V_{w_2}}{u_2}$$

$$\therefore \quad \text{Work reqd./kg} = \mu u_2^2 = 0.82 \times 518.36^2 \text{ N-m/kg}$$

$$= \mathbf{220.331\ kN/kg}$$

Also, work reqd./kg of air = $C_p\left(T_{o_2} - T_{o_1}\right)$

$$220.331 = 1.005\left(T'_{o_2} - T_{o_1}\right)$$

$$\therefore \quad \left(T'_{o_2} - T_{o_1}\right) = \frac{220.331}{1.005} = 219.23°\text{C}$$

And Isentropic efficiency = $\dfrac{T_{o_2} - T_{o_1}}{T'_{o_2} - T_{o_1}}$

$$\therefore \quad \boldsymbol{T_{o_2} = T_{o_1} + 0.76(219.36) = 466.6\ \text{K}}$$

And, Static pressure ratio $\left(\dfrac{P_2}{P_1}\right) = \left(\dfrac{T_{o_2}}{T_{o_1}}\right)^{\frac{\gamma-1}{\gamma}} = \mathbf{4.69}$

Example 8.10 An axial flow compressor has 8 stages and with 50 % degree of reaction compresses air with a pressure ratio of 4 : 1. Air enters the compressor at 20°C and flows through it with a constant speed of 90 m/s. The mean blade speed is 180 m/s. The isentropic efficiency is 82 %.

Calculate:

(i) Work required/kg of air

(ii) Blade angles.

Assume $\gamma = 1.4$ and $C_p = 1.005$ kJ/kg-K

Solution

No. of stages = 8

$$\text{D. R.} = 0.50, \qquad \therefore \alpha_1 = \beta_2 \text{ and } \beta_1 = \alpha_2$$

$$r_p = 4.0, \qquad T_1 = 293 \text{ K}$$

$$u = 180 \text{ m/s}$$

$$\eta_{\text{isent}} = \frac{T_2' - T_1}{T_2' - T_1} \quad \text{But } T_2 = T_1\left(r_p\right)^{\frac{\gamma-1}{\gamma}}$$

$$= 293 \times 4^{\frac{0.4}{1.4}}$$

$$0.82 = \frac{435.4 - 293}{T'_2 - 293} = 435.4 \text{ K}$$

$$T_2' = 466.6 \text{ K}$$

$$\text{Total work reqd./kg} = C_p\left(T_2' - T_1\right) = 1.005\ (466.6 - 293) \text{ kJ/kg}$$

$$= \mathbf{174.47\ kJ/kg}$$

$$\text{Total work reqd./kg} = \text{No. of stages} \times u\left(V_{w_2} - V_{w_1}\right)$$

$$= 8 \times u \times V_f\left(\tan\alpha_2 - \tan\alpha_1\right)$$

$$\therefore \quad 174.47 = 8 \times 180 \times 90\left(\tan\alpha_2 - \tan\alpha_1\right) \times 10^3$$

$$\therefore \quad \tan\alpha_2 - \tan\alpha_1 = 1.346 \qquad \text{(i)}$$

With 50% degree of reaction,

$$\alpha_1 = \beta_2 \qquad \text{and } \beta_2 = \alpha_1$$

$$\text{Also,} \quad \tan\alpha_1 + \tan\beta_1 = \frac{u}{V_f} = \frac{180}{90} = 2 \qquad \text{(ii)}$$

Solving Eqs (i) and (ii)

$$\boldsymbol{\beta_1 = \alpha_2 = 59.1^\circ}$$

$$\boldsymbol{\alpha_1 = \beta_2 = 18.1^\circ}$$

Example 8.11 Determine the pressure rise and work done by a rotary axial flow compressor for which the following data is available.

u = 200 m/s, $V_w = V_f$ = 186 m/s, α_1 = 45°, α_2 = 14°,

and ρ = 1.0 kg/m³.

Solution

Pressure rise through a ring of axial flow compressor

$$\Delta p = \rho V_w^2\left(\tan^2\alpha_1 - \tan^2\alpha_2\right)$$

$$= \frac{1 \times 186^2}{2 \times 10^5}\left(\tan^2 45 - \tan^2 14\right)$$

$$= \mathbf{0.1622\ bar}$$

Work required /kg of air on the blades

$$= uV_w\left(\tan\alpha_1 - \tan\alpha_2\right)$$

$$= \frac{200 \times 186}{10^3}(\tan 45 - \tan 14)$$

$$= \mathbf{27.92\ kW}$$

Example 8.12 An eight stage axial flow compressor admits air at 1.0 bar and 20°C. It has an overall pressure ratio of 5:1 and overall isentropic efficiency of 90%. The work is equally divided between the stages. The mean blade speed is 175 m/s and 50% reaction design is used. The axial velocity of flow through the compressor is contant at 100 m/s.

Calculate the power required and blade angles.

Data: Degree of reaction = 50%

$P_1 = 1.0$ bar $\qquad T_1 = 293$ K

$\eta_{isent} = 0.90, \qquad r_p = 5.0$

$u = 175$ m/s $\qquad V_f = 100$ m/s

$\alpha_1 = \beta_2$ and $\alpha_2 = \beta_1$

Solution

Let sufix z denote the number of stages.

$$T_{o_2} = T_{o_1} \times r_p^{\frac{\gamma-1}{\gamma}} = 293 \times 5^{\frac{1.4}{1.4}}$$

$$= 464 \text{ K}$$

But
$$\eta_{isent} = \frac{T_{o_2} - T_{o_1}}{T'_{o_2} - T_{o_1}}$$

$$\therefore \quad 0.90 = \frac{464 - 293}{T'_{o_2} - 293} \quad \therefore \quad T'_{o_2} = 483 \text{ K}$$

$$\text{Work reqd./kg} = C_p\left(T'_{o_2} - T_{o_1}\right) = \left(V_{w_2} - V_{w_1}\right) u \times z$$

$$= V_f\left(\tan\alpha_2 - \tan\alpha_1\right) u \times z$$

$$\therefore \quad \left(\tan\alpha_2 - \tan\alpha_1\right) = \frac{1.004(483 - 293) \times 10^3}{100 \times 175 \times 8}$$

$$= 1.3639 \qquad \text{(i)}$$

Referring to velocity triangles,

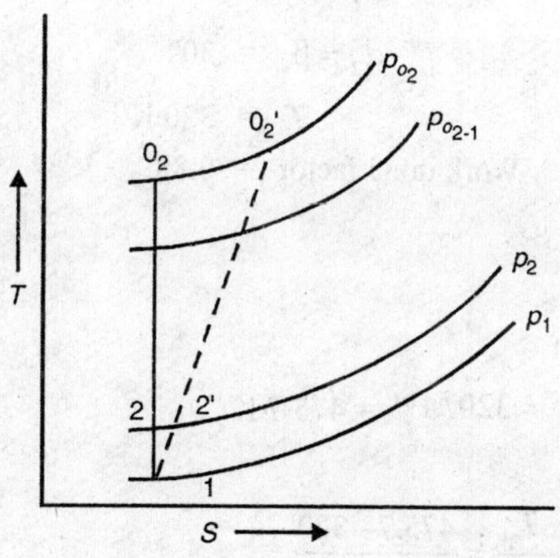

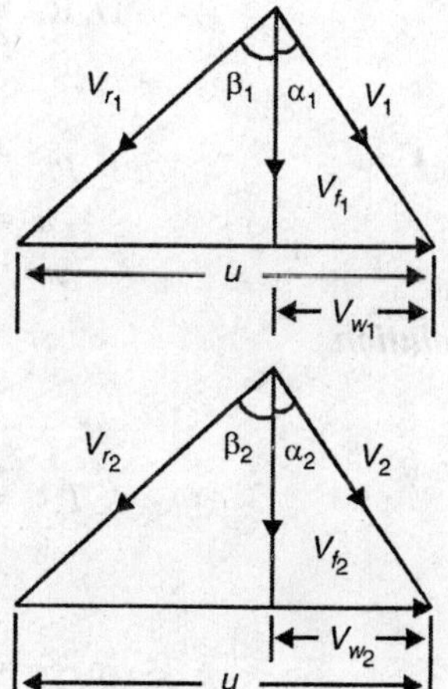

Fig. Ex. 8.12

$$\frac{u}{V_f} = \tan\alpha_1 - \tan\beta_1 = \frac{175}{100}$$

$$= 1.75$$

From Eqs (i) and (ii), we get

$$\tan\beta_1 = \frac{1.365 + 1.75}{2}$$

$$= 1.5569$$

$$\therefore \quad \beta_1 = \alpha_2 = 57.4^\circ$$

and substituting in Eq. (i),

$$\alpha_1 = \beta_2 = 10.95^\circ$$

And power rquired to drive the compressor

$$= \dot{m}C_p\left(T'_{o_2} - T_{o_1}\right)$$

$$= 1.005\ (483 - 293)/1$$

$$= \mathbf{190.93\ kW}$$

Example 8.13 An axial flow compressor operates with an overall stagnation pressure ratio of 4 and overall stagnation isentropic efficiency 86 %. The inlet stagnation pressure and temperature are 1.0 bar and 320 K. The mean blade speed is 190 m/s. The degree of reaction is 0.5 at the mean radius with relative air angles of 30° and 10° at inlet and outlet of rotor respectively. The work done factor is 0.88.

Determine the stagnation temperature. If the hub : tib ratio is 0.4 and mass flow rate is 20 kg/s, calculate the blade height in the first stage and stagnation polytropic efficiency.

Data: $\frac{p_{o_2}}{P_{o_1}} = 4,$ $\quad \eta_{isent} = 0.86$

D. R. = 0.50, $\quad \alpha_2 = \beta_2 = 10^\circ$

$\alpha_2 = \beta_1 = 30^\circ$

$p^1 = 1.0$ bar, $\quad T_1 = 320$ K

$u = 190$ m/s $\quad$ Work done factor = 0.88

$\dot{m} = 20$ kg/s

Solution

$$T_{o_2} = T_{o_1}\left(\frac{P_{o_2}}{P_{o_1}}\right)^{\frac{\gamma-1}{\gamma}} = 320/4^{\frac{0.4}{1.4}} = 475.7\ \text{K}$$

$$\eta_{isent} = 0.86 = \frac{T_{o_2} - T_{o_1}}{T'_{o_2} - T_{o_1}} = \frac{475.7 - 320}{T_{o_2} - 320}$$

$$T'_{o_2} = 501 \text{ K}$$

Stagnation polyteropic efficiency, $\eta_p = \dfrac{\ln\left(\dfrac{P_{o_2}}{P_{o_1}}\right)^{\frac{\gamma-1}{\gamma}}}{\ln\left(\dfrac{T_{o_2}}{T_{o_1}}\right)}$

$$= \frac{\ln(4)^{\frac{0.4}{1.4}}}{\ln\left(\dfrac{501}{302}\right)}$$

$$= \mathbf{88.4\%}$$

Referring to velocity triangles shown in Fig. Ex. 8.12.

$$\frac{u}{V_f} = \tan\alpha_1 + \tan\beta_1 = \tan 10° + \tan 30°$$

$$= 0.7536$$

$$\therefore \quad V_f = \frac{190}{0.7536} = 252.12 \text{ m/s}$$

Also $\quad V_{w_2} = V_f \tan\alpha_2 = 252.12 \tan 30° = 145.56 \text{ m/s}$

$$V_{w_1} = V_f \tan\alpha_1 = 252.12 \tan 10° = 44.45 \text{ m/s}$$

$\therefore$ Work required/kg of air per stage

$$= u\left(V_{w_2} - V_{w_1}\right) \times \lambda \times 10^{-3} \text{ kJ/kg}$$

$$= \frac{190 \times 0.88(145.56\text{-}44.45)}{1000}$$

$$= 16.906 \text{ kJ/kg per stage}$$

Total work required/kg for all stages

$$= C_p\left(T'_{o_2} - T_{o_1}\right)$$

$$= 1.005\,(501 - 320)$$

$$= 181.9 \text{ kJ/kg}$$

$\therefore$ No. of stages $z = \dfrac{181.9}{16.906} = 10.76$

$$= 11 \text{ stages}$$

Now, $\quad V_1 = \dfrac{V_f}{\cos\alpha_1} = \dfrac{252.12}{\cos 10°} = 256 \text{ m/s}$

and $$T_1 = T_{o_1} - \frac{V^2}{2C_p} = 320 - \frac{190^2}{2 \times 1.005 \times 1000} = 302.03 \text{ K}$$

$$\frac{P_1}{P_{o_1}} = \left(\frac{T_1}{T_{o_1}}\right)^{\frac{\gamma}{\gamma-1}} \quad \therefore \quad \frac{P_1}{P_{o_1}} = \left(\frac{302.03}{320}\right)^{\frac{1.4}{0.4}}$$

$$\therefore \quad P_1 = 0.816 \text{ bar}$$

and $$\rho_1 = \frac{P_1}{RT_1} = \frac{0.816 \times 10^2}{0.287 \times 302.03}$$

$$= 0.941 \text{ kg/m}^3$$

From the continuity equation,

$$\rho_1 A_1 V = \dot{m} = 20 \text{ kg/s}$$

$$\therefore \; 0.941 \times \pi r_1^2 \left[1 - 0.4^2\right] \times 252.12 = 20$$

$$r_1 = 0.1787 \quad m = 17.87 \text{ cm}$$

But, $$\frac{r_2}{r_1} = 0.4$$

$$\therefore \quad r_1 = 17.87 \times 0.4 = 7.1 \text{ cm}$$

∴ Hence height of blade in the first stage

$$z = r_2 - r_1$$
$$= 17.87 - 7.148$$
$$= 10.722 \text{ cm}$$

PROBLEMS FOR PRACTICE

1. Air at 101.3 kPa and 288 K is to be compressed at a rate of 5.6 m³/min to 11.75 bar. Two compressors are considered (i) Roots blower and (ii) Sliding vane rotary compressor. Compare the power required assuming that for vane type compressor, 75% of the pressure rise occurs due to internal compression before the delivery takes place and that compressor is uncooled.

   [**Ans.** (i) 6.88 kW, (ii) 5.75 kW]

2. Free air of 30 m³/min. is compressed from 101.3 kPa to 2.23 bar. Determine the power required (i) if the roots blower is used, (ii) if vane blower is used. Assume that there is 25% reduction in volume before the back flow occurs. Also find the isentropic efficiency in each case.

   [**Ans.** (i) 60.85 kW, (ii) 48.46 kW, (iii) 73.69%, (iv) 92.53%]

3. The design pressure ratio of a centrifugal compressor is 3.5 : 1. The diameter of the impeller eye is 30 cm. The axial velocity of flow at inlet is 130 m/s and the

mass flow rate of air is 5 kg/s. The velocity in the delivery duct is 115 m/s. The tip speed of the impeller is 450 m/s and runs at 16000 r.p.m. with a total head isentropic efficiency of 78%. The pressure coefficient is 0.72.

Determine:

(i) Static pressure ratio

(ii) Static pressure and temperature at the inlet and outlet of the compressor

(iii) Work required/kg of air

(iv) Theoretical power required.

Assume ambient conditions as 1.013 bar and 15°C.

[**Ans.** (i) 4.21, (ii) 0.917 bar, 279.6 K, 3.86 bar, 461.67 K, (iii) 180.29 kJ/kg (iv) 901.45 kW]

4. A centrifugal compressor runs at 20000 r.p.m. draws in air at 290 K. It has slip factor 0.80, total head isentropic efficiency 0.75 and outer diameter of blade tip 50 cm.

Calculate,

(i) temperature rise of air flowing through the compression

(ii) static pressure ratio.

Assume that velocities of air at inlet and outlet to remain constant.

[**Ans.** (i) 218.62°C, (ii) 4.8]

5. Air at a temperature of 17°C flows axially into the eye of the impeller of an uncooled centrifugal compressor running at 18000 r.p.m. and leaves the impeller with a velocity of whirl 0.88 times the velocity of the blade tip. The outer diameter of the blade tip is 46 cm and the kinetic energy of air at the outlet from the impeller is equal to that at the inlet to the impeller. Assuming isentropic efficiency of 0.75, determine

(i) Rise in temperature of the air compressed

(ii) Static head pressure ratio.

(**Ans.** Actual Δt = 165°C, Pressure ratio = 3.48)

6. A single sided centrifugal compressor of a gas turbine plant has to deliver 8.2 kg/s of air with a pressure ratio of 4.4 : 1 at 18000 r.p.m. The intake conditions are static pressure 1.013 bar and 15°C. Assuming isentropic efficiency of 78%, ratio of whirl speed to tip speed 0.94. Calculate rise in stagnation temperature, tip speed, external diameter of impeller, external diameter of the eye for which the internal diameter is 12.5 cm and the axial velocity is to be 140 m/s with no pre-whirl.

(**Ans.** Outer dia. of impeller D_2 = 48.5 cm, Eye external dia. = 28.6 cm, ΔT_o, = 195 °C)

7. An eight stage axial flow compressor takes in air at a temperature of 20°C at a rate of 3 kg/s. The pressure ratio is 6 and the isentropic efficiency is 0.89. The compression process is adiabetic. All stages of the compressor are similar and operate with 50 reaction. In each stage, the mean blade speed is 180 m/s. The constant axial velocity of flow of air is 105 m/s.

Determine the power required and the direction of air at the entry to and exit from the rotor and stator. Assume air as an ideal gas.

(**Ans.** Power = 660.3 kW, $\alpha_1 = \beta_2 = 7.4°$. $\alpha_2 = \beta_1 = 57.8°$)

8. The mean blade speed in the second stage of an uncooled axial flow compressor is 230 rn/s and the axial velocity of flow is 150 m/s. The shape of the fixed and moving blades is same and the difference between the tangents of the angles at inlet and outlet is 0.56 If the adiabetic efficiency of the stage is 85% and the air inlet temperature to the stage is 290 K, find the number of required for a pressure rise from 1 bar to 3.8 bar. Also find the power required to compress a mass flow of 1000 kg/min. Assume C_p = 1.005, mechanical efficiency 99%. Neglect the difference in the kinetic energy entering successive stages is negligible.

(**Ans.** W = 19.32 kJ/kg, ΔT = 19.22°C, r_p = 1.211,
No. of stages = 7, power = 2276 kW)

9. An axial flow compressor compresses air from an inlet condition of 1 bar and 290 K so a delivery pressure of 5 bar with an overall isentropic efficiency 87%. The degree of reaction is 0.5 and the blade angles at inlet and outlet are 44° and 13° respectively. The mean blade speed and axial velocity of flow are constant throughout the compressor. Assuming a blade velocity of 180 m/s and work done factor of 0.85.

Determine:

(i) Number of stages,

(ii) Change in entropy.

[**Ans.** (i) 12, (ii) 0.4898 kJ/kg-K]

10. The first stage of an axial flow compressor is to be designed for an axial velocity of flow 133 m/s, mass flow rate of air 22.7 kg/s and for ambient conditions of 1 bar an 10°C.

At the mean radius, both moving and fixed blade sections are to be suitable for relative air angles of 50° at inlet and 30° at outlet. A row of guide vanes giving an outlet air angle of 30° is to be fitted before each stage. Draw the triangles of velocity assuming the ratio of the hub to tip diameters as 0.68, free vortex conditions.

Calculate:

(i) Speed

(ii) Theoretical temperature rise across the stage

(iii) Number of similar stages theoretically required to give a total head delivery, pressure of 8.4 bar if the stage isentropic efficiency is 85%.

[**Ans.** (i) 9000 r.p.m., (ii) 16°C, (iii) 13]

11. A six stage axial flow compressor take air at a temperature of 25°C at a rate of 5 kg/s. The pressure ratio is 5 and isentropic efficiency 0.85. The compressor stag are similar and has 50% degree of reaction. In each stage, the mean blade speed is 190 m/s and axial velocity of flow is 100 m/s. Calculate the power required and blade angles.

(**Ans.** 1074.675 kW, $\beta_1 = 72.23°$, $\beta_2 = 62.72°$)

12. In an xial flow air compressor has a pressure ratio 6 : 1. Air enters at 20°, mean velocity of rotor blades is 200 m/s and the inlet and exit angles of both moving and fixed blades are 45° and 15° respectively. The degree of reaction is 50%, the work done factor is 0.86 throughout. There are 12 stages and axial velocity remains constant throughout.

    Calculate the isentropic efficiency of the compressor. Explain why the overall pressure ratio is not equal to the stage pressure ratio raised to the power of *n*, is the number of stages.

    (Ans. $\eta_{isent} = 0.83$)

13. Sketch the velocity diagram for an axial flow compressor and derive the relation,

$$\frac{p_{o_2}}{p_{o_1}} = \left[1 + \eta_c \frac{V_b (V_{w_2} - V_{w_1})}{C_p T_{o_1}}\right]^{\frac{\gamma}{\gamma - 1}}$$

ꕤ ꕤ

9

Steam Generation and Properties, Processes of Steam

CHAPTER OBJECTIVES

After reading this chapter you will be able to learn the following

- Steam Generation Process at Constant Pressure.
- Condition of Steam—Wet, Dry and Superheated.
- Steam Tables.
- Steam Processes.
- Phase Diagram and Tripple Point.
- Vapour Processes.
- Determination of Dryness Fraction of Steam with Steam Calorimeter.

9.1 INTRODUCTION

Upto now we have considered a fluid, which under normal conditions remains in a gaseous state. Such substances are oxygen, nitrogen, air etc. Now we consider fluids, which under normal working conditions may be either in a liquid state or in a gaseous state. Such substances are known as vapours. Examples of vapours, employed in engineering practices are steam, ammonia, freon, mercury etc. The behaviour of vapours is typically represented by steam and water.

With vapours, the simple laws of perfect gas does not even approximately apply and therefore sometimes the definition of vapour is given as: "A vapour is any substance in the gaseous state which does not even approximately follow the general gas laws". The reason for this departure of the performance of vapours from the general gas law is that the vapours from the general gas law is that the vapours when heated or cooled experience some desegregation work while the general gas laws are based on a perfect gas, which would experience no desegregation work when heated or cooled. It should be remembered that all known actual gases do experience a small amount of desegregation work. So we can use the formula derived for perfect gas, for vapour calculations. For this purpose we use values, which have been determined by experiments and listed in tables.

Highly superheated vapours approximately follow the perfect gas laws. Vapour obeys its own Laws.

9.2 FORMS OF MATTER

There are three forms of matter, viz. (i) Solid, (ii) Liquid and (iii) Gaseous.

In order to explain the changes in form of matter, a hypothetical experiment may be described with the help of Fig. 9.1.

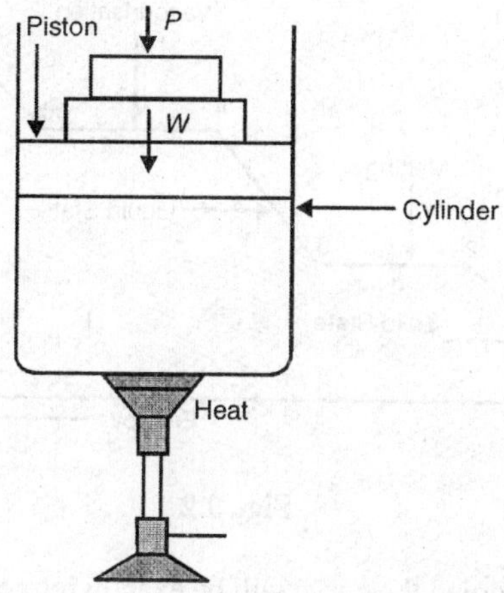

Fig. 9.1

As substance in the solid state is placed in a cylinder, which is fitted with a frictionless right fitting piston being loaded by a weight as shown in Fig. 9.1. Heat is supplied to the cylinder. The substance may change in volume during the heating process but the pressure is kept constant by the piston and weights on it. The change in temperature of the substance will be recorded by a thermometer (Not shown).

At first, when the heat is added, the temperature of the solid rises and a certain value of temperature will be reached after which a considerable amount of heat is added, without a change in temperature. Through a sight window provided in the cylinder wall it can be seen that the solid is melting. When the solid has been completely transformed into liquid, further addition of heat again causes a rise in temperature. This rise in temperature continues till a point is reached where further addition of heat does not change the temperature of the substance but on observation we see that the liquid is boiling. When the liquid has been completely transformed into a vapour, the temperature again rises when the heat is added. The relation between the heat added, entropy change and the temperature of the substance is shown in Fig. 9.2.

1–2 Shows the stage during which the rise in temperature of a substance in solid state takes place.

2–3 Shows the stage during which the substance is being transformed from a solid state to a liquid state without rise in temperature.

3–4 Shows the stage during which rise in temperature of the substance in liquid state takes place.

4–5 Shows the stage during which the substance is being transformed from a liquid state to a vapour state without rise in temperature.

5–6 Shows the stage, during which rise in temperature of the substance may take place in a vapour state.

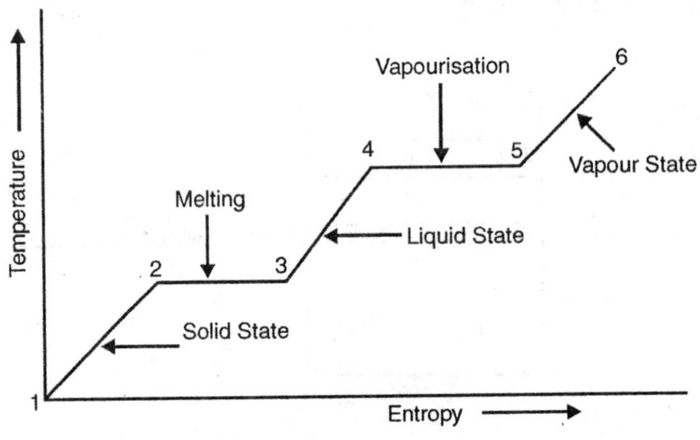

Fig. 9.2

If the hot vapour is cooled the curve will be exactly retraced with all changes taking place in reverse manner. If the weight W is increased and the experiment is repeated the same general behaviour is obtained but the temperature will be higher than that in the first case. If the weight is reduced still the behaviour will be the same but temperature will be lower than that in first case.

The above hypothetical experiment brings out certain important facts.

- When a solid melts the temperature remains constant and is always the same for a given substance at a given pressure. When a liquid freezes the temperature remains constant. This temperature is called the freezing point. The quantity of heat removed or added during the change of state is constant and is known as *latent heat of fusion.* The latent heat of fusion depends upon pressure.
- If heating of water is continued after total melting, temperature increases till it starts boiling. The amount of heat added to one kg of water from freezing point to boiling point is called *sensible heat.*
- When the liquid boils the temperature remains constant and is always the same for a given substance at a given pressure. If the pressure is increased, the temperature at which the change of state takes place also increases. Similarly when the vapour

condenses the temperature also remains constant. This temperature is called the *boiling point* or saturated temperature of the liquid. The definite quantity of heat removed or added during the change of state is called the *latent heat of vaporisation.* The latent heat of vaporisation depends upon pressure.

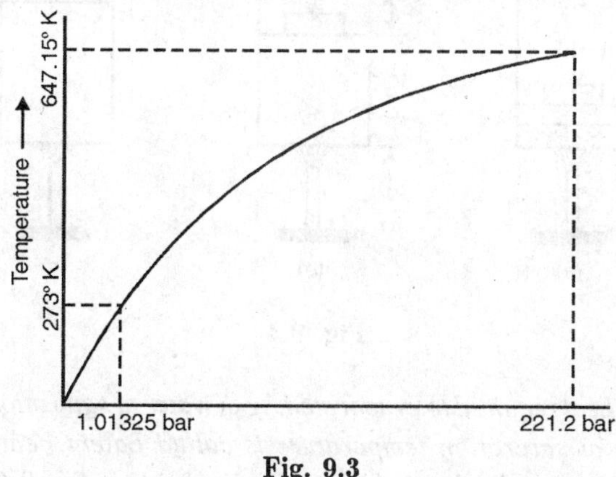

Fig. 9.3

The freezing point, the latent heat of fusion, the boiling point and the latent heat of vaporisation vary widely with different substances.

When the pressure is very low, an interesting phenomenon is noticed. A solid when heated directly transforms into a vapour without passing through an intermediate liquid state. Such a phenomenon is known as *sublimation.* A familiar example is the conversion of solid CO_2 (dry ice) directly into vapour. Figure 10.3 shows a temperature pressure curve.

In practice, steam is generated in a device called steam boiler.

9.3 GENERATION OF STEAM AT A GIVEN PRESSURE

Consider one kg of water at 0°C. Also let us decide upon the pressure under which one kg of water is to be heated. Consider a cylinder with a frictionless piston loaded such that the pressure P under consideration is acting on water. See Fig. 9.1.

Heat is added to the water so that its temperature increases from 0°C to t_{sat} (saturation temperature) at which the water boils at pressure P. Volume of water increases to V_f. This is shown in Fig. 9.4(a).

Heat supplied or added to 1 kg water to raise its temperature from 0°C to saturation temperature t_{sat} is called sensible heat because it can be detected by the sense of touch and produces a rise in temperature to be seen on a thermometer. Generally this sensible heat is denoted by h_f.

Note. V_f means volume of fluid which is taken as liquid.

Heating is continued after the water reaches saturation temperature. Evaporation of water takes place and steam is generated at pressure P. This heating is continued till all the water is evaporated and the temperature of the steam is saturation temperature. V_g or V_{sat} denotes the volume of steam at this condition. The external work done during evaporation is given by $W = P(V_{sat} - V_f)$ or $p(V_g - V_f)$ with proper units.

Because the heat during this stage of evaporation cannot be recorded by a rise in temperature on the thermometer, it is called Latent heat (hidden heat).

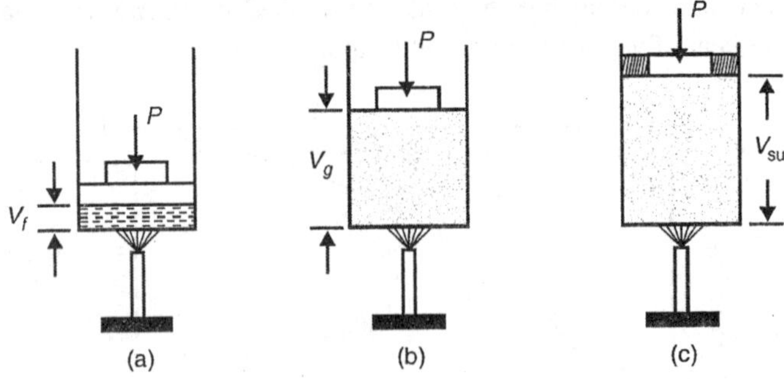

Fig. 9.4

The amount of heat required to evaporate 1 kg of water at saturation temperature to steam or vapour at saturation temperature is called Latent heat or enthalpy of evaporation and is generally denoted by h_{fg}. This is shown in Fig. 9.4(b).

Once all the water is evaporated to saturated steam, if heat is added further, the steam behaves approximately as an ideal gas and the heating of saturated steam increases its temperature. *The steam whose temperature is greater than saturation temperature at the given pressure, is called a superheated steam*; and let its temperature be denoted by t_{sup}. Its volume will also increase to V_{sup}.

The amount of heat required to raise the temperature of 1 kg steam from t_{sat} *to* t_{sup} *is called superheat* and the temperature difference called *degrees of superheat.*

$$\therefore \quad \text{Superheat} = (\text{Sp. heat of steam}) \times (\text{Rise in temperature})$$

$$= C_{pv}(t_{sup} - t_{sat})$$

Degree of superheat $= t_{sup} - t_{sat}$

The value of C_{pv} varies from 2.1 to 2.3 kJ/kg-K.

Similarly, $$\frac{T_{sup}}{T_{sat}} = \frac{V_{sup}}{V_{sat}}$$

This is according to Charles's law of $p =$ Const. (volumes are proportional to absolute temperatures)

$$\therefore \quad V_{sup} = V_{sat} \times \frac{T_{sup}}{T_{sat}}$$

This is shown in Fig. 9.4(c).

Note. V_{sat} and V_g also means volume of gas which is in this case, steam

$$h_f = \text{sensible heat} = Cp_w(t_{sat} - 0)\,\text{kJ/kg}$$

where Cp_w = specific heat of water which is taken as 4.187 kJ/kg-K if not given.

$$h_g = h_f + h_{fg} \text{ for 1 kg of dry and saturated steam.}$$

Total heat or enthalpy of superheated steam

$$h = h_g + Cp_v(t_{sup} - t_{sat})$$

9.4 CONDITIONS OF STEAM

Steam may occur in any one of the following three conditions:

(a) Saturated steam, which may be either dry or wet
(b) Superheated steam
(c) Supersaturated steam

(a) Saturated Steam. Saturated steam is a vapour at the temperature corresponding to the boiling point of the liquid at the given pressure. A substance, which is in the vapour state in a confined space and is in contact with some of the same substance in liquid state is always at the same temperature as the liquid and is saturated vapour.

If saturated steam does not contain any water, it is known as dry and saturated steam. It contains just sufficient heat energy to maintain all the water in a gaseous state.

If saturated steam contains liquid particles, it is known as wet steam. It does not contains sufficient heat energy to maintain all water in a gaseous state.

If some of the heat energy is absorbed from the dry saturated steam, the steam becomes wet. In practice it is difficult to get an absolutely dry saturated steam if it is produce by boiling water, because some of water particles are carried away with the steam.

(b) Superheated Steam. If the temperature of the steam is greater than that of the boiling point or saturation temperature corresponding to the pressure of steam generation, the steam is known as superheated steam.

The amount of superheat in the steam is given in terms of the difference between its temperature and that of the saturated steam at the pressure of steam generation. If the temperature of the steam is 100 degrees higher than the saturation temperature corresponding to the pressure of steam generation, we say that the steam has 100 degrees of superheat.

(c) Supersaturated Steam. Supersaturated steam at a particular saturation pressure has temperature less and density greater than the corresponding values given in the steam tables. This condition of steam is obtained when it is cooled by its own expansion until it contains less heat energy than the saturated steam under the same conditions. This state of steam is varying, unstable and the steam soon returns to the saturation condition. Such a state of steam occurs in expansion in a nozzle.

9.5 PROPERTIES OF STEAM

The properties of steam are interrelated. If we know certain properties, the other properties may be found out or derived. For example, if the pressure of saturated steam is observed by the pressure gauge, its temperature can be found from the steam tables in which the results of the various experiments have been tabulated.

(a) Dryness Fraction of Saturated Steam. We have seen that steam in contact with water contains liquid particles in suspension. Thus the steam consists of dry saturated and water particles in suspension.

The dryness fraction of steam is defined as the ratio of the weight of dry steam in a certain quantity of steam to the weight of total wet steam. It is generally denoted by the letter x. If W is the weight of dry steam in the total steam W_t, then $x = \frac{W_s}{W_t}$.

The dryness fraction of a wet steam may also be defined as the amount of dry steam in unit amount of the steam.

When $x = 1$, the steam is said to be dry and saturated.

(b) Sensible heat (h_f): Refer Sec.12.3

(c) Latent heat or enthalpy of evaporation h_{fg} Refer Sec. 12.3

(d) Superheat and degrees of superheat: Refer Sec. 12.3

(e) Volume of wet steam: $V = xV_g$

(f) Volume of dry and saturated steam: $V = V_g$

(g) Volume of superheated steam: $V_{sup} = V_{sat} \times \frac{T_{sup}}{T_{sat}} = V_g \times \frac{T_{sup}}{T_{sat}}$

9.6 STEAM TABLES

Experimentally, accurately determined, values are tabulated in tables that are called as Steam Tables. These values form the basis for many calculations concerned with steam engineering. These Tables are to be used because vapours do not obey general gas laws. The values given in the Tables are for one kg of dry and saturated steam but these values can also be employed for wet steam calculations. Properties of dry and saturated steam and superheated steam at various pressures and temperatures for SI units are given at the end of this chapter.

In order to determine the properties of steam at some pressure between those given in Tables, we interpolate assuming the linear relation between these values. The method is very simple and at the same time accurate.

Steam Table: *Properties of Dry and Saturated Steam.*

Pressure bar	Sat. temp. °C	Volume m³ /kg		Enthalpy kJ/kg K			Entropy kJ/kg K		
		V_f	V_g	H_f	H_{fg}	H_g	S_f	S_{fg}	S_g
10.0	179.88	0.0011274	0.19429	762.61	2013.6	2776.2	2.1382	4.4443	6.5828
10.5	182.02	0.0011303	0.18545	772.03	2005.9	2778.0	2.1588	4.4071	6.5659

Method of linear interpolation to find, say, enthalpy at 10.2 bar pressure.

Difference between two values at 10 and 10.5 is 772.03 – 762.61= 009.42 kJ/kg.

This is the difference for 0.5 bar pressure.

∴ For 0.2 bar difference, value is $\left(\frac{0.2}{0.5}\times 9.42\right) = 3.768$ kJ/kg.

Now, this is to be added to the value at 10 bar.

(This addition should be done if the value is increasing with pressure. If the value is decreasing, it should be deducted. This is seen in case of volumes.)

∴ Enthalpy at 10.2 bar pressure

$$= 762.61 + 3.768$$

$$= 766.378 \text{ kJ/kg.}$$

9.7 INTERNAL ENERGY AND ENTROPY OF STEAM

9.7.1 Internal Energy of Steam

As in case of ideal gases, internal energy of steam is given by ($u = h - pV$).

∴ For wet steam $\quad u = (h_f + xh_{fg}) - p.xVg$

For dry and saturated steam $u = h_g - p.V_g$

and for superheated steam $u = [h_g + C_{pv}(t_{sup} - t_{sat})] - \left[p.V_g \cdot \frac{T_{sup}}{T_{sat}}\right]$

9.7.2 Entropy of Steam

The change of entropy during any reversible process is defined as the ratio of heat added or rejected to the absolute temperature at which the heat is added or rejected.

We can apply the definition for getting entropy of steam above 0°C.

Consider one kg of water at 0°C and pressure p. Let heat be added so that the temperature increases to saturation temperature corresponding to pressure p. Here the temperature changes. Let during the heating process, T be the temperature and an infinitesimal heat δQ is added and for the process 1–2; where the temperature changes from 0°C (273K) to T_{sat} (saturation temperature at p).

$$\therefore \quad \delta Q = Cp_w dT \text{ and } dS = \frac{\delta Q}{T}$$

$$\therefore \quad dS = \frac{Cp_w \cdot dT}{T}$$

∴ Change of entropy during sensible heating of water from 273 K to T_{sat},

$$\int_1^2 dS = S_2 - S_1 = \Delta S_{1-2} = Cp_w \int_{273}^{T_{sat}} \frac{dT}{T}$$

This in denoted by s_f and is given in Steam Tables.

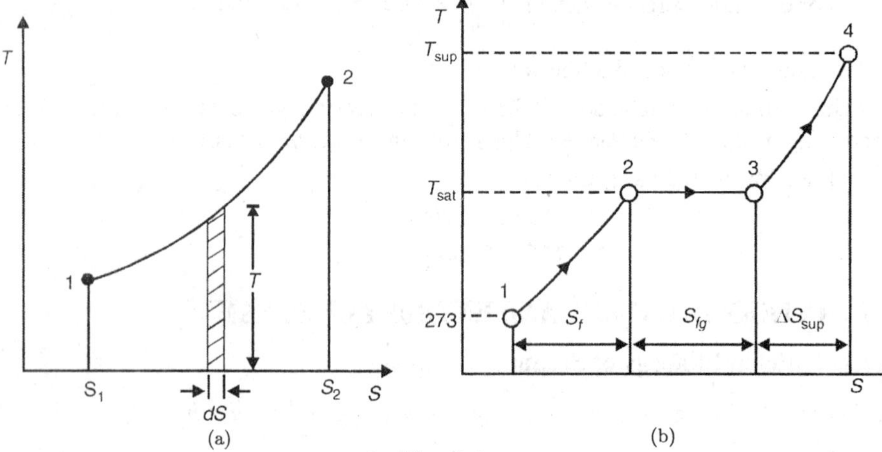

Fig. 9.5.

$$\therefore \qquad s_f = Cp_w \ln . \frac{T_{sat}}{273}$$

Water, at saturation temperature, if heated further, evaporation takes place. Here the heat added is latent heat or enthalpy of evaporation and temperature is saturation temperature; heat added is h_{fg}, and the temperature is constant.

$\therefore$ Change of entropy during evaporation 2–3 is given by

$$\Delta S_{2-3} = \frac{h_{fg}}{T_{sat}} = s_{fg} \text{ and is given in Steam Tables.}$$

and the entropy of dry and saturated steam is given by

$$\Delta S_{1-3} = \left[Cp_w \ln \frac{T_{sat}}{273} + \frac{h_{fg}}{T_{sat}} \right] = s_g .$$

It is given in Steam Tables.

Similarly, if the steam is further heated so that its temperature is T_{sup} (temperature of superheated steam), then the entropy change during superheating is given by

$$\Delta S_{3-4} = Cp_v \ln . \frac{T_{sup}}{T_{sat}}$$

$\therefore$ Total change of entropy of superheated steam or simply entropy of superheated steam is given by

$$\Delta S_{1-4} = S = Cp_w \ln . \frac{T_{sat}}{273} + \frac{h_{fg}}{T_{sat}} + Cp_v \ln . \frac{T_{sup}}{T_{sat}} .$$

$$= S_f + S_{fg} + Cp \ln . \frac{T_{sup}}{T_{sat}} .$$

$$= S_g + C_{pv} \ln \frac{T_{sup}}{T_{sat}}$$

If the steam is wet having a dryness fraction x, then entropy of steam is given by

$$S = S_f + x.S_{fg}.$$

Temperature – Entropy diagram (T–S diagram) is shown in Fig. 9.5.

9.8 PHASE DIAGRAM

Pure substance is a substance consisting only of a single molecular species such as He, H_2, H_2O or NH_3, phases of these substances being different namely solid, liquid and gaseous.

We have considered the phase change for the matter in Secs 9.2 and 9.3.

We will consider a pressure temperature co-ordinates to study the constant pressure process discussed in Secs 9.2 and 9.3.

Figure 9.3 shows such a diagram where the phase changes are shown. Such a diagram is called a phase diagram for pure substance.

For a particular pressure, the horizontal line *a, b, c, d, e* is drawn as shown in Fig. 9.6.

a = Solid state
b = Melting point of ice
c = Boiling or saturation temperature point, *C* does not apply to substance that expand on freezing
d = dry and saturation point
de = Superheated vapour
bd = Liquid vapour region

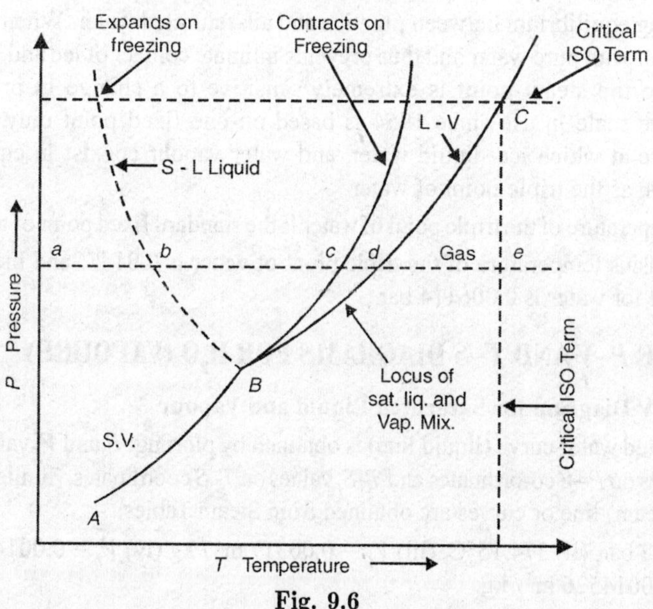

Fig. 9.6

This process is repeated at numerous pressure and the point analogous to b and d duly noted, say down to point B. A smooth curve through all the d points is the phase boundary between liquid and vapour; it is the locus of all points representing saturated mixtures. The smooth curve through all points b is the phase boundary for solid and liquid; points on this line represent all equilibrium mixtures of solid and liquid.

The states B and C in Fig. 9.6 are particularly significant. Point B, the junction of the three phase boundaries, is called the triple point because it represents all equilibrium mixtures of solid, liquid and vapour.

The critical point C where the curve BC ends is a state where liquid and vapour are indistinguishable. As this point is approached the latent heat of evaporation decreases. At C it is zero. The temperature T_C here is called the critical temperature and the corresponding pressure is the critical pressure (P_C). The specific volume at P_C and T_C is the critical volume.

Density of water and vapour at the point C is same.

For water, P_C = 221.2 bar

T_C = 374.15°C

V_C = 0.00317 m³/kg

9.9.1 Triple Point

We know that to determine the temperature of any system, the thermometer must be brought to thermal equilibrium with an arbitrarily chosen standard system in an easily reducible state. The temperature of the standard system in the chosen state is called the fixed point. Before 1954 there were two fixed points : (i) the ice point and (ii) steam point.

The use of two fixed points was found unsatisfactory, partly because of the difficulty of achieving equilibrium between pure ice and air-saturated water. When ice melts, it sounds itself with pure water and thus prevents intimate contact of ice and air saturated water. Also the steam point is extremely sensitive to a change in pressure. The temperature scale in use since 1954 is based on one fixed point only. This is the temperature at which ice, liquid water, and water vapour coexist in equilibrium, a state known as the triple point of water.

The temperature of the triple point of water is the standard fixed point of thermometry.

The Celsius temperature of the triple point of water is 0.01 °C and the pressure at triple point for water is 0.006114 bar.

9.9 THE P–V AND T–S DIAGRAMS FOR H_2O (VAPOURS)

9.9.1 P–V Diagram for Saturated Liquid and Vapour

The saturated water curve (liquid line) is obtained by plotting: P and V_f values form the steam tables on P–V co-ordinates and T–S_f values on T–S coordinates. Similarly saturated vapour (steam) line or curves are obtained from Steam Tables.

(i) 221.2 bar, (ii) 374.15°C, (iii) V_C = 0.00317 m³ / kg (iv) V_b = 0,0014526 m³ / kg (v) V_a = 0.0014526 m³ / kg.

Notice here that saturated vapour line flattens out towards the right, indicating that the volume increases at an increasing rate at low pressure.

In Fig. 9.7, *m* represents the state where $V_f = 0.0010434$ m^3/kg at 1 bar pressure. *C* is the critical point where pressure is 221.2 bar, saturation temperature is 374.15°C and volume is 0.317 m^3/kg.

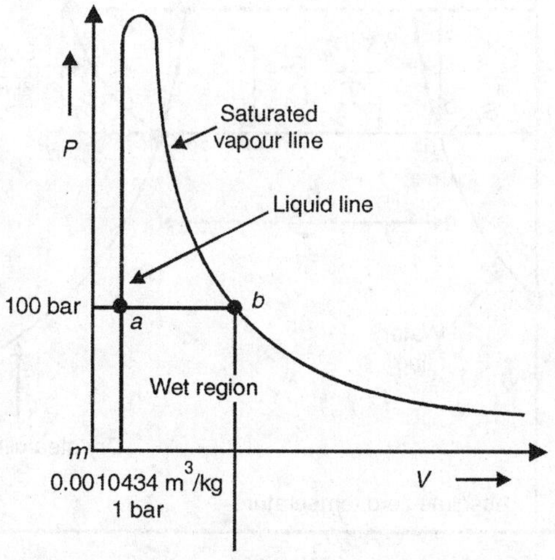

Fig. 9.7

This diagram is not used in practice but is useful in general to study the property relations and the vapour behaviour.

9.9.2 The Temperature–Entropy Diagram for Steam

It is possible to plot on a Temperature Entropy diagram, a chart which gives the entropy of water and steam at all temperatures and pressures. Such a chart is very convenient for solving problems containing an adiabatic (reversible) expansion of steam. A Temperature Entropy Chart (*T–S* Chart) is plotted in Fig. 9.8 with co-ordinates representing the absolute temperature and entropy. The units of entropy are reckoned from the freezing point of water.

9.9.3 Construction of T–S Diagram

Consider 1 kg steam at absolute pressure P_1, the absolute temperature of formation at this pressure will be T_1; the value of T_1 can be obtained from the Steam Tables.

Then $$S_{f1} = \log_e \frac{T_1}{273}$$

$$= ab \text{ on the diagram.}$$

Let this value of S_{f1} be represent by *ab* to a convenient scale, and plot this distance on *T–S* diagram of Fig. 9.8. Repeat this for other pressure of steam and draw a curve through the points obtained. This curve is known as the saturation liquid line or simply liquid line. Some times it is also called water line and will give the values of S_f at all temperatures and pressure.

Again consider the steam at pressure P_1. During evaporation the change of entropy of this steam will be

$$S_{fg} = \frac{h_{fg}}{T_1}$$

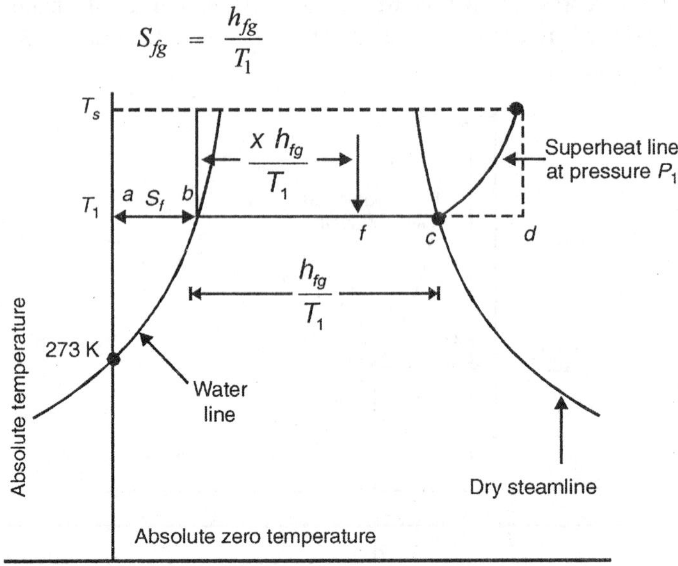

Fig. 9.8

Let this be represented by the distance *bc* and *be* plotted on the *T–S* chart. The point *C* will represent dry steam at pressure P_1. Repeat this for other pressures and temperatures and draw a curve through the points obtained. This curve is known as the dry steam line, and its horizontal distance from the vertical ordinate represents the total entropy of steam at any pressure.

If the steam at pressure P_1 had been wet, with a dryness fraction of *x*, the entropy of evaporation would be

$$\frac{x.h_{fg}}{T_1}$$

Let this be represented by *bf* and *be* plotted to the same scale. Then *af* is the total entropy of the wet steam, and the dryness fraction is the ratio of *bf* to *bc*.

$$\therefore \qquad x = \frac{bf}{bc}$$

Again, consider 1 kg of dry steam at condition *c*. Let steam be now superheated by increasing the absolute temperature from T_1 to T_S. Then, increase in entropy due to superheating is

$$= Cp_w \log_e \frac{T_s}{T_1}$$

$$= Cp_w \log_e \frac{T_{sup}}{T_{sat}}$$

Let this increase of entropy be represented by the horizontal distance *cd*. Draw a horizontal through T_S and a vertical through point *d* let these lines intersect at *e*. Then the point *e* represents the condition of the superheated steam at pressure P_1. Repeat this for other valves of T'_s the temperature T_1 remaining the same, as the pressure is constant.

Draw the curve *ce* through the points so obtained. Then this curve represents the entropy of superheated steam at pressure P_1, for all temperatures for superheat. Actually, the curve *ce* is a constant pressure line. Similar constant pressure lines can be drawn for superheated steam at other pressures.

It should be noticed that the area of the chart between the water line and the dry steam line represents wet steam. The area of the chart to the right of the dry steam represents superheated steam, and is known as the area of superheat, or superheat region.

Thus a complete *T–S* diagram for pressure P_1 is shown in Fig. 9.9.

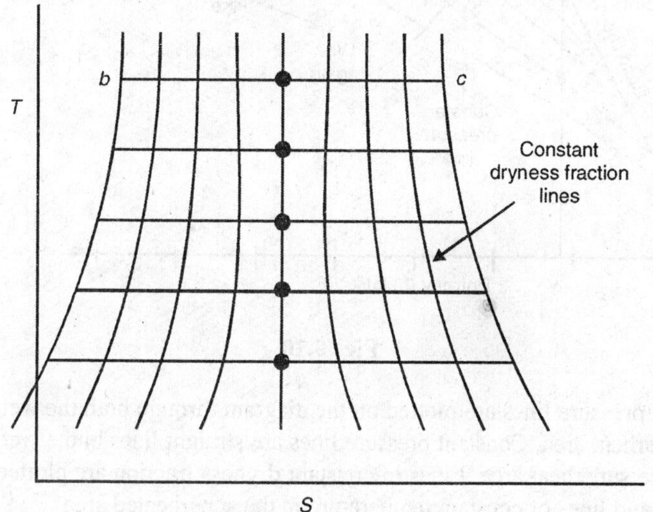

Fig. 9.9

In this diagram, the evaporation lines *bc* at all pressure can be divided into ten equal parts. Curves can then be drawn through each corresponding tenth part of these lines. These curves represent the dryness fraction of the steam at any temperature and pressure. These curves or lines are known as constant dryness fraction lines. Thus, any condition of steam can be represented by a corresponding point on the *T–S* diagram. Such a point is known as state point.

Dryness fraction lines are shown on *T–S* diagram of Fig. 9.9.

9.10 MOLLIER DIAGRAM OR TOTAL-HEAT-ENTROPY OR ENTHALPY-ENTROPY DIAGRAM (h–s DIAGRAM)

Another type of entropy diagram which is used by engineers is the Mollier diagram or Enthalpy – Entropy (*h–s*) diagram. This diagram represents the entropy and total heat or enthalpy of steam. It is shown in Fig. 9.10.

The base of the diagram represents the entropy of the steam and the vertical ordinate represents the total heat or enthalpy of steam. The lower portion of the diagram is the wet steam area, whilst the upper portion of the diagram represents the superheated condition of the steam.

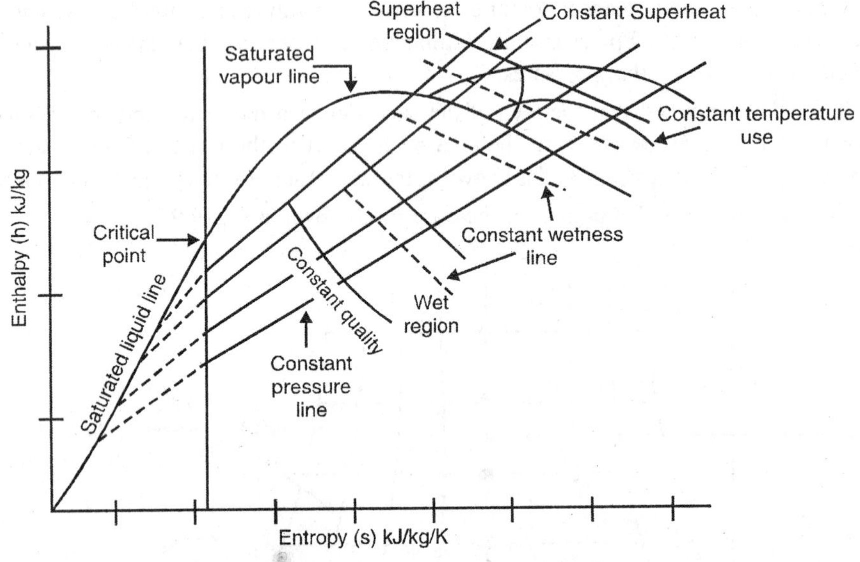

Fig. 9.10

Constant pressure lines are plotted on the diagram through both the wet steam area and the superheat area. Constant pressure lines are straight lines in the wet region and curved in the superheat area. Lines of constant dryness fraction are plotted in the wet steam area, and lines of constant temperature in the superheated area.

As the base of the diagram represents entropy of the steam, an adiabatic (isentropic) expansion will be represented by a vertical line.

Again a throttling expansion will be represented by a horizontal line on this diagram, as throttling process is a constant enthalpy (isenthalpic) or total heat process.

The main use of this diagram is for finding the drop in total heat or enthalpy of steam during an isentropic expansion. The final condition of the steam after expansion can also be read off from the diagram. The drop in enthalpy during an adiabatic or isentropic expansion. The final condition of the steam after expansion can also be read off from the diagram. The drop in enthalpy during an adiabatic or isentropic expansion of steam is known as isentropic enthalpy drop.

9.11 NON-FLOW AND STEADY FLOW PROCESSES

In the previous course of thermodynamics we have studied about the different thermodynamic systems. Here we will first revise to certain extent about the system and the process.

Thermodynamic system is classified as:

1. Closed or Non-Flow system
2. Open onflow system and
3. Isolated system

In closed or non-flow system the boundary allows the energy to flow in or flow out i.e. the energy crosses the boundary of the system. Secondly, the boundary does not allow the mass to cross the boundary of the system. When the energy crosses the boundary of the system, the boundary may or may not move. The process that will take place in this closed or Non-flow system is called *Non-Flow Process.*

The energy equation for this Non-Flow Process is, with usual relations, as

$$Q = \Delta U + W_{nf}$$

where Q = Heat exchange during the process

ΔU = Change of internal energy during the process

W_{nf} = Work of non-flow process

$$= \int_1^2 p \cdot dv$$

In Open or Flow system, both energy and mass will cross the boundary of the system. This open or flow system can further be classified as *Steady-Flow System* or *Unsteady-Flow-System.*

At this stage, we are required to consider steady flow system. In this *steady flow* system the rate of mass flow and energy flow into the system is equal to the rate of flow of mass and energy out of the system. i.e. $m_1 = m_2$, and $E_1 = E_2$. This means that there is no accumulation of mass as well as energy, within the system.

The process that takes place in a Steady-Flow-System is called *Steady Flow Process.*

Steady flow energy equation can be written as

$$Q = \Delta P + \Delta K + \Delta H + W_{sf}$$

where Q = Heat exchange during the process

ΔP = change in potential energy

ΔK = change in kinetic energy

ΔH = change in enthalpy

W_{sf} = work of steady flow process

$= h_1 - h_2$ for isentropic process when $\Delta K = 0$

and $h_1 - h_2 = K_1 - K_2$ for isentropic process and $W_{sf} = 0$

9.12 PROCESSES OF VAPOURS

The basic energy relations for the processes as defined for perfect gases also hold for vapours all previous equations in terms of the general symbols *W, Q, H, h, U, u, K, P* apply to any substance under the circumstances specified. The equations derived from the assumption of an ideal gas do not hold.

Remember that the areas on the *P–V* diagram under the curve at an internally reversible process represent $\int p.dv$, and that this area is the work of a non-flow process. The area behind the same curve is the $-\int v.dp$.

The vapour processes that are to be studied here are:

(i) Constant pressure process (Isobaric process)
(ii) Constant volume process (Isochoric process)
(iii) Reversible adiabatic process (Isentropic process)
(iv) Irreversible adiabatic process (Throttling process)
(v) Isothermal process
(vi) Polytropic process (Reversible)
(vii) Hyperbolic process

(i) Constant Pressure Process

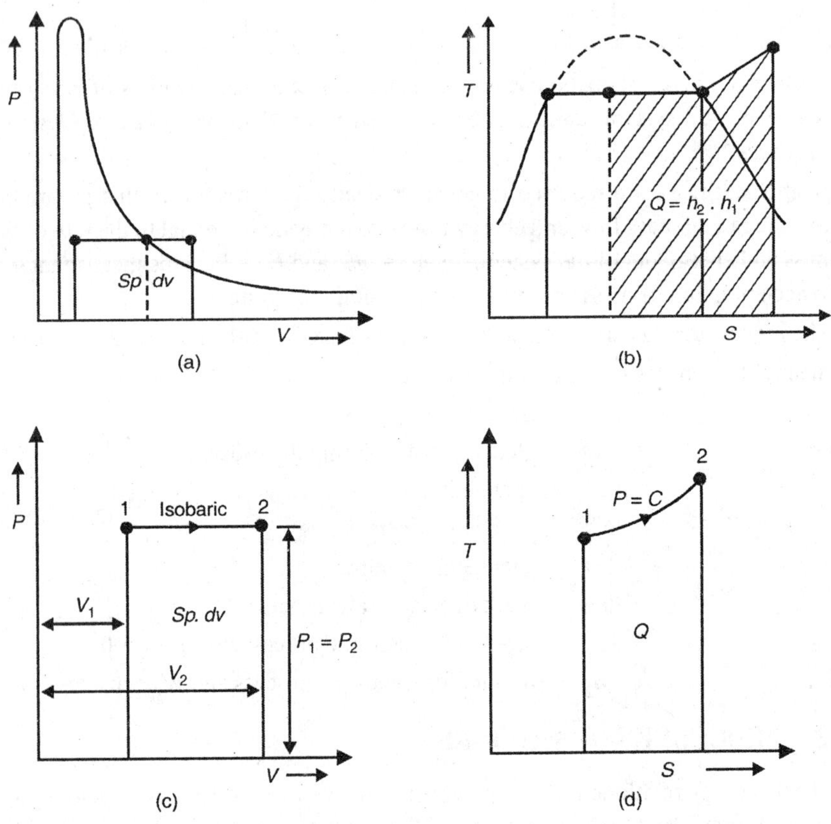

Fig. 9.11

A constant pressure, also called an isobaric process, is a change of state during which the pressure remains constant. On the P–V plane, (Fig. 9.11), the process is represented by a horizontal line, and on the T–S plane the process is represented by a horizontal line in the wet region and by a curve obtained from the indefinite integral of

$$ds = \frac{\delta q}{T}$$

In addition to defining relation, such as $p = c$, other quantities defining the limits of the process are needed. Suppose, in this case that point 1 is in the wet region and point 2 in the superheat region. With the pressure known, a point in the wet region is generally defined by giving its quality x or percentage moisture y. Similarly, a point 2 in the superheat region is generally, but not necessarily, defined by giving its temperature.

The work of reversible non-flow process at $p = c$ is

$$W_{nf} = \text{work of a non-flow system}$$

$$= \int p \cdot dV$$

$$= p(V_2 - V_1)$$

Also

$$Q = \text{Heat} = \Delta u + \int (p.dV)$$

$$= u_2 - u_1 + P_2V_2$$

$$= (P_2V_2 + u_2) - (u_1 + bV_1)$$

$$= (P_2V_2 + U_2) - (P_1V_1 + U_1)$$

$$= (h_2 - h_1)$$

(ii) Constant Volume Process

A constant volume process, also called an *isometric process* is a change of state during which the volume remains constant. This process is represented by a vertical line 1–2 in Fig. 9.12.

For non-flow, constant volume process we have

$$Q = \Delta U + \int p \cdot dV$$

$$Q = U_2 - U_1 \text{ Non-flow process } \int p.dv = 0$$

For a steady for isometric process, we have,

$$Q = \Delta U + \Delta W_f + \Delta K + \Delta P + W_{sf}$$

$$= \Delta P + \Delta K + \Delta H + W$$

For a simple for process, (like nozzle), $W_f = 0$ and assuming $\Delta P = 0$, we have

$$Q = \Delta K + \Delta H$$

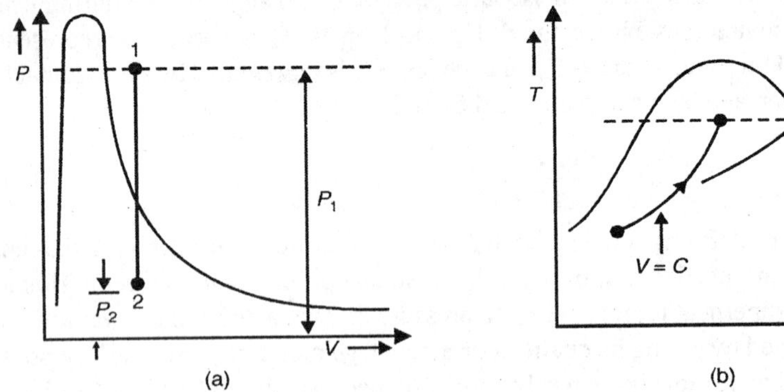

Fig. 9.12

For isentropic process, $Q = 0$

$$\therefore \quad O = \Delta K + \Delta H$$

or

$$\Delta H = \Delta K$$

$$H_2 - H_1 = K_1 - K_2 \text{ or } (H_1 - H_2) = (K_2 - K_1)$$

$$H = U - PV$$

(iii) Reversible Adiabatic Process (Isentropic Process)

Reversible adiabatic process is that process during which heat exchange is not taking place and $Q = 0$. This process is also called Isentropic process during which heat is zero and entropy change is also zero. i.e. $\Delta S = 0$.

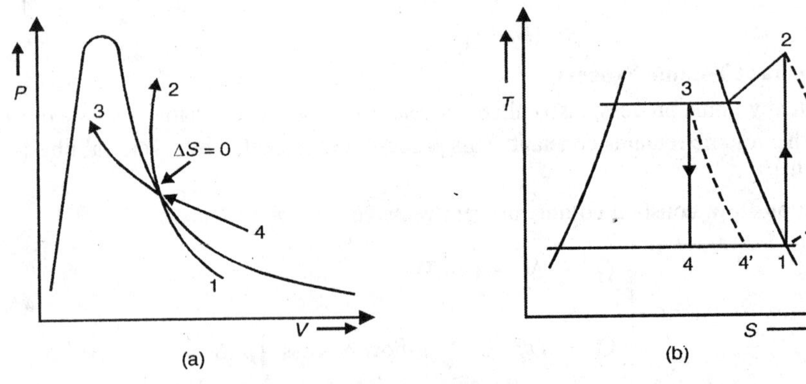

Fig. 9.13

This process is represented on PV plane, 1–2 Isentropic Compression, 3–4 Isentropic expansion and $T-S$ Plane of Fig. 9.13.

The isentropic and irreversible adiabatic can either be non-flow or steady flow. Let the second state along isentropic be designated along an irreversible adiabatic 2'.

Then, from the non-flow energy equations with $Q = 0$ for isentropic process, we write $Q = \Delta U + W_{nf}$

$\therefore\ 0 = \Delta U + W_{nf}$ or $W_{nf} = U_1 - U_2$ or $W = U_1 - U_2$ (Reversible) and $W = U_1 - U_2'$ (Irreversible)

For steady flow process we get, in general,

$$W = (h_3 - h_4) \text{ when } \Delta K = 0 \text{ (Reversible)}$$

or

$$W = h_3 - h_4 \text{ when } \Delta K = 0 \text{ (Irreversible)}$$

An example of steady flow to which this equation applies is a steam turbine. If the process reversible (Isentropic),

$S_1 = S_2 = (S_f + xS_{fg})_2$ (Reversible non-flow or steady flow)

In which it is presumed that state 2 is in the wet region.

(iv) Irreversible Adiabatic or Throttling Process

This term is generally reversed for an irreversible steady flow process, but the kinetic energies are not necessarily zero. This adiabatic process ($Q = 0$) occur from the extreme of no work being done, called a throttling process. This is a simple flow process with work $W = 0$. For an ideal gas, this throttling process is at constant temperature process, therefore when $W = 0$, $\Delta P = 0$ and $Q = 0$, the energy equation is

$$Q = \Delta P + \Delta K + \Delta h + W$$

$$O = O + \Delta K + \Delta H + O$$

or

$$h_1 + K_1 = h_2 + K_2$$

If velocity changes are negligible, then $K_1 = K_2$ and $h_1 = h_2$. This is a constant enthalpy process or *Isenthalpic Process.*

Thus the throttling process is defined by $h_1 - h_2$ and it is commonly used in steam power practice to determine the quantity of steam (dryness fraction of steam). Generally this process is shown by a dotted line on *T–S* plane as shown in Fig. 9.14 and by a vertical line on *h–s* diagram of Fig. 9.15.

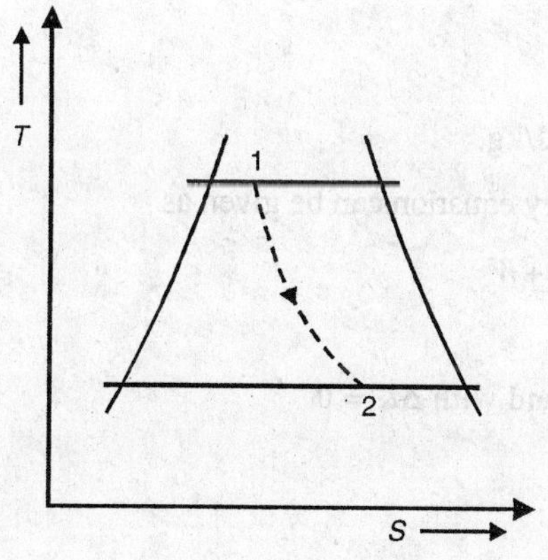

Fig. 9.14

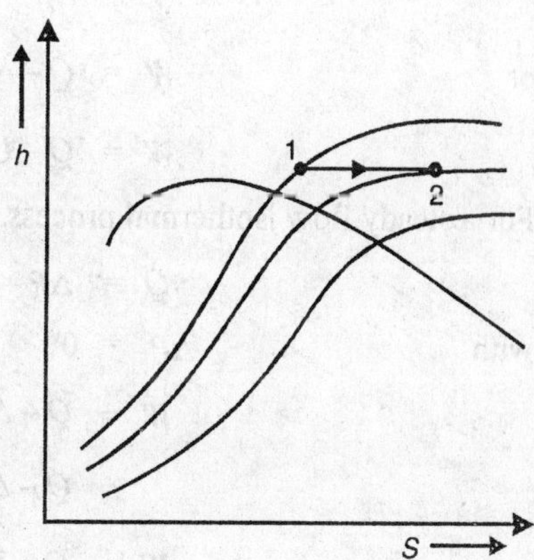

Fig. 9.15

(v) Isothermal Process

An Isothermal process is one carried out at constant temperature $T = C$. Unless stated otherwise, always it will be taken as a reversible process. For an ideal gas with $T = C$, we think immediately of Boyle's Law, PV = Constant or $P_1V_1 = P_2V_2$. For steam, an 'internally reversible' isothermal process from a point in the superheat region to a point in the wet region is shown in Fig. 9.16.

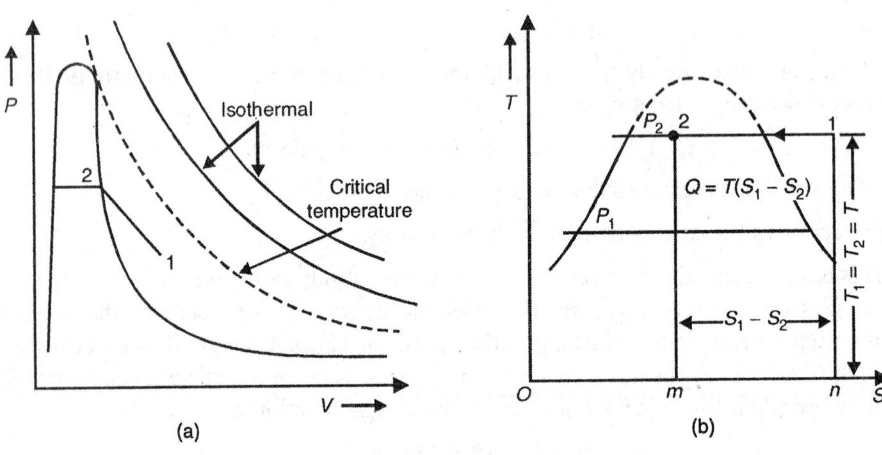

Fig. 9.16

The area m–2-1-n, $\int T.ds = T\int ds$, represents heat

$$\text{or} \qquad Q = T(S_2 - S_1)\text{kJ/kg}$$

where $(S_2 - S_1)$, being a negative term or number indicates that heat is passing from the substance. Then non-flow energy equation will be

$$Q = \Delta U + \int p.dv = \Delta U + W$$

$$\text{or} \qquad W = Q - \Delta U$$

$$W = Q - U_1 - U_2 \text{ kJ/kg.}$$

For a steady flow isothermal process, the energy equation can be given as

$$Q = \Delta P + \Delta K + \Delta h + W$$

$$\text{with} \qquad \Delta P = 0$$

$$W = Q - \Delta K - \Delta h \text{ and with } \Delta K = 0$$

$$= Q - \Delta h$$

$$W = Q + h_1 - h_2$$

(vi) Polytropic Process

A polytropic process is an internally reversible process, which conforms to the relation $pv^n = 0$ or $P_1V_1^n = P_2V_2^n$ which n is any constant.

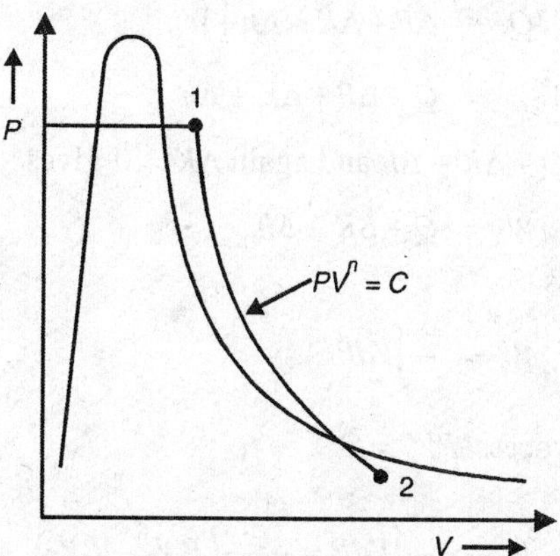

Fig. 9.17

If a process is known or assumed to be polytropic and if two state points are defined (say, pressures and volumes are known), the value of n can be found by use of logarithms.

$$\therefore \quad P_1V_1^n = P_2V_2^n \text{ or } \frac{P_1}{P_2} = \left(\frac{V_2}{V_1}\right)^2$$

or $$\log(P_1/P_2) = n.\log\left(\frac{V_2}{V_1}\right)$$

or $$n = \frac{\log_e(P_1/P_2)}{\log(V_2/V_1)}$$

Using $pv^n = C$, we get.

$$\int_1^2 p.dv = \frac{P_2V_2 - P_1V_1}{(1-n)} \frac{\text{N-m}}{\text{kg}}$$

Which is the work of non-flow process.

$$\therefore \quad W_{nf} = \frac{P_2V_2 - P_1V_1}{(1-n)}$$

The process is shown on P–V diagram as in Fig. 9.17.

Non-flow energy equation can be written as

$$Q = \Delta U + \int P \cdot dv$$

$$= u_2 - u_1 + \frac{P_2V_2 - P_1V_1}{1-n}$$

steady flow energy equation for the polytropic equation gives

$$Q = \Delta P + \Delta K + \Delta h + W_{sf}$$

or $$W_{sf} = Q - \Delta P + \Delta K + \Delta h$$

when $\Delta P = 0$, $W = Q - \Delta K - \Delta h$ and again $\Delta K = 0$ gives

$$W = Q - \Delta K - \Delta h.$$

Also for flow process

$$W = -\int VdP$$

and for polytropic process $PV^n = C$

$$W = -\int_1^2 VdP = \frac{n}{n-1}(P_2V_2 - P_1V_1)$$

Besides constant volume, constant pressure, constant temperature, adiabatic (reversible and irreversible) processes the following are the other important operations:

(i) Hyperbolic and (ii) Free expansion.

(i) Hyperbolic Process

When a gas or steam expands or is compressed in such a manner that the product of pressure and volume remains constant during the process of expansion or compression, the process is known as hyperbolic process because such a process on the *P–V* diagram will give a rectangular hyperbola. Hence a hyperbolic expansion or compression process follows the law,

$$\text{Pressure} \times \text{Volume} = \text{Constant}$$

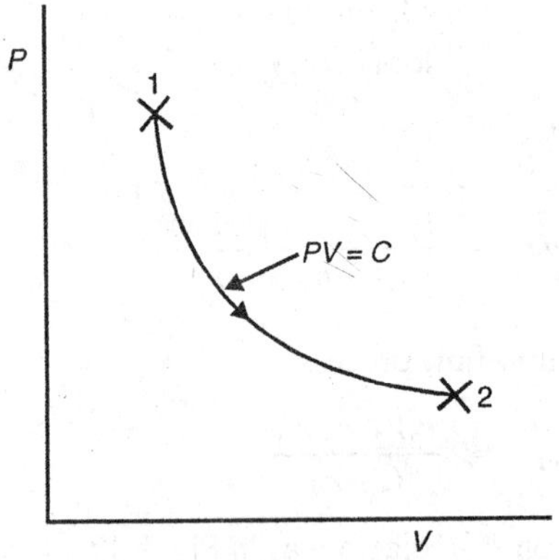

Fig. 9.18

$$\therefore \quad PV = \text{Constant}$$

or $$P_1V_1 = P_2V_2$$

For ideal gas, this process is similar to isothermal process

For such a process,

$$\int_1^2 P \cdot dV = \text{work done of non-flow process}$$

$$= P_1V_1 \log_e\left(\frac{P_1}{P_2} = \frac{V_1}{V_2}\right)$$

$$-V.dP = P_1V_1 \log_e\left(\frac{P_1}{P_2} = \frac{V_1}{V_2}\right)$$

For steam, we can find the volume of steam after the process. Thus

$$P_1V_1 = P_2V_2$$

$$\therefore \quad V_2 = \frac{P_1}{P_2} - V_1$$

If $V_2 > V_{g_2}$, the steam is superheated.

$V_2 < V_{g_2}'$, the steam is wet.

$V_2 = V_{g_2}'$

the steam is dry and saturated.

when $V_2 > V_{g_2}'$

$$V_2 = V_{sat_2} \times \frac{T_{sup2}}{T_{sat2}}$$

$\therefore \quad T_{sup}$ is determined.

$$\therefore \quad h_2 = \text{enthalpy} = h_{g_2}^T Cpv(T_{sup} - T_{sat})$$

and $\quad U_2 = h_2 - P_2V_2$

and hence Q heat can be calculated for non-flow process. Similarly,

when $\quad V_2 < V_{g_2}$

Then $\quad V_2 = x_2V_{g_2'}$

$$\therefore \quad x_2 = \frac{V_2}{V_{g_2}}$$

$$\therefore \quad h_2 = h_{f_2} + x_2hf_{g_2}$$

$$\therefore \qquad U_2 = h_2 - P_2V_2$$

and hence Q heat can be calculted. Again,

when
$$V_2 = V_{g_2}$$
$$x_2 = 1$$
$$h_2 = h_{g_2}$$
$$U_2 = h_{g_2} - P_2V_2$$

and thus Q can be calculated.

Note. Remember we will have to calculate h_1, U_1 etc. at the condition of steam before the process commences.

(ii) Free Expansion

When a fluid is allowed to expand suddenly into a vacuum chamber through an orifice of large dimensions, the free expansion of the fluid occurs. During this operation no heat has been supplied or rejected and no external work has been done; hence it follows that the total enthalpy of the fluid remains constant. This type of expansion is also called the constant total heat expansion.

After studying the different vapour processes, we can conclude that for any process, it will be required to determine or calculate

(i) Heat during the process (Q)

(ii) Change in Internal energy (ΔU)

(iii) Change in Enthalpy (ΔH)

(iv) Work done (W)

(v) $\int p.dV$

(vi) $-\int V.dP$

(vii) Change in entropy (ΔS)

9.13 STEAM DRYNESS FRACTION AND ITS DETERMINATION WITH STEAM CALORIMETERS

We have defined the dryness fraction of steam as the ratio of (dry steam) to (total steam containing it) and is represented by x.

$$\therefore \qquad x = \frac{\text{Dry steam (kg)}}{\text{Total steam (kg)}}$$

Dryness fraction of steam is determined with the help of steam calorimeters.

These calorimeters are as follows.

(i) Barrel Calorimeter

(ii) Separating Calorimeter

(iii) Throlling Calorimeter and

(iv) Combined separating and throtling Calorimeter

Any one combination these Calorimeters can be used for finding the dryness fraction of steam. Each of these methods will be dealt within term.

(i) Barrel Calorimeter. To get the approximately dryness fraction of steam, a very simple appe called Barrel Calorimeter is used. This consists of a wooden barrel containing cold water.

Steam is taken from the pipe through which it is flowing and is condensed in the cold water.

Temperatures of water before and after condensation are recorded. Knowing the quantities of steam and water, we can very easily calculate the dryness fraction steam. The calorimeter is shown in Fig. 9.19.

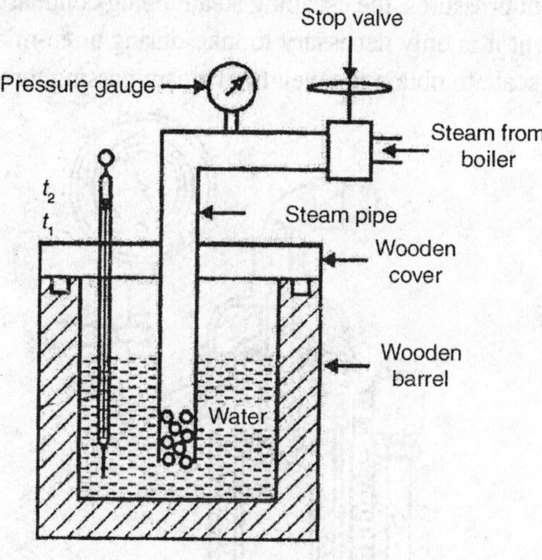

Fig. 9.19

Let, ms = weight of steam condensed

m_w = weight of cold water in barrel

C_{pw} = sp. heat of water

x = Dryness fraction of steam

h_{fg} = Latent heat of the steam at pressure of steam P

t_1 = Initial temperature of water before mixing steam

t_2 = Final temperature of steam after mixing of steam.

t = Saturation the temperature of steam at P.

Then we can find the dryness fraction of steam by using the equations given below.

[Heat lost by steam] = [Heat gained by coal water]

$$Ms\left[x.h_{fg} + C_{pw}(t - t_2)\right] = M_w C_{pw}(t_2 - t_1)$$

From this equation, we can find the dryness fraction of steam (x)

(ii) Separating Calorimeter. The separating calorimeter is quite simple in its action and it may be used for testing steam of practically any degree of wetness. Figure 9.20 shows a design of separating calorimeter by **Prof Carpenter.**

The sample of steam to be tested enters the calorimeter at D and passes down through the central passage DF into the perforated metal cup H. The current of steam then reverses and flows as shown by the arrows in the jacket J surrounding the inner chamber C. From the jacket the steam escapes into the atmosphere through the small orifice at the bottom and exhaust pipe E.

The water in the steam is thrown out in the cup H and collected in the chamber C. A gouge G has two scale on its dial, the inner one showing the pressure of the steam in the jacket while the outer one shows the weight of steam passing through the jacket for each pressure in ten minutes. The graduations on the outer scale are determined by experiment with steam of different pressures, the escaping steam being condensed and weighted, but in using the instrument it is only necessary to take during a ten-minute test the average reading on the outer scale to obtain the weight of steam passing through the jacket.

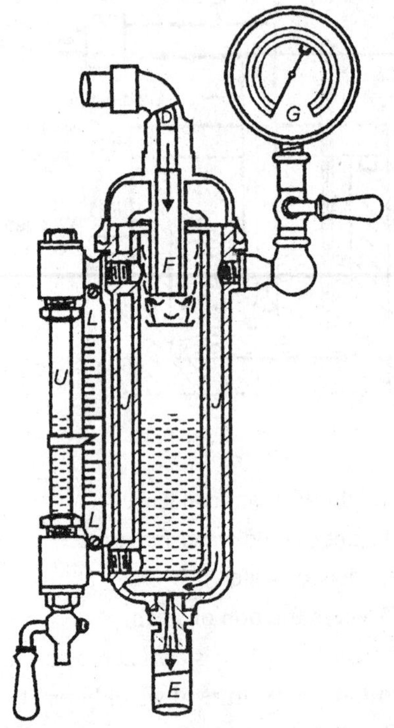

Fig. 9.20

The water separated from the steam is shown by means of the glass gauge U. A pointer P having a frictional grip of the glass tube is set to the water level at the beginning of the taste. The change in level of the separated water during a test is read off on the scale L. Which is graduated in fractions of mols.

Let W be the weight of steam passing through the jacket during a test and W is the weight of water separated in the same time, then x, the dryness fraction of the steam tested is $x = \frac{W}{W + W}$.

The weight of steam passing through the jacket may also be readily determined by passing the escaping steam into a bucket of water where it is condensed, the increase in the weight of the bucket and its contents giving the value of W. In that case gauge G is not required, but it used its reading may be taken as a check.

(iii) Throttling Calorimeter. The throttling calorimeter was first introduced by professor Peabody. Figure 9.21 shows Professor Carpenter's modification of the Peabody throttling calorimeter.

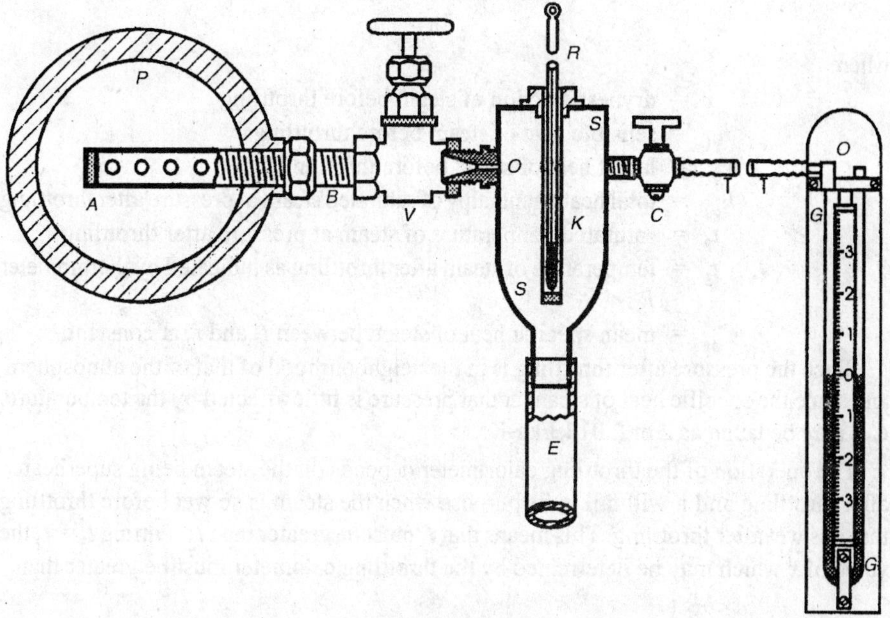

Fig. 9.21

P is the pipe carrying the steam to be tested. AB is the sampling tube through which the sample of steam flows to the vessels S which is the principle part of the calometer. V is the stop valve. The sample of steam enters the vessel S through a small orifice O and expands to a pressure a little above that of atmosphere into which it escapes freely through the exhaust pipe E.

The pressure of the steam in S is measured by means of the manometer or siphon gauge G which is a glass U-tube containing mercury. G is connected to S by 2a rubber tube T. Steam to the gauge may be cut-off by means of the stop cock C. A deep pocket K, containing cylinder oil, takes the thermometer R which indicates the temperature of the steam in S. The vessel S, the stop valve V, and the exposed part of the sampling tube AB must be carefully lagged to prevent undue loss of heat by radiation.

In using this instrument four observations have to be made

(1) The pressure P_1 or temperature t_1 of the steam in the pipe P.

(2) The pressure of the steam in the vessel S as shown by the gauge G in inches/centimeters of mercury.

(3) Reading of thermometer R

(4) The height of the barometer in inches/centimeters of mercury.

The principle upon which the operation of the throttling calorimeter depends is that the total heat (enthalpy) in the steam after throttling is the same as the total heat (enthalpy) in the steam before throttling. This is given by the equation

$$h_{f1} + xh_{fg1} = h_{g_2} + Cp_v(t_3 - t_2)$$

$$\therefore \qquad x = \frac{h_{g_2} - h_{f1} + C_{pv}(t_3 - t_2)}{h_{fg1}}$$

where

x = dryness fraction of steam before throttling

h_{f1} = sensible heat of steam before throttling

h_{fg1} = latent heat of steam before throttling

h_{g2} = total heat or enthalpy of saturated steam at pressure after throttling

t_2 = saturated temperature of steam at pressure after throttling

t_3 = temperature of steam after throttling as indicated by thermometer R.

C_{pv} = mean specific heat of steam between t_3 and t_2 at constant.

Since the pressure after throttling is in the neighbourhood of that of the atmosphere, and since the specific heat of steam at that pressure is little affected by the temperature, C_{PV} may be taken as 2 or 2.01 kJ/kg-K.

The operation of the throttling calorimeter depends on the steam being superheated after throttling and it will fail in its purpose when the steam is so wet before throttling that it is wet after throttling. This means that t_3 must be greater than t_2. Putting $t_3 = t_2$ the value of x which may be determined by the throttling calometer must be greater than

$$= \frac{h_{g_2} - h_{f1}}{h_{fg1}}$$

9.14 LIMITATIONS OF CALORIMETERS

(a) Barrel Calorimeter. Basically, this calorimeter is used when we want to know the approximate dryness fraction of steam, or to have a rough idea about the dryness fraction. This calorimeter gives better results if the dryness fraction of steam is greater than 0.95.

(b) Separating Calorimeter. This calorimeter will give better results when the dryness fraction of steam to be determined is above 0.95.

(c) Throttling Calorimeter. Basically the condition of steam after throttling must be known so that the unknown x-dryness fraction before throttling can be calculated. For this purpose the steam after throttling must be separated and its temperature should be more than the saturated temperature of steam corresponding to the pressure after throttling . To get such a result dryness fraction before throttling should be more than 0.9.

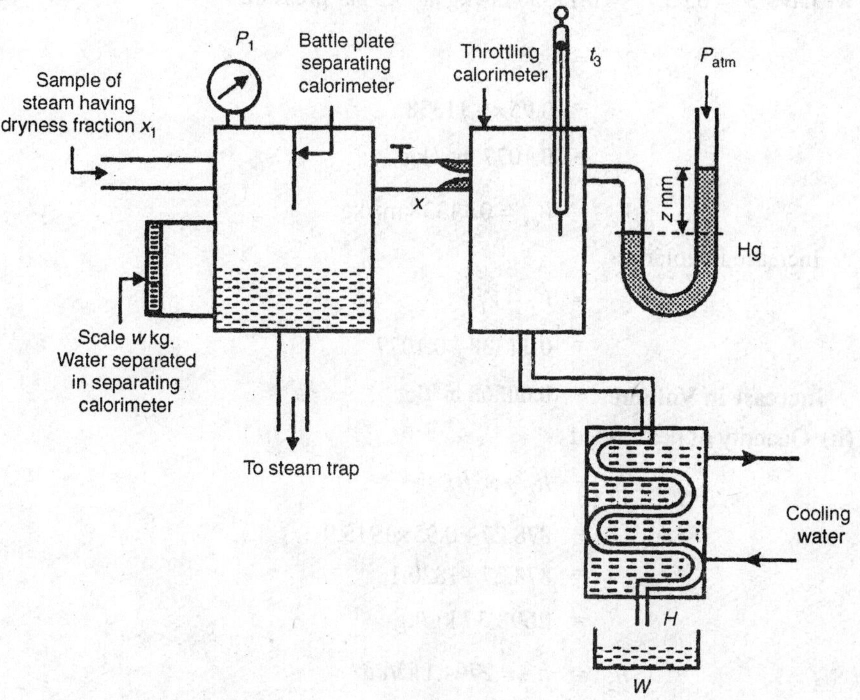

Fig. 9.22

If the dryness fraction of steam is less than 0.9, then the steam will not be superheated after throttling and the dryness before throttling can not be found.

In such a case, a combined separating and throttling calorimeter is used. In this, steam is first admitted to separating calorimeter where some portion of the water particles are removed and the quality of steam is improved to the required level for throttling calorimeter.

A Schematic diagram of a combined separating and throttling calorimeter is shown in Fig. 9.22 and now is self explanatory.

SOLVED EXAMPLES

(I) Constant Pressure Process

Example 9.1 One kg of steam at a pressure of 17.5 bar and dryness 0.95 is heated at constant pressure, until it is completely dry. Determine

(i) Increase in volume. (ii) Quantity of heat added. (iii) Change in entropy.

Solution

(i) Increase in volume

We have $x_1 = 0.95, V_{g_1} = 0.11338\ m^3/kg$ at 17.5 bar pressure

$$\therefore \quad V_1 = x_1 V_{g_1}$$

$$= 0.95 \times 0.11338$$

$$= 0.1077\ m^3/kg$$

$$V_2 = V_{g2} = 0.11338\ m^3/kg$$

$\therefore$ Increase in volume

$$= V_2 - V_1$$

$$= 0.11338 - 0.1077$$

Increase in Volume $= \mathbf{0.00568\ m^3/kg}$

(ii) Quantity of heat added :

$$h_1 = h_{f_1} + x_1 h_{fg_1}$$

$$= 878.27 + 0.95 \times 1915.9$$

$$= 878.27 + 1820.1$$

$$= 2698.37\ kg/kg$$

$$h_2 = h_{g_2} = 2794.1\ kJ/kg$$

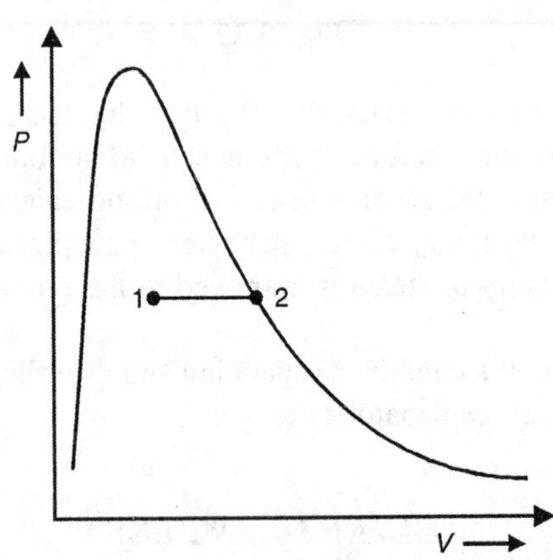

Fig. Ex. 9.1

$\therefore$ Heat added during constant pressure process is the difference in enthalpy $h_2 - h_1$

$$\therefore \quad Q = h_2 - h_1$$

$$= 2794.1 - 2698.37$$

$$\mathbf{Q = 95.73\ kJ/kg}$$

(iii) Change in entropy :

$$\Delta S = S_2 - S_1$$

$$= S_{g_2} - (s_{f_1} + x_1 S_{fg_1})$$

$$= (S_{f_2} + S_{fg_2}) - (Sf_1 + x_1 S_{fg_1})$$

But $S_{f_1} = S_{f_2}$ and $S_{fg_1} = S_{fg_2}$ for $P = C$

$$\therefore \quad \Delta S = S_{fg_1}(1 - x_1)$$

$$= 0.05 \times 4.0007$$

$$\mathbf{\Delta S = 0.200035\ kJ/kg\text{-}K.}$$

Example 9.2 A piston cylinder arrangement is filled with a wet steam of quality 0.8 at a pressure of 0.1 MPa. Energy is added at constant pressure till the temperature of steam rises to 300°C. Calculate the heat added and the work done.

Solution

0.1 MPa = 1 bar

Volume of one kg before heat is added

$$= X V_g$$

$$= 0.8 \times 1.6938$$

$$= 1.35504\ \text{m}^3/\text{kg}$$

Heat is added at constant pressure. The temperature of steam after heat added is while saturated temperature is 170.17°C

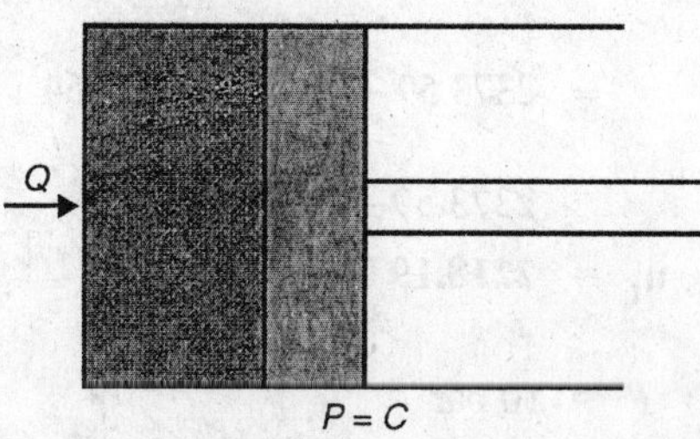

Fig. Ex. 9.2

∴ Steam is superheated.

$$V_2 = \text{Volume of 1 kg steam} = 2.639\ \text{m}^3/\text{kg}$$

$$\therefore \quad \text{Work done/kg} = P(V_2 - V_1)$$

$$= 100 \times 1(2.639 - 1.35504)$$

$$= 100 \times 1.28396$$

Work done/kg $=$ **128.396 kJ/kg**

For constant pressure process,

$$Q = \text{Heat} = h_2 - h_1 \qquad \text{(for non-flow process)}$$

$$= 3074.5 - \left(h_{f_1} + x_1 h_{fg_1}\right)$$

$$= 3074.5 - (417.5 + 0.8 \times 2257.9)$$

$$= 3074.5 - 2223.82$$

$$\mathbf{Q = 850.68\ kJ/kg}$$

Example 9.3 Steam having a dryness fraction 0.8 at a pressure of 10 bar changes its state so that the temperature of steam become 275°C. The pressure of steam remains the same. Calculate per kg of steam, the change in total heat, change in volume and change in internal energy.

Solution

Initial Condition:

$$\text{Enthalpy of steam/kg} = h_1 = h_{f_1} + x_1 h_{fg_1}$$

$$= 762.61 + 0.8 \times 2013.6$$

$$= 762.61 + 1610.88$$

$$= 2373.59 \text{ kJ/kg}$$

Volume

$$V_1 = x_1 v_{g1}$$

$$= 0.8 \times 0.19429$$

$$\mathbf{V_1 = 0.1554\ cu.m/kg}$$

Internal energy

$$u_1 = h_1 - P_1 V_1$$

$$= 2373.59 - \frac{10^5}{1000} \times 10 \times 0.1554$$

$$= 2373.59 - 155.4$$

$$\mathbf{u_1 = 2218.19\ kJ/kg}$$

Final condition

$$P = 10 \text{ bar}$$

$$\text{Temperature} = 275°\text{C}$$

$$\text{Enthalpy of 1 kg steam} = hg + Cp_V\left(T_{sup} - T_{sat}\right)$$

$$h_2 = \frac{3052.1 + 2943}{2} \qquad \text{(From Table Interpolation)}$$

$$= \frac{5995}{2}$$

$$= 2997.5 \text{ kJ/kg}$$

$$V_2 = \frac{0.2327 + 0.2580}{2} \quad \text{(Volume from Tables)}$$

$$= \frac{0.4907}{2}$$

$$= 0.24535 \text{ m}^3$$

$\therefore$ Internal energy $u_2 = h_2 - P_2 V_2$

$$= 2997.5 - 100 \times 0.24535 \times 10$$

$$= 2997.5 - 245.35$$

$$= 2752.15 \text{ kJ/kg}$$

$\therefore$ Change in volume / kg $= V_2 - V_1$

$$= 0.24535 - 0.1554$$

Change in volume/kg = 0.18995 m³/kg

Change in internal energy / kg

$$= 2752.15 - 2218.19$$

Change in internal energy/kg = 2533.96 kJ/kg

Example 9.4 (a) The internal energy of 4.5 kg of wet steam at 7 bar is 11388 kJ. What is the dryness fraction of steam?

(b) By how many heat units, the internal energy of the above steam shall increase if it is superheated at constant pressure to a temperature of 250°C ?

Assume specific heat of steam as 2.1 kJ/kg-K.

Solution

(a) Dryness Fraction

Let x be the dryness fraction of steam

$$\text{Internal energy/kg} = \frac{11388}{4.5}$$

$$= 2530.67 \text{ kJ}$$

$$= h - PV$$

$$= \left(h_f + xh_{fg}\right) - p \times xvg$$

$$\therefore \quad 2530.67 = 697.06 + x \times 2064.9 - 100 \times 7 \times x \times 0.27268$$

$$\therefore \quad 2530.67 - 697.06 = 2064.9\,x - 190.88\,x$$

$$\therefore \quad 1833.61 = 1874.02\,x$$

$$x = \frac{1833.61}{1874.02}$$

x = 0.9784

Dryness fraction of steam

(b) Increase in Internal energy

After superheating, enthalpy of 1 kg steam is given by

$$h = h_g + C_{pv}\left(t_{sup} - t_{sat}\right) \text{ This equation used } C_p \text{ is given}$$

$$h = 2762.0 + 2.1(250 - 164.96)$$

$$= 2762.0 + 2.1 \times 85.04$$

$$= 2762.0 + 178.584$$

$$= 2940.584 \text{ kJ/kg}$$

$$V = v_{sat} \times \frac{T_{sup}}{T_{sat}}$$

$$= 0.27268 \times \frac{250 + 273}{273 + 16496}$$

$$= \frac{0.27268 \times 523}{437.96}$$

$$= 0.32563 \text{ m}^3\text{/kg}$$

$$u = \text{Internal energy}$$

$$= h - PV$$

$$= 2940.584 - 100 \times 7 \times 0.32563$$

$$= 2940.584 - 227.94 = 2712.644 \text{ kJ/kg}$$

∴ Rise in heat units of Internal energy

$$= u_2 - u_2$$

$$= 2712.644 - 2530.67 = 181.974 \text{ kJ/kg}$$

Rise in heat units of Internal energy = 818.883 kJ

Example 9.5 1.5 cu.m of steam at a pressure 9 bar and having dryness fraction 0.95 loses 9420 kJ at constant pressure. Calculate dryness fraction of the steam at the end of heat loss.

Solution

We will have calculations on kg basis and therefore first we will find mass of steam in 1.5 cu.m at 9 bar and 0.95 dry

$$= x \,.\, V_g \quad \text{(Volume of 1 kg steam at 9 bar and 0.95 dry)}$$

$$= 0.95 \times 0.21481$$

$$= 0.2041 \text{ cu.m}$$

∴ Mass of steam in 1.5 cu.m

$$= \frac{\text{Total volume}}{\text{Sp. volume}}$$

$$= \frac{1.5}{0.2041} = 7.35 \text{ kg}$$

$$\therefore \quad \text{1 kg steam loses} = \frac{9420}{7.35} \text{ kJ} = 1281.63 \text{ kJ at constant pressure}$$

$$\text{Initial enthalpy/kg steam} = h_1 = h_{f1} + x_1 h_{fg1}$$

$$= 742.64 + 0.95 \times 2029.5$$

$$= 742.64 + 1924 = 2670.64 \text{ kJ}$$

$$\therefore \quad \text{Final enthalpy/kg} = 2670.64 - 1281.63$$

$$= 1389.01 \text{ kJ}$$

$$\therefore \quad 1389.01 = h_{f2} + x_2 h_{fg2} = 742.64 + x_2 h_{fg2}$$

$$= 742.64 + x_2 \times 2029.5$$

$$\therefore \quad x_2 = \frac{1389.01 - 742.64}{2029.5}$$

$$\mathbf{x_2 = 0.3185}$$

Dryness fraction of steam at the end of heat loss.

Example 9.6 A vessel contains 4 kg of steam at 10 bar and 220°C. Find the volume of the vessel. If the vessel is cooled till the steam pressure drops to 3 bar. Find the final condition of steam and the heat transfer during cooling.

Solution

Volume of 1 kg steam at 10 bar and 220°C (from superheated Steam Tables and by interpolation)

$$V_1 = 0.2059 + 0.4(0.2328 - 0.2059)$$

$$= 0.21666 \text{ m}^3/\text{kg}$$

$$\text{Total volume of vessel} = \text{Mass of steam} \times \text{Specific volume}$$

$$= 4 \times 0.21666$$

$$\textbf{Total volume of vessel} = \mathbf{0.86664\ m^3}$$

Volume of 1 kg steam at 3 bar after cooling

$$V_{g_2} = 0.60533 \text{ m}^3/\text{kg}$$

This is greater than that of V_1

∴ Steam is wet

$$\therefore \quad 0.60533\, x_2 = V_1 = 0.21666$$

$$x_2 = \frac{0.21666}{0.60533}$$

$$\mathbf{x_2 = 0.3579}$$

For constant volume process, we have

$$Q = \Delta u + \int P \cdot dv$$

$$= \Delta U \text{ as } \int P.dV = 0 \qquad \text{(Constant volume)}$$

$$= u_2 - u_1$$

$$u_2 = h_2 - P_2V_2$$

$$= h_{f_2} + x_2 h_{fg_2} - P_2 V_2$$

$$= (561.5 + 0.3579 \times 2163.2) - \frac{3 \times 10^5}{1000} \times 0.21666$$

$$= 1335.7 - 65$$

$$= 1270.7 \text{ kJ/kg}$$

$$u_1 = h_1 - p_1 v_1$$

$$= 2873.28 - \frac{10 \times 10^5}{1000} \times 0.21666$$

$$= 2873.28 - 216.66$$

$$= 2656.62 \text{ kJ/kg}$$

$$\therefore \quad Q = u_2 - u_1$$

$$= 1270.7 - 2656.62$$

$$= -1385.92 \text{ kJ/kg}$$

$\therefore$ Total heat transfer during cooling

$$= -4 \times 1385.92$$

Total heat transfer during cooling = –5543.68 kJ

– ve sign indicates that the heat is rejected by the system.

Example 9.7 1 kg of steam at initial condition of 6 bar and 0.2 dry is heated at constant volume until the pressure is 20 bar. Determine the final state of steam, heat added and change of energy.

Solution

Initial volume of 1 kg steam

$$V_1 = x_1 V_{g_1} = 0.2 \times 0.31546$$

$$= 0.063092 \text{ m}^3/\text{kg} = V_2$$

After cooling pressure is 20 bar

$$V_{g_2} = 0.09955 \text{ and is greated than } V_1$$

$\therefore$ Steam after cooling is wet.

$$\therefore \quad 0.063092 = x_2 \times 0.09955$$

$$x_2 = \frac{0.063092}{0.09955}$$

$$\mathbf{x_2 = 0.634}$$

Initial entropy per kg is given by

$$S_1 = S_{f_1} + x_1 S_{fg_1}$$

$$= 1.931 + 0.2 \times 4.827$$

$$= 1.931 + 0.9654$$

$$= 2.8964 \text{ kJ/kg°K}$$

Similarly $\quad S_2 = S_{f_2} + x_2 S_{fg_2}$

$$= 2.447 + 0.634 \times 3.890$$

$$= 2.447 + 2.4663$$

$$= 4.9133 \text{ kJ/kg/K}$$

$$\Delta S = S_2 - S_1$$

$$= \text{Change of entropy}$$

$$= 4.9133 - 2.8964$$

$$\mathbf{\Delta S = 2.0169 \text{ kJ/kg-K}}$$

Now $\quad u_1 = h_1 - P_1 V_1 = h_{f_1} + x_1 h_{fg_1} - P_1 V_1$

$$= 670.4 + 0.2 \times 2085.1 - 100 \times 6 \times 0.063092$$

$$= 1087.42 - 37.86$$

$$= 1049.56 \text{ kJ/kg} \qquad \text{(I)}$$

$$u_2 = h_2 - P_2 V_2$$

$$= h_{f_2} + x_2 h_{fg_2} - P_2 V_2$$

$$= 908.5 + 0.634 \times 1888.7 - 100 \times 20 \times 0.063092$$

$$= 2105.94 - 126.184$$

$$= 1979.756 \text{ kJ/kg} \qquad \text{(II)}$$

For constant volume process

$$Q = \Delta u + \int P.dV = \Delta u \text{ as } \int P.dV = 0$$

$$= u_2 - u_1 + 0$$

$$= 1979.756 - 1049.56$$

$$\mathbf{Q = 930.196\ kJ/kg}$$

Example 9.8 A closed vessel of 0.24 cu. m capacity, constant steam at a pressure of 11 bar and at a temperature of 200°C.

Find the quantity of steam in the vessel.

At what pressure in the vessel will the steam be dry and saturated if the vessel is cooled? The cooling is further continued till the temperature of steam in the vessel becomes 116.93°C. Find out the pressure and condition of the steam at the end of operation.

Solution

The capacity of the vesel is 0.24 m³. Saturation temperature of steam corresponding to pressure of 11 atm. bar is 184.07°C. Therefore steam is superheated steam can be found out by Charle's Law steam behaves as a perfect gas and as the pressure of steam is constant, the volume of superheated steam can be found out by the formula

$$v_{sup} = v_{sat} \times \frac{T_{sup}}{T_{sat}}$$

$$= \frac{0.17738 \times 473}{184.07 + 273} = \frac{0.17738 \times 473}{457.07}$$

$$= 0.1836\ m^3/kg$$

As the vessel is closed, the volume of steam/kg will always remain constant whatever the pressure will be

$$\therefore \text{Quantity of steam in vessel} = \frac{\text{Total volume}}{\text{Sp. volume}}$$

$$= \frac{0.24}{0.1836} = 1.307\ kg$$

When the steam is cooled, its quality will not be improved and the saturated steam having volume of steam = 0.1836 will have a pressure $\approx$ 10.5 bar.

Now the steam is further cooled so that its temprature is 116.93°C. The sturated pressure is 1.8 bar.

At 1.8 bar, volume of dry and saturated steam

$$= 0.97723\ m^3/kg$$

$$\therefore \quad 0.97723\,x = 0.1836$$

$$\therefore \quad \text{Condition of steam} = x = \frac{0.1836}{0.97723}$$

Condition of steam = 0.188

∴ Quantity steam in vessel = 1.307 kg

Pressure after cooling = 10.5 bar

Pressure after further cooling = 1.8 bar

Condition after further cooling D.F. = 0.188

Example 9.9. Calculate the total energy and internal energy of 1 kg of steam at a pressure of bar and dryness 0.95. One cu.m of this steam is cooled at constant volume until its pressure has fallen to 5 4 bar. Determine the heat abstracted from steam.

Solution

Total energy is enthalpy of steam

$$\therefore \quad h_1 = h_{f_1} + x_1 h_{fg_1}$$

$$= 742.64 + 0.95 \times 2029.5$$

$$= 742.64 + 1928.02$$

$$\mathbf{h_1 = 2670.66\ kcal/kg}$$

Internal energy kg $u_1 = h_1 - P_1V_1$

$$= 2670.66 - 100 \times 9 \times 0.95 \times 0.21481$$

$$= 2670.66 - 183.33$$

$$\mathbf{u_1 = 2497.00\ kJ/kg}$$

$$\text{Mass of one cu.m steam} = \frac{\text{Total volume}}{\text{Sp. volume}}$$

$$= \frac{1}{0.95 \times 0.21481} = 4.99 \text{ kg}$$

After cooling, the pressure is 5.4 bar at 5.4 bar, $V_{g_2} = 0.34846\ m^3/kg$

$$\text{Actual volume / kg} = xv_{g_2} = 0.95 \times 021481$$

$$\therefore \quad x_2 = \text{Dryness fraction of steam at 5.4 bar}$$

$$= \frac{0.95 \times 0.21481}{0.34846} = 0.586$$

$$\therefore \quad h_2 = h_{f_2} + x_2 h_{fg_2}$$

$$= 652.76 + 0.586 \times 2098.1$$

$$= 652.76 + 1229.5 = 1882.26 \text{ kJ/kg}$$

$$\therefore \quad u_2 = h_2 - p_2 v_2$$

$$= 1882.26 - 100 \times 5.4 \times 0.95 \times 0.21481$$
$$= 1882.26 - 110.2$$
$$= 1672.06$$

$$\therefore \quad u_2 - u_1 = \text{Change in internal energy} = \text{Heat}$$

$$\therefore \quad Q = 1672.03 - 2497.00 = -824.94 \text{ kJ/kg}$$

Total heat $Q = -499 \times 824.94$

$$\mathbf{Q = -4116.45 \text{ kJ}}$$

Example 9.10 A closed vessel of 0.2 cu.m capacity contains steam at a pressure of 8 bar and a temperature of 200°C.

(a) What is the weight of steam contained in the vessel ?

(b) The vessel is cooled and the steam becomes just dry and saturated. What will be the pressure of steam at this state? Estimate the specific entropy.

(c) The vessel is further cooled till the temperature drops to 160.8°C. Find the pressure and condition of steam at this state.

Solution

(a) Weight of Steam

Specific volume of superheated steam

Here pressure = 8 bar

$$V_{sup} = V_{sat} \times \frac{T_{sup}}{T_{sat}}$$

$$V_{sat} = 0.24026 \text{ m}^3$$

$$T_{sat} = 273 + 170.41$$
$$= 443.41°\text{K}$$

$$\therefore \quad V_{sup} = 0.24026 \times \frac{200 + 273}{170.41 + 273}$$

$$= \frac{0.24026 \times 473}{443.41} = 0.2563 \text{ m}^3/\text{kg}$$

$\therefore$ Weight of steam in the vessel

$$= \frac{\text{Actual volume}}{\text{Sp. volume}} = \frac{0.2}{0.2563}$$

Weight of steam in the vessel = 0.78 kg

(b) Pressure and Entropy after Cooling

After cooling, the steam becomes dry and saturated.

∴ Volume of 1 kg dry and saturated steam $= 0.2563\ \text{m}^3$

∴ **Saturation pressure of steam ≈ 7.5 bar**

(c) Pressure and Entropy

Saturation temperature of steam = 160.8°C

∴ Saturation pressure corresponding to this temperature ≈ 6.3 bar

= 6.3 bar

$$V_g = \frac{0.30585 + 0.29681}{2}$$

$$= 0.30133\ \text{m}^3$$

$$\therefore \quad 0.2563 = x \times 0.30133$$

$$x = \frac{0.2563}{0.30133}$$

$$\mathbf{x = 0.85}$$

= Condition of steam (final)

(II) Isothermal Process

Example 9.11 One kilogram of 0.3 dry steam at 10 bar is heated isothermally so as to make it dry and saturated. Calculate the mechanical work, change in internal energy, enthalpy and entropy during the process.

Solution

Note. Here the heating process is taking place in the wet region of *T–S* diagram. During this heating process, both temperature and pressure are constant

$$\therefore \quad P_1 = P_2 = 10\ \text{bar};$$

$$x_1 = 0.3 \text{ and } x_2 = 1$$

$$\therefore \quad V_1 = \text{Sp. volume of 1 kg steam before heating}$$

$$x_1 V_{g_1} = 0.3 \times 0.1946$$

$$= 0.05838\ \text{m}^3/\text{kg}$$

∴ Mechanical work available

$$W = P(V_2 - V_1)$$

$$= 100(0.1946 - 0.05838) \times 10$$

$$= 1000 \times 0.13622$$

$$\mathbf{w = 136.22\ kJ/kg}$$

$$h_1 = \text{Enthalpy of 1 kg}$$

$$= h_{f_1} + x_1 h_{fg_1}$$

$$= 762.2 + 0.3 \times 2013.8$$

$$= 1366.34\ \text{kJ/kg}$$

$$h_2 = h_{g_2} = 2776.1\ \text{kJ/kg}$$

$$h_2 - h_1 = \text{Change in enthalpy}$$

$$= 2776.1 - 1366.4$$

$$\mathbf{h_2 - h_1 = 1409.7\ kJ/kg}$$

$$u_1 = h_1 - P_1 V_1$$

$$= 1366.34 - 100 \times 10 \times 0.05838$$

$$= 1366.34 - 58.38$$

$$= 1307.96\ \text{kJ/kg}$$

$$u_2 = h_2 - P_2 V_2$$

$$= 2776.1 - 100 \times 10 \times 0.1946$$

$$= 2776.1 - 194.6$$

$$= 258.5\ \text{kJ/kg}$$

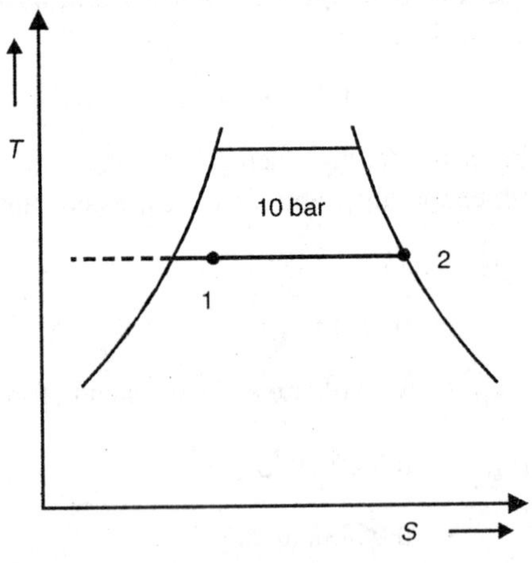

Fig. Ex. 9.11

$$\therefore \quad u_2 - u_1 = \text{Change in internal energy}$$

$$= 2581.5 - 1307.96$$

$$\mathbf{u_2 - u_1 = 1273.54\ kJ/kg}$$

Again
$$S_1 = S_{f_1} + x_1 S_{fg_1}$$
$$= 2.138 + 0.3 \times (6.582 - 2.138)$$
$$= 2.138 + 1.3332$$
$$= 3.4712\ \text{kJ/kg-K}$$
$$S_2 = S_{g_2} = 6.582\ \text{kJ/kg-K}$$
$$\therefore \quad S_2 - S_1 = \text{Change in entropy}$$
$$= 6.582 - 3.4712$$
$$\mathbf{S_2 - S_1 = 3.1108\ kJ/kg\text{-}K}$$

Example 9.12 2 kg steam at 7 bar and 190°C is compressed isothermally to 11 bar. Find the change in entropy, heat added or removed and final condition of steam. Assume $C_p = 2.3$ kJ/kg-K.

Solution

Here saturation temperature at 7 bar is 164.92°C and that at 11 bar 184.06°C. Actual temperature is 190°C. Therefore, steam is superheated before and after isothermal compression. The process is shown on *T–S* diagram.

∴ Final condition of steam is superheated steam with 5.94°C of superheat

$$S_1 = S_{g_1} + C_P \log_e \frac{T_{\text{sup}}}{T_{\text{sat}}}$$
$$= 6.705 + 2.3 \log_e \frac{273 + 190}{273 + 164.92}$$
$$= 6.705 + 2.3 \log_e \frac{463}{437.92}$$
$$= 6.705 + 0.1281$$
$$= 6.8330\ \text{kJ/kg-K}$$
$$S_2 = S_{g_2} + C_P \log_e \frac{T_{\text{sup}}}{T_{\text{sat}}}$$
$$= 6.55 + 2.3 \log_e \frac{463}{457.06}$$
$$= 6.55 + 0.0297 = 6.5797\ \text{kJ/kg-K}$$

∴ Change in entropy

$$= S_2 - S_1 = 6.5797 - 6.8331$$

$$= -0.2534 \text{ kJ/kg-K}$$

$$= -0.5068 \text{ kJ/K.}$$

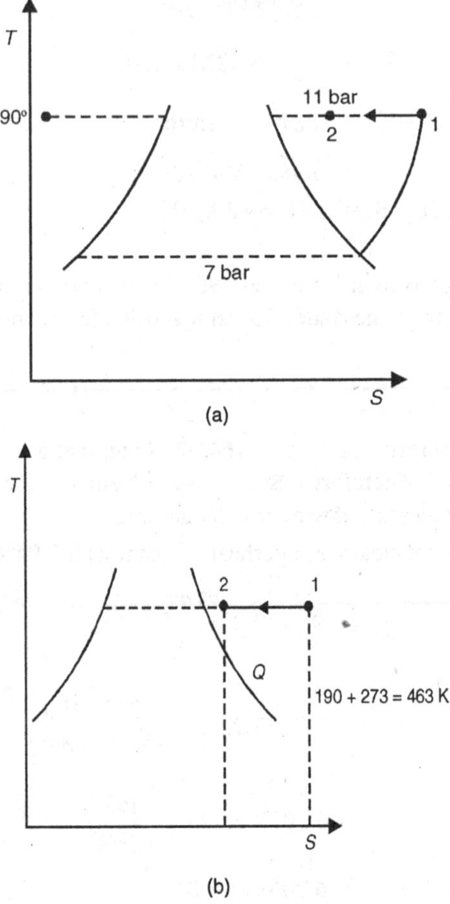

Fig. Ex. 9.12

Heat during the process

$$Q = T \times (S_2 - S_1)$$

$$= (273 + 190) \times (-0.50680)$$

$$= -463 \times 0.5068$$

$$\mathbf{Q = -234.65 \text{ kJ}}$$

– ve sign indicates that heat is rejected or removed.

Example 9.13 Dry and saturated steam at 6.5 bar is expanded isothermally to 0.5 bar in a special apparatus. Calculate the external work available from 2 m³ of steam initially taken, as well as the change in entropy during the process.

Solution

The process is shown in Fig. Ex. 10.13

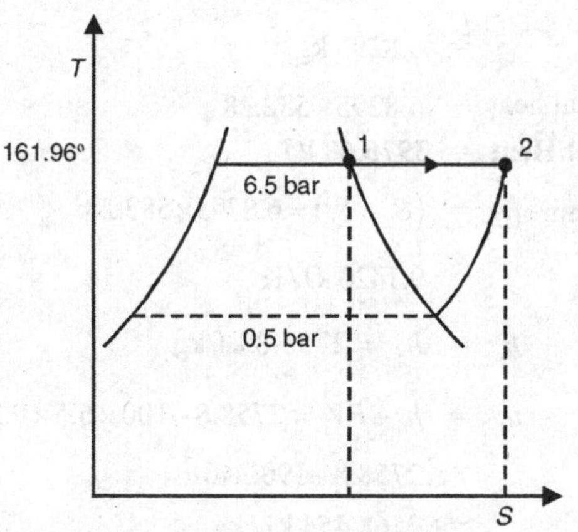

Fig. Ex. 9.13

$$T_1 = T_2 = 161.96 + 273 = 434.96 \text{ K}$$

$$S_1 = \text{Entropy before expansion}$$

$$= S_{g_1} = 6.730 \text{ kJ/kg-K.}$$

$$S_2 = S_{g_2} + C_P \ln \frac{T_{sup}}{T_{sat}}$$

$$= 7.597 + 2.3 \log_e \frac{434.96}{273 + 81.31}$$

$$= 7.597 + 2.3 \ln \frac{434.96}{354.31}$$

$$= 7.597 + 0.4717$$

$$= 8.0687 \text{ kJ/Kg-K}$$

$$\therefore \quad Q = \text{Heat during the expansion process}$$

$$= T_1 \times (S_2 - S_1)$$

$$= 434.96 \times (8.0687 - 6.730)$$

$$= 434.96 \times 1.3387$$

$$= 582.28 \text{ kJ/kg}$$

It is positive. i.e. heat is added.

$$\text{Initial volume of steam} = V_1 = 0.292847 \text{ m}^3/\text{kg}$$

$$\therefore \quad \text{Mass of steam processed} = \frac{2}{0.292847}$$

$$= 6.8295 \text{ kg}$$

$$\therefore \quad \text{Total heat} = 6.8295 \times 582.28$$

$$\textbf{Total Heat} = \textbf{3976.68 kJ}$$

$$\text{Also change in entropy} = (S_2 - S_1) = 6.8295 \times 582.28$$

$$= 9.1426 \text{ kJ/K}$$

$$h_1 = h_{g_1} = 2758.8 \text{ kJ/kg}$$

$$\therefore \quad u_1 = h_1 - P_1V_1 = 2758.8 - 100 \times 6.5 \times 0.29284$$

$$= 2758.8 - 190.346$$

$$= 2568.454 \text{ kJ/kg}$$

$$h_2 = h_{g_2} + C_P\left(t_{\text{sup}} - t_{\text{sat}}\right)$$

$$= 2645.3 + 2.3(161.96 - 81.31)$$

$$= 2645.3 + 185.495$$

$$= 2830.795 \text{ kJ/kg}$$

$$V_2 = V_{g_2} \times \frac{T_{\text{sup}}}{T_{\text{sat}}} = 3.24827 \times \frac{434.96}{354.31}$$

$$= 3.9877 \text{ m}^3/\text{kg}$$

$$\therefore \quad u_2 = h_2 - P_2V_2$$

$$= 2830.795 - 100 \times 0.5 \times 3.9877$$

$$= 2830.795 - 199.383$$

$$= 2631.412 \text{ kJ/kg}$$

$$\therefore \quad u_2 - u_1 = 2631.412 - 2568.454$$

$$= 62.958 \text{ kJ/kg}$$

$$\therefore \quad Q = \Delta u + \text{Work}$$

$$3976.68 = 6.8295 \times 62.958 + \text{Work}$$

$$\therefore \quad \text{Work} = 3976.68 - 429.97$$

$$\textbf{Work} = \textbf{3546.71 kJ}$$

Example 9.14 A quantity of dry and saturated steam occupies 0.2876 cu.m at 14 bar. Determine the final condition of steam if it is compressed until its volume is halved, if the compression is carried out isothermally. Determine the heat rejected during compression.

Solution

$$\text{Mass of steam} = \frac{\text{Total steam volume}}{\text{Sp. volume}}$$

$$= \frac{0.2876}{0.14072}$$

Consider 1 kg steam. Then we have

$$P_1 = 14 \text{ bar}; \quad V_1 = 0.14072 \text{ m}^3;$$

$$t_1 = 195.04°\text{C}; \quad V_2 = 0.07036 \text{ m}^3;$$

$$t_2 = 195.04°\text{C}$$

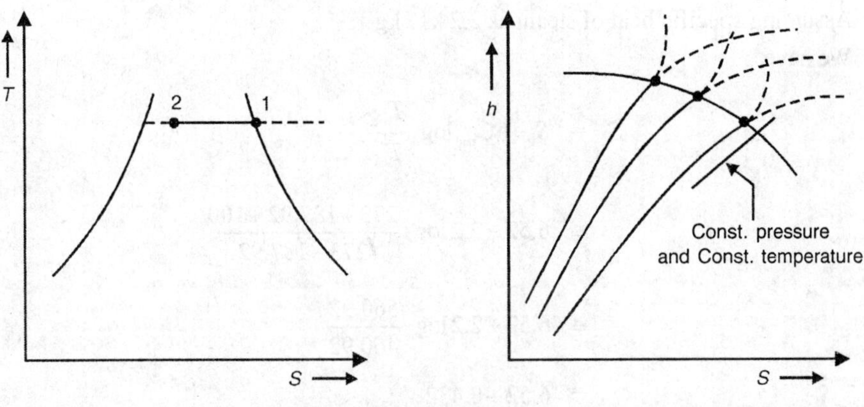

Fig. Ex. 9.14

As the temperature remains same and as the steam is initially dry and saturated, the compression of steam is at the same pressure because 195.04°C is going to be the saturation temperature at p_2 or $p_1 = p_2$.

Now as the volume is halved, the dryness fraction is given by

$$V_2 = x_2 V_{g_2}$$

$$\therefore \quad 0.07036 = x_2 \times 0.14072$$

$$\mathbf{x_2 = 0.5}$$

Compression takes place in the wet region and therefore that is also a constant pressure (in addition to constant temperature process) process.

$$\therefore \quad \text{Heat rejected} = (1 - x_2) h_{fg_2}$$

$$= (1-0.5)\times 1957.7$$
$$= 0.5\times 1957.7$$
$$= 978.85 \text{ kJ/kg}$$

$\therefore$ Total heat rejected $= 2.044\times 978.85$

Total heat rejected = 2000 kJ

(III) Isentrpoic Process

Example 9.15 Steam at 12 bar and 100°C superheat is expanded adiabatically and reversibly till becomes dry and saturated. Calculate the lower pressure: Calculate also the work done if the process is non-flow type.

Solution

For reversible adiabatically process, entropy before and after process is unchanged.

$$\therefore \quad S_1 = S_2$$

Assuming specific heat of steam as 2.2 kJ / kg-K

We have,

$$S_1 = S_{g_1} + C_{pv}\log_e \frac{T_{\text{sup}}}{T_{\text{sat}}}$$

$$= 6.52 + 2.2\log_e \frac{273+187.92+100}{273+187.92}$$

$$= 6.52 + 2.2\log_e \frac{560.92}{460.92}$$

$$= 6.52 + 0.432$$

$$= 6.952 \text{ kJ/kg-K}$$

This can also be found from superheat steam tables

$$\therefore \quad S_2 = 6.952 = S_{g_2}$$

as steam is dry and saturated after expansion.

$\therefore$ Pressure corresponding to S_{g_2} = 6.952 is approximately 3.4 bar.

$$\therefore \quad \mathbf{P = 3.4 \text{ bar}}$$

For non-flow process, $Q = \Delta u + \int P.\,dV$

$$\therefore \quad O = \Delta u + \int P.\,dV$$

$$\therefore \quad \text{Work done/kg} = -\Delta u = u_1 - u_2$$

$$h_1 = \text{Enthalpy of 1 kg steam before expansion}$$

$$= h_{g_1} + Cp_V \left(\text{Degree of superheated}\right)$$

$$= 2782.6 + 2.2 \times 100$$

$$= 2782.6 + 220$$

$$= 3002.6 \text{ kJ/kg}$$

$$V_1 = \text{Volume of 1 kg steam before expansion}$$

$$= V_{g_1} \times \frac{T_{\text{sup 1}}}{T_{\text{sat 1}}}$$

$$= 0.163446 \times \frac{5609.92}{460.92}$$

$$= 0.1989 \text{ m}^3\text{/kg}$$

$$\therefore \quad u_1 = \text{Internal energy of 1 kg steam before expansion}$$

$$= h_1 - P_1V_1$$

$$= 3002.6 - 100 \times 12 \times 0.1989$$

$$= 3002.6 - 238.68$$

$$= 2763 \text{ kJ/kg}$$

$$\therefore \quad h_2 = h_{g_2} = \text{Enthalpy of 1 kg steam after expansion}$$

$$= 2730.4 \text{ kJ/kg}$$

$$\therefore \quad u_2 = h_2 - P_2V_2$$

$$= 2730.4 - 100 \times 3.4 \times 0.539$$

$$= 2730.4 - 183.26$$

$$= 2547.14 \text{ kJ/kg}$$

$$\therefore \quad \text{Work done/kg} = u_1 - u_2$$

$$= 2763.92 - 2547.14$$

Work done/kg = 216.78 kJ/kg

Example 9.16 Steam flows into a steam nozzle with a pressure of 14 bar when it is dry and saturated. If the expansion is isentropic to 2 bar, find the final condition of steam and change in internal energy, taking $C_P = 2.2$ kJ/kg-K.

Find also the velocity of steam leaving the nozzle.

Solution

This is an example of steady flow system. Flow is isentropic.

$P_1 = 14$ bar, $P_2 = 2$ bar

$$S_1 = S_{g_1} = 6.465 \text{ kJ/kg-K}$$

$$= S_2 = S_{f_2} + x_2 S_{fg_2}$$

$$= 1.529 + x_2 (7.127 - 1.529)$$

$$= 1.529 + 5.598\, x_2$$

$\therefore \quad x_2 =$ Find dryness fraction of steam

$$= \frac{6.465 - 1.529}{5.598} = \frac{4.936}{5.598}$$

$$\mathbf{x_2 = 0.8817}$$

Now $\quad h_1 = h_{g_1} = 2787.9$ kJ/kg

$$h_2 = h_{f_2} + x_2 h_{fg_2}$$

$$= 504.5 + 0.8817 \times 2201.8$$

$$= 504.5 + 1941.33$$

$$= 2445.83 \text{ kJ/kg}$$

$$V_1 = V_{g_1} = 0.1409 \text{ m}^3\text{/kg}$$

$$V_2 = x_2 V_{g_2} = 0.8817 \times 0.8872$$

$$= 0.7822 \text{ m}^3\text{/kg}$$

$\therefore$

$$u_1 = h_1 - P_1 V_1$$

$$= 2787.9 - 100 \times 14 \times 0.1409$$

$$= 2787.9 - 197.26$$

$$= 2590.64 \text{ kJ / kg}$$

$$u_2 = h_2 - P_2 V_2$$

$$= 2445.83 - 100 \times 2 \times 0.7822$$

$$= 2445.83 - 156.44$$

$$= 2289.39 \text{ kJ / kg}$$

$\therefore \quad \Delta u = u_2 - u_1 =$ change in internal energy

$$= 2289.39 - 2590.64$$

$$\Delta \mathbf{u} = \mathbf{-301.25\ kJ/kg}$$

Assuming change in potential energy and initial velocity, we have

$$Q = \Delta P + \Delta K + \Delta H + W_f$$

$$O = O + \Delta K + \Delta H + O$$

$$\therefore \quad \Delta K = -\Delta H = H_1 - H_2$$

$$\frac{C_2^2 - C_1^2}{2000} = H_1 - H_2$$

$$= 2787.9 - 2445.83$$

$$= 342.07$$

$$C_2 = \sqrt{2000 \times 342.07}$$

$$\mathbf{C_2} = \mathbf{827.13\ m/s}$$

$$= \text{Velocity at exit}$$

Example 9.17 Steam at 15 bar and 250°C expands reversibly and adiabatically to 3 bar in a steam turbine, find the condition of steam after expansion and the work done per kg steam admitted to the turbine. Assume ΔP and ΔK to be zero.

Solution

This is an other example of steady flow system. Here $P_1 = 3$ bar, $T_1 = 523$ K

$$h_1 = 2923.8\ \text{kJ/kg}$$

$$S_1 = 6.711\ \text{kJ/kg-K} = S_2$$

$$\therefore \quad 6.711 = s_{f_2} + x_2 s_{fg_2}$$

$$\therefore \quad 6.711 = 1.671 + x_2(6.991 - 1.671)$$

$$x_2 = \frac{6.711 - 1.671}{5.320}$$

$$= \frac{5.040}{5.320}$$

$$\mathbf{x_2} = \mathbf{0.9474}$$

$$= \text{Final condition of steam}$$

$$\therefore \quad h_2 = \text{Enthalpy of steam after expansion}$$

$$= h_{f_2} + x_2 f_{fg_2}$$

$$= 561.3 + 0.9474 \times 2163.3$$

$$= 561.3 + 2049.5$$

$$= 2610.8\ \text{kJ/kg}$$

For steady flow system, we have

$$Q = \Delta P + \Delta K + \Delta H + W_{sf}$$

$$O = O + O + \Delta H + W_f$$

$$\therefore \quad W_{sf} = \text{work of steady flow system}$$

$$= h_1 - h_2$$

$$= 2923.8 - 2610.8$$

$$\mathbf{W_{sf} = 313.0\ kJ/kg}$$

Example 9.18 Find the internal energy of steam 0.9 dry at 16 bar reckoned from water at 0° C. If the steam is expanded isentropically in a cylinder to 0.9 bar, find the final internal energy of the steam.

Solution

$$h_1 = h_{f_1} + x_1 f_{fg_1}$$

$$= 858.3 + 0.9 \times 1933.4$$

$$= 858.3 + 1740.06$$

$$= 2598.36 \text{ kJ/kg}$$

$$v_1 = \text{Initial volume of steam}$$

$$= X_1 V_{g_1} = 0.9 \times 0.123843$$

$$= 0.1115 \text{ m}^3\text{/kg}$$

Neglecting volume of water at 0°C,

$$u_1 = h_1 - P_1 V_1 = 2598.36 - 100 \times 16 \times 0.1115$$

$$= 2598.36 - 178.33$$

$$\mathbf{u_1 = 2420.03\ kJ/kg\text{-}K}$$

$$= \text{Internal energy reckoned from water at 0°C}$$

Now $$S_1 = s_{f_1} + x_1 s_{fg_1} = 2.343 + 0.9(6.417 - 2.343)$$

$$= 2.343 + 3.6666$$

$$= 6.0096 \text{ kJ/kg-K}$$

$$S_2 = S_{f_2} + x_2 S_{fg_2} = 1.269 + x_2 (7.396 - 1.269)$$

$$= 1.269 + x_2 \times 6.127$$

$$\therefore \quad S_1 = S_2 \text{ for isentropic process}$$

$$6.0096 = 1.269 + 6.127$$

$$x_2 = \frac{6.0096 - 1.2690}{6.127} = \frac{4.7406}{6.127}$$

$$= 0.7737$$

$$\therefore \quad h_2 = h_{f_2} + x_2 h_{fg_2} = 405.1 + 0.7737 \times 2265.8$$

$$= 405.1 + 1753.1 = 2158.2 \text{ kJ/kg}$$

$$V_2 = x_2 v_{g_2} = 0.7737 \times 1.87135$$

$$= 1.4479 \text{ m}^3\text{/kg}$$

$$\therefore \quad u_2 = \text{Final internal energy above 8°C}$$

$$= h_2 - P_2 V_2$$

$$= 2158.2 - 100 \times 0.9 \times 1.4479$$

$$= 2158.2 - 130.31$$

$$\mathbf{u_2 = 2027.89 \text{ kJ/kg}}$$

Example 9.19 Calculate the index of adiabatic expansion if steam expanded adiabatically from 6 bar and 200°C to 1 bar has dryness fraction 0.93 at the end of expansion.

Take $C_p = 2.3$ kJ/kg-K ; $p_1 = 6$ bar; $p_2 = 1$ bar; $x_2 = 0.93$

Solution

$$V_1 = V_{g_1} \times \frac{T_{sup}}{T_{sat}}$$

$$= 0.3159 \times \frac{200 \times 273}{158.81 + 273} = \frac{0.3159 \times 473}{431.81}$$

$$= 0.346 \text{ m}^3\text{/kg}$$

$$V_2 = x_2 V_{g_2}$$

$$= 0.93 \times 1.6949$$

$$= 1.5762 \text{ m}^3\text{/kg}$$

Let n be the index of adiabatic expansion.

$$\therefore \quad P_1 V_1^n = P_2 V_2^n$$

or

$$\left(\frac{V_2}{V_1}\right)^n = \frac{P_2}{V_2}$$

or $$n \ln \frac{V_2}{V_1} = \ln \frac{p_1}{p_2}$$

$$N = \frac{\ln P_1 / P_2}{\ln V_2 / V_1} = \frac{\ln \frac{6}{1}}{\ln . \frac{1.5762}{0.346}}$$

$$= \frac{1.7918}{1.5163}$$

$$\mathbf{n = 1.1817}$$

Example 9.20 1 kg of steam at 100 bar and 375° C expands isentropically upto 38 bar. At the end of expansion steam is dry and saturated. For both non flow and steady flow process with $\Delta K = 0$, find the work done.

Data: $p_1 = 100$ bar; $t_1 = 375°$ C; $p_2 = 38$ bar; $x_2 = 1$.

Solution

From superheated steam tables

$$V_1 = \frac{0.02242 + 0.2641}{2} = \frac{0.04883}{2} = 0.024415$$

$$h_1 = \frac{2925.8 + 3099.9}{2} = \frac{6025.7}{2} = 3012.85 \text{ m}^3/\text{kg}$$

$$\mathbf{h_1 = 3012.85}$$

$$V_2 = V_{g_2} = 0.052438 \text{ cu.m/kg}$$

$$h_2 = 2801.1 \text{ kJ/kg}$$

Non-flow process :

$$u_1 = h_1 - P_1 V_1 = 3012.85 - 100 \times 0.024415$$

$$= 3012.85 - 244.15$$

$$= 2768.70 \text{ kJ/kg}$$

$$u_2 = h_2 - P_2 V_2 = 2801.1 - 100 \times 0.052438$$

$$= 2801.1 - 199.26$$

$$= 2601.84 \text{ kJ/kg}$$

$\therefore$ For non-flow process $= Q = \Delta u + \text{Work}$

$$0 = (u_2 - u_1) + w_{nf}$$

$$\therefore \quad W_{nf} = \text{Work done} = u_1 - u_2$$

$$= 2768.70 + 2601.84$$

$$W_{nf} = 166.86 \text{ kJ/kg}$$

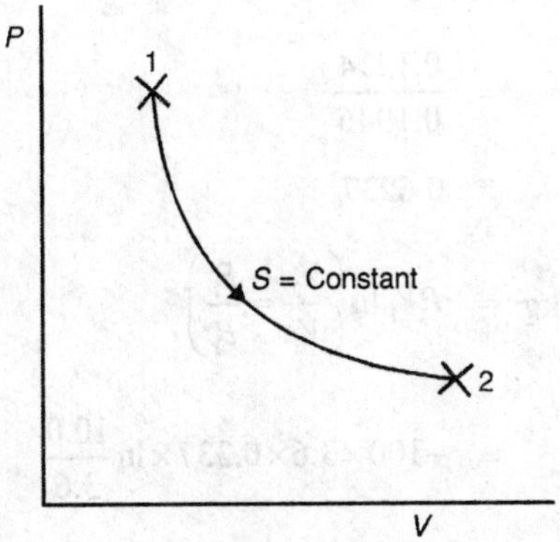

Fig. Ex. 9.20

For steady flow process

with $\Delta K = 0$

We have $\text{Work} = -\Delta H = h_1 - h_2$

$$= 3012.85 - 2801.1$$

$$\textbf{Work} = \textbf{277.75 kJ/kg}$$

(IV) Hyperbolic Process

Example 9.21 Steam at 3.6 bar and 0.66 dry is compressed hyperbolically to 10 bar. Determine the heat removed from 2 m³ of steam.

Solution

$P_1 = 3.6$ bar; $x_1 = 0.66$

$$\therefore \quad V_1 = x_1 V_{g_1} = 0.66 \times 0.5108$$

$$= 0.3371 \text{ m}^3\text{/kg.}$$

$$P_2 = 10 \text{ bar}$$

Law of expansion is $PV = C$

$$\therefore \quad P_1 V_1 = P_2 V_2$$

$$V_2 = V_1 \times \frac{P_1}{P_2} = 0.337 \times \frac{3.6}{10}$$

$$= 0.1214 \text{ m}^3/\text{kg}$$

This is less than V_{g_2}. Therefore the steam is wet after compression.

$$\therefore \quad 0.1214 = x_2 \times 0.1946$$

$$x_2 = \frac{0.1214}{0.1946}$$

$$= 0.6237$$

$$\therefore \quad \text{Work done/kg} = P_1V_1 \ln\left(\frac{V_2}{V_1} = \frac{P_1}{P_1}\right)$$

$$= -100 \times 3.6 \times 0.337 \times \ln\frac{10.0}{3.6}$$

$$= -367.8 \text{ kJ/kg}$$

Now

$$h_1 = h_{f_1} + x_1 h_{fg_1}$$

$$= 588.4 + 0.66 \times 2144.5$$

$$= 2003.77 \text{ kJ/kg}$$

$$\therefore \quad u_1 = h_1 - P_1 V_1$$

$$= 2003.77 - 100 \times 3.6 \times 0.3371$$

$$= 2125.126 \text{ kJ/kg}$$

Similarly

$$h_2 = h_{f_2} + x_2 h_{fg_2}$$

$$= 762.2 + 0.6237 \times 2013.8$$

$$= 2018.2 \text{ kJ/kg}$$

$$\therefore \quad u_2 = h_2 - P_2V_2$$

$$= 2018.2 - 121.356$$

$$= 1896.844 \text{ kJ/kg}$$

$$\therefore \quad u_2 - u_1 = 1896.844 - 2125.126$$

$$= -228.282 \text{ kJ/kg}$$

$$\therefore \quad \text{Heat } Q = \Delta u + \int P \cdot dV$$

$$= -228.282 - 367.8$$

$$\textbf{Heat Q} = \mathbf{-596.082 \text{ kJ/kg}}$$

$$= \text{Heat removed.}$$

$$= -596.082 \times \frac{2}{0.3371} = -3536.53 \text{ kJ}$$

Example 9.22 Steam at 8 bar and 175°C is throttled down to 4 bar and then expanded hyperbolically to 1.2 bar. Determine

(i) change in entropy during throttling

(ii) change in enthalpy during expansion and

(iii) work done

Solution

(i) Change in entropy during throttling : (1 – 2)

$$h_1 = \text{Enthalpy before throttling}$$

$$= h_{g_1} + C_p\left(t_{sup} - t_{sat}\right)$$

$$= 2676.3 + 2.3\left(175 - 170.38\right)$$

$$= 2767.3 + 10.626$$

$$= 2777.923 \text{ kJ/kg}$$

$$S_1 = S_{g_1} + C_p \ln\frac{T_{sup}}{T_{sat}} = 6.659 + 2.3\ln\frac{175+273}{170.38+273}$$

$$= 6.659 + 0.0238$$

$$= 6.6828 \text{ kJ/kg-K}$$

For throttling process,

$$h_1 = h_2 > h_{g_2}$$

$$\therefore \quad 2777.926 = 2737 + 2.3\left(t_{sup_2}\right) - 143.6$$

$$\therefore \quad t_{sup_2} = \frac{2777.962 - 2737.7}{2.3} + 143.6$$

$$= \frac{40.226}{2.3} + 143.6$$

$$= 17.49 + 143.6 = 161.09°\text{C}$$

$$\therefore \quad S_2 = S_{g_2} + C_p \ln\frac{T_{sup}}{T_{sat}}$$

$$= 6.894 + 2.3 \ln \frac{273 + 161.09}{273 + 143.6}$$

$$= 6.894 + 2.3 \ln \frac{434.09}{416.6}$$

$$= 6.894 + 0.0946$$

$$= 6.9886 \text{ kJ/kg}$$

$\therefore$ Change in entropy during throttling

$$= S_2 - S_1 = 6.9886 - 6.6828$$

$$= 0.3058 \text{ kJ/kg-K}$$

(i) Change in enthalpy during expansion (2 – 3)

$$h_2 = 2777.926 \text{ kJ/kg}$$

$$V_2 = V_{g_2} \times \frac{T_{sup}}{T_{sat}}$$

$$= 0.462626 \times \frac{434.09}{416.6}$$

$$\mathbf{V_2 = 0.482\ m^3/kg}$$

$P_2 = 4$ bar; $P_3 = 1.2$ bar

$$\therefore \quad P_2 V_2 = P_3 V_3$$

$$\therefore \quad V_3 = V_2 \times \frac{P_2}{P_3} = \frac{0.482 \times 4}{1.2}$$

$$= 1.607 \text{ m}^3/\text{kg}$$

$$> v_{g_3}$$

$\therefore$ Steam is superheated after expansion

$$\therefore \quad 1.607 = 1.43016 \times \frac{T_{sup_3}}{104.77 + 273}$$

$$\therefore \quad T_{sup_3} = \frac{1.607}{1.43016} \times \frac{377.77}{1}$$

$$= 424.436 \text{ K}$$

$$\therefore \quad h_{g_3} = h_{g_3} + C_P \left(T_{sup3} - T_{sat3}\right)$$

$$= 2684.9 + 2.3(424.436 - 377.77)$$

$$= 2684.9 + 2.3 \times 46.666$$

$$= 2684.9 + 107.33$$

$$= 2792.23 \text{ kJ/kg}$$

∴ Change in enthalpy during expansion

$$h_3 - h_2 = 2792.23 - 2777.926$$

$$\mathbf{h_3 - h_2 = 14.304 \text{ kJ/kg}}$$

(iii) Work done

$$\text{Work done} = P_2 V_2 \ln\left(\frac{V_3}{V_2} = \frac{P_2}{P_3}\right)$$

$$= 100 \times 4 \times 0.482 \ln \frac{4}{1.2}$$

$$\textbf{Work done} = \mathbf{232.126 \text{ kJ/kg}}$$

Example 9.23 The steam in the steam engine cylinder expands according to the hyperbolic law from 8.5 bar and 0.85 dry to 0.7 bar.

Find the final condition of steam and work done per kilogram assuming $C_p = 2.01$ kJ/kg-K.

Solution

We have $P_1 = 8.5$ bar; $P_2 = 0.7$ bar; $x_1 = 0.85$ dry

$$\therefore \quad V_1 = x_1 V_{g_1} = 0.85 \times 0.227037$$

$$= 0.193 \text{ m}^3/\text{kg}.$$

$$\therefore \quad P_1 V_1 = P_2 V_2$$

$$V_2 = \frac{P_1}{P_2} \cdot V_1 = \frac{8.5}{0.7} \times 0.193$$

$$= 2.3433 \text{ m}^3/\text{kg}$$

$$< V_{g_2}$$

∴ The steam is wet

$$\therefore \quad 2.3433 = x_2 \times 2.36795$$

$$\therefore \quad x_2 = \frac{2.3433}{2.36795}$$

$$\mathbf{x_2} = \mathbf{0.9896}$$

$$\text{Work done/kg} = P_1 V_1 \log_e \left(\frac{V_2}{V_1} = \frac{P_1}{P_2} \right)$$

$$= 100 \times 8.5 \times 0.193 \ln \frac{8.5}{0.7}$$

$$\textbf{Work done/kg} = \mathbf{409.59\ kJ/kg}$$

Example 9.24 10 kg of steam at 8 bar and 0.9 dry expanded in a steam engine cylinder hyperbolically to 0.4 bar. Find the final condition of steam and the external work done during the process.

Solution

$$P_1 = 8 \text{ bar}; \quad P_2 = 0.4 \text{ b ar};$$

$$V_1 = xV_{g_1} = 0.9 \times 0.2405 = 0.21645 \text{ m}^3/\text{kg}$$

$$\therefore \quad P_1 V_1 = P_2 V_2$$

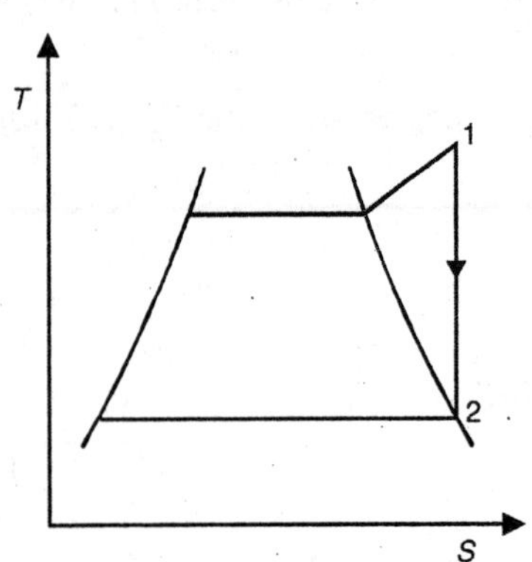

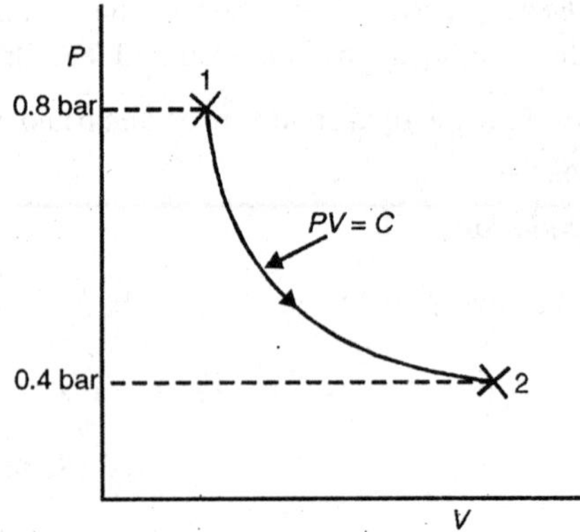

Fig. Ex. 9.24

$$\therefore \quad V_2 = \frac{P_1}{P_2} \times V_1 = \frac{8}{0.4} \times 0.21645$$

$$= 4.329 \text{ m}^3/\text{kg}$$

$$> V_{g_2}(3.993)$$

∴ Steam is superheated.

$$\therefore \quad 4.329 = V_{g_2} \times \frac{T_{\text{sup}\,2}}{T_{sat\,2}} = \frac{3.993 \times T_{\text{sup}\,2}}{75.9 + 273}$$

$$\therefore \quad T_{sup\,2} = \frac{4.329}{3.993} \times 348.9$$

$$= 378.26°K$$

$$\mathbf{T_{sup\,2} = 105.26° \, C}$$

$$\therefore \quad \text{Work done/kg} = P_1 V_1 \ln \frac{P_1}{P_2}$$

$$= 100 \times 8 \times 0.21645 \ln \frac{8}{0.4}$$

$$= 518.74 \text{ kJ/kg}$$

$$\textbf{Work done/kg} = \textbf{5187.4 kJ}$$

Example 9.25 Dry and saturated steam of 3.5 bar pressure is expanded hyperbolically to 0.5 bar. Using steam tables, determine

(i) Final condition of steam

(ii) Change of specific entropy during the process

Solution

(i) Hyperbolic process is defined by PV = Const. or $P_1V_1 = P_2V_2$

Here $P_1 = 3.5$ bar and $P_2 = 0.5$ bar; $x_1 = 1$

$$\therefore \quad V_1 = V_{g1} = 0.524 \text{ m}^3\text{/kg}$$

$$\therefore \quad V_2 = \text{volume after expansion}$$

$$= \frac{P_1 V_1}{P_2} = \frac{3.5 \times 0.524}{0.5}$$

$$= 3.668 \text{ cu.m/kg}$$

$$V_{g_2} = 3.2402 \text{ m}^3\text{/kg}$$

∴ Actual volume is greater than volume of dry and saturated steam at the pressure.

∴ Steam after expansion is superheated.

$$\therefore \text{ We write,} \quad \frac{V_2}{V_{g_2}} = \frac{T_2}{T_{sat_2}}$$

$$\text{or} \quad T_2 = \text{Temp. of superheated steam}$$

$$= \frac{3.668}{3.2402} \times \frac{273 + 81.343}{1}$$

$$T_2 = \mathbf{401.13\ K}$$

(ii) Change in specific entropy:

S_2 = Entropy of 1 kg at 3.5 bar and $x_1 = 1 = S_{g1} = 6.9392$ kJ/kg-K

Entropy of 1 kg steam at 0.5 bar

$$S_2 = S_{g_2} + C_{pv} \ln \frac{T_{sup}}{T_{sat}}$$

$$= 7.5947 + 2 \ln \frac{401.13}{354.343}$$

$$= 7.5947 + 2 \times 0.1240$$

$$= 7.5947 + 248$$

$$= 7.8427 \text{ kJ/kg-K}$$

$$\therefore \quad \text{Change in entropy} = S_2 - S_1$$

$$= 7.8427 - 6.9392$$

Change in entropy = 0.9035 kJ/kg-K

Example 9.26 A quantity of dry and saturated steam occupies 0.2876 cu.-m at 14 bar. Determine the final condition of the steam if it is compressed until its volume is halved, and if the compression is carried out hyperbolically.

Determine the heat rejected during compression.

Solution

Volume of 1 kg dry and saturated steam at 14 bar = 0.14072 cu.-m.

$$\therefore \quad \text{Mass of steam} = \frac{0.2876}{0.14072} = \frac{\text{Total volume}}{\text{Sp. volume}}$$

$$= 2.044 \text{ kg}$$

Hyperbolic process follows the law $PV = C$

$$\therefore \quad P_1V_1 = P_2V_2 \text{ Consider 1 kg of steam}$$

$$\therefore \quad \frac{V_1}{V_2} = \frac{P_2}{P_1} = \frac{0.14072}{0.14072/2} = 2$$

$$\therefore \quad P_2 = P_1 \times 2 = 14 \times 2 = 28 \text{ bar}$$

$$V_2 = 0.14072/2 = 0.07036 \text{ cu.-m/kg}$$

$$< V_{g_2}(0.071389)$$

$\therefore$ Steam after compression is wet.

$$\therefore \quad V_2 = X_2 V_{g_2} = x_2 \times 0.071389$$

$$\therefore \quad x_2 = \text{Dryness fraction after compression}$$

$$= \frac{0.07036}{0.071389}$$

$$x_2 = 0.9856$$

Calculation for heat $h_1 = 2787.8$ kJ/kg

$$\therefore \quad u_1 = h_1 - P_1V_1 = 2787.8 - \frac{10^5 \times 14 \times 0.14072}{1000}$$

$$= 2787.8 - 197$$

$$= 2590.8 \text{ kJ/kg}$$

Similarly, $h_2 = h_{f_2} + x_2 h_{fg_2}$

$$= 990.48 + 0.9856 \times 1811.5$$

$$= 990.48 + 1785.41$$

$$= 2775.89 \text{ kJ / kg}$$

$$\therefore \quad u_2 = h_2 - P_2 V_2 (= P_1 V_1)$$

$$= 2775.89 - 197$$

$$= 2578.89 \text{ kJ / kg}$$

$$\text{Work done kg} = P_1 V_1 \ln \frac{V_2}{V_1}$$

$$= \frac{10^5 \times 14 \times 0.14072}{1000} \times \ln \frac{1}{2}$$

$$= 197 \ln \frac{1}{2}$$

$$= -136.55 \text{ kJ / kg}$$

$$\therefore \quad Q = \Delta u + \int P \cdot dV$$

$$= (u_2 - u_1) + (-136.55)$$

$$= (2578.89 - 2590.8) - 136.55$$

$$= 11.91 - 136.55$$

$$= -148.46 \text{ kJ / kg}$$

$\therefore$ Total heat rejected $= -148.46 \times 2.044$

$\therefore$ **Total heat rejected= –303.45 kJ**

(V) Polytropic Process

Example 9.27. Steam originally at 16 bar and 0.9 dry expands in a cylinder according to law $pV^{1.1}$ = const. Until the pressure is 1.6 bar. Determine the amount of heat put into or extracted from the steam per kg during process stating the direction of heat transfer.

Given: $P_1 = 16$ bar; $x_1 = 0.9$ dry; $P_2 = 1.6$ bar; $PV^{1.1}$ = constant.

Solution

From steam table V_1 = Initial volume of steam/kg

$$= x_1 V_{g_1}$$

$$= 0.9 \times 0.12369$$

$$= \frac{0.1113}{n} \text{m}^3/\text{kg}$$

We have, $P_1 V_1^n = P_2 V_2$

$$\therefore \quad \left(\frac{V_2}{V_1}\right)^n = \frac{P_1}{P_2}$$

or

$$V_2 = V_1 \cdot \left(\frac{P_1}{P_2}\right)^{1/n}$$

$$= 0.1113\left[\left(\frac{16}{1.6}^{1/1.17}\right)\right]$$

$$= 0.1113 \times (10)^{0.9091}$$

$$= 0.1113 \times 8.1115$$

$$= 0.9028 \text{ m}^3/\text{kg}$$

This is less than $V_{g_2} = 1.0911$

Therefore, steam after expansion is wet and

$$x_2 = \frac{0.9028}{1.0911}$$

$$= 0.8274$$

$$\therefore \quad h_1 = \text{Enthalpy of 1 kg steam before expansion}$$

$$= h_{f1} + x_1 h_{fg_2}$$

$$= 858.56 + 0.9 \times 1933.2$$

$$= 858.56 + 1739.88$$

$$= 2598.44 \text{ kJ/kg}$$

$$\therefore \quad u_1 = h_1 - P_1 V_1$$

$$= 2598.44 - 100 \times 16 \times 0.1113$$

$$= 2598.44 - 178.08$$

$$= 2420.36 \text{ kJ/kg}$$

Similarly,

$$h_2 = \text{Enthalpy of 1 kg steam after expansion}$$

$$= hf_2 + x_2 h_{fg_2}$$

$$= 475.38 + 0.8274 \times 2220.9$$

$$= 2302.95 \text{ kJ/kg}$$

$$\therefore \quad u_2 = h_2 - P_2 V_2$$

$$= 2302.95 - 100 \times 1.6 \times 0.9028$$

$$= 2158.50 \text{ kJ/kg}$$

$$\text{Work done/kg} = \int P.\,dV$$

$$= \frac{P_1 V_1 - P_2 V_2}{n-1}$$

$$= \frac{178.08 - 144.45}{1.1 - 1}$$

$$= \frac{33.63}{0.1}$$

$$= 336.3 \text{ kJ/kg}$$

Energy equation gives

$$Q = \Delta u + \int P.\,dV$$

$$= (u_2 - u_1) + 336.3$$

$$= (2158.5 - 2420.36) + 336.3$$

$$\mathbf{Q = 74.44 \text{ kJ/kg}}$$

Here Q is positive. This means that heat is added to the system i.e. steam.

Example 9.28 0.9 kg steam initially at a pressure if 15 bar and temperature of 250°C expands reversibly and polytropically to 1.5 bar. Find (i) the final condition (ii) work done (iii) heat transferred and (iv) change in entropy if the index of expansion is 1.25.

Show the process on *P–V* and *T–S* diagrams.

Given: $P_1 = 15$ bar; $T_1 = 250°C$; $P_2 = 1.5$ bar; $n = 1.25$

Solution

The process is shown on *P–V* and *T–S* diagram in Fig. Ex. 9.28.

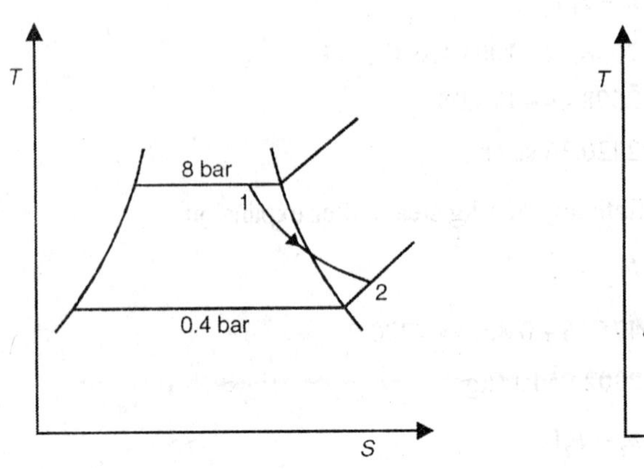

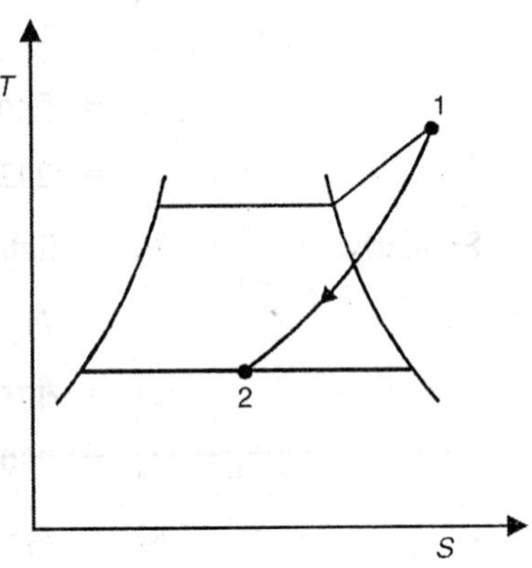

Fig. Ex. 9.28

First we will calculate on kg basis.

(i) Final condition :

$V_1 = 0.1520$ m³/kg; $h_1 = 2923.5$ kJ/kg; $S_1 = 6.7099$ kJ/kg-K

$$\therefore \quad P_1 V_1^n = P_2 V_2^n$$

or

$$V_2 = V_1 \times \left(\frac{P_1}{P_2}\right)^{1/n}$$

$$= 0.1520 \times \left(\frac{15}{1.5}\right)^{1/1.25}$$

$$= 0.152 \times 6.3096$$

$$= 0.9591 \text{ m}^3/\text{kg}$$

$$V_{g_2} = 1.1590 \text{ m}^3/\text{kg}$$

$\therefore$ Steam is wet

$\therefore \quad x_2$ = Dryness fraction of steam after expansion

$$= \frac{0.9591}{1.159}$$

$$\mathbf{x_2} = \mathbf{0.8275}$$

(ii) Work done :

Work done for non-flow process = $\int P.\,dV$

$$= \frac{P_1V_1 + P_2V_2}{n-1}$$

$$= \frac{100}{0.25}\left[\frac{15\times 0.152 - 1.5\times 0.9591}{1}\right]$$

$$= \frac{228 - 143.865}{0.25}$$

$$= \frac{84.135}{0.25}$$

$$= 336.54 \text{ kJ/kg}$$

$\therefore$ Work done for 0.9 kg steam

$$= 0.9\times 336.54 = 302.886 \text{ kJ.}$$

(iii) Heat transferred

$$h_1 = 2923.5 \text{ kJ/kg}$$

$$\therefore \quad u_1 = h_1 - P_1V_1$$

$$= 2923.5 - 228$$

$$= 2695.5 \text{ kJ / kg}$$

$$h_2 = \text{enthalpy after expansion}$$

$$= h_{f2} + x_2 h_{fg_2}$$

$$= 467.13 + 0.8275\times 2226.2$$

$$= 467.13 + 1842.18$$

$$= 2309.31 \text{ kJ / kg}$$

$$\therefore \quad u_2 = h_2 - P_2V_2 = 2309.31 - 143.865$$

$$= 2165.445 \text{ kJ / kg}$$

$\therefore$ For non-flow process,

$$Q = \Delta u + \int P.dV = u_2 - u_1 + \int P.dV$$

$$= (2165.445 - 2695.5) + 336.54$$

$$= -530.055 + 336.54$$

$$= -193.575 \text{ kJ/kg}$$

$\therefore$ For 0.9 kg $\quad Q = -0.9 \times 193.515$

$$\mathbf{Q = -174.1635 \text{ kJ/kg}}$$

– ve sign indicates that heat is rejected by the steam.

(iv) Change in entropy

$$S_1 = \text{Entropy before expansion}$$

$$= 6.7099 \text{ kJ/kg-K}$$

$$S_2 = \text{Entropy after expansion}$$

$$= S_{f_2} + x_2 S_{fg_2}$$

$$= 1.4336 + 0.8275 \times 5.7898$$

$$= 6.2247 \text{ kJ/kg-K}$$

$\therefore$ Change in entropy

$$\Delta S = S_2 - S_1 = 6.2247 - 6.7099$$

$$= -0.4852 \text{ kJ/kg-K}$$

$\therefore$ Total change in entropy

$$= -(0.9 \times 0.4852)$$

$$= -0.43668 \text{ kJ-K}$$

– ve sign indicates that entropy is decreasing.

Example 9.29 1 kg of steam, initially saturated at 11 bar, expands non flow according to $pV^{1-13} = C$ to 1 bar. Determine: (a) final volume v_2 (b) x_2 (c) work (d) Δu and (e) Q. (f) If the process is steady flow which $\Delta K = +4$ kJ / kg, what is the work?

Data: $p_1 = 11$ bar; $x_1 = 1$; $P_2 = 1$ bar; $PV^{1.3} = C$

Solution

(a) Final volume V_2

From tables, $\quad V_1 = 0.17738 \text{ m}^3\text{/kg}$

$$P_1 V_1^n = P_1 V_2^n$$

$$\therefore \quad V_2 = \left(\frac{P_1}{P_2}\right)^{1/n} \cdot V_1$$

(b) Final dryness fraction x_2

$$V_{g_2} = 1.6937 \text{ m}^3/\text{kg}$$

$\therefore$ Steam is wet

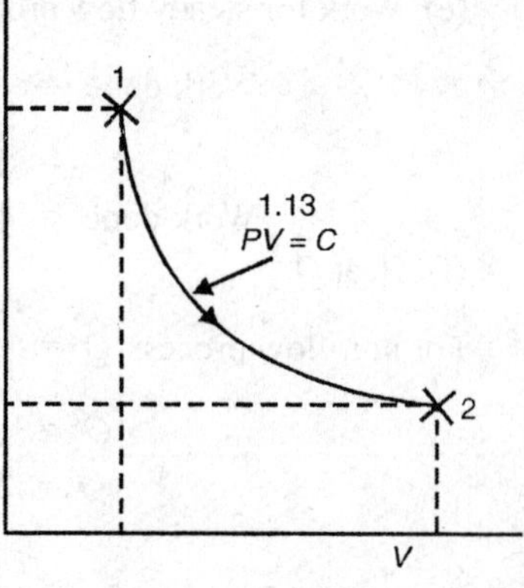

Fig. Ex. 9.29

$$\therefore \quad 1.481 = x_2 \times 1.6937$$

$$\therefore \quad x_2 = \frac{1.481}{1.6937}$$

$$\mathbf{x_2 = 0.8743}$$

(c) Work done :

Work of non-flow process $= \int P.dV$

$$= \frac{P_1 V_1 - P_2 V_2}{n-1}$$

$$= \frac{100}{1.13-1}[11 \times 0.17738 - 1 \times 1.481]$$

$$= \frac{100}{0.13}[1.9512 - 1.481]$$

$$= 769.231 \times 0.4702$$

$$= 361.69 \text{ kJ/kg}$$

(d) Change in internal energy

$$h_1 = h_{f_1} + x_2 h_{fg_1} = h_{g_1}$$

$$= 2779.7 \text{ kJ/kg}$$

$$\therefore \quad u_1 = h_1 - P_1 V_1 = 2779.7 - 195.12$$

$$= 2574.58 \text{ kJ/kg}$$

$$h_2 = h_{f_2} + x_2 h_{fg_2} = 417.51 + 0.8743 \times 2257.9$$

$$= 417.51 + 1974.08$$

$$= 2391.59 \text{ kJ/kg}$$

$$\therefore \quad u_2 = h_2 - P_2 V_2$$

$$= 2391.59 - 100(1.481) = 2391.59 - 148.1$$

$$= 2243.49 \text{ kJ/kg}$$

$$\therefore \quad \Delta u = u_2 - u_1 = 2243.49 - 2574.58$$

$$\mathbf{\Delta u = -331.09 \text{ kJ/kg}}$$

(e) Work for steady flow process

$$\text{Work done} = -\int P.dV + \Delta K = 1.13\int P.dV - 4$$

$$= 1.13 \times 361.69 - 4 = 408.71 - 4$$

$$\text{Work done} = 404.71 \text{ kJ/kg}$$

(f) Heat Q

$$\text{For non-flow process, } Q = \Delta u + \int P.dV$$

$$Q = -331.09 + 361.69$$

$$\mathbf{Q = 30.6 \text{ kJ/kg}}$$

Example 9.30 Four kg of steam at 13 bar and 260°C expand into wet region to 95°C in a polytropic process where $PV^{1.21} = C$. Find (a) y_2, ΔH, Δu, ΔS. (b) $\int P.dV$ (c) $-\int V.dP$ (d) work for nonflow process and (e) work for steady flow process if $\Delta K = -10$ kJ/ kg (f). Find Q for both non-flow and steady flow processes.

Solution

(a) Wetness $y_2, \Delta H, \Delta U, \Delta S$:

At 95°C, saturation pressure $= P_2 = 0.8453$ bar; $P_1 = 13$ bar;

$$V_1 = 0.1769 + 0.2(0.1969 - 0.1769)$$

$$= 0.1769 + 0.2 \times 0.0200$$

$$= 0.1769 + 0.004$$

$$= 0.1809 \text{ cu.m / kg}$$

$$\therefore \quad V_2 = \left(\frac{P_1}{P_2}\right)^{1/n} \times V_1 = 0.1809 \times \left(\frac{13}{0.845}\right)^{1/1.20}$$

$$= 0.1809 \times (15.38)^{0.8264}$$

$$= 1.7312 \text{ m}^3\text{/kg}$$

But $\quad V_{g_2} = 1.982 \text{ m}^3\text{/kg}$

Actual volume is less than v_{g_2}. Therefore steam is wet

$$\therefore \quad 1.7312 = x_2.V_{g_2}$$

$$x_2 = \frac{1.7312}{1.982}$$

$$= 0.8734$$

$$\therefore \quad y_2 = \text{Wetness} = 0.1216$$

$$\therefore \quad h_1 = 2931.5 + 0.2(3044.3 - 2931.5)$$
$$= 2931.2 + 0.2 \times 112.8$$
$$= 2931.5 + 22.56 = 2954.06 \text{ kJ/kg}$$

$$h_2 = h_{f_2} + x_2 h_{fg_2} = 397.99 + 0.8734 \times 2270.2$$
$$= 397.99 + 1982.79$$
$$= 2380.78 \text{ kJ/kg}$$

$$\therefore \quad \Delta H = h_2 - h_2 = 2380.78 - 2954.06$$

$$\mathbf{\Delta H = -573.28 \text{ kJ/kg}}$$

$$S_2 = S_{f_2} + x_2 S_{fg_2} = 1.2501 + 0.8734 \times 6.1665$$
$$= 1.2581 + 5.3858$$
$$= 6.6359 \text{ kJ/kg-K}$$

$$S_1 = 6.7878 + 0.2(6.9938 - 6.7878)$$
$$= 6.7878 + 0.2 \times 0.1060 = 6.7878 + 0.02012$$
$$= 6.80792 \text{ kJ/kg-K}$$

$$\therefore \quad \Delta S = S_2 - S_1 = 6.63590 - 6.80792$$

$$\mathbf{\Delta S = 0.17202 \text{ kJ/kg-K}}$$

$$u_1 = h_1 - P_1 V_1 = 2954.06 - 100 \times 13 \times 0.1809$$
$$= 2954.06 - 235.17$$
$$= 2718.89 \text{ kJ/kg}$$

$$u_2 = h_2 - P_2 V_2 = 2380.78 - 100 \times 0.8453 \times 1.1712$$
$$= 2380.78 - 99.0$$
$$= 2281.78 \text{ kJ/kg}$$

$$\therefore \quad \Delta u = u_2 - u_1 = 2281.78 - 2718.89$$

$$\mathbf{\Delta u = -437.11 \text{ kJ/kg}}$$

(b) $\int P.dV$:

$$\int P.dV = \frac{P_1 V_1 - P_2 V_2}{n-1} = \frac{100[13 \times 0.1809 - 0.8453 \times 1.7312]}{0.21}$$

$$= \frac{1}{0.21}[295.17 - 99] = \frac{136.17}{0.21}$$

$$\mathbf{\int P.dV = 648.43 \text{ kJ/kg}}$$

(c) $-\int V.dP$: $-\int V.dP \;=\; n\int P.dV = 1.21\times 648.43$

$$-\int \mathbf{V.dP = 784.6\ kJ/kg}$$

(d) Work for non-flow process

$$W_{nf} \;=\; \int P.dV = 648.43 \text{ kJ/kg}$$

(e) Work for steady flow process

$$W_{sf} \;=\; -\int V.dP + \Delta k = 784.6 + 10 = 794.6 \text{ kJ/kg}$$

(f) Q-Heat

For non-flow process $Q \;=\; (u_2 - u_1) + \int P.dV$

$$= -437.11 + 648.43 = 211.32 \text{ kJ/kg}$$

For steady flow process

$$Q \;=\; \Delta K + W_{sf} + \Delta H \cdot 10 + 794.6 - 573.28$$

$$\mathbf{Q = 784.6\ kJ/kg}$$

Example 9.31 One kg per second of steam expands in a polytropic process from P_1 = 10 bar and V_1 = 0.2059 cu-m to 1 bar and 1.58 cu-m. Find (a) n in $PV^n = C$ (b) $\int P.dV$ (c) $-\int V.dP$ (d) W and Q for non-flow process (e) w and Q for steady flow process with ΔK = + 9 kJ/kg.

Data: P_1 =10 bar; V_1= 0.2059 m³/ kg, P_2 =1 bar , V_2 = 1.58 m³/ kg

Solution

(a) Index n in $PV^n \;=\; C$

We have $P_1 V_1^n \;=\; PV_2^n$

$$\therefore \quad \left(\frac{V_2}{V_1}\right)^n = \frac{P_1}{P_2} = \frac{10}{1} = 10$$

$$\therefore \quad \left(\frac{1.58}{0.2059}\right)^n = (7.6736)^n = 10$$

$$\therefore \quad n \log_{10} 7.6736 = \log_{10} 10$$

$$\therefore \quad n = \frac{1}{0.885}$$

$$\mathbf{n = 1.13}$$

(b) $\int P.dV$:

$$\int P.dV = \frac{P_1 V_1 - P_2 V_2}{n-1}$$

$$= \frac{100}{0.13}[10 \times 0.2059 - 1 \times 1.58]$$

$$= \frac{100}{0.13}[2.059 - 1.580]$$

$$= \frac{100}{0.13} \times 0.479$$

$$\int \mathbf{P.dV = 368.46\ kJ / kg / sec}$$

(c) $-\int V.dP$:

$$-\int V.dP = n\int P.dV$$

$$= 1.13 \times 368.46$$

$$-\int \mathbf{V.dP = 416.36\ kJ / kg / sec}$$

(d) Work and Heat for non-flow process :

For non-flow process

$$\text{Work} = \int P.dV$$

$$\textbf{Work} = \textbf{368.46 kJ/kg/sec}$$

$$V_1 = 0.2059 \text{ m}^3\text{/kg}$$

$$V_{g_1} = 0.19429$$

∴ Steam is superheated

$$\therefore \quad 0.2059 = 0.19429 \times \frac{T_{sup}}{T_{sat}}$$

$$= \frac{0.19429 \times T_{sup}}{179.88 + 273}$$

$$\therefore \quad T_{sup1} = \frac{0.2059}{0.19429} \times 452.88$$

$$= 480°K = 207.°C \approx 200°C$$

$$\therefore \quad h_1 = 2826.8 \text{ kJ / kg}$$

$$\therefore \quad u_1 = h_1 - P_1 V_1 = 2826.8 - 100 \times 10 \times 0.2059$$

$$= 2826.8 - 205.9$$

$$= 2620.9 \text{ kJ/kg/sec}$$

$$V_2 = 1.58 \text{ m}^3/\text{kg}$$

$$V_{g2} = 1.6937$$

$$\therefore \quad x_2 = \frac{1.58}{1.6937} = 0.9329$$

$$\therefore \quad h_2 = h_{f_2} + x_2 h_{fg_2} = 417.51 + 0.9329 \times 2257.9$$

$$= 417.51 + 2106.32 = 2523.83 \text{ kJ/kg/sec}$$

$$\therefore \quad u_2 = h_2 - P_2 V_2 = 2523.83 - 158$$

$$= 2365.83 \text{ kJ/kg/sec}$$

$$\therefore \quad Q = \Delta u + \int P.dV$$

$$= (u_2 - u_1) + 368.46$$

$$= (2365.83 - 2620.9) + 368.46$$

$$\mathbf{Q = 113.39 \text{ kJ/kg/sec}}$$

(e) W and Q for steady flow process :

$$\text{Work} = -\int V.dP + \Delta K$$

$$= 416.36 + 9$$

$$\mathbf{Work = 425.36 \text{ kJ/kg/sec.}}$$

$$Q = \Delta k + \Delta H + W_{sf}$$

$$= 9 + (2523.83 - 2826.8) + 425.36$$

$$= 9 + 425.36 - 302.97$$

$$\mathbf{Q = 131.39 \text{ kJ/kg/sec}}$$

Example 9.32 Steam at 20 bar and 300°C is expanded to four times its initial volume according to the law $PV^{1.3}$ = constant. Find the heat during the process.

Data: $P_1 = 20$ bar; $t_1 = 300°C$; $V_2 = 4V_1$; $PV^{1.3}$ = constant.

Solution

$$\therefore \quad P_1 V_1^{1.3} = P_2 V_2^{1.3}$$

$$\therefore \quad \left(\frac{V_2}{V_1}\right)^{1.3} = \frac{P_1}{P_2} = \left(\frac{4V_1}{V_1}\right)^{1.3} = 4^{1.3} = 6.063$$

$$\therefore \quad P_2 = \frac{20}{6.063} = 3.3 \text{ bar}$$

From tables, $V_1 = 0.126 \text{ m}^3/\text{kg}$

$$\therefore \quad V_2 = 4 \times 0.126 = 0.504 \text{ m}^3/\text{kg}$$

$$\therefore \quad \text{Work done/kg} = \frac{P_1 V_1 - P_2 V_2}{n-1} \qquad \text{(For non-flow)}$$

$$= \frac{100}{1.3-1}[20 \times 0.126 - 3.3 \times 0.504]$$

$$= \frac{100}{0.3}[2.52 - 1.6632]$$

$$= \frac{100 \times 0.8568}{0.3}$$

$$= 285.6 \text{ kJ/kg}$$

$$h_1 = 3025.2 \text{ kJ/kg}$$

$$u_1 = h_1 - P_1 V_1 = 3025.2 - 100 \times 20 \times 0.126$$

$$= 3025.2 - 252$$

$$= 2773.2 \text{ kJ/kg}$$

$$V_2 = 0.504 \text{ m}^3/\text{kg}$$

$$V_{g_2} = \frac{0.570627 + 0.539025}{2}$$

$$= \frac{1.1099652}{2}$$

$$= 0.554826$$

$$< V_{g_2},$$

$$= \frac{1.109652}{2} = 0.554826$$

$\therefore$ Steam is wet after expansion

$$\therefore \quad x_2 = \frac{0.504}{0.554826} = 0.910$$

$$\therefore \quad h_2 = h_{f_2} + x_2 h_{fg_2}$$

where
$$h_{f_2} = \frac{570.8+579.8}{2}$$
$$= \frac{1150.6}{2} = 575.3 \text{ kJ/kg}$$
$$h_{fg_2} = \frac{2156.9+2150.5}{2} = \frac{4307.4}{2}$$
$$= 2153.7 \text{ kJ/kg}$$

$\therefore$
$$h_2 = 575.3+0.910\times 2153.7$$
$$= 575.3+1959.87$$
$$= 2535.17 \text{ kJ/kg}$$

$\therefore$
$$u_2 = h_2 - P_2V_2 = 2535.17-100\times 3.3\times 0.504$$
$$= 2535.17-166.32$$
$$= 2368.85 \text{ kJ/kg}$$

Applying the non-flow energy equation,
$$Q = \Delta u + \int P.dV$$
$$= \text{Heat during the process}$$
$$= (u_2 - u_1) + W_n$$
$$= (2368.85-2773.2)+285.6$$
$$= -404.35+285.6$$
$$\mathbf{Q = -118.75 \text{ kJ/kg}}$$

– ve sign indicates that heat is rejected.

Example 9.33 Steam expands from the condition 16 bar and 250°C to 0.9 bar according to the law $pV^{1.1}$ = constant. Find for 1 kg of steam, the expansive work and change in internal energy of steam.

Data: P_1= 16 bar; P_2 = 0.9 bar; T_1 = 523 K and $pV^{1.1} = C$.

Solution

$$V_1 = \text{volume of 1 kg steam before expansion}$$
$$= V_{g1}\frac{T_{sup_1}}{T_{sat_1}} = 0.123843\times\frac{523}{474.33}$$
$$= 0.1366 \text{ m}^3/\text{kg}$$

Similarly, h_1 = Enthalpy before expansion

$$= h_{g1} + C_P\left(T_{sup} - T_{sat}\right)$$

$$= 2791.7 + 2.3(250 - 201.33)$$

$$= 2791.7 + 111.941 = 2903.641 \text{ kJ/kg}$$

$\therefore$ u_1 = Internal energy

$$= h_1 - P_1V_1$$

$$= 2903.641 - 100 \times 16 \times 0.1366$$

$$= 2903.641 - 218.56$$

$$= \frac{2685.081}{1.1} \text{ kJ/kg}$$

Now we have $P_1V_1 = P_2V_2$

or

$$V_2 = \left(\frac{P_1}{P_2}\right)^{1/1.1} \times V_1$$

$$= \left(\frac{16}{0.9}\right)^{1/1.1} \times 0.1366$$

$$= (17.78)^{0.9091} \times 0.1366$$

$$= 13.69 \times 0.1366$$

$$= 1.87 \text{ m}^3/\text{kg}$$

$$\approx V_{g_2} \text{ approx.}$$

$\therefore$ The steam is dry and saturated after expansion.

$\therefore$ $h_2 = h_{g_2} = 2670.9$ kJ/kg

$$v_2 - 1.87 \text{ m}^3/\text{kg}$$

$\therefore$ $u_2 = h_2 - P_2V_2 = 2670.9 - 100 \times 0.9 \times 1.87$

$$= 2670.9 - 168.3 = 2502.6 \text{ kJ/kg}$$

$\therefore$ Change in internal energy / kg

$$= u_2 - u_1 = 2502.6 - 2685.081$$

$$= -182.481 \text{ kJ/kg}$$

Work done/kg (non-flow)

$$W = \frac{P_1V_1 - P_2V_2}{n-1} = \frac{218.56 - 168.3}{0.1}$$

$$= \frac{50.26}{0.1}$$

$$\mathbf{W = 502.6\ kJ/kg}$$

Example 9.34 At a particular instant during expansion stroke of the steam engine, the volume of the steam was 0.05 m^3, the pressure and condition of steam being 8 bar and 0.9 dry respectively. Determine (i) the index of expansion (ii) the work done per kg and (iii) the change in entropy during the process.

Data: $P_1 = 8$ bar; $P_2 = 2$ bar; $V_1 = 0.05$ m^3; $V_2 = 0.15$ m^3; $x_1 = 0.9$

Solution

(i) Index of expansion:

Let n be the index of expansion.

$$\therefore \quad P_1V_1^n = P_2V_2^n$$

$$\therefore \quad \left(\frac{V_2}{V_1}\right)^n = \frac{P_1}{P_2} = \frac{8}{2} = 4$$

$$\therefore \quad \left(\frac{0.15}{0.05}\right)^n = 4 = 3^n$$

$$\therefore \quad n \log 3 = \log 4$$

$$n = \frac{\log 4}{\log 3} = \frac{0.6021}{0.4771}$$

$$\mathbf{n = 1.262}$$

(ii) The work done :

$$\text{Work done} = \frac{P_1V_1 - P_2V_2}{n-1} = \frac{100}{0.262}[8\times0.05 - 2\times0.15]$$

$$= \frac{40-30}{0.262} = \frac{10}{0.262} = 38.168 \text{ kJ}$$

$$\text{Specific volume} \quad V_1 = x_1 V_{g1} = 0.9\times0.240506$$

$$= 0.2165 \text{ kJ/kg}$$

$$\therefore \quad \text{Mass of steam} = \frac{0.05}{0.2165} = 0.231 \text{ kg}$$

$\therefore$ Work done/kg $= \dfrac{38.168}{0.231}$

Work done/kg = 165.23 kJ/kg

(iii) Change in entropy:

Initial entropy $S_1 = S_{f_1} + x_1 S_{fg_1}$

$= 2.045 + 0.9(6.659 - 2.045) = 6.1976$

Final volume/kg $= \dfrac{0.15}{0.231}$

$= 0.6493 \text{ m}^3/\text{kg}$

$< V_{g_2}(0.8872)$

$\therefore$ $x_2 = \dfrac{0.6493}{0.8872}$

$= 0.732$

Final entropy $S_2 = S_{f_2} + x_2 S_{fg_2}$

$\therefore$ $S_2 = 1.529 + 0.732(7.127 - 1.529)$

$= 1.529 + 4.098$

$= 5.627 \text{ kJ/kg-K}$

$\therefore$ Change in entropy/kg $\Delta S = S_2 - S_1$

$= 5.627 - 6.1976$

$= -5706 \text{ kJ/kg-K}$

$\therefore$ Change in entropy $\Delta S = -0231 \times 0.5706 \text{ kJ/K}$

ΔS = –0.1318 kJ/K

Example 9.35 Dry and saturated steam at 20 bar passes through an expander where it is expanded to a pressure of 500 mm Hg (vacuum) (barometer reads 725 mm Hg) according to the law $pV^{1.07}$ = constant. Calculate

(i) Condition of steam after expansion

(ii) $\int P.dV$

(iii) $-\int V.dP$

(iv) Work of non-flow process

(v) Work of steady flow system with $\Delta P = \Delta K = 0$

(vi) Change in internal energy

(vii) Change in internal entropy.

(viii) Heat Q.

Solution

(i) Condition of steam expansion:

Abs. pressure of steam after expansion

$$P_2 = \text{barometer reading vacuum} = 725 - 500 = 225 \text{ mm Hg}$$

In bar
$$P_2 = \frac{225 \times 1.01325}{760} = 0.3 \text{ bar}$$

$$P_1 = 20 \text{ bar and } V_1 = 0.099623 \text{ m}^3/\text{kg}$$

$$\therefore \quad P_1 V_1^n = P_2 V_2^n \text{ or } V_2 = \left(\frac{P_1}{P_2}\right)^{1/n} \cdot V_1$$

$$\therefore \quad V_2 = \left(\frac{20}{0.3}\right)^{1/1-0.7} \times 0.099623$$

$$= \left(\frac{20}{0.3}\right)^{0.9346} \times 0.099623$$

$$= 5.0464 \text{ m}^3/\text{kg}$$

$$< V_{g2}(5.229)$$

$$\therefore \quad x_2 = \frac{5.0464}{5.229}$$

$$\mathbf{x_2 = 0.9651}$$

(ii) $\int P.dV$:
$$\int P.dV = \frac{P_1V_1 - P_2V_2}{n-1}$$

$$= \frac{100}{0.07}[20 \times 0.099623 - 0.3 \times 5.0464]$$

$$= \frac{199.246 - 151.392}{0.07}$$

$$= \frac{47.854}{0.07}$$

$$\int \mathbf{P.dV = 683.63 \text{ kJ/kg}}$$

(iii) $-\int V.dP$: $-\int V.dP = n\int P.dV$

$$= 1.07 \times 683.63$$

$$-\int \mathbf{V.dP} = \mathbf{731.48\ kJ/kg}$$

(iv) Work for non-flow process

$$W_{nf} = \text{Work of non-flow process}$$

$$= \int P.dV$$

$$\mathbf{W} = \mathbf{683.63\ kJ/kg}$$

(v) Work of steady flow process

We have $W_{sf} = -\int V.dP$

$$\mathbf{W_{sf}} = \mathbf{731.48\ kJ/kg}$$

(vi) Change in internal energy

$$h_1 = h_{g_1} = 2797.3 \text{ kJ/kg}$$

$$\therefore \quad u_1 = h_1 - P_1 V_1 = 2797.3 - 199.246$$

$$= 2598.054 \text{ kJ/kg}$$

$$h_2 = h_{f_2} + x_2 h_{fg_2}$$

$$= 289.3 + 0.9651 \times 2336.1$$

$$= 289.3 + 2254.57 = 2543.87 \text{ kJ/kg}$$

$$\therefore \quad u_2 = h_2 - P_2 V_2$$

$$= 2543.87 - 151.392$$

$$= 2392.478 \text{ kJ/kg}$$

$$\therefore \quad \Delta u = u_2 - u_1 \text{ change in internal energy}$$

$$= 2392.478 - 2598.054$$

$$\Delta u = -205.576 \text{ kJ/kg}$$

(vii) Change in entropy

$$S_1 = S_{g1} = 6.337 \text{ kJ/kg-K}$$

$$S_2 = S_{f_2} + x_2 S_{fg_2}$$

$$= 0.944 + 0.9651(7.767 - 0.944)$$

$$= 0.944 + 0.9651 \times 6.823$$

$$= 0.944 + 6.5849$$

$$= 7.5289 \text{ kJ/kg-K}$$

$$\therefore \quad S_2 - S_2 = \text{Change in entropy}$$

$$= 7.5289 - 6.3370$$

$$\mathbf{S_2 - S_2 = 1.1919 \text{ kJ/kg-K}}$$

(viii) Heat Q

Heat during the non-flow process is given by

$$Q = \Delta u + \int p.dv$$

$$= -205.576 + 683.63$$

$$\mathbf{Q = 478.054 \text{ kJ/kg-K}}$$

For Steady Flow Process

$$Q = \Delta P + \Delta K + \Delta H + W_{sf}$$

$$= 0 + 0 + (2543.87 - 2797.3) + 731.48$$

$$\mathbf{Q = 478.05}$$

(VI) Combined Processes

Example 9.36 Steam at 2 MN/m^2 and 250°C expands isentropically to 3.5 bar. It is then further expanded hyperbothically to 0.5 bar. Using steam tables, find out

(i) condition of steam after each process.

(ii) work done kg/of steam in each process.

Data: $P_1 = 2 \text{ MN/m}^2 = 20$ bar; $T_1 = 250°C$; $P_2 = 3.5$ bar; $P_3 = 05$ bar.

Solution

The process are shown in Fig. Ex. 9.36

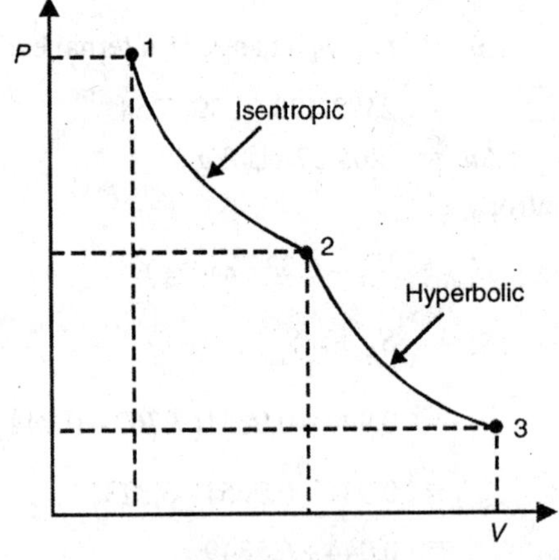

Fig. Ex. 9.36

Isentropic Process (1 – 2) :

From steam tables : $h_1 = 2902.4$ kJ / kg

$S_1 = 6.5454$ kJ / kg-K

For isentropic process, $S_1 = S_2 = S_{f_2} + x_2 S_{fg_2}$

$$\therefore \quad 6.5454 = 1.7273 + x_2 \times 5.2119$$

$$\therefore \quad x_2 = \frac{6.5454 - 1.7273}{5.2119}$$

$$x_2 = 0.9244$$

$$\therefore \quad h_2 = h_{f2} + x_2 h_{fg2}$$

$$= 584.27 + 0.9244 \times 2147.4$$

$$= 2569.32 \text{ kJ/kg}$$

Assuming non-flow process,

$$\text{Work done / kg} = u_1 - u_2$$

$$= (h_1 - P_1 V_1) - (h_2 - P_2 V_2)$$

$$= \left[290234 \frac{-10^5 \times 0.1114}{1000}\right]$$

$$- \left[2569.32 \frac{3.5 \times 10^5 \times 0.9255 \times 0.524}{1000}\right]$$

Work done/kg = 279.81 kJ/kg

For steady flow process with $\Delta P = \Delta K = 0$

$$\text{Work done/kg} = h_1 - h_2$$

Work done/kg = 2902.4 – 2569.42 = 333.08 kJ/kg

Hyperbolic Expansion (2 – 3): The law of expansion is pV = Const.

$$\therefore \quad P_2 V_2 = P_3 V_3$$

where $P_2 = 3.5$ bar, $P_3 = 0.5$ bar.

$$V_2 = x_2 V_{g_2} = 0.9244 \times 0.524 = 0.4844 \text{ m}^3\text{/kg}$$

$$\therefore \quad V_3 = \frac{P_2 V_2}{P_3} = \frac{3.5 \times 0.4844}{0.5} = 3.391 \text{ m}^3\text{/kg}$$

From tables, $V_3 = 3.391 > (V_{g3} = 3.2402)$

$$\therefore \quad \frac{V_3}{V_{g3}} = \frac{T_3}{T_{sat_3}}$$

$$\therefore \quad \frac{3.391}{3.2402} = \frac{T_3}{273+81.345}$$

$$\therefore \quad T_3 = \frac{3.391 \times 354.345}{3.2402}$$

$$\mathbf{T_3 = 370.8\ K = 97.8^\circ C}$$

Superheated Condition

For pV = Const., the work done of non-flow process and steady flow process with

$$\Delta P = \Delta K = 0, \text{ is same}$$

$$\text{Work done/kg} = P_2 V_2 \ln \frac{P_2}{P_3}$$

$$= \frac{3.5 \times 105 \times 0.4844}{1000} \ln \frac{3.5}{0.5}$$

$$\textbf{Work done/kg} = \mathbf{329.91\ kJ/kg}$$

Example 9.37 Two boilers of equal evaporative capacities generate steam at the same pressure of 16 bar to a common pipe line. One boiler produces steam with 50°C superheat and other produces wet steam. If the steam in the main pipe is just dry and saturated, find the dryness fraction of wet steam.

Solution

The flow diagram is shown in Fig. Ex. 9.37. Consider 1 kg steam from each boiler

$$h_A + h_B = 2hC$$

$$h_{gA} + C_{P_v}\left(t_{sup} - t_{sat}\right) + \left(h_{fB} + x_B h_{fgB}\right) = 2h_{gc}$$

$$\therefore \quad (2791.76 + 2.3 \times 50) + (858.56 + x_B \times 1933.2)$$

$$= 2 \times 2791.76$$

$$2906.76 + 858.56 + 1933.2\, x_B = 5583.52$$

$$\therefore \quad x_B = \frac{5583.56 - 2906.76 - 858.56}{1933.2}$$

$$= \frac{1818.2}{1933.2}$$

$$\mathbf{x_B = 0.9405}$$

This is Dryness fraction of wet steam.

Example 9.38 A closed vessel of 0.3 cu.m capacity contains steam at a pressure of 1.2 bar and 0.8 dry. It is connected to a steam pipe supplying steam of 0.9 dryness fraction at a pressure of 4.5 bar. Neglecting the volume of moisture in suspension and thermal capacity of the vessel, determine, (a) the weight of steam which will pass through the steam pipe to the vessel, when the valve in the pipe is opened and (b) the final dryness of the steam in the vessel when the flow ceases.

Solution

When the valve is opened, the steam will flow into the vessel till the pressure in the vessel pressure increases to 4.5 bar and then the flow will cease.

Initially,

Sp. vol. of steam in vessel= $V_1 = x_1\, V_{g_1}$

$$= 0.8 \times 1.4281$$

$$= 1.14248 \text{ m}^3/\text{kg}$$

∴ Initial wt. of steam in the vessel

$$W_1 = \frac{0.3}{0.14248} = 0.2626 \text{ kg}$$

After the steam is admitted, the pressure is 4.5 bar

Let W_2 = wt. of steam dryness

x_2 = Dryness fraction of steam admitted

x_3 = Dryness fraction of after mixing

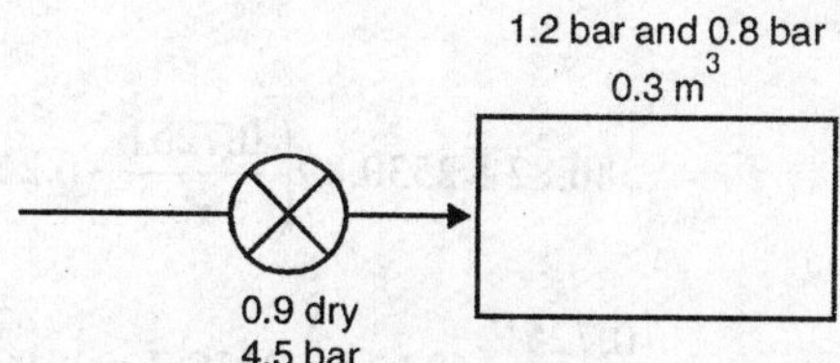

Fig. Ex. 9.38

Total heat of steam in vessel before mixing.

$$h_1 = 0.2626\left[h_{f_1} + x_1 h_{fg_1}\right]$$

$$= 0.2626[439.36 + 0.8 \times 2244.1]$$

$$= 586.82 \text{ kJ} \quad (1)$$

Total heat of steam admitted

$$h_2 = W_2\left[h_{f_2} + x_2 h_{fg_2}\right]$$

$$= W_2[623.16 + 0.9 \times 2119.7]$$

$$= 2530.89\, W_2 \text{ kJ} \quad (2)$$

Total heat of steam after mixing

$$h_3 = (W_1 + W_2)\left[h_{f_3} + x_3 h_{fg_3}\right]$$

$$= (0.2626 + W_2)[623.16 + x_3 \times 2119.7] \quad (3)$$

$\therefore$ 1 + 2 + 3

$$586.82 + 2530.89\, W_2 = (0.2626 + W_2)(623.16 + x_3 \times 2110.7) \quad (4)$$

Again volume of mixture is 0.3 m^3

$$\therefore \quad = (W_1 + W_2)x_3\, V_{g3} = 0.3$$

$$= (0.2626 + W_2)x_3 \times 0.41375 = 0.3$$

$$(2626 + W_2)x_3 = \frac{0.3}{0.41375} = 0.7251 \quad (5)$$

$$\therefore \quad (2626 + W_2) = \frac{0.7251}{x_3} \quad (6)$$

$$\therefore \quad W_2 = \frac{0.7251}{x_3} - 0.2626 \quad (7)$$

$\therefore$ Eq. (4) gives

$$= 586.82 + 2530.89\left(\frac{0.7251}{x_3} - 0.2626\right)$$

$$= \frac{0.7251}{x_3}(623.16 + 2119.7\, x_3)$$

$$= 586.82 + \frac{1835.15}{x_3} - 664.61$$

$$= \frac{451.85}{x_3} + 1537$$

$$\frac{1835.15 - 451.85}{x_3} = 1537 + 664.61 - 586.82$$

$$\frac{1383.30}{x_3} = 2201.61 - 586.82 = 1614.79$$

$\therefore \quad x_3 = 0.8566$

$$\therefore \quad W_2 = \frac{0.7251}{0.8566} - 0.2626$$

$$= 0.5839 \text{ kg}$$

(a) The weight of steam admitted = 0.5839 kg

(b) Final dryness fraction = 0.8566

Example 9.39 Steam initially at 14 bar and having specific volume of 0.12 m³/ kg is throlled to 7 bar and then expanded hyperbolically to 1.1 bar. For each process determine change in entropy.

Solution

Throttling Process

Volume of steam at 14 bar and dry and saturated is 0.14073 m³/kg. Actual volume is 0.12 m³/kg. Therefore the steam is wet before throttling.

$$\therefore \quad x_1 = \frac{0.12}{0.14073} = 0.8527$$

$$\therefore \quad h_1 = \text{Enthalpy of 1 kg steam before throttling}$$

$$= h_{f_1} + x_1 h_{fg_1}$$

$$= 830 + 0.8527 \times 1957.7$$

$$= 2799.32 \text{ kJ/kg}$$

This enthalpy is less than h_{g_2}. Therefore steam is wet after throttling.

$$\therefore \quad h_2 = 2499.32 = h_{f_2} + x_2 h_{fg_2} \text{ at 7 bar}$$

$$= 697.1 + x_2 \times 2064.9$$

$$\therefore \quad x_2 = \frac{2499.32 - 697.1}{2064.9} = \frac{1802.22}{2064.9}$$

$$= 0.8728$$

Entropy of wet steam before throttling is

$$S_1 = S_{f_1} + x_1 S_{fg_1}$$

$$= 2.2837 + 0.8527 \times 4.1814$$

$$= 5.8492 \text{ kJ/kg-K}$$

Similarly $\quad S_2 = \text{Entropy after throttling}$

$$= S_{f_2} + x_2 S_{fg_2}$$

$$= 1.9918 + 0.8728 \times 4.7134$$

$$= 6.1056 \text{ kJ/kg-K}$$

$\therefore$ Change of entropy during throttling

$$\Delta S = S_2 - S_1 = 6.1056 - 5.8492$$

$$= 0.2564 \text{ kJ/kg-K}$$

Hyperbolic Process

$P_2 = 7$ bar; $V_2 =$ volume of 1 kg steam before hyperbolic expansion

$$V_2 = x_2 V_{g_2}$$

$$= 0.8728 \times 0.27268 = 0.2379 \text{ m}^3\text{/kg}$$

$$P_3 = 1.1 \text{ bar} = \text{Pressure after hyperbolic expansion}$$

$$\therefore \quad P_2 V_2 = P_3 V_3$$

$$\therefore \quad V_3 = \frac{P_2 V_2}{P_3}$$

$$= \frac{7}{1.1} \times 0.2379 = 1.5147 \text{ m}^3\text{/kg}$$

This is less than $V_{g_3} = 1.5492$

$\therefore$ Steam is wet after hyperbolic expansion.

$$\therefore \quad 1.5147 = x_3 \times 1.5492$$

$$x_3 = \frac{1.5147}{1.5492}$$

$$= 0.9774 \text{ m}^3\text{/kg}$$

$$\therefore \quad S_3 = S_{f_3} + x_3 S_{fg_3}$$

$$= 1.3330 + 0.9774 \times 5.9947$$

$$\mathbf{S_3 = 7.1922 \text{ kJ/kg-K}}$$

Example 9.40 Two boilers discharge equal amounts of steam into the same main. The steam from one boiler is at 17 bar and 270°C and other at 17 bar and 0.95 quality. What is (a) the equilibrium condition after mixing. ($\Delta K = 0$) (b) the change of entropy by the higher temperature steam, (c) the change of entropy of the lower temperature steam? (d) is there a net increase or decrease of entropy?

Solution

Energy diagram is shown in Fig. Ex. 9.40 $\Delta K = 0$;

$$h_A + h_B = 2\,h_C$$

From table
$$h_A = \frac{2915.3 + 3033.5}{2}$$

$$= \frac{5948.8}{2}$$

$$= 2974.4 \text{ kJ/kg}$$

$$S_{A1} = \frac{6.6398 + 6.8557}{2} = \frac{13.4955}{2}$$

$$= 6.74775 \text{ kJ/kg-K}$$

$$h_B = h_f + xh_{fg}$$

$$= 871.84 + 0.95 \times 1921.5$$

$$= 2697.26 \text{ kJ/kg}$$

$$S_{B1} = S_f + xS_{fg}$$

$$= 2.3713 + 0.95 \times 4.0244$$

$$= 6.1945 \text{ kJ/kg.K}$$

Fig. Ex. 9.40

After mixing in the main pipe

$$2h_C = h_A + h_B = 2974.4 + 2697.26$$

$$= 5671.66$$

$$\therefore \quad h_C = \frac{5671.66}{2} = 2835.83 \text{ kJ/kg}$$

The steam is superheated in the main pipe.

Assuming $Cp_V = 2.3$ kJ/kg-K

$$2835.83 = h_g + C_{pv}\left(t_{sup} - t_{sat}\right)$$

$$= 2793.4 + 2.3\left(t_{sup} - 204.31\right)$$

$$\therefore \quad t_{sup} = \text{Temp of steam in main pipe}$$

$$= \frac{2835.83 - 2793.4}{2.3} + 204.31$$

$$= 222.76°\text{C}$$

∴ Equilibrium condition is

$$P = 17 \text{ bar.}$$
$$t = 222.76°C$$

Entropy in main pipe is

$$S_c = S_{gc} C_{pv} \log_e \frac{T_{sup}}{T_{sat}}$$

$$= 6.3957 + 2.3 \log_e \frac{222.76+273}{204.31+273}$$

$$\mathbf{S_c = 6.48293 \text{ kJ/kg-K}}$$

(b) Change in entropy of A stream :

$$\Delta S_A = S_C - S_A$$

$$= 6.48293 - 6.74775$$

$$\mathbf{\Delta S_A = -0.26482 \text{ kJ/kg-K}}$$

(c) Change in entropy of B stream

$$\Delta S_B = S_C - S_B$$

$$= 6.48293 - 6.19450$$

$$\mathbf{\Delta S_B = 0.28843 \text{ kJ / kg-K}}$$

(d) Net change

Net change in entropy

$$= 0.28843 + (-0.26482)$$

$$= 0.02361 \text{ kJ / kg-K}$$

Net increase is there.

(VII) Separating and Throttling Calorimeter

Example 9.41 Discuss under what circumstances, a throttling calorimeter is capable of being used as a device for measuring the steam dryness.

Find the maximum degree of wetness that can be determined by a throttling calorimeter if the inlet pressure of steam to it is 10.5 bar and exit pressure is 1.2 bar.

Solution

Condition of the use of throttling calorimeter for finding the dryness fraction of steam is that the condition of steam after throttling must be known. For this purpose the steam should be slighlty superheated so that temperature can be measured by the thermometer. In the limit, the steam should be dry and saturated after throttling.

Let x, be the dryness fraction of steam to be found.

For throttling process, enthalpies are same before and after throttling.

$$\therefore \quad h_1 = h_2$$

$$\text{or} \quad h_{f1} + x_1 h_{fg1} = h_{g2}$$

$$\therefore \quad 772.03 + x_1 \times 2005.9 = 2683.4$$

$$\therefore \quad x_1 = \frac{2683.4 - 772.03}{2005.9}$$

$$= \frac{1911.37}{2005.9}$$

$$= 0.9529$$

$$\therefore \quad \text{Maximum wetness} = 1 - 0.9529 = 0.0471 \text{ or } 4.71\,\%$$

Example 9.42 The following, data were obtained with a separating and throttling calorimeter:

Pressure in the pipe line 1.5 MPa
Condition after throttling 0.1 MPa and 110°C
Moisture collected in separating
Calorimeter /5 min 0.15 lit at 70°C
Steam condensed after throttling 3.24 kg
Find the quality of steam in the pipe.

Solution

If w is the wt. of steam separated in separating calorimeter and w the wt. of condensed steam after throttling, then dryness fraction of steam is given by

$$x = \frac{Wx_1}{W + w}$$

where x_1 is the dryness fraction of steam entering the throttling calorimeter

$$\therefore \quad x = \frac{3.24\, x_1}{3.24 + 0.15}$$

$$= \frac{3.24\, x_1}{3.39} \quad (1)$$

For throttling calorimeter,

[Enthalpy before] = [Enthalpy after throttling]

$$h_{f_1} + x_1 h_{fg_1} = h_{g2} + C_V\left(t_{\text{sup}} - t_{\text{sat}}\right)$$

$$844.66 + x_1 \times 1945.2 = 2676.2 + 2.0[110 - 99.632]$$

$$= 2676.2 + 2 \times 10.368$$
$$= 2676.2 + 20.736$$
$$= 2696.936 \text{ kJ/kg}$$

$$\therefore \quad x_1 = \frac{2696.936 - 844.66}{1945.2}$$

$$= \frac{4852.276}{1945.2} \quad (2)$$

$$= 0.9522$$

$\therefore$ Dryness fraction of steam in the pipe

$$x = \frac{3.24}{3.39} \times 0.9522$$

$$\mathbf{x = 0.91}$$

Example 9.43 (a) Explain the influence of pressure of the following properties of steam with the help of *T–S* diagram.

(i) Enthalpy of evaporation
(ii) Temperature of saturation
(iii) Enthalpy of dry saturated steam

(b) With a separating and throttling calorimeter arranged in series, the following observations, were taken. Estimate the quality of steam sampled.

Water separated = 2 kg
Steam discharged from throttling calorimeter = 20.5 kg
Temperature of steam after throttling = 110° C
Initial pressure = 11 bar (gauge)
Barometer = 760 mm of Hg
Final pressure = 28 mm of Hg

Solution

(a) The *T–S* diagram for steam is shown in Fig. Ex. 9.43. Effect of pressure on

(i) Enthalpy of evaporation: Horizontal distance on *T–S* diagram represents to scale, the enthalpy of evaporation.
If the pressure increases, i.e. saturation temperature increases, enthalpy of evaporation decreases and vice-versa.

(ii) Temperature of saturation: As the pressure increases, saturation temperature increases and vice-versa.

(iii) Enthalpy of dry saturated steam: Enthalpy of dry saturated steam decreases with the increase in pressure and vice-versa. This is beyond certain pressure only otherwise it increases.

(b) The dryness fraction of steam sampled is given by

$$x = \frac{W_1}{W_1 + W_2} . x$$

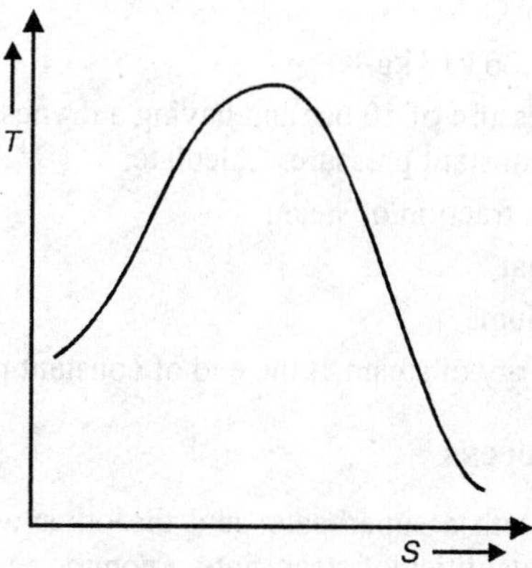

Fig. Ex. 9.43

here W_1 = 20.5 kg, W_2 = 2 kg, x_1 = Dryness fraction of steam leaving the separating calorimeter and entering the throttling calorimeter Atm. pressure = 760 mm Hg = 1.01325 bar

$$\therefore \text{ Abs. pressure of steam sampled} = 11 + 1.01325 \approx 12.01325 \text{ bar}$$

$$\text{Similarly,} \quad 28 \text{ mm Hg} = \frac{28}{700} \times 1.01325 = 0.03733 \text{ bar}$$

$$\text{Abs. pressure} = 0.03733 + 1.01325 = 1.05058 \text{ bar, considering } 1.1 \text{ bar}$$

For throttling calorimeter, we have

$$h_1 = h_2$$

$$\therefore \; 798.43 + x_1 \times 1984.3 = \frac{2679.6 + 2.1(110 - 102.32)}{1984.3}$$

$$= \frac{1897.298}{1984.3} = 0.9561$$

$$x = \frac{09561 \times 20.5}{22.5} = 0.871$$

PROBLEMS FOR PRACTICE

1. The internal energy of 2 kg of steam at a pressure of 9 bar is 4605 kJ. Calculate the dryness fraction of the steam. By how many heat units the internal energy of the above steam will increase if it is superheated, at constant pressure, to a temperature of 225°C.

   Take $C_p = 2.26$ kJ / kg-K

2. 1 kg steam at a pressure of 10 bar and having a dryness fraction of 0.2 receives 1046 kJ heat at a constant pressure. Calculate:
   (a) The dryness fraction of steam
   (b) The total heat
   (c) Specific volume
   (d) Internal energy of steam at the end of constant pressure heat supply.

Constant Pressure Process

3. Two boilers one with a superheater and the other without a superheater are delivering equal quantities of steam into a common main. The pressure in the boilers and the main is 14 bar absolutely. The temperature of steam from the boiler with superheater is 325°C and that of the steam in the main is 250°C. Estimate the quality of steam supplied by other boiler.

   (**Ans.** 0.9785 dry)

4. In a Babcock Wilcox boiler, the wet steam at 20 bar and 0.93 dry is taken, from the steam drum to the superheater, where it is superheated to 150°C superheat at constant pressure. Find the heat during the process, work done and the change of entropy during the process. Take C_p for superheated steam as 2.3 kJ / kg-K.

   [**Ans.** 477.2 kJ / kg; 75.5 kJ / kg; 0.8922 kJ / kg-K]

Constant Volume Process

5. A pressure vessel contains 1.25 kg of steam at 5 bar and 0.7 dry. The vessel is cooled until the steam attains the final dryness fraction of 0.5.

   Find (a) The final pressure of steam and (b) Heat transfer during the process.

   [**Ans.** (i) 2.676 bar (ii) –1034.25 kJ]

6. A closed vessel of 0.2 m³ capacity contains steam at a pressure of 8 bar and a temperature of 200°C.
   (i) What is the weight of steam contained in the vessel ?
   (ii) The vessel is cooled till the steam becomes just dry and saturated. What will be the pressure of steam at this state ? Estimate the specific entropy.
   (iii) The vessel is further cooled till the temperature drops; to 160.8°C. Find the pressure and condition of steam at this stage.

   [**Ans.** (i) 0.78 kg (ii) 7.5 bar; 6.686 kJ/ kg-K (iii) 6.32 bar, (iv) 0.849 dry]

Isentropic Process

7. Steam initially at 1.5 mPa and 300°C expands reversibly and adiabatically in a steam turbine to 40°C. Determine:
   (i) condition of steam after expansion
   (ii) work done by the turbine per kg of steam

   [**Ans.** (i) 0.818 (ii) 870 kJ / kg-K]

8. Steam at 2 MN/m² and 250°C is expanded isentropically to 0.35 MN/m². Using the steam tables, determine
   (a) condition of steam after expansion (b) work done for non-flow process

   [**Ans.** (a) 0.915 (b) 297.38 kJ/ kg]

9. In a steam engine cylinder steam expands hyperbolically from 8 bar pressure and dryness fraction 0.9 to 1 bar. Find the final condition of steam and change in specific volume. Find also the non-flow work during-the process.

   [**Ans.** (i) 107.61C; (ii) 1.51515 m³/kg; (iii) 360 kJ/kg]

10. Five kilogram steam having a pressure of 16 bar and 0.9 dry is expanded in a cylinder to a pressure of 0.8 bar. Determine the final condition of steam mechanical work done and the heat transfer during the process if the process is hyperbolic.

    [**Ans.** (1) 117.82C; 534.055 kJ/kg (2670.276 kJ) ; (2) 657.445 kJ]

11. 1 kg of wet steam at 250°C and with enthalpy of 2500 kJ is confined in a rigid vessel. Heat is supplied until the steam becomes saturated. Determine
    (a) initial pressure of steam
    (b) final pressure of steam
    (c) heat supplied

    [**Ans.** (a) 39.7894 bar (b) 48 bar (c) 262.15 kg]

12. A sample of steam at 15 bar has an internal energy, of 2500 kg. Find the condition of steam.

    (**Ans.** $x = 0.946$)

13. A rigid vessel of volume 0.01 m³ contains 90% (by volume) water and 10% steam at a pressure of 2 bar. Deterrnine
    (i) temperature of water
    (ii) total mass of water
    (iii) steam in the vessel

    [**Ans.** (i) 120.2°C (ii) 9 kg (iii) .0:00113 kg]

14. Determine the specific volume, enthalpy of steam under the following conditions
    (i) 4 MPa pressure 360°C
    (ii) 150 kPa, 10% quality
    (iii) 350 kPa, 10% moisture

    [**Ans.** (i) 0.062 m³, 3053.31 kJ/kg (ii) 0.1159 m³, 689.6 kJ/kg
    (iii) 0.4717 m³;1517.2 kJ kg]

15. A closed vessel of 0.75 cu.m capacity contains dry and saturated steam at 5 bar. The vessel is cold until the pressure is reduced to 2 bar. Determine,

(i) mass of steam present in the vessel

(ii) final dryness fraction

(iii) amount of heat transferred during the process.

[**Ans.** (i) 2.001 kg (ii) 0.4232 (iii) –23.78.52 kg]

16. 0.125 cu.-m of wet steam at 10 bar has an enthalpy of 1800 kJ/kg. Calculate mass and dryness fraction of steam.

(**Ans.** m = 1.249 kg, x = 0.5152)

17. 1 kg of steam at an initial condition of 6 bar and 0.2 dry is heated at constant volume until the pressure is 20 bar. Determine the final state of steam, heat added.

(**Ans.** x_2 = 0.635, Q = 932 kJ/kg)

18. Three boilers A, B, C working at 12 bar absolute supply steam in equal quantities to a plant through a common steam pipe. The boiler A supplies, dry saturated steam while the steam supplied by boiler B is superheated. to 350°C. The steam from boiler C is of dryness fraction 0.96. Calculate the condition of steam in common pipe. Assume that there is no loss of heat in pipes and specfic heat of superheated steam is 2.1 kJ/kg°K.

(**Ans.** 229.2 °C)

19. Steam is stored in a container 0.45 m³ capacity. The pressure of steam is 8 bar and temperature is 200°C. The container is cooled and pressure drops to 25 bar. Find

(i) initial internal energy in kJ

(ii) quality of steam after cooling

(iii) heat lost to atmosphere due to cooling in kJ

Neglect volume of water.

[**Ans.** (i) 4533.77 kJ, (ii) 0.3625, (iii) – 2360.42 kJ]

20. 1 kg of steam at 20 bar and 300°C is cooled at constant pressure till the dryness fraction of steam becomes 0.5. Find the change in enthalpy work done and the heat transfer during the process.

(**Ans.** ΔH = – 1172 kJ, W = –151.5 kJ, Q = –1172 kJ)

ꕤ

STEAM TABLES

Introduction

1. The following steam tables are based on the Mollier tables and give all the necessary data with regard to saturated and superheated steam required by Engineering students.

2. The various tables are arranged as follows:

(*a*) *Saturated Steam*

Table 1 : It provides data with reference to varying *temperatures.*

Table 2 : It provides data with reference to varying *pressures.*

(*b*) *Superheated Steam*

Table 3: It provides *Enthalpy* of steam superheated to various temperatures at different pressures.

Table 4: It provides *Entropy* of steam superheated to various temperatures at different pressures.

Table 5: It provides *Volume* of steam superheated to various temperatures at different pressures.

3. All heat energy quantities expressed in these tables are in terms of kJ/kg of steam.

4. All pressures are in terms of N/m^2.

Important Relationships

(i) 1 J is equivalent of 1 Nm or 0.10197 m kg_f of work.

(ii) $T = t + 273$, where T = Temperature expressed in centrigrade absolute scale, (K).

(iii) Absolute pressure = Gauge pressure + Atmospheric pressure.

(iv) Volume of wet steam $= x \times V$,

where x = dryness fraction.

V_s = Specific volume of saturated Steam.

(v) Enthalpy of wet steam = Enthalpy of liquid + (Latent heat × dryness fraction).

or $H = h_L \times xL$

(This also applies to *Entropy* and *Internal Energy*)

(vi) $1\ N/m^2 = 10^{-5}$ bar $= 0.10197\ kg_f/m^2$ of pressure.

Table 1. *Saturated Steam (Temperature Table)*

Temperature		Pressure N/m^2 p × 10^{-5}	Volume m^2/kg		Density of Vapour kg/m^2	Enthalpy kJ/kg		Latent heat L kJ/kg	Entropy kJ\kg-K		L/T kJ/kg-K
°C	K		Liquid v_i	Vapour v_i		Liquid h_i	Vapor h_s		Liquid s_i	Vapour s_v	
1	2	3	4	5	6	7	8	9	10	11	12
0	273	0.00610	0.001000	206.3	0.004847	0.00	2501.6	2501.6	0.00	9.1577	9.1577
1	274	0.00656	0.001000	192.6	0 005192	4.17	2503.4	2499.2	0.0152	9.1311	9.1159
2	275	0.00705	0.001000	179.9	0.005558	8.39	2505.2	2496.8	0.0306	9.1047	9.0741
3	276	0.00757	0.001000	168.2	0.005946	12.60	2507.1	2494.5	0.0459	9.0785	9.0326
4	277	0.00813	0.001000	157.3	0.006358	16.80	2508.9	2492.1	0.0611	9.0526	8.9915
5	278	0.00872	0.001000	147.2	0.006795	21.01	2510.7	2489.7	0.0762	9.0269	8.9507
6	279	0.00934	0.001000	137.8	0.007258	25.21	2512.6	2487.4	0.0913	9.0015	8.9102
7	280	0.01001	0.001000	129.1	0.007748	29.41	2514.4	2485.0	0.1063	8.9762	8.8699
8	281	0.01072	0.001000	121.0	0.008267	33.60	2516.2	2482.6	0.1213	8.9513	8.8300
9	282	0.01147	0.001000	113.4	0.008816	37.80	2518.1	2480.3	0.1361	8.9265	8.7904
10	283	0.01227	0.001000	106.4	0.009396	41.99	2519.9	2477.9	0.1510	8.9020	8.7510
11	284	0.01311	0.001000	99.91	0.01001	46.19	2521.7	2475.5	0.1658	8.8776	8.7118
12	285	0.01401	0.001000	93.84	0.01066	50.38	2523.6	2473.2	0.1805	8.8536	8.6731
13	286	0.01496	0.001000	88.18	0.01134	54.57	2525.4	2470.8	0.1952	8.8297	8.6345
14	287	0.01597	0.001000	82.90	0.01206	58.75	2527.2	2468.5	0.2098	8.8060	8.5962
15	288	0.01703	0.001000	77.98	0.01282	62.94	2529.1	2466.1	0.2243	8.7826	8.5583
16	289	0.01817	0.001001	73.38	0.01363	67.13	2530.9	2463.8	0.2388	8 7593	8.5205
17	290	0.01936	0.001001	69.09	0.01447	71.31	2532.7	2461.4	0.2533	8.7363	8.4830
18	291	0.02062	0.001001	65.09	0.01447	75.50	2534.5	2459.0	0.2677	8.7135	8.4453
19	292	0.02196	0.001001	61.34	0.01630	79.68	2536.4	2456.7	0.2820	8.6908	8.4088
20	293	0.02337	0.001002	57.84	0.01729	83.86	2538.2	2454.3	0.2963	8.6684	8.3721
21	294	0.02485	0.001002	54.56	0.01833	88.04	2540.0	2452.0	0.3105	8.6462	8.3357
22	295	0.02642	0.001002	51.49	0.01942	92.23	2541.8	2449.6	0.3247	8.6241	8.2994

(Contd.)

Table 1. *Saturated Steam (Temperature Table) Contd.*

1	2	3	4	5	6	7	8	9	10	11	12
23	296	0.02808	0.001002	48.62	0.02057	96.41	2543.6	2447.2	0.3389	8.6023	8.2634
24	297	0.02982	0.001002	45.93	0.02177	100.59	2545.5	2444.8	0.3530	8.5806	8.2276
25	298	0.03166	0.001003	43.40	0.02304	104.77	2547.3	2442.5	0.3670	8.5592	8.1922
26	299	0.03360	0.001003	41.03	0.02437	108.95	2549.1	2440.2	0.3810	8.5379	8.1569
27	300	0.03564	0.001003	38.81	0.02576	113.13	2550.9	2437.8	0.3949	8.5168	8.1219
28	301	0.03778	0.001004	36.73	0.02723	117.31	2552.7	2435.4	0.4088	8.4959	8.0871
29	302	0.04004	0.001004	34.77	0.02876	121 48	2554.5	2433.1	0.4227	8.4751	8.0524
30	303	0.04241	0.001004	32.93	0.03037	125.66	2556.4	2430.7	0.4363	8.4546	8.0181
31	304	0.04491	0.001005	31.20	0.03205	129.84	2558.2	2428.3	0.4503	8.4342	7.9339
32	305	0.04753	0.001005	29.57	0.03382	134.02	2560.0	2425.9	0.4640	8.4140	7.9500
33	306	0.05029	0.001005	28.04	0.03566	138.20	2561.8	2423.6	0.4777	8.3939	7.9162
34	307	0.05318	0.001006	26.60	0.03759	142.38	2563.6	2421.2	0.4913	8.3740	7.8827
35	308	0.05622	0.001006	25.24	0.03961	146.56	2565.4	2418.8	0.5049	8.3543	7.8494
36	309	0.05940	0.001006	23.97	0.04172	150.74	2567.2	2416.4	0.5184	8.3348	7.8164
37	310	0.06274	0.001007	22.76	0.04393	154.91	2569.0	2414.1	0.5319	8.3155	7.7835
38	311	0.06624	0.001007	21.63	0.04624	159.04	2570.8	2411.7	0.5453	8.2962	7.7509
39	312	0.06991	0.001007	20.56	0.04865	163.27	2572.6	2409.3	0.5588	8.2772	7.7184
40	313	0.07375	0.001008	19.55	0.05116	167.45	2574.4	2406.9	0.5721	8.2583	7 6862
41	314	0.07777	0.001008	18.59	0.05379	171.63	2576.2	2404.5	0.5854	8.2395	7.6541
42	315	0.08198	0.001009	17.69	0.05652	175.81	2577.9	2402.1	0.5987	8.2209	7.6222
43	316	0.08639	0.001009	16.84	0.05938	179.99	2579.7	2399.7	0.6120	8.2025	7.5305
44	317	0.09100	0.001009	16.04	0.06236	184.17	2581.5	2397.3	0.6252	8.1342	7.5590
45	318	0.09582	0.001010	15.28	0.06546	188.35	2583.3	2394.9	0.6383	8.1661	7.5278
46	319	0.10086	0.001010	14.56	0.06869	192.53	2585.1	2392.5	0.6514	3.1481	7.4967
47	320	0.10612	0.001011	13.88	0.07206	196.71	2586.9	2390.1	0.6645	3.1302	7.4737
48	321	0.11162	0.001012	13.23	0.07557	200.89	2588.6	2387.7	0.6776	8.1125	7.4349
49	322	0.11736	0.001012	12.62	0.07922	205.07	2590.4	2385.3	0.6906	8.0950	7.4044

(Contd.)

Table 1. *Saturated Steam (Temperature Table) Contd.*

1	2	3	4	5	6	7	8	9	10	11	12
50	323	0.12335	0.001012	12.05	0.08302	209.26	2592.2	2382.9	0.7035	8.0776	7.5741
55	328	0.15741	0.001014	9.579	0.1044	230.17	2601.0	2370.8	0.7677	7.9926	7.2249
60	333	0.1992	0.001017	7.679	0.1302	251.09	2609.7	2358.6	0.8310	7.9108	7.0798
65	338	0.2501	0.001020	6.202	0.1612	272.02	2618.4	2346.3	0.8933	7.8322	6.9389
70	343	0.3116	0.001023	5.046	0.1982	292.97	2626.9	2334.0	0.9548	7.7565	6.8017
75	348	0.3855	0.001026	4.134	0.2419	313.94	2635.4	2321.5	1.0154	7.6835	6.6681
80	353	0.4736	0.001030	3.409	0.2933	334.92	2643.8	2308.8	1.0753	7.6132	6.5379
85	358	0.5780	0.001032	2.829	0.3535	355.92	2652.0	2296.5	1.1343	7.5454	6.4111
90	363	0.7011	0.001036	2.361	0.4235	376.94	2660.1	2283.2	1.1925	7.4799	6.3074
95	368	0.8453	0.001040	1.982	0.5045	397.99	2668.1	2270.2	1.2501	7.4166	6.1665
100	373	1.0133	0.001043	1.673	0.5977	419.06	2676.0	2256.9	1.3069	7.3554	6.0485
105	378	1.2080	0.001050	1.419	0.7096	440.17	2683.7	2243.6	1.3630	7.2962	5.9332
110	383	1.4327	0.001052	1.210	0.8265	461.32	2696.3	2230.0	1.4185	7.2388	5.8203
115	388	1.6906	0.001056	1.036	0.9650	482.50	2698.7	2216.2	1.4733	7.1832	5.7099
120	393	1.9854	0.001060	0.8915	1.122	503.72	2706.0	2202.2	1.5276	7.1293	5.6017
125	398	2.3210	0.001065	0.7702	1.298	524.99	2713.0	2188.0	1.5813	7.0769	5.4956
130	403	2.7013	0.001070	0.6681	1.497	546.31	2719.9	2173.6	1.6344	7.0261	5.3917
135	408	3.131	0.001075	0.5818	1.719	567.68	2726.6	2158.9	1.6869	6.9766	5.2897
140	413	3.131	0.001080	0.5085	1.967	589.10	2733.1	2144.0	1.7390	6.9284	5.1894
145	418	4.155	0.001085	0.4460	2.242	610.60	2739.3	2128.7	1.7906	6.8815	5.0919
150	423	4.760	0.001090	0.3924	2.548	632.15	2745.4	2113.2	1.8416	6.8358	4.9942
155	428	5.433	0.001096	0.3464	2.886	653.78	2751.2	2097.4	1.8923	6.7911	4.8988
160	433	6.181	0.001100	0.3068	3.260	675.47	2756.7	2081.3	1.9425	6.7475	4.8050
165	438	7.008	0.001108	0.2724	3.671	697.25	2762.0	2064.8	1.9923	6.7048	4.7125
170	443	7.920	0.001110	0.2426	4.123	719.12	2767.1	2047.9	2.0416	6.6630	4.6214
175	448	8.924	0.001120	0.2165	4.618	741.07	2771.8	2030.7	2.0906	6.6221	4.4715
180	453	10.027	0.001127	0.1938	5.160	763.12	2776.3	2013.1	2.1393	6.5813	4.4426

(Contd.)

Table 1. *Saturated Steam (Temperature Table) Contd.*

1	2	3	4	5	6	7	8	9	10	11	12
185	458	11.233	0.001134	0.1739	5.752	785.26	2780.4	1995.2	2.1876	6.5424	4.3548
190	463	12.551	0.001141	0.1563	6.397	807.52	2784.3	1976.7	2.2356	6.5036	4.2680
195	468	13.987	0.001149	0.1408	7.100	829.88	2787.8	1957.9	2.2833	6.4654	4.1821
200	473	15.549	0.001565	0.1272	7.864	852.37	2790.9	1938.6	2.3307	6.4278	4.0971
205	478	17.243	0.001164	0.1150	8.694	874.99	2793.8	1918.8	2.3778	6.3906	4.0128
210	483	19.077	0.001173	0.1042	9.593	897.74	2796.2	1898.5	2.4247	6.3539	3.9292
215	488	21.060	0.001181	0.0946	10.57	920.63	2798.3	1877.6	2.4713	6.3176	3.8463
220	493	23.198	0.001190	0.0860	11.62	943.67	2799.9	1856.2	2.5178	6.2817	3.7639
225	498	25.501	0.001199	0.0783	12.76	966.89	2801.2	1834.3	2.5641	6.2461	3.6820
230	503	27.976	0.001209	0.0714	14.00	990.26	2802.0	1811.7	2.6102	6.2107	3.5995
235	508	30.632	0.001218	0.0652	15.33	1013.8	2802.3	1788.5	2.6562	6.1756	3.5194
240	513	33.478	0.001229	0.0596	16.76	1037.6	2802.3	1764.6	2.7020	6.1406	3.4386
245	518	36.523	0.001240	0.0546	18.31	1061.6	2801.6	1740.0	2.7478	6.1057	3.3579
250	523	39.776	0.001251	0.0500	19.99	1085.5	2800.4	1714.6	2.7935	6.0708	3.2773
255	528	43.246	0.001263	0.0459	21.79	1110.2	2798.7	1688.5	2.8392	6.0359	3.1967
260	533	46.943	0.001276	0.04213	23.73	1134.9	2796.4	1661.5	2.8848	6.0010	3.1162
265	538	50.877	0.001288	0.03871	25.83	1159.9	2793.5	1633.6	2.9306	5.9658	3.0352
270	543	55.058	0.001302	0.03559	28.10	1185.2	2789.9	1604.6	2.9763	5.9304	2.9541
275	548	59.496	0.001317	0.03274	30.55	1210.9	2785.5	1574.7	3.0223	5.8947	2.8724
280	553	64.202	0.001332	0.03013	33.19	1236.8	2780.4	1543.6	3.0683	5.8536	2.7903
285	558	69.186	0.001348	0.02773	36.06	1263.2	2774.5	1511.3	3.1146	5.8220	2.7074
290	563	74.461	0.001366	0.02554	39.16	1290.0	2767.6	1477.6	3.1611	5.7848	2.6237
295	568	80.037	0.001384	0.02351	42.53	1317.3	2759.8	1442.6	3.2079	5.7469	2.5390
300	573	85.927	0.001404	0.02165	46.19	1345.0	2751.0	1406.0	3.2552	5.7081	2.4529
305	578	92.144	0.001425	0.01993	50.18	1373.4	2741.1	1367.7	3.3029	5.6685	2.3656
310	583	98.700	0.001448	0.01833	54.54	1402.4	2730.0	1327.6	3.3512	5.6278	2.2766
315	588	105.67	0.001472	0.01686	59.33	1432.1	2717.6	1285.6	3.4002	5.5858	2.1856

(Contd.)

Table 1. *Saturated Steam (Temperature Table) Contd.*

1	2	3	4	5	6	7	8	9	10	11	12
320	593	112.89	0.001499	0.01548	64.60	1492.6	2703.7	1241.1	3.4500	5.5423	2.0923
325	598	120.56	0.001529	0.01419	70.45	1490.0	2688.0	1194.0	3.5008	5.4969	1.9961
330	603	128.63	0.001561	0.01299	76.99	1526.5	2670.2	1143.6	3.5528	4.4490	1.8962
335	608	137.12	0.001597	0.01185	84.36	1560.3	2649.7	1089.5	3.6063	5.3979	1.7916
340	613	146.05	0.001638	0.01078	92.76	1595.5	2626.2	1030.7	3.6616	5.3427	1.6811
345	618	155.45	0.001685	0.00976	102.4	1632.5	2598.9	966.4	3.7193	5.2828	1.5635
350	623	165.35	0.001741	0.00880	113.6	1671.9	2567.7	895.7	3.7800	5.2177	1.4377
355	628	175.77	0.001808	0.00786	127.2	1716.6	2530.4	813.8	3.8489	5.1442	1.2953
360	633	186.75	0.001896	0.00694	144.1	1764.2	2485.4	712.3	3.9210	5.0600	1.1390
365	638	198.33	0.002016	0.00601	166.3	1818.0	2428.0	610.0	4.0021	4.9579	0.9558
370	643	210.54	0.002136	0.00497	201.1	1890.2	2342.8	452.6	4.1108	4.8144	0.7036
374.15	647.15	221.20	0.00317	0.00317	315.5	2107.4	2107.4	0.0	4.4429	4.4429	0.0

Table 2. *Saturated Steam (Pressure Table)*

Pressure N/m^2 $p \times 10^{-5}$	Temperature °C	Volume, v m^2/kg		Density of Vapour, p kg/m^2	Enthalpy, h, kJ/kg		Latent heat L kJ/kg	Entropy, kJ/kg-K		L/T kJ/kg-K
		Liquid v_i	Vapour v_s		Liquid h_i	Vapour h_s		Liquid s_i	Vapour s_v	
1	2	3	4	5	6	7	8	9	10	11
0.010	6.983	0.001000	129.2	0.00774	29.34	2514.4	2485.0	0.1060	8.9767	8.8707
0.015	13.036	0.001000	87.98	0.01137	54.71	2525.5	2470.7	0.1957	8.8288	8.6331
0.020	17.513	0.001001	67.01	0.01492	73.46	2533.6	2460.2	0.2607	8.7246	8.4639
0.025	21.096	0.001002	54.26	0.01843	88.45	2540.2	2451.7	0.3119	8.6440	8.3321
0.030	24.100	0.001003	45.67	0.02190	101.00	2545.6	2444.6	0.3544	8.5785	8.2241
0.035	26.694	0.001003	39.48	0.02533	111.85	2550.4	2438.5	0.3907	8.5232	8.1325
0.040	28.983	0.001004	34.80	0.02873	121.41	2554.5	2433.1	0.4225	8.4755	8.0530
0.045	31.035	0.001005	31.14	0.03211	129.99	2558.2	2428.2	0.4507	8.4335	7.9828
0.050	32.898	0.001005	28.19	0.03547	137.77	2561.6	2423.8	0.4763	8.3960	7.9197
0.055	34.605	0.001006	25.77	0.03880	144.91	2564.7	2419.8	0.4995	8.3621	7.8626
0.060	36.183	0.001006	23.74	0.04212	151.50	2567.5	2416.0	0.5209	8.3312	7.8103
0.065	37.651	0.001007	22.02	0.04542	157.64	2570.2	2412.5	0.5407	8.3029	7.7622
0.070	39.025	0.001007	20.53	0.04871	163.38	2572.6	2409.2	0.5591	8.2767	7.7176
0.075	40.316	0.001008	19.24	0.05198	168.77	2574.9	2406.2	0.5763	8.2523	7.6760
0.080	41.534	0.001008	18.10	0.05523	173.86	2577.1	2403.2	0.5925	8.2296	7.6371
0.085	42.689	0.001009	17.10	0.05848	178.69	2579.2	2400.5	0.6079	8.2082	7.6003
0.090	43.787	0.001009	16.20	0.06171	183.28	2581.1	2397.9	0.6224	8.1881	7.5657
0.095	44.833	0.001010	15.40	0.06493	187.65	2583.0	2395.3	0.6361	8.1691	7.5330
0.10	45.833	0.001010	14.67	0.06814	191.83	2584.8	2392.9	0.6493	8.1511	7.5018
0.15	53.997	0.001014	10.02	0.09977	225.97	2599.2	2373.2	0.7549	8.0093	7.2544
0.20	60.086	0.001017	7.650	0.1307	251.45	2609.0	2358.4	0.8321	7.9094	7.0773
0.25	64.992	0.001019	6.204	0.1612	271.99	2618.3	2346.4	0.8932	7.8323	6.9391
0.30	69.124	0.001023	5.229	0.1912	289.30	2625.4	2336.1	0.9441	7.7695	6.8254

(Contd.)

Table 2. *Saturated Steam (Pressure Table) Contd.*

1	2	3	4	5	6	7	8	9	10	11
0.35	72.651	0.001024	4.529	0.2210	304.20	2631.5	2327.3	0.9880	7.7168	6.7288
0.40	75.886	0.001026	3.993	0.2504	317.65	2636.9	2319.2	1.0261	7.6709	6.6448
0.45	78.743	0.001028	3.576	0.2796	329.64	2614.7	2312.0	1.0603	7.6307	6.5704
0.50	81.345	0.001030	3.240	0.3086	340.56	2646.0	2305.4	1.0912	7.5947	6.5035
0.55	83.737	0.001031	2.964	0.3374	350.61	2649.9	2299.3	1.1194	7.5623	6.4429
0.60	85.954	0.001033	2.732	0.3661	359.93	2653.6	2293.6	1.1454	7.5327	6.3873
0.65	88.021	0.001035	2.535	0.3945	368.62	2656.9	2288.3	1.1696	7.5055	6.3359
0.70	89.959	0.001036	2.365	0.4229	376.77	2660.1	2283.3	1.1921	7.4804	6.2883
0.75	91.785	0.001037	2.217	0.4511	384.45	2663.0	2278.6	1.2131	7.4570	6.2439
0.80	93.512	0.001038	2.087	0.4792	391.72	2665.8	2274.0	1.2330	7.4352	6.2022
0.85	95.152	0.001040	1.972	0.5071	398.63	2668.4	2269.8	1.2518	7.4147	6.1629
0.90	96.713	0.001041	1.869	0.5350	405.21	2670.9	2265.6	1.2696	7.3954	6.1258
0.95	98.204	0.001042	1.777	0.5627	411.49	2673.2	2261.7	1.2865	7.3771	6.0906
1.00	99.632	0.001043	1.694	0.5904	417.51	2675.4	2257.9	1.3027	7.3598	6.0571
1.50	111.37	0.001053	1.159	0.8628	467.13	2693.4	2226.2	1.4336	7.2234	5.7898
2.00	120.23	0.001060	0.8854	1.129	504.17	2706.3	2201.6	1.5301	7.1268	5.5967
2.50	127.43	0.001067	0.7184	1.392	535.34	2716.4	2181.0	1.6071	7.0520	5.4449
3.00	133.54	0.001073	0.6056	1.651	561.43	2724.7	2163.2	1.6716	6.9906	5.3190
3.50	138.87	0.001079	0.5240	1.908	584.27	2731.6	2147.4	1.7273	6.9392	5.2119
4.00	143.62	0.001084	0.4622	2.163	604.67	2737.6	2133.0	1.7764	6.8943	5.1179
4.50	147.92	0.001088	0.4132	2.417	623.16	2742.9	2119.7	1.8204	6.8647	5.0343
5.00	151.84	0.001092	0.3747	2.669	640.12	2747.5	2107.4	1.8604	6.8192	4.9588
5.50	155.46	0.001097	0.3425	2.920	655.66	2751.7	2096.1	1.8969	6.7869	4.8900
6.00	158.84	0.001101	0.3155	3.170	670.42	2755.5	2085.0	1.9308	6.7575	4.8267
6.50	162.00	0.001104	0.2923	3.419	684.16	2758.8	2074.7	1.9623	6.7304	4.7681
7.00	164.96	0.001108	0.2727	3.667	697.06	2762.0	2064.9	1.9918	6.7302	4.7134
7.50	167.76	0.011116	0.2554	3.916	709.30	2764.8	2055.5	2.0191	6.6812	4.8621

(Contd.)

Table 2. *Saturated Steam (Pressure Table) Contd.*

1	2	3	4	5	6	7	8	9	10	11
8.00	170.41	0.001115	0.2403	4.162	720.94	2767.5	2046.5	2.0457	6.6596	4.6139
8.50	172.95	0.001118	0.2268	4.410	732.05	2769.9	2037.9	2.0670	6.6388	4.5718
9.00	175.36	0.001121	0.2148	4.655	742.64	2772.1	2029.5	2.0941	6.6192	4.5251
9.50	177.66	0.001125	0.2040	4.890	752.82	2774.2	2021.4	2.1164	6.6078	4.4914
10.00	179.88	0.001127	0.1943	5.147	762.61	2776.2	2013.6	2.1382	6.5828	4.4446
11.00	184.07	0.001133	0.1774	5.637	781.13	2779.7	1998.5	2.1786	6.5497	4.3711
12.00	187.96	0.001138	0.1632	6.127	798.43	2782.7	1984.3	2.2161	6.5194	4.3033
13.00	191.61	0.001144	0.1511	6.617	814.70	2785.7	1970.7	2.2510	6.4913	4.2403
14.00	195.04	0.001149	0.1407	7.106	830.08	2787.8	1957.7	2.2837	6.4651	4.1814
15.00	198.29	0.001154	0.1317	7.596	844.67	2789.9	1945.7	2.3145	6.4406	4.1261
16.00	201.37	0.001159	0.1237	8.085	858.56	2791.7	1933.2	2.3436	6.4175	4.0739
17.00	204.31	0.001163	0.1166	8.575	871.84	2793.4	1921.5	2.3713	6.3957	4.0244
18.00	207.11	0.001168	0.1103	9.065	884.58	2794.8	1910.3	2.3976	6.3751	3.9775
19.00	209.80	0.001172	0.1047	9.555	896.81	2796.1	1899.3	2.4228	6.3554	3.9326
20.00	212.37	0.001176	0.0995	10.05	908.59	2797.2	1886.6	2.4469	6.3367	3.8898
21.00	214.85	0.001180	0.09489	10.54	919.96	2798.2	1878.2	2.4700	6.3187	3.8487
22.00	217.24	0.001185	0.09065	11.03	930.95	2799.1	1868.1	2.4922	6.3015	3.8093
23.00	219.55	0.001189	0.08677	11.52	941.60	2799.8	1858.2	2.5136	6.2849	3.7713
24.00	221.78	0.001193	0.0832	12.02	951.93	2800.4	1848.5	2.5343	6.2690	3.7347
25.00	223.94	0.001197	0.07991	12.51	961.96	2800.9	1839.0	2.5543	6.2536	3.6993
26.00	226.04	0.001201	0.07686	13.01	971.72	2801.4	1829.6	2.5736	6.2387	3.6651
27.00	228.07	0.001205	0.07402	13.51	981.22	2801.7	1820.5	2.5924	6.2244	3.6320
28.00	230.05	0.001209	0.07139	14.01	990.48	2802.0	1811.5	2.6106	6.2104	3.5998
29.00	231.97	0.001212	0.06893	14.51	999.53	2802.2	1802.6	2.6283	6.1969	3.5686
30.00	233.84	0.001216	0.06663	15.01	1000.84	2802.3	1793.9	2.6455	6.1837	3.5382
31.00	235.67	0.001220	0.06447	15.51	1017.0	2802.3	1785.4	2.6623	6.1709	3.5086
32.00	237.45	0.001223	0.06244	16.02	1025.4	2802.3	1776.9	2.6786	6.1585	3.4799

(Contd.)

Table 2. *Saturated Steam (Pressure Table) Contd.*

1	2	3	4	5	6	7	8	9	10	11
33.00	239.18	0.001227	0.06053	16.52	1033.7	2802.3	1768.6	2.6945	6.1463	3.4518
34.00	240.88	0.001231	0.05873	17.03	1041.8	2802.1	1760.3	2.7101	6.1344	3.4243
35.00	242.54	0.001234	0.05703	17.54	1049.8	2802.0	1752.2	2.7253	6.1228	3.3975
36.00	244.16	0.001238	0.05541	18.05	1057.6	2801.7	1744.2	2.7401	6.1115	3.3714
37.00	245.75	0.001241	0.05389	18.56	1065.2	2801.4	1736.2	2.7545	6.1004	3.3457
38.00	247.31	0.001245	0.05244	19.07	1072.7	2800.1	1728.4	2.7689	6.0896	3.3207
39.00	248.84	0.001248	0.05106	19.58	1080.1	2800.8	1720.6	2.7829	6.0789	3.2960
40.00	250.33	0.001252	0.04975	20.10	1087.4	2800.3	1712.9	2.7965	6.0885	3.2720
41.00	251.80	0.001255	0.04850	20.62	1094.6	2799.9	1705.3	2.8099	6.0583	3.2484
42.00	253.24	0.001258	0.04731	21.14	1101.6	2799.4	1697.8	2.8231	6.0482	3.2251
43.00	254.66	0.001262	0.04617	21.66	1108.5	2798.9	1690.3	2.8360	6.0863	3.2023
44.00	256.05	0.001265	0.04508	22.18	1115.4	2798.3	1682.9	2.8487	6.0286	3.1799
45.00	257.41	0.001269	0.04404	22.71	1122.1	2797.7	1675.6	2.8612	6.0191	3.1579
46.00	258.75	0.001272	0.04304	23.24	1128.8	2797.0	1668.3	2.8735	6.0097	3.1362
47.00	260.07	0.001276	0.04208	23.76	1135.3	2796.4	1661.1	2.8855	6.0004	3.1149
48.00	261.37	0.001279	0.04116	24.29	1141.8	2795.7	1653.9	2.8974	5.9913	3.0939
49.00	262.65	0.041282	0.04028	24.83	1148.2	2794.9	1646.8	2.9091	5.9824	3.0733
50.00	263.91	0.001286	0.03943	25.36	1154.5	2794.2	1639.7	2.9206	5.9735	3.0529
52.00	266.37	0.001292	0.03782	26.44	1166.8	2792.6	1625.7	2.9431	5.9561	3.0130
54.00	268.76	0.001299	0.03633	27.52	1178.9	2790.8	1611.9	2.9650	5.9392	2.9742
56.00	271.09	0.001305	0.03495	28.62	1190.8	2789.0	1598.2	2.9863	5.9227	2.9364
58.00	273.35	0.001312	0.03365	29.72	1202.3	2787.0	1584.7	3.0071	5.9066	2.8995
60.00	275.55	0.001319	0.03244	30.83	1213.7	2785.0	1571.3	3.0273	5.8908	2.8635
62.00	277.70	0.001325	0.03130	31.95	1224.8	2782.9	1558.0	3.0471	5.8753	2.8282
64.00	279.79	0.001331	0.03023	33.08	1235.7	2780.6	1544.9	3.0664	5.8601	2.8057
66.00	281.84	0.001338	0.02922	34.22	1246.5	2778.3	1531.9	3.0853	5.8452	2.7601
68.00	283.84	0.001345	0.02827	35.37	1257.0	2775.9	1518.9	3.1038	5.8306	2.7268
70.00	285.79	0.001351	0.02737	36.53	1267.4	2773.5	1506.0	3.1219	5.8162	2.6943

(Contd.)

Table 2. *Saturated Steam (Pressure Table) Contd.*

1	2	3	4	5	6	7	8	9	10	11
72.00	287.70	0.001358	0.02652	37.70	1277.6	2770.9	1493.3	3.1397	5.8020	2.6637
74.00	289.57	0.001364	0.02572	38.89	1287.7	2768.3	1480.5	3.1571	5.7880	2.6311
76.00	291.41	0.001371	0.02495	40.08	1297.6	2765.5	1467.9	3.1742	5.7742	2.6000
78.00	293.21	0.001377	0.02422	41.29	1307.4	2762.8	1455.3	3.1911	5.7606	2.5695
80.00	294.97	0.001384	0.02353	42.51	1317.1	2759.9	1442.8	3.2076	5.7471	2.5395
82.00	296.70	0.001390	0.02286	43.74	1326.6	2757.0	1430.3	3.2239	5.7338	2.5099
84.00	298.39	0.001397	0.02223	44.98	1336.1	2754.0	1417.9	3.2399	5.7207	2.4808
86.00	300.06	0.001404	0.02163	46.24	1345.4	2750.9	1405.5	3.2558	5.7076	2.4518
88.00	301.70	0.001411	0.02105	47.51	1354.6	2747.8	1393.2	3.2713	5.6948	2.4235
90.00	303.31	0.001418	0.02050	48.79	1363.7	2744.6	1380.9	3.2867	5.6820	2.3953
92.00	304.89	0.001425	0.01996	50.09	1372.8	2741.4	1368.6	3.3018	5.6694	2.3676
94.00	306.44	0.001432	0.01945	51.40	1381.7	2738.0	1356.3	3.3168	5.6568	2.3400
96.00	307.97	0.001438	0.01897	52.73	1390.6	2734.7	1344.1	3.3315	5.6444	2.3129
98.00	309.48	0.001445	0.01849	54.07	1399.3	2731.2	1331.9	3.3461	5.6321	2.2860
100.00	310.96	0.001453	0.01804	55.43	1408.0	2727.7	1319.7	3.3605	5.6198	2.2593
102	312.42	0.001460	0.01760	56.80	1416.7	2724.2	1307.5	3.3748	5.6076	2.2328
104	313.86	0.001467	0.01718	58.19	1425.2	2720.6	1295.3	3.3889	5.5955	2.2134
106	315.27	0.001474	0.01678	59.60	1433.7	2716.9	1283.1	3.4029	5.5835	2.1806
108	316.67	0.001481	0.01639	61.03	1442.2	2713.1	1270.9	3.4167	5.5715	2.1548
110	318.05	0.001488	0.01601	62.48	1450.6	2709.3	1258.7	3.4304	5.5595	2.1291
112	319.40	0.001496	0.01564	63.94	1458.9	2705.4	1246.5	3.4440	5.5476	2.1036
114	320.74	0.001503	0.01528	65.43	1467.2	2701.4	1234.3	3.4575	5.5358	2.0783
116	322.06	0.001511	0.01494	66.93	1475.4	2697.4	1222.0	3.4708	5.5239	2.0531
118	323.36	0.001519	0.01461	68.46	1483.6	2693.3	1209.7	3.4840	5.5121	2.0281
120	324.65	0.001526	0.01428	70.01	1491.8	2689.2	1197.4	3.4972	5.5002	2.0030
122	325.91	0.001534	0.01397	71.59	1499.9	2684.9	1185.0	3.5102	5.4484	1.9382
124	327.17	0.001542	0.01366	73.19	1508.0	2680.6	1172.6	3.5232	5.4765	1.9533

(*Contd.*)

Table 2. *Saturated Steam (Pressure Table) Contd.*

1	2	3	4	5	6	7	8	9	10	11
126	328.40	0.001550	0.01337	74.81	1516.0	2676.1	1160.1	3.5361	5.4646	1.9285
128	329.62	0.001559	0.01308	76.46	1524.0	2671.6	1147.6	3.5488	5.4527	1.9039
130	330.83	0.001567	0.01280	78.14	1532.0	2667.0	1135.0	3.5616	5.4408	1.8792
132	332.02	0.001576	0.01252	79.85	1540.0	2662.3	1122.3	3.5742	5.4288	1.8546
134	333.19	0.001584	0.01226	81.59	1547.9	2657.4	1109.5	3.5868	5.4168	1.8300
136	334.36	0.001593	0.01200	83.36	1555.8	2652.5	1096.7	3.5993	5.4047	1.8054
138	335.51	0.001602	0.01174	85.16	1563.7	2647.5	1083.8	3.6118	5.3925	1.7807
140	336.64	0.001610	0.01150	86.09	1571.6	2642.4	1070.7	3.6242	5.3803	1.7561
142	337.76	0.001620	0.01125	88.86	1579.5	2637.1	1057.6	3.6366	5.3679	1.7313
144	338.87	0.001630	0.01102	90.77	1587.4	2631.8	1044.4	3.6490	5.3555	1.7065
146	339.97	0.001638	0.01079	92.71	1595.3	2626.3	1031.0	3.6613	5.3431	1.6818
148	341.06	0.001648	0.01056	94.69	1603.1	2620.7	1017.6	3.6736	5.3305	1.6569
150	342.13	0.001658	0.01034	96.71	1611.0	2615.0	1004.0	3.6859	5.3178	1.6319
152	343.19	0.001668	0.01012	98.77	1618.9	2609.2	990.3	3.6981	5.3051	1.6070
154	344.24	0.001678	0.00991	100.9	1626.8	2603.3	976.5	3.7104	5.2922	1.5818
156	345.28	0.001688	0.00970	103.0	1634.7	2597.3	962.6	3.7226	5.2793	1.5567
158	346.31	0.001699	0.00950	105.2	1642.6	2591.1	948.5	3.7348	5.2663	1.5315
160	347.33	0.001710	0.00931	107.4	1650.5	2584.9	934.3	3.7471	5.2531	1.5060
162	348.34	0.001722	0.00912	109.7	1658.5	2578.5	920.0	3.7594	5.2399	1.4805
164	349.34	0.001733	0.00892	112.0	1666.5	2572.1	905.6	3.7717	5.2267	1.4550
166	350.32	0.001748	0.00874	114.4	1674.5	2565.5	891.0	3.7842	5.2132	1.4290
168	351.30	0.001757	0.00855	116.9	1683.0	2558.6	875.6	3.7974	5.1994	1.4020
170	352.26	0.001770	0.00837	119.5	1691.7	2551.6	859.9	3.8107	5.1855	1.3748
172	353.22	0.001782	0.00819	122.1	1700.4	2544.4	844.1	3.8240	5.1713	1.3473
174	354.17	0.001796	0.00801	124.8	1709.0	2537.1	828.1	3.8372	5.1570	1.3198
176	355.11	0.001810	0.00784	127.6	1717.6	2529.5	811.9	3.8504	5.1425	1.2921
178	356.04	0.001825	0.00767	130.4	1726.2	2521.8	795.6	3.8635	5.1278	1.2643

(Contd.)

Table 2. *Saturated Steam (Pressure Table) Contd.*

1	2	3	4	5	6	7	8	9	10	11
180	356.96	0.001840	0.00749	133.4	1734.8	2513.9	779.1	3.8765	5.1128	1.2363
182	357.87	0.001856	0.00733	136.4	1743.4	2505.8	762.3	3.8896	5.0975	1.2079
184	358.77	0.001872	0.00716	139.6	1752.1	2497.4	745.3	3.9028	5.0820	1.1792
186	359.67	0.001890	0.00700	142.8	1760.9	2488.8	727.9	3.9160	5.0661	1.1501
188	360.55	0.001907	0.00684	146.2	1769.7	2479.8	710.1	3.9294	5.0498	1.1204
190	361.43	0.001926	0.00668	149.8	1778.7	2470.6	692.0	3.9429	5.0332	1.0903
192	362.30	0.001945	0.006517	153.4	1787.8	2461.1	673.3	3.9566	5.0160	1.0594
194	363.16	0.001966	0.006358	157.3	1797.0	2451.1	654.1	3.9706	4.9983	1.0277
196	364.02	0.001988	0.006198	161.3	1806.5	2440.7	634.2	3.9849	4.9801	0.9952
198	364.86	0.002012	0.006038	165.6	1816.3	2429.8	613.5	3.9996	4.9610	0.9624
200	365.70	0.002037	0.005877	170.2	1826.5	2418.4	591.9	4.0149	4.9412	0.9263
202	366.53	0.002064	0.005714	175.0	1837.0	2406.2	569.2	4.0308	4.9204	0.8896
206	368.17	0.002125	0.005379	185.9	1859.9	2379.3	519.5	4.0651	4.8750	0.8099
210	369.78	0.002201	0.005023	199.1	1886.3	2347.6	461.3	4.1048	4.8223	0.7175
214	317.37	0.002306	0.004624	216.3	1919.0	2307.4	388.4	4.1543	4.7569	0.6026
218	379.92	0.002483	0.004115	243.0	1967.2	2248.0	280.8	4.2276	4.6622	0.4346
221.20	374.15	0.00317	0.00317	315.5	2107.4	2107.4	0.0	4.4429	4.4429	0.0

Table 3. *Enthalpy of Superheated Steam*

Pressure $p \times 10^{-5}$ N/m^2	Saturated Temp. °C	Saturated Enthalpy kJ/kg	Enthalpy of Superheated System at t°C										
			T = 100	150	200	250	300	350	400	450	500	550	600
1	2	3	4	5	6	7	8	9	10	11	12	13	14
0.01	69.83	2514.4	2688.6	2783.7	2808.1	2977.7	3076.8	3177.5	3279.7	3383.6	3489.2	3596.5	3705.6
0.05	32.90	2561.6	2688.1	2783.4	2879.9	2977.6	3076.7	3177.4	3279.7	3383.6	3489.2	3596.5	3705.6
0.10	45.83	2584.8	2687.5	2783.1	2879.6	2977.4	3076.6	3177.3	3279.6	3383.5	3489.1	3596.5	3705.6
0.20	60.09	2609.9	2686.3	2782.3	2879.2	2977.4	3076.4	3177.1	3279.4	3383.4	3489.0	3596.4	3705.4
0.30	69.12	2625.4	2685.1	2781.6	2878 7	2976.8	3076.1	3176.9	3279.3	3383.3	3488.9	3596.3	3705.4
0.40	75.89	2636.9	2683.8	2780.9	2878.2	2976.5	3075.9	3176.8	3279.1	3383.1	3488.8	3596.2	3705.1
0.50	81.35	2646.0	2682.6	2780.1	2877.7	2976.1	3075.7	3176.6	3279.0	3383.0	3488.7	3596.1	3705.2
0.60	85.95	2653.6	2681.3	2779.4	2877.3	2975.8	3075.4	3176.4	3278.8	3382.9	3488.6	3596.0	3705.2
0.70	89.96	2660.1	2680.0	2778.6	2876.8	2975.5	3075.2	3176.2	3278.7	3382.7	3488.5	3596.9	3705.0
0.80	93.51	2665.8	2678.8	2778.8	2876.3	2975.2	3075.0	3176.0	3278.5	3382 6	3488.4	3596.9	3705.0
0.90	96.71	2670.9	2677.5	2771.1	2875.8	2974.8	3074.7	3175.8	3278.4	3382.5	3488.3	3595.7	3704.9
1.0	99.63	2675.4	2676.2	2776.3	2875.4	2974.5	3074.5	3175.6	3278.2	3382.4	3488.1	3595.6	3704.8
1.2	104.81	2683.4	–	2774.8	2874.4	2973.9	3074.0	3175.3	3277.9	3382.1	3487.9	3595.4	3704.6
1.4	109.32	2690.3	–	2773.2	2873.4	2973.2	3073.5	3174.9	3277.6	3381.8	3487.7	3595.2	3704.5
1.6	113.32	2696.2	–	2771.7	2872.5	2972.5	3073.0	3174.5	3277.3	3381.6	3487.5	3595.0	3704.5
1.8	116.93	2701.5	–	2770.1	2871.5	2971.9	3072.6	3174.1	3777.0	3381.3	3487.3	3594.9	3704.1
2.0	120.23	2706.3	–	2768.5	2870.5	2971.2	3072.1	3173.8	3276.7	3381.1	3487.0	3594.7	3704.0
2.5	127.43	2716.4	–	2764.5	2868.0	2969.6	3070.9	3172.8	3275.9	3380.4	3486.5	3594.2	3703.6
3.0	133.54	2724.7	–	2760.4	2865.5	2967.9	3069.7	3171.9	3275.2	3379.8	3486.0	3593.7	3703.2
3.5	138.87	2731.6	–	2756.2	2862.9	2966.2	3068.5	3170.9	3274.4	3379.1	3485.4	3593.2	3702.7
4.0	143.62	2737.6	–	2752.0	2860.4	2964.5	3067.2	3170.0	3273.6	3378.5	3484.9	3592.8	3702.3
4.5	147.92	2742.9	–	2747.6	2857.8	2962.7	3065.9	3169.0	3272.8	3377.8	3484.3	3592.3	3701.9
5.0	151.84	2747.5	–	–	2855.1	2961.1	3064.8	3168.1	3272.1	3377.2	3483.8	3591.8	3701.5
5.5	155.47	2751.7	–	–	2852.5	2959.3	3063.5	3167.2	3271.3	3376.6	3483.2	3591.4	3700.1
6.0	158.84	2755.5	–	–	2849.7	2957.6	3062.3	3166.2	3270.6	3376.0	3482.7	3590.9	3700.7

(*Contd.*)

Table 3. *Enthalpy of Superheated Steam (Contd.)*

1	2	3	4	5	6	7	8	9	10	11	12	13	14
6.5	161.99	2758.9	–	–	2847.0	2955.8	3061.0	3165.3	3269.8	3375.3	3482.1	3590.4	3700.3
7.0	164.96	2762.0	–	–	2844.2	2954.0	3059.8	3164.3	3269.0	3374.7	3481.6	3589.9	3699.9
7.5	167.76	2764.8	–	–	2841.4	2952.2	3058.5	3163.4	3268.3	3374.0	3481.0	3589.5	3699.5
8.0	170.41	2767.5	–		2838.6	2950.4	3057.3	3162.4	3267.5	3373.4	3480.5	3589.0	3699.1
8 5	172.94	2769.9	–	–	2835.7	2948.6	3056.0	3161.4	3266.7	3372.7	3479.9	3538.5	3698.6
9.0	175.36	2772.1	–		2832.7	2946.8	3054.7	3160.5	3266.0	3372.1	3479.4	3588.1	3698.2
9.5	177.67	2774.2	–	–	2829.8	2944.9	3053.4	3159.5	3265.2	3371.5	3478.8	3587.6	3697.8
10.0	179.88	2776.2	–	–	2826.8	2943.0	3052.1	3158.5	3264.4	3370.8	3478.3	3587.1	3697.4
11.0	184.07	2779.7	–	–	2820.7	2939.3	3049.6	3156.6	3262.9	3369.5	3477.2	3586.2	3696.6
12.0	187.96	2782.7	–	–	2814.4	2935.4	3046.9	3154.6	3261.3	3368.2	3476.1	3585.2	3695.8
13.0	191.61	2785.4	–	–	2808.0	2931.5	3044.3	3152.7	3259.7	3366.9	3475.0	3584.3	3695.5
14.0	195.04	2787.8	–	–	2801.4	2927.6	3041.6	3150.7	3258.2	3365.6	3473.9	3383.3	3694.0
16.0	201.37	2791.7	–	–	–	2919.4	3036.2	3146.7	3255.0	3363.0	3471.7	3581.4	3692.1
18.0	207.11	2794.8	–	–	–	2911.0	3030.7	3142.7	3251.4	3360.4	3469.5	3579.5	3690.6
20.0	212.37	2797.2	–	–	–	2902.4	3025.0	3138.6	3248.7	3357.8	3467.3	3577.6	3689.2
22.0	217.24	2799.1	–	–	–	2893.4	3019.3	3134.5	3245.5	3355.2	3465.1	3575.7	3687.6
24.0	221.78	2800.4	–	–	–	2884.2	3013.4	3130.4	3242.3	3452.6	3562.9	3573.8	3685.9
26.0	226.04	2801.4	–	–	–	2874.7	3007.4	3126.1	3239.0	3349.0	3460.6	3571.9	3684.3
28.0	230.05	2802.0	–	–	–	2864.9	3001.3	3121.9	3235.8	3347.3	3458.4	3570.0	3682.5
30.0	233.84	2802.3	–	–	–	2854.8	2995.1	3117.5	3232.5	3344.6	3456.2	3568.1	3681.0
35.0	242.54	2802.0	–	–	–	2828.1	2979.0	3106.5	3224.2	3338.0	3450.6	3563.4	3676.9
40.0	250.33	2800.3	–	–	–	–	2962.0	3095.1	3215.7	3331.2	3445.0	3558.6	3672.8
45.0	257.41	2797.7	–	–	–	–	2944.2	3083.3	3324.4	3324.4	3439.3	3553.8	3668.6
50.0	263.91	2794.2	–	–	–	–	2925.5	3071.2	3198.3	3317.5	3433.7	3549.0	3664.5
55.0	269.93	2789.9	–	–	–	–	2905.7	3058.7	3189.3	3310.5	3430.3	3544.1	3660.3
60.0	275.55	2785.0	–	–	–	–	2885.0	3045.8	3180.1	3303.5	3422.2	3539.3	3656.2
65.0	280.82	2779.5	–	–	–	–	2862.9	3032.4	3170.8	3296.3	3416.4	3534.4	3652.0

(*Contd.*)

Table 3. *Enthalpy of Superheated Steam (Contd.)*

1	2	3	4	5	6	7	8	9	10	11	12	13	14
70.0	285.79	2773.5	–	–	–	–	2839.4	3018.7	3161.2	3289.1	3410.6	3529.6	3647.9
75.0	290.50	2766.9	–	–	–	–	2814.0	3004.5	3151.6	3281.8	3404.7	3524.7	3643.7
80.0	294.97	2759.9	–	–	–	–	2786.8	2989.9	3141.6	3274.3	3398.8	3519.7	3639.5
85.0	299.23	2752.5	–	–	–	–	2758.2	2974.7	3131.4	3266.8	3392.8	3514.8	3635.3
90.0	303 31	2744.6	–	–	–	–	–	2959.0	8121.2	3259.2	3386.8	3509.8	3631.1
95.0	307.21	2736.4	–	–	–	–	–	2942.7	3110.7	3251.8	3380.7	3504.9	3626.9
100.0	310.96	2727.7	–	–	–	–	–	2925.8	3099.9	3243.6	3374.6	3499.8	3622.7
110.0	318.05	2709.3	–	–	–	–	–	2889.6	3077.8	3227.7	3362.2	3489.7	3614.2
			300	350	400	450	500	550	600	650	700	750	800
120.0	324.65	2689.2	–	2849.7	3054.8	3211.4	3349.6	3479.6	3605.2	3730.2	3854.3	3978.3	4102.7
130.0	330.83	2667.0	–	2805.0	3030.7	3194.6	3336.8	3469.3	3597.1	3722.9	3618.0	3972.9	4098.0
140.0	336.64	2642.4	–	2754.2	3005.6	3177.4	3323.8	3458.8	3588.5	3715.6	3841.7	3967.5	4093.3
150.0	342.13	2615.0	–	2694.8	2979.1	3159.7	3310.6	3448.3	3579.8	3708.3	3835.4	3962.1	4088.6
160.0	347.33	2584.9	–	2620.8	2951.3	3141.6	3297.1	3437.7	3571.0	3700.0	3829.1	3956.7	4084.0
170.6	352.26	2551.6	–	–	2921.7	3123.1	3283.5	3427.0	3557.8	3689.8	3819.7	3998.5	4077.0
180.0	356.96	2513.9	–	–	2890.3	3104.0	3269.6	3416.1	3553.4	3686.1	3816.5	3945.8	4074.6
190.0	361.43	2470.6	–	–	2856.7	3084.4	3255.4	3405.2	3544.5	3678.6	3810.2	3940.4	4070.0
200.0	365.70	2418.4	–	–	2820.5	3064.3	3241.1	3394.1	3535.5	3671.1	3803.8	3935.0	4065.3
210.0	369.78	2347.6	–	–	2781.3	3043.6	3226.5	3382.9	3526.5	3663.6	3797.5	3925.5	4060.6
220.0	373.69	2195.6	–	–	2738.8	3022.3	3211.7	3371.6	3517.6	3656.1	3791.1	3924.1	4055.9
221.20	374.15	2107.4	–	–	2732.5	3012.3	3202.7	3363.7	3510.4	3650.2	3786.1	3920.1	4052.9
250.0	–	–	–	–	2582.0	2954.3	3165.9	3337.0	3489.9	3633.4	3771.9	3907.7	4041.9
300.0	–	–	–	–	2161.8	2825.6	3085.0	3277.4	3443.0	3595.0	3739.7	3880.3	4018.5
350.0	–	–	–	–	–	2676.4	2998.3	3215.4	3395.1	3556.1	3707.3	3852.9	3995.1
400.0	–	–	–	–	–	2515.6	2906.8	3151.6	3346.4	3517.0	3674.8	3825.5	3971.7
450.0	–	–	–	–	–	2384.2	2813.5	3086.5	3297.4	3477.8	3642.4	3798.1	3948.4
500.0	–	–	–	–	–	2293.2	2723.0	3021.1	3248.3	3438.9	3610.2	3770.9	3925.3

(*Contd.*)

Table 4. *Entropy of Superheated Steam (Contd.)*

1	2	3	4	5	6	7	8	9	10	11	12	13	14
500.0	–	–	–	–	4.0083	4.6026	5.1782	5.5525	5.8207	6.0331	6.2138	6.3749	6.5222
550.0	–	–	–	–	3.9705	4.5000	5.0470	5.4421	5.7321	5.9558	6.1436	6.3095	6.4600
600.0	–	–	–	–	–	4.4246	4.9374	5.3463	5.6477	5.8827	6.0775	6.2483	6.4031
700.0	–	–	–	–	–	4.3182	4.7688	5.1748	5.4931	5.7480	5.9562	6.1361	6.2979
800.0	–	–	–	–	–	4.2434	4.6488	5.0382	5.3595	5.6270	5.8470	6.0354	6.2034
900.0	–	–	–	–	–	4.1852	4.5602	4.9276	5.2468	5.5195	5.7479	5.9439	6.1179
1000.0						4.1373	4.4913	4.8393	5.1505	5.4267	5.6579	5.8600	6.0397

Table 5. *Volume of Superheated Steam*

Pressure $p \times 10^{-5}$ N/m²	Saturated Temp. °C	Saturated Volume m³/kg	Volume of Superheated Steam at t°C										
			t = 100	150	200	250	300	350	400	450	500	550	600
1	2	3	4	5	6	7	8	9	10	11	12	13	14
0.01	6 983	129.20	172.20	195.30	218.40	241.40	264.50	287.60	310.70	333.70	356.80	379.90	403.00
0.05	32.90	28.19	34.420	39.04	43.66	48.28	52.90	57.51	62.13	66.74	71.36	75.98	80.59
0.10	45.83	14.67	17.200	19.51	21.83	24.14	26.45	28.75	31.06	33.37	35.68	37.99	40.29
0.20	60.09	7.650	8.585	9.748	10.907	12.064	13.219	14.374	15.529	16 684	17.838	18.992	20.146
0.30	69.12	5.229	5.714	6 493	7.268	8.040	8.811	9.581	10.351	11.121	11.891	12.661	13.430
0.40	75.89	3.993	4.279	4.865	5.447	6.027	6.606	7.184	7.7625	8.340	8.917	9.494	10.072
0.50	81.35	3.240	3.418	3.889	4.356	4.821	5.284	5.747	6.209	6.671	7.133	7.595	8.057
0.60	85.95	2.732	2.844	3.238	3.628	4.016	4.402	4.788	5.174	5.559	5.944	6.329	6.714
0.70	89.96	2.365	2.434	2.773	3.108	3.441	3.772	4.103	4.434	4.764	5.095	5 425	5.425
0.80	93.51	2.087	2.126	2.425	2.718	3.010	3.300	3.590	3.879	4.168	4.457	4.746	5.035
0.90	96.71	1.869	1.887	2.153	2.415	2.674	2.933	3.190	3.448	3.705	3.962	4.219	4.475
1.0	99.63	1.694	1.696	1.936	2.172	2.406	2.639	2.871	3.102	3.334	3.565	3.797	4.028
1.2	104.81	1.428	—	1.611	1.808	2.004	2.198	2.391	2.585	2.778	2.971	3.163	3.356
1.4	109.32	1.236	—	1.378	1.548	1.716	1.883	2.049	2.215	2.380	2.546	2.711	2.876
1.6	113.32	1.091	—	1.204	1.353	1.501	1.647	1.792	1.937	2.082	2.227	2.372	2.517
1.8	116.93	0.972	—	1.068	1.202	1.333	1.463	1.593	1.722	1.851	1.979	2.108	2.237
2.0	120.23	0 8854	—	0.9595	1.0804	1.1989	1.3162	1.4328	1.5492	1.6653	1.7812	1.8971	2.0129
2.5	127.43	0.7185	—	0.7641	0.8620	0.9574	1.0516	1.1452	1.2385	1.3315	1.4244	1.5172	1.6099
3.0	133.54	0.6056	—	0.6337	0.7164	0.7964	0.8753	0.9535	1.0314	1.1090	1.1865	1.2639	1.3412
3.5	138.87	0.5240	—	0.5410	0.6120	0.6810	0.7495	0.8172	0.8848	0.9506	1.0181	1.0830	1.1504
4.0	143.62	0.4622	—	0.4707	0.5343	0.5952	0.6549	0.7139	0.7725	0:8309	0.8892	0.9474	1.0054
4.5	147.92	0.4138	—	0.4167	0.4738	0.5284	0.5816	0.6341	0.6862	0.7386	0.7908	0.8425	0 8939
5.0	151.84	0.3747	—	—	0.4250	0.4744	0.5226	0.5701	0.6172	0.6640	0.7108	0.7574	0.8039
5.5	155.47	0.3425	—	—	0.3852	0.4305	0.4745	0.5178	0.5607	0.6034	0.6459	0.6883	0.7307
6.0	158.84	0.3155	—	—	0.3520	0.3939	0.4344	0.4742	0.5136	0.5528	0.5918	0.6308	0.6696
6.5	161.99	0.2925	—	—	0.3240	0.3629	0.4005	0 4373	0.4738	0.5100	0.5461	0 5821	0.6180

(Contd.)

Table 5. *Volume of Superheated Steam (Contd.)*

1	2	3	4	5	6	7	8	9	10	11	12	13	14
7.0	164.96	0.2727	—	—	0.2999	0.3364	0.3744	0.4057	0.4396	0.4733	0.5069	0.5403	0.5737
7:5	167.76	0.2554	—	—	0.2791	0.3134	0.3462	0.3783	0 4100	0.4416	0.4729	0.5041	0.5353
8.0	170.41	0.2403	—	—	0.2608	0.2932	0.3241	0.3543	0.3842	0 4137	0.4432	0 4725	0.5017
8 5	172.94	0.2268	—	—	0.2447	0.2754	0.3047	0.3332	0.3613	0.3892	0.4169	0.4446	0.4721
9.0	175.36	0.2148	—	—	0.2303	0.2596	0.2874	0.3144	0.3410	0.3674	0.3936	0.4197	0.4458
9.5	177.67	0.2040	—	—	0.2175	0.2455	0.2719	0.2976	0 3228	0.3479	0.3727	0.3975	0.4222
10.0	179.94	0.1943	—	—	0.2059	0.2327	0.2580	0.2824	0.3065	0.3303	0.3540	0.3775	0.4010
11.0	184.07	0.1774	—	—	0.1859	0.2107	0.2339	0.2563	0.2782	0.3000	0.3215	0.3430	0.3643
12.0	—	0.1632	—	—	0.1692	0.1924	0.2139	0.2345	0.2547	0.2747	0.2945	0.3142	0.3338
13.0	191.61	0.1511	—	—	0.1551	0.1769	0.1969	0.2160	0.2348	0.2533	0.2716	0.2898	0.3080
14 0	195.06	0.1407	—	—	0.1429	0.1635	0.1823	0.2002	0.2177	0.2349	0.2520	0.2690	0.2859
16.0	201.37	0.1237	—	—	—	0.1419	0.1587	0.1745	0.1900	0.2051	0.2202	0.2351	0.2499
18.0	207.11	0.1103	—	—	—	0.1250	0.1402	0.1546	0.1684	0.1820	0.1954	0.2087	0.2219
20.0	212.37	0.0995	—	—	—	0.1114	0.1255	0.1386	0.1511	0.1634	0.1756	0.1876	0.1995
22 0	217.24	0.0906	—	—	—	0.10035	0.11343	0.12547	0.13700	0.14825	0.15934	0.17030	0.18119
24.0	221.78	0.08320	—	—	—	0.09107	0.10336	0.11455	0.12522	013561	0.14582	0.15591	0.18991
26.0	226.04	0.07680	—	—	—	0.08320	0.09483	0.10532	0.11526	0.12491	0.13438	0.14374	0.15301
28.0	230.05	0.07139	—	—	—	0.07644	0.08751	0.09740	0.10671	0.11574	0.12458	0.13330	0.14194
30.0	233.84	0.06663	—	—	—	0.07055	0.08116	0.09053	0.09931	0.10779	0.11608	0.12426	0.13234
35.0	242.54	0.05703	—	—	—	0.05869	0.06842	0.07678	0.08449	0.09189	0.09909	0.10617	0.11315
40.0	250.33	0.04975	—	—	—	—	0.05883	0.06645	0.07338	0.07996	0.08634	0.09260	0.09876
45.0	257.41	0.04404	—	—	—	—	0.05134	0.05840	0.06472	0.07068	0.07643	0.08204	0.08757
50.0	263.91	0.03943	—	—	—	—	0.04530	0.05194	0.05779	0.06325	0.06342	0.07360	0.07862
55.0	269.93	0.03563	—	—	—	—	0.04033	0.04665	0.05212	0.05719	0.06202	0:06666	0.07131
60.0	275.55	0.03244	—	—	—	—	0.03614	0.04222	0.04738	0.05210	0.05659	0.06094	0.06518
65.0	280.82	0.02972	—	—	—	—	0.03225	0.03850	0.04340	0.04783	0.05203	0.05598	0.06012
70.0	285.79	0.02737	—	—	—	—	0.02946	0.03523	0.03992	0.04413	0.04809	0.05189	0.05559
75.0	290.50	0.02533	—	—	—	—	0.02672	0.03242	0.03694	0.04095	0.04469	0.04825	0.05323
80.0	294.97	0.02353	—	—	—	—	0.02426	0.02995	0.03431	0.03431	0.03814	0.04170	0.04510

(*Contd.*)

Table 5. *Volume of Superheated Steam (Contd.)*

1	2	3	4	5	6	7	8	9	10	11	12	13	14
85.0	299.23	0.02193	—	—	—	—	0.02200	0:02774	0.03199	0.03567	0.03909	0.04232	0.04544
90.0	303.31	0.02050	—	—	—	—	—	0.02579	0.02993	0.03348	0.03674	0.03982	0.04280
95.0	307.21	0.01921	—	—	—	—	—	0.02402	0.02805	0.03151	0.03465	0.03759	0.04045
100.0	310.96	0.01804	—	—	—	—	—	0.02244	0.02641	0.02974	0.03276	0.03560	0.03832
110.0	318.05	0.01601	—	—	—		—	0.01961	0.02351	0.02668	0.02950	0.03214	0.03466
				t = 350	**400**	**450**	**500**	**550**	**600**	**650**	**700**	**750**	**800**
120.0	324.65	0.01428	—	0.01721	0.02108	0.02412	0.02679	0.02926	0.03160	0.03387	0.03607	0:03822	0.04033
130.0	330.83	0.01280	—	0.01510	0.01902	0.02194	0.02440	0.02682	0.02902	0.03114	0 03319	0.03519	0.03716
140.0	336.64	0.01150	—	0.01321	0.01723	0.02008	0.02251	0.02472	0.02680	0.02880	0.03072	0.03260	0.03444
150.0	342.13	0.01034	—	0.01146	0.01566	0.01845	0.02080	0.02291	0.02488	0.02677	0.02859	0.03036	0.03209
160.0	347.33	0.009308	—	0.009764	0.01427	0.01703	0.01929	0.02132	0.02320	0.02499	0.02672	0.02839	0.03002
170.0	352.26	0.008371	—	—	0.01303	0.01576	0.01797	0.01992	0.02104	0.02271	0.02431	0.02587	0.02738
180.0	356.96	0.007498	—	—	0.01191	0.01464	0.01678	0.01867	0.02040	0.02204	0.02360	0.02512	0.02659
190.0	361.43	0.006677	—	—	0.01089	0.01362	0.01573	0.01755	0.01922	0.02079	0.02229	0.02374	0:02515
200.0	365.70	0.005877	—	—	0.009947	0.01271	0.01477	0.01655	0.01816	0.01967	0.02111	0.02250	0.02385
210.0	369.78	0.005023	—	—	0.009071	0.01187	0.01391	0.01564	0.01720	0.01866	0.02004	0.02138	0.02267
220.0	373 69	0.003728	—	—	0:008251	0.0111	0.01312	0.01481	0.01633	0.01774	0.01907	0.02036	0.02160
221.20	374.15	0.00317	—	—	0.008041	0.009878	0.01212	0.01395	0.01553	0.01708	0.01847	0.0200	0.01987
250.0	—	—	—	—	0.006014	0.009171	0.01113	0.01272	0.01413	0.01542	0.01663	0.01779	0.01891
300.0	—	—	—	—	0.002831	0.006735	0.008681	0.01017	0:01144	0.01258	0.01365	0.01465	0.01562
350.0	—	—	—	—	0.002179	0.004950	0.006925	0.008364	0.009519	0.01056	0.01152	0.01242	0.01327
400.0	—	—	—	—	0.001909	0.003675	0.005616	0.006982	0.008088	0.009053	0.009935	0.01075	0.01152
450.0	—		—	—	0.001801	0.002913	0.004625	0.005934	0.006984	0.007886	0.008699	0.009452	0.01016
500.0	—	—	—	—	0 001729	0 002492	0 003882	0.005113	0.006111	0.006960	0.007720	0.008420	0.00907
550.0	—	—	—	—	0.001672	0.002246	0.003320	0.004484	0.005454	0 006200	0.006925	0.007581	0.008195
600.0	—	—	—	—	0.001632	0.002084	0.002952	0.003947	0.004835	0.005596	0.006269	0:006885	0.007460
700.0	—	—	—	—	—	0.001890	0.002467	0.003922	0.003972	0.004652	0.005257	0.005808	0 006321
800.0	—	—	—	—	—	0.001772	0.002188	0.002764	0.003379	0.003974	0.004519	0.005017	0.005481
900.0	—	—	—	—		0.001691	0.002013	0.002458	0.002967	0.003476	0.003964	0.004419	0.004841
1000.0	—	—	—	—		0.001629	0.001893	0.002246	0.002668	0.003106	0.003536	0.003952	0.004341

10

Steam Generators

CHAPTRER OBJECTIVES

After reading this chapter you will be able to learn the following

- Historical development of boilers.
- Classification of boilers.
- Landcashire, Cornish, Locomotive, Cochran, Simple Vertical, Babcock and Wilcox Stirling boiler etc.
- Mounting and accessories of boiler.
- Boiler Thermal Efficiency and Equivalent of Evaporation.
- Chimney height, Equivalent height of hot gas column.
- Energy Balance for the Boiler Plant.
- Introduction to IBR.

10.1 INTRODUCTION

Amongst a number of power plants fulfilling the power requirements of the world, major share is contributed by thermal power plants. As this type of power plants use steam as the working fluid, the generator of the steam can be said to be heart of it.

The steam generator is also known as 'Boiler'. The steam produced in the boiler is used in the industries for different purposes viz:

1. For generation of electrical energy as in power plants.
2. For generating mechanical energy in steam turbines.
3. For process heating purpose as heating of ground nuts in Oil Mills.
4. For sizing and bleaching in textile industries.

Normally a boiler produces steam at high pressure. This steam is first used to drive turbines for generating power. Then the steam exhausted from turbines, which is at lower pressure is used for process heating purpose.

A typical boiler has two different parts known as pressure part and combustion chamber. Pressure part is closed vessel, strongly fabricated out of steel, this contains water in it. The combustion chamber is constructed around pressure part with the help of fire bricks. The fuel undergoes combustion in this combustion chamber producing high temperature gases. These gases are also known as flue gases, hot furnace gases, or products of combustion. When flue gases come in contact with the water container, they give their heat to water, causing generation of steam. Afterwards flue gases are exhausted high into the atmosphere through the chimney.

10.2 HISTORICAL DEVELOPMENT OF THE BOILER

As we know boiler is steam generator, or it is a closed vessel in which water is heated and converted into steam. The boiler was initially developed in the very first century by Mr. Hero of Alexandria. However it was too small and it was used as toy only.

Consider a tea kettle on stove or burner, which has water or tea boiling in it, we can see that steam spreads as it leaves the spout. If the spout is closed by means of a cork, after some time the cork will be blown off.

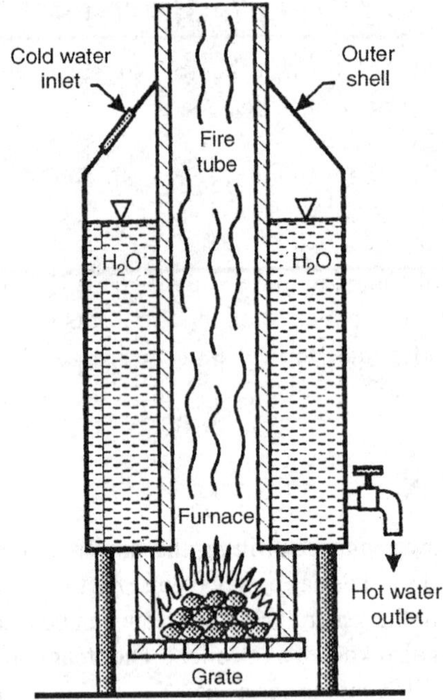

Fig. 10.1. *Small tank boiler used for heating water.*

This shows that the steam generated in a closed vessel keeps on the developing pressure. Such steam at high pressure is utilised to do work in expanding in a steam engine.

At the beginning a small closed vessel was used to generate steam by the application of heat. At the beginning the quantity of Steam generated and subsequently used in an engine was not sufficient. Therefore larger-vessel called **Boilers** were developed for giving a sufficient quantity of steam at the required pressure. The pressure was maintained constant in the boiler by opening the outlet valve such that the steam generated is the same as the steam supplied to the engine.

At the initial stages boilers used were of the fire tube type or tank boilers.

In our everyday life we used to have hot water for bathing purpose a device similar to a tank boiler but on a very smaller scale as shown Fig. 10.1.

Taking into consideration the safety, the development of simple boiler took place over the years. Till 17th century, there was no much development in the boiler design and manufacturing. Mr. Denis Papin of France designed and developed the first boiler with safety valve in A.D. 1679.

The major developments took place after A.D. 1769, when James Watt of Scotland developed the practical steam engine.

Afterwards taking into consideration the quantity of steam generated with respect to time, water tube boilers were developed. Subsequently high pressure boilers and supercritical boilers for power generation were developed.

The steam generated in the boilers is utilised for 2-purposes viz:

(i) Power Generation. The boilers generating steam for power-generation purpose are bigger in size, generate steam at high pressure and this high pressure steam is first expanded in the turbines for generating power. Then the exhausted steam is utlised for process heating purpose.

(ii) Process Heating Purpose. For example, heating of the oil seeds in the oil mills before they are being crushed.

10.3 CLASSIFICATION

Boilers are mainly classified according to the following factors.

1. By the Relative Position of Flue Gases and Water (or Tube Contents)

(*a*) *Fire tube.* In this case flue gasses flow through the tubes, which are surrounded by the water which is to be evaporated. Example: Lancashire, Cornish, Cochran, Locomotive, Simple vertical, Scotch marine boiler, etc.

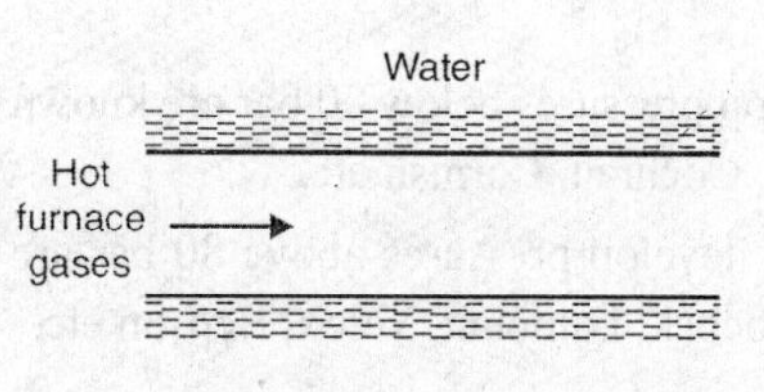

Fig. 10.2

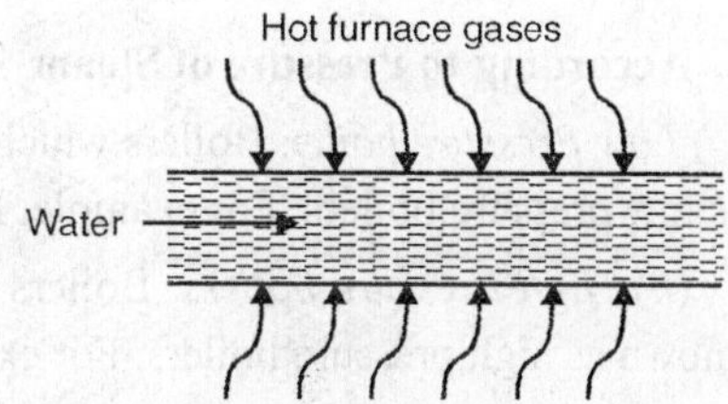

Fig. 10.3

(*b*) *Water tube.* In this case water flows through the tubes and hot gases heat the tubes from out side. For example, Babcock Wilcox, Stirling Boiler, etc.

2. By the Method of Firing

(*a*) *Internally fired.* In this case the furnace is located inside the boiler shell as in case of Lancashire boiler.

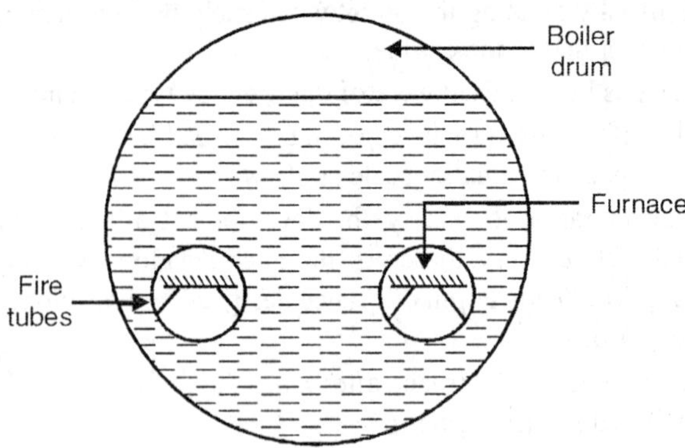

Fig. 10.4. *Internal fired.*

(*b*) *Externally fired.* The furnace is located outside the boiler shell as in case of Babcock Wilcox boiler Fig. 10.5.

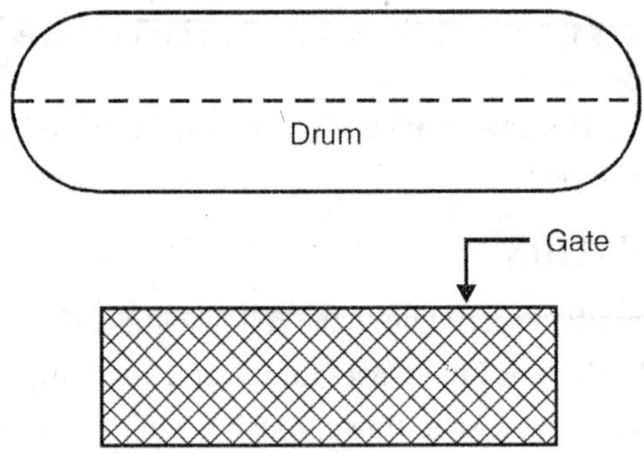

Fig. 10.5. *External fired.*

3. According to Pressure of Steam

(*a*) *Low Pressure boiler.* Boilers which can develop pressures below 80 bar are known as low pressure boilers. For example, Lancashire, Cochran, Cornish etc.

(*b*) *High Pressure boilers.* Boilers which can develop pressures above 80 bar are known as high pressure boilers. For example, Babcock, Lamount, Velox, Benson etc.

4. Nature of Draught

Draught is the pressure difference, which is necessary to draw the sufficient quantity of it for combustion. It is of the following types:

(*a*) *Natural draught.* In this case the draught is produced by mean of chimney only. The amount of draught directly depends upon the height of chimney.

(*b*) *Artificial draught.* Artificial draught is produced by means of fans.

(i) Induced Draught. Induced Draught is produced by sucking the flue gases from chamber. It is obtained by providing a fan between the boiler and the chimney. This fan is known as I.D. Fan.

(ii) Forced Draught. Forced draught is produced by forcing the air into the combustion chamber with the help of a fan, known as F.D. fan.

(iii) Balanced Draught. Balanced draught is produced by means of I.D. fan and F.D. fans.

5. Method of Circulation of Water

(a) *Natural circulation.* In which the circulation of water is due to gravity.

(b) *Forced circulation.* In which the circulation of water is obtained by a centrifugal pump.

6. By the Use

(a) *Land type.* Boilers used with stationary plants.

(b) Marine type

(c) Locomotive boilers.

7. By the Design of Flue Gas Passages

The flue gas may follow a single pass, return pass or multipass.

8. By the Number of Drums

These are of two types, (a) Single drum (b) Multidrum boilers.

9. According to Energy Source (fuel) Used

The heat energy may be derived from,

(a) Combustion of fossile fuels such as coal, wood, oil or natural gas etc.

(b) Electric or Nuclear energy.

(c) Hot waste gases of other chemical reactions.

10. According to the Material of Construction of the Boiler Shell

(a) Steel Boilers: Power boilers are generally fabricated out of steel plates.

(b) C.I. Boilers: Generally low pressure boilers are, sometimes built out of castings.

10.4 LANCASHIRE BOILER

It is fire tube, stationary, horizontal straight tube, internally fired, natural circulation boiler. Normal working pressure of this boiler is about 15 bar and steam generation capacity is upto 8T/hr.

Constructional Details

The main parts of the boiler are as follows.

(1) Boiler Drum. This boiler consists of a very large boiler drum. The size of the boiler drum is approximately 7–9 metres in length and 2–3 metres in diameter. This is a very old type of boiler, in this boiler, the meeting edges of the plate are connected by means of rivetted joints. Steel plates are rolled to form shell and the meetings edges of the shell are connected by means of rivetted joints. The alternate shells are made smaller in a diameter, so that the

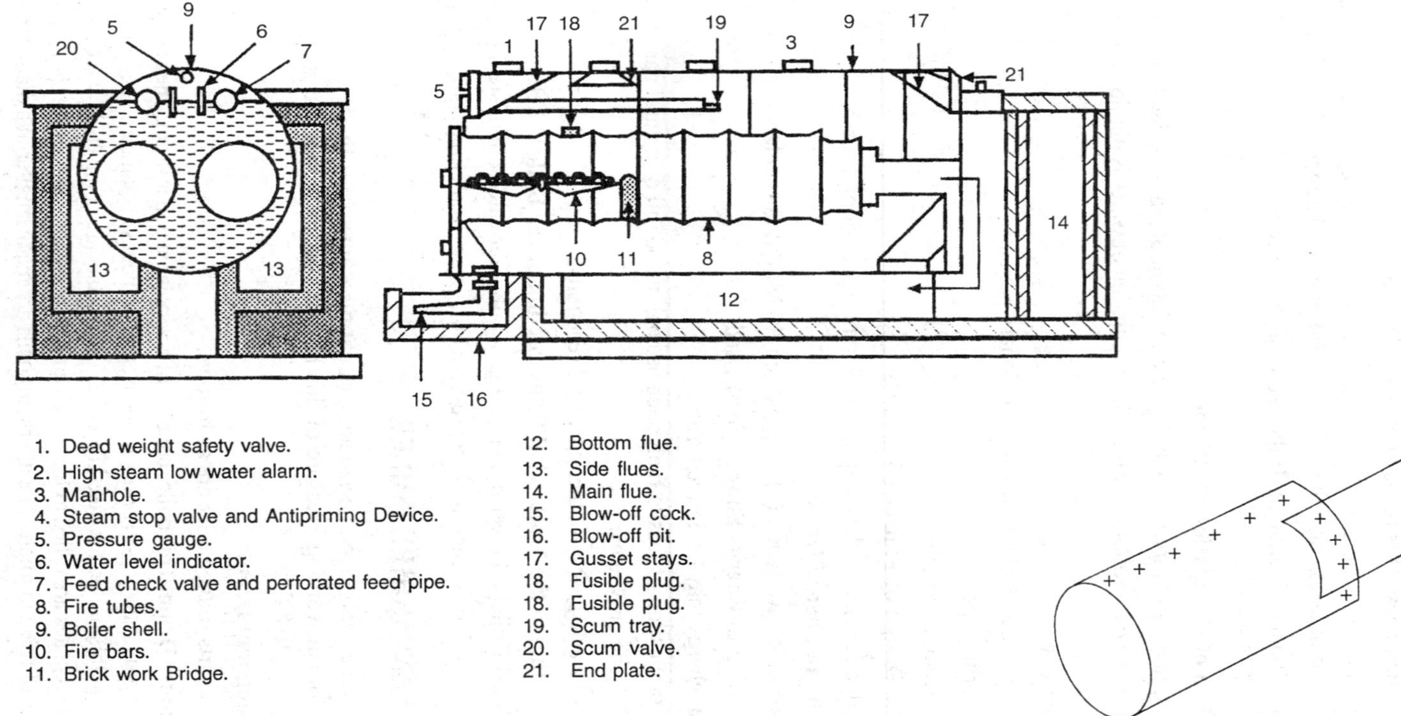

Fig. 10.6. **(a)** *Lancashire boiler.*

Fig. 10.6. (b)

longer shells of required length can be obtained by inserting one shell into the other and rivetting over the circumference.

The end plates of the boiler are flat. These flat end plates, when they are subjected to steam pressure from inside, they will have a tendency to bulge out. In order to prevent them from bulging, they are strengthened by means of stays. There are two types of stays.

(*i*) *Longitudinal stay.* It consist of round bar which extends from one end plate to the other and are connected to the end plates by means of lock nuts.

(*ii*) *Diagonal gusset stay.* Consists of a triangular plate, which is rivetted to the shell plate and end plate as shown in Fig. 10.6(a).

(2) Fire Tubes or Flue Tubes. Through the boiler drum passes two fire tubes.

Part of each of fire tubes is corrugated, so as to take up expansion and contraction when the boiler gets hot and cold. Also by providing corrugations surface area increases. These tubes are also made of number of shells. The ends of each of the shell are flanged outwards and the flanges of the two adjacent shells are connected by means of rivets, so as to get shells of required length.

It is to be noted that, the centre line of fire tubes is slightly below that of the centre line of boiler shell. Because fire tubes are provided approximately in the centre of water, so that heat can be transferred uniformly in all directions.

(3) Grate, Brickwork Bridge etc. Combustion of fuel takes place over the Grate, Top surface of the grate is made up of fire-bars. Below the grate Ash-pit is provided.

When the combustion of fuel takes place, hot gases are generated and the Brickwork bridge deflects the gases to move upwards.

(4) Manhole. It gives access to the inside of the boiler drum. It will be sufficiently large enough for a man to pass through it, for cleaning and inspection purpose.

(5) Mountings. The boiler is fitted with various mountings. Mountings are necessary for the safe and efficient operation of the boiler.

The term mountings refers to the items such as,

(*i*) *Safety valves*

(a) Dead weight safety valve.

(b) Spring loaded safety valve.

(c) High steam low water alarm.

When the pressure of steam in the boiler drum increases beyond safe value, then these valves will open and allow the surplus steam to escape to the atmosphere, till the pressure of steam in the boiler drum comes to normal working pressure.

(*ii*) *Pressure gauge.* It indicates pressure inside the boiler drum.

(*iii*) *Feed check valve.* Water enters the boiler drum through feed check valve. On the inner side of the feed check valve, perforated feed pipe is fitted for feeding the water uniformly in the boiler drum. Below the feed pipe scum tray is fitted which separates scum and is removed through the scum valve.

(*iv*) *Water level indicator.* It indicates the water level in the drum.

(*v*) *Blow-off valve.* It blows out impurities collected due to gravity.

(*vi*) *Steam stop valve.* Through this steam will be taken out for use.

(*vii*) *Antipriming device.* It is always desirable that the steam should leave the boiler as dry as possible, for that purpose an antipriming device or steam collecting pipe is provided in the steam space. It consists of a pipe and rectangular openings are provided in its bottom surface. It is fitted by means of baffles from inside. When the wet steam (steam and water particles) enters the pipe, the water particles being heavier than steam, after striking the baffles fall back to water space and steam almost dry will be taken out through the steam stop valve.

(6) Accessories. Accessories are used for improving the efficiency of boiler plant. Generally Super heater and Economiser are used as accessories.

(7) Path of Flue Gases. When the combustion of fuel takes place, hot gases are generated and the brick work bridge deflects these gases to move upwards.

During their first pass, these gases flow from the front end of the boiler to the rear end through central tubes. During their I-pass, they pass through the superheater, then they enter the common bottom flue at the rear end. Now their II-pass starts, during this gases flow from rear end to front end through the common bottom flue. At the end of II-pass, the gases are bifurcated into two side flues, when the flue gases enter side flues, their III-pass starts end during their III-pass, they flow from the front end to rear end through side flues and during this pass they heat the water from sides.

After their third pass they meet in main common flue, from where they pass through the economiser and then they are exhausted through the chimney.

10.5 CORNISH BOILER

The only difference between Cornish boiler and Lancashire boiler is that, in case of

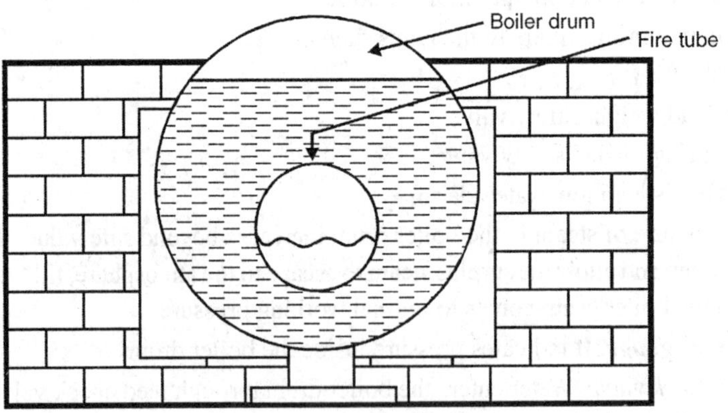

Fig. 10.7. *Cornish boiler.*

Cornish boiler there is only one fire tube, whereas in case of Lanchashire boiler there are two fire tubes, rest all the arrangement is same.

10.6 LOCOMOTIVE BOILER

It is a Fire tube, Horizontal, Multitubular, Internally fired, Natural circulation boiler.

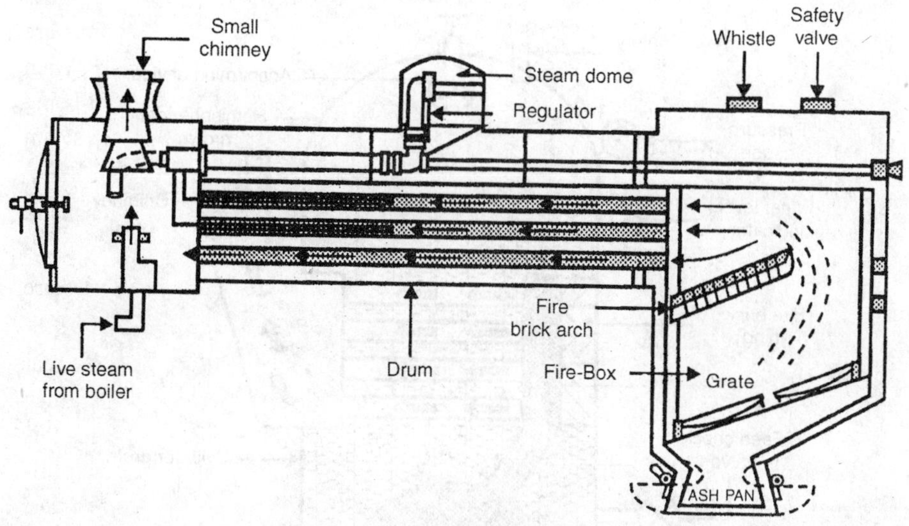

Fig. 10.8. *Locomotive boiler.*

As shown in Fig. 10.8 it consist of a large cylindrical shell. This shell is fitted to a rectangular fire box at one end and smoke box at other end. In between the smoke box and fire box a number of tubes are fitted.

When the combustion of fuel (coal) takes place over the grate, hot gases are generated. These hot gases rising from the grate are deflected by the Fire brick arch, so that they come in contact with the complete heating area of the fire box. The dempers as shown, control the flow of air for combustion. These hot gases then flow through the tubes, heat will be transferred to water and water will be converted into steam and accumulates in the steam dome and gases afterward are exhausted to the atmosphere through the small chimney.

The regulator can be operated from the cabin by turning the lever. In order to have superheated steam, the steam is passed through the superheater tubes. These superheater tubes are of smaller diameter and are laid in the large diameter fire tubes as shown. Then this superheated steam is supplied to the engine. This boiler is fitted with various mountings as shown in Fig. 10.8.

10.7 COCHRAN BOILER

It is a Fire tube, Stationary, Multitubular, Internally fired, Vertical boiler.

Cochran boiler is made in many sizes capable of producing steam upto 3T/hour and working pressures upto 17 bar. This boiler gives Thermal efficiency (η_{th}) of 70% with coal firing and about 75% with oil firing.

The boiler consists of a cylindrical shell, with its crown having hemispherical shape. Such a shape of the crown plate, gives sufficient strength to withstand the bulging effect.

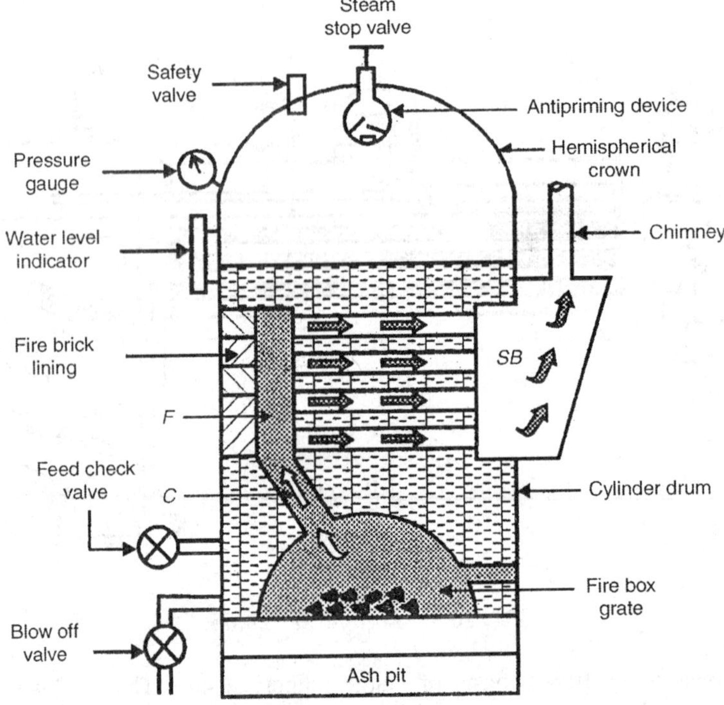

Fig. 10.9. *Cochran boiler.*

The furnace or fire box is also crown shaped or dome shaped, which deflects the unburnt fuel back to the grate. Combustion of fuel takes place over the grate. Hot gases generated in the fire box, enter into the chamber *C*, though a short flue pipe *F* and strikes the boiler shell plate or Back plate which is lined by fire bricks. The back plate directs the gases into the smoke or fire tubes. The gases after passing through the horizontal tubes enter the smoke box *SB* and from where they are exhausted through the chimney.

The hot gases passing through the horizontal tubes, give their heat to the water and in doing so, convert water into steam, which gets accumulated in the steam space.

10.8 BABCOCK WILCOX BOILER

It is a water tube boiler. It mainly consists of three parts.

(i) A horizontal Steam and Water Drum. This is the main part of the Babcock Wilcox boiler. The size of the boiler drum is very much smaller, when compared with boiler drums of fire tube boiler of same capacity. Hence these are simpler from transportation point of view.

(ii) A bundle of Steel Tubes. These are provided in the combustion chamber.

(iii) Combustion Chamber. It is the space above the grate and below the drum where the combustion of fuel takes place. This complete chamber is divided into three separate chambers. These chambers are so arranged. that, the gases can flow from one

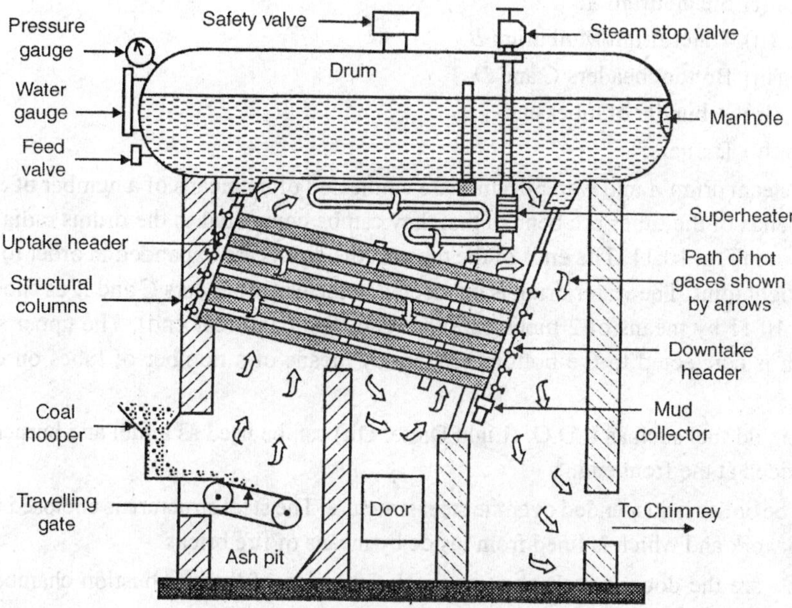

Fig. 10.10. *Babcock wilcox boiler.*

chamber to the other and give out heat in each chamber gradually. Hence the first chamber is hottest and the last chamber is at lowest temperature.

The front end of the boiler is connected to the uptake header by a short tube and the rear end is connected to the downtake header by a long tube. In between the headers a number of tubes are fitted. The ends of the tubes in the headers are either expanded or welded in order to have gas tight joint. These tubes are arranged in zig-zag manner, so that more surface area of the tubes is exposed to hot gases. These tubes are sloping, downwards towards the rear end. The inclination of the tubes with the horizontal, will be in between 5–15°. A mud collector is provided at the bottom most position of down take header for collecting and disposing the mud or any other suspended impurities. The entire boiler except the furnace in suspended on steel structure, so that the boiler can expand or contract freely without producing the cracks into the brick work, which encloses the boiler and furnace. This brickwork is lined from inside by means of fire bricks. Doors provided give access for cleaning inspection and repair purpose. Dampers provided are used for regulating draught.

This boiler is fitted with various mountings viz. Safety valve, Steam stop valve, Pressure gauge, Water level indicator, Fusible plug, Feed check valve, Blow-off valve etc. This boiler is also fitted with superheater as shown in Fig. 10.10.

10.9 STERLING BENT TUBE BOILER

It is (a) water tube, (b) multi drum and (c) externally fired boiler.

The boiler mainly consists of,

(i) Steam drum *A*.

(ii) Water drum/Mud drum *B*.

(iii) Bottom beaders *C* and *D*.

(iv) A bundle of tubes and

(iv) Burners.

The steam drum *A* and water drum *B* are connected of by means of a number of tubes. The ends of the tubes are bent so that they can be connected to the drums radially as shown in Fig. 10.11. The ends of the tubes in the drums are expanded in order to have gas tight joint. The water drum is connected to the water headers *C* and *D* as shown in Fig. 10.11 by means of 2 pipes on either end (front and rear end). The upper steam drum is connected to the bottom headers by means of a number of tubes on either side.

Liquid fuel such as L.D.O. (Light Diesel Oil) can be used as a fuel and burners are provided at the front end.

The boiler is suspended over the steel structure. The steel structure is enclosed in red brickwork and which is lined from inside by means of fire bricks.

E—are the doors which give access to the inside of the combustion chamber for inspection and repair purpose.

F—are the baffles which divide the combustion chamber into 3 separate chambers (I, II and III). These 3-chambers are so connected that the hot gases can flow from the I-chamber to III-chamber and they are made to pass through economiser and finally they are exhausted through chimney. The water in the tubes will absorb heat in each chamber, so the temperature of flue gases will go on decreasing towards the rear end.

The boiler is fitted with various mountings viz. :

(a) Spring loaded safety valves.

(b) Pressure gauges.

(c) Thermometers.

(d) Water level indicators.

(e) Feed check valve.

(f) Steam stop valve.

(g) Blow off valve.

Also superheater and economiser are connected in the boiler circuit as accessories.

Water will be pumped by means of a centrifugal multi-stage feed pump. Water first enter the bottom header of the economiser and when it rises to through the tubes, it recovers some of the heat of the flue gases and becomes hot. The hot water will then be

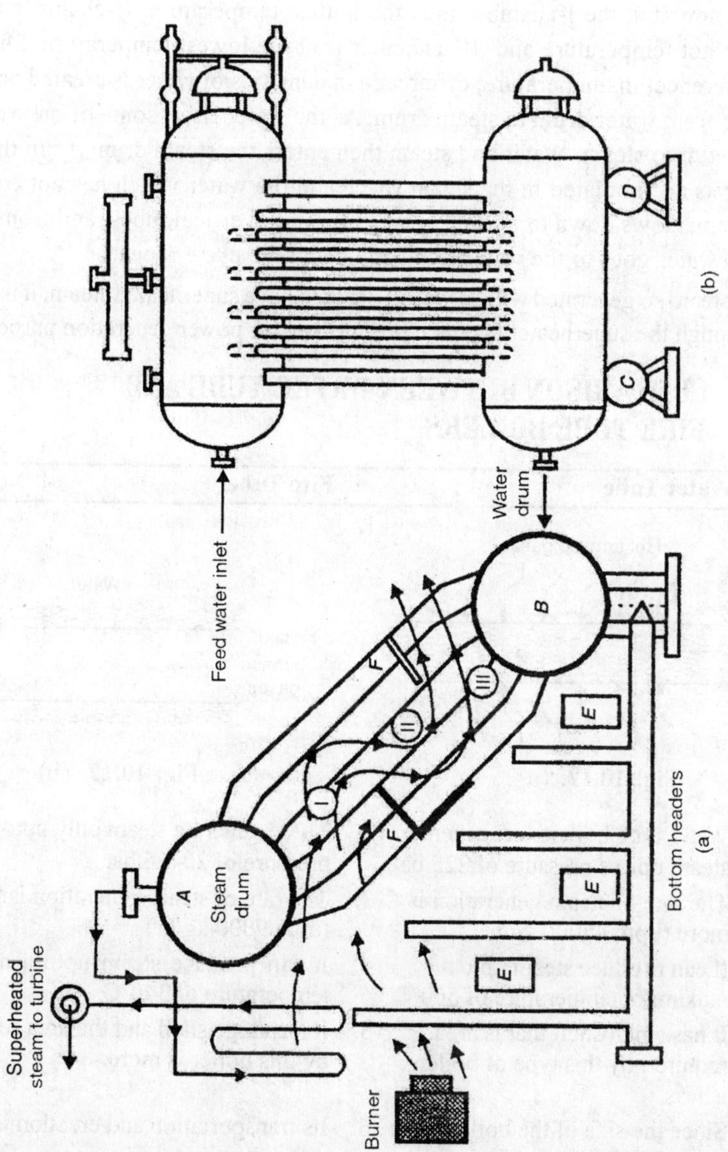
Superheated steam to turbine
A
Steam drum
Burner
F
I
II
III
F
B
E
E
E
Bottom headers
(a)
Feed water inlet
Water drum
C
D
(b)

Fig. 10.11.

collected into the upper header of the economiser and from where it goes to the upper drum.

Water is filled in the drum upto a pre-specified level, so that tubes, water drum, headers and vertical tubes are flooded by means of water.

We know that the I-chamber is at the hottest temperature, II-chamber is at the medium hot temperature and III-chamber is at the lowest temperature. So, due to this difference, in temperature, difference in densities of water is created and water will rise from water drum to steam drum. As the water rises, some of the water gets converted into steam. Water and steam then enters the steam drum A. In this drum steam gets accumulated in the steam space and the water which has not converted into steam; flows down to bottom headers through vertical tubes, and from bottom headers water goes to the water drum and again the cycle repeats.

The steam so generated will be wet, in order to have superheated steam, it is made to pass through the superheater and then it is utilised for power generation purpose.

10.10 COMPARISON BETWEEN WATER TUBE AND FIRE TUBE BOILERS

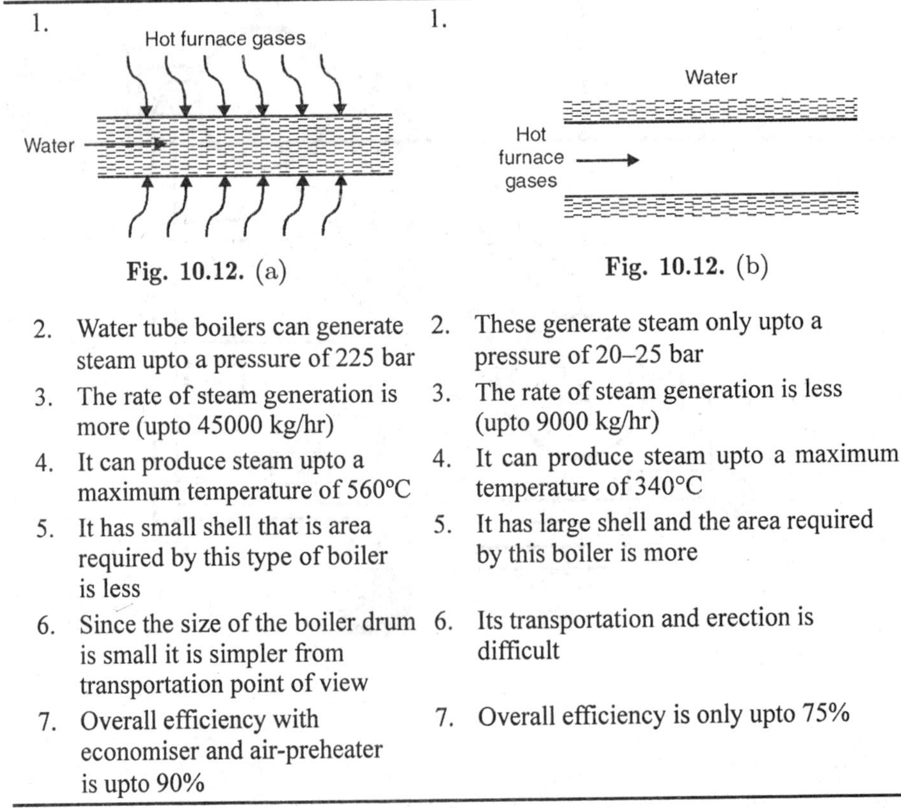

Water Tube	Fire Tube
1. **Fig. 10.12.** (a)	1. **Fig. 10.12.** (b)
2. Water tube boilers can generate steam upto a pressure of 225 bar	2. These generate steam only upto a pressure of 20–25 bar
3. The rate of steam generation is more (upto 45000 kg/hr)	3. The rate of steam generation is less (upto 9000 kg/hr)
4. It can produce steam upto a maximum temperature of 560°C	4. It can produce steam upto a maximum temperature of 340°C
5. It has small shell that is area required by this type of boiler is less	5. It has large shell and the area required by this boiler is more
6. Since the size of the boiler drum is small it is simpler from transportation point of view	6. Its transportation and erection is difficult
7. Overall efficiency with economiser and air-preheater is upto 90%	7. Overall efficiency is only upto 75%

	Water Tube		Fire Tube
8.	Construction is very much complex, hence more skill is required for operation	8.	Construction is relatively simpler and hence less skill is required
9.	Operating costs are more	9.	Operating costs are less
10.	Steam generation capacity is more, hence large power plants use these boilers	10.	It is not suitable for power plants, but is used in Dairy, Laundry and some process industries
11.	Boiler won't get blown off, even if some	11.	Danger of explosion of complete drum of the tubes fail

10.11 ESSENTIALS OF A GOOD BOILER

1. With minimum fuel consumption, the boiler should be capable of generating steam at the required quality.
2. It should be light and occupy less space.
3. It should be capable of quick starting.
4. The initial cost, installation cost and maintenance cost of the boiler should be minimum.
5. The different parts of the boiler should be easily approachable for repairs.
6. The boiler should have minimum number of joints and those too should be away from direct flames.
7. There should be no deposition of mud and other deposits on the inner surface of the boiler drum.
8. Boiler should conform to safety regulations, laid down in Boiler act [Indian Boiler Regulations (IBR) Act].

10.12 BOILER MOUNTINGS AND ACCESSORIES

Mountings. Mountings are the fittings mounted on the pressure part of the boiler. Mountings are necessary for the safe and efficient operation of the boiler. Without mountings boiler cannot work safely. Special provisions are made on the pressure part to mount the mountings.

The term mountings refers to the items such as,

(1) Safety valves

   (a) Dead weight safety valve

   (b) Spring loaded safety valve

   (c) High steam and low water alarm.

(2) Water level indicators

(3) Fusible plug

(4) Pressure gauges

(5) Steam stop valve

(6) Feed check valve

(7) Blow-off valve etc.

The above mountings are usually installed in accordance with IBR.

Accessories. Accessories are used to improve the efficiency of the boiler plant. Without accessories boiler can work safely. These are not mounted on the boiler, but these are connected in the boiler circuit.

Usually following accessories are provided with the boiler.

1. Super heater
2. Economiser
3. Air-preheater
4. Steam injector
5. Feed pump
6. Steam trap
7. Induced draught fan
8. Forced draught fan etc.

10.12.1 Dead Weight Safety Valve

As the name implies, this valve is used to assure safe working of the boiler. When the pressure of steam inside the boiler drum increases beyond the safe value, it may cause damage to the pressure part, or even a severe accident. It can be prevented by allowing the surplus steam to escape to the atmosphere, by means of the safety valve.

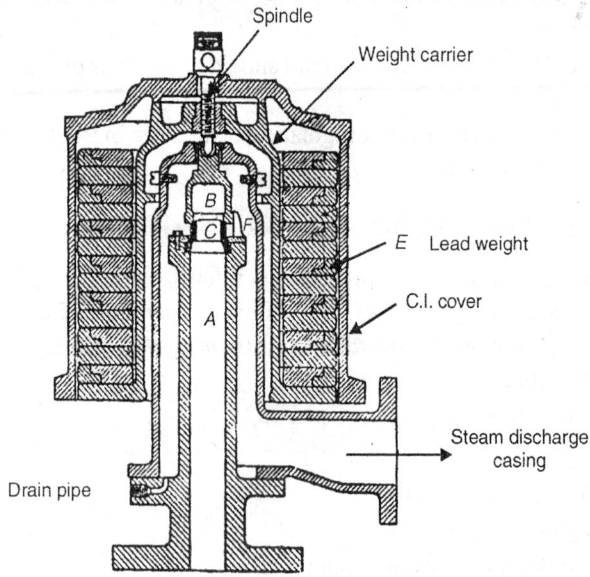

Fig. 10.13. *Dead-weight safety valve.*

This valve was first introduced by M/s. J.J. Hopkinson and Co. Ltd. It consists of a vertical pipe *A*, bottom end of which is flanged and through this flanged end it is mounted

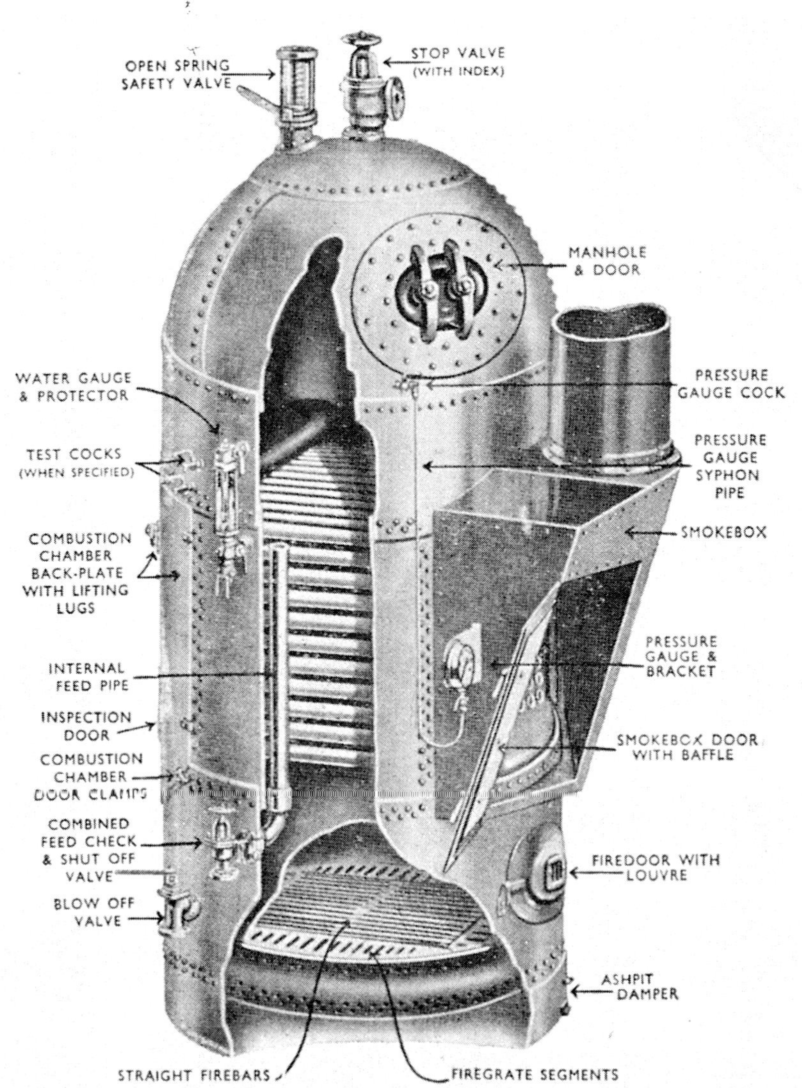

Fig : Cochran Boiler

Sterling Bent tube boiler.

Fire Tube, 6 TPH, LIPI Make Boiler

Water Tube, 90TPH, LIPI Make Boiler

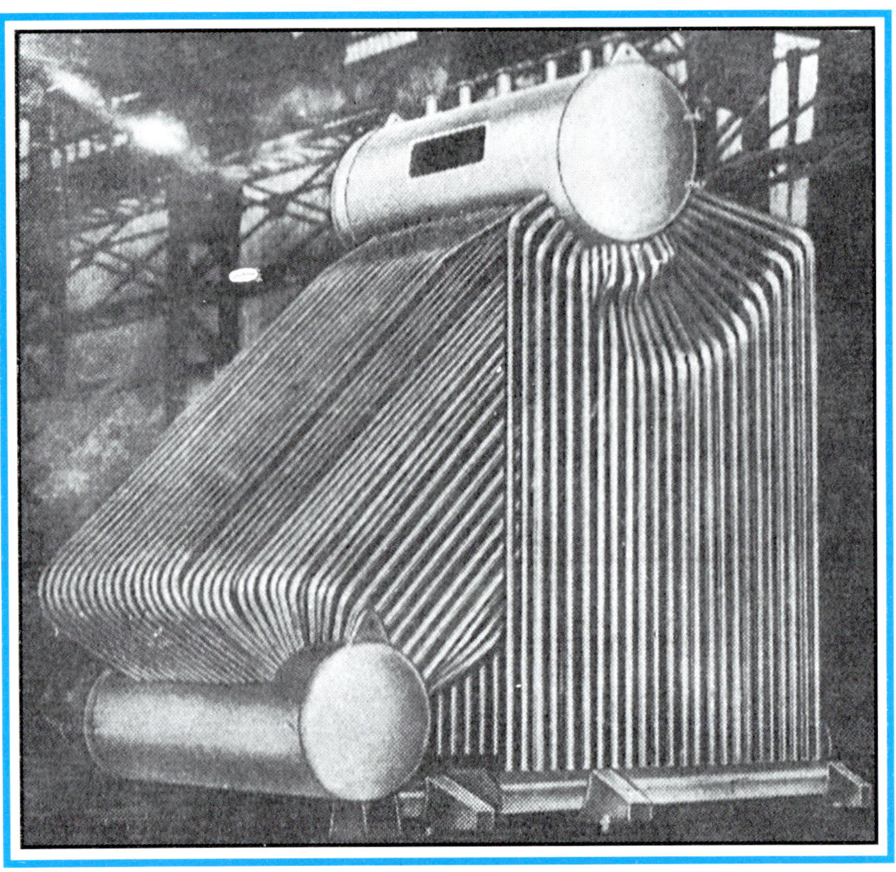

Fig : Sterling Bent tube boiler.
Courtesy : LIPI BOILERS LTD. PUNE-1

on the mounting blocks of the boiler. *B* is the Gun metal valve, resting over the valve seat *C*. Valve seat *C* is fixed on the top of the pipe *A* by means of a securing ring and screw. *F* is the feather which guides the valve *B*. Valve *B* is also guided by a ring in steam discharge casing. The weight carrier is suspended from the top of the valve *B*. Dead weights are placed on the weight carrier and these are covered by the Cast Iron cover.

Weight of the valve *B*, spindle, carrier, dead weights and cover will be acting downwards. During normal working conditions, this downward acting load will be balanced by the steam pressure from below. When the steam pressure exceeds pre-specified or safe value, then the valve will be lifted up and the surplus steam will be exhausted to the atmosphere.

For example Dead weight safety valve used on pressure cookers.

10.12.2 High Steam Low Water Alarm

This valve was also introduced by M/s Hopkinson and Co. Ltd. This is a two-in-one valve, i.e. it consists of two valves.

(i) High steam valve. During high steam conditions this valve opens, and allows surplus steam to escape to the atmosphere.

(ii) Low water valve. When the water level in the boiler drum falls below the safe level, then this valve opens and allows the steam to escape. When the steam is allowed to escape, it makes whistling sound i.e. it gives warning to the boiler attendant. The whistling sound for high steam conditions will be louder than low water conditions, so that one can distinguish between the two. Since it makes whistling sound it is known as alarm.

As shown in Fig. 10.13 *A* is the high steam valve, resting over the valve seat *B*. Valve seat *B* in secured in position by means of a ring and feather *S* 'is cost on it. This feather guides the valve *A*. *C* is the low water valve, which is resting over the inner edge of the valve *A* as shown. *R* is the rib which guides the valve *C*. Low water valve. *C* will be held in position by means of the Dead weigh, *D* which is connected to the rod *E*, which position by means of the weight *D* and partly by means of weight and counter weight outside the steam discharge casing.

In the steam space lever *FK* is provided. A large fire brick float *L* is suspended from the end *K*.

During normal working conditions, the float will be submerged in water and the weight of this float in partly balanced by means of counter weight *M* and partly by means of buoyant force due to water, which is acting in the upward direction. But when the water level goes below the safe level, the float gets uncovered by means of water, then there won't be buoyant force and counter, weight. *M* alone will not be sufficient to balance the float. So, the lever *FK* tilts in the clockwise direction. Then the knife edge projection of the lever *FK* comes in contact with the collar *N* and in turn lifts the hemispherical valve *C*. When the valve *C* is lifted, steam escapes and makes whistling sound.

During high steam conditions, both the valves are lifted since the low water valve is resting over inner edge of the high steam valve C is lifted, steam escapes and makes whistling sound.

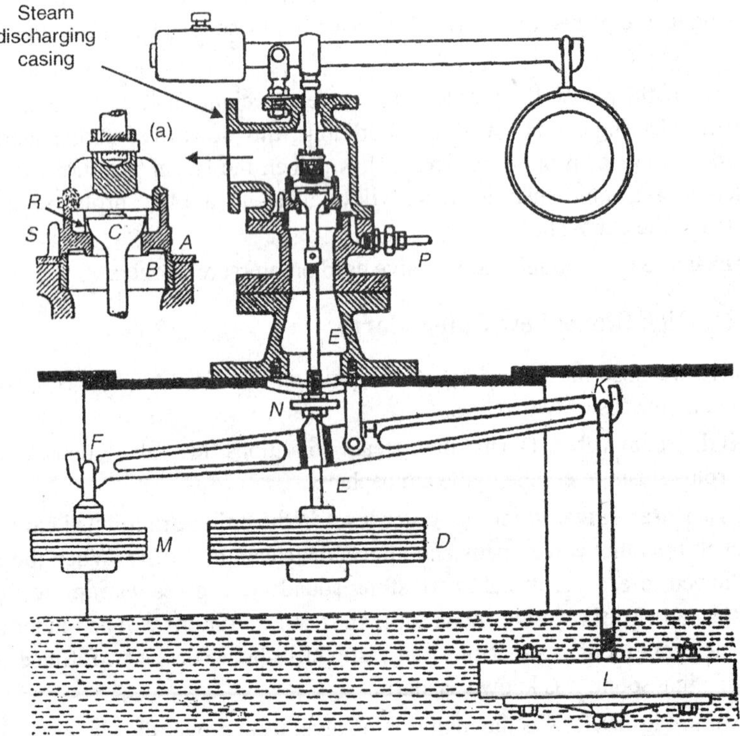

Fig. 10.14. *High steam low water alarm.*

During high steam conditions, both the valves are lifted since the low water valve is resting over inner edge of the high steam valve, and surplus steam will be exhausted.

10.12.3 Water Level Indicator

It indicates the water level inside the boiler drum. This water level can be made visible from the boiler room floor by using mirrors. This unit was also introduced by M/s Hopkinson and Co. Ltd.

The unit consist of two hollow gun-metal stuffing boxes. One end of each of the stuffing box will be flanged and through this flanged end it is connected to the front end plate of the boiler. Upper stuffing box is connected to the steam space and lower stuffing box is connected to the water space. In between the two stuffing boxes a strong glass tube is fitted. (In between the glass and metal, rubber packing is provided in order to have gas tight joint).

Thus two valves A and B which control the passage of steam and water between the boiler and the glass tube. When these valves are open, the handles are vertical and the water level in the glass tube will be the water level in the boiler drum. A third valve *C* called as Blow-off valve is generally closed. When this valve is closed, its handle is vertical and to blow out sediments and impurities handle is to be made horizontal.

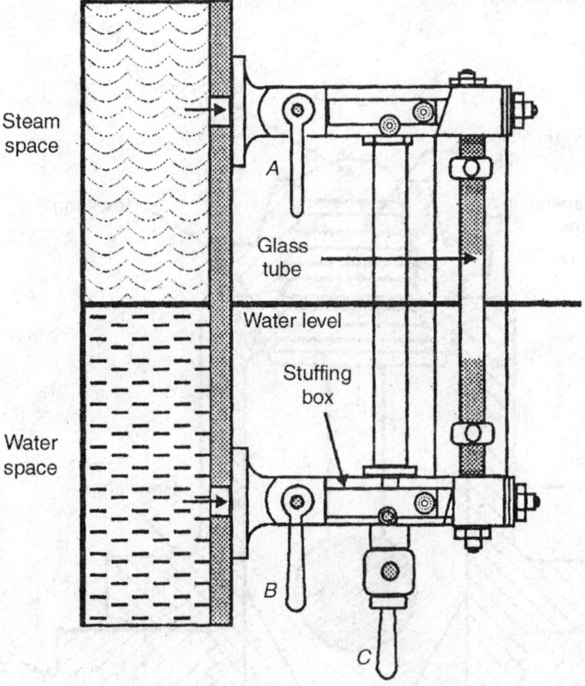

Fig. 10.15. *Water level indicator.*

An arrangement is also provided, in order to shut off automatically the supply of steam and water to the glass tube, if the glass tube breaks due to any reason. This arrangement consists of a hollow gun-metal column *D* and two steel balls as shown.

During normal working conditions these balls will be in position shown by full circles if the glass tube breaks the rush of water in the bottom passage carries the ball into the position shown by dotted circle and shuts off the supply of water. At the same time the steam rushing through the upper passage aided by the water rushing upwards through the column *D* carries the ball into the position shown by dotted circle and shuts off the steam supply. Then the boiler attendant can safely close the valves *A* and *B* and can replace the tube.

Screw plugs provided at the end of stuffing boxes, give access for cleaning of passages. The glass tube is generally covered from the front and from sides by Guard glass in order to protect the boiler attendant from the flying fragments of glass tube, if it breaks.

10.12.4 Fusible Plug

It actuates itself as the last resort, when the water level in the boiler drum falls below the minimum permissible water level. As a result of which, fusible metal melts by the heat of steam, the plug drops out and the steam rushes into the combustion chamber and extinguishes the fire. Thus it safeguards the combustion chamber crown from burning and also warns the boiler attendant about the fall of water level in the drum below minimum permissible level.

Function:

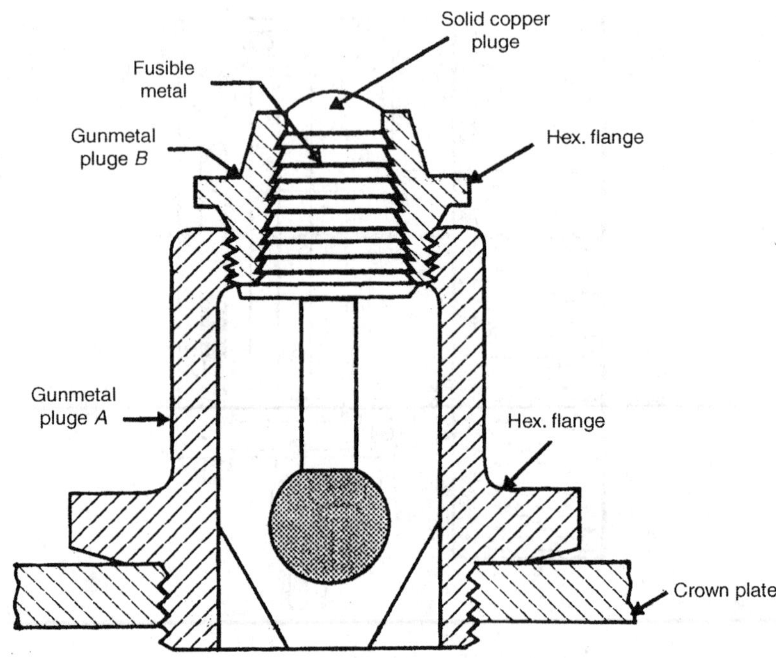

Fig. 10.16. *Fusible plug.*

Fusible plug is fitted on the combustion chamber crown, but in the drum. As shown in Fig. 10.16, *A* is the hollow gun metal plug screwed into the crown plate. *B* is fhe second hollow gun metal screwed into the plug *A* and *C* is the third solid copper conical plug. The inner surface of *B* and *C* for removal and refitting purpose by using spanners.

Under normal working conditions, fusible plug will be covered by means of water in the drum. This water keeps the temperature of the fusible metal below its melting point. But when due to any reason, water level in the drum falls below the minimum permissible water level, the plug gets uncovered from water and is exposed to steam. Due to heat of steam, fusible metal melts and copper plug drops and the steam rushes through the access into the combustion chamber. Before referring the boiler, the attendant removes and refits the plug after interposing the fusible metal between the plugs *B* and *C*.

10.12.5 Bourdon Pressure Gauge

This is the most common type of pressure gauge. This is used to measure pressure inside the containers and pipe lines etc.

The basic element of this gauge is the tube *A*. It is elliptical in cross section and is bent into and arc of circle as shown in Fig. 10.17.

End *B* of this tube is sealed, while open end *C* is connected to the connecting union *D*. Through the connecting union the gauge will be mounted over the containers or pipes of whose pressures are to be measured.

When it is mounted on the systems and if the pressure of the system is more than atmosphere (i.e. + ve pressure) then the tube will tend to curl out. Conversely if the pressure of the system is less than the atmosphere, (i.e. –ve pressure) the tube will tend to curl in. Thus the change of pressure will therefore appear as a movement of the end *B*. This end *B* is connected by means of a link *E* to a quadrant gear *F*. The quadrant gear engages with a small gear *G* on to which a point *H* is attached.

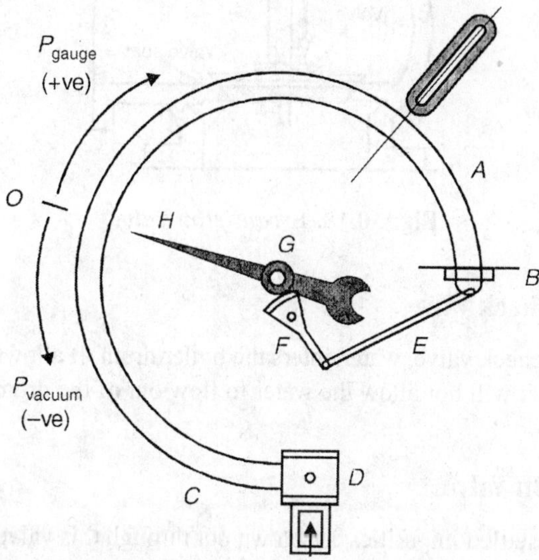

Fig. 10.17. *Bourdon pressure guage.*

Thus then change of pressure will appear as the movement of the end *B*, which will be transmitted to the quadrant gear, which in turn rotate the pinion and hence the pointer. The pointer rotates over the calibrated scale, which directly gives the pressure readings.

10.12.6 Steam Stop Valve

Through the steam stop valve, steam will be taken out for power generation purpose.

As shown in Fig. 10.18 valve body is made up of C.I. Through its bottom flange *A* it is mounted in the boiler drum, and its side flange *F* is connected to the pipe through which steam goes to the steam engine or steam turbine.

It is clear from full figure that when the hand wheel is rotated in clockwise or anticlockwise direction, the valve will move up and down. And the required quantity of steam can be taken out for power generation.

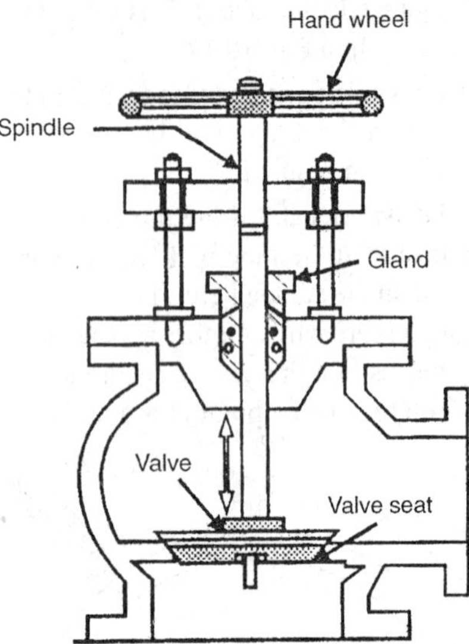

Fig. 10.18. *Steam stop valve.*

10.12.7 Feed Check Valve

Through the feed check valve, water enters the boiler drum. It allows the water to flow into the drum and it will not allow the water to flow out of the drum, so it is called as check valve.

10.12.8 Blow Off Valve

Mud or any other settled impurities are blown out through this valve.

10.12.9 Superheater

The steam generated from a simple low pressure boiler is generally wet. So, for superheating the steam, superheaters are used. In the superheater, wet steam is first dried at the same temperature and pressure and then it is superheated at constant pressure.

Generally, same heat of the flue gases is used for superheating purpose and hence superheaters are placed in the path of flue gases, However in bigger installations superheaters are provided with independent furnaces. And such superheaters with independent furnaces are known as Independently fired Superheaters.

Advantages of Superheating the Steam

(1) Increase in amount of work output, for the same amount of steam and hence increase in cycle efficiency.

(2) Loss due to condensation of steam in the pipe line connecting the boiler to the steam engine or steam turbine is reduced.

(3) Loss due to condensation of steam engine/steam turbine is reduced.

(4) When the superheated steam is used, moisture will be absent, hence it will reduce corrosion and erosion of steam engine and steam turbine parts.

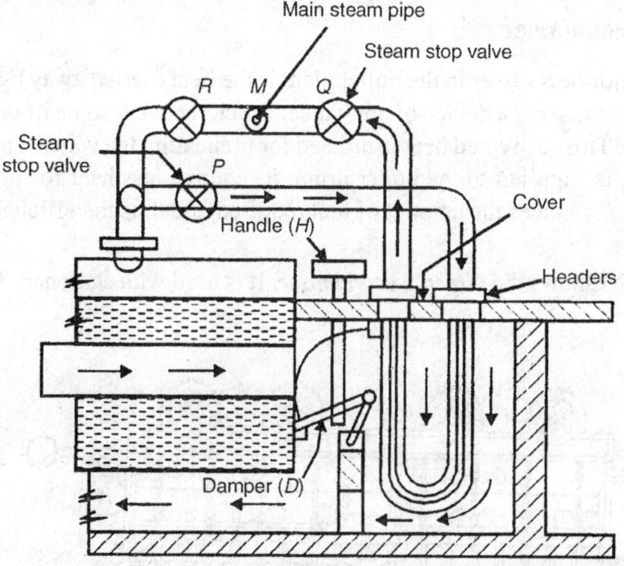

Fig. 10.19. *Superheater.*

Figure 10.19 shows *Sudgen's Hair pin type of superheater* arranged with Lancashire/ Cornish boiler. The superheater is placed at the rear end of the boiler and the flue gases after their first pass, they pass through the superheater. Here the temperature of the flue gases is not less than 550°C. This superheater gives a superheat of 40 – 95°C.

This superheater consists of two Mild Steel boxes or headers and a number of tubes bent to U-shape. The ends of the tubes, in the headers will be either expanded or welded in order to have gas tight joint. Between front end rear header some space is provided and this space is covered by means of a cover. This gives access for cleaning, inspection and repair purpose.

As long as temperature of the flue gases is less than 550°C then there is no danger of burning of tubes. But when the temperature increases beyond this and also when the delivery of steam from the boiler to the steam turbine through the superheater is suspended for along time, then the following arrangements are to be made so as to protect the tubes from burning out.

(1) **Flooding the superheater tubes.** This water is to be drained before the supply of steam through the superheater again starts.

(2) **Diverting the flow of flue gases.** This is done by Damper *D* which is operated by means of handle *H*.

An arrangement is also provided so as to pass the steam through superheater or directly to the main steam pipe as may be required *M* is the main steam pipe, *PQ* and *R* are stop valves. When the superheater is in action valves *P* and *Q* are open *R* is closed.

When steam is to be taken out directly from the boiler to the steam pipe, then valves *P* and *Q* are closed and *R* is opened.

10.12.10 Economiser

One of the major heat losses in the boiler plant is the heat carried away by the exhaust gases. An economiser is a device or appliance, which recovers some of the heat of the exhaust gases. This recovered heat in utilised for preheating the water. This preheated water when it is supplied to the boiler drum, it requires less heat for its conversion into steam. This reduces the amount of fuel required and thus the efficiency of boiler plant increases.

Figure 10.20 shows the *Green's economiser*. It is used with stationary low pressure boilers.

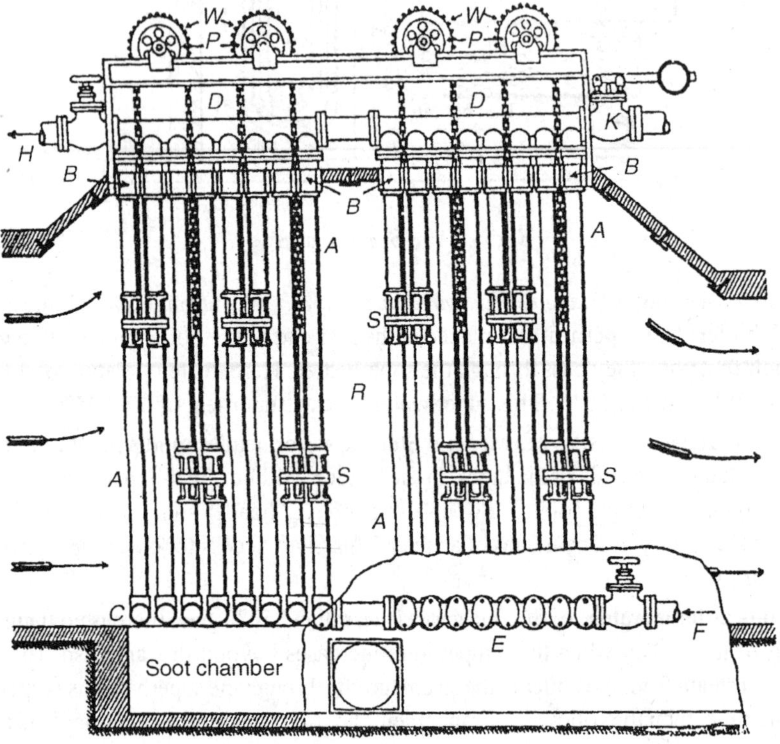

Fig. 10.20. *Green's economiser.*

It consists of a number of vertical pipes *A*. These pipes are arranged in groups of 4, 8, 12,16 etc. Upper ends of the pipes are connected to the upper boxes *B*: These boxes *B* are inturn connected to the common header *D*. Lower ends of the pipes are connected to lower boxes. These lower boxes are connected to the horizontal pipe *C* through the cross pipes. All the tubes are enclosed in the brick work of the economiser except header *D* and pipe *C*.

Water will be pumped by means of a feed pump, water enters the economiser through the stop valve at the inlet F and passes into the pipe C, from where it flows through the cross pipe and then rises through the vertical tubes A. Since the economiser is provided in the path of flue gases the water rising through the tubes recovers heat and becomes hot. Hot water will be then collected into the top common header.

This hot water from the top header is taken out through the stop valve H, from where it goes to the boiler drum. K is the safety valve for the economiser.

Since the economiser is provided in the path of flue gases, soot (or ash) is likely to be collected over the tubes. It retards the rate of heat transfer from the flue gases to the water inside the tubes. In order to avoid this, soot is scrapped by means of scrapers. Scrapers of the two-adjacent tubes are coupled to form one pair and two such pairs of scrapers are connected by means of a chain, which passes over pulley P. Pulley P is connected to worm shaft, in such a way that when one pair of scrappers comes to top most position, another pair will be at the bottom most position, then it automatically reverses the direction of motion. The scrapers are kept in motion as long as economiser is in action.

By Pass Arrangement. Figure 10.21 by pass arrangements are provided for the flue gases and for feed water, so that the economiser may be put out of action when not required. Figure 10.21 shows such an arrangement for two Lancashire boilers fitted

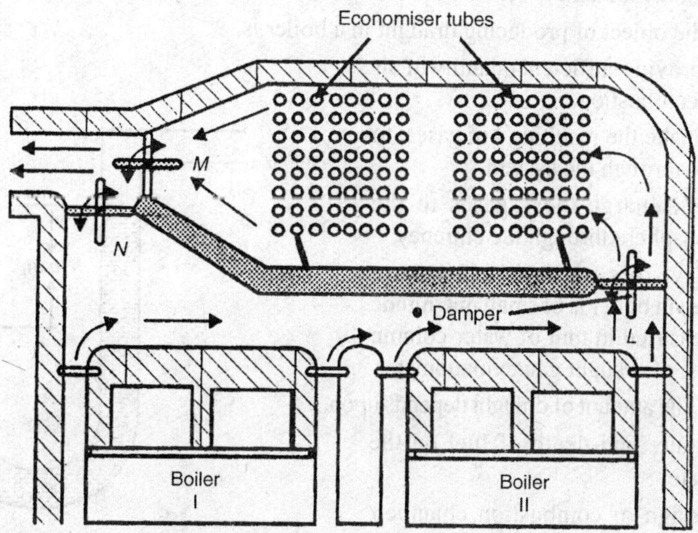

Fig. 10.21. *By pass arrangement.*

with an economiser. When. the economiser is in service, then dampers L and M are open and the damper N is closed. For isolating the economiser damper N is opened and dampers L and M are closed.

10.12.11 Air Preheater

The air is heated, by using the heat of flue gases in a device called preheater, which otherwise would have been lost to the atmosphere. By using Air preheater the overall efficiency will increase by about 10%.

10.12.12 Feed Pump

It is an appliance used to pump the feed water into the boiler drum. The feed pump may be either rotary or reciprocating type.

10.12.13 Steam Injector

When the feed pump fails due to any reasons, then steam injector, injects or pumps or delivers feed water to the boiler drum, by the use of steam.

10.12.14 Steam Trap

Steam trap automatically collects and returns to the boiler, the water resulting from partial condensation of steam, without allowing any steam to escape.

10.13 BOILER DRAUGHT

Draught is the pressure difference which is necessary to draw the required quantity of air for combustion and to remove the flue gases out of the system.

Thus the object of producing draught in a boiler is,

(i) To provide sufficient quantity of air for combustion.

(ii) To make the resulting hot gases, to flow through the system.

(iii) To discharge these gases to the atmosphere through the chimney.

Usually this drought (pressure difference) in boiler is of small magnitude and is measured in mm of water column by means of draught gauge/manometer.

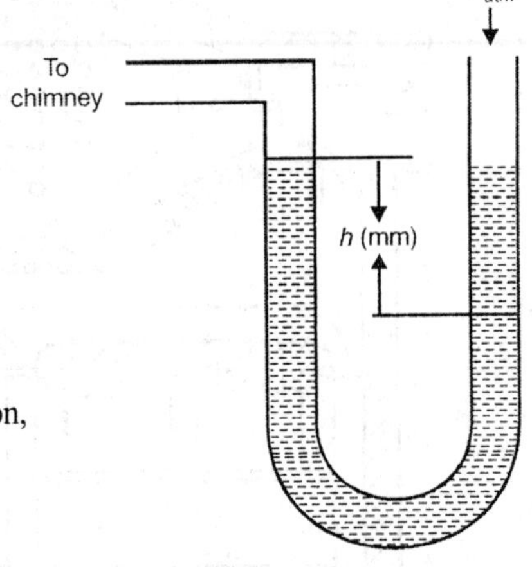

Fig. 10.22

Note. The amount of draught depends upon,

(i) Nature and depth of fuel on the grate.

(ii) Design of combustion chamber/ firebox.

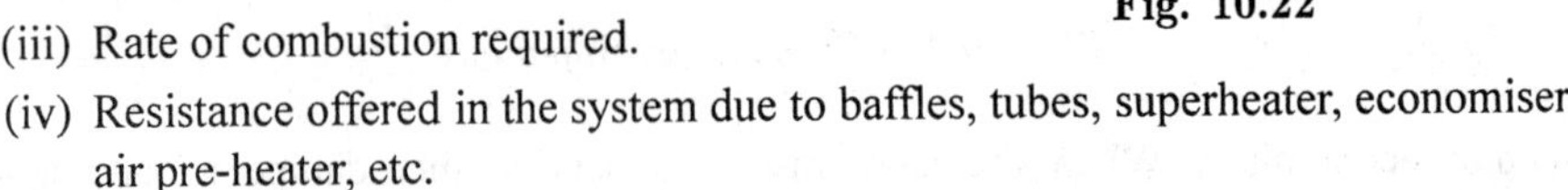

(iii) Rate of combustion required.

(iv) Resistance offered in the system due to baffles, tubes, superheater, economiser, air pre-heater, etc.

10.14 CLASSIFICATION

Draught is broadly classified into 2-types,

(1) Natural or chimney draught

(2) Artificial draught

  (a) Fan Drought (Produced by Mechanical Fans)

    (i) Forced draught

    (ii) Induced draught

    (iii) Balanced draught

  (b) Steam jet draught (Produced by steam jet)

    (i) Induced draught

    (ii) Forced draught

1. Natural or Chimney Draught. In this case the amount of draught directly depends upon the height of chimney. It is produced due to the difference in densities between the column of hot gases in the chimney and a similar column of cold air out side the chimney.

Let us first consider the case when fires are not lighted.

Let, the atmospheric pressure at grate level be P_1 and P_2 be the atmospheric pressure at an altitude H. The pressure P_2 is lower than the pressure P_1 because with the altitude pressure goes on decreasing.

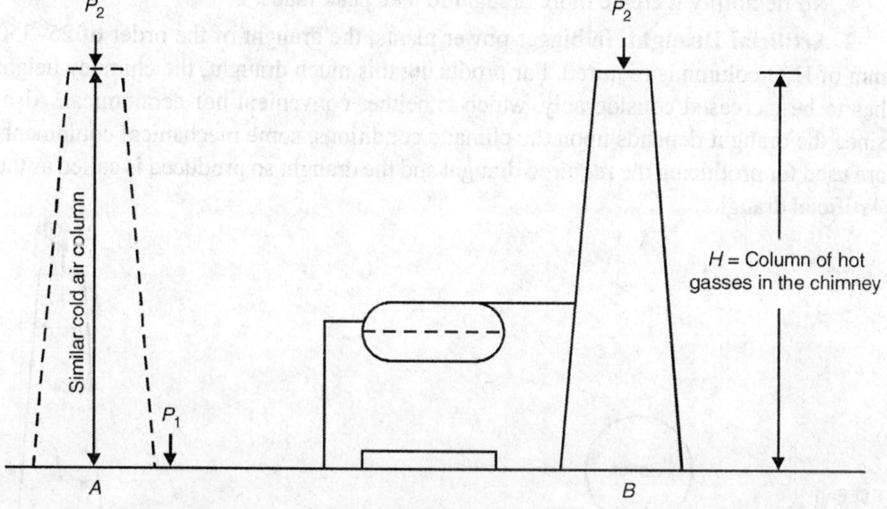

Fig. 10.23

Now let us consider the case when fires are lighted and the chimney is full of hot gases. Under these circumstances, the pressure at the base of the chimney is the sum of. pressure P_2 at the top and the pressure due to hot gas column H. But preŝsure P_1 at grate is the sum of pressure P_2 and the pressure due to similar cold column of air H.

Since, ρ cold air > ρ hot gases

i.e. $P_A > P_B$

∴ P_2 + Pressure due to cold column $H > P_2$ + Pressure due to hot column H.

∴ Pressure at grate due to cold column > Pressure at the chimney base due to hot column H.

This difference is called **static draught** and because of the pressure difference, (draught) air will rush to the combustion chamber, where combustion of air and fuel takes place and hot gases are generated. Then these hot gases because of draught, flow through the system and finally they are exhausted to the atmosphere through the chimney.

Advantages of Natural Draught

1. Easy to construct.
2. No power is required for producing the draught.
3. Long life of chimney.
4. No maintenance is required.

Disadvantages

1. Tall chimney is required.
2. Poor efficiency.
3. Decreases with increase in outside temperature.
4. No flexibility it create more draught to take peak loads.

2. Artificial Draught. In bigger power plants, the draught of the order of 25–350 mm of H_2O column is required. Far producing this much draught, the chimney height has to be increased considerably, which is neither convenient nor economical. Also, since the draught depends upon the climatic conditions, some mechanical equipments are used for producing the required draught and the draught so produced is called as the Artificial draught.

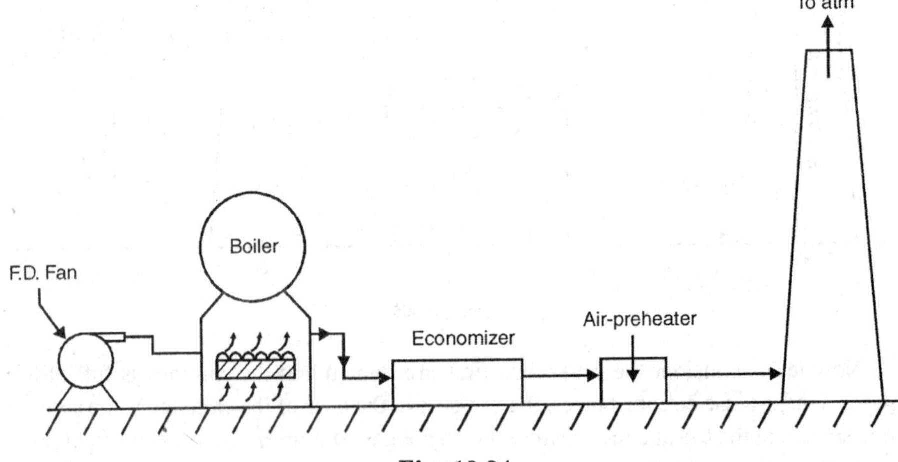

Fig. 10.24

(a) (i) Forced Draught. In a Forced draught system, a Fan or Blower is provided is shown in figure which forces the air in the combustion chamber. In the combustion chamber combustion of air and fuel takes place and hot gases generated. These gases are forced to pass through the flues, economiser, air pre-heater and then they are exhausted after recovering heat of flue gases. This draught system is known positive draught system, since the pressure of gases throughout the system is above atmospheric pressure.

It is to be noted that, the function of chimney use is to discharge the gases high in the atmosphere to reduce air pollution and it is not much significant for producing draught.

(ii) Induced Draught. In this system, the Blower or Induced Draught fan is located near the base of chimney. The air is sucked in the system, by reducing the pressure through the system below atmosphere. The flue gases, generated after combustion are drawn through the system and after recovering heat in the economiser, air-preheater, they are exhausted through the chimney to the atmosphere.

Here it is to be noted that the draught produced is independent of the temperature of hot gases, so the gases may be discharged as cold as possible after recovering as much heat as possible.

Advantage of Forced Draught (F.D.) over Induced Draught (I.D.)

1. The size and power required by I.D. fan is more because this fan handles more gases.
2. Since the I.D. fan handles hot gases, water cooled or air cooled bearings are to be used.
3. F.D. fan consumes less power and normal bearing can be used.

(iii) Balanced Draught. It is always preferable to use combinations of I.D. and F.D. instead of Forced or Induced draught alone.

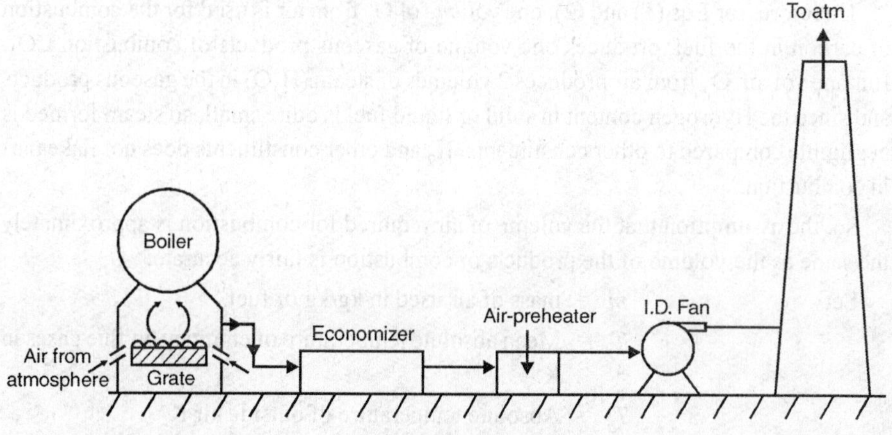

Fig. 10.25

If Forced Draught alone is used then the furnace cannot be opened for firing or for inspection. Because the high pressure air/gases inside the furnace will try to blowout, and there is every chance of blowing out of the fire completely and the furnace may stop.

If induced Draught fan alone is used, then also furnace cannot be opened either for firing or for inspection. Because the cold air will try to rush into the furnace, which reduces the effective draught.

To overcome both these difficulties Balanced Draught is used. In this case I.D. fan and F.D. fans are provided as shown.

(b) Steam Jet Draught. It may be Forced or Induced type. (Refer Locomotive boiler Fig. 10.8) If the steam jet is directed into the smoke box, near, the chimney, the air is sucked through the system, into the smoke box. If the jet is Located before grate, air is forced through the system. Induced type is favoured as forced draught increases heat losses.

Induced draught is produced by steam jet in case of Locomotive boiler. When the Locomotive in stationary, steam from the boiler may be supplied to the smoke box through the nozzle to create draught. While locomotive is running, due to motion, the air flows to the furnace.

10.15 CALCULATION OF CHIMNEY HEIGHT

We know that,

$$C + O_2 = CO_2 \quad (1)$$

$$1\,\text{vol} + 1\,\text{vol} = 1\,\text{vol}$$

and

$$2H_2 + O_2 = 2H_2O$$

$$2\,\text{vol} + 2\,\text{vol} = 2\,\text{vol} \quad (2)$$

Therefore, for Eqs (1) and (2), one *volume* of O_2 from air is used for the combustion of carbon in the fuel, produces one volume of gaseous products of combustion CO_2. But one vol. of O_2 from air produces 2 volumes of steam (H_2O) in the gaseous products and since the Hydrogen content in solid or liquid fuel is quite small, so steam formed is negligible compared to other constituents. N_2 and other constituents does not flake part in combustion.

So, the assumption that the volume of air required for combustion is approximately the same as the volume of the products of combustion is fairly accurate.

Let m = mass of air used in kg/kg of fuel.

T_f = Mean absolute temperature of chimney on flue gases in K.

T_a = Absolute temperature of outside air K

∴ Mass of chimney gases = $(m + 1)$ kg/kg of fuel

Volume per kg of flue gases = 273 K or 0°C

= Volume per kg of air at 273 K

From Perfect gas equation, $PV = RT$.

Volume at standard conditions,

$$V_N = \frac{RT}{P} = \frac{287 \times 273}{101325}$$

$$= 0.7734 \, m^3/kg$$

The pressure difference being very small, of the order of 25 mm of H_2O column, hence the pressure can be considered as constant at furnace and chimney base for the purpose of calculation of volumes at high temperatures.

∴ Volume per kg of air at T_a K,

Since we know that

$$\frac{P_N V_N}{273} = \frac{P_a V_a}{T_a}$$

$$V_a = V_N \times \frac{T_a}{273} \times \frac{P_N}{P_a}$$

$$\therefore \quad V_a = 0.7734 \times \frac{T_a}{273} \times \frac{P_N}{P_a}$$

∴ Volume of m kg of air at

$$T_a K = 0.7734 \times m \times \frac{T_a}{273} \times \frac{P_N}{P_a} \text{ and}$$

Volume of $(m + 1)$ kg of chimney gases at $T_f K$,

$$= \frac{0.7734 \times m \times T_f}{273} \times \frac{P_N}{P_a} \text{ and}$$

∴ Density of air at

$$T_a K = \frac{\text{Mass}}{\text{Volume}} = \frac{m \times 273}{0.7734 \times m \times T_a} \times \frac{P_a}{P_N}$$

$$= \frac{273}{0.7734 \times T_a'} \times \frac{P_a}{P_s}$$

$$\rho_a = 1.293 \left[\frac{273}{T_a}\right] N \frac{kg}{m^3} \times \frac{P_a}{P_N}$$

And density of flue gases at T_f at K,

$$= \frac{273}{0.7734 \times T_a} \times \frac{P_a}{P_s}$$

$$\rho_a = 1.293 \frac{273}{T_a} \frac{\text{kg}}{\text{m}^3} \times \frac{P_a}{P_N}$$

And density of flue gases at $T_f K$

$$= \frac{(m+1) \times 273}{0.7734 \times m \times T_f} \times \frac{P_a}{P_N}$$

$$\rho_g = 1.293 \left[\frac{m+1}{m}\right] \times \frac{273}{T_f} \times \frac{P_a}{P_N}$$

Let H = chimney height in metre from grate level.

Then pressure exerted by the column of hot gases.

$$H = \rho_g . g . H$$

$$P_f = 1.293 \left[\frac{m+1}{m}\right] \times \frac{273}{T_1} \times g \times H \ \text{N/m}^2 \tag{1}$$

Pressure exerted due to cold air column,

$$\rho_a \times g \times H = 1.293 \times \frac{273}{T_a} \times g \times H \ \text{N/m}^2 \tag{2}$$

Draught pressure, in terms of static pressure difference between hot gas column in chimney and a similar cold column.

$$P_d = \text{Eq. (2)} - \text{Eq. (1)}$$

$$= g \times H [P_a - P_f]$$

$$= 1.293 \times 273 \times g \times H \left[\frac{1}{T_a} - \left[\frac{m+1}{m}\right] \times \frac{1}{T_f}\right] \text{N/m}^2$$

$$P_d = 353 \times g \times H \left[\frac{1}{T_a} - \left[\frac{m+1}{m}\right] \times \frac{1}{T_f}\right] \text{N/m}^2$$

If P_a and P_N are different, then

$$P_d = 353 \times g \times H \times \frac{P_a}{P_N} \left[\frac{1}{T_a} - \left[\frac{m+1}{m}\right] \times \frac{1}{T_f}\right] \text{N/m}^2$$

To express in mm of H_2O column

$$P_d = \rho_w . g . h_w \frac{\text{N}}{\text{m}^2}$$

where ρ_w = Density of water in kg/m^3

g = Acceleration due to gravity in m/sec^2

h_w = water column in mm

Then, P_d = Pressure difference wil be in N/m^2

i.e.
$$P_d = \frac{\text{kg}}{\text{m}^3}\times\frac{\text{m}}{\text{sec}^2}\times\text{m}$$

$$= \left[\text{kg}\times\frac{\text{m}}{\text{sec}^2}\right]\times\frac{1}{\text{m}^2} = \text{Force}\times\frac{1}{\text{m}^2}$$

$$= \frac{\text{N}}{\text{m}^2}\ \text{resultant unit.}$$

$$= h_a = \frac{P_d}{P_w.g}$$

$$= \frac{353\times g\times H\times\frac{P_a}{P_N}\left[\frac{1}{T_a}-\left[\frac{m+1}{m}\right]\times\frac{1}{T_f}\right]}{1000\times g}\ \text{meters of } H_2O$$

$$\therefore\quad h_w = 353\times H\times\frac{P_a}{P_w}\left[\frac{1}{T_a}-\left[\frac{m+1}{m}\right]\times\frac{1}{T_f}\right]\ \text{mm of water}$$

If $P_a = P_N$ then,

$$h_w = 353\times H\left[\frac{1}{T_a}-\left[\frac{m+1}{m}\right]\times\frac{1}{T_f}\right]\ \text{mm of } H_2O$$

To express, P_d in meters of hot flue gas

$$\therefore\quad P_d = \rho_f\cdot g\cdot h_f$$

where ρ_f = Density of hot flue gas

h_f = height of flue gas column

$$\therefore\quad P_d = \rho_f\cdot g\cdot H$$

$$H = \frac{P_d}{\rho_f\cdot g}$$

$$= \frac{353\times g\times H\times\frac{P_a}{P_N}\left[\frac{1}{T_a}\left[\frac{m+1}{m}\right]\times\frac{1}{T_f}\right]}{g\times 353\times\left[\frac{m+1}{m}\right]\times\frac{1}{T_f}\times\frac{P_d}{P_N}}$$

On simplification we get,

$$H' = H\left[\frac{m}{m+1}\times\frac{T_f}{T_a}-1\right]\ \text{mt. of hot gas column}$$

Condition for maximum discharge through the chimney.

Mass of the flue gases discharged through the chimney is proportional to the density, cross sectional areas and velocity.

i.e. $M \alpha$ density × cross section area × velocity

Note. Cross section areas is constant because of construction,

And we know that,

$$\rho_f = \frac{1}{T_f}$$

$$M = \text{Constant} \times \frac{1}{T_f} \times \text{Velocity}$$

$$= k \times \frac{1}{T_f} \times \sqrt{2gH}$$

$$= \text{Constant} \times \frac{1}{T_f} \times \sqrt{H}$$

Mass of gas

$$= \frac{1}{T_f} \times \left[H \times \left[\frac{m}{m+1} \times \frac{T_f}{T_a} - 1 \right] \right]^{1/2} \times \text{Constant}$$

$$= \frac{1}{T_f} \times \left[\frac{m}{m+1} \times \frac{T_f}{T_a} - 1 \right]^{1/2} \times \text{Constant (Since } H = \text{Constant)}$$

$$M = \left[\frac{m}{m+1} \times \frac{1}{T_f . T_a} - \frac{1}{T^2{}_f} \right]^{1/2} \times \text{Constant}$$

For maximum discharge,

$$\frac{dM}{dT_f} = 0$$

$$\frac{dM}{dT_f} = \frac{1}{2} \left[\frac{m}{m+1} \times \frac{1}{T_f . T_a} - \frac{1}{T^2{}_f} \right]^{1/2}$$

$$\times \left[\frac{m}{m+1} \times \frac{1}{T_a} \times \frac{1}{T_f^2} - \frac{1}{T_f^3} \right] = 0$$

$$\therefore \quad \left[-\frac{m}{m+1} \times \frac{1}{T_a . T_f^3} \right] = 0$$

$$\frac{m}{m+1} \times \frac{1}{T_a . T_f^{\,2}} = \frac{2}{T_f^{\,3}}$$

$$T_f = 2\left[\frac{m}{m+1}\right] \times T_a$$

We know that, $$H = \left[\frac{m}{m+1} \times \frac{T_f}{T_a} - 1\right] - H$$

Substituting T_f in above equation we get,

$$H' = H\,[2-1] = H$$

$$\therefore \quad H' = H$$

10.16 DRAUGHT LOSSES

The actual or available draught will be less than the theoretical Draught obtained by calculations, because of the draught losses viz.,

(1) Head loss due to fuelled resistance h_{fb}: Fuel bed resistance depends upon the size of the fuel, depth of fuel on grate and combustion rate.

(2) Head loss due to frictional resistance to the flow of flue gases because of the equipments like superheater, economiser, air preheater, etc. (h_e)

(3) Head loss due to friction in bends and curves in the flue gas passages (h_b).

Thus, total draught loss = $h_{\text{Total}} = h_{fb} + h_e + h_b$

10.17 INTRODUCTION TO IBR OR REGULATIONS FOR CHIMNEY (WITH REFERENCE TO IBR/MSNC MUMBAI)

To keep control on boiler accidents, Indian Government after independence made following rules and regulations, which the boiler should conform. They are as under:

1. A plan of the boiler house and chimney is to be approved by the chief inspector of Indian Boiler Regulations (IBR)/Maharashtra Chief Inspector of Nuisance Commission (MSNC)—Mumbai.
2. Chimney fabrication drawings are to be approved before fabrication.
3. A plan of 1 m =1 cm in blue print showing the exact position of furnace, flues, chimney, duly certified by the chartered engineer /owner should be submitted.
4. Substantial structure certificate for the chimney stating that the chimney in self supported/supported by wire ropes and with stands a wind load of 207 kg/m^2 is to be submitted (Note this should be certified by the chartered engineer).
5. Chimney height minimum required as per IBR/MSNC is 100 feet that i.e. 30.48 m from the firing floor of the boiler.
6. An undertaking for the height of chimney is to be submitted by the owner to MSNC.
7. An undertaking for the use of specific fuel is to be sumitted by the owner.
8. No objection certificate (NOC) is to be obtained from Air Force Deptt.

9. Smoke generated should conform to Bombay Smoke Nuisance Act 1912.
10. Chimney should be day painted. (i.e. Red and white strips of equal width in total 7 strips are to be painted).
11. Anti collision lights are to be fitted, for sighting in the night and poor visibility.
12. Following details are to be submitted,
    (a) Height of own factory building/boiler house.
    (b) Height of tallest building within a radius of 300 m around ground level.
    (c) Capacity of boiler.
    (d) Working pressure.
    (e) Type of fuel.
    (f) Quantity of fuel/hour.
    (g) Type of chimney self-supported/supported by wire ropes.
    (h) Material of construction.

10.18 BOILER CALCULATIONS

In case of boilers it is very important to find,

(a) Heat transfer required to form steam. We have studied steam generation process,

Let h_2 = Specific enthalpy of steam formed in kJ/kg

h_1 = Specific enthalpy of feed water in kJ/kg

Then, since the steam is generated ($P = C$) at constant pressure,

And for a constant pressure process $\delta Q = dH$

i.e. Heat supplied = Change of enthalpy,

∴ Heat transfer required to form 1 kg of steam in the boiler,

$$= (h_2 - h_1) \text{ kJ/kg}$$

If m_s is the mass of steam formed in kg, then heat transfer required to form total steam.

$$= m_s.(h_2 - h_1)\left[\text{kg} \times \frac{\text{kJ}}{\text{kg}} = \text{kJ}\right] \qquad (1)$$

(b) Energy received from fuel. This is obtained from the mass of fuel burnt and its calorific value of fuel,

If mass of fuel used = m_f kg

And calorific value of fuel = C.V. kJ/kg

Then,

$$\text{Energy received} = m_f \times \text{C.V.}\left[\text{kg} \times \frac{\text{kJ}}{\text{kg}} = \text{kJ}\right] \qquad (2)$$

(c) Boiler thermal efficiency. It is the ratio of the energy received by the steam to the energy supplied by the fuel to produce steam.

Thus, Boiler Thermal Efficiency;

$$\eta_{th} = \frac{\text{Energy to steam}}{\text{Energy from fuel}}$$

$$\eta_{th} = \frac{\text{Energy received by the steam}}{\text{Energy supplied by the fuel}}$$

From Eqs (1) and (2)

$$\textbf{Boiler Thermal Efficiency } \eta_{th} = \frac{m_s(h_2 - h_1)}{m_f \times C.V.} \times 100\%$$

(d) Boiler performance. As we have studied, the main function of the boiler is to produce steam from feed water, with the use of heat liberated by the combustion of fuel. It is to be noted the number of kg of water evaporated per hour is not an exact measure of the performance of the boiler. Since different boilers may be operating at different at different conditions, so their comparison may be made on the basis of some reference condition known as Equivalent of Evaporation.

Equivalent of Evaporation from and at 100° C

"The evaporation which would be obtained if the feed water were supplied at 100°C and converted into dry and saturated steam at 100°C, at standard atmospheric pressure, is known as equivalent of evaporation from and at 100°C".

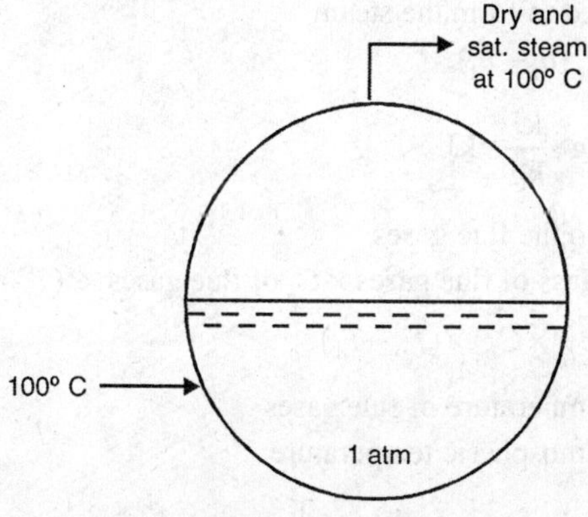

Fig. 10.27

We know that from Eq. (1) energy received by the steam

$$= m_s(h_2 - h_1)\,\text{kJ}$$

It is to be noted that, at 100°C, it is the enthalpy of evaporation which is supplied and the specific enthalpy of evaporation at 100°C, = 2256.9 kJ/kg. (From steam tables h_{fg} at 100° C and 1 atm.)

The equivalent of evaporation of a boiler from and at 100°C,

$$= \frac{m_s(h_2 - h_1)}{2256.9} \text{ kg/kg of coal.}$$

Evaporative Capacity

"It is defined as the amount of water and water evaporated into steam per hour". The evaporative capacity may be expressed in kg/hr. or kg/kg, of fuel or kg/hr-m^2 of heating surface area.

Actual Evaporation

It is defined as the amount of water evaporated into steam, at actual working conditions, per kg of fuel burnt.

10.19 ENERGY BALANCE

Energy released due to the combustion of fuel will be utilised for various purposes and is tabulated in heat balance sheet.

Heat balance sheet gives the account of heat energy released and heat energy utilised

(1) Energy released due to the combustion of fuel

$$= m_f \times C.V. = \text{kg} \times \frac{\text{kJ}}{\text{kg}} = \text{kJ}$$

(2) This energy released will be utilised for various purposes, viz.

(i) Heat energy utlised to form the steam

$$= m_s (h_2 - h_1)$$

$$\text{Units} = \text{kg} \times \frac{\text{kJ}}{\text{kg}} = \text{kJ}$$

(ii) Heat energy lost to the flue gases

$$= \text{Mass of flue gases} \times C_p \text{ of flue gases} \times (T_f - T_a)$$

$$= m_{f_g} \times C_{pg} \times (T_f - T_a)$$

where T_f = temperature of flue gases

and T_a = atmospheric temperature

$$\left[\text{Units} = \frac{\text{kg}}{\text{kg of coal}} \times \frac{\text{kJ}}{\text{kg.K}} \times \text{K} = \frac{\text{kJ}}{\text{kg of coal}}\right]$$

(3) Energy lost due to the unburnt fuel falling through the grate bars—When solid fuels are used, some of the fuel falls through the grate bars and is lost with ash and this is

$$= m_{uf} \times C.V_{uf}$$

(4) Energy lost due to the water vapour formed due to the combustion of H_2 in the flue gases

$$= m_1 (h'_2 - h'_1)$$

where m_1 = mass of steam formed due to the combustion of H_2 in kg/kg of coal burnt.

h'_2 = Specific enthalpy of steam formed due to the combustion of H_2 in the flue gases at the temperature of flue gases and at 1 atmospheric pressure or, at 1 bar pressure or at the given partial pressure.

h'_1 = Specific enthalpy of water at the boiler room temperature

(5) Energy lost due to radiation

Note. In order to reduce radiation losses, Asbestos lagging is to be provided on the exposed boiler surface and is to be painted by a black paint.

All the above items may be expressed in the balance sheet as follows.

Energy supplied	**kJ per minutes or Per hr**	**%**	**Energy utilised**	**kJ**	
	say X	100	(i) To form steam	x_1	.....
			(ii) Energy lost to, flue gases	x_2	.....
			(iii) Energy lost unburnt fuel	x_3	.....
			(iv) Energy lost to water vapour	x_4	.....
			(v) Every unaccounted by difference	x_5	.....
Total	X	100	Total	X	100

FORMULAE

(1) Draught produced in mm of H_2O

$$h_W = 353 \times H \left\{ \frac{1}{T_a} - \left[\frac{m+1}{m} \right] \times \frac{1}{T_f} \right\} \frac{P_a}{P_N}$$

where H = Height of chimney in metre

P_a = Actual pressure at boiler location.

P_N = Standard atmospheric pressure

T_a = Actual atmospheric temperature K

T_f = Temperature of flue gases in K

m = Mass of air in kg to burn 1 kg of fuel

∴ $m + 1$ = Mass of flue gases in kg/kg of fuel.

(i.e. when 1 kg of fuel is burnt with m kg of air, then total flue gases produced = $m + 1$)

(2) Density of air (ρ_a)

$$\rho_a = 1.293\left[\frac{273}{T_a}\right]\frac{P_a}{P_N} \text{ in } \frac{\text{kg}}{\text{m}^3}$$

(3) Density of flue gases

$$\rho_f = 1.293\left[\frac{m+1}{m}\right].\frac{273}{T_f}.\frac{P_a}{P_N}$$

or

$$\rho_f = \frac{m+1}{m}\times\frac{353}{T_f}\times\frac{P_a}{P_N} \text{ kg/m}^3$$

(4) Draught pressure in terms of static pressure difference between hot gas column and a parallel cold air column.

$$P_d = g\times H\times[P_\alpha - P_f]\frac{\text{N}}{\text{m}^2}$$

and in terms of mm of water column

$$h_w = H[P_\alpha - P_f] \text{ mm of } H_2O$$

(5) Velocity of flue gases = $C=\sqrt{2gH'}$

(6) Equivalent height of hot gas column H_{gases} or H'

$$H' = H\left[\frac{m}{m+1}\times\frac{T_f}{T_a}-1\right] \text{ m of hot gas column.}$$

(7) Condition for maximum discharge through chimney $H' = H$ and during maximum discharge conditions, Maximum temperature of flue gases,

$$T_f = 2\left[\frac{m+1}{m}\right]\times T_a$$

SOLVED EXAMPLES

(Problems on Boiler Draught)

Example 10.1 Find the minimum height of chimney required to produce a draught of 16 mm of H_2O, if 19 kg of air is required per kg of fuel burnt on the grate. The mean temperature of flue gases inside the chimney is 330°C.

Solution

Data: H = ? h_w = 16 mm of H_2O

m = 19 kg/kg of fuel $\qquad T_f = 330 + 273 = 603$ K

$T_a = 30 + 273 = 303$ K

We know that, the drought produced in mm of H_2O column is given by,

$$h_w = 353 \times H\left\{\frac{1}{T_a} - \frac{m+1}{m} \times \frac{1}{T_f}\right\}\frac{P_a}{P_N} \quad (1)$$

Since in the problem they have not mentioned anything about P_a and P_N

∴ Assuming $P_a = P_N$

Now substituting the values in the Eq. (1) we get,

$$h_w = 353 \times H\left\{\frac{1}{303} - \frac{20}{19} \times \frac{1}{603}\right\} \text{mm of } H_2O$$

$$16 = 353 \times H\{0.00330033 - 0.0017456577\}$$

$$\mathbf{H = 29.155\ m}$$

Example 10.2 Readings from a boiler trial,

(i) Height of chimney : 100 m

(ii) Ambient temperature : 32 °C

(iii) Mass of air : 18 kg/kg of fuel

(iv) Flue gas temperature : 300°C

Find natural draught in mm of H_2O

Data: H = 100 m $\qquad T_a = 32 + 273 = 305$ K

m = 18 kg/kg of fuel $\qquad T_f = 300 + 273 = 573$ K

h_w = ?

Solution

We know that, $$h_w = 353 \times H\left\{\frac{1}{T_a} - \left[\frac{m+1}{m}\right] \times \frac{1}{T_a}\right\} \text{ mm of } H_2O$$

Assuming $P_a = P_N$

Now $$h_w = 353 \times 100\left\{\frac{1}{305} - \frac{19}{18} \times \frac{1}{573}\right\}$$

$$h_w = 35300 \times \{0.0014365322\}$$

$$\mathbf{h_w = 50.7096\ m}$$

Example 10.3 A boiler is equipped with a chimney of 30.48 m height. Atmospheric conditions are 755 mm of Hg and 35°C. Average temperature of flue gases in the Chimney is 300°C. If the boiler is supplied with 20 kg of air kg of fuel.

Calculate theoretical draught created in mm of H_2O. Take standard atmospheric pressure as 760 mm of Hg.

Solution

We know that

$$h_w = 353 \times H \times \frac{P_a}{P_N}\left\{\frac{1}{T_a} - \left[\frac{m-1}{m}\right] \times \frac{1}{T_f}\right\}$$

$$= 353 \times 30.48 \times \frac{755}{760}\left\{\frac{1}{308} - \frac{21}{20} \times \frac{1}{573}\right\}$$

$$h_w = \mathbf{15.1168 \text{ mm of } H_2O}$$

Example 10.4 Find the amount of air used in a boiler/kg of fuel, when chimney height is 50 m. Draught produced is 25 mm of H_2O. The temperature of flue gases is 50 m. Draught produced is 25 mm of H_2O. The temperature of flue gases is 353°C and that of room is 35°C.

Also determine the draught produced in meters of hot gas column and flue gas temperature for maximum discharge conditions.

Data:

$m = ?$	$H = 50$ m
$h_w = 25$ mm of H_2O	$T_f = 353 + 273 = 626$ K
$T_a = 35 + 273 = 308$ K	$H = ?$
$T_f = ?$ for max. discharge conditions	

Solution

We know that,

$$h_w = 353 \times H\left\{\frac{1}{T_a} - \left[\frac{m+1}{m}\right] \times \frac{1}{T_f}\right\} \quad \text{mm of } H_2O$$

$$25 = 353 \times 50\left\{\frac{1}{308} - \left[\frac{m+1}{m}\right] \times \frac{1}{626}\right\} \quad \text{mm of } H_2O$$

$$\therefore \quad \mathbf{m = 6.8595 \text{ kg} = 6.86 \text{ kg}}$$

(ii) And we know that Equivalent height of hot gas column,

$$H' = H\left\{\frac{m}{m+1} \times \frac{T_f}{T_a} - 1\right\} m \text{ of hot gas column}$$

$$= 50\left\{\frac{6.86}{7.86} \times \frac{626}{308} - 1\right\}$$

$$\mathbf{H' = 38.6942 \text{ m of hot gas column}}$$

(iii) We know that for max discharge conditions

T_f is given by, $T_f = 2\times\frac{(m+1)}{m}\times T_a = 2\times\frac{7.86}{6.86}\times 308$

$$\mathbf{T_f = 705.7959\ K}$$

Example 10.5 Find the minimum temperature of the flue gases required to produce a draught of 1.5 cm of water by a chimney of 30 m height, when air-fuel ratio required in the combustion is 20. Atmospheric air temperature is 27°C.

Data: $T_f = ?$ $h_w = 15$ mm

$H = 30$ m $\frac{\text{Air}}{\text{Fuel}} = \frac{20}{1}$

Solution

∴ Mass of air, $m = 20$ kg

and mass of flue gases $m + 1 = 21$ kg/ kg of fuel

$$T_a = 27 + 273 = 300\ K$$

We know that, $h_w = 353\times H\times\left[\frac{1}{T_a}-\frac{(m+1)}{m}\times\frac{1}{T_f}\right]$ mm of H_2O

$$15 = 353\times 30\times\left[\frac{1}{300}-\frac{21}{20}\times\frac{1}{T_f}\right]$$

∴ $\mathbf{T_f = 547.75\ K}$

Example 10.6 How much air is used/Kg of coal burnt in a boiler having a chimney of 35 m height to create a draught of 20 mm of water, when the temperature of flue gas in the chimney is 370° C and the boiler house temperature is 34°C. Does this chimney satisfy the condition of max. discharge ?

Data: $m = ?$ $H = 35$ m

$h_w = 20$ mm of H_2O $T_f = 643$ K

$T_a = 307$ K

Solution

We know that, $h_w = 353\times H\times\left\{\frac{1}{T_a}-\frac{m+1}{m}\times\frac{1}{T_f}\right\}$ mm of H_2O

$$20 = 353\times 35\left\{\frac{1}{307}-\frac{m+1}{m}\times\frac{1}{643}\right\}$$

∴ $\mathbf{m = 18.66\ kg/kg\ of\ coal}$

For max, discharge, $$T_f = 2\times\left[\frac{m+1}{m}\right]\times T_a$$

$$= 2\times\frac{19.66}{18.66}\times 307$$

$$\mathbf{T_f = 646.905\ K}$$

Whereas actual T_f given is 643 K which is less than T_f max

∴ Condition is satisfied.

Example 10.7 The draught produced in a certain boiler is 18.5 mm of water when the temperature of the flue gases in the chimney is 370°C and the temperature of the boiler house is 30°C. Estimate the air fuel ratio. Also state whether chimney operates under the maximum discharge condition or not. The height of the chimney is 32 m.

Data: h_w = 18.5 mm of H_2O T_f = 370 + 273 = 643 K

T_a = 30 + 273 = 303 K $m = ?$ in $\frac{\text{kg of air}}{\text{kg of fuel}}$

H = 32 m

Solution

Check T_f for maximum discharge conditions.

We know that, the draught produced in mm of H_2O column is given by,

$$h_w = 353\times H\left\{\frac{1}{T_a}-\left[\frac{m+1}{m}\right]\times\frac{1}{T_f}\right\}$$

$$18.5 = 353\times 32\left\{\frac{1}{303}-\left[\frac{m+1}{m}\right]\times\frac{1}{643}\right\}$$

$$\mathbf{m = 14.52\frac{kg\ of\ air}{kg\ of\ fuel}}$$

∴ Air fuel ratio is 14.52:1

Now to check for T_f maximum discharge conditions,

$$T_f = 2\left[\frac{m+1}{m}\right]T_a = 2\times\left[\frac{14.52+1}{14.52}\right]\times 303$$

$$\mathbf{T_f = 647.76\ K}$$

Since T_f actual given is 643 which is less than T_f for maximum discharge condtitions, So the chimney operates under maximum discharge conditions.

Example 10.8 In a certain boiler installation a chimney of 42.5 m height produces a natural drought equivalent of 33.25 m of hot gas column. Boiler house temperature is 33°C. The mass of air supplied per kg of fuel is 19 kg, determine,

(i) Temperature of gas leaving the chimney.

(ii) Temperature of flue gases and corresponding draught for maximum discharge.

Data: $H = 42.5$ m $\quad H' = 33.25$ m of hot gas column

$T_a = 33 + 273 = 306$ K $\quad m = 19$ kg/kg of fuel

$m + 1 = 20$ kg of gases/kg of fuel $\quad T_f = ?$

and T_f for max discharge consitions = ?

Solution

We know that, equivalent height of hot gas column,

$$H' = H\left\{\frac{m}{m+1}\times\frac{T_f}{T_a}-1\right\} \text{ m of hot gas column}$$

$$33.25 = 42.5\left\{\frac{19}{20}\times\frac{T_f}{306}-1\right\} \text{ m of hot gas column}$$

$$33.25 = 42.5\{3.1\times10^{-3}T_f-1\}$$

$$T_f = 547.105 \text{ K}$$

We know that T_f for maximum discharge conditions,

$$T_f = 2\left[\frac{m+1}{m}\right]\times T_a$$

$$T_f = \left[\frac{20}{19}\right]\times 306$$

$$\mathbf{T_f = 644.211 \text{ K}}$$

Example 10.9 Determine the height of chimney required for a boiler having an average coal consumption of 1500 kg/hr and producing 15 kg of dry flue gases per kg of coal fired. The pressure drop in mm of water at various stages in a boiler flue passes are (i) fuelbed 4.9 mm, (ii) Boiler flue 7.8 mm, (iii) Boiler tube passage 2.7 mm. (iv) Through bends, dampers etc. 0.6 mm.

Actual draught may be taken as 25% greater than the theoretical draught. The flue gases have a mean temperature of 277°C and ambient temperature is. 27°C.

Does this chimney fulfill the condition of maximum discharge?

Data: $H = ?$ $m_f = 1500$ kg/hr

$$m_{fg} = (m+1) = \frac{15 \text{ kg of gases}}{\text{kg of coal}}$$

Solution

Total pressure drop or Draught = 4.9 + 7.8 + 2.7 + 0.6 =16 mm of H_2O

Since Actual draught in 25% more than the theoretical draught,

$$\therefore \quad h_w = 16 \times 1.25 = 20 \text{ mm of } H_2O$$

$$T_f = 277°C + 273 = 500 \text{ K (Actual)}$$

$$T_a = 27 + 273 = 300 \text{ K} \quad \text{(Mean)}$$

Now we know that the draught,

$$h_w = 353H\left\{\frac{1}{T_a} - \left[\frac{m+1}{m}\right] \times \frac{1}{T_f}\right\}$$

$$20 = 353 \times H\left\{\frac{1}{300} - \frac{15}{14} \times \frac{1}{550}\right\}$$

$$H = 40.91 \text{ m}$$

And the condition for max discharge is,

$$T_f = 2\left\{\frac{m+1}{m}\right\} \times T_a$$

$$= 2\left\{\frac{15}{14}\right\} \times 300$$

$$\mathbf{T_f = 642.8571\ K}$$

As T_f given in 550 K which is within T_f maximum. So the chimney satisfies the condition for maximum discharge.

Example 10.10 Find the draught produced in mm of water by a chimney of 40 m height. Let the mass of flue gas be 20 kg/kg of fuel burnt in the combustion chamber. The temperature of flue gases and the ambient air are 270° C and 23° C. Assuming the diameter of the chimney is 150 cm and 30% of the theoretical draught is lost in friction, find the mass of the flue gases passing through the chimney/min.

Data: $h_w = ?$ $h = 40$ m

$m + 1 = 20$ kg of flue gases/kg of fuel

$T_f = 543$ K $T_a = 296$ K

Also mass of flue gases passing through the chimney/min = ?

Solution

We know that, $$h_w = 353H\left\{\frac{1}{T_a} - \left[\frac{m+1}{m}\right] \times \frac{1}{T_f}\right\}$$

$$= 353 \times 40\left\{\frac{1}{296} - \frac{20}{19} \times \frac{1}{543}\right\}$$

$$h_w = 20.33 \text{ mm}$$

For the 2nd part from the continuity equation,
Mass of flue gases discharged

$$= \text{c/s area} \times P_f \times \text{velocity of flue gases} \quad (1)$$
$$= A \times P_f \times C$$

$$\text{Units} = \text{m}^2 \times \frac{\text{kg}}{\text{m}^3} \times \frac{\text{m}}{\text{sec}} = \frac{\text{kg}}{\text{sec}} \quad (2)$$

We also know that,

$$A = \frac{\pi}{4}D^2 = \frac{\pi}{4} \times 1.5^2 = 1.766 \text{m}^2$$

$$P_f = \frac{m+1}{m} \times \frac{353}{T_f} = \frac{20}{19} \times \frac{353}{543}$$

$$P_f = 0.6843 \frac{\text{kg}}{\text{m}^3} \quad (3)$$

and $$c = \sqrt{2gH'}$$

To find H', we know that,

$$H' = H\left\{\frac{m}{m+1} \times \frac{T_f}{T_a} - 1\right\} \text{ metres of hot gas column.} \quad (4)$$

$$= 40\left\{\frac{19}{20} \times \frac{543}{296} - 1\right\} = 40 \times 0.7427364$$

$$H' = 29.7095 \text{ m}$$

Since 30% of draught is lost frcition only 70% of the draught will be available

$\therefore$ Available draught $= 0.7 \times 29.7095$

H' available $= 20.7966$ m

$\therefore$ Substitute this in Eq. (4) we get,

$$C = \sqrt{2 \times g \times H'} = \sqrt{2 \times 9.81 \times 20.7966}$$

$$C = 20.199745 \text{ m/sec}$$

Now mass of the flue gases discharged from Eq. (1)

$$= 1.766 \times 0.6843 \times 20.199745$$
$$= 24.4109 \text{ kg/sec}$$

Mass of flue gases = 1464.65 kg/min

Example 10.11 In a boiler trial following data were noted,

Height of chimney : 25 m

Temperature of flue gases : 320°C

Ambient temperature : 22°C

Mass of air : 17 kg/kg of coal

Atmospheric pressure : 720 mm of Hg

Calculate $h_w = ?$

Solution

We know that draught produced in mm of H_2O column is given by,

$$h_w = 353 \times H\left\{\frac{1}{T_a} - \frac{m+1}{m} \times \frac{1}{T_f}\right\}\frac{P_a}{P_N} \text{ mm of } H_2O$$

$$= 353 \times 25\left\{\frac{1}{295} - \frac{18}{17} \times \frac{1}{593}\right\}\frac{720}{760}$$

$$h_w = 13.40 \text{ mm of } H_2O$$

This problem can also be solved by using fundamental equations.

We know that, $h_w = H[P_\alpha - P_f]$

Volume of 17 kg of air at NTP

$$P_0V_0 = mRT_0$$

$$V_0 = \frac{17 \times 287 \times 273}{101325}$$

$$V_0 = 13.145 \text{ m}^3$$

Note. $P_0 = 760$ mm of Hg $= 101325$ N/m^2 and $T_0 = 273$

Volume of air at T_a = 22° C = 295 K

and pressure of 720 mm of Hg is 95992.105 N/m^2

$$\therefore \quad \frac{P_0V_0}{T_0} = \frac{P_aV_a}{T_a}$$

$$V_a = \frac{101325\,\text{N/m}^2 \times 13.145\,\text{m}^3 \times 295\,\text{K}}{273\,\text{K} \times 95992.105\,\text{N/m}^2}$$

$$V_a = 15\ \text{m}^3$$

$$\therefore \quad \rho_a = \frac{\text{Mass of air}}{\text{Volume of air at 295 K}}$$

$$\rho_a = \frac{17}{15} = 1.133\,\text{kg/m}^3$$

Note volume of flue gas at NTP =13.145 m^3
and to find volume of flue gases at 320°C.

$$\frac{P_0V_0}{T_0} = \frac{P_fV_f}{T_f}$$

$$V_f = \frac{101325 \times 13.145 \times 593}{273 \times 95992.105}$$

$$V_f = 30.137\ \text{m}^3$$

$$\rho_f = \frac{\text{Mass of flue gases}}{\text{Volume of flue gases}}$$

$$\rho_f = \frac{18}{30.137} = 0.5972\,\text{kg/m}^3$$

$$h_w = H[\rho_\alpha - \rho_\tau]25[1.133 - 0.5972]$$

$$h_w = 13.40\ \text{mm of } H_2O$$

FORMULAE

1. Heat transfer required to form the steam

$$= m_s(h_2 - h_1)\,\text{kJ.}$$

m_s = Mass of steam generated in kg

h_2 = Specific enthalpy of steam formed in kJ/kg

h_1 = Specific enthalpy of feed water in kJ/kg

2. Energy received from fuel

$$= (m_f \times \text{C.V.})\text{kJ}$$

where m_f = mass of fuel in kg.

C.V. = Calorific value in kJ/kg

3. Boiler thermal efficiency,

$$\eta_{th} = \frac{m_s(h_2 - h_1)}{m_f \times \text{C.V.}}$$

4. Equivalent of Evaporation from and at 100°C

$$= \frac{\dot{m}_s(h_2 - h_1)}{2256.9} \frac{\text{kg}}{\text{kg of coal}}$$

Note. Here $\dot{m}_s$ should be in kg/kg of fuel burnt and it is written as $\dot{m}_s$ or m_a.

(Problems on Boiler Performance)

Example 10.12 A boiler working at a pressure of 14 bar evaporates 8 kg of water / kg of coal burnt from the feed water entering at 39°C. The steam at the steam stop valve is 0.95 dry. Determine the equivalent of evaporation from and at 2009°C.

Data: P = 14 bar $\quad \frac{m_s}{m_f}$ = 8 kg of water/kg of coal = m_s

Feed water temperature = 39°C $\quad x = 0.95$

Equivalent of evaporation?

Solution

We know that the equivalent of evaporation form and at 100°C

$$= \frac{\dot{m}_s(h_2 - h_1)}{2256.9} \qquad (1)$$

where h_2 is specific enthalpy if steam formed at 14 bar pressure and 0.95 dry and the steam formed is wet,

$$\therefore \quad h_2 = h_f + x \cdot h_{fg}$$

From steam tables at 14 bar pressure

$$h_f = 830.07 \text{ kJ/kg}$$

$$h_{fg} = 1957.7 \text{ kJ/kg}$$

$$\therefore \quad h_2 = 830.07 + 0.95 \times 1957.7$$

$$h_2 = 2689.885 \text{ kJ/kg} \qquad (2)$$

and h_1= specific enthalpy of feed water at 39°C from steam tables,

$$h_1 = h_f = 163.27 \text{ kJ/kg} \qquad (3)$$

Also $h_1 = h_f = m.C_p.\Delta t = 1 \times 4.187 \times 39°C$

$h_1 = 163.29$ kJ/kg

∴ From Eq. (1),

Now, Equivalent of evaporation $= \dfrac{\dot{m}_s \times (h_2 - h_1)}{2256.9}$

m_s should be in kg/kg of fuel burnt

$$= \frac{8 \times [2689.885 - 163.27]}{2256.9}$$

Equivalent of Evaporation = 8.9560548 kg of steam/kg of coal

Example 10.13 A boiler plant supplies 5400 kg of steam /hr at 7502 and 0.98 dry from feed water at 41.5°C, when using 670 kg of coal / hr having a calorific value of 31000 kJ/kg.

Determine,

(i) The efficiency of the boiler;

(ii) The equivalent of evaporation from and at 100°C.

Solution

We know that the thermal efficiency of the boiler is given by

$$\eta_{th} = \frac{m_s(h_2 - h_1)}{(m_f \times \text{C.V.})} \times 100\%$$

Now Actual Evaporation

or Steam raised/kg of coal = m_a or $m_s = \dfrac{m_s}{m_f} = \dfrac{5400}{670}$

$$m_a \text{ or } \dot{m}_s = 8.06 \text{ kg/kg of coal}$$

and h_2 = specific enthalpy of steam formed in kJ/kg at 7.5 bar and 0.98 dry,

$$\therefore \quad h_2 = h_f + x \,.\, h_g$$

$$= 709.3 + 0.98 \times 2055.55$$

$$= 2723.74 \text{ kJ/kg}$$

Note h_f and h_{fg} are to be found at a pressure of 7.5 bar by interpolation, and h_1 = Specific enthalpy of feed water 41.5°C = 173.7 kJ/kg

$$\therefore \quad \eta_{th} = \frac{8.06 \times (2723.74 - 173.7)}{31000} \times 100$$

$$\eta_{th} = \mathbf{66.30\ \%}$$

(ii) Equivalent of evaporation from and at 100°C

$$= \frac{\dot{m}_s \times (h_2, h_1)}{2256.9}$$

$$= \frac{8.06 \times (2723.74 - 173.7)}{2256.9}$$

$$= \mathbf{9.11\ kg/kg\ of\ coal.}$$

Example 10.14 Duration of a boiler trial the following data was obtained,

During of trial = 8 hrs
Pressure of steam = 14 bar
Dryness fraction = 0.973
Feed Water evaporated = 26700 kg
Temperature of water at inlet = 50°C
Coal used = 4260 kg
Calorific value of coal = 28900 kJ/kg
Air used/kg of coal = 17 kg
Temperature of flue gases = 344°C
Boiler room temperature = 21 °C
C_p of flue gases = 1.1 kJ / Kg-K
Determine,

1. Boiler efficiency
2. Equivalent of evaporation
3. Heat lost to flue gases in kJ/kg of coal and in percentage

Solution

(i) Boiler efficiency

We know that, $$\eta_{th} = \frac{m_s [h_2 - h_1]}{m_f \times \text{C.V.}} \times 100\%$$

And steam raised/kg of coal = Actual evaporation

$$\dot{m}_s = \frac{m_s}{m_f} = \frac{26700}{4260}$$

$$= 6.2676 \frac{\text{kg}}{\text{kg of coal}}$$

Specific enthaply of steam formed at a pressure of 14 bar and 0.973 dry,

$$h_2 = h_f + x\, h_{fg}$$

$$= 830.3 + 0.973 \times 1957.7$$

$$h_2 = 2735.14 \text{ kJ/kg}$$

and h_1 = specific enthaply of feed water at 50°C

$$h_1 = 209.26 \text{ kJ/kg}$$

$$\therefore \quad \eta_{th} = \frac{6.2676 \times [2735.14 - 209.26]}{28900} \times 100$$

$$\boldsymbol{\eta_{th} = 54.77\%}$$

(ii) Equivalent of evaporation

$$= \frac{\dot{m}_s [h_2 - h_1]}{2256.9}$$

$$\left[\text{here } m_s \text{ should be in } \frac{\text{kg}}{\text{kg of coal}} \text{ i.e. } \dot{m}_s \text{ is to be used} \right]$$

$$= \frac{6.2676 \times [2735.14 - 209.26]}{2256.9}$$

$$= 7.01 \text{ kg/kg of coal}$$

(iii) Heat to flue gases = Mass of flue gases × C_p of flue gases × temperature rise

$$= m_g \times C_{pg} \times (T_f - T_a)$$

$$\left[\textbf{Note.} \text{ Units} = \frac{\text{kg}}{\text{kg of coal}} \times \frac{\text{kJ}}{\text{kg-K}} = \frac{\text{kJ}}{\text{kg of coal}} \right]$$

$$= 18 \times 1.1 \times (344 - 21)$$

$$= 6395.4 \frac{\text{kJ}}{\text{kg of coal}}$$

Now to get heat to flue gases in % divide by C.V. i.e.

$$= \frac{m_g \times C_{pg} (T_f - T_a)}{\text{C.V.}} \times 100$$

$$= \frac{6395.4}{28900} \times 100 = 22.129\ \%$$

Example 10.15 During a boiler trial following readings were noted,

Chimney height = 90 m

Feed water = 1400 kg/hr

Temperature of feed water = 30°C

Boiler working pressure of dry and saturated steam = 20 bar
Coal used = 200 kg/hr
Calorific value of coal = 28000 kJ/kg
Flue gas temperature in chimney = 300°C
Flue gas formed = 18 kg/kg of coal burnt
Ambient temperature = 32° C
C_p of products of combustion = 1.028 kJ/kg-K
Find

(1) η_{th}
(2) Equivalent of evaporation
(3) Heat carried away by flue gases
(4) Natural draught produced in mm of H_2O

Solution

(1) η_{th}

We know that, $$\eta_{th} = \frac{m_s(h_2 - h_1)}{m_f \times \text{C.V.}} \times 100$$

From Steam Tables,
Specific enthalpy of dry and saturated steam at 20 bar
i.e. $h_2 = h_g = 2797.2 \text{ kJ / kg}$
And specific enthalpy of fluid at 30° C.

$$= h_1 = h_f \text{ at } 30°\text{C} = 125.7 \text{ kJ/kg}$$

And steam generator per kg of fuel

$$\frac{m_s}{m_f} = \frac{1400}{200} = 7\frac{\text{kg}}{\text{kg of fuel}} = m_a = \dot{m}_s$$

$$\therefore \quad \eta_{th} = \frac{7 \times (2797.2 - 125.7)}{28000} \times 100$$

$$\boldsymbol{\eta_{th} = 66.781\%}$$

(2) Equivalent of evaporation

$$= \frac{\dot{m}_s(h_2 - h_1)}{2256.9}$$

$$\text{Equivalent of evaporation} = \frac{(2797.2 - 125.7)}{2256.9}$$

$$= 8.2859\frac{\text{kg}}{\text{kg of coal}}$$

$$= 8.2859 \times 200 \text{ kg/hr}$$

Equivalent of evaporation = 1657.18 kg/hr

(3) Heat to flue gases $= m_g \times C_{pg} \times (T_f - T_a)$

$$= 18\frac{\text{kg}}{\text{kg of coal}} \times 1.028\frac{\text{kJ}}{\text{kg K}} \times (300-32)\text{K}$$

$$= 4959.072\frac{\text{kJ}}{\text{kg of coal}}$$

or Heat of the flue gases $= 4959.072\frac{\text{kJ}}{\text{kg of coal}} \times 200\frac{\text{kg of coal}}{\text{hr}}$

$$= 991814.4\frac{\text{kJ}}{\text{hr}}$$

or $= 17.71\ \%$

(4) Chimney draught

$$h_w = 353 \times H\left\{\frac{1}{T_a} - \frac{m+1}{m} \times \frac{1}{T_f}\right\}$$

$$= 353 \times 90\left\{\frac{1}{305} - \frac{18}{17} \times \frac{1}{573}\right\}$$

$$h_w = 45.46 \text{ mm of } H_2O$$

Note. m = Mass of air supplied / kg of fuel

$m + 1$ = Mass of flue gases in kg / kg of fuel

Example 10.16. In a boiler trial, following readings were noted,

Mean temperature of feed water : 15°C

Mean boiler working pressure : 1.2 MPa

Mean dryness fraction : 0.95

Mass of coal burnt/hr : 250 kg

C. V. of coal burnt : 32400 kJ /kg

Mass of water supplied to the boiler in 7 hrs and 14 minutes = 16500 kg.

Mass of water in the boiler at the end of the test was less than at the commencement by 1000 kg.

Calculate,

(i) Actual evaporation/kg of coal

(ii) Equivalent of evaporation from and at 100°C

(iii) η_{th}

Solution

(i) Water will be filled into the boiler, upto the prespecified level before firing.

Then the water is supplied continuously during the test period of 7 hrs and 14 minutes = 16500 kg and at the end of the trial water in the drum was less by 1000 kg.

∴ Total mass of steam raised in 7 hrs and 14 minutes

$$= 16500 + 1000 = 17500 \text{ kg}$$

$$\text{Mass of steam raised/hr} = \frac{17500}{7\frac{14}{60}} = 2419.35 \text{ kg/hr (Note)}$$

∴ Actual evaporation or steam raised/kg of coal

$$m_a = \frac{m_s}{m_f} = \dot{m}_s = \frac{2419.35}{250} = 9.677 \text{ kg/kg of coal}$$

(ii) Equivalent of evaporation

$$= \frac{m_s(h_2 - h_1)}{2256.9}$$

h_2 at a pressure of 12 bar and 0.92 dry from steam tables

$$h_2 = h_f + x h_{fg} = 798.43 + 0.95 \times 1984.3$$

$$h_2 = 2683.515 \text{ kJ /kg}$$

And h_1 at 15°C from steam tables

$$h_1 = h_f$$
$$= 62.94 \text{ kJ/kg}$$

∴ Equivalent of evaporation

$$= \frac{9.667[2683.515 - 62.94]}{2256.9}$$

$$= 11.2363 \text{ kg/kg of coal}$$

(iii) Boiler thermal efficiency,

$$\eta_{th} = \frac{m_s(h_2 - h_1)}{m_f \times C.V.} \times 100$$

$$= \frac{9.677 \times [2683.515 - 62.94]}{32400} \times 100$$

$$\eta_{th} = \mathbf{78.2695\%}$$

Example 10.17 A boiler is supplied with 15,900 kg of water at 15°C during a trial for a period of 8 hours 20 minutes. The mass of water at the end of trial was 680 kg less than that of the commencement of trial. The steam is produced at a pressure of 125 N/cm^2 and is 0.9 dry. The coal having calorific value of 30,000 kJ/kg was fired at the rate of 275.8 kg per hour. Calculate

(i) The actual evaporation
(ii) The equivalent evaporation
(iii) The thermal efficiency of boiler.

Data: $P = 125\frac{N}{cm^2} = 12.5$ bar

Solution

Note.
$$\left[\begin{array}{l} 100\,cm \times 100\,cm = 1000\,cm^2 = 1\,m^2 \\ P = 125\frac{N}{cm^2} = 125 \times 10000 \\ P = 1250000\frac{N}{m^2} - 12.5\,bar \end{array}\right]$$

m_s = 15900 kg of water in 8 hours 20 min.

Mass of water at the end of test was less by 680 kg. So, this much water is also converted into steam . So Total m_s = 15900 + 680 = 16500 kg of water in 8 hr and 20 min.

$$C.V. = 30000\frac{kJ}{kg}$$

$$m_f = 275.8\frac{kg}{h_r}$$

Calculate :

(i) Actual evaporation $= m_a = \frac{m_s}{m_f} = \dot{m}_s$

(ii) Equivalent of evaporation

(iii) η_{th}

Now, mass of steam produced/hr

$$= \frac{16500}{8\frac{20}{60}}$$

$$m_s = 1980\,kg/hr$$

(i) Now actual evaporation

$$m_a = \frac{m_s}{m_f} = \dot{m}_s$$

$$m_a = \frac{1980}{275.8}$$

$$m_a = 7.1791 \frac{\text{kg of steam}}{\text{kg of fuel}}$$

(ii) Equivalent of evaporation

$$= \frac{\dot{m}_s(h_2 - h_1)}{2256.9}$$

and
$$h_2 = h_f + x h_{fg} \text{ at 12.5 bar and 0.9 dry}$$
$$= 806.7 + 0.9 \times 1977.5$$
$$h_2 = 2586.45 \text{ kJ/ kg}$$

and
$$h_1 = h_f \text{ at } 15°\text{C} = 62.9 \text{ kJ/kg}$$

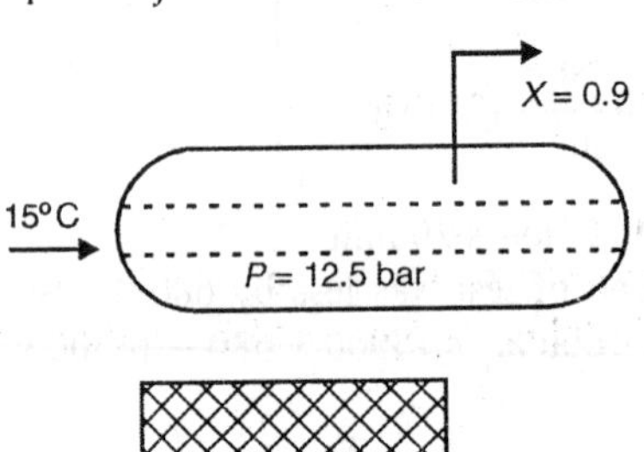

Fig. Ex. 10.17

Now from Eq. (1)
Equivalent of evaporation

$$= \frac{1980}{275.8} \times \frac{2586.45 - 62.9}{2256.9}$$

(iii) Thermal efficiency η_{th}

$$= \frac{m_s(h_2 - f_1)}{m_f \times \text{C.V.}} \times 100\%$$

$$= \frac{1980 \times (2586.45 - 62.9)}{275.8 \times 30000} \times 100$$

$$\eta_{th} = \mathbf{60.389\%}$$

Example 10.18 Following readings were obtained during a trial on two boilers.

Boiler	Pressure	Quality and steam	Evaporation rate	Feed water temp.
I	10 bar	0.9 dry	8.5 kg/kg of fuel	40°C
I	15 bar	300°C	8.0 kg/kg of fuel	60°C

Fuel used has a clorific value of 30,000 kJ/kg. Compare these boilers in respect of equivalent evaporation and thermal efficiency.

Data. $\frac{m_s}{m_f} = 8.5\frac{\text{kg of steam}}{\text{kg of fuel}} = m_a = \dot{m}_s$

$$\text{C.V.} = 30000\frac{\text{kJ}}{\text{kg}}$$

Equivalent of Evaporation = ?

$\eta_{\text{th I}}$ = ?

Solution

I –Boiler

We know that equivalent of evaporation from and at 100°C

$$= \frac{\dot{m}_s(h_2 - h_1)}{2256.9} \quad ..(1)$$

$$h_2 = h_f + xh_{fg} \text{ at 10 bar and 0.98 dry}$$
$$= 762.6 + 0.98 \times 2013.6$$
$$h_2 = 2735.928 \text{ kJ/kg}$$

and $h_1 = h_f$ at 40°C = 167.5 kJ/kg

$$\text{Now from Eq. (1)} = \frac{8.5\frac{\text{kg of steam}}{\text{kg of fuel}} \times (2735.928 - 167.5)}{2256.9}$$

$$= \frac{21831.638}{2256.9}$$

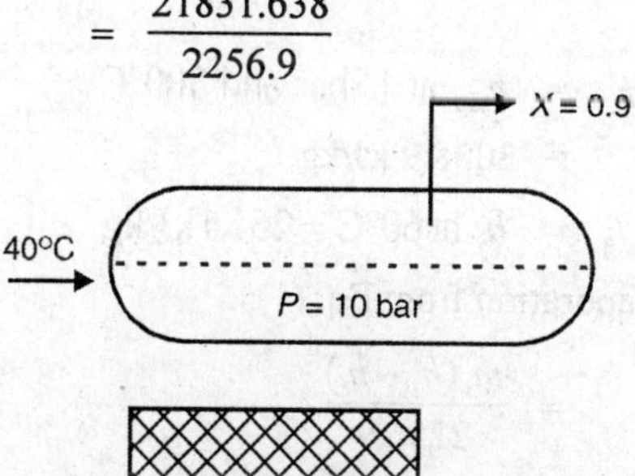

Fig. Ex. 10.18 (a)

Equivalent of evaporation $= \mathbf{9.673\frac{kg\ of\ steam}{kg\ of\ fuel}}$

And Boiler thermal efficiency,

$$\eta_{th} = \frac{m_s(h_2 - h_1)}{m_f \times \text{C.V.}} \times 100$$

$$= \frac{m_s}{m_f} \times \frac{[h_2 - h_1]}{\text{C.V.}} \times 100\%$$

$$= \frac{8.5 \times (2735.928 - 167.5)}{30000} \times 100$$

$$\eta_{th} = \mathbf{72.772\%}$$

II – Boiler $\quad \dfrac{m_s}{m_f} = \dfrac{8 \text{ kg of steam}}{\text{kg of fuel}} = m_a = \dot{m}_s$

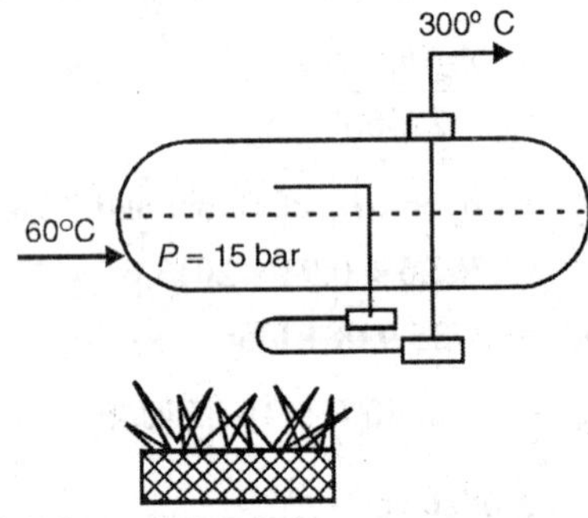

Fig. Ex. 10.18 (b)

At 15 bar pressure T_{sat} is 198.3, and the given temperature of steam is 300° C, so the steam is superheated.

Now $\quad h_2 = h_{sup}$ at 15 bar and 300°C

$$= 3038.9 \text{ kJ/kg}$$

and $\quad h_1 = h_f$ at $60°\text{C} = 251.1 \text{ kJ/kg}$

Now equivalent of evaporation from Eq. (1),

$$= \frac{\dot{m}_s (h_2 - h_1)}{2256.9}$$

$$= \frac{8 \frac{\text{kg of steam}}{\text{kg of fuel}} \times (3038.9 - 251.1)}{2256.9}$$

$$= \frac{22302.4}{2256.9}$$

$$\textbf{Eq. of eva.} = \mathbf{9.8818 \frac{kg\ of\ steam}{kg\ of\ fuel}}$$

and Thermal efficiency from Eq. (2),

$$\eta_{th} = \frac{\dot{m}_s(h_2 - h_1)}{m_f \times \text{C.V.}} \times 100\%$$

i.e.
$$\eta_{th} = \frac{m_s}{m_f} \times \frac{(h_2 - h_1)}{\text{C.V.}} \times 100\%$$

$$= \frac{8 \times (3038.9 - 251.1)}{30000} \times 100$$

$$\eta_{th} = \mathbf{74.34\%}$$

So, second boiler is more efficient.

Example 10.19 Design for principal dimensions i.e., determine the height and diameter of a chimney of a boiler which uses 18 kg of air/kg of coal burnt, when the coal is consumed at the rate of 1800 kg/hr. Actual draught required is 22 mm of H_2O taking into account all losses viz.

(i) Head loss due to fuel bed resistance

(ii) Head loss due to frictional resistance because of equipments and various bends/curves

Take, surrounding air temperature = 37°C

Tempereture of flue gases = 294°C

And the velocity of gas flow to be 35% of the theoretical velocity because of roughness at the interior surface of chimney.

Data: m = 18 kg/kg of coal

Fuel consumed 1800 kg/hr.

h_w = 22 mm of H_2O $\qquad T_a = 37 + 273 = 310$ K

$T_f = 294 + 273 = 567$ K $\qquad C_{\text{actual}} = 0.35 \times C_{\text{theoretical}}$

H = ? and D = ?

Solution

We know that
$$h_w = 353 \times H\left\{\frac{1}{T_a} - \frac{m+1}{m} \times \frac{1}{T_f}\right\}$$

$$22 = 353 \times \left\{\frac{1}{310} - \frac{19}{18} \times \frac{1}{567}\right\}$$

$\therefore$
$$\mathbf{H = 45.68607\ metres}$$

We also know that,
$$\rho_f = \frac{m+1}{m} \times \frac{353}{T_f}$$

$$\rho_f = \frac{19}{18} \times \frac{353}{567} = 0.6572 \text{ kg/m}^3$$

And note mass flow rate of flue gases = $(m + 1) \times$ Mass of coal burnt/hr

$$\left[\textbf{Note.}\ \text{Units} = \frac{\text{kg}}{\text{kg of coal}} \times \frac{\text{kg of coal}}{\text{hr}} = \frac{\text{kg}}{\text{hr}} = \frac{\text{kg}}{\text{hr}} \times \frac{1}{3600} = \text{kg/sec}\right]$$

$\therefore$ Mass flow rate of the flue gases

$$= 19 \times \frac{1800}{3600} = 9.5\,\text{kg/sec}$$

We also know that, $C = \sqrt{2\,g\,H'}$

To find H' $= H\left\{\frac{m}{m+1} \times \frac{T_f}{T_a} - 1\right\}$

$$= 45:687\left\{\frac{18}{19} \times \frac{567}{310} - 1\right\}$$

$$\mathbf{H' = 33.479\ m\ of\ hot\ gas\ column}$$

Since $C_{\text{actual}} = 0.35$

$$C_a = 0.35\sqrt{2 \times 9.81 \times 33.4799}$$

$$C_a = 8.97008\ \text{m/sec}$$

From continuity equation, we know that,

$$\text{Mass flow rate} = \rho_f \times A \times C_f$$

$$9.5\frac{\text{kg}}{\text{sec}} = 0.6572\frac{\text{kg}}{\text{m}^3} \times A \times 8.97\frac{\text{m}}{\text{sec}}$$

$$A = 1.6115122\ \text{m}^2 = \frac{\pi}{4}D^2$$

$$\mathbf{D = 1.4324\ m}$$

Example 10.20. Following data was available in a boiler trial.

Feed water = 2400 kg/hr
Ambient temperature = 33° C
Temperature of feed water = 42°C
Fuel used = 205 kg/hr
Constituents of fuel,
Carbon 84%, Hydrogen 9.27% and Oxygen = 6.73%
Calorific value of fuel = 39500 kJ/kg
Average chimney gas temperature = 307°C
Height of chimney = 32 m
Steam pressure = 11.054 bar (Gauge)
Barometric pressure = 710 mm of Hg
Air supplied = 50% in excess of stoichiometric (theoretical) air,
Steam condition = 0.96 dry.

Determine

(i) Boiler efficiency

(ii) Equivalent of evaporation from and at 100° and

(iii) Draught produced in min of H_2O.

Solution

Since in the problem Gauge pressure is given, we know that

$$P_{abs} = P_{atm} + P_g$$

Since 760 mm of Hg = 1.01325 bar

710 mm of Hg = ?

$$\therefore \quad \frac{710}{760} \times 1.01325 = P_{atm}$$

$$\therefore \quad P_{atm} = 0.9465888 \text{ bar}$$

$$\therefore \quad P_{abs} = 12.00 \text{ bar i.e. } (P_{atm} + P_g)$$

(i) We know that $\eta_{th} = \dfrac{m_s(h_2 - h_1)}{m_f \times \text{C.V.}} \times 100\%$

Now specific enthalpy of steam formed, h_2 at 12 bar and 0.96 dry

$$\therefore \quad h_2 = h_f + x h_{fg}$$

$$= 798.43 + 0.96 \times 1984.3$$

$$h_2 = 2703.3 \text{ kJ/kg}$$

And specific enthalpy of feed water at 42°C

$$h_1 = 175.81 \text{ kJ/kg}$$

$$\text{Actual evaporation } ma = \dot{m}_s = \frac{m_s}{m_f} = \frac{2400}{205}$$

$$= 11.7073 \frac{\text{kg of steam}}{\text{kg of fuel}}$$

$$\therefore \quad \eta_{th} = \frac{11.7073(2703.3 - 175.81)}{39500} \times 100$$

$$\boldsymbol{\eta_{th} = 74.912\%}$$

(ii) Equivalent of evaporation from and at 100°C

$$= \frac{\dot{m}_s(h_2 - h_1)}{2256.9} = \frac{11.7073(2703.3 - 175.81)}{2256.9}$$

$$= 13.1 \text{ kg/kg of coal}$$

(iii) Mass of theoretical air required to burn 1 kg of fuel. From fuels and combustion chapter,

$$= \frac{100}{23}\left[\frac{8}{3}C + 8H - O_2\right]$$

$$= \frac{100}{23}\left[\frac{8}{3}\times 0.84 + 8\times 0.0927 - 0.0673\right]$$

$$= 12.6709 \text{ kg/kg of coal}$$

∴ As given in the example

Actual air supplied = 1.5 × theoretical air

$$m = 12.6709 \times 1.5$$

$$m = 19.006 \text{ kg/kg of fuel}$$

$$\therefore \quad h_w = 353\times H\left\{\frac{1}{T_a} - \frac{m+1}{m}\times\frac{1}{T_f}\right\}$$

$$= 353\times H\left\{\frac{1}{306} - \frac{20.006}{19.006}\times\frac{1}{580}\right\}$$

$$\mathbf{h_w = 16.415 \text{ mm of } H_2O}$$

Example 10.21 Equivalent of evaporation = 10.4 kg/kg of fuel. Boiler products 15 T/hr of steam at 2.00 MN/m^2. Feed water temperature is 40°C. Fuel consumption is 1650 kg/hr find boiler efficiency and condition of steam produced. Take calorific value = 29800 kJ/kg.

Data: Equivalent of evaporation = 10.4 kg/kg of coal

m_s = 1500 kg/hr

Pressure = 20 bar

Temperature of feed water at the inlet = 40°C

η_{th} = ? $\quad x$ = ?

C.V. = 29800 kJ/kg

Solution

We know that,

$$\text{Equivalent of evaporation} = \frac{\dot{m}_s(h_2 - h_1)}{2256.9}\frac{\text{kg}}{\text{kg of coal}}$$

$$10.4 = \frac{1500}{1650}\times\frac{(h_2 - h_1)}{2256.9}$$

$$(h_2 - h_1) = 2581.894\frac{\text{kJ}}{\text{kg}} \qquad (1)$$

$$\text{We also know that, } \eta_{th} = \frac{m_s(h_2 - h_1)}{m_f \times \text{C.V.}}\times 100$$

$$\eta_{th} = 78.76\%$$

To determine the condition of steam.

We have already found that

$$(h_2 - h_1) = 2581.894$$

and h_1 is the specific enthalpy of feed water at 40°C = 167.45 kJ/kg from steam tables and

$$\therefore \quad (h_2 - h_1) = 2581.894$$

$$h_2 = 2581.894 + 167.45$$

$$h_2 = 2749.334 \text{ kJ/kg}$$

At a pressure of 20 bar

$$h_2 = 908.6 \text{ kJ/kg}$$

$$h_{fg} = 1.888.7 \text{ kJ/kg}$$

$$\therefore \quad h_2 = h_f + xh_{fg}$$

$$2749.334 = 908.6 + \times 1887.7 \text{ kJ/kg}$$

$$\mathbf{x = 0.9746}$$

Example 10.22 The observation recorded during a boiler trial are given below,

Duration : 60 min

Steam generated : 5250 kg

Coal burnt : 695 kg

Calorific value of coal burnt : 30200 kJ/kg

Boiler pressure : 12 bar

Dryness fraction : 0.94

Temperature of steam leaving the superheater : 250 °C

Temperature of hot well : 45°C

Calculate without superheater

(i) Equivalent of evaporation

(ii) η_{th} of boiler

Also calculate with superheater

(i) Equivalent of evaporation in kg/kg of coal

(ii) η_{th} of boiler

(iii) Heat supplied by the superheater

Solution

(i) Without superheater

We know that equivalent of evaporation,

$$= \frac{\dot{m}_s (h_2 - h_1)}{2256.9} \frac{\text{kg of steam}}{\text{kg of coal}}$$

Also steam raised/kg of coal = $\frac{m_s}{m_f} = m_a = \dot{m}_s$

$$= \frac{5250}{695} = 7.554 \text{ kg/kg of coal}$$

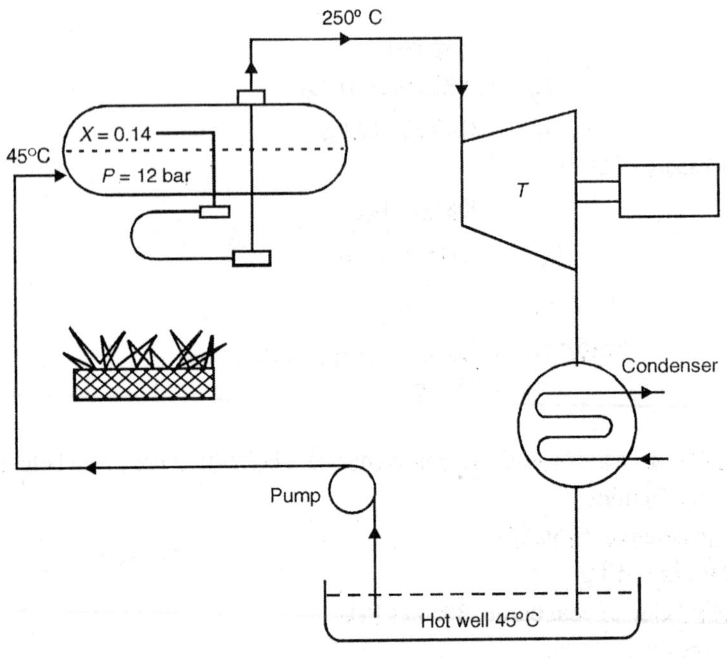

Fig. Ex. 10.22

h_2, corresponding to a pressure of 12 bar and 0.94 dry

$$h_2 = h_f + x \cdot h_{fg}$$

$$= 798.43 + 0.94 \times 1984.3$$

$$h_2 = 2663.672 \text{ kJ/kg}$$

And h_1 corresponding to a temperature of hot well 45°C = 188.35 kJ/kg (Since from the hot well water goes to the boiler drum)

∴ Equivalent of evaporation

$$= \frac{7.554 \times [2663.672 - 188.35]}{2256.9}$$

$$= 8.2851 \text{ kg/kg of coal}$$

(ii) Boiler Thermal Efficiency,

$$\eta_{th} = \frac{m_s(h_2 - h_1)}{m_f \times \text{C.V.}} \times 100$$

$$= \frac{7.554 \times [2663.67 - 188.35]}{30200} \times 100\%$$

$$\eta_{th} = 61.91\ \%$$

(ii) With superheater

Here h_2 should be corresponding to a pressure of 12 bar and temerature of 250°C from superheated steam tables,

$$h_2 = 2935.4 \text{ kJ/kg}$$

and Note h_1 will be the same.

∴ Now,

(i) Equivalent of evaporation,

$$= \frac{7.554 \times [2935.4 - 188.35]}{2256.9}$$

$$= 9.1945 \text{ kg/kg of coal}$$

(ii) $$\eta_{th} = \frac{m_s (h_2 - h_1)}{m_f \times \text{C.V.}} \times 100\%$$

$$= \frac{7.554 \times [2935.4 - 188.35]}{30200} \times 100$$

$$\eta_{th} = 68.71\ \%$$

(iii) Heat added in the superheater

= Enthalpy of superheated steam – Enthalpy of wet steam

$$= h_{sup} - h_{wet}$$

$$= (2935.4 - 2663.672) \text{ kJ/kg}$$

$$= 271.728 \frac{\text{kJ}}{\text{kg of steam}}$$

∴ Total heat added in the superheated/hr

$$= 271.728 \frac{\text{kJ}}{\text{kg of steam}} \times 5250 \frac{\text{kg of steam}}{\text{hr}}$$

$$= 1426572 \text{ kJ / hr}$$

Example 10.23 The observation recorded during a boiler trial are as under:

Duration : 60 min

Steam generated : 35,500 kg

Steam pressure : 12 bar

Steam temperature : 250°C

Temperature of water entering economiser = 17°C

Temperature of water leaving economiser = 77°C

Fuel burnt = 3460 kg

Calorific value of fuel used = 39500 kJ / kg

Calculate:

(1) Equivalent of evaporation from and at 100°C per in kg per kg of fuel.

(2) Thermal efficiency of the plant.

(3) The percentage of heat energy of the fuel utilised by the economiser.

Solution

We know that,

(i) Equivalent of evaporation from and at 100°C

$$= \frac{\dot{m}_s (h_2 - h_1)}{2256.9} \frac{\text{kg of steam}}{\text{kg of fuel}}$$

Here h_2 is the specific enthalpy of steam formed at 12 bar and 250°C. The steam is superheated so h_2 from superheated steam tables = 2935.4 kJ/kg

And h_1 at the feed water inlet temperature of economiser i.e.

17 °C = 71.31 kJ / kg

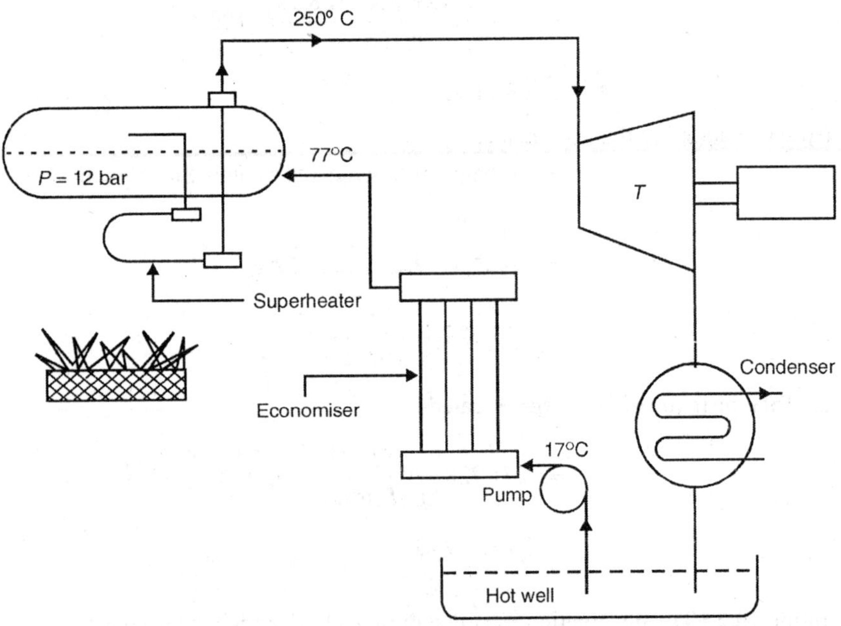

Fig. Ex. 10.23

And h_1 at the economiser outlet temperature i.e. inlet temperature of the water to the boiler

i.e. 77 °C = 322.33 kJ/kg

Now steam raised/kg of coal = $\dfrac{m_s}{m_f} = \dfrac{35500}{3460} = 10.26 = \dot{m}_s = m_a$

Equivalent of evaporation= $\dfrac{\dot{m}_s(h_2 - h_1)}{2256.9}$

$$= \frac{10.26\times(2935.4-322.33)}{2256.9}$$

$$= 11.879 \text{ kg/kg of coal}$$

Note. Here h_1 should be the enthaply of water entering the drum, from the definition of equivalent of evaporation.

(ii) Thermal efficiency of the plant

$$\eta_{th} = \frac{m_s(h_2 - h_1)}{m_f \times \text{C.V.}} \times 100$$

i.e.

$$\eta_{th} = \frac{m_s(h_2 - h_1)}{m_f \times \text{C.V.}} \times 100\%$$

$$= \frac{10.26\times(2935.4-71.31)}{39500}$$

$$= 74.395\%$$

Note. Since η_{th} of the plant is to be found, h_1 should correspond to the temperature of water entering the economiser

(iii) The % of used in the economiser

= Mass of water passing through Economiser/kg of coal

$$\times \frac{C_p \text{ of water} \times \text{temperature rise}}{\text{C.V.}} \times 100\%$$

$$\left[\textbf{Note.}\ \text{Unit} = \frac{\dfrac{\text{kg}}{\text{kg of coal}} \times \dfrac{\text{kJ}}{\text{kg K}} \times \text{K}}{\dfrac{\text{kJ}}{\text{kg of coal}}} = \frac{\text{kJ/kg of coal}}{\text{kJ/kg of coal}}\right]$$

$$= \frac{10.26\times 4.187\times(77.17)}{39500} = 6.52\%$$

Example 10.24 During a boiler trial, following readings were noted

Steam pressure = 15 bar

Amount of water evaporated = 5400 kg/hr

Condition of steam = 0.92

Feed water is supplied at 50°C

Determine the rate of coal consumption in kg/hr if η_{th} = 70%

Assume C.V. of coal = 40000 kJ/kg.

Also determine the equivalent of evaporation from and at 100°C in kJ/hr

If the superheater is used then the temperature of steam leaving the super heater is 250° C. Find η_{th} = ? Assume C_p for superheated steam as 2.27 kJ/kg-K.

Solution

We know that,

$$\eta_{th} = \frac{m_s (h_2 - h_1)}{m_f \times \text{C.V.}} \times 100 \qquad (1)$$

Now, h_2 = specific enthaply of steam formed at 15 bar and 0.92 dry.

$$= h_f + x \cdot h_{fg}$$

$$= 844.66 + 0.92 \times 1945.2$$

$$h_2 = 2634.244 \text{ kJ/kg}$$

$$\text{And } h_1 \text{ at } 50°\text{C} = 209.26 \text{ kJ/kg} = h_f$$

$$\therefore \text{ From Eq. (1)}, \quad \eta_{th} = \frac{\dot{m}_s (h_2 - h_1)}{m_f \times \text{C.V.}} \times 100$$

$$\therefore \quad m_f = \frac{5400 \text{ kg/hr} (2634.24 - 209.26)}{0.7 \times 40000}$$

$$m_f = 467.675 \text{ kg/hr}$$

(ii) Equivalent of evaporation,

$$= \frac{\dot{m}_s (h_2 - h_1)}{2256.9}$$

Note. Here m_s should be in kg/hr so as to get Equivalent of Evaporation in $\frac{\text{kg of steam}}{\text{hr}}$

$$= \frac{5400(2634.24 - 209.26)}{2256.9} = 5802.159 \text{ kg/hr}$$

(iii) With superheater, since C_p of superheated steam is given,

$$\therefore \quad h_2 = h_g + c_p (T_{\text{sup}} - T_{sat}) \text{ at } 15 \text{ bar}$$

$$h_2 = 2789.9 + 2.27 (250 - 198.29)$$

$$h_2 = 2907.282 \text{ kJ/kg}$$

$$\eta_{th} = \frac{m_s(h_2 - h_1)}{m_f \times \text{C.V.}} \times 100 \; \frac{5400(2907.28 - 209.26)}{467.675 \times 40000}$$

$$= 77.88\ \%$$

Example 10.25 During a boiler trial on fire tube boiler, following readings were noted,

Duration of trial = 24 hrs

Amount of water evaporated = 20 T

Feed water temperature = 50°C

Condition of steam = 0.92

Absolute pressure = 15 bar

Coal burnt = 6T

Grate area = 4 m^2

Heating surface = 3200 m^2

C.V. of coal = 20000 kJ / kg

Calculate:

(1) Mass of coal burnt in kg per m^2 of grate area per hr $\left[\text{i.e.} \frac{\text{kg}}{\text{m}^2 - \text{hr}}\right]$

(2) Equivalent of evaporation from and at 100° C in kg/kg of coal.

(3) η_{th}

(4) Equivalent of evaporation in kg/ m^2 of heating surface per hour.

Solution

Now, mass of coal burnt /hr $= \frac{6000\,\text{kg}}{24\,\text{hrs}} = 250\,\text{kg/hr}$

And steam roduced/hr $= \frac{20000}{24}$

$= 833.33$ kg/hr

Therefore,

(1) Mass of coal burnt in kg/m^2 of grate area/hr

$$= \frac{250\ \text{kg/hr}}{4\text{m}^2} = 62.5\,\text{kg/m}^2\text{-hr.}$$

(2) Mass of steam produced /kg of coal

$$= \frac{20000}{6000} = 3.333\,\text{kg/kg of coal}$$

∴ Equivalent of evaporation from and at 100°C in kg/kg of coal

$$= \frac{\dot{m}_s(h_2 - h_1)}{2256.9} \quad (1)$$

Now from Eq. (1),

And h_2 = specific enthalpy of steam formed at 15 bar and 0.92 dry

$$= h_f + x \cdot h_{fg}$$

$$= 844.66 + 0.92 \times 1945.2 = 2634.244 \text{ kJ/kg}$$

$$h_1 \text{ at } 50°\text{C} = 209.26 \text{ kJ/kg}$$

Now from Eq. (1), Equivalent of Evaporation,

$$= \frac{3.333(2634.24 - 209.26)}{2256.9}$$

$$= 3.5815795 \text{ kg/kg of coal}$$

(3) $$\eta_{th} = \frac{m_s(h_2 - h_1)}{m_f \times \text{C.V.}} \times 100$$

$$= \frac{3.33(2634.24 - 209.26)}{20000} \times 100$$

$$= 40.41\ \%$$

(4) Mass of steam produced per m^2 of heating surface/hr

$$= \frac{833.33 \text{ kg/hr}}{3200 \text{ m}^2 \text{ of heating surface area}}$$

$$= 0.2604156 \frac{\text{kg}}{\text{hr - m}^2}$$

∴ Equivalent, of evaporation in $\frac{\text{kg}}{\text{hr - m}^2}$

$$= \frac{m_s(h_2 - h_1)}{2256.9}$$

∴ Equivalent, of evaporation in $\frac{\text{kg}}{\text{hr - m}^2}$

$$= \frac{0.2604156(2634.24 - 209.26)}{2256.9}$$

Equivalent of Evaporation = 0.2798097 kg / hr – m^2

(Problems on Energy Balance)

Example 10.26 The following results were obtained in a boiler trial,

Feed water/hr : 700 kg

Feed water inlet temperature : 27°C

Steam produced at a pressure : 8 bar
Dryness fraction : 0.97
Coal used : 100 kg/hr
C.V. of coal : 25000 kJ/kg
Ash and unburnt coal collected from beneath the grate bars = 7.25 kg/hr
C. V. of unburnt fuel = 2000 kJ/kg
Flue gases formed/kg of fuel = 17.3 kg
Flue gas temperature = 325° C
Temperature of air in the room = 16° C
C_p of flue gases = 1. 025 kJ/kg-K
Draw up energy balance on minute basis and find the boiler efficiency.

Solution

1. Energy supplied / released due to the combustion of fuel

$$= m_f \times \text{C.V. of fuel}$$
$$= 100 \text{ kg/hr} \times 25000 \text{ kJ/kg}$$
$$= 25{,}00{,}000 \text{ kJ/hr}$$
$$= 41666.667 \text{ kJ/min} = 100\% \text{ (Say)}$$

2. Energy utilised.

(i) Energy utilised to form the steam

$$= m_s (h_2 - h_1)$$

where,

h_2 = specific enthalpy of steam formed in kJ/kg corresponding to a pressure of 8 bar and since the steam is wet.

$$h_2 = h_f + x \cdot h_{fg} = 720.9 + 0.97 \times 2046.5$$
$$\therefore \quad h_2 = 2706.005$$

And h_1 specific enthaply of feed water at 27°C

$$= 113.13 \text{ kJ/kg}$$

∴ Energy or Heat to form steam

$$= m_s (h_2 - h_1)$$
$$= 700 \text{ kg/hr} (2706.005 - 113.13) \text{ kJ/kg}$$
$$= 1815012.5 \text{ kJ/hr}$$
$$= 30250.21 \text{ kJ/min}$$

Total energy released $= 41666.67 \text{ kJ/min} = 100\%$

for 30250.21 kJ/min = ?

$$= \frac{30250.21}{41666.67} \times 100 = 72.60\% = \eta_{th}$$

(ii) Heat to flue gases

$$= \text{Mass of flue gases} \times C_p \text{ of flue gases} \times (T_f - T_a)$$
$$= 17.3 \times 1.025 \times (325 - 16)$$
$$= 5479.3425 \text{ kJ/kg of coal}$$

Since 100 kg of coal is burnt/hr

$$\therefore \qquad = 5479.3425 \frac{\text{kJ}}{\text{kg of coal}} \times 100 \frac{\text{kg of coal}}{\text{hr}}$$
$$= 547934.25 \text{kJ/hr}$$
$$= 9132.2375 \text{kJ/min}$$

$$\therefore \qquad \text{In \%} = \frac{9132.24}{41666.67} \times 100$$

(iii) Heat lost to unburnt fuel

$$= m_{uf} \times \text{C.V. of unburnt fuel}$$
$$= 7.25 \text{kg/hr} \times 2000 \text{kJ/kg}$$
$$= 14500 \text{kJ/hr} = 241.667 \text{kJ/min}$$
$$= 0.58\%$$

(iv) Heat unaccounted by difference

$$= \text{Total heat generated} - \text{Total heat utilised}$$
$$= 41666.67 - (30250.21 + 9132.24 + 241.67)$$
$$= 2024.55 \text{ kJ/min}$$
$$= 4.1\%$$

Energy Supplied	kJ/min	%	Energy utilised	kJ/min	%
	41666.67	100%	(i) To steam (ii) To flue gases (iii) To unburnt fuel (iv) To unaccounted	30250.21 9132.24 241.67 2042.55	72.60 21.92 0.58 4.10
	41666.67	100%		41666.67	100%

Example 10.27 The following data were obtained in a boiler trial,

Feed water supplied : 680 kg/ hr

Temperature of feed water 20° C

Steam pressure : 15 bar

Steam temperature : 300° C

Coal used : 95 kg l hr

C.V. of coal used : 34000 kJ/kg

Ash and unburnt coal collected in the ash pit = 4 kg/hr

C.V. of unburnt coal = 2200 kJ/kg

Flue gas formed = 18 kg/kg of coal burnt
Flue gas temperature = 343° C
Ambient temperature = 28° C
Mean C_p of flue gases = 1.025 kJ/kg-K
Determine:

(1) η_{th}
(2) Equivalent of evaporation from and at 100°C
(3) % of heat unaccounted
(4) Drought produced in mm of H_2O, if the chimney height is 50 m.

Solution

1. We know that,

$$\eta_{th} = \frac{m_s(h_2 - h_1)}{m_f \times \text{C.V.}} \times 100\%$$

At 15 bar and 300°C, from superheated Steam Tables,

$$h_2 = 3038.9\,\text{kJ/kg}$$

and $$h_1 \text{ at } 20°\text{C} = 83.86\,\text{kJ/kg} = h_f$$

$$\eta_{th} = \frac{680 \times (3038.9 - 83.86)}{95 \times 34000} \times 100$$

$$\boldsymbol{\eta_{th} = 62.21\%}$$

2. Equivalent of evaporation from and at 100°C

$$= \frac{\dot{m}_s(h_2 - h_1)}{2256.9}$$

$$= \frac{680}{95} \times \frac{(3038.9 - 83.86)}{2256.9}$$

$$\textbf{Eq. of Evaporation} = \mathbf{9.3721 \frac{kg}{kg\,of\,coal}}$$

3. (a) Heat supplied from coal

$$= m_f \times \text{C.V.}$$

$$= 95\,\text{kg/hr} \times 34000\,\text{kJ/kg}$$

$$= 32{,}30{,}000\,\text{kJ/hr}$$

$$= 53833.333\,\text{kJ/min}$$

$$= 100\%\,(\text{Say})$$

(b) (i) Heat to steam $= m_s(h_2 - h_1)$

$$= 680(3038.9-83.89)\,\text{kJ/hr}$$

$$= 2009406.8\,\text{kJ/hr}$$

$$= 33490.113\,\text{kJ/min}$$

$$= 62.210\%$$

(ii) Heat carried away by the flue gases,

$$= m_{\text{flue gas}} \times C_{p \text{ of flue gases}} \times (T_f - T_a)$$

$$= 18 \times 1.025(343-28)$$

$$= 5811.77 \frac{\text{kJ}}{\text{kg of coal}}$$

$$= 5811.75\,\text{kJ/kg of coal} \times 95\,\text{kJ of coal/hr}$$

$$= 552116.25\ \text{kJ/hr}$$

$$= 9201.9375\,\text{kJ/min}$$

$$= 17.09\%$$

(iii) Heat lost to unburnt fuel

$$= m_{\text{uf}} \times \text{C.V.}_{\text{uf}}$$

$$= 4\frac{\text{kg}}{\text{hr}} \times 2200\frac{\text{kJ}}{\text{kg}}$$

$$= 8800\,\text{kJ/hr}$$

$$= 146.667\,\text{kJ/min}$$

$$= 0.2724\%$$

(iv) Heat unaccounted by difference

$$= 53833.333 - [33490.113 + 9201.938 + 146.667]$$

$$= 10994.615\,\text{kJ/min}$$

$$= 20.423\%$$

(4) Chimney height = 50 m

Mass of flue gases = 18 kg/kg of coal

∴ Mass of flue air $= 18 - 1 = 17$ kg/kg of coal

$$T_a = 28 + 273 = 301\,\text{K}$$

$$T_f = 343 + 273 = 616\,\text{K}$$

∴
$$h_w = 353 \times H \times \left\{\frac{1}{T_a} - \frac{m+1}{m} \times \frac{1}{T_f}\right\} \text{mm of } H_2O$$

$$h_w = 353 \times 50 \left\{ \frac{1}{301} - \frac{18}{17} \times \frac{1}{616} \right\}$$

$$\mathbf{h_w = 28.3 \text{ mm of } H_2O}$$

Example 10.28 Following observation were made during a boiler trial,

Feed water = 1500 kg/hr

Inlet temp. of feed water = 20°C

Steam generated = 15 bar and 300°C

Coal consumed = 200 kg/hr

C.V. = 34000 kJ/kg

Ash in the ash pit contain unburnt products of 10 kg/hr of C.V. 2000 kJ/kg.

Flue gases formed = 18 kg/kg of coal burnt.

Flue gas temperature = 343° C

Atmospheric temp. = 28° C

Mean specific heat of fuel gases = 1.025 kJ/kg-K

Determine

(i) Boiler thermal efficiency

(ii) Equivalent of evaporation in $\frac{\text{kg of steam}}{\text{kg of coal}}$

(iii) Heat balance in kJ/min for the boiler and

(iv) Drought produced in mm of H_2O if chimney height H = 100 m.

Solution

$$m_s = 1500 \text{ kg/hr}$$

$$m_f = 200 \text{ kg/hr}$$

$$\text{C.V.} = 34000 \text{ kJ/kg}$$

$$m_{uf} = 10 \text{ kJ/kg}$$

$$\text{C.V.}_{uf} = 2000 \text{ kJ/kg}$$

$$m_{fg} = (m+1) = 18 \frac{\text{kg}}{\text{kg of coal}}$$

$$T_f = 343 + 273 = 616 \text{K}$$

$$T_a = 28 + 273 = 301 \text{K}$$

Mean $Cp_g = 1.025 \text{ kJ/kg-K}$

At 15 bar pressure T_{sat} – 198.3°C and the steam temperature given is 300°C so the steam is superheated.

Now $h_2 = h_{sup}$ at 15 bar and 300°C from superheated steam tables

$$= 3038.9 \text{ kJ/kg}$$

and $$h_1 = h_f \text{ at } 20°\text{C} = 83.9 \text{ kJ/kg}$$

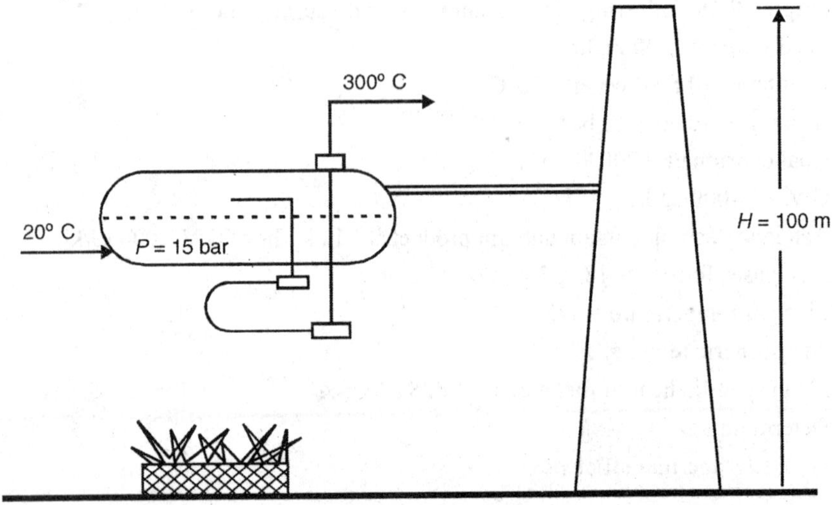

Fig. Ex. 10.28

(i) Now we know that Boiler thermal efficiency,

$$\eta_{th} = \frac{m_s(h_2 - h_1)}{m_f \times \text{C.V.}} \times 100\%$$

$$= \frac{1500 \frac{\text{kg}}{\text{hr}} (3038.9 - 83.9)}{200 \frac{\text{kg}}{\text{hr}} \times 34000}$$

$$= \frac{4432500}{200 \times 34000}$$

$$\eta_{th} = \mathbf{65.1838\%}$$

(ii) Equivalent of evaporation

$$= \frac{\dot{m}_s(h_2 - h_1)}{2256.9}$$

$$= \frac{1500}{200} \times \frac{(3038.9 - 83.9)}{2256.9}$$

Equivalent of evaporation = 9.8198 kg of st/kg of coal

(iii) Heat Balance sheet

(A) Heat energy released due to the combustion of fuel

$$= m_f \times \text{C.V.}$$

$$= 200\,\text{kg/hr} \times 34000\,\text{kJ/kg}$$

$$= 6800000\,\text{kJ/hr}$$

$$= 113333.33\,\text{kJ/min}$$

$$= 100\%\,(\text{say})$$

(B) Heat Energy Utilised

(a) To form steam $= m_s(h_2 - h_1)$

$$= 1500\,\text{kg/hr}\,(3038.9 - 83.9)\,\text{kJ/kg}$$

$$= 4432500\,\text{kJ/hr}$$

$$= 73875.0\,\text{kJ/min}$$

(b) Heat to flue gases $= m_{fg} \times C_{Pg} \times (T_f - T_a)$

$$= 18\frac{\text{kg of gases}}{\text{kg of coal}} \times 1.025\frac{\text{kJ}}{\text{kg-K}}(616 - 301)\,\text{K}$$

$$= 5811.75\,\text{kJ/kg of coal}$$

$$= 5811.75\frac{\text{kJ}}{\text{kg of coal}} \times 200\frac{\text{kg of coal}}{\text{hr}}$$

$$= 1162350\,\text{kJ/hr}$$

Heat to flue gases $= 19372.5\,\text{kJ/min}$

(C) Heat lost due to unburnt fuel

$$= m_{uf} \times \text{C.V.}_{uf}$$

$$= 10\,\text{kg/hr} \times 2000\text{kJ/kg}$$

$$= 2000\,\text{kJ/hr}$$

$$= 333.333\,\text{kJ/min}$$

(d) Heat unaccounted by difference

$$= \{113333.33\} - \{73875 + 5811.75 + 19375.5 + 333.33\}$$

$$= (113333.33 - 99392.58)\,\text{kJ/min}$$

$$= 13940.75\,\text{kJ/min}$$

(iv) Since $H = 100$ m

We know that,
$$h_w = 353\times H\left\{\frac{1}{T_a}-\frac{m+1}{m}\times\frac{1}{T_f}\right\}$$
$$= 353\times 100\{353\times 100\}\left\{\frac{1}{301}-\frac{18}{17}\times\frac{1}{616}\right\}$$
$$\mathbf{h_w = 56.61\text{ mm of }H_2O}$$

Example 10.29 A boiler generates 8 kg of steam/kg of coal at 30 bar and 350° C. Water is fed at 40°C. The mass of dry flue gas and water vapour/kg of coal burnt are 18.75 kg and 0.7 kg respectively. Boiler house temperature is 27°C and flue gas temperature leaving the boiler is 127°C.

Draw-up heat balance sheet on kJ/kg of coal burnt basis and % basis.

Assume C_p of gases = 1.1 and C_p for the steam formed due to the combustion of H_2 = 2.1 kJ/kg K. Partial pressure of steam as 0.07 bar and C.V. of coal = 32000 kJ/kg.

Solution

(I) Heat energy liberated due to the combustion of 1 kg of fuel
$$= m_f\times \text{C.V.}$$
$$= 1\times 32000$$
$$= 32000\text{ kJ/kg of coal} = 100\%\text{ (say)}$$

(II) Energy utilised

(a) Heat to steam
$$= m_s(h_2 - h_1)$$
$$= 8\text{ kg/kg of coal}\times(3117.5-167.45)\text{ kJ/kg}$$
$$= 23600.4\text{ kJ/kg of coal} = 73.75\%$$

(b) Heat lost to the flue gases
$$= m_g\times C_{pg}\times(T_f - T_a)$$
$$= 18.75\times 1.1(127-27)$$
$$= 2062.5\text{ kJ/kg of coal}$$
$$= 6.45\ \%$$

(c) Heat lost due to the water vapour formed due to the combustion of H_2
$$= m_1(h_2' - h_1')$$

Here h_2' – is the specific enthalpy of steam formed due to the combustion of H_2 at flue gas temperature and at the given partial pressure.

Since C_p of steam formed is given, from steam tables corresponding to 0.07 bar pressure,

$$\therefore \quad h_2' = h_g + C_p(T_{sup} - t_{sat})$$

$$= 2572.6 + 2.1(127 - 39)$$

$$h_2' = 2557.4 \text{ kJ/kg}$$

and h_1' = specific enthalpy of water at room temperature

i.e. at 27°C = 113.13 kJ / hr.

Heat lost to water vapour

$$= m_1(h_2' - h_1')$$

$$= 0.7\{2757.4 - 113.13\}$$

$$= 1850.99 \text{ kJ/kg}$$

$$= 5.79\%$$

(d) Heat unaccounted by difference

$$= 32000 - \{23600.4 + 2062.5 + 1850.99\}$$

$$= 4486.11 \text{ kJ/kg}$$

$$= 14.02\ \%$$

THEORY QUESTIONS

1. What is a boiler? Explain how they are classified.
2. Explain the following boilers with the help of neat sketches.
   (a) Landcashire boiler
   (b) Cornish boiler
   (c) Locomotive boiler
   (d) Cochran boiler
   (e) Bacock Wilcox boiler
   (f) Stirling boiler
3. Compare water tube and fire tube boilers.
4. What are the essentials of good boiler?
5. What is the difference between the Mountings and Accessories ? Mention the various Mountings and Accessories.
6. Explain with neat sketches the working of the following Mountings and Accessories
   (a) Dead weight safety valve
   (b) High steam low water alarm
   (c) Water level indicator
   (d) Fusible Plug
   (e) Bourdon pressure gauge
   (f) Steam top valve

(g) Super heater
(h) Economiser

7. What is boiler draught ?
8. Explain various types of draught used in a boiler.
9. Compare Artificial draught versus Natural (Chimney) draught.
10. Derive, $h_w = 353H\left[\frac{1}{T_a} - \frac{m+1}{m} \times \frac{1}{T_f}\right]$ mm of H_2O .
11. Derive the condition for maximum discharge of flue gases through the chimney.
12. What is a Steam Jet Draught ?
13. Explain,
    (i) Equivalent evaporation from and at 100°C
    (ii) Boiler Efficiency
14. What are the draught losses ?
15. Write a note on IBR regulation.
16. State the essentials of a good boiler.
17. Draw a neat sketch of any one type of boiler you have studied. Label the parts and describe its working.
18. State the function and location of the following :
    (a) Fusible plug
    (b) Feed check valve
    (c) Blow-offcock
    (d) Pressure gauge
19. Staff the factors considered for selection of a boiler, for a specific application.
20. Describe with a neat indicator:
    (a) Water level indicator
    (b) Superheater
    (c) Fusible plug
    (d) Economiser
    (e) Spring loaded safety valve
    (f) Feed check valve
21. Describe briefly the advantages of incorporating an economiser, air-preheater and superheater in a steam boiler. Indicate the position of these in the passage of flue gases by a line diagram.
22. Define
    (i) Equivalent evaporation
    (ii) Boiler accessories
23. What is a boiler draught ? Compare natural draught versus artificial draught.

24. Define :
   (i) Boiler efficiency,
   (ii) Equivalent evaporation and explain the working parameters which influence them.

25. Define the term 'equivalent evaporation' and explain its significance.

26. Write a short note on 'classification of boiler'.

27. Write a short note on 'regulations for chimney height'.

28. State the working principle of
   (a) Fusible plug,
   (b) Water level indicator,
   (c) Air preheater.

29. Prove that under maximum discharge condition of chimney, the draught produced in terms of the hot gas column is same as the chimney height.

30. State the advantage of high pressure boilers.

31. Explain Evaporative capacity of a boiler.

32. Explain Function and working of the following:
   (a) Superheater
   (b) Spring loaded safety valve
   (c) Air preheater

33. What is IBR? What are the provisions made in IBR?

34. Compare water tube boilers with five tube boilers.

35. Explain:
   (i) Grate
   (ii) Shell
   (iii) Mounting
   (iv) Heating surface
   (v) Manhole
   (vi) Damper

36. Make a list of any five boiler mountings and write down their functions in brief.

37. Make 'a list of any five boiler accessories and write down their functions in brief.

38. With the help of neat sketch explain working of Benson boiler.

39. What is a boiler? How they are classified?

40. Explain the following brifers with the help of neat sketches.
   (a) Lancashire Boiler
   (b) Cornish Boiler
   (c) Locomotive Boiler
   (d) Cochram Boiler
   (e) Babcock Wilcox Boiler
   (f) Stirling Boiler

41. Compare water tube and fire tube brifers.

42. Whar are the essentials of a good brifer.

43. What is the difference between mountings and accessories mountings and accessories.

44. Explain with neat sketches working of following mountings and accessories.

(a) Dead water safety valve

(b) High-steam low water alarm

(c) Water level indicator

(d) Fusible plug

(e) Bourden pressure gauge

(f) Steam stop valve

(g) Superheater

(h) Economiser

45. What is boiler draught?

46. Explain various types of draught used in boiler?

47. Compare Artificial draught verses Natural draught.

48. Derive $h_w = 353 \times H\left[\frac{1}{T_a} - \frac{me1}{m} \times \frac{1}{T_f}\right]$ mm of H_2O.

49. Derive the condition for maximum discharge of the gases through the chimney?

50. What is a steam jet draught?

51. Explain

(i) Equivalent of evaporation from and @ 100°C

(ii) Boiler efficiency

52. What are the draught losses?

53. Write a note on IBR Regulations.

54. Explain the function of

(i) Scum valve

(ii) Steam stop valve

(iii) Steam trap

55. Boiler accessories.

56. Explain

(i) Feed check valve

(ii) Blow off cock

57. State the difference between

(i) Externally fired internally fired

(ii) Forced circulation and Natural circulation

(iii) High pressure and Lower pressure

(iv) Stationary Portable

(v) Single tube and multi tube.

Give at least one example of each type.

58. What do you understand by steam jet draught? Where is it generally employed. State advantages of it.

59. Draw neat pencil sketch of any one type of boiler you have studied. Label the parts and describe its working

60. Explain the function and working of

(i) Superheater

(ii) Spring loaded safety valve

(iii) Air preheater

61. Bring out a comprehensive comparison between FT and WT Boilers.

62. Explain the following terms an applied, to a St. generator

(i) Grate

(ii) Shell

(iii) Mounting

(iv) Heating surface

(v) Marble

(vi) Demper.

63. What is IBR? Explain some of its provisions.

64. Explain in detail classification of boilers.

65. Draw-neat sketches of following and explain their working

(i) Feed check valve

(ii) Water level indicator

(iii) Economiser

66. Enumerate the factors to be considered in selections of a boiler.

67. Write the location and function of the following components used in a boiler

(i) Fusible plug

(ii) Blow off cock

(iii) Superheater

(iv) St. stop valve

68. What is equivalent of evaporation of a boiler? What is its importance?

69. What ave the factors involved in the selection of a boiler?

In what circumstances W.T boilers are used in preferance to F.T. boilers? Give reasons.

Suggest a boiler for a small chemical industry which is in need of 500 kg/hrof steam at 12 atm pressure.

70. S.N

(i) Classification of boilers

(ii) IBR

PROBLEMS FOR PRACTICE

1. A boiler generates steam at 18 bar and 325° C feed water is supplied at 41.45° C. The thermal efficiency of the boiler is 80%. It uses furnace oil of C.V. 45500 kJ/kg. The steam is supplied to a turbine producing 100 kW power, having a specific steam consumption of 10 kg/hr-kW. Calculate the furnace oil consumption in kg/hr and equivalent of evaporation.

   (**Ans.** 395.71 kg/hr, 6382.55 kg/hr)

2. A boiler has a chimney of 35 m height. It produces a draught of 20 mm of water column. The mean temperature of flue gases is 370°C and the boiler house temperature is 34° C. Estimate the air-fuel ratio. Does this boiler satisfy the condition for maximum discharge? Also find the height of hot gas column under maximum discharge condition.

   [**Ans.** (i) The chimney nearly fulfils the condition for maximum discharge, (ii) $H' = H = 35$ m of hot gas columb]

3. A boiler generates 5000 kg/hr of steam at 16 bar and 300°C from feed water at 30°C. Coal used in 600 kg/hr of C.V. 30,000 kJ/kg. Find
   (a) Equivalent of evaporation.
   (b) Boiler efficiency

   [**Ans.** (a) 10.75 kg/kg of coal (b) 80.9%]

4. The equivalent of evaporation of a boiler from and a 100°C is 10.4 kJ/kg of fuel. The C.V. of fuel is 29800 kJ/kg.

   Determine efficiency of boiler. If the boiler produces 15000 kg of steam per hour at 20 bar from a feed water at 40° C and the fuel used 1650 kg/hr. Determine the condition of steam produced.

   [**Ans.** (a) 78.76 % (b) $x = 0.9734$]

5. The following readings were noted during an hour trial on a boiler.

   Coal burnt : 750 kg

   Feed water : 7000 kg

   C.V. of coal : 33500 kJ/kg

   Feed water temperature entering the economiser : 30°C

   Leaving the economiser : 90°C

   Boiler pressure : 10 bar

   Steam dryness : 0.96

   Temperature of steam leaving superheater : 250°C

   Determine the thermal efficiency of the plant and also calculate the enthalpy received by feed water in various components as a % of total enthalpy increase.

   [**Ans.** (1) $\eta_{th} = 77.92\%$ (ii) Enthalpy received in economizer = 230.0 kJ/kg i.e. 8.23%
   (iii) Enthalpy received in boiler = 2138.8 kJ/kg i.e. 82.91%
   (iv) Enthalpy received in superheater = 247.4 kJ/kg i.e. 8.86%]

6. The following observations were obtained in a boiler trial:
   Steam condition = 15 bar and 300°C
   Coal used = 95 kg/hr having C.V. = 34,000 kJ/kg
   Feed water supplied = 680 kg/hr at 20°C.
   Ash and unburnt coal collected in ash pit = 4 kg/hr.
   C.V. of unburnt coal = 2200 kJ/kg.
   Flue gas formed = 18 kg/kg of coal burn
   Flue gas formed = 18 kg/kg of coal burnt
   Flue gas temperature = 343°C
   Boiler room temperature = 28°C
   Mean specific heat of flue gases = 1.025 kJ/kg-K
   Determine:
   (i) Boiler efficiency (**Ans.** 62.21%)
   (ii) Heat carried away by the flue gases in kJ/min (**Ans.** 9201.9375 kJ/min)
   (iii) Heat unaccounted in kJ/min and in percentage. (**Ans.** 10994.615 kJ/min, 20.4%)

ꕤ

11

Simple Vapour Power Cycle

CHAPTER OBJECTIVES

After reading this chapter you will be able to learn the following

- Carnot Cycle for Vapours.
- Practical Vapour Cycle – Rankine Cycle for Thermal Power Plant.
- Effect of Boiler Pressure, Back Pressure, and Superheat on Rankine Efficiency.

11.1 INTRODUCTION OR IDEAL STEAM PLANT CYCLES

Figure 11.1 shows the simple steam power plant. The heat energy released due to the combustion of fuel will be utilised in the boilers for converting water into steam (i.e.

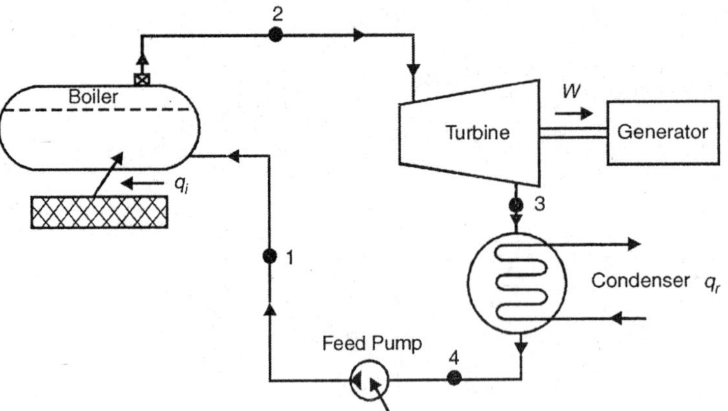

Fig. 11.1. *General arrangement of steam power plant.*

vapour) and this steam is then expanded into the steam engine/steam turbines to obtain useful work. The steam after producing work output, is generally condensed in the condensers. If the condensate is pure, then it is pumped as feed water to the boiler by means of a feed pump.

11.2 CARNOT CYCLE AND STEAM PLANT

As we know, the Carnot cycle has the greatest possible efficiency between any two given temperature limits T_1 and T_2 and is given by $\eta_{th} = \frac{T_1 - T_2}{T_1}$. It is important therefore to see whether Carnot cycle can be successfully applied to the steam plant. Referring the *T–S* diagram as shown in Fig. 11.2.

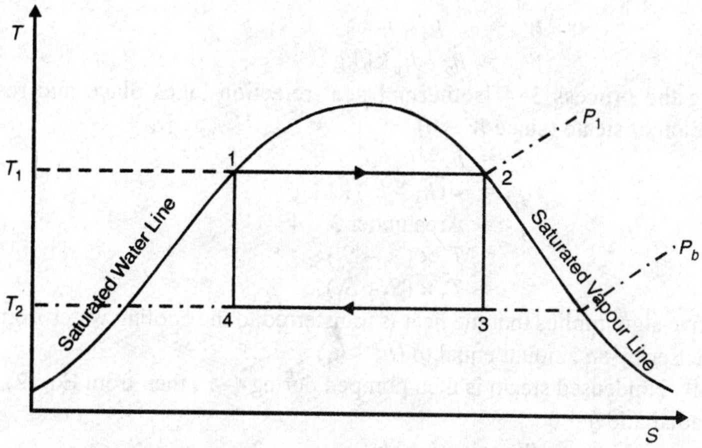

Fig. 11.2. *Carnot cycle on T–S diagram.*

Process 1–2:

This is an isothermal heat addition process and during this, water in the boiler gets converted into steam at constant temperature and pressure.

At point 2, steam is dry saturated.

Process 2–3:

The steam is expanded adiabatically isentropically in the steam turbine or steam engine from pressure P_1 to P_b

Process 3–4:

The steam after producing work output in the turbine enters the condenser. During this process isothermal heat rejection q_r takes place and results into condensation of steam.

Process 4–1:

Partially condensed steam at 4 is fed into the feed pump, which pumps/compresses the mixture of wet steam and condensate isentropically to boiler pressure.

Since all the devices in Steam Power Plant are steady flow ones, from steady flow energy equation (SFEE) for unit mass,

$$q - w = \Delta h + \Delta KE + \Delta PE \quad (1)$$

Since the change in KE and PE are small, on neglecting them Eq. (1) reduces to

$$q - w = \Delta h \qquad (2)$$

We can write the energy equation for all the devices of Steam Power Plant considering that, boiler produces no work, turbine is adiabatic, condenser produces no work and pump is adiabatic.

∴ Heat added in the boiler q_1 during 1 – 2

$$q_i = h_2 - h_1 \qquad (\because W = 0 \text{ for boiler})$$

i.e.

$$q_i = h_2 - h_1 \text{ kJ/kg}$$
$$= \text{Area under } 1 - 2$$
$$= T_1 \times (S_2 - S_1) \qquad (3)$$

Work produced by the turbine W_T during the expansion process 2–3 and since expansion adiabatic.

$$q = 0$$

$$\therefore \quad 0 - W_T = h_3 - h_2$$
$$W_T = h_2 - h_3 \text{ kJ/kg} \qquad (4)$$

During the process 3–4 isothermal heat rejection takes place and results into condensation of steam (since $W = 0$)

$$q_r = h_4 - h_3$$
$$q_r = -(h_3 - h_4) \text{ kJ/kg} \qquad (5)$$
$$= \text{Area under } 3 - 4$$
$$= T_2 \times (S_3 - S_4)$$
$$= T_2 \times (S_2 - S_1)$$

Negative sign implies that the heat is transferred to the cooling water from steam in the condenser by an amount equal to $(h_3 - h_4)$.

Partially condensed steam is then pumped during 4–1 , then from Eq. (2), since the pump is adiabatic $q = 0$

$$\therefore \quad -W_p = h_1 - h_4$$

or

$$W_p = -(h_1 - h_4)$$

Negative sign implies that work is to be supplied to the pump for pumping by amount equal to $(h_1 - h_4)$

$$\eta_{th} = \frac{\text{Net work output}}{\text{Heat supplied}} = \frac{\text{Shaft work } (W_s)}{\text{Heat supplied } q_i}$$

$$\eta_{th} = \frac{\text{Difference between turbine and pump work}}{q_i}$$

$$= \frac{W_T - W_p}{q_i}$$

$$\eta_{th} = \frac{[h_2 - h_3] - [h_1 - h_4]}{[h_2 - h_1]} \qquad (7)$$

$$= \frac{(h_2 - h_1) - (h_3 - h_4)}{h_2 - h_1}$$

$$= \frac{T_1(S_1 - S_2) - T_2(S_1 - S_1)}{T_1(S_2 - S_1)}$$

$$= \frac{T_1 - T_2}{T_1} \text{ as for ideal gas.}$$

Even though, the Carnot efficiency is maximum for a given value T_1 and T_2. But because of some inherent practical difficulties, as discussed below, Carnot cycle cannot be applied for the steam plant.

Carnot cycle when applied to S.P.P is practical upto a point that, Isothermal heat addition and the adiabatic expansion of steam in the turbine, which is quite reasonable.

The impractical part is in handling of steam in the condenser and pump. In the condenser, the steam is only partially condensed and the condensation must be stopped at 4. Also the feed pump must be capable of handling both wet steam and condensate.

A slight modification is made to this Carnot cycle which however produces a more practical cycle, called Rankine cycle. This cycle is accepted as, the practical but ideal cycle for the steam plant even though it has reduced thermal efficiency.

11.3 RANKINE CYCLE

To obtain Rankine cycle, modification made to Carnot cycle, is that, instead of stopping the condensation at some intermediate condition, the condensation is completed as shown in Fig. 11.3.

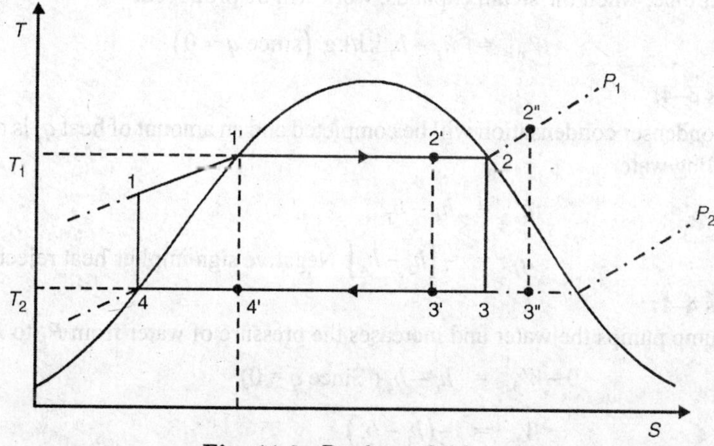

Fig. 11.3. *Rankine Cycle.*

On *T–S* diagram Carnot cycle, would be 1'–2–3–4'. For Rankine cycle, however, condensation is continued until it is complete at 4. At this point there is all water. This can be successfully dealt with by the feed pump.

Process 4–1:

Pumping process, where water of condenser pressure P_b is raised to boiler pressure P_1. The pumping compression process is isentropic. At point (1) temperature is lower than T_{sat}. Subcooled state:

Process 1–2: (1 – 1' and 1' – 2)

During 1 – 1' feed water is heated to T_{sat} (may occur in economiser/boiler itself.) During 1' – 2 heat added q_t at constant temperature and pressure P_1 convert water into steam in the boiler.

Process 2–3:

Isentropic expansion of steam in the turbine from boiler pressure P_1 to back pressure P_b, which results into turbine work W_T.

Process 3–4:

The steam is condensed at constant pressure P_b in the condenser. The steam rejects heat q_r to the cooling water.

It should be noted that the steam generated in the boiler or steam entering the turbine may be wet, dry-saturated and superheated. Correspondingly the expansion process in the turbine will be 2'–3', 2–3, 2"–3" respectively.

Analysis of Rankine Cycle (Assuming 1 kg of steam)

From Eq. (5) we get $q - w = \Delta h$

Process 1–2:

In the boiler heat q_i converts water into steam

$$q_i = h_2 - h_1 \ (\text{since } q = 0)$$

Process 2–3:

In the turbine, when the steam expands, work will be produced.

$$W_T = h_2 - h_3 \text{ kJ/kg } (\text{since } q = 0)$$

Process 3–4:

In the condenser condensation will be completed and an amount of heat q_r is rejected to the cooling water.

$$q_r = h_4 - h_3$$

i.e.

$$q_r = -[h_3 - h_4] \text{ Negative sign implies heat rejection.}$$

Process 4–1:

Feed pump pumps the water and increases the pressure of water from P_b to P_1.

$$0 - W_p = h_1 - h_4 \ (\text{Since } q = 0)$$

or

$$W_P = -(h_1 - h_4)$$

Negative sign implies work to be supplied to pump.

Also since pump is a steady flow equipment, we know that work done in a steady flow system.

$$= \int v dP$$

$$\therefore \quad W_P = -\int_{P_b}^{P_1} v dP$$

The specific volume of water remains constant since the effect of pressure is negligible on liquids.

$$\therefore \quad W_P = -V\int_{P_b}^{P_1} dP$$

$$\therefore \quad \text{Magnitude of } W_P = V\left[P_1 - P_b\right]$$

We also know that

$$\text{Specific volume} = \frac{1}{\rho}$$

and for water $\rho_w = 1000 \text{ kg/m}^3$

$\therefore$ Specific volume $= 0.001 \text{ m}^3/\text{kg}$. If the pressure are measured in bar then,

$$W_P = 0.001 \text{m}^3/\text{kg}\left[P_1 - P_b\right]10^5 \text{N/m}^2$$

$$= 0.001\left(P_1 - P_b\right)10^5 \text{Nm/kg or J/kg}$$

$$= \frac{0.001\left(P_1 - P_b\right)10^5 \text{kJ}}{1000} \text{kJ/kg}$$

$$W_P = \left[\frac{P_1 - P_b}{10} \text{kJ/kg}\right] \quad (8)$$

And the efficiency of Rankine cycle,

$$\eta_R = \frac{\text{Shaft work } W_s}{\text{Heat supplied } q_i} = \frac{W_T - W_P}{q_i}$$

$$\therefore \quad \eta_R = \frac{\left[h_2 - h_3\right] - \left[h_1 - h_4\right]}{\left[h_2 - h_1\right]}$$

Also we can write this as,

$$\therefore \quad \eta_R = \frac{\left[h_2 - h_3\right] - \left[h_1 - h_4\right]}{\left[h_2 - h_4\right] - \left[h_1 - h_4\right]}$$

Since the pump work $W_p = h_1 - h_4$ is very small quantity, it can be neglected,

$$\therefore \quad \eta_{th} = \frac{\left[h_2 - h_3\right]}{\left[h_2 - h_4\right]}$$

Note.: For solving the problems we can read the value of $h_4 = h_f$ from steam tables at back pressure P_b, h_2 and h_3 can be determined with the help of Mollier diagram or by calculations from steam tables. Since $W_P = h_1 - h_4$ the volume of h_1 can be determined by using Eq. (8) i.e. by finding W_p.

It should be noted that $\eta_R < \eta_{carnot}$ because the mean temperature of heat addition process in Rankine cycle is lower than the mean temperature of heat addition process in Carnot cycle. Whereas the mean temperature of heat rejection process in both the cycles is same.

In order to improve the efficiency of Rankine cycle, superheated steam has to be used, as represented by the process 2–2". It improves the cycle efficiency, as it increases the mean temperature of heat addition process in the boiler. Also when the superheated steam is used, it reduces the corrosion of turbines.

The Rankine efficiency has the same form as before, from Fig. 11.3 on similar lines

$$\eta_R = \frac{h_2'' - h_3''}{h_2'' - h_4''}$$

and when the steam is wet, then the cycle would be 1 – 1' – 2' – 3' – 4 –1

and $$\eta_R = \frac{h_2' - h_3'}{h_2' - h_4'}$$

11.4 EFFECT OF INLET PRESSURE (BOILER PRESSURE) AND BACK PRESSURE (EXHAUST PRESSURE) AND SUPERHEAT ON RANKINE EFFICIENCY

The effect of inlet pressure, back pressure and the superheat on the performance of Rankine cycle can be studied with the help of *T–S* diagram as shown in Fig. 11.4.

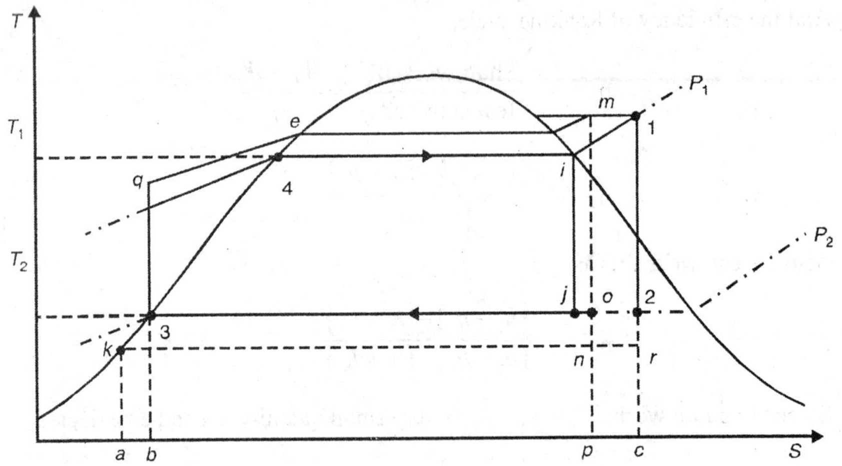

Fig. 11.4. *Rankine cylce varying working conditions.*

(a) Boiler or inlet pressure. If the boiler pressure is raised, the evaporation line is raised as the saturation temperature is increased from 4–1 to *em*. The maximum temperature t_1 is kept constant i.e. $t_{m'}; = t_1$. 1 – 2 – 3 – 4 –1 is the basic cycle taken for comparison.

With increased boiler pressure, keeping the condenser pressure and temperature constant, the new cycle will be m 03 *qem*. The amount of heat rejected is decreased given by C20 pc and difference in the heat added –*b*341*cb* and *bqempb* is small and the thermal efficiency increases. It is experimentally found that at p_1 = 170 bar and p_2 = 0.14 bar, the efficiency is maximum and decrease with increase in p_1.

(b) Condenser pressure and temperature is lowered. The new cycle wilt be 1*dk* 41. Work done is increased and heat rejected will be less so that the efficiency will be increased. Nature, however, precisely defines the limit of improvement that may be obtained by this means. Atmospheric temperature and therefore temperature of the cooling water puts the limits of the steam temperature and pressure.

(c) Superheat. If the steam is dry and saturated at the inlet of turbine, and if the steam is superheated then the work is increased and the proportional rise of work is more than the heat supplied so that the efficiency of the Rankine cycle increases.

It is interesting to note that condensing plants commonly operate more efficiently in the winter than in summer because of the lower cooling water temperature in the winter.

SOLVED EXAMPLES

Example 11.1 A steam turbine operates on the Rankine cycle. It receives steam from the boiler at 30 bar, 250°C and it is exhausted into a condenser at 0.5 bar. Condensate is returned to the boiler by a feed pump. Calculate,

(1) Heat supplied in the boiler drum.

(2) Dryness fraction of steam entering the condenser.

(3) Work done.

(4) Steam rate/kWh and

(5) Efficiency of the cycle.

Solution

Since it operates on Rankine Cycle,

Refer to Rankine cycle 1–2–3–4, (Fig. Ex. 14.1)

h_4 from tables at 0.5 bar condenser pressure $h_4 = h_f$ = 340.56 kJ/kg.

We know that,

$$W_P = \frac{[P_1 - P_b]}{10}\frac{\text{kJ}}{\text{kg}} = \left[\frac{30-0.5}{10}\right]\frac{\text{kJ}}{\text{kg}}$$

$$W_P = 2.95 \text{ kJ/kg}$$

Also, $$W_P = [h_1 - h_4]\text{kJ/kg}$$

or $$h_1 = W_p + h_4 = 2.95 + 340.56 \text{ kJ/kg}$$

$$h_1 = 343.51 \text{ kJ/kg}$$

From steam tables, h_2 at 30 bar and 250°C

$$h_2 = 2855 \text{ kJ/kg } (2854.8)$$

and $$s_2 = 6.28 \text{ kJ/kg-K}$$

For isentropic process (2–3)

Entropy before expansion s_2 = Entropy after expansion s_3

$$6.28 = s_{f3} + x_3.s_{fg3} = 1.0912 + x_3.(6.5035)$$

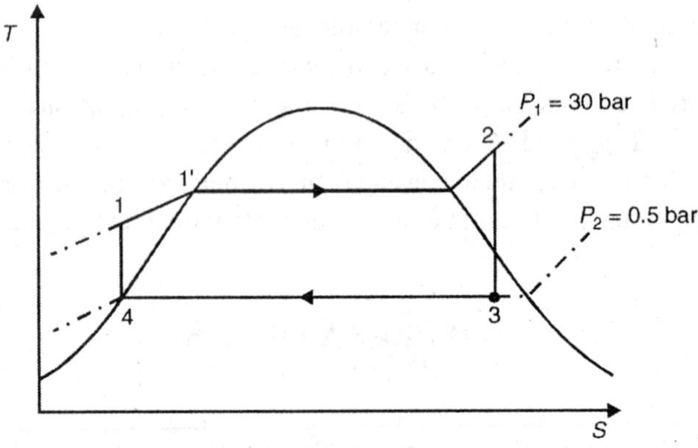

Fig. Ex. 11.1

∴ Dryness fraction of steam entering the condenser

$$x_3 = 0.797 \qquad \text{(A)}$$

∴ $$h_3 = h_{f3} + x_3.h_{fg3} \text{ corresponding to condenser pressure}$$

$$= 340.56 + 0.797 \times 2305.4$$

$$h_3 = 2177.9 \text{ kJ/kg}$$

Now,

(1) Heat supplied in the boiler drum

$$q_i = h_2 - h_1 = 2855 - 343.51$$

$$q_i = 2511.49 \text{ kJ/kg}$$

(2) Dryness fraction of steam entering condenser

$$x_3 = 0.797 \text{ as above}$$

(3) W.D. or shaft work

$$W_s = W_T - W_P = [h_2 - h_3] - [h_1 - h_4]$$

$$= (2855 - 2177.9) - 2.95 = 674.15 \text{ kJ/kg}$$

(4) Since 1 kWh = 3600 kJ

$$\text{Steam rate} = \frac{3600}{W_s\,(\text{kJ/kg})} = \frac{3600}{674.15}$$

$$= 5.34 \frac{\text{kg}}{\text{kWh}}$$

(5) $$\eta_R = \frac{W_s}{q_1} = \frac{674.15}{2511.49}$$

$$\boldsymbol{\eta_R = 26.84\ \%}$$

Example 11.2 Steam at 20 bar and 360°C expands in a steam turbine to 0.08 bar. It is then condensed in a condenser to saturated water. The pump feeds back the water to the boiler. Assume ideal Rankine cycle and determine.

(i) The net work done/kg of steam

(ii) η_R

Solution

Refer Fig. Ex. 11.2. From steam tables corresponding to condenser pressure 0.08 bar $h_4 = h_f = 173.86$ kJ/kg.

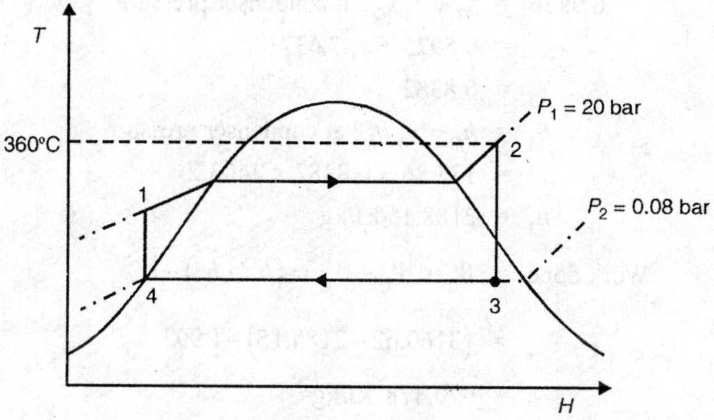

Fig. Ex. 11.2

We know that, $$W_P = \left[\frac{P_1 - P_b}{10}\right]\frac{\text{kJ}}{\text{kg}} = h_1 - h_4$$

$$W_P = \left[\frac{20 - 0.08}{10}\right]\frac{\text{kJ}}{\text{kg}}$$

$$W_P = 1.992 \frac{\text{kJ}}{\text{kg}} = h_1 - h_4$$

or $$h_1 = W_P + h_4 = 1.992 + 173.86$$

$$\mathbf{h_1 = 175.852\ kJ/kg}$$

or corresponding to 20 bar and 360° C, $h_2 = 3160.62$

At 350° C, $h = 3138.6$ kJ/kg and 400° C = h 3248.7 kJ/kg

$$\text{Difference is} \frac{110.1\,\text{kJ/kg}}{50^\circ\text{C}} = 22.02\,\text{kJ/kg}$$

$$\therefore \quad 2.202 \times 10 = 22.02\,\text{kJ/kg}$$

$$\therefore \quad \text{At } 360^\circ\text{C}, h_2 = 3138.6 + 22.02 = 3160.62\,\text{kJ/kg}$$

This method of finding the values at intermediate points is known as Interpolation.

To find h_s:

For isentropic process 2–3

Entropy before expansion s_2 = Entropy after expansion s_3

To find s_2 = 7.1296 at 400° C

6.9596 at 350° C

$$0.17/50 = 0.0034 \text{ kJ/kg-K}$$

$$0.0034 \times 10 = 0.034 \text{ kJ/kg-K}$$

$$\therefore \quad 6.9596 + 0.034 = 6.9936 \text{ kJ/kg-K}$$

$$6.9936 = f_{f3} + x_3.s_{fg3} \text{ at condenser pressure}$$

$$= 0.5925 + x_3.7.6371$$

$$\therefore \quad x_3 = 0.8382$$

$$h_3 = h_{f3} + x_3\,h_{fg3} \text{ at condenser pressure}$$

$$= 173.86 + 0.8382 \times 2403.2$$

$$h_3 = 2188.15 \text{ kJ/kg}$$

Now,
$$\text{Work done} = W_s = W_T - W_P = (h_2 - h_3) - W_P$$

$$= (3160.62 - 2188.15) - 1.992$$

$$= 970.478 \text{ kJ/kg}$$

and
$$\eta_R = \frac{W_S}{q_i} = \frac{970.478}{h_2 - h_1} = \frac{970.478}{3160.62 - 175.852}$$

$$\eta_R = \mathbf{32.51\%}$$

Example 11.3 A steam turbine operates on Rankine cycle. Steam entering the turbine has a pressure and temperature of 50 bar and 500°C, expands to a condenser pressure of 0.05 bar. Steam is condensed completely. Find heat supplied, turbine work, pump work, dryness fraction of steam entering the condenser and Rankine cycle efficiency.

Solution

Refer Fig. Ex. 11.3. Since the steam is superheated,

From steam tables corresponding to 0.05 bar.

$$h_4 = h_f = 137.8\,\text{kJ/kg}$$

Also pump work $W_P = \frac{P_1 - P_b}{10} = \frac{50 - 0.05}{10}$

$W_P = 4.995\,\text{kJ/kg} = h_1 - h_4$

$\therefore \quad h_1 = 4.995 + 137.8$

$h_1 = 142.795\,\text{kJ/kg}.$

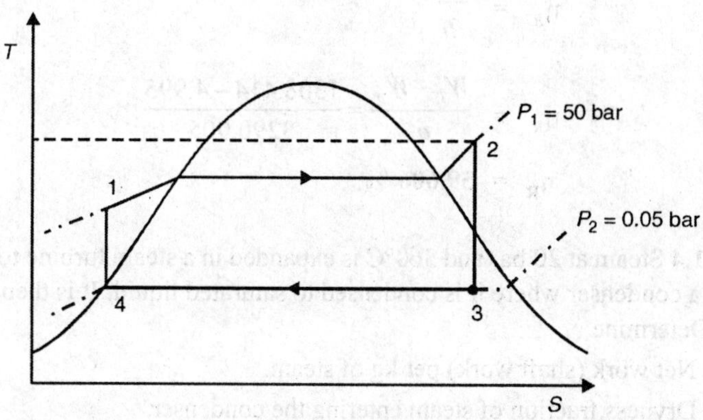

Fig. Ex. 11.3

Now, h_2 corresponding to 50 bar and 500° C from superheated Steam Tables,

$h_2 = 3433.7$ kJ/kg

To find h_3; from the isentropic process 2–3

Entropy before expansion = Entropy after expansion.

$s_2 = s_3$

$$s_2 = s_{f_3} + x_3 \cdot s_{fg_3} \qquad \text{(A)}$$

Now s_2 at 50 bar and 500° C = 6.977 kJ/kg-K

$\therefore$ From (A), $\quad 6.977 = 0.4763 + x_3 \times (7.92)$

(At condenser pressure 0.05 bar)

$\Rightarrow \quad x_3 = 0.8208 = 0.82$

$h_3 = h_{f_3} + x_3 . h_{fg_3} = 137.77 + 0.82 \times 2423.8$

$\mathbf{h_3 = 2125.286\,kJ/kg}$

Now,

(i) Heat supplied, $q_i = h_2 - h_1 = 3433.7 - 142.795$

$\mathbf{q_i = 3290.905\,kJ/kg}$

(ii) Turbine work $W_T = h_2 - h_3 = 3433.7 - 2125.286$

$$\mathbf{W_T = 1308.414\,kJ/kg}$$

(iii) Pump work $W_P = 4.995$ kJ/kg, already found in the beginning.

(iv) Dryness fraction of steam entering the condenser.

$$x_3 = 0.8208 \text{ already found}$$

(v) Rankine cycle efficiency

$$\eta_R = \frac{W_s}{q_i}$$

$$\eta_R = \frac{W_T - W_P}{q_i} = \frac{1308.414 - 4.995}{3290.905}$$

$$\eta_R = \mathbf{39.606\ \%}$$

Example 11.4 Steam at 20 bar and 360°C is expanded in a steam turbine to 0.08 bar. It then enters a condenser where it is condensed to saturated liquid. It is then fed back to the boiler. Determine,

(i) Net work (shaft work) per kg of steam.

(ii) Dryness fraction of steam entering the condenser.

Solution

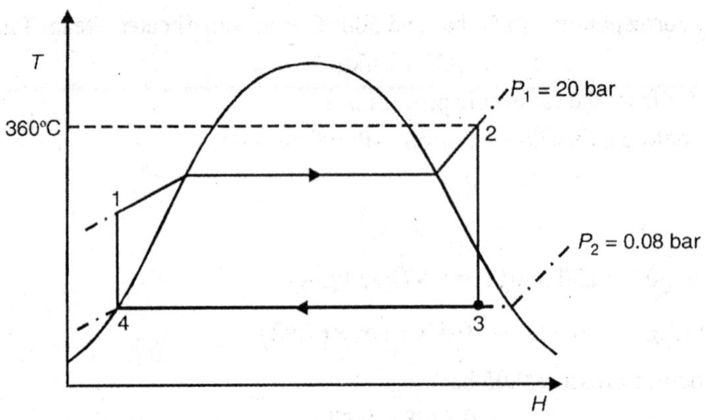

Fig. Ex. 11.4

From steam tables at 0.08 bar

$$h_4 = h_f = 173.9 \text{ kJ/kg}$$

And pump work $$W_P = \left[\frac{P_1 - P_b}{10}\right] = \left[\frac{20 - 0.08}{10}\right]$$

$$W_P = 1.992\,\text{kJ/kg} = h_1 - h_4$$

$$h_1 = 1.992 + h_4$$

$$h_1 = 1.992 + 173.9$$

$$h_1 = 175.892 \text{ kJ/kg}$$

h_2 corresponding to 20 bar and 360° C from superheated steam tables = 3160.62 kJ/kg (By interpolation)

To find h_3 from the isentropic process 2–3

Entropy before expansion = Entropy after expansion

$$s_2 = s_3$$

$$s_2 = s_{f_3} + x_3 \cdot s_{fg_3} \qquad \text{(A)}$$

Now s_2 at 20 bar and 360° C = $6.994 \dfrac{\text{kJ}}{\text{kg K}}$ (By interpolation)

Now from (A) $\quad 6.994 = 0.593 + x_3 \times 7.637$ at 0.08 bar

$$6.994 - 0.593 = 7.637\, x_3$$

$$0.8381 = x_3$$

$$\therefore \quad h_3 = h_{f_3} + x_3 \,.\, h_{fg_3}$$

at $\quad$ 0.08 bar $= 173.9 + 08381 \times 2403.2$

$$\mathbf{h_3 = 2188.157 \text{ kJ/kg}}$$

(i) ∴ Net work or shaft work = $W_s = W_T - W_P$

$$W_s = [h_2 - h_3] - [h_1 - h_4]$$

$$= (3160.62 - 2188.157) - (175.892 - 173.9)$$

$$= (972.4626) - (1.992)$$

$$\mathbf{W_s = 970.4706 \text{ kJ/kg}}$$

(ii) Dryness fraction of steam entering the condenser $x_3 = 0.8381$ as found earlier.

Example 11.5 A steam power plant separating on Rankine cycle, receives steam from boiler at 3.5 MPa and 350°C. It is exhausted to condenser at 10 kPa. Calculate,

(i) Energy supplied per kg of steam generated in boiler.

(ii) Quality of steam entering the condenser.

(iii) Rankine cycle efficiency considering feed pump work.

(iv) Specific steam consumption.

Data: $P = 3.5$ MPs = 35 bar, P_b =10 kPa = 0.1 bar

Solution

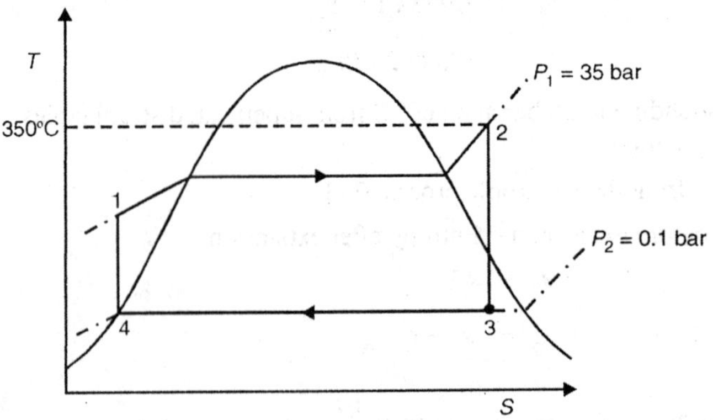

Fig. Ex. 11.5

As $100 \dfrac{\text{kN}}{\text{m}^2} = 1\,\text{bar}$

(i) $q_i = ?$ (ii) $x_3 = ?$

(iii) $\eta_R = ?$ (iv) $SR = ?$

From steam tables corresponding to 0.1 bar,

$$h_4 = h_f = 191.8 \text{ kJ/kg}$$

Also pump work $$W_p = \frac{P_1 - P_b}{10}$$

$$= \frac{35 - 0.1}{10} = 3.490\,\text{kJ/kg} = h_1 - h_4$$

i.e.
$$3.499 + h_4 = h_1$$
$$3.490 + 191.8 = h_1$$
$$195.29 = h_1$$

And h_2 corresponding to 35 bar and 350° C from superheated steam tables = 3106.45 kJ/kg by interpolation.

To find h_3 from the isentropic process 2–3

$$s_2 = s_3$$

$$s_2 = s_{f_3} + x_3 \cdot s_{fg_3} \qquad \text{(A)}$$

Now s_2 at 35 bar and 350° C = $6.663 \dfrac{\text{kJ}}{\text{kg-K}}$ by interpolation.

$\therefore$ From (A) $6.663 = 0.649 + x_3 \times 7.502$ at 0.1 bar

$$\frac{6.663 - 0.649}{7.502} = x_3$$

$$0.8016 = x_3$$

$$\therefore \quad h_3 = h_{f_3} + x_3.h_{fg_3} \text{ at } 0.1 \text{ bar}$$

$$= 191.8 + 0.8016 + 2392.9$$

$$h_3 = 2110.075 \text{ kJ/kg}$$

Now,

(i) Heat supplied $\quad q_i = h_2 - h_1 = 3106.45 - 195.29$

$$q_i = 2911.16 \text{ kJ/kg}$$

(ii) $x_3 = 0.8016$ dryness fraction of steam which is already found.

(iii) Rankine cycle efficiency η_R condidering feed pump work,

$$\eta_R = \frac{W_s}{q_i} = \frac{W_T - W_P}{q_i}$$

$$= \frac{[h_2 - h_3] - 4.90}{2911.16}$$

$$= \frac{(3106.45 - 2110.0) - 4.9}{2911.16}$$

$$\eta_R = 0.34226$$

$$\eta_R = 34.226\%$$

(iv) Steam Rate $= \dfrac{3600}{W_S} = \dfrac{3600}{[W_T - W_P]}$

$$= \frac{3600}{[h_2 - h_3] - 4.9} = \frac{3600}{996.375 - 4.9}$$

$$= \frac{3600}{991.475}$$

S.R. = 3.630 kg/kWh

Example 11.6. Find heat supplied, turbine work, pump work, shaft work, efficiencies, steam rate of Carnot and Rankine cycle using steam between 30 bar and 0.04 bar.

Steam is dry and saturated at 30 bar. Compare the results.

Data: $q_i = ?$ $\quad W_T = ?$

$W_P = ?$ $\quad W_S = ?$

η_c = ? η_R = ?

S.R. = ?

P_1 = 30 bar P_b = 0.04 bar.

Solution

Refer Fig. Ex. 11.6

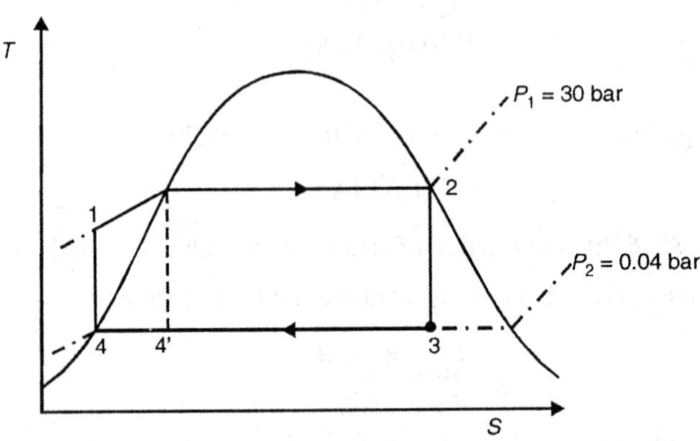

Fig. Ex. 11.6

First consider the Carnot cycle 1' – 2 – 3 – 4'

From steam tables at 30 bar,

$$h_{1'} = h_f = 1008.35 \text{ kJ/kg}$$

$$s_{1'} = s_f = 2.6455 \text{ kJ/kg-K}$$

and

$$h_2 = h_g \text{ at } 30 \text{ bar} = 2802.3 \text{ kJ/kg}$$

$$s_2 = s_g = 6.1837 \text{ kJ/kg-K}$$

For the isentropic process 2–3

$$s_2 = s_{f_3} + x_3 . s_{fg_3} \text{ at } 0.04 \text{ bar}$$

$$6.1837 = 0.423 + x_3 \times 8.053$$

$$x_3 = 0.7154$$

$\therefore$

$$h_3 = h_{f_3} + x_3 h_{fg_3}$$

$$= 121.41 + 0.7154 \times 2433.1$$

$$h_3 = 1862.04 \text{ kJ/kg}$$

Now for Isentropic Compression process 4'–1'

$$s_1' = s_4' = S'_{f_4} + x_4' s'_{fg_4}$$

$$2.6455 = 0.4225 + x'_4 \times 8.0530$$

$$x'_4 = 0.276 \qquad \text{Nearer to liquid line}$$

$$\therefore \quad h_4' = h'_{f_4} + x'_4 h'_{fg_4}$$

$$= 121.41 + 0.276 \times 2433.1$$

$$h_4' = 792.94 \text{ kJ/kg}$$

Now,

(i) Heat supplied $q_i = h_2 - h_{1'}$

$$= 2802.3 - 1008.35$$

$$q_i = 1793.95 \text{ kJ/kg}$$

(ii) Turbine work $= W_T = h_2 - h_3 = 2802.3 - 1862.04$

$$W_T = 940.26 \text{ kJ/kg}$$

(iii) Pump work $W_P = h_{1'} - h_{4'} = 1008.35 - 792.94 = 215.41 \text{ kJ/kg}$

(iv) Shaft work $W_s = W_T - W_P = (940.26) - (215.41) = 724.85 \text{ kJ/kg}$

(v) Carnot cycle efficiency

$$\eta_{carnot} = \frac{W_s}{q_i} = \frac{724.85}{1793.95} = 0.4040$$

$$\boldsymbol{\eta_{carnot} = 40.40\%}$$

(iii) Steam Rate $= \dfrac{3600}{W_s} = \dfrac{3600}{724.85}$

$$\mathbf{SR = 4.96654 \frac{kg}{kWh}}$$

II. Rankine Cycle 1 – 2 – 3 – 4

From Fig. Ex. 11.6 it is to be noted that

h_2 and h_3 will remain same.

$$\therefore \quad h_2 = 2802.3 \text{ kJ/kg}$$

$$h_3 = 1862.04 \text{ kJ/kg}$$

and $h_4 = h_f$ at 0.04 bar $= 121.41 \dfrac{\text{kJ}}{\text{kg}}$

Now to find h_1,

We also know that, pump work

$$W_P = \frac{P_1 - P_b}{10} = \frac{30 - 0.04}{10} = 2.996$$

$$\cong 3.0 \text{ kJ/kg}$$

Also,

$$W_p = h_1 - h_4$$

$$W_p + h_4 = h_1$$

$$3 + 121.41 = h_1$$

$$\Rightarrow \quad h_1 = 124.41 \text{ kJ/kg}$$

Now,

(a) Heat supplied $q_i = h_2 - h_1 = 2802.3 - 124.41$

$$q_i = 2677.89 \text{ kJ/kg}$$

(b) Turbine work $W_T = h_2 - h_3 = 940.26 \dfrac{\text{kJ}}{\text{kg}}$ · It will remain same.

(c) Pump work $W_P = h_1 - h_4 = 124.41 - 121.41$

$$\mathbf{W_p = 3 \text{ kJ/kg}}$$

(d) Shaft Work $W_S = W_T = h_2 - h_3 = 940.26 - 3 = 937.26 \text{ kJ/kg}$

(e) Rankine Cycle Efficiency

$$\eta_R = \frac{W_S}{q_i} = \frac{937.26}{2677.89}$$

$$\eta_R = 0.34999$$

$$\eta_R = 34.999\%$$

(f) Steam Rate,

$$\text{SR} = \frac{3600}{W_s} = \frac{3600}{937.26}$$

SR = 3.8409 kg/kWh

Comparison:

(1) Pump work for Carnot Cycle is more, since it has to handle large amount of mixture of wet steam and condensate.

(2) Turbine work in both the cases is same as the state points are same.

(3) Carnot cycle efficiency is more than Rankine cycle efficiency.

Example 11.7 Calculate the cycle efficiency and steam consumption in kg/kWh for Carnot cycle and Rankine cycle using steam between two pressure of 20 bar and 0.1 bar. The steam is dry and saturated at 20 bar.

Find: $\eta_{\text{carnot}} = ?\ \ \eta_R = ?\ \ \text{SR} = ?$

Solution

Refer Fig. Ex. 11.7

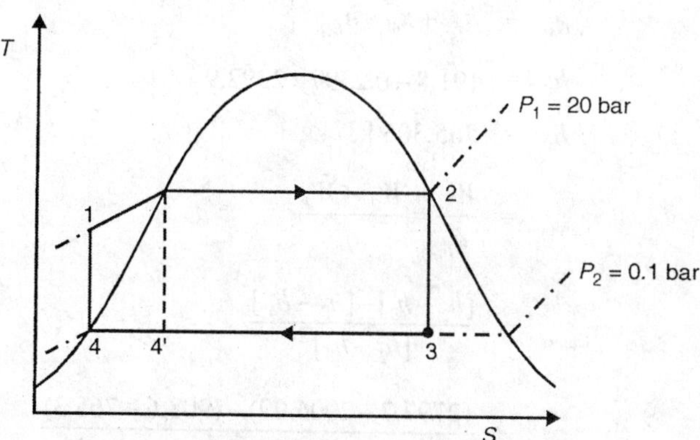

Fig. Ex. 11.7

(I) Consider carnot Cycle 1' – 2 – 3 – 4'

$$h_{1'} = h_f \text{ at 20 bar} = 908.6\frac{\text{kJ}}{\text{kg}}$$

$$s_{1'} = s_f = 2.447\frac{\text{kJ}}{\text{kg-K}}$$

and

$$h_2 = h_g \text{ at 20 bar} = 2797.2\frac{\text{kJ}}{\text{kg}}$$

$$s_2 = s_g \text{ at 20 bar} = 6.337\frac{\text{kJ}}{\text{kg-K}}$$

For Isentropic Process 2–3

$$s_2 = s_3$$

$$s_2 = s_{f_3} + x_3 \cdot s_{fg_3}$$

$$6.337 = 0.649 + x_3 \times 7.502$$

$$x_3 = 0.75819 \cong 0.76$$

$$\therefore \quad h_3 = h_{f_3} + x_3 . s_{f_{g_3}} = 191.8 + 0.76 \times 2392.9$$

$$h_3 = 2006.09 \text{ kJ/kg}$$

Now for Isentropic Compression process 4' – 1'

$$s_{1'} = s_{4'} = s_{f4'} + x_{4'} \times s_{fg_{4'}}$$

$$2.447 = 0.649 + x_{4'} 7.502 \text{ at } 0.1 \text{ bar}$$

$$0.2397 = x_{4'}$$

$$h_{4'} = h_{f_{4'}} + x_{4'} \times h_{fg_{4'}}$$

$$h_{4'} = 191.8 + 0.2397 \times 2392.9$$

$$h_{4'} = 765.305 \text{ kJ/kg}$$

Now $$\eta_{\text{carnot}} = \frac{W_s}{q_i} = \frac{W_T - W_P}{q_i}$$

$$\eta_{\text{carnot}} = \frac{[h_2 - h_3] - [h_{1'} - h_{4'}]}{[h_2 - h_{1'}]}$$

$$= \frac{(2797.2 - 2006.09) - (908.6 - 765.3)}{2797.2 - 908.6}$$

$$= \frac{797.11 - 143.3}{1888.6} = 034618$$

$$\boldsymbol{\eta_{\text{carnot}} = 34.618\ \%}$$

And Steam Rate $$= \frac{3600}{W_S} = \frac{3600}{W_T - W_P}$$

$$= \frac{3600}{[h_2 - h_3] - [h_{1'} - h'_{4'}]}$$

$$\text{S.R} = \frac{3600}{653.81} = 5.506 \text{ kg/kWh}$$

(II) Now Consider Rankine Cycle = 1 – 2 – 3 – 4

$$h_4 = h_f \text{ at } 0.1 \text{ bar} = 191.8 \frac{\text{kJ}}{\text{kg}}$$

From Fig. Ex. 11.7 h_2 and h_3 will remain same.

i.e. $$h_2 = 2797.2 \text{ kJ/kg}$$

$$h_3 = 2006.09 \text{ kJ/kg}$$

Now to find h_1, we also know that,

$$W_P = \frac{P_1 - P_b}{10} = \frac{20 - 0.1}{10}$$

$$W_P = 1.99 \text{ kJ/kg} = h_1 - h_4$$

$\therefore$ $$h_1 = 1.99 + 191.8$$

$$h_1 = 193.79\,\text{kJ/kg}$$

$$\text{Now, } \eta_g = \frac{W_s}{q_i} = \frac{W_T - W_P}{q_i}$$

$$\eta_R = \frac{[h_2 - h_3] - W_P}{[h_2 - h_1]}$$

$$= \frac{(2797.2 - 2006.09) - 1.99}{(2797.2 - 193.79)}$$

$$= \frac{791.11 - 1.99}{2603.41} = \frac{789.12 (i.e. W_s)}{2603.41}$$

$$= 0.30311$$

$$\eta_R = 30.311\%$$

and

$$\text{Steam Rate} = \frac{3600}{W_s} = \frac{3600}{[h_2 - h_3] - W_p} = \frac{3600}{789.12}$$

$$\mathbf{S.R = 4.562\,kg/kWh}$$

Example 11.8. A steam plant operating on Rankine cycle generates superheated steam at 10 bar and 380°C. Condensation occurs at 0.06 bar. Calculate the thermal efficiency of the plant and compare this with the thermal efficiency of Carnot cycle working between the same temperature limits.

Find: $\eta_R = ?$ $\eta_{car} - ?$

Solution

Refer Fig. Ex. 11.8.

I. Rankine Cycle (1 – 2 – 3 – 4)

From Steam Tables, at 0.06 bar,

$$h_4 = h_f = 151.5 \text{ kJ/kg}$$

Also Pump work

$$W_P = \frac{P_1 - P_b}{10} = \frac{10 - 0.06}{10}$$

$$W_P = 0.994 \text{ kJ/kg} = h_1 - h_4$$

$$0.994 + 151.5 = h_1$$

$$h_1 = 152.494\,\text{kJ/kg}$$

And h_2 at 10 bar and 380°C from superheated steam tables = 3222.04 kJ/kg by interpolation.

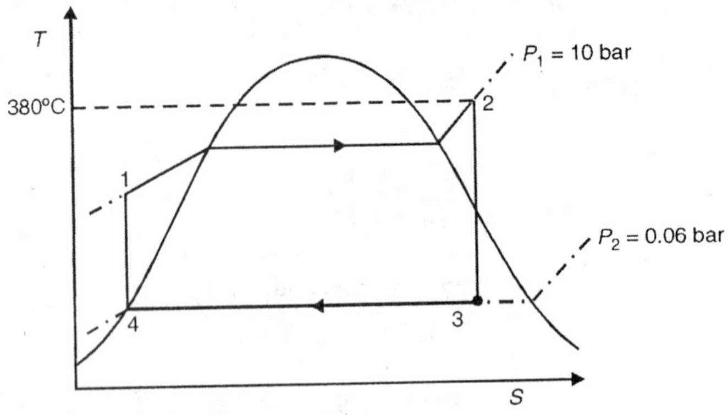

Fig. Ex. 11.8

To find h_3; from the Isentropic process 2–3

$$s_2 = s_3$$

$$s_2 = s_{f_3} + x_3 s_{f_{g3}} \qquad \text{(A)}$$

Now s_2 at 10 bar and 380°C = 7.4.014 kJ/kg-K

Now from Eq. (A),

$$7.4014 = 0.521 + x_3 \times 7.81 \text{ at } 0.06 \text{ bar}$$

$$\frac{7.4014 - 0.521}{7.81} = x_3$$

$$0.8809 = x_3$$

Now $$h_3 = h_{f_3} + x_3 . h_{f_{g3}} \text{ at } 0.06 \text{ bar}$$

$$= 151.5 + 0.8809 \times 2416$$

$$h_3 = 2279.93 \text{ kJ/kg}$$

Now, (i) $$\eta_R = \frac{W_S}{q_i}$$

$$\eta_R = \frac{W_T - W_p}{q_i}$$

$$\eta_R = \frac{[h_2 - h_3] - W_p}{[h_2 - h_1]}$$

$$\eta_R = \frac{(3222.04 - 2279.93) - (0.994)}{(3222.04 - 152.494)}$$

$$\eta_R = \frac{(942.108 - 0.994)}{3069.546}$$

$$\eta_R = \frac{941.1149}{3069.546}$$

$$\eta_R = 0.30659.$$

$$\boldsymbol{\eta_R = 30.659\%}$$

II. Carnot Cycle 1'–2–3–4'

From Steam Tables at 10 bar

$$h_{1'} = h_f = 762.6 \text{ kJ/kg}$$

$$s_1 = s_f = 2.138 \text{ kJ/kg-K}$$

And $h_2 = 3222.04$ and $h_3 = 2279.93$

These will remain same.

Now for Isentropic process 4'–1'

$$s_{1'} = s_{4'} = s_{f4'} + x_{4'} \times s_{fg_{4'}}$$

$$2.138 = 0.521 + x \times 7.81 \text{ at } 0.06 \text{ bar}$$

$$\frac{2.138 - 0.521}{7.81} = x_{4'}$$

$$0.2070 = x_{4'}$$

$$h_{4'} = h_{f_{4'}} + x_{4'} \times h_{fg_{4'}} \text{ at } 0.06 \text{ bar}$$

$$= 151.5 + 0.2070 \times 2416$$

$$h_{4'} = 651.714 \text{ kJ/kg}$$

Now,
$$\eta_{\text{carnot}} = \frac{W_s}{q_i} = \frac{W_T - W_P}{q_i}$$

$$= \frac{[h_2 - h_3] - [h_{1'} - h_{4'}]}{h_2 - h_{1'}}$$

$$= \frac{(3222.04 - 2279.93) - (762.6 - 651.714)}{3222.04 - 762.6}$$

$$= \frac{(942.11) - (110.886)}{(2459.44)}$$

$$= 0.33797$$

$$\boldsymbol{\eta_{\text{carnot}} = 33.797\%}$$

Hence Carnot Cycle efficiency is more than Rankine cycle for an engine working between same temperature limits.

THEORY QUESTIONS

1. Derive the expression for efficiency of Carnot cycle.
2. Draw Carrot cycle on *P–V* and *T–S* diagram.
3. Explain Rankine cycle with *P–V* and *T–S* diagram.
4. Derive an expression for efficiency of Rankine cycle.
5. Compare Carnot cycle with Rankine cycle.
6. Define the terms : (i) Work ratio (ii) SSC.
7. Write a short note on factors affecting performance of Rankine cycle.
8. Explain why Rankine cycle and not Carnot cycle is considered as the thermodynamic cycle for the steam power plant.
9. Discuss the effects of the following parameters on the performance of Rankine cycle:
   (i) Steam pressure at turbine inlet.
   (ii) Steam temperature at turbine inlet.
   (iii) Steam pressure at turbine exhaust.
10. Draw a neat sketch of Rankine cycle of steam plant, label the parts and describe the working of each component, giving the formula for energy exchange for the same.
11. Enumerate the advantages of Rankine cycle over the Cannot cycle for steam power plant.
12. Explain the drawbacks of Cannot cycle for a steam power plant. How are they overcome in a Rankine cycle ?
13. In what respect does the Rankine cycle differ from the Carnot cycle. Draw Carnot and Rankine cycle on *T–S* plane.
14. Enumerate the advantages of Rankine cycle over the Carnot cycle for steam power plant.
15. Why is Carnot cycle not practicable for a steam power plant.
16. Draw a neat sketch of Rankine cycle steam plant, label the part and describe the working of each component, giving formula for energy exchange for the same.
17. Explain clearly how the following affects the performance of a Rankine cycle.
    (i) Lowering the condensation pressure.
    (ii) Superheating the steam to high temperature.
    (iii) Increasing the boiler pressure.
18. Parameters affecting the performance of Rankine cycle.
19. Explain the factors affecting the efficiency of Rankine cycle.
20. Explain with *T–S* diagram, factors affecting performance of Rankine cycle.
21. Explain the effect of following on the efficiency of Rankine cycle.
    (i) Increasing the average temperature at which heat is supplied.
    (ii) Decreasing the temperature at which heat is rejected.

22. Show the Rankine cycle on
    (i) Wet
    (ii) Dry saturated
    (iii) Superheated

PROBLEMS FOR PRACTICE

1. A steam power plant operating on Rankine cycle recieves steam from boiler at 35 bar and 350°C. It is exhausted to condenser at 0.1 bar, calculate
    (i) Energy supplied/kg of steam generated in boiler
    (ii) Quality of steam entering the condenser
    (iii) Rankine cycle efficiency considering feed pump work.

    [**Ans.** (i) 2911.16 kJ/kg, (ii) 0.8016, (iii) 34.226%]

2. Consider a steam power plant operating on the ideal Rankine cycle. The steam enters the turbine at 3 MPa and 350°C and is condensed in the condenser at a pressure of 10 kPa. Determine
    (i) The thermal efficiency of this power plant.
    (ii) The thermal efficiency if the steam is superheated to 600°C instead of 350°C.
    (iii) The thermal efficiency if the boiler pressure is raised to 15 MPa while the turbine inlet temperature is maintained at 600° C.

ꟸ ꟹ

12

Fuels and Combustion

CHAPTER OBJECTIVES

After reading this chapter you will be able to learn the following

- Classification of fuels and alternative fuels, such as bio-gas, bagasse, natural gas, LPG, LNG etc.
- Proximate and Ultimate analysis of coal.
- Calorific values of fuels.
- Stoichiometric Air, Excess air.
- Mass Fraction and Mole Fraction.
- Combustion Equations.
- Flue Gas Analysis.
- Determination of Gravimetric and Volumetric Analysis of the Products of Combustion and their Conversions.
- Orsats Apparatus, Bomb Calorimeter, Bogs Gas Calorimeter etc.
- Determination of Actual Air Supplied.

12.1 INTRODUCTION

The fuel is material which when once raised to its ignition temperature continues to burn if sufficient oxygen or air is available. The main constituents of any fuel are carbon and hydrogen. These constituents are called combustibles. Small traces of sulphur are also present in certain fuels.

12.1.1 Classification of Fuels

Fuels may be classified as given follows:

(i) According to origin (mineral and organic)
(ii) According to mode of occurrence (natural, synthetic)
(iii) According to state (solid, liquid and gaseous)
(iv) According to condition (pulversied, briquetted, cooked)
(v) According calorific value (high and low)
(vi) According to combination of C and H (paraffin based, aromatic, benzoic)

Another way to classifying the fuels is as given on next page.

12.2 ALTERNATIVE FUELS

12.2.1 Biogas

Biogas is produced by anaerobic fermentation of wet organic waste like animal dung, municipal garbage etc. Biogas is mostly (CH_4(50 – 70%) and CO_2 (remaining). Biogas can be conveniently used as alternate fuel. India has much scope to tap the large animal waste available in rural areas. India has very large animal population and its dung can be used not only as manure that is a byproduct in Gobar gas plant but also to produce biogas.

Another way of classifying the fuels is as follows.

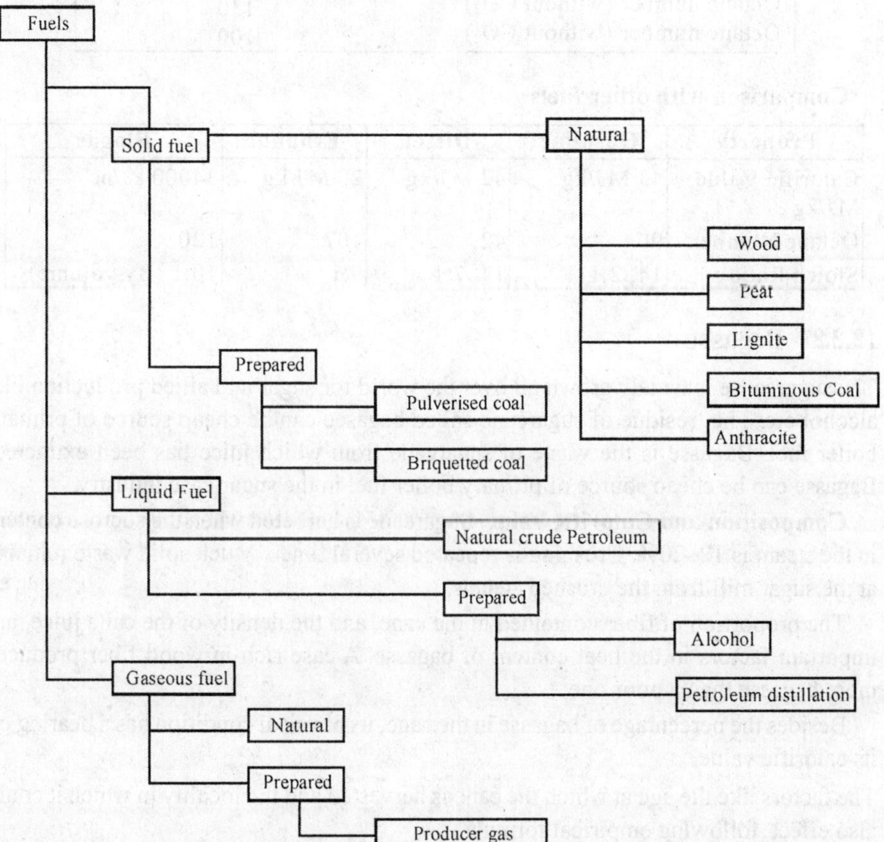

The biogas being lighter than air will rise up in case of any leak without any harm. Its ignition temperature is also higher than many fuels that make it quite safe. Recent advance in storage and transportation also make it quite useful as attractive fuel for IC engines apart from domestic use.

The biogas is produced usually in a gobar gas plant. Factors affecting Biogas production are:

(a) Animal dung and organic waster should contain adequate nitrogen. Best carbon to nitrogen (C/N) ratio is about 30.

(b) Best temperature is about 35°C.

(c) pH ratio of the slurry should be between 7.2 to 8.2.

(d) The percentage of solids in the slurry should not exceed 10%.

The calorific value of Biogas is low but it has excellent knock resistance. In a typical Biogas composition and Octane No. is as follows.

Component	% by volume
CH_4	50–60
CO_2	30–45
H_2	5–10
Octane number (without CO_2)	130
Octane number (without CO_2)	100

Comparison with other fuels

Property	Gasoline	Diesel	Ethanol	Biogas
Calorific Value MJ/kg	43 MJ/kg	42 MJ/kg	20 MJ/kg	34000 kJ/m^3
Octane Number	90	42	107	120
Stoich Ratio	14, 2:1	14, 7:1	9:1	10:1 (By volume)

12.2.2 Bagasse

The sugarcanse is widely grown all over the world for sugar and allied production like alcohol etc. The residue of sugarcane called bagasee can be cheap source of primary boiler fuel. Bagasse is the waste of sugarcane from which juice has been extracted. Bagasse can be cheap source of primary boiler fuel in the sugar case industry.

Composition and Calorific Value. Sugarcane is harbeted when the sucrose content in the steam is 12–20%. Crushing is repeated several times. Much solid waste remains at the sugar mill from the crushed steams.

The proportion of fiber contained in the cane, and the density of the cane juice, are important factors in the heat content of bagasse. A case rich in wood fiber produces more bagasse than a poor one.

Besides the percentage of bagasse in the cane, its physical condition has a bearing on its calorific value.

The factors like the age at which the cane is harvested and the locality in which it could also effect following empirical formula

Heat content of bagasse in kJ/kg = 200 F + 165 S + 157 G – 22.5 W

F = Percentage of fiber in the sample
S = Percentage of sucrose in the sample
G = Percentage of glucose in the sample
W = Percentage o£ water in the sample

This formula gives the total available heat per pound of bagasse, that is the heat generated per pound less the heat required to evaporate its moisture and superheat the steam thus formed to the temperature of the stack gasses.

Three bagasses samples yield the following results

F = 50 S and G = 50 W = 42.5 9710 kJ/kg
F = 33.3 S and G = 6.0 W = 51.0 7780 kJ/kg
F = 33.3 S and G = 7.0 W = 56.7 6500 kJ/kg

The bagasses with a high moisture constant that will not burn properly but which smolders and produces a large quantity of products of destructive distillation mainly heavy hydrocarbons, which escape unburned.

The Energy Aspect. The steam is raised in bagasse fired boilers which usually have a secondary fuel to accommodate imbalances in bagasse supply and steam or power demand. The factory designer attempts to balance the site such that bagasse is neither left over not insufficient any secondary fuel costs money and a large surplus of bagasse may cost money to dispose. Today, more and more factories are considering power export as another by product of sugar production. To do this they are improving the efficiency of their thermodynamic cycles and converting equipment drives to optimise power output.

Factories are establishing frequently in very undeveloped places and have no connection to an external power supply. This requires special techniques to start the factory and means that any breakdown in the powerhouse impacts on the entire neighbourhoods.

12.2.3 Natural Gas

It is the most important fuel belonging to gaseous fuels and is found in the vicinity of coal mines or oil reservoirs. Natural gas is a mixture of methane, ethane, propane, butane, pentane, nitrogen and carbondioxide.

Natural gas is obtained from oil wells as underground reservoirs, with or without petroleum oil, hence, it is either dry or wet. When there is no oil but only gas in the petroleum well, the natural gas is said to be dry whereas if natural gas occurs along with petroleum in oil wells, it is called wet gas. The wet contains gaseous hydrocarbons like methane and its mixture with higher hydrocarbons like propane, butane, isopentane pentane whereas the dry gas consists of methane, ethane with small amount of CO, CO_2, H_2S, N_2, H_2 and inert gases. The calorific value of wet natural gas is higher than that of dry one because of higher percentage of heavier unsaturated molecules. Dry gas does not form a liquid phase during production conditions whereas a gas is said to be wet if a liquid phase is produced at the surface without retrograde condensation in the reservoir. The wet gas is suitably treated to remove propane, propene, butane and butene and is called as LPG. For using the natural gas commercially it is necessary to remove, H_2S (toxic and corrosive), CO_2 (corrosive and has no heating value), Mercury (toxic), water (forms hydrates), heavy hydrocarbons (condenses in the transport system), Nitrogen (with no heating value).

Composition of Natural Gas. The approximate composition of natural gas is $CH_4 = 70 - 90\%$, $C_2H_6 = 5 - 10\%$, $C_3H_8 = 1 - 2\%$, $C_4H_{10} = 0.55 - 1\%$, Pentaye = 5 – 1%, $H_2 = 3\%$, $CO + CO_2$ = Rest and the calorific value varies from 1200 – 1400 kcal/m³. This composition of natural gas varies widely as it depends on the underground conditions methods used for winning the gas etc. Figure 12.1 shows the temperature scale of Natural gas.

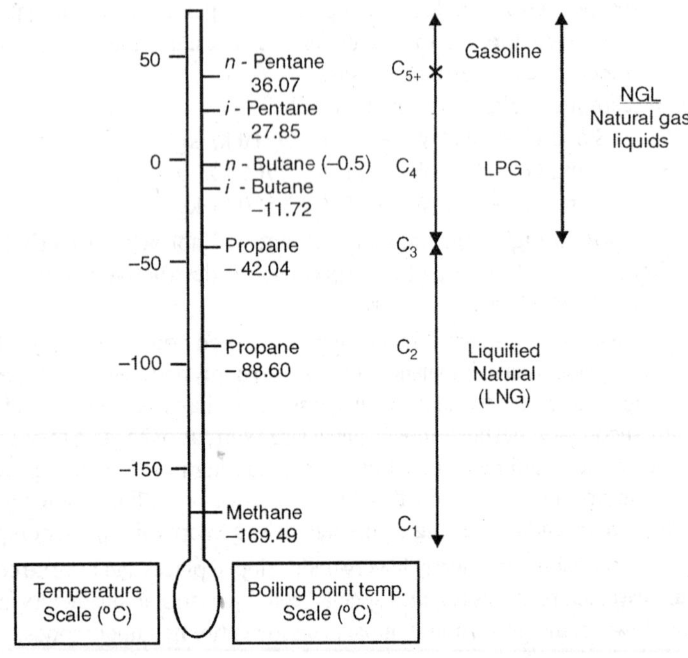

Fig. 12.1. *Temperature scale of natural gas.*

Application of Natural Gas. Natural gas has got wide application as it is used not only for domestic and industrial purpose but also as a chemical raw material for various synthesis.

(i) It is an excellent domestic fuel, for example LPG.

(ii) Various byproducts are obtained from raw natural gas like methane, ethane, propane, butane, natural gasoline etc.

(iii) It is used in manufacturing of carbon black (used as a filler) and hydrogen (used in ammonia synthesis).

(iv) Microbiological fermentation of methane yields synthetic proteins which is used asanimal feed.

(v) Natural gas and its fractions can be transported in various forms like CNG, LNG, LPG and chemicals like methanol, amonia, area.

(i) Liquified Petroleum Gas (LPG). It is generally manufactured from natural gas or the gas from the cracking units of petroleum refineries. LPG in general contains hydrocarbons of such volatility that they can exist as gases under atmospheric pressure, but can be readily liquified under pressure. The mixture of lower hydrocarbons is condensed under high pressure in refrigeration units and passed through de-ethaniser and debutaniser. The gas is finally passed through washing column and mixed with mercaptans (which acts as a warning gas on leaking). It is then supplied under pressure in containers. The pressure of the gas at room temperature within the cooking cylinder is about 3 bar. Figure 12.2 gives a schematic diagram showing processing operation of natural gas.

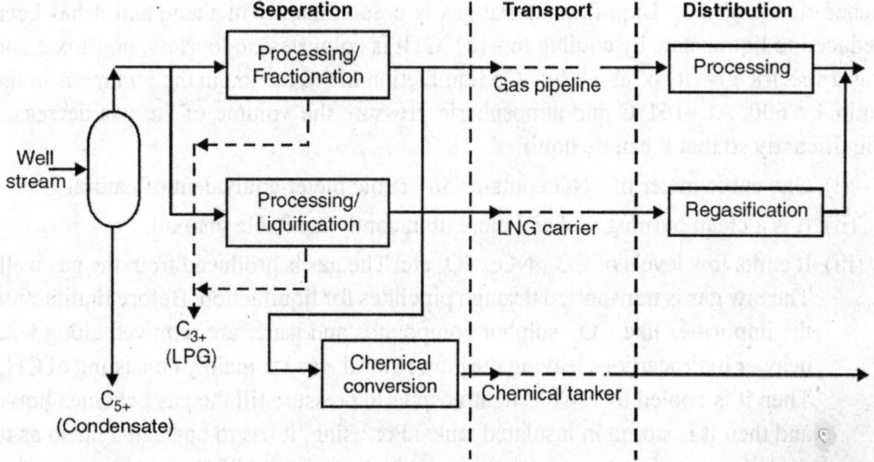

Fig. 12.2. *Processing operation of natural gas.*

Composition of LPG. LPG contains mainly paraffinic hydrocarbons upto C_4 and its main constituents are *n*-butane, isobutane, butylene and propane with little or no propylene or ethane. The exact composition in *n*- butane = 27%, 2-Butane = 25% Butane = 43%, Propylene = 2.5%, Propane = 2.5 %. LPG is available under the trade name Bharat Gas, Hindustan Petroleum etc.

Uses. LPG (Liquid butane) is used largely as a domestic fuel and liquid propane as an industrial fuel. Research is on to use LPG as a motor fuel. It is also a source of petrochemicals.

Advantages of LPG over Gaseous Fuels

1. The calorific value is 3 times that of natural gas and 7 times that of coal gas
2. Smoke problem is not there
3. Low maintenance
4. Easy transportation in cylinders or pipelines
5. High efficiency and heating rate

Advantages of LPG over Gasoline as a Motor Fuel

1. LPG is cheaper than gasoline
2. It mixes easily with air as compared to gasoline
3. It burns cleanly without giving any residue

Disadvantages of LPG over Gasoline as a Motor Fuel

1. Its octane number is quite low
2. It has to be handled under pressure
3. It cannot be blended with other fuels
4. Due to its faint colour, one cannot easily detect its leakage.

(ii) Liquid Natural Gas (LNG). It is mostly used as an industrial fuel and in power generation projects. Liquified natural gas is predominantly methane and it has been reduced to liquid state by cooling to –161°C. It is colourless, odourless, non-toxic and has a specific gravity of about 0.5. On liquifaction the reduction in the volume is in the ratio 1 : 600. At –161°C and atmospheric pressure the volume of the gas decreases significantly so that it can be liquified.

(i) One cubic meter of LNG contains 584 cubic meter equivalent of natural gas.

(ii) It is a clean burning fuel and more abundantly available than oil.

(iii) It emits low levels of CO_2, NO_2, SO_2 etc. The gas is produced from the gas well. The raw gas is transported through pipelines for liquifaction. Before liquifaction the impurities like CO_2 sulphur compounds and water are removed along with heavier hydrocarbons leaving the pure natural gas i.e. mainly consisting of CH_4. Then it is cooled to –161°C at atmospheric pressure till the gas becomes liquid and then it is stored in insulated tanks. For using, it has to be heated up so as to regasify it and then transport it through pipelines. Figure 12.2 gives its processing and transportation scheme.

Applications

1. LNG can be used as a fuel where pollution control, and efficient heat control are major requirements.
2. Used in Industries like fertilizer units, glass industry, ceramic industry, and general transport.
3. Used in Boiler fuel for steam generation.

(iii) Compressed Natural Gas (CNG). This is another alternative fuel obtained from natural gas which is much more fuel efficient and requires less maintenance.

Compared to Diesel and Petrol, CNG cause lower pollution without affecting the performance of the vehicle. The exhaust level of CO_2 in CNG vehicle is about 40% lower than that of gasoline or diesel engine. Since CNG does not operate on fuel containing high lead and sulphur hence these are not emitted in CNG emission. Even though CNG has a higher self ignition temperature as compared to petrol but because of its low inflammability it is much safer than petrol. CNG is regarded as one of the cleanest fuel as it emits less CO and has lower potential of ozone formation as compared to petroleum gasoline. Costwise also it is less expensive than Diesel and petrol. The carcinogenic potential of even the best quality diesel in vehicles is 5 times that of CNG. CNG operated vehicles emit less particulate matter than Diesel operated vehicles, Catalytic converters which are used to reduce the emissions from automobiles cannot function effectively because of high sulfur content in both petrol and diesel.

Composition. CNG is compressed natural gas i.e. natural gas is compressed at high pressure (100 – 200 bar). Composition is same as natural gas.

Applications

1. As a substitute for petrol/diesel.
2. On blending with diesel (15% diesel and rest CNG) the fuel can be more cost effective.
3. Less polluting than gasoline.
4. Least contribution towards smog as compared to gasoline.

12.3 COAL ANALYSIS

Coal as found in nature is neither a pure substance nor of uniform composition. A definite chemical formula cannot be therefore written for a coal found in coal mines. Consequently two different methods of analysis are employed to know the composition of coal. They are known as *Ultimate analysis* and *Proximate analysis.*

In ultimate analysis a complete chemical breakdown of coal into its chemical constituents is carried out by chemical process. This analysis is required when important large scale trials are being performed. The analysis serves the basis of calculations of the amount of air required for complete combustion of the kg of fuel. The analysis gives the percent content on mass basis of carbon, hydrogen, nitrogen, oxygen, sulphur and ash, and their sum is taken as equal to 100%. Moisture is expressed as a separate item. The analysis also enables us to determine calorific value or heating value of coal.

Proximate analysis is the separation of coal into its physical components and can be made without the knowledge of analytical chemistry. The analysis is made by means of a chemical balance and a temperature controlled furnace. The sample of fuel is heated in the furnace or oven. The component in the analysis we fixed carbon, volatile matter, moisture and ash. These components are expressed in percent on mass basis and their sum if taken as 100%. Sulphur is determined separately. This analysis also enables us to determine the C.V. of coal. Compacted vegetation, in the absence of air and under the influence of pressure and temperature, is converted into coal. The various stages of conversion are as follows:

Wood → Peat → brown → coal → lignite → sub-bituminous coal → bituminous coal → anthracite.

This conversion takes 10 – 300 million years — the peat being the youngest and the anthracite the oldest one.

12.3.1 Proximate Analysis

(1) Powdered coal – *m* gm (weighted) → heated at 110°C for 20 min. Sample is again weighted, (m_1)

$$(m - m_1) = \text{moisture evaporated}$$

$$\% \text{ of moisture } (M) = \frac{m - m_1}{m} \times 100$$

(2) m_1 is then heated in a close container at 954°C for seven (7) minutes. Sample is again weighed. Volatile matter evolves at about 600°C and leaves the sample of m_2.

$$\% \text{ of volatile volume } (MV) = \frac{m_1 - m_2}{m} \times 100$$

(3) m_2 is heated at 732°C in an open crucible. It is completely burnt. The residue is the ash fraction $\rightarrow m_3$

$$\% \text{ Ash (A)} = \frac{m_3}{m} \times 100$$

12.3.2 Ultimate Analysis

The ultimate analysis consists of determining the percentages of the ultimate constituents

(a) Carbon
(b) Hydrogen
(c) Oxygen
(d) Sulphur
(e) Nitrogen
(f) Ash

Ultimate analysis is used for calculative and scientific work.

12.4 CALORIFIC VALUE OF FUELS : (HCV AND LCV)

The calorific value of a fuel is amount of heat liberated by its complete combustion. For solid and liquid fuels, calorific value is expressed in kJ/kg, whereas for gaseous fuels it is expressed as kJ/m^3 where m^3 is normal cubic metre measured at NTP conditions i.e. at 0°C temperature and 760 mm Hg barometric pressure (1.01325 bar). Sometimes, calorific value of gaseous fuels can also be expressed as kJ per cubic meter expressed at STP conditions. STP (standard temperature and pressure) conditions are taken as 15°C and 760 mm Hg barometric pressure (1.01325 bar).

A fuel consists of one or more combustible components like carbon, hydrogen, carbon monoxide, hydrocarbons, sulphur etc. Of the above mentioned combustibles, sulphur is not a desirable ingredient due to corrosive properties of SO_2 formed by its combustion. The above mentioned combustibles liberate heat by their combustion, but to evaporate the water formed by combustion of hydrogen or hydrocarbons as well as to evaporate the moisture content of the fuel, its sensible heat from temperature to the saturation temperature at combustion pressure and its latent heat and superheat, if any, have to be supplied to bring it to the temperature of the products of combustion. This heat is naturally taken from the heat of combustion of the fuel. Thus the total heat of combustion of the fuel is not available to do external work. The heat liberated by the combustion of the fuel, neglecting the heat required to evaporate the water is called the *Higher Calorific Value* of the fuel (HCV). Higher calorific value is the maximum heat energy liberated by the complete combustion of the fuel. This is also called as the gross calorific value of the fuel. If we subtract from the higher calorific value, an amount of heat required to evaporate the water formed, we get *Lower Calorific Value* (LCV) or *Net Calorific Value* (LCV) or *Net Calorific Value* of the fuel.

No definite agreement is to be found in the literature on fuel as to whether the lower calorific value shall be found simply by subtracting latent heat of steam or both the latent heat and sensible heat in cooling from 100°C, from the gross calorific value of the fuel; in the latter case it would be necessary to fix the temperature to which the products are finally reduced. In literature the net calorific value of the fuel is obtained by subtracting from higher calorific value the amount 2466 kJ/kg (latent heat of dry and saturated steam at STP (15°C) conditions) (amount of water formed because of the combustion of 1 kg fuel.)

Calorific values of solid and liquid fuels are found experimentally with the help of **Bomb Calorimeter.**

12.4.1 Theoretical Determination of Calorific Value (Provided Composition by Weight is Known)

The calorific value of a fuel is the heat liberated from the combustion of 1 kg of the fuel. This value can be calculated from the analysis of the fuel if the calorific value of its constituents are known. Any oxygen contained in the fuel is assumed to be already combined with the hydrogen; the heat of this hydrogen is not, therefore, available for further combustion.

The calorific values of the individual combustible in the fuel are given below:

Carbon C burns to CO_2 → 33915 kJ/kg

Carbon C burns to CO_2 → 10200 kJ/kg

Hydrogen H_2 burns to H_2O → 1444515 kJ/kg

Sulphure S burns to SO_2 → 9630 kJ/kg

Now let C, H, O and S% be the carbon, hydrogen, oxygen and sulphur contents of the fuel respectively.

∴ Assume or consider 100 kg of fuel.

∴ The quantity of heat evolved or liberated due to combustion of carbon is C × 33915 kJ when burnt to CO_2.

The quantity of heat evolved due to combustion of hydrogen is $\left[H-\frac{0}{8}\right]\times 144451.5$ kJ.

Here the assumption is made that any oxygen present in the fuel is already wholly in combination with hydrogen. Since it is known that 8 parts of oxygen by weight are combined with 1 part of hydrogen, it is customary to deduct from the total hydrogen an amount equal to 1/8 of the oxygen present, calling the remainder available as hydrogen.

The quantity of heat evolved due to combustion of sulphur is $S\times 9630$ kJ

Therefore, total heat evolved due to combustion of 100 kg of fuel

$$= 33915C + 144452.5\left(H - \frac{0}{8}\right) + 9630\ S\ kJ$$

∴ Higher Calorific Value of the fuel is

$$HCV = \frac{33915C + 144452.5\left(H - \frac{0}{8}\right) + 9630\ S}{100}\ kJ/kg$$

The amount of steam formed = $\frac{9H}{100}$ kg/kg fuel because from 1 kg of hydrogen 9 kg of water vapour is formed.

∴ Lower Calorific Value of the fuel is

$$LCV = \frac{33815C + 144451.5\left(H - \frac{0}{8}\right) + 9635S - 9H \times 2466}{100}\ kJ/kg$$

$$\left[HCV = \frac{33815C + 144451.5\left(H - \frac{0}{8}\right) + 9635S}{100}\right] \text{ is } \textit{Dulong farmula}$$

This higher value is based on ultimate analysis. Similarly, for gaseous combustibles we have

H_2 — 12800 kJ/m^3

CO_2 — 12620 kJ /m^3

CH_2 — 40070 kJ/m^3

Goutel suggested the following formula form calculating the higher calorific value when the percentage proximate analysis of fuel is known. The formula is, cal. value = 343.3 × fixed carbon %+ α × % volatile matter kJ/kg.

The value of the factor α depends on the percent amount of volatile matter on the dry ash free basis. The value of α decreases with the rise in volatile matter as can be seen from the table below.

Volatile matter % (dry ash free)	5	10	15	20	25	30	35	40
α	607	544	490	456	431	410	394	335

Goutel formula is unreliable for fuels having high percentage in oxygen.

12.4.2 Experimental Determination of Calorific Value of Fuel

An apparatus which is used for determining the calorific value of a fuel is known as a fuel calorimeter.

The principle of all the calorimeters is the transference of heat of combustion of the given weight of fuel to water and the vessel. From the observed rise of temperature of the water and the container the calorific value of the fuel can be determined by equating the heat given out by the fuel to the heat taken by the water and the container. In order to know the heat taken by the container, water equivalent of the container should be known.

In this method of determining the calorific value of the fuel the following conditions should be satisfied:

(i) the combustion of the fuel must be complete

(ii) the heat must be transferred completely to the water

(iii) cooling losses from the calorimeter must be corrected

(iv) the rise of temperature of water must be correctly determined because the mass of the fuel is very small in comparison with the quantity of the water heated.

(a) Boy's Gas Calorimeter. This is the standard instrument used for measuring the calorific value of gas. The sectional view of the calorimeter is shown in Fig. 12.3 given on next page. It consists of two burners in which a known volume of gas is burnt.

The hot gases produced by combustion pass up the copper chimney. This is surrounded by a double coil of metal tubing through which a known weight of cooling water is circulating. After passing upwards the hot gas are deflected downwards through the space containing the inner coil. By the time the gases reach the top of this passage, practically the whole of their heat has been absorbed by the circulating water. The gases are then passed away into the atmosphere.

The cooling water enters the outer coil. After circulating through both coils it leaves at the water box. Its temperature is measured at inlet and outlet by thermometers at inlet and outlet by thermometers shown. These can be read to a fraction of a degree by the reading leases. For this purpose special thermometers giving least count of 0.01 can also be used. The temperature of the exhaust gases is measured by another thermometer. Any water formed by the combustion of hydrogen is condensed on the coils and is collected at the base of the instrument. From here it is drained through the pipe and is collected in a measuring glass.

In order to increase the cooling surface of the coils, the outsides of the tubes are ribbed or finned as shown in the cross-sectional view. The outer cylinder of the instrument is lined with a felt jacket to prevent any leakage of heat. For measuring the volume of the gas consumed, a meter is used reading 1/100 cu. meter for one revolution of the hand.

The test is carried out after the gas has been burning under uniform conditions for 45 minutes, the rate of gas flow is obtained by timing the rotating hand of the gas meter with a stop watch. The cooling water is collected in a measuring vessel during the test. The cold water supply to the coils should not be less than 5°C, below the room temperature. The barometer reading, gas temperature and room temperature should be taken during the test. The thermometers should be read at regular intervals whilst the test is in progress and the average values obtained from these readings.

Let V = Volume of the gas burnt during the test, reduced to NTP

m_w = Weight or mass of cooling water used during the test

t_1 = Average inlet water temperature

t_2 = Average outlet water temperature

w = Weight or mass of condensed water collected

h_{fg} = Latent heat of steam at its partial pressure

= 2466 kJ/kg (assumed)

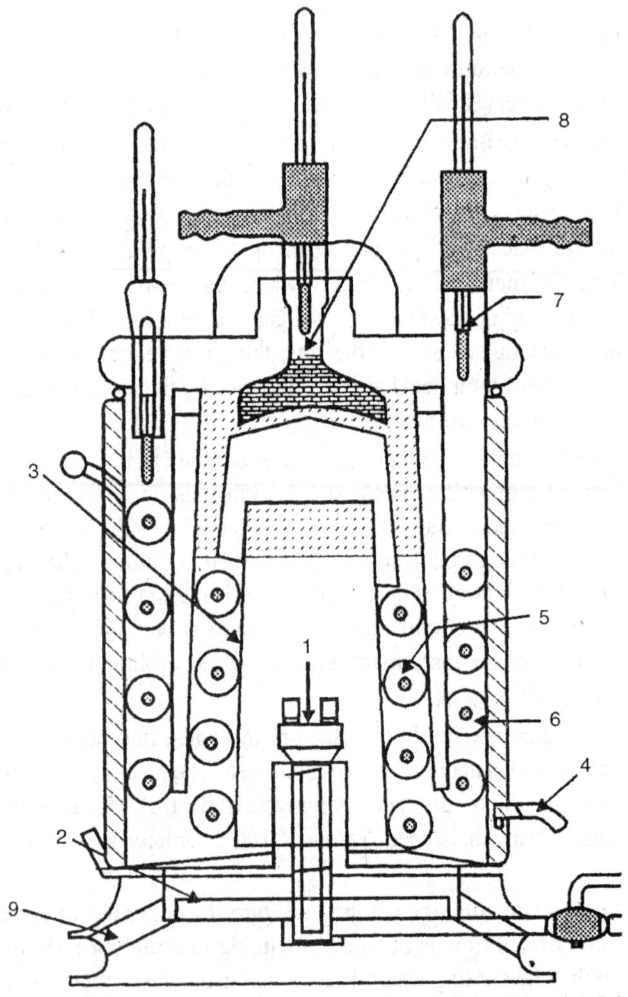

1. Gas burners, 2. Metal vessel, 3. Copper chimney, 4. Condensed water outlet, 5. Inner coil, 6. Outer coil, 7. Inlet water box, 8. outlet water box, 9. Holes in base

Fig. 12.3. *Boy's gas calorimeter.*

$\therefore$ Higher calorific volume $\text{HCV} = \dfrac{m_w C_{pw}(t_2 - t_1)}{V}$ kJ/m^3 and lower calorific value $\text{LCV} = \text{HCV} - w.h_{fg}$ kJ/m^3.

Generally mass flow rate of cooling water is adjusted such that the gas exhaust temperature is same as that of atmosphere.

(b) Bomb Calorimeter. One of the best instruments for measuring the calorific value of powdered and liquid fuels is the bomb calorimeter. The fuel is burnt in a strong steel chamber, known as a bomb which is immersed in a known mass of water. The fuel is placed in a crucible inside the bomb which is filled with oxygen under a pressure of 25–30 atmospheres. It is then electrically ignited by a platinum or magnesium wire. The heat liberated by the rise in temperature of water surrounding the bomb and then calorific value of the fuel is determined.

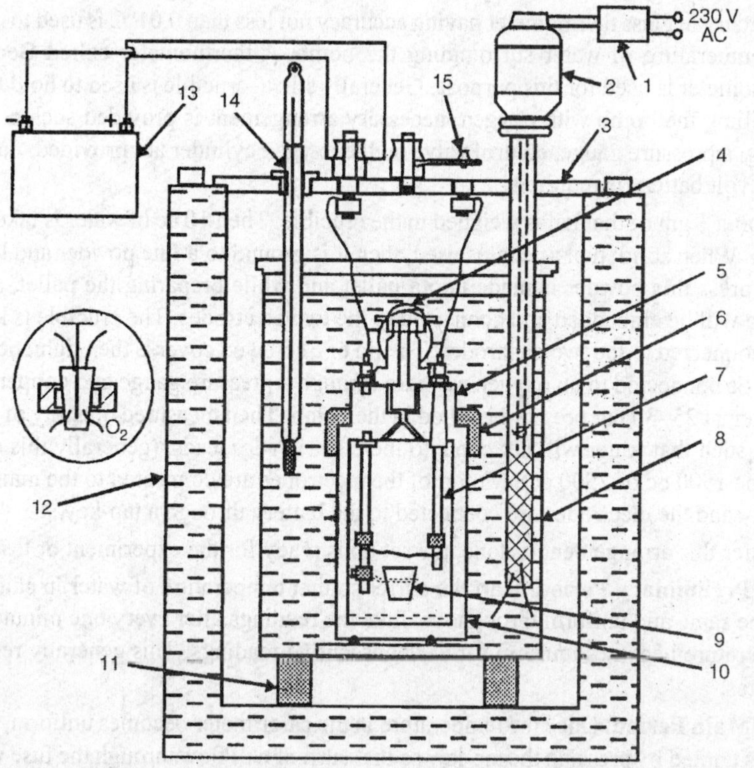

1. Rheostat, 2. Motor, 3. Cover, 4. Cap, 5. Lid, 6. Union Nut, 7. Electrode, 8. Calorimeter vessel, 9. Crucible, 10. Bomb, 11. Support, 12. Outer vessel, 13 Battery, 14. Thermometer, 15. Key

Fig. 12.4. *Bomb calorimeter.*

The advantages of using such a procedure for the determination of calorific value are as given below:

(a) Due to high pressure of oxygen supplied, large excess oxygen is provided and therefore the combustion is complete.

(b) All heat liberated will be given to the surrounding water as the combustion is at constant volume.

(c) The addition of oxygen and very small amount of water (generally 10 cc) in the bomb do not affect the combustion.

(d) The test procedure and the arrangement employed favour the accurate computation of temperature loss correction (cooling correction).

The general arrangement of the Bomb calorimeter is shown in Fig. 12.2. The calorimeter consists of a stainless steel vessel called the bomb which is placed in a calorimeter vessel. The bomb is also made of monel metal. The calorimeter is placed in a double walled chromium plated jacket vessel containing water. The calorimeter is closed on top with ebonite cover. This is because of the reduction of radiation losses to the surroundings. An electrically driven stirrer is provided to agitate water in the vessel. A mercury in glass thermometer having accuracy not less than 0.01°C is used to measure the temperature of water surrounding the bomb. A thermometer called Beckman's thermometer is used for this purpose. Generally silica–crucible is used to hold the fuel. For filling the bomb with oxygen, necessary arrangement is provided such as copper tubing, a pressure gauge, control valve on the oxygen cylinder are provided. For test 4 or 12 volt battery is required.

About 1 gm of the fuel is weighed in the crucible. Then 10 cc of water is taken in the bomb. When solid fuel as coal is used, then it is ground to a fine powder and then in a hand press this powder is made into a pallet and while preparing the pallet, a cotton thread will be embedded and connected to the two electrodes. The crucible is kept in a ring connected to the two electrodes. The lid or a screwed cover is then tightened. Then bomb is connected to an oxygen cylinder through a pressure gauge and copper tubing. Oxygen at 25–30 bar pressure is filled in the bomb. Then measured quantity of water is taken such that water will not come to the electric wire leads, (generally this quantity may be 1500 cc or 2500 cc). Motor of the electrodes are connected to the main power supply and the electrodes are connected to the battery through a tap-key.

After this arrangement is done, the set up is ready for the experiment or test.

1. Preliminary Period. Start the stirrer so that temperature of water in calorimeter will be same and uniform throughout. Take the readings after every one minute till the temperature becomes uniform for 2-3 consecutive readings. This generally requires 5 minutes.

2. Main Period. Once the temperature in the calorimeter becomes uniform, then the fuel is ignited by pressing the tap-key so that current will flow through the fuse wire and it will glow. A cotton thread is connected to the wire so that fuel pallet will be ignited giving out heat. Continue to take readings of temperature at the interval of 1 minute till the maximum temperature is reached.

3. After Period. After the maximum temperature is reached, the temperature will start decreasing. Continue to take temperature readings at the interval of one minute. Take 10–15 readings, and then stop the test.

Take the bomb out of the calorimeter and release the pressure by opening the valve. After removing the lid of the bomb, measure the quantity of water in the bomb. The difference in this water will be the weight of H_2O formed because of the combustion of hydrogen in the fuel. This weight is very much essential to measure or determine the lower calorific value of the fuel. Ash content of the fuel is also obtained after weighing the crucible.

Observation Table :

1. Weight of empty crucible — W_1 gm
2. Weight of crucible + fuel — W_2 gm
3. Weight of the fuel taken — $(W_2 - W_1)$ gm
4. Weight of water in the bomb — 10 gm (generally)
5. Weight of water after test — $10 + x$ gm
6. Weight of moisture condensed — x gm
7. Weight of water in the calorimeter — m_w gm
8. Weight of water equivalent of calorimeter — m_e gm
9. Weight of fuse wire — mf_w gm
10. Calorific value of fuse-wire — (C.V.)FW
11. Weight of crucible + ash
12. Weight of ash — $W_4 = W_3 - W_1$ gm
13. Weight of ash.
14. Observed temperature rise during main period — $\Delta t = (T_n - T_0)$

Temperature Record

Initial Period		Main Period		After or Find	
Time	**Temperature**	**Time**	**Temperature**	**Time**	**Temperature**
0 time (min)		6 (min)		11	
:		:		$\}t_{0_1}$	
:		:		12	
:		:		$\}t_{0_2}$	
:		:		13	
5	To	10 say		:	
				:	
				20	

Cooling correction by Whitaker method

Average rate of change of temperate in after period

$$DT_2 = \frac{t_{0_1} + t_{0_2} + \ldots}{\text{No. of reading taken}}$$

Then cooling correction $T_c = (DT_2) \times$ (No. of reading in main period)

$\therefore$ Corrected rise in temperature $= (T_n - T_0) + T_c + T_t$

where T_t = Minimum temperature that can be read on thermometer
= 0.01 with Beckman thermometer.

Calculations for Calorific Value of Fuel

Total heat liberated = Total heat absorbed

Heat released by fuel + Heat released by fuse wire + Heat released by cotton thread = Total heat absorbed by (Water + Bomb + Calorimeter)

$(m_f + \text{C.V})_{\text{fuel}} + (m \times \text{C.V})_{\text{fuese wire}} + (m \times \text{CV})_{\text{Thread}}$ = (Total weight of water and water equivalent of calorimeter) × C_{Pw} × (Total rise of temperature of water)

$$(g_m \times \text{J/gm})_{\text{fuel}} + (g_m \times \text{J/gm})_{\text{fuse wire}} + (g_m \times \text{J/gm})_{\text{thread}} = g_m \times \text{J/kg°C} \times \text{°C}$$

Joules/gm = Joules/gm

or kJ/kg = kJ/kg

Calorific value obtained with observation taken on Bomb Calorimeter Test, is the higher calorific value of the fuel (HCV).

Lower calorific value (LCV) of the fuel is then obtain as

LCV = HCV – (weight of moisture formed per kg fuel × 2466)

For liquid fuels, the same procedure is followed except of making a pallet of a coal powder. One end of the cotton thread will be connected to the fuse wire and the other end is dipped in the liquid fuel.

12.5 FUELS AND COMBUSTION

The fuel is a material which when once raised to its ignition temperature continues to burn if sufficient oxygen or air is available. The principal constituents of any fuel are carbon and hydrogen. The materials which evolve heat after burning, are called combustibles. Carbon and hydrogen are combustibles. Sulphur is also a combustible material.

When anything slowly combines chemically with oxygen, the process is called *oxidation*. When the same process occurs with a considerable swiftness and exotherm chemical reaction, it is called *combustion*; whereas such a process with almost instantaneous action is called *detonation*.

The mechanical engineers are interested in combustion. Chemists are interested in oxygenation while designers of arms, ammunition, missiles etc. are interested in the process of detonation.

Main priorities of fuels that in general interest to an engineer are

1. Heat liberated by complete combustion i.e. the calorific value of the fuel
2. Combustion temperature
3. Air required for complete combustion and
4. Volume and composition of the products of combustion.

Combustion Temperature

The combustion temperature indicates the degree of calorific intensity of the fuel. Quicker the rate of burning of a fuel, larger is the calorific intensity. The combustion temperature is the temperature of the burning fuel as measured on the products of combustion. If C_{Pm} is the mean specific heat in kJ/kg-K of the products, W_P kg is the mass of the products of combustion per kg fuel burnt, Q kJ the amount of heat released by combustion, t_a the temperature of atmosphere, then the combustion temperature t_c is given by the equation

$$Q = m_p C_{Pm} (t_c - t_a)$$

The amount of air supplied per kg fuel, as well as the completeness of combustion influence the combustion temperature of fuels.

12.5.1 Chemical Equations (Combustion Equations) Involved in Combustion

The process of combustion of elements and compounds can be represented by the chemical equations. These equations are also called as combustion equations. The equations can be treated as those dealing with molar weights or molar volumes (kmol). These dealings can further be simplified by the fact, that equal volumes of different gases at same pressure and at any temperature sufficiently above the liquifying temperature of the respective gas, always contain the same number of molecules. Thus the relative weights of equal volumes of gases will be the same as their molar weights. Thus, at constant pressure, in case of complete combustion on carbon, is represented by

$$C + O_2 = CO_2$$

So that, 1 kmol of carbon + 1 kmol of oxygen = 1 kmol of carbon dioxide

This equation represents molecules.

Similarly, (negligible volume of C) + (1 volume of O_2) = (1 vol of CO_2)

This equation represents volumes.

Again, (1 molar weight of carbon) + (1 molar weight of oxygen) = (1 molar weight of carbndioxides)

This equation represents weights or masses.

Thus, (12 kg of carbon) + (32 kg of oxygen) = (44 kg of carbon dioxide)
or (1 kg of carbon) + (8/3 kg of oxygen) = (11/3 kg of carbon dioxide)

From this, we see that 1 kg of carbon requires 8/3 = 2.667 kg of oxygen to procedure 11 /3 3.667 kg of carbon dioxide or negligible volume of carbon combines with 1 cu.m (m^3) oygen and produces 1 m^3 of carbon dioxide. In these two cases, both total weight and total volume of the gases before combustion have remained the same even after combustion.

Let us now consider the case of hydrogen chemical or combustion equation is

$$2H_2O + O_2 = 2H_2O$$

i.e.	2 kmol + 1 kmol	=	2 kmol
or	$2\ m^3 + 1\ m^3$	=	$2\ m^3$ by volume
or	4 kg + 32 kg	=	36 kg by weight
or	1 kg + 8 kg	=	9 kg by weight

Thus, the total volume before combustion was ($2 + 1 = 3\ m^3$) and became $2\ m^3$ after combustion. Thus contraction took place during combustion. According to the principle of conservation of mass, the total weight before and after combustion is unaltered. Similarly, in case of some other combustion process expansion takes place during the process.

The above type of equations can also be made to show the amount of heat liberated or absorbed during the process, depending upon whether the process is exothermic or endothermic Thus, for example,

$$C + O_2 = CO_2 + 406976 \text{ kJ}$$

This means that complete combustion of 1 kmol (12 kg) of carbon produces 1 kmol (44 kg) of carbondioxide, thereby liberating 406976 kJ of heat. Similarly,

$$2H_2 + O_2 = 2H_2O + 573619 \text{ kJ}$$

means that two kmols (4 kg) of hydrogen liberate 573619 kJ heat during its complete combustion.

We will now, consider some more examples of combustions.

If carbon burns in insufficient supply of oxygen, the product of combustion will be carbon monoxide instead of carbon dioxide, and the process will be represented by the chemical equation

$$2C + O_2 = 2\,CO$$

$$\text{Carbon + Oxygen} = \text{Carbon monoxide}$$

i.e.

$$2 \times 12 + 32 = 56$$

$$1\,kg(C) + \frac{4}{3}kg(O_2) = \frac{7}{3}kg(CO)$$

The carbon monoxide produced during the above combustion is a fuel and therefore it should not be allowed to escape otherwise it would be a loss. It is highly poisonous and has no smell, therefore it is dangerous if it escapes.

Also, combustion of carbon monoxide

$$2CO + O_2 = 2CO_2$$

$$2 \times 28 + 32 = 2 \times 44 \text{ by weight}$$

$$1 + \frac{4}{7} = \frac{11}{7} \text{ kg by weight}$$

Sulphur is also one of the combustibles. It is of very minor importance in contributing to the calorific value of a fuel because only small quantities are present and its calorific value is very low being 9260 kJ. It has got injurious effects because it forms, on burning, sulphuric acid which corrodes the metal and therefore fuel having higher percentage of sulphar is undesirable particularly for steam raising purposes.

When sulphur burns with oxygen, it produces sulphur dioxide. The combustion of sulphur can be represented by a chemical equation

$$S + O_2 = SO_2$$

i.e. Sulphur + Oxygen = Sulphur dioxide

32 kg + 32 kg = 64 kg

1 kg + 1 kg = 2 kg

∴ Sulphur + Oxygen = Sulphur dioxide

12.6 MINIMUM QUANTITY OF AIR REQUIRED FOR THE COMPLETE COMBUSTION OF 1 KG OF SOLID OR LIQUID FUEL

It will be seen from the above discussion that an adequate supply of oxygen is essential for complete combustion. If the combustion is complete, then and then only maximum heat is available from the given fuel. The theoretically exact amount of oxygen required can be calculated with the help of various chemical or combustion equations discussed above.

These calculations will give us the minimum or theoretical quantity of oxygen if we know the ultimate analysis of the fuel.

The oxygen required for combustion is taken from the atmospheric air although in some cases a certain amount of oxygen is a constituent of the fuel. Air is a mixture of oxygen, nitrogen, a very amount of carbon dioxide and very small traces of rare gases such as neon, argon, krypton etc. For all practical purposes we assume that air is made up of 23% by weight of oxygen and the remaining 77% by weight of nitrogen. If considered by volume, air contains 21% of oxygen and 79% of nitrogen. Once we know the amount of oxygen necessary for complete combustion of 1 kg of fuel, we can determine the weight of air necessary for complete combustion of 1 kg fuel. This minimum or theoretical amount of air required for complete combustion of 1 kg, 1 kg fuel is also known as *stoichiometric air.*

Let C, H and S be the weights of carbon, hydrogen and sulpher contained in 1 kg of fuel

∴ O_2 required for C $= \frac{8}{3}\text{C kg} = 2.667 \text{ kg}$

O_2 required for H $= 8 \text{ H kg}$

and O_2 required for S $= \text{S kg}$

Therefore, total oxygen required for 1 kg fuel,

$$= \left(\frac{8}{3}\text{C} + 8\text{H} + \text{S}\right) \text{kg}$$

∴ Minimum quantity of air required for complete combustion of 1 kg fuel

$$= \frac{100}{23}\left(\frac{8}{3}\text{C} + 8\text{H} + \text{S}\right) \text{kg/kg fuel}$$

$$\therefore \qquad \text{Minimum air} = \frac{100}{23}\left(\frac{8}{3}C + 8H + S\right) \text{ kg/kg fuel}$$

If fuel contains 0 kg of oxygen then total minimum oxygen required/kg fuel

$$= \left(\frac{8}{3}C + 8H + S - 0\right)$$

$$= \left[\frac{8}{3}C + 8\left(H - \frac{0}{8}\right) + S\right]$$

Here $\left(H - \frac{0}{8}\right)$ is called the available hydrogen for combustion.

∴ Minimum amount of air required for complete combustion of 1 kg fuel.

$$(\text{Air})_{\min} = \frac{100}{23}\left[\frac{8}{3} + \left(H - \frac{0}{8}\right) + S\right]$$

12.7 EXCESS AIR

In the previous section we have calculated the minimum quantity of air required for complete combustion of one kg fuel or 1 cu.m of gaseous fuel. In actual practice we supply air more than the theoretical minimum amount in order to ensure the complete combustion of the fuel because all of the air supplied does not come into intimate contact with the particles of the fuel. Weight of air supplied over and above the theoretical air supplied, is called excess air. Generally this is given as a percentage of the theoretical air.

A large amount of excess air has a cooling effect on the process of combustion and represents a loss, and in order to avoid this cooling effect air is preheated before it enters the furnace of a boiler. With natural draught system the excess air is more in comparison with the artificial draught system. The excess air may approach 100% but the modern practice is to use 20–25% or even 50%. When the Lancashire type boiler uses hand firing with natural draught system 10 to 12% CO_2 in the flue gases would be considered good practice. With mechanical stokers and artificial draught it would be quite reasonable to expect 12 to 15% CO_2.

In case of I.C. engines all the air taken in during the induction, or suction stroke will not come in contact with the fuel particles. Therefore excess air is supplied and it should be reduced to minimum to get more specific output.

If the supply of air is less than the minimum required then the mixture is rich and if the supply of air is more than 30% in excess of the theoretical minimum, the mixture is known as lean or weak mixture.

12.8 MASS FRACTION AND MOLE FRACTION

Let *A*, *B*, and *C* be three gases in a mixture. Let each of these gases having a mass of m_a, m_b and m_c.

Then,

$$m = \text{Total mass of the mixture considered}$$

Then, mass fraction of the mass of the constituent is defined as the ratio of the mass of the constituent to that of the mixture and is denoted by m_f.

Thus, $$m = m_a + m_b + m_c$$

and mass fraction of gas A is given by

$$m_{fa} = \frac{m_a}{m}$$

Similarly $$m_{fb} = \frac{m_b}{m}$$

$$m_{fc} = \frac{m_c}{m}$$

$$\therefore \quad \sum m_{fa} = \frac{m_a}{m} + \frac{m_b}{m} + \frac{m_c}{m} = \frac{m_a + m_b + m_c}{m}$$

Mole is an unit used to express a quantity of gas whose mass is equal to its molecular weight in magnitude. Thus mass is given in kg, the one mole is represented by K-mol. Similarly, when the molecular weight is expressed in gm or pound, mole will be represented by gram-mole or g mole and pound-mole or p.mol. Thus one g mol of O_2 is 32 gm and p.mol, if H_2O is 18.0156 pounds and so on. The number of moles n, is obtained by dividing the mass m, in grams or pound–mass or kilograms, by the molecular weight M.

$$\therefore \quad n = \frac{m}{m}$$

Thus in a mixture of gases A, B and C if n_a, n_b and n_c are the moles, then the total number of moles is $n_a + n_b + n_c = n$ sag.

Then, mole fraction of a component in a gas mixture is defined as the ratio of number of moles of the component to the total number of moles of the mixture.

$$n_{fa} = \frac{n_a}{n}, n_{fb} = \frac{n_b}{n} \text{ and so on.}$$

12.9 FLUE GAS (EXHAUST GAS) ANALYSIS

When the combustion of the fuel takes place, the products of combustion will be carbon dioxide, (some times carbon monoxide may also be present) sulphur dioxide, vapour and nitrogen all the while, oxygen also will be present in the flue gases.

Generally, the composition of solid or liquid fuels is expressed by weight, whereas that of the gaseous fuels is expressed by volume. Products of combustion are in the gaseous state and the analysis of the products of combustion will be in volumes or volumetric analysis will be obtained.

In order to get the analysis of the exhaust gases by volume, first the analysis of the exhaust gases is found by weight. i.e. the weights of the individual products are first determined and then the percentage analysis of the gas is found. When the weight analysis is known, it can be converted into volumetric analysis by the help of Avogadro's hypothesis.

12.9.1 Analysis by Weight or Gravimetric Analysis

(a) Theoretical Air Supplied As before, let C, H, S and O be the weight of carbon, Hydrogen sulphur and oxygen per kg of fuel. Then the chemical equations can be written as

$$C + O_2 = CO_2$$

$$1\,\text{kg} + \frac{8}{3}\,\text{kg} = \frac{11}{3}CO_2$$

i.e. 1 kg carbon produces $\frac{11}{3}$ kg of CO_2

Similarly $$2H + O_2 = 2H_2O$$

$$1\text{ kg} + 8\text{ kg} = 9\text{ kg}$$

or 1 kg hydrogen produces 9 kg of H_2O

$$S + O_2 = O_2$$

$$1\text{ kg} + 1\text{ kg} = 2\text{ kg}$$

or 1 kg of sulphur produces 2 kg of SO_2

$$\text{Theoretical air required} = \frac{100}{23}\left[\frac{8}{3}C + 8H + S - O\right]$$

∴ Nitrogen present in the gas will be kg of fuel

$$N = \frac{77}{23}\left[\frac{8}{3}C + 8H + S - O\right]\text{kg/kg fuel}$$

∴ The total weight of the products per kg of fuel

$$= \frac{11}{3}C + 9H + 2S + \frac{77}{23}\left[\frac{8}{3}C + 8H + S - O\right]$$

$$= \text{P says}$$

∴ Gravimetric analysis of the products of combustion of 1 kg fuel is

$$CO_2 \rightarrow \frac{11}{3}C \times \frac{100}{P}$$

$$H_2O \rightarrow \frac{9H \times 100}{P}$$

$$SO_2 \rightarrow \frac{2S \times 100}{P}$$

$$N_2 \rightarrow \frac{77 \times 100}{23P}\left[\frac{8}{3}C + 8H + S - O\right]$$

All these quantities can be tabulated as given below.

Constituent of fuel	Wt. per kg/fuel	O_2 required	Product	% by weight of product
Carbon	C	$\frac{8}{3}C$	$CO_2 = \frac{11}{3}C$	$\frac{11}{3}C \times \frac{100}{P}$
Hydrogen	H	8 H	$H_2O = 9H$	$\frac{9H \times 100}{P}$
Sulphur	S	S	$SO_2 = 2H$	$2S \times \frac{100}{P}$
Oxygen	O	O	–	
Nitrogen	–	–	$N_2 = \frac{77}{23}\sum O_2$	$\frac{77}{23} \times \frac{100}{P}\sum O_2$

When H_2O vapours are included in the exhaust or flue gases, then the products of combustion are wet.

When H_2O vapours are excluded from the analysis, then the products of combustion are known as Dry Products of Combustion and in practice we consider the dry products of combustion. With the exclusion of the H_2O vapours, the table will be slightly modified accordingly. Note here that hydrogen has to be considered for finding the total oxygen required. N_2 will depend on this total oxygen required.

(b) Excess Air Supplied. Let x be the excess air supplied per kg of fuel. Then as before, consider 1 kg fuel to contain carbon (C), hydrogen (H), sulphur (S) and oxygen (O) and the products of combustion will be

$$CO_2 \rightarrow \frac{11}{3}C$$

$$H_2O \rightarrow 9H$$

$$SO_2 \rightarrow 2S$$

$$O_2 \rightarrow \frac{23}{100}x$$

$$N_2 \rightarrow \frac{77}{23}\left[\frac{8}{3}C + 8H + S - O + \frac{23}{100}x\right]$$

Total weight of wet products = P

∴ Percentage analysis by weight of the wet products of combustion will be

$$CO_2 = \frac{11}{3}C \times \frac{100}{P}$$

$$H_2O = 9H \times \frac{100}{P}$$

$$SO_2 = \frac{2S \times 100}{P}$$

$$O_2 = \frac{23}{100} x \times \frac{100}{P}$$

$$N_2 = \frac{77}{23}\left[\frac{8}{3}C + 8H + S - O + \frac{23}{100}x\right]$$

Similarly, the percentage analysis by weight, of the dry products of combustion will be

$$CO_2 = \frac{11}{3}C$$

$$SO_2 = 2S$$

$$O_2 = \frac{23}{100}x$$

$$N_2 = \frac{77}{23}\left[\frac{8}{3}C + 8H + S - O + \frac{23}{100}x\right]$$

Total weight of dry products = P_1

∴ Percentage analysis is given as

$$CO_2 = \frac{11}{3}C \times \frac{100}{P_1}$$

$$SO_2 = 2S \times \frac{100}{P_1}$$

$$O_2 = \frac{23}{100} \times \frac{100}{P_1}$$

$$N_2 = \frac{77}{23}\left[\frac{8}{3}C + 8H + S - O + \frac{23}{100}x\right] \times \frac{100}{P_1}$$

12.9.2 Analysis by Volume or Volumetric Analysis

Once we have the analysis of the products of combustion by weight, we can convert this into volumetric analysis of the products of combustion. Remember, for this purpose we make use of Avagadro's law or hypothesis (molecular weights of all gases at NTP occupy equal volumes). Therefore, weight or gravimetric analysis can be converted into volumetric analysis by dividing the weight of the gas by its molecule weight to get the volume of the gas.

Now consider the percentage analysis of the wet products.

Product Weight	Molecular Weight	Relative Volumes V
$CO_2 = \frac{11}{3}C \times \frac{100}{P}$	44	$\frac{11}{3}C \times \frac{1}{44}$
$H_2O = 9H \times \frac{100}{P}$	18	$9H \times \frac{1}{18}$
$SO_2 = 2S \times \frac{100}{P}$	64	$2S \times \frac{1}{64}$
$O_2 = \frac{23}{100} x \times \frac{100}{P}$	32	$\frac{23}{100} x \times \frac{1}{32}$
$N_2 = \frac{77}{23}\left[\frac{8}{3}C + 8H + S - O + \frac{23}{100}x\right] \times \frac{100}{P_1}$	28	$\frac{N_2}{28}$
		Total $= \Sigma V$

∴ % analysis by volume of the wet products of combustion is given as

$$CO_2 \rightarrow \frac{C}{12} \times \frac{100}{\Sigma V'} H_2O \rightarrow \frac{H}{2} \times \frac{100}{\Sigma V'} SO_2 \rightarrow \frac{S}{32} \times \frac{100}{\Sigma V}$$

$$O_2 \rightarrow \frac{23}{100} \times \frac{100}{32 \Sigma V}$$

$$N_2 \rightarrow \frac{77}{23}\left[\frac{8}{3}C + 8HS - O + \frac{23}{100}x\right]\frac{1}{28} \times \frac{100}{\Sigma V}$$

Similarly, volumetric analysis of the dry products of combustion will be as

	% analysis
$CO_2 \rightarrow \frac{11}{3}C \times \frac{1}{44}$	$\frac{C}{12} \times \frac{100}{\Sigma V_1}$
$SO_2 \rightarrow \frac{2S}{64}$	$\frac{S}{32} \times \frac{100}{\Sigma V_1}$
$O_2 \rightarrow \frac{23}{100} x \times \frac{1}{32}$	$\frac{23}{100} \times \frac{x}{32} \times \frac{100}{\Sigma V_1}$
$N_2 \rightarrow \frac{77}{23}\left[\frac{8}{3}C + 8HS - O + \frac{23}{100} \times\right] \times \frac{1}{28}$	$\frac{77}{23}\left[\frac{8}{3}C + 8HS - O + \frac{23x}{100}\right]\frac{100}{28x} \div \Sigma V_1$

Now, when the volumetric analysis of the products of combustion is known, we can convert the volumetric analysis into gravimetric analysis in converting volume into weight of the gas by multiplying the volume with the molecular weight of the gas. Thereby, weights of all constituents of the flue or exhaust gases are determined and therefore total weight of the gas and then the percentage composition of the gas can be determined. Table given below will show this procedure of converting volumetric analysis into gravimetric analysis.

The illustrative examples will give the reader idea about the conversion of gravimetric into volumetric analysis and vice-versa.

12.9.3 Conversion between Gravimetric and Volumetric Analysis

For converting the gravimetric analysis of the products of combustion into volumetric analysis and vice-versa, two tables are given below.

(a) Gravimetric to Volumetric Analysis

Sr. No.	Name of the gas	Molecular weight (a)	% composition by weight (b)	Relative volume $c = \frac{b}{a}$	% composition volume $= \frac{c}{\Sigma c} \times 100$
1					
2					
			$\Sigma b = 100$	Σc	100

(b) Volumetric to Gravimetric Analysis

Sr. No.	Name of the gas	Molecular weight (a)	% composition by weight (b)	Relative volume $c = a \times b$	% composition volume $= \frac{c}{\Sigma c} \times 100$
1					
2					
				Σc	100

12.9.4 Flue Gas Analysis (Using Orsat Appartus)

A simple and convenient apparatus used for analysis of dry flur gases by volume is called an Orsat Appartus (Fig. 12.5). It consists of a graduated measuring bottle also called as eudiometer, an aspirating bottle, three double reagant pipettes to absorb CO_2, O_2 and CO. Eudiometer is connected to aspirating bottle by means of rubber tube.

The first reagent pipette next to eudiometer contains the KOH solution (33% KOH + 67% water by weight) to absorb CO_2. The second pipette contains alkaline solution of pyrogallic acid (5 gm of pyrogallic acid in 15 cc of water + 33 gm KOH in 67 gm H_2O) to absorb O_2 and the third pipette contains acidic solution of cuprous chloride (5 cc of cuo + 1000 cc of HCl) to absorb CO.

Each pipette is made of two pipettes joined together by a glass tube at the bottom as shown. Front row of pipette is connected to the front flue gas tube and rear row of pipette are connected to another tube which runs parallel to flue gas sample tube. Each front pipette is provided with number of glass tubes inside to increase the wetted surface area, hence it accelerates action of the gas absorption.

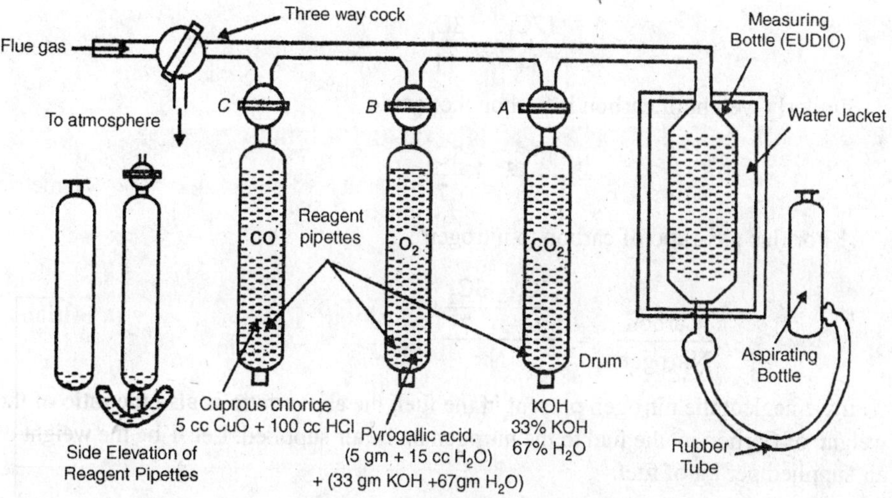

Fig. 12.5. *Orsat apparatus.*

Firstly, the existing air or the gas is expelled from the eudiometer by raising the aspirating bottle and keeping three way valve open to atmosphere. The three cocks *a, b, c* and a three way cock are closed. The aspirating bottle is lowered so that the eudiometer reads zero on graduated scale. The three way cock is open and the sample of flue gas is drawn in the measuring bottle. The flue gas is expelled by raising the aspirating bottle. Now the three way cock is kept in the closed position.

Now the cock *a* is opened and the sample of the gas is forced into the first reagent pipette containing KOH by raising the aspirating bottle. Here the CO_2 component of the flue gas is absorbed. Aspirating bottle is moved up and down several times to ensure the complete absorption of CO_2 by KOH solution in reagent pipette. The sample of gas is returned to eudiometer and cock *a* is closed by keeping the reagent to its original level of solution. The eudiometer reading is taken and the difference of two readings gives the percentage of CO_2 by volume in the gas sample.

The same procedure is repeated with the pipettes *b* and *c* to find the percentage analysis of O_2 and CO respectively in the flue gas. The remainder is N_2 gas.

Thus, Orsat apparatus gives analysis of the dry gas.

12.9.5 Determination of the Quantity of Air Supplied per kg of Fuel from the Analysis of Flue Gas given in Weight (Gravimetric Analysis Given)

Let C be the percentage of carbon in one kg of fuel and the percentage analysis of the dry flue gas by weight is as given below.

$$\text{Carbon dioxide} = C_1\%$$
$$\text{Carbon monoxide} = C_2\%$$
$$\text{Oxygen} = 0\,\%$$
$$\text{Nitrogen} = N_2$$

Weight of carbon present in carbon dioxide

$$= \frac{12C_1}{44} = \frac{3C_1}{11}$$

Similarly weight of carbon in carbon monoxide

$$= \frac{12C_2}{28} = \frac{3C_2}{7}$$

∴ For flue gas-ratio of carbon to nitrogen

$$\frac{\text{Carbon}}{\text{Nitrogen}} = \frac{\frac{3C_1}{11} + \frac{3C_2}{7}}{N}$$

If we neglect the nitrogen present in the fuel, the above ratio is also the ratio of the weight of Carbon in the fuel to the nitrogen in the air supplied. Let W be the weight of air supplied per kg of fuel.

∴ Weight of carbon in fuel = C/100

Weight of nitrogen in air = 0.77W

$$\therefore \quad \text{Ratio} = \frac{C/100}{0.77W}$$

$$\therefore \quad \frac{C}{77W} = \frac{\frac{3C_1}{11} + \frac{3C_2}{7}}{N}$$

$$\therefore \quad W = \frac{NC}{77\left[\frac{21C_1 + 33C_2}{77}\right]}$$

$$= \frac{NC}{21C_1 + 33C_2}$$

12.9.6 Determination of Actual Air Supplied per kg Fuel when Volumetric Analysis of the Dry Flue Gases is Given

In the power plants the amount of fuel used is very large and consequently the amount of air supplied to a boiler furnace is also great and so it cannot be measured directly. Its indirect measurement can be done if we know the volumetric analysis of the dry flue gas. The main constituents of the flue gases are carbon dioxide, carbon monoxide, oxygen and nitrogen. Quantity of sulphur dioxide is negligibily small and the water vapours are condensed and do not enter the flue gas.

Let C be the percentage by weight of carbon in the fuel. And the percentage volumetric analysis of the dry products of combustion is given as

$$\text{Carbon Dioxide} = C_1$$
$$\text{Carbnon Monoxide} = C_2$$
$$\text{Oxygen} = 0$$
$$\text{Nitrogen} = N$$

We will convert this volumetric analysis into gravimetric analysis and then the procedure is same as the above article.

∴ To convert volumetric analysis into gravimetric analysis, multiply volume by its molecular weights to get relative weights.

∴ Relative weights of these constituents are

$$\text{Carbon dioxide} = 44C_1$$
$$\text{Carbon monoxide} = 28C_2$$

∴ Carbon content of carbon dioxide

$$= \frac{12C_1}{44} \times \frac{44}{1} = 12C_1$$

Carbon contents of carbon monoxide

$$= \frac{12 \times 28C_2}{28} = 12C_2$$

$$\therefore \text{Total carbon contents} = 12(C_1 + C_2)$$

∴ In exaust or flue gases,

$$\text{Ratio} \frac{\text{Carbon}}{\text{Nitrogen}} = \frac{12(C_1 + C_2)}{28N} \qquad \text{(I)}$$

Let W be the weight of air supplied/kg fuel

$$\therefore \quad \text{Ratio} \frac{\text{Carbon}}{\text{Nitrogen}} = \frac{C/100}{0.77W} \qquad \text{(II)}$$

$$\therefore \quad I = II$$

$$\therefore \quad \frac{12(C_1 + C_2)}{28N} = \frac{C}{77W}$$

$$\therefore \quad W = \frac{C}{77} \times \frac{28N}{12(C_1 + C_2)}$$

$$= \frac{NC}{33(C_1 + C_2)}$$

$$\therefore \quad W = \text{Wt. of air supplied / kg fuel} = \frac{NC}{33(C_1 + C_2)}$$

Note: All values are in percentage

To find excess air supplied, first we have to consider the oxygen required for combustion of carbon monoxide to carbon dioxide and after deducting this oxygen from the oxygen present in the flue gases, the excess oxygen is obtained. Once this excess oxygen is obtained, then we can find the weight of excess air supplied.

The illustrative example given below will give you an procedure for accounting oxygen for the combustion of carbon monoxide to carbon dioxide.

Approximate equation for excess air supplied is

$$\text{Excess air supplied/kg fuel} = \frac{79°C}{21 \times 33(C_1 + C_2)}$$

This equation does take into consideration the complete combustion of carbon to carbon dioxide and hence approximation.

Note: All values are in percentages.

12.10 DETERMINATION OF AIR FUEL RATIO WITH THE HELP OF DRY FLUE GAS ANALYSIS

When the volumetric analysis of dry gas is given, then the amount of air supplied per kg of fuel can be obtained by any one of the following methods (Carbon content of the fuel is not given directly)

1. Carbon balance
2. Hydrogen balance
3. Carbon-hydrogen balance
   (i) with known composition of fuel
   (ii) with unknown composition of fuel
4. Oxidized product method

Of these method (2) and (3) required that three should be sufficient proportion of hydrogen in the fuel and hence they are not applicable for boiler fuel as coal.

We can write the chemical equation by considering a certain moles of fuel and coresponding moles of air such that the analysis of dry products given will be obtained. For balancing purpose, we have the consider the moles of H_2O produced.

(i) Carbon Balance In this method, carbons in the reactants (fuel) and the products are balanced and thereby getting the moles of fuel taken. The condition required for this method is that the formation of free carbon does not take place.

(ii) Hydrogen Balance After writing the combustion equation, we can proceed to balance the hydrogen also and the unknown.

(iii) Carbon-Hydrogen Balance This method is very much useful when the composition of the fuel is now known.

These methods will be better understood with the help of illustrative examples.

SOLVED EXAMPLES

(Theoretical Method)

Example 12.1 A boiler uses anthracite coal containing 8% moisture by weight. The higher calorific value of the coal is observed to be 26000 kJ/kg. Calculate the heat available from complete combustion of the fuel to the boiler flue gases, if the rate of firing in the boiler is 1500 kg per hour. Neglect the hydrogen content of the coal.

Solution

In case of complete combustion of a fuel, its lower calorific value is available as heat to the products of combustion. In the problem here, the moisture content of the coal is 8%.

$\therefore$ Weight of moisture/kg coal = 0.08 kg

$\therefore$ Lower calorific value of the coal

$$\begin{aligned} LCV &= HCV - 2466 \times 0.08 \\ &= 26000 - 197.28 \\ &= 25802.72 \text{ kJ/kg} \end{aligned}$$

$\therefore$ Total heat available to the flue gases per hour

$$\begin{aligned} &= \frac{\text{Coal}}{\text{Hour}} \times LCV \\ &= 1500 \times 25802.72 \\ &= 38.7041 \times 10^6 \text{ kJ/kg} \end{aligned}$$

Example 12.2 The composition of motor car petrol was found to be 86%C, 12% H_2, 0.1 % O_2 and 0.5% moisture by weight. Calculate the lower and higher calorific values of petrol.

Solution

Higher calorific value (HCV) given by,

$$HCV = 33915 \times C + 144452.5\left(H - \frac{0}{8}\right) + 9630\,S$$

$$= 33915 \times 0.86 + 144451.5\left[0.12 - \frac{0.001}{8}\right] + 0.0001 \times 9630$$

$$= 29166.9 + 17332.374 + 0.963$$

$$\mathbf{HCV = 46500.237 \text{ kJ/kg}}$$

$$\simeq 46500 \text{ kJ/kg Approx.}$$

$$\text{Moisture formed} = 0.005 + 9 \times 0.12$$
$$= 0.005 + 1.08 = 1.085 \text{ kg}$$

$\therefore$ Lower calorific value is given by

$$\text{LCV} = \text{HCV} - 2466 \times 1.085$$
$$= 46500 - 2675.61$$
$$\textbf{LCV} = \textbf{43924.39 kJ/kg}$$

Example 12.3 An analysis of a simple Diesel oil fuel is given by H_2 = 12.4%, carbon = 84.6%, sulphur =1.3% and incombustible substances 1.7%. Calculate the lower and higher calorific values of this fuel.

Solution

Higher calorific value of fuel is given by

$$\text{HCV} = 33915\,\text{C} + 144451.5\left(\text{H} - \frac{0}{8}\right) + 9630\,\text{S}$$
$$= 333915 \times 0.846 + 144451.5 \times 0.124 + 9630 \times 0.013$$
$$= 28692.09 + 17911.986 + 125.19$$
$$\textbf{HCV} = \textbf{46729.266 kJ/kg}$$
$$\text{Moisture formed} = 9\text{H} = 9 \times 0.124 = 1.116 \text{ kg}$$

Lower calorific value is given by

$$\text{LCV} = \text{HCV} - 2466\,(H_2O)$$
$$= 46729.266 - 2466 \times 1.116$$
$$= 46729.266 - 2752.056$$
$$\textbf{LCV} = \textbf{43977.210 kJ/kg}$$

Example 12.4 If the higher calorific value of carbon and hydrogen are 36000 kJ/kg and 142300 kJ/kg calculate the lower calorific value of methane, CH_4 kJ/kg.

Solution

One 1 kmol of CH_4 contains $12 \times 1 = 12$ kg of carbon and $1 \times 4 = 4$ kg of hydrogen.

$\therefore$ Wt. of 1 kmol of CH_4 = $12 + 4 = 16$ kg

$$\text{Carbon in 1 kg } CH_4 = \frac{12}{16} = 0.75 \text{ kg}$$

$$\text{Hydrogen in 1 kg } CH_4 = \frac{04}{16} = 0.25 \text{ kg}$$

Moisture or H_2O formed per kg of CH_4 is

$$9\text{H} = 9 \times 0.25 = 2.25 \text{ kg}$$

Now, heat produced by 0.75 kg carbon

$$= 0.75 \times 36000$$

$= 27000$ kJ

Heat produced by 0.25 kg hydrogen

$= 0.25 \times 142300$

$= 35575$ kJ

$\therefore$ Total heat produced by combustion of 1 kg

$= 2700 + 35575$

$= 62575$ kJ/kg

$\therefore$ LCV $=$ HCV $- 9H \times 2466$

$= 62575 - 2.25 \times 2466$

LCV = 57026.5 kJ/kg

Example 12.5 Calculate the higher calorific value of the fuel whose proximate analysis is given below as moisture 5% volatile matter 29%, fixed carbon 59%, ash 7%.

Solution

Goutel formula is to be used when the proximate analysis of the fuel is known.

$$\text{Percentage of fixed carbon} = \frac{59 \times 100}{59 + 29} = \frac{59}{88} \times 100 = 67\%$$

$\therefore$ Percentage of volatile matter on dry ash free basis = 100 – 67 = 33%

For dry as free basis volatile matter % of 33 will be

$\alpha = 410 - (410 - 394) \times 0.6$........ (Refer data Given)

$= 410 - 0.96$

$= 400.04$

Calorific value of fuel $= 343.3 \times 59 + 400.4 \times 29 = 20254.7 + 11611.6$

Calorific value of fuel = **31866.3 kJ/kg**

Example 12.6 Estimate the higher and lower calorific values of a gaseous fuel whose volumetric composition is CH_4 = 77%, C_2H_6 – 22.5%, nitrogen = 0.5%. The higher calorific values of the gases are given below – CH_4 = 40,000 kJ/m^3, C_2H_6 = 695000 kJ/m^3

Solution

Let us consider one cu-meter of gas. The volume of methane is 0.77 cubic metre and heat evolved = 0.77×40000 = 30800 kJ. The volume of ethane is 0.225 cubic metre and the heat evolved = 0.225×69500 = 15637.5 kJ. Nitrogen does not evolve any heat. Therefore the higher calorific value of gaseous fuel is 30800 + 15637.5 = **46437.5 kJ/Nm^3**.

Volume of water vapour formed due to combustion of 0.77 cu-m of CH_4 is 2×0.77 = 1.54 cu-m.

Volume of water vapour formed due to combustion of 0.225 cu-m of C_2H_6 = 0.225 × 3 = 0.675 cu-m

∴ Total volume of water vapour formed due to combustion of 1 cu-m gaseous fuel

$$= 1.54 + 0.675$$

$$= 2.215 \text{ cu-m.}$$

Wt. of one cu-m of water vapour $= \dfrac{18}{22.41}$ (1 molar volume = 22.41cu-m. (This ways 18 Kg (H_2O))

$$\therefore \quad \text{Wt. of 1 cu-m} = \frac{18}{22.41} = 0.803$$

∴ Total wt. of water vapour = 2.215 × 0.803 = 1.7786 kg

∴ Heat of water vapour formed/cu-m of gas fuel

$$= 1.7786 \times 2466$$

$$= 4386.14 \text{ kJ}$$

∴ Lower calorific value of the gaseous fuel is given by

$$\text{LCV} = \text{HCV} - 4386.14$$

$$= 46437.5 - 4386.14$$

$$\textbf{LCV} = \mathbf{42051.36\ kJ/Nm^3}$$

Example 12.7 The analysis of a sample producer gas shows 60 % CO, 20% H_2, 10% N_2, 1 % O_2, 2% CO_2 and rest moisture, by weight. If the higher calorific value of the gas was 1465 kJ/m³ and if the density of the gas was 1.4 kg/m³, calculate the lower calorific value of the producer gas. Neglect the steam formed by the combustion of hydrogen.

Solution

1 m³ gas weighs 1.3 kg/m³ and its moisture content is 7% by difference i.e.

$$1.3 \times 0.07 = 0.091 \text{ kg/m}^3$$

∴ Heat required to evaporate this moisture is

$$= 2466 \times 0.091$$

$$= 224.4 \text{ kJ}$$

∴ Lower calorific value of 1 Nm³ of producer gas is given by LCV = 1465 – 224.4

$$\text{LCV} = \mathbf{1240.6\ kJ/m^3}$$

Example 12.8 Estimate the higher and lower calorific value of a gas whose volumetric analysis is 75% CH_4, 24.5 % C_2H_2, and 0.5% N_2. Corresponding higher calorific values of CH_4, $C_2H_2$and $N_2$are 43000, 48600 and 0 kJ/m³ respectively. Heat required to evaporate 1 kg of water is 2466 kJ/kg for the condition after combustion.

Solution

Let us consider 1m^3. The volume of $CH_4 = 0.75$ m^3 and the heat evolved by its combustion is $0.75 \times 43000 = 32250$ kJ.

The volume of C_2H_2 is 0.245 m^3 and heat evolved by its combustion is $0.245 \times 48600 = 11907$ kJ. Nitrogen evolves no heat. Hence the higher calorific value of the gas is given by

$$HCV = 32250 + 11907$$
$$= 44157 \text{ kJ/m}^3$$

Similarly, volume of water vapour by burning of 0.75 m^3 of CH_4 is given by

$$CH_4 + 2O_2 = 2H_2O + CO_2$$
$$1 \text{ vol} = 2 \text{ vol}$$
$$(\text{vol})_{CH_4} = 2 \times 0.75 = 1.5 \text{ Nm}^3$$

Also, volume of water vapour formed by burning 0.245 m^3 of C_2H_2 is given by

$$C_2H_2 + 3O_2 = CO_2 + H_2O$$
$$1 \text{ vol} = 1 \text{ vol}$$
$$(\text{Vol})_{C_2H_2} = 0.245 \text{ m}^3$$

∴ Total water vapour in the product of combustion

$$= 1.5 + 0.245 = 1.745 \text{ Nm}^3$$

∴ Weight of this vapour is

$$= 1.745 \times \frac{18}{22.41}$$
$$= 1.4 \text{ kg}$$

∴ Heat required to evaporate this water

$$= 1.4 \times 2466$$
$$= 3456.36 \text{ kJ}$$

∴ Lower calorific value is given by,

$$LCV = 44157 - 3456.36$$
$$\mathbf{LCV = 40700.64 \text{ kJ/Nm}^3}$$

(Practical Methods)

Example 12.9 In a bomb calorimeter test following observations were recorded:

Weight of coal tested 2 gm

Weight of water in the calorimeter 1.2 kg

Water equivalent of the calorimeter 0.8 kg

Rise in temperature of jacket water 8.125°C

If the coal contains 2% moisture by weight, the room temperature is 20°C and 1 kg moisture at 0°C requires 2466 kJ to evaporate to form dry and saturated steam, calculate calorific value of the coal.

Solution

Total water equivalent of the calorimeter is

$$= 1.2 + 0.8$$

$$= 2 \text{ kg}$$

∴ Heat received by water

$$= m_w C_p \Delta T = 2 \times 4.187 \times 8.125$$

$$= 68.04 \text{ kJ} \quad (1)$$

Again heat given by coal

$$= \text{HCV} \times \text{wt. of fuel} \quad (2)$$

$$= \text{HCV} \times 0.002$$

∴ Therefore by balancing these two equations we get

$$\text{HCV} \times 0.002 = 68.04$$

or

$$\text{HCV} = \frac{68.04}{0.002}$$

$$\textbf{HCV} = \textbf{34020 kJ/kg}$$

Now the heat that would be consumed to evaporate the moisture per kg coal

$$= 2466 \times 0.02$$

$$= 49.32 \text{ kJ}$$

∴ The lower calorific value of the coal is given by

$$\text{LCV} = \text{HCV} - 49.32$$

$$\textbf{LCV} = \textbf{34020} - \textbf{49.32} = \textbf{33970.68 kJ/kg}$$

Example 12.10 During bomb calorimeter test on diesel oil according to the BSS specifications, following data were recorded:

Room temperature 25°C
Weight of the crucible 8.116 gm
Weight of the crucible and oil 8.702 gm
Weight of can 1.051 kg
Water equivalent of the can 0.559 kg
Weight of can and water 3.492 kg
Rise in temperature of the can and water 2.305°C
Fuse correction for the heat supplied by the fuse to the bomb 0.12 kJ
Find the higher calorific value of diesel oil.

Data:

$$\text{Weight of oil taken} = 8.702 - 8.116$$
$$= 0.586 \text{ gm}$$
$$= 0.000586 \text{ kg}$$

Solution

$\therefore$ Heat released by the combustion of oil

$$= \text{HCV} \times 0.000586 \text{ kJ}$$

Heat received by the bomb from fuse = 0.12 kJ

weight of water in the can

$$= 3.492 - 1.051$$
$$= 2.441 \text{ kg}$$

$\therefore$ Total equivalent of can and water

$$= 2.441 + 0.559$$
$$= 3.000 \text{ kg}$$

$\therefore$ Heat received by can and water

$$= m_w C_{P_w} \times \text{Rise in temp.}$$
$$= 3 \times 4.187 \times 2.305$$
$$= 28.9531 \text{ kJ}$$

$\therefore$ Equating heat evolved and received by water we get

$$\text{HCV} \times 0.000586 + 0.12 = 28.9531$$

$$\therefore \quad \text{HCV} = \frac{28.9531 - 0.12}{0.000586} = \frac{28.8331}{0.000586}$$

$$\therefore \quad \textbf{HCV} = \textbf{49203.242 kJ/kg}$$

Example 12.11 The following particulars refer to an experimental determination of the calorific value of a sample of dry coal containing 88% carbon and 4.2 % hydrogen by weight.

Weight of coal taken 0.78 gm

Weight of fuse wire 0.02 gm

Calorific value of the fuse 6.7 kJ/gm

Weight of water in the calorimeter 1.9 kg

Water equivalent of calorimeter 0.35 kg

Observed temperature rise 2.98°C

Cooling correction 0.016°C

Room temperature 20°C

Find the lower calorific value of the sample. Assume the hydrogen content of the fuel requires, 2600 kJ heat to convert 1 kg water at 0°C to dry and saturated steam.

Solution

$$\text{Observed temp. rise} = 2.98°\text{C}$$

$$\therefore \text{ Corrected temp. rise} = 2.98 + 0.16$$

$$= 2.996°\text{C}$$

Similarly corrected weight of water

$$= 1.9 + 0.35$$

$$= 2.25 \text{ kg}$$

$\therefore$ Heat received by water and calorimeter

$$= \text{weight} \times \text{sp. ht} \times \text{Rise in temp.}$$

$$= 2.25 \times 4.187 \times 2.996$$

$$= 28.255 \text{ kJ}$$

$$\text{Heat supplied by fuse wire} = 0.02 \times 6.7 = 0.134 \text{ kJ}$$

$$\text{Heat evolved by coal} = 0.78 \times \text{HCV}$$

Then equating the heat evolved and heat gained by water, we get

$$0.78 \times \text{HCV} + 0.134 = 28.225$$

or

$$\frac{0.78\text{HCV}}{1000} = 28.225 - 0.134$$

$$= 28.091$$

$$\therefore \quad \text{HCV} = \frac{28.091}{0.78} \times 1000$$

$$\therefore \quad \text{HCV} = 36014 \text{ kJ/kg}$$

Heat required to evaporate water above 20°C

$$= 2466 \times \text{Mass of water}$$

$$= 9 \times 0.042 \times [2466 - (20 \times 4.18)]$$

$$= 9 \times 0.042 \times [2466 - 83.74]$$

$$= 900.49 \text{ kJ}$$

$\therefore$ Lower calorific value of coal is given by

$$\text{LCV} = \text{HCV} - 900.49$$

$$= 36014 - 900.49$$

$$\textbf{LCV} = \textbf{3513.51 kJ}$$

Example 12.12 In a test to determine the calorific value of a produce gas, the following observations were made.

Volume of gas burned per hour at 18°C and 100 mm H_2O gas pressure above atmosphere pressure of 750 mm Hg = 500 litres.

Volume of water flowing per minute =2360 cm^3

Temperature of incoming water = 15°C

Temperature of outgoing water = 16.5°C

Rate of condensation water = 4.2 cm^3/min

Calculate the lower calorific value of the gas at NTP. Allow 2466 kJ/kg steam evaporated at 0°C.

Solution

Heat taken or received by water/min

$$= \frac{2360}{1000} \times 4.18 \times (16.5 - 15)$$

$$= 2.36 \times 4.187 \times 1.5$$

$$= 14.822 \text{ kJ/min}$$

Now we will convert actual conditions of gas to NTP conditions

$$\text{Absolute pressure of gas} = 750 + \frac{100}{13.6}$$

$$= 750 + 5.353$$

$$= 757.353 \text{ mm Hg}$$

$\therefore$ We can write $$\frac{P_0 V_0}{T_0} = \frac{PV}{T}$$

$\therefore$ Volume of gas NTP $= V_0 = \frac{PV}{T} \times \frac{T_0}{P_0}$

$$= \frac{757.353 \times 500}{273 + 18} \times \frac{273}{760}$$

$$V_0 = 467.44 \text{ litres/hour}$$

$\therefore$ Heat supplied by the fuel/hour

$$= \frac{\text{HCV} \times 497.44}{1000} = \text{HCV} \times 0.46744 \text{ kJ/hr}$$

$\therefore$ Heat given by gas = Heat received by water

$$\text{HCV} \times 0.46744 = 14.822 \times 60$$

$$\text{HCV} = \frac{14.822 \times 60}{0.46744}$$

Again heat supplied by fuel to steam/min

$$= \left(\frac{4.2}{106} \times 1000\right) \times (2466 - 18 \times 4.187)$$

$$= 0.0042 \times (2466\ 75.366)$$

$$= 0.0042 \times 2390.634 = 10.041 \text{ kJ/min}$$

$$\text{LCV} = \frac{(14.822 - 10.041) \times 60}{0.46744}$$

$$= \frac{4.781 \times 60}{0.46744}$$

$$\textbf{LCV} = \textbf{613.68 kJ/Nm}^3$$

Example 12.13 In a Boy's Gas calorimeter 500 litres of water gas at 27°C, 75 mm of Hg about the atmosphere pressure of 750 mm Hg, was burned during a test conducted to determine the lower calorific value of water gas. During the test 0.3 kg of water was collected from the products of combustion. The circulating was admitted at 30°C and left at 47°C, the rate of its flow being 250 kg during the test.

Calculate the higher and lower calorific values of water gas. Assume that 2466 kJ of heat of required to convert 1 kg water at 0°C to dry and saturated steam.

Solution

$$\text{Pressure of gas} = 750 + 75 = 825 \text{ mm Hg}$$

From the relation $\dfrac{P_0 V_0}{T_0} = \dfrac{PV}{T}$

$$V_0 = \text{Volume of gas at NTP}$$

$$= \frac{PV}{T} \times \frac{T_0}{P_0}$$

$$= \frac{825}{760} \times \frac{273}{303} \times \frac{500}{1000} \text{ Nm}^3$$

$$= 0.489 \text{ Nm}^3$$

Heat given by gas is given by

$$= \text{HCV } 0.489$$

$\therefore$ HCV = Higher calorific value of gas

$$= \frac{177794.95}{0.489}$$

$$\textbf{HCV} = \textbf{36390 kJ}$$

Condensed water from the products of combustion leaving the calorimeter is at 27°C

$\therefore$ Heat rejected by 1 kg steam in condensing to water at 27°C is

$$= 2466 - 4.187 \times 27$$

$$= 2466 - 113.049$$

$$= 2352.951 \text{ kJ}$$

$\therefore$ Total heat recovered from condensate during the test

$$= 2352.951 \times 0.3$$
$$= 705.885 \text{ kJ}$$

Therefore heat supplied by gas to water

$$= 177794.75 - 705.885$$
$$= 17088.865 \text{ kJ}$$

$$\text{LCV} = \text{Lower calorific value}$$
$$= \frac{17088.865}{0.489}$$
$$\textbf{LCV} = \textbf{34946 kJ/Nm}^3$$

Example 12.14 Write the equations of combustion for CH_3, C_6H_{14} and CH_3OH. Find the amount of oxygen required in terms of kg/kg fuel for complete combustion in each case.

Assume constant pressure. Find also air to fuel ratio in each case.

Solution

(a) CH_3: We will require 4 K moles of O_2

$$4CH_3 + 7O_2 = 4CO_2 + 6H_2O$$
$$4\,(12 + 3) + 7 \times 32 = 4\,(44) + 6 \times 18$$

or $$60 \text{ kg} + 224 \text{ kg} = 176 + 108$$

∴ Required ratio of oxygen to fuel

$$= \frac{224}{60} = 3.733$$

∴ Air to fuel ratio

$$A/F = \text{Mass of Air/kg fuel}$$
$$= \frac{100}{23} \times 3.733$$
$$\textbf{A/F} = \textbf{16.232 kg/kg fuel}$$

(b) C_6H_{14}: Chemical equation will be as given

$$2C_6H_{14} + 19O_2 = 12CO_2 + 14H_2O$$

$$\therefore 2\,(12 \times 6 + 1 \times 14) + 19 \times 32 = 12 \times 44 + 14 + 18$$
$$\therefore \quad 2\,(72 \times 14) + 608 = 528 + 252$$
$$\therefore \quad 2 \times 84 + 608\,(O_2) = 780$$

∴ Oxygen to Fuel ratio

$$= \frac{608}{172} = 3.535$$

$$\therefore \quad \text{A/F ratio} = \frac{100}{23} \times 3.535$$

$$\textbf{A/F ratio} = \textbf{15.37 kg/kg fuel}$$

(c) CH_3OH : Chemical equation is as given

$$2CH_3OH + 3O_2 = 2CO_2 + 4H_2O$$

$$2(12 + 4 + 16) + 3 \times 32 = 2 + 44 + 4 \times 18$$

$$(64) + (96) = 88 + 72$$

$$(\text{fuel}) \; (O_2)$$

$$\text{Required } \frac{\text{Oxygen}}{\text{Fuel}} = \frac{96}{64}$$

$$= 1.5$$

$$\therefore \quad \text{A/F} = 1.5 \times \frac{100}{23}$$

$$\textbf{A/F} = \textbf{6.522}$$

Example 12.15. Calculate the volume of oxygen required for complete combustion of $1 m^3$ of C_2H_2, C_2H_4 and CO at constant pressure.

Solution

(a) C_2H_2

The chemical equation will be

$$2C_2H_2 + 5O_2 = 4CO_2 + 2H_2O$$

i.e. $\quad 2 \text{ K mol} + 5 \text{ K mol} = 4 \text{ K mol} + 2 \text{ K mol}$

or $\quad 1 \text{ m}^3 + 2.5 \text{ m}^3 = 4 \text{ m}^3 + 2 \text{ m}^3$

$\therefore$ Volume of O_2 required is 2.5 m^3/m^3 of C_2H_2

(b) CO

Again the chemical equation will be

$$2CO + O_2 = 2CO_2$$

i.e. $\quad 2 \text{ K mol} + 1 \text{ K mol} = 2 \text{ K mol}$

$$1\text{m}^3 + \frac{1}{2}\text{m}^3 = 2\text{m}^3$$

$\therefore$ Volume of O_2 required per 1 m^3 of CO is 0.5 m^3.

Example 12.16 Calculate the amount of air required for combustion of producer gas containing 25% CO, 10% CO_2, 17% H_2, 5% O_2 and 43%,N_2 by volume. Assume that the combustion takes place at constant pressure and air contains 21% oxygen by volume.

Solution

We will write chemical or combustion equations for individual constituent of the produced gas.

CO_2: $\quad 2CO + O_2 = 2CO_2$

$1\ m^3 + 0.5\ m^3$

H_2: $\quad 2H + O_2 = 2H_2O$

$$1\ m^3 + 0.5\ m^3 = 2\ m^3$$

∴ Total volume of O_2 required for 1 m^3 gas

$$= 0.25 \times 0.5 + 0.17 \times 0.5$$
$$= 0.125 + 0.085$$
$$= 0.210\ m^3 \text{ of } O_2$$

Oxygen present in the gas is 0.05 m^3.

∴ Net volume of O_2 required

$$= 0.210 - 0.05$$
$$= 0.160\ m^3$$

∴ Air required for complete combustion

$$= \frac{0.16}{0.21} = 0.762\ m^3 \text{ per cu-m of producer gas}$$

Air required for complete combustion = 0.762 cu-m of air/m³ gas

Example 12.17 A diesel fuel contains 70% C, 10% H_2, 5% O_2, 1% S and rest incombustible by weight. If air contains 23% oxygen by weight, find the amount of air required for complete combustion of 1 kg fuel.

Solution

Total oxygen required for complete combustion of 1 kg fuel is given by the relation

$$= \left(\frac{8}{3}C + 8H + S - O\right)$$

$$= \frac{8}{3} \times 0.7 + 8 \times 0.1 + 0.01 - 0.05$$

$$= 1.867 + 0.8 + 0.01 + 0.05 = 2.627 \text{ kg}$$

∴ Weight of air required for complete combustion of 1 kg fuel

$$W_a = \frac{100}{23} \times 2.627$$

$$\mathbf{W_a = 11.422\ kg/kg\ fuel}$$

Example 12.18 Determine the theoretical weight of air required for complete combustion of 1 kg of coal whose analysis by weight is given as under:

Carbon – 83%, Hydrogen – 5%, Oxygen – 2%, Sulphur – 0.2% remainder being incombustible.

Solution

In one kg of coal, the constituents by weight are

$$\begin{aligned} \text{Carbon} &= 0.83 \text{ kg} \\ \text{Hydrogen} &= 0.05 \text{ kg} \\ \text{Oxygen} &= 0.02 \text{ kg} \\ \text{Sulphur} &= 0.002 \text{ kg} \end{aligned}$$

∴ Minimum quantity of air required for complete combustion of 1 kg fuel

$$= \frac{100}{23}\left[\frac{8}{3}C + 8H + S - O\right]$$

$$= \frac{100}{23}\left[\frac{8}{3} \times 0.83 + 8 \times 0.05 + 0.002 - 0.02\right]$$

$$= \frac{100}{23}[2.2133 + 0.4 + 0.002 - 0.02]$$

$$= \frac{100}{23} \times 2.5953 = 11.284 \text{ kg/kg fuel}$$

Example 12.19 Determine the theoretical weight of air required for the complete combustion of one litre of petrol; given that the composition of petrol by weight is 86% carbon, and 14% hydrogen specific gravity of petrol being 0.8. If one kg or air at atmospheric temperature and pressure occupies 0.77 cu-m. Find also the volume of air supplied.

Solution

Constituent of 1 kg petrol are 0.86 kg carbon, 0.14 kg of hydrogen.

∴ Minimum quantity of air required /kg fuel

$$= \frac{100}{23}\left[\frac{8}{3}C + 8H\right]$$

$$= \frac{100}{23}\left[\frac{8}{3} \times 0.86 + 8 \times 0.14\right]$$

$$= \frac{100}{23}[2.2933 + 1.12] = \frac{100}{23} \times 3.4133$$

$$= 14.84 \text{ kg/kg fuel}$$

$$\text{1 litre of petrol} = 0.8 \text{ kg}$$

∴ Minimum quantity of air required/litre

$$= 0.8 \times 14.84$$

Minimum quantity of air = 11.87 kg

∴ Minimum volume of air required/litre

$$= 11.87 \times 0.77$$

Minimum quantity of air = 9.142 m^3/litre fuel

Example 12.20 If the analysis by volume of a product of combustion is 20% CO_2, 3.8% CO 70% N_2 and 6.2% O_2 find, the analysis by weight.

Solution

This example can be solved by tabulating the various quantities.

Substance	Volume/m^3 (b)	Mol. Weight (b)	Proportional wt =(a) × (b)	Wt/kg of gas × 100
CO_2	0.2	44	$0.2 \times 44 = 88$	27.983
CO	0.038	28	$0.038 \times 28 = 1.064$	3.383
N_2	0.7	28	$0.7 \times 28 = 19.6$	62.32
O_2	0.062	32	$0.062 \times 32 = 1.984$	6.31
			$\sum P = 31.448$	99.996

Thus the percentage composition of the products by weight is

$$CO_2 = 27.983\%$$
$$CO = 3.383\%$$
$$N_2 = 62.32\%$$
$$O_2 = 6.31\%$$

Example 12.21 A sample of water-gas contains 38% H_2, 23% CO, 17% CH_4, 9% C_2H_6, 6% H_2O, 4% CO_2 and 3% N_2. Convert this to analysis by weight.

Solution

We will make a table as in Ex. 12.21

S. No.	Substance	Vol/m^3 of gas (a)	Mol. Weight (b)	Proportional weight (a)× (b)	% by weight = $\frac{(a)\times(b)}{\sum(a\times b)}\times 100$
1	H_2	0.38	2	0.76	4.6626
2	CO	0.23	28	6.44	39.5100
3	CH_4	0.17	16	2.72	16.6781
4	C_2H_6	0.09	30	2.70	16.5644
5	H_2O	0.06	18	1.08	6.6258
6	CO_2	0.04	44	1.76	10.7975
7	N_2	0.03	28	0.84	5.1534
				$\sum(a\times b)$ =163	100.00

∴ Percentage analysis of water gas by weight is

H_2 = 4.6626%, CO = 39.5100%, CH_4 = 16.6871 %, C_2H_6 = 165644%, H_2O = 6.6258%, CO_2 = 10.7975 %, N_2 = 5.1534%.

Example 12.22. The composition of a dry flue gas, as obtained by using Orsat apparatus was 10% CO_2, 1.8 % CO, 7% O_2, and 81.2 % N_2 by volume. What will be the percentage composition of the flue gas by weight ?

Solution

Relative weight of CO_2 = $0.1 \times 44 = 4.4$ kg

Relative weight of CO = $0.018 \times 28 = 0.504$ kg

Relative weight of O_2 = $0.07 \times 32 = 2.24$ kg

Relative weight of N_2 = $0.812 \times 28 = 22.736$ kg

∴ Total weight of the gas = 29.88 kg

$$\therefore \text{Percentage of } CO_2 \text{ in gas} = \frac{4.4 \times 100}{29.88} = 1477$$

$$\therefore \text{Percentage of CO in gas} = \frac{0.504 \times 100}{29.88} = 1.687$$

$$\therefore \text{Percentage of } O_2 \text{ in gas} = \frac{2.24 \times 100}{29.88} = 7.497$$

$$\therefore \text{Percentage of } N_2 \text{ in gas} = \frac{22.736 \times 100}{29.88} = 76.091$$

Total = 100

Example 12.23 The diesel oil used in an engine consisted of 82% C, 7% O_2, 8% H_2 and 3% H_2O by weight. Convert this to analysis by volume.

Solution

First convert weight into volume by dividing weight of the substance by its molecular weight. Thereby relative volumes will be obtained.

$$\therefore \quad 82\% \text{ of carbon} = \frac{82}{44} = 1.8636 \text{ volume}$$

$$7 \text{ kg of } O_2 = \frac{7}{32} = 0.2188 \text{ volume}$$

$$8 \text{ kg } H_2 = \frac{8}{2} = 4.0000 \text{ volume}$$

$$3 \text{ kg } H_2O = \frac{3}{18} = 0.1667 \text{ volume}$$

Total volume = 6.2491

∴ Volumetric analysis will be

$$\text{Carbon} = \frac{1.8636 \times 100}{6.2491} = 29.822\%$$

$$\text{Oxygen} = \frac{0.2188 \times 100}{6.2491} = 3.501\%$$

$$\text{Hydrogen} = \frac{4 \times 100}{6.2491} = 64.009\%$$

$$H_2O = \frac{0.1667 \times 100}{6.2491} = 2.668\%$$

Example 12.24 In a boiler trial the fuel analysis was carbon 88%, hydrogen 3%, sulphur 0.5% by weight and the remainder being ash. Determine the weight of air required per kg of fuel for complete combustion. If the actual supply of air is 50% in excess of this, estimate the percentage analysis of the dry fuel gas by

(a) by weight

(b) by volume

Solution

We can write the chemical equations for the minimum, theoretical air required for complete combustion of 1 kg fuel.

(i) $C + O_2 = CO_2$

1 kg of carbon requires 8/3 kg of carbon dioxide.

$$\therefore \quad O_2 \text{ required} = \frac{8}{3} \times 0.88 = 2.347 \text{ kg}$$

$$\text{and produces } \frac{11}{3} \times 0.88 = 3.227 \text{ kg}$$

(ii) $H_2 + O = H_2O$

1 kg of hydrogen requires 8 kg of xygen and produces 9 kg H_2O.

$$\therefore \quad O_2 \text{ required} = 8 \times 0.03 = 0.24 \text{ kg}$$

$$\text{and } H_2O \text{ produced} = 9 \times 0.03 = 0.27 \text{ kg}$$

(iii) $S + O_2 = SO_2$

∴ 1 kg of sulphur requires 1 kg of oxygen

∴ Oxygen required for 0.005 kg of sulphur

$$0.005 \times 1 = 0.005 \text{ kg and produces}$$

$$2S = 2 \times 0.005 = 0.01 \text{ kg of sulphurdioxide.}$$

∴ Total minimum oxygen required for complete combustion of 1 kg coal

$$= 2.347 + 0.24 + 0.005$$

$$= 2.592 \text{ kg}$$

Now 50% of the minimum air is supplied as excess air

∴ Actual weight of air supplied / kg fuel

$$= 1.5 \times 11.261$$

$$= 16.89 \text{ kg}$$

With this actual air supplied, the weights of the dry products of combustion are

$$\text{Carbon dioxide} = 3.227 \text{ kg}$$

$$\text{Sulphur dioxide} = 0.010 \text{ kg}$$

$$\text{Oxygen} = 0.5 \times 2.592$$

$$= 1.296 \text{ kg}$$

$$\text{Nitrogen} = 0.77 \times 16.89$$

$$= 13.005 \text{ kg}$$

Total weights of the dry products

$$= 3.227 + 0.01 + 1.296 + 13.005$$

$$= 17.5383 \text{ kg}$$

∴ Percent analysis by weight of the dry products is

$$\text{Carbon Dioxide} = \frac{3.227}{17.5383} \times 100 = 18.402$$

$$\text{Sulphur Dioxide} = \frac{1}{17.5383} = 0.057$$

$$\text{Oxygen} = \frac{129.6}{17.5383} = 7.389$$

$$\text{Nitrogen} = \frac{1300.5}{17.5383} = \frac{74.152}{100.00}$$

We can get relative volumes by dividing weights by molecular weights.

∴ Relative volumes are

$$\text{Carbon dioxide} = \frac{18.402}{44} = 0.4182$$

$$\text{Sulphar dioxide} = \frac{1}{64} = 0.0156$$

$$\text{Oxygen} = \frac{7.389}{32} = 0.2309$$

$$\text{Nitrogen} = \frac{74.152}{28} = \frac{2.6483}{\Sigma = 3.313}$$

Percent analysis by volume of the products

$$\text{Carbon dioxide} = \frac{0.4182}{3.13} \times 100 = 12.6230$$

$$\text{Sulphur dioxide} = \frac{0.0156}{313} \times 100 = 0.4708$$

$$\text{Oxygen} = \frac{0.2309}{313} \times 100 = 6.9695$$

$$\text{Nitrogen} = \frac{2.6483}{313} \times 100 = \frac{79.9367}{1.00}$$

Example 12.25 A sample of coal has the following analysis by weight : Carbon 84%, Hydrogen 5%, incombustible material 11%. This coal is supplied to a furnace and burnt with a volume of air 50% in excess of that theoretically required for complete combustion. Estimate the volumetric composition of the dry flue gas.

Solution

We now can write directly the equation for theoretical air required for complete combustion of 1 kg fuel.

$$\text{Theoretical air/kg fuel} = \frac{100}{23}\left[\frac{8}{3}C + 8H\right]$$

$$= \frac{100}{23}\left[\frac{8}{3} \times 0.84 + 8 \times 0.05\right]$$

$$= 11.4783 \text{ kg}$$

Here excess air supplied is 50 % by volume. Volume of theoretical air can be determined and 50 % of that can be obtained. This excess volume again will be converted into weight which will be 50 %, by weight, of the theoretical air.

∴ We will consider 50 % by weight, excess air.

∴ Actual weight of air supplied / kg fuel

$$= 1.5 \times 11.4783$$

$$= 17.2174 \text{ kg}$$

Now first find the flue gas analysis by weight and then this is to be converted into volumetric analysis.

$$\text{Weight of Carbon Dioxide} = \frac{11}{3}C = \frac{11}{3} \times 0.84$$

$$= 3.08 \text{ kg}$$

$$\text{Weight of oxygen} = 0.5 \times 2.64 = 1.32 \text{ kg}$$

$$\text{Weight of nitrogen} = 0.77 \times 17.2174$$

$$= 13.26 \text{ kg}$$

$\therefore$ Total weight of flue gases produced per kg of fuel

$$= 3.08 + 1.32 + 13.26$$

$$= 17.66 \text{ kg}$$

$$\text{Volume of carbon dioxide} = \frac{3.08}{44} = 0.07$$

$$\text{Weight of oxygen} = \frac{1.32}{32} = 0.04125$$

$$= 1.32 \text{ kg}$$

$$\text{Weight of nitrogen} = \frac{13.26}{28} = 0.4736$$

$$= 13.26 \text{ kg}$$

$$\text{Total} = \Sigma = 0.58485$$

$\therefore$ Percentage analysis of the dry products of combustion is given as

$$\text{Carbon dioxide } CO_2 = \frac{0.07 \times 100}{0.58485} = 11.9$$

$$\text{Oxygen } O_2 = \frac{0.04125 \times 100}{0.58485} = 7.053$$

$$\text{Nitrogen } N_2 = \frac{0.4738 \times 100}{0.58485} = 81.0122$$

Example 12.26 A petrol has a composition by weight of 86% carbon and 14% hydrogen. When used in an engine the air supply is 90% of that theoretically required for complete combustion. Assuming that all the hydrogen is burnt and that the carbon burns to carbon monoxide and carbon dioxide so that there is no free carbon left. Calculate the percentage analysis of dry exhaust gases by volume.

Solution

Theoretically air required for complete combustion of 1 kg fuel

$$= \frac{100}{23}\left[\frac{8}{3}C + 8H\right] = \frac{100}{23}\left[\frac{8}{3}\times 0.86 + 8\times 0.14\right]$$

$$= \frac{100}{23}\times[2.2933 + 1.12]$$

$$= \frac{100}{23}\times 3.4133$$

$$= 14.84 \text{ kg}$$

∴ Actual air supplied/kg fuel

$$= 0.9\times 14.84$$
$$= 13.3564 \text{ kg}$$

∴ Actual oxygen supplied/kg fuel

$$= 0.9\times 3.4133$$
$$= 3.072 \text{ kg}$$

∴ Actual nitrogen supplied/kg fuel

$$= 0.77\times 13.3564$$
$$= 10.2844 \text{ kg}$$

Oxygen used for burning of hydrogen

$$8H = 8\times 0.14$$
$$= 1.12 \text{ kg}$$

∴ Oxygen left for combustion of carbon

$$= 3.072 - 1.12$$
$$= 1.952 \text{ kg}$$

Let C be weight of carbon burnt to CO_2

∴ (0.86 – C) will be weight of carbon burnt carbon monoxide.

∴ Total oxygen required for this combustion

$$= \frac{8}{3}C + \frac{4}{3}(0.86 - C)$$

$$= \frac{8}{3}C + \frac{3.44 - 4C}{3} = \frac{4C + 3.44}{3}$$

$$\therefore \quad 1.952 = \frac{4C + 3.44}{3}$$

$$\therefore \quad 4C = 3\times 1.952 - 3.44$$

$$\therefore \quad = 5.86 - 3.44 = 2.416$$

$$\therefore \quad C = \frac{2.416}{4} = 0.604 \text{ kg}$$

∴ Weight of carbon burnt carbon monoxide.

$$= 0.86 - 0.604 = 0.256 \text{ kg}$$

Weight of carbon dioxide produced / kg fuel

$$= \frac{11}{3} \times 0.604 = 2.215 \text{ kg}$$

Weight of carbon monoxide produced / kg fuel

$$= \frac{7}{3} \times 0.256 = 0.597 \text{ kg}$$

Weight of nitrogen = 10.2844 kg

∴ Relative volumes produced are

$$CO_2 = \frac{2.215}{44} = 0.0503$$

$$CO = \frac{0.597}{28} = 0.0213$$

$$N_2 = \frac{10.2844}{28} = 0.3673$$

Total relative volumes = 0.4389

∴ Percentage analysis of dry products of combustion is as

$$CO_2 = \frac{0.0503}{0.4389} \times 100 = 11.46\%$$

$$CO = \frac{0.0213}{0.4389} \times 100 = 4.85\%$$

$$N_2 = \frac{0.3673}{0.4389} \times 100 = 86.69\%$$

Total = 100.00

Example 12.27 A sample of coal burnt in a Lancashire boiler has the following composition by weight: carbon 80%, hydrogen 5%, oxygen 2%, sulphar 1%, and the remainder ash. The coal burns with 60% of excess air over the minimum quantity of air required for the complete combustion of the fuel. Estimate the volumetric composition of the dry flue gas. If the mean temperature of the gas leaving the boiler plant is 315°C and that of the boiler house be 27°C, estimate the heat carried away by flue gases leaving the boiler plant. Assume C_p of gas to be 1.01 kg/kg-K and total heat of water vapour in the gases to be 2658 kJ/kg.

Solution

Theoretical or minimum air required for complete combustion of 1 kg fuel is similarly carbon content of CO_2 / kg gas

$$= \frac{22}{44} \times 0.1618 = 0.04413 \text{ kg}$$

$\therefore$ Total carbon/kg gas $= 0.441 + 0.00802$

$= 0.05215$ kg

$\therefore$ Weight of gas produced per kg fuel

$$= \frac{0.81}{0.05215} = 15572 \text{ kg}$$

$\therefore$ Excess air/kg fuel $= 15.532 \times 0.193$

Example 12.28 In a boiler trial, dry fuel as burnt contained 84% carbon and 3% free hydrogen. The flue gas analysis by volume gave CO_2 – 11.5% , oxygen – 8.4% and nitrogen – 80.1%. Calculate per kg of dry fuel the weight of air supplied and weight of excess air.

Solution

We will solve this problem by tabulating the data as

Constituent	% Volume (a)	Mol. Wt (b)	Rel. wt (a x b)	Wt. per kg gas	Carbon Content kg gas
Carbon Dioxide	C_1=11.5	44	506	0.1677	0.04574
Oxygen	0=8.4	32	268.8	0.0891	
Nitrogen	N=80.1	28	2242.8	0.7432	
			Σ3017.6	1.0000	

$\therefore$ Weight of gas/kg fuel $= \frac{0.84}{0.04574} = 18.36$ kg

$$\text{Theoretical air per kg} = \frac{100}{23} \times \left[\frac{8}{3}\text{C} + 8\text{H} + \text{S} - \text{O}\right]$$

$$= \frac{100}{23} \times \left[\frac{8}{3} \times 0.8 + 8 \times 0.05 + 0.01 - 0.02\right]$$

$= 10.97$ kg

$\therefore$ Actual weight of air supplied/kg fuel

$= 1.6 \times 10.97 = 17.55$ kg

Dry products of combustion by weight are

$$\text{Carbon dioxide} = \frac{11}{3}\text{C} = \frac{11}{3} \times 0.08 = 2.933 \text{ kg}$$

Sulphur dioxide $= 2\text{S} = 2 \times 0.01 = 0.02$ kg

Oxygen $= 0.6 \times 2.523 = 1.5138$ kg

$$\text{Nitrogen} = 0.77 \times 17.55 = 13.5135 \text{ kg}$$

∴ Relative volumes of the dry products are

$$\text{Carbon dioxide} = \frac{2.933}{44} = 0.0667$$

$$\text{Sulphur dioxide} = \frac{0.02}{64} = 0.0003125$$

$$\text{Oxygen} = \frac{1.5138}{32} = 0.04731$$

$$\text{Nitrogen} = \frac{13.5135}{28} = 0.4826$$

$$\text{Total} = 0.59691$$

Percentage composition of dry products of combustion by volume is

$$\text{Carbon dioxide} = \frac{0.0667}{0.59691} \times 100 = 11.174\%$$

$$\text{Sulphur dioxide} = \frac{0.0003}{0.59691} \times 100 = 0.050\%$$

$$\text{Oxygen} = \frac{0.04731}{0.59691} \times 100 = 7.926\%$$

$$\text{Nitrogen} = \frac{0.4826}{0.59691} \times 100 = 80.850\%$$

$$\text{Total} = 100.00$$

Heat carried away by the flue gases leaving the boiler plant consists of

(i) Heat with dry gases.

(ii) Heat with vapours.

Heat carried by dry gas = $mg\, C_{pg}\,(T_g - T_R)$

where mg = Weight of the gases/kg fuel

= Air + Fuel – H_2O formed – ash.

= 17.55 + 1 – 9 × 0.05 –12

= 18.43 – 0.45 = 17.98 kg

∴ Heat carried by gases = 17.98 × 1.01(315 – 27)

= 5230 kJ/kg fuel

Heat carried by vapours

= 9H × 2658

= 0.45 × 2658 = 1196.1 kJ

∴ Total heat carried by exhaust or flue gases per kg of fuel

$$= 5230 + 1196.16 = 6426.1 \text{ kJ / kg fuel}$$

Example 12.29 The analysis of fuel used in an engine is carbon 50%, hydrogen 16.6% and oxygen 33.4% by weight. It is burnt with twice the theoretical minimum quantity of dry air required for complete combustion. Give the analysis of the exhaust gases (a) by weight and (b) by volume.

Solution

Minimum weight of air required for complete combustion of 1 kg fuel is

$$= \frac{100}{23}\left[\frac{8}{3}C + 8H - O\right]$$

$$= \frac{100}{23}\left[\frac{8}{3} \times 0.5 + 8 \times 0.166 - 0.334\right]$$

$$= 10.12 \text{ kg/kg fuel}$$

∴ Weight of air supplied/kg

$$= 10.12 \times 2 = 20.24 \text{ kg}$$

∴ The weight of the products/kg fuel are

$$\text{Carbon dioxide} = \frac{11}{3} \times 0.5 = 1.8333 \text{ kg}$$

$$\text{Water vapour} = 9H = 9 \times 0.166 = 1.494 \text{ kg}$$

$$\text{Oxygen} = 2.3273 \text{ kg}$$

$$\text{Nitrogen} = 0.77 \times 20.24 = 15.5848$$

$$\text{Total Weight} = 21.2394$$

∴ Percentage analysis by weight of the products of combustion is as

$$\text{Carbon dioxide} = \frac{1.8333}{21.2394} \times 100 = 0.6317$$

$$\text{Water vapour} = \frac{1.494}{21.2394} \times 100 = 7.0341$$

$$\text{Oxygen} = \frac{2.3273}{21.2394} \times 100 = 10.9575$$

$$\text{Nitrogen} = \frac{15.5848}{21.2394} \times 100 = 73.3768$$

Relative volumes of the products are

$$CO_2 = \frac{1.8333}{44} = 0.04167$$

$$H_2O = \frac{1.494}{18} = 0.08300$$

$$O_2 = \frac{2.3273}{32} = 0.07273$$

$$N_2 = \frac{15.5848}{28} = 0.55660$$

Total = 0.754

∴ Percentage analysis of the products of combustion by volume is

$$\text{Carbon dioxide} = \frac{0.04167}{0.754} \times 100 = 5.5265\%$$

$$\text{Water vapour} = \frac{0.083}{0.754} \times 100 = 9.6459\%$$

$$\text{Oxygen} = \frac{0.07273}{0.754} \times 100 = 9.6459\%$$

$$\text{Nitrogen} = \frac{0.5566}{0.754} \times 100 = 73.8196\%$$

Total = 100.0000

Example 12.30 In a boiler trail the composition of dry flue gases by volume was as under : CO_2 = 11%, CO = 2%, O_2 = 4%, N_2 = 83 %. Determine the quantity of air passing to the furnace per kg of fuel burnt if the carbon in the fuel was 81% by weight. Also determine the weight of excess air per kg of gas.

Solution

$$\text{Actual air supplied per kg fuel} = \frac{NC}{33(C_1 + C_2)} \text{ with usual notations}$$

$$\therefore \quad \text{Air/kg fuel} = \frac{83 \times 81}{33(11+2)} = \frac{83 \times 81}{33 \times 13}$$

Air/kg fuel = **15.67 kg**

Oxgen present in l kg flue gas can be found as

Gas	Quantity m^3	Mol. Wt.	Rel Wt.	Wt/kg gas
CO_2	11	44	44 × 11 = 484	$\frac{484}{\Sigma} = 0.16176$
CO	2	28	28 × 2 = 056	$\frac{56}{\Sigma} = 0.01872$

We have $CO + O = CO_2$

$\therefore$ Oxygen required to burn 0.01872 kg CO to carbondioxide

$$= \frac{16}{28} \times 0.01872$$

$$= 0.01067 \text{ kg}$$

$\therefore$ Excess oxygen, after accounting for CO,

$$= 0.04278 - 0.0107$$

$$= 0.03208 \text{ kg/kg gas}$$

$$\therefore \quad \text{Excess air} = \frac{100}{23} \times 0.03208 = \mathbf{0.1395 \text{ kg}}$$

Example 12.31 During a boiler trial, the average volumetric analysis of the flue gases gave the following results.

CO_2 = 10.7%, CO = 1.1 %, O_2 = 6.85 % and N_2 = 81.35 %. The chemical analysis of coal as fired gave, C = 87.6 %, H_2 = 3.4 %, O_2 = 4.5 %, ash etc. = 4.5 %. Determine per kg of coal

(i) The weight of flue gas.

(ii) The weight of air supplied.

Solution

(i) The weight of flue gas

CO_2	10.7 × 44	=	470.8	0.1570	C = 0.04282
CO	1.1 × 28	=	30.8	0.0103	C = 0.00421
O_2	6.85 × 32	=	219.2	0.0731	0.04723
N_2	81.35 × 28	=	2277.8	0.7596	
			2998.6	0.9997 = 1 kg gas	

$$\therefore \quad \text{Weight of gases/kg fuel} = \frac{0.876}{0.04723}$$

$$\text{Weight of gases/kg fuel} = \mathbf{18.59 \text{ kg}}$$

$$\text{Wt. of wet gases} = 18.55 + 9 \times 0.034$$

$$= 18.55 + 0.306 = 19.906$$

$$\therefore \quad \text{Weight of flue gas/kg fuel} = \mathbf{19.906 \text{ kg}}$$

(ii) Weight of air supplied

Weight of air supplied /kg fuel

$$= \frac{NC}{33(C_1 + C_2)} = \frac{8.135 \times 87.6}{33(10.7 + 1.1)}$$

$$= \mathbf{18.3 \text{ kg/kg fuel}}$$

Example 12.32 A SI engine uses octane (C_8H_{18}) as the fuel and the exhaust gas analysis gave the following composition. $CO_2 = 9.9\%$, $CO = 7.2\%$, $H_2 = 3.3\%$, $CH_4 = 0.3\%$, $N_2 = 79.3\%$. Calculate the air fuel ratio.

Solution

For volumetric analysis of air 1 mol of O_2 will be accompanied by 3.76. mol. of N_2

$$\left(\frac{79}{21} = 3.76\right)$$

Now we can write the chemical equation for combustion of this fuel. Thus,

$$a\,(C_8 + H_8) + b(O_2 + 3.76N_2) = 9.9\,CO_2 + 7.2\,CO + 3.3\,H_2 + 0.3\,CH_4 + 79.3\,N_2 + DH_2O$$

Last term in this equation is because of the combustion of hydrogen in the fuel.

Values of a and b in the above equation should be such that we will get the analysis as given on the right hand side.

First we will balance carbon.

$$8a = 9.9 + 7.2 + 03 = 17.4$$

$$\therefore \quad a = \frac{17.4}{8} = 2.175$$

Again, we can balance nitrogen. Then

$$3.76\,b = 79.3$$

$$b = \frac{79.3}{3.76} = 21.09$$

Then we will balance hydrogen

$$18a = 6.6 + 1.2 + 2D = 7.8 + 2D$$

$$\therefore \quad 18 \times 2.175 = 7.8 + 2D = 39.15$$

$$\therefore \quad D = \frac{39.15 - 7.8}{2} = \frac{31.35}{2}$$

$$= 15.675$$

Thus all unknowns a, b and D in the equation are known. Therefore, the combustion equation is

$$2.175(C_8H_{18}) + 21.09(O_2 + 3.76N_2)$$
$$= 9.9\,CO_2 + 7.2\,CO + 3.3\,H_2 + 0.3\,CH4 - 79.3\,N_2 + 15.675\,H_2O$$

$$\therefore \quad \frac{\text{Air}}{\text{Fuel}} = \frac{21.09(32 + 3.76 \times 28)}{2.175(96 + 18)} = \frac{2.109 \times (32 + 105.78)}{2.175 \times 114}$$

$$\frac{\text{Air}}{\text{Fuel}} = \textbf{11.68 kg air/kg fuel}$$

Example 12.33 Volumetric analysis of products of combustion of a hydrocarbon fuel is as under : $CO_2 = 12.5\,\%$, $CO = 0.3\%$, $O_2 = 5\%$ N_2 is found by difference. Determine (a) air : fuel ratio (b) carbon and hydrogen percentage by mass (c) percentage of excess air.

Solution

Let the fuel be denote by C_xH_y. Nitrogen present in the exhaust gases = 100 – 17.8 = 82.2

The combustion equation is

$$C_xH_y + b(O_2 + 3.76N_2) = 12.5CO_2 + 0.3CO + 5O_2 + 82.2N_2 + DH_2O$$

∴ Nitrogen balance gives

$$3.76\,b = 82.2$$

$$\therefore \quad b = \frac{82.2}{3.76} = 21.862$$

Carbon balance gives

$$x = 12.5 + 0.3 = 12.8$$

Oxygen balance gives

$$b = 12.5 + \frac{0.3}{2} + 5 + \frac{D}{2}$$

$$21.862 = 17.65 + \frac{D}{2}$$

$$\text{or} \quad D = \left(\frac{21.862 - 17.654}{1}\right)2 = 4.208 \times 2$$

$$= \mathbf{8.416}$$

Hydrogen balance gives

$$y = D$$

$$= 8.416$$

∴ Actual combustion equation is given by

$$(C_{12.8}H_{8.416}) + 21.862O_2 + 21.862 \times 3.76N_2$$

$$= 12.5CO_2 + 0.3CO + 5O_2 + 82.2N_2 + 8.416H_2O$$

(i) Air fuel ratio

$$\frac{\text{Air}}{\text{Fuel}} = \frac{(21.862 \times 32) + (21.862 \times 3.76 \times 28)}{12.8 \times 12 + 8.416 \times 1}$$

$$= \frac{699.584 + 2301.631}{153.6 + 8.416} = \frac{3001.215}{162.016}$$

$$\frac{\textbf{Air}}{\textbf{Fuel}} = \textbf{18.524 kg/kg of fuel}$$

(ii) Composition of fuel

$$\begin{aligned} \text{Carbon weight} &= 153.6 \text{ kg} \\ \text{Hydrogen weight} &= 8.416 \\ \text{Total weight} &= 162.016 \text{ kg} \end{aligned}$$

$$\therefore \quad \% \text{ of carbon in fuel} = \frac{153.6}{162.016} \times 100 = 94.8$$

$$\therefore \quad \textbf{\% of hydrogen in fuel} = \textbf{5.2}$$

(iii) Percentage of excess air

$$\text{Theoretical air/kg fuel} = \frac{100}{23}\left[\frac{8}{3} \times 0.9012 + 8 \times 0.0988\right]$$

$$= 13.885 \text{ kg/kg fuel}$$

$$\text{Actual air supplied/kg fuel} = 18.524 \text{ kg/kg fuel}$$

$$\therefore \quad \text{Excess air} = 18.524 - 13.885 = 4.639 \text{ kg}$$

$$\therefore \quad \% \text{ of excess air} = \frac{4.639}{13.885} \times 100 = \textbf{33.41 \%}$$

Example 12.34 (a) Derive an equation for minimum quantity of air required for complete combustion of 1 kg of solid or liquid fuel, assuming that the fuel contains carbon, hydrogen, sulphur and oxygen as constituents.

(b) The analysis by mass of the coal fired in a boiler is 82 % carbon, 8% hydrogen, 3% oxygen, 7% ash. The boiler uses 0.19 kg/sec of air and is supplied to the furnace by a fan. The air supplied to the furnace by a fan. The air supplied is 30% in excess of that require for theoretical correct, combustion. Calculate,

(i) Volume of air taken in by the fan per sec. when the pressure and temperature of the air at the fan intake are 100 kN/m^2 and 18°C respectively. R for air 0.287 kJ/kg-K.

(ii) Percentage composition by mass of the dry flue gases. Air contains 23% oxygen by mass.

Solution

(b) (i) Volume of air sec

$$\text{Volume of air supplied/sec} = \frac{mRT}{p} = \frac{0.19 \times 287 \times 291}{100 \times 1000}$$

$$= 0.1587 \text{ cu-m/sec}$$

(ii) Gravimetric analysis of dry products:

Minimum air required/kg fuel

$$= \frac{100}{23}\left[\frac{8}{3}C + 8H + S - O\right]$$

$$= \frac{100}{23}[2.187 + 0.64 \times 0.03]$$

$$= 12.16 \text{ kg/kg fuel}$$

∴ Actual air supplied /kg fuel

$$= 1.3 \times 12.16 = 15.81 \text{ kg/kg fuel}$$

Dry products of combustion

$$CO_2 = \frac{11C}{3} = \frac{11}{3} \times 0.82 = 3.001 \text{ kg}$$

$$N_2 = 0.77 \times 15.81$$

$$= 12.1737 \text{ kg}$$

$$O_2 = (0.3 \times 12.16) \times 0.23$$

$$= 0.839 \text{ kg}$$

Total mass of products (assuming ash remain in the furnace only)

$$= 3.001 + 12.1737 + .839$$

$$= 16.014 \text{ kg}$$

∴ Percentage of CO_2 in products $= \frac{3.001 \times 100}{16.014} = 18.74\%$

Percentage of O_2 in products $= \frac{0.839}{16.014} \times 100 = 5.24\%$

Percentage of N_2 in products $= \frac{12.1737}{16.014} \times 100 = 76.02\%$

100.00 %

Example 12.35 (a) Describe briefly a method of obtaining the higher calorific value of a sample of fuel oil, showing how the result is calculated from the data obtained. State the possible sources of inaccuracy in the result.

(b) A fuel consists of the following percentage analysis by mass

$C = 84\ \%$, $H_2 = 10\%$, $O_2 = 2\%$, $S = 1\ \%$ and $N_2 = 3\%$.

Find out the amount of air required to completely burn 1 kg of this fuel and determine the mass of products of combustion.

Solution

Minimum amount of air required for complete combustion of 1 kg fuel is given by

$$m_a = \frac{100}{23}\left[\frac{8}{3}C+8H+S-O\right]$$

$$= \frac{100}{23}\left[\frac{8}{3}\times 0.84+8\times 0.1+0.1+0.01-0.02\right]$$

$$= \frac{100}{23}[2.24+0.84-0.01]$$

$$= \frac{100}{23}\times 3.03$$

$$\mathbf{m_a = 13.174\ kg/kg\ fuel}$$

Products of combustion are

$$CO_2 = \frac{11}{3}C = \frac{11}{3}\times 8.84 = 3.08\text{ kg}$$

$$H_2O = 9H = 9\times 0.1 = 0.90\text{ kg}$$

$$O_2 = 2S = 2\times 0.01 = 0.02\text{kg}$$

$$N_2 = 0.03+\frac{77}{23}\times 0.03 = 10.144$$

$$= 10.174\text{ kg}$$

$\therefore$ Total mass $= 14.174$ kg

$$\therefore \quad CO_2 = \frac{3.08\times 100}{14.174} = 21.73\%$$

$$H_2O = \frac{0.9\times 100}{14.174} = 6.35\%$$

$$SO_2 = \frac{0.02\times 100}{14.174} = 0.14\%$$

$$\mathbf{N_2 = \frac{10.174\times 100}{14.174} = 71.78\%}$$

Example 12.36 (a) Derive an equation for minimum quantity of air required for complete combustion of 1 kg of solid or liquid fuel, assuming that the fuel contains carbon, hydrogen, sulphur and oxygen as constituents.

(b) The analysis by mass of the coal fired in a boiler is 82% carbon, 8% hydrogen, 3% oxygen, 7% ash. The boiler uses 0.19 kg/s of air and is supplied 30% in excess of that required for theoretically correct combustion. Calculate,

(i) Volume of air taken in by the fan per sec. when the pressure and temperature of the air at the fan intake are 100 kN/m^2 and 18°C respectively. R for air 0.287 kJ/kg-K.

(ii) Percentage composition by mass of the dry flue gases. Air contains 23% oxygen by mass.

Solution

(i) Volume of air/sec:

$$\text{Volume of air supplied/sec} = \frac{mRT}{P} = \frac{0.19 \times 287 \times 291}{100 \times 1000}$$

Volume of air supplied/sec = 0.1587 cu.m/sec

(ii) Gravimetric analysis of dry products:

$$\text{Minimum air required/kg fuel} = \frac{100}{23}\left[\frac{8}{3}C + 8H + S - O\right]$$

$$= \frac{100}{23}\left[\frac{8}{3} \times 0.82 + 8 \times 0.08 - 0.03\right]$$

$$= 12.16 \text{ kg/kg fuel}$$

$\therefore$ Actual air supplied/kg fuel = 1.3×12.16

$$= 15.81 \text{ kg/kg fuel}$$

Dry products of combustion

$$CO_2 = \frac{11C}{3} = \frac{11}{3} \times 0.82 = 3.001 \text{ kg}$$

$$N_2 = 0.77 \times 15.81 = 12.1737 \text{ kg}$$

$$O_2 = (0.3 \times 12.16) \times 0.23$$

$$= 0.839 \text{ kg}$$

Total mass of products (assuming ash remain in the furnace only)

$$= 3.001 + 12.1737 + 0.839$$

$$= 16.014 \text{ kg}$$

Total mass of products assuming ash remain in the furnace only

$$\textbf{\% of } \mathbf{CO_2} \textbf{ in exhaust by mass} = \frac{\mathbf{3.001}}{\mathbf{16.014}} \times \mathbf{100} = \mathbf{18.73\%}$$

$$\textbf{\% of } \mathbf{O_2} \textbf{ in exhaust by mass} = \frac{\mathbf{0.839}}{\mathbf{16.014}} \times \mathbf{100} = \mathbf{5.24\ \%}$$

$$\textbf{\% of } \mathbf{N_2} \textbf{ in exhaust by mass} = \frac{\mathbf{12.1737}}{\mathbf{16.014}} \times \mathbf{100} = \mathbf{76.02\ \%}$$

Example 12.37 (a) Define excess air, stoichiometric air fuel ratio.

(b) Methods of coal analysis

(c) The following data was obtained during experimental determination of calorific value of fuel by bomb calorimeter:

Mass of fuel = 0.8 gm

Mass of fuse wire = 0.03 gm

Calorific value of fuse wire = 7 kJ/gm

Mass of water in calorimeter = 2 kg

Water equivalent of calorimeter = 0.4 kg

Rise in temperature of calorimeter = 3.2°C

Cooling correction = 0.02°C

Determine HCV and LCV of coal. Given the coal contains 90% of carbon and 5% of hydrogen.

Solution

We can use the energy balance principle.

Heat given by combustion of fuel + Heat of fuse wire = Heat taken by water

$$\therefore \frac{0.8}{1000} \times CV + \frac{0.03}{1000} \times \frac{7 \times 1000}{1} = \left[(2+0.4)(3.2+0.02) \times 4.187\right]$$

$$\therefore \quad \frac{0.8 \times CV}{1000} + \frac{210}{1000} = 2.4 \times 4.187 \times 3.22$$

$$= 32.36$$

$$\therefore \quad 0.8 \times CV + 210 = 32.36 \times 1000$$

$$\therefore \quad CV = \frac{32.36 \times 1000 - 210}{0.8}$$

$$= 40187.5 \text{ kJ/kg}$$

Bomb calorimeter readings or observations give us high calorific value HCV.

$$\therefore \quad \mathbf{HCV = 40187.5 \text{ kJ/kg}}$$

In one kg coal, 5% is hydrogen

$$\therefore \quad H_2O \text{ formed/kg coal} = 0.05 \times 9 = 0.45 \text{ kg}$$

$$\therefore \quad LCV = \text{Lower calorific value}$$

$$= HCV - \frac{H_2O}{\text{kg coal}} \times 2466$$

$$= 40187.5 - 0.45 \times 2466$$

$$\mathbf{LCV = 39077.8 \text{ kJ/kg}}$$

Example 12.38 (a) Define the following terms:

(i) Higher calorific value of a fuel

(ii) Lower calorific value of a fuel

(b) A fuel has the following percentage composition by mass : C = 85 %, H_2 = 15%
Determine:

(i) The stoichiometric mass of air required for complete combustion of 1 kg of this fuel.

(ii) The percentage composition by mass of the dry products of combustion air contain 23.3% oxygen by mass.

(c) The dry exhaust gas from an oil engine had the following composition by volume CO_2 8.85%, CO = 1.2%, O_2 = 6.8 %, N_2 = 83.15.

Convert the volumetric analysis of the exhaust gas into a mass analysis.

Solution

(i) Stoichiometric air

Mass of air required for complete combustion of 1 kg of fuel

$$= \left[\frac{8}{3}C + 8H - O + S\right]\frac{100}{23.2}$$

$$= \left[\frac{8}{3}\times 0.85 + 8\times 0.15\right]\frac{100}{23.2}$$

$$= \textbf{14.94 kg/kg of fuel}$$

(ii) Gravimetric analysis: (Dry products)

$$CO_2 \rightarrow \frac{11}{3}C = \frac{11}{3}\times 0.85 = 3.117 \text{ kg}$$

$$N_2 \rightarrow \frac{76.7}{23.3}\times 3.4667 = 11.476 \text{ kg}$$

$$\text{Total mass of products} = 3.117 + 11.476$$

$$= \textbf{14.593 kg/kg fuel}$$

$$\% CO_2 \rightarrow \frac{3.117}{14.593}\times 100 = 21.3658\ \%$$

$$\% N_2 \rightarrow \frac{12.8}{14.593}\times 100 = 78.64\ \%$$

(c) Volumetric to gravimetric analysis

			Rel. wts	%
CO_2	8.85×44	=	089.4	13.116
CO	1.2×28	=	33.6	1.132

O_2	6.8×32	=	217.6	7.330
N_2	83.15×28	=	2328.2	78.442
			2968.8	100.00

Example 12.39 A sample of dry anthracite has the following composition by mass:

C = 90%, H = 3%, O =2.5%, N = 1%, S = 0.5%, Ash = 3%.

Calculate:

(i) The stoichoimetric A/F ratio

(ii) The A/F ratio and the dry and wet analysis of the products of combustion by mass and by volume, when 20% excess air is supplied.

Solution

(i) Stoichoimetric A/F ratio:

Theoretical or stoichiometric air supplied per kg of fuel

$$= \frac{100}{23}\left[\frac{8}{3}C + 8H + S - O\right]$$

$$= \frac{100}{23}\left[\frac{8}{3} \times 0.9 + 8 \times 0.03 + 0.005 - 0.025\right]$$

$$= \mathbf{11.3913\ kg}$$

(ii) Product analysis:

Products are $CO_2 = \frac{11}{3} \times 0.9 = 3.3\text{ kg} \div 44 = 0.075\text{ mol}$

$$H_2O = 9H = 9 \times 0.03 = 0.27\text{ kg} \div 18 = 0.015\text{ mol}$$

$$SO_2 = 2S = 2 \times 0.005 = 0.01\text{ kg} = 0.001563\text{ mol}$$

$$O_2 = 0.2 \times 2.62 = 0.524\text{ kg} = 0.01638\text{ mol}$$

$$N_2 = 1.2 \times 0.77 \times 11.3913 = 10.5256 + 0.01$$

$$= 10.5356\text{ kg} = 0.376\text{ mol}$$

Gravimetric Analysis (Dry Products)

$$\text{Total} = 3.3 + 0.01 + 0.524 + 10.5356$$

$$= 14.3696\text{ kg}$$

$$CO_2\% \rightarrow \frac{3.3}{14.3696} \times 100 = 22.965\%$$

$$O_2\% \rightarrow \frac{0.524 \times 100}{14.3696} = 3.6466\%$$

$$SO_2\% \rightarrow \frac{0.01 \times 100}{14.3696} = 0.0696\%$$

$$N_2\% \rightarrow \frac{10.5356 \times 100}{14.3696} = 73.32\%$$

Wet Products

Total products are 14.3696 + 0.27 = 14.63 kg

$$\% CO_2 \rightarrow \frac{330}{14.6396} = 22.5416\%$$

$$O_2\% \rightarrow \frac{52.4}{14.6396} = 3.5793\%$$

$$SO_2\% \rightarrow \frac{1}{14.6396} = 0.06830\%$$

$$N_2 \rightarrow \frac{1053.56}{14.6396} = 71.964\%$$

$$H_2 \rightarrow \frac{27}{14.6396} = 1.8443\%$$

Volumetric Analysis

Dry products

$$\text{Total mol} = 0.0752 + 0.3763 + 0.0001563 + 0.01638$$

$$\cong 0.46784$$

$$\therefore \quad CO_2 \rightarrow \frac{0.075 \times 100}{0.46784} = 16.031\%$$

$$SO_2 \rightarrow \frac{0.1563}{0.46784} = 0.0334\%$$

$$O_2 \rightarrow \frac{1.638}{0.46784} = 3.501\%$$

$$N_2 \rightarrow \frac{37.63}{0.46784} = 80.4335\%$$

Wet Products

$$\text{Total} = 0.46784 + 0.015 = 0.48284$$

$$\therefore \quad CO_2 \rightarrow \frac{7.5}{0.48284} = 15.5331\%$$

$$SO_2 \rightarrow \frac{0.01563}{0.46284} = 0.0324\%$$

$$O_2 \rightarrow \frac{1.638}{0.48284} = 3.3924\%$$

$$N_2 \rightarrow \frac{37.63}{0.48284} = 77.9342\%$$

$$H_2O \rightarrow \frac{1.5}{0.48284} = 3.1066\%$$

PROBLEMS FOR PRACTICE

1. A sample of coal when analyzed, gave following composition by weight—Carbon 85%, hydrogen 12%, ash 1% and moisture 2%. Find the higher and lower calorific values of the coal.

(**Ans.** 46141 kJ/kg, 43498 kJ/kg)

2. Find the higher and lower calorific value of a gaseous fuel whose volumetric composition is 7% CO_2, 25% CO, 15% H_2, 4% CH_4 and 49% N_2. The latent heat of water may be assumed to be 2466 kJ/kg at the condition after combustion. The higher calorific value of CO_2, CO, H_2 CH_4 and N_2 is zero, 12600, 12800, 43000 and zero kJ/Nm³ respectively.

3. A sample of lignite coal contains 6% hydrogen and 3% moisture by weight. If the higher calorific value of the sample was found to be 29300 kJ/kg, calculate its lower calorific value. Assume 2466 kJ of heat is required to evaporate water at room temperature to dry and saturated steam.

4. A producer gas contains 16% H_2, 20% CO, 6% CO_2 and 58% N_2 by volume. Determine the HCV and LCV of the gas in kJ/Nm³.

Assume calorific value of H_2 = 144000 kJ/Nm²

Assume calorific value of CO = 10000 kJ/Nm²

5. In a test carried on diesel oil in a bomb calorimeter following observations were recorded

Wt. of crucible = 2.5 gm

Wt. of crucible + oil = 3.079 gm

Wt. of can = 1 kg

Wt. of can + water = 2.4 kg

Wt. of cotton used in igniting oil = 0.005 gm

Calorific value of cotton = 16750 kJ/kg

Room temperature = 20°C

Sp. heat of the can = 2.1 kJ/kg-K

Temp. rise of water from 25°C to 27.912°C

Temp. correction due to radiation = 0.012C

If the oil contains 14% H_2 and 3% moisture, calculate HCV and LCV of the oil.

6. During an engine trial, the gas as used was tested in a Boy's calorimeter with the following results

   Gas supply 18°C and 1.04 bar

   Gas burnt 0.5 m^3

   Temp. rise of water 16°C to 24.5°C

   Wt. of cooling water 24.5 kg

   Condensate from the products 0.004 kg and 18°C

   Calculate the higher and lower calorific value of the gas.

7. The dry exhaust gases from an Internal combustion engine burning a pure hydrocarbon fuel were found to contain CO_2 =125%, CO = 2.7 %, N_2 = 83.4 % with small quantities of other gases. The fuel consumption was observed to be 11 kg/hour.

   Estimate the ratio of carbon to hydrogen by weight in fuel. Find the air to fuel ratio and thus the air consumption of the engine.

   (Ans. A/F = 14.12, $\frac{C}{H}$ = 5.4672, 155.32 kg/hr)

ꙮ ꙮ

Index